KT-133-059

Third Edition
Introduction to Organic Chemistry

William H. Brown
Beloit College

Willard Grant Press
Boston, Massachusetts

PWS PUBLISHERS
Prindle, Weber & Schmidt · Willard Grant Press · Duxbury Press ·
Statler Office Building · 20 Providence Street · Boston, Massachusetts 02116

PWS Publishers is a division of Wadsworth, Inc.

Printed in the United States of America.
10 9 8 7 6 5 4 3 — 86 85 84

Library of Congress Cataloging in Publication Data

Brown, William Henry
Introduction to organic chemistry.

Includes index.
1. Chemistry, Organic. I. Title.
QD251.2.B76 1981 547 81-10477
ISBN 0-87150-747-1 AACR2

ISBN 534-98022-8 (international)

ISBN 0-87150-747-1
ISBN 534-98022-8

Photograph credits not given in text: page 116e, Dow Chemical Co.; 116h top and bottom, Sharon A. Bazarian; 251k, H. C. Brown; 284g, Ellis Herwig, Stock, Boston; 388a, Joslin Diabetes Center, Inc.; 429a, World Health Organization.

Cover: "Oil Streamers" A thin film of lubricating oil separates into uniform streamers as it exits from the region of contact between two surfaces in relative motion. This phenomenon was observed on an experimental apparatus used to study the mechanics of lubricant flow in very thin gaps — a condition typically found in engine bearings. Theoretical and experimental investigations of this type have led to the design of bearings with low friction and high durability. (Courtesy of General Motors Research Laboratories)

Text designers: Nancy Ross McJennett and John A. Servideo
Art Studio: J & R Services
Composition: Composition House Ltd
Text Printing and binding: Halliday Lithograph Corp.
Cover Printer: Lehigh Press Lithographers
Production coordinator: John A. Servideo

PREFACE

The objective of this third edition is to provide an introduction to organic chemistry for those aiming toward careers in science. I recognize, however, that few students enrolled in the short course intend to become professional chemists. Rather, they are preparing for careers that require a grounding in the fundamentals of organic chemistry. For this reason, I have made an effort throughout the text to show the interrelationship between organic chemistry and other areas of science.

There are several changes in the third edition that recur throughout the text. One of the most significant is the addition of example problems within the chapters. Students often have trouble in applying chemical principles when solving problems, and I hope examples will help bridge this gap. The worked-out solutions show students how to formulate strategies and approach problem solving in a systematic way. Following every example is a similar problem for the student to solve. In addition to the intrachapter problems, many new exercises have been added at the ends of chapters. In all, the number of problems has been increased by one-third over the second edition.

The number of illustrations has been increased greatly to help students visualize organic molecules as three-dimensional substances. Ball-and-stick models have been added to complement the structural drawings included throughout the book. In addition, space-filling models are used to illustrate the shapes of proteins, lipids, and nucleic acids.

Summaries of important reactions have been added at the ends of appropriate chapters. Included in the summaries are references to the sections in the chapter where each reaction is covered in detail.

This text can be divided roughly into two parts. Chapters 1 to 10 lay the foundation in organic chemistry by discussing the structures and typical reactions of the important functional groups with which the student must be familiar. Chapters 11 to 14, the second part, provide a comprehensive coverage of the structure and function of the four key classes of biomolecules: carbohydrates, amino acids and proteins, lipids, and nucleic acids.

The discussion of chemical bonding in Chapter 1 is expanded to include a review of electronegativity and its relationship to the polarity of bonds. Also added is a description of the valence-shell electron-pair repulsion model and its application to predicting bond angles and the shapes of molecules. The concept of a functional group is treated more thoroughly and concentrates on alcohols, ethers, aldehydes, ketones, and carboxylic acids, groups that are encountered repeatedly throughout the first seven chapters. Emphasis is on the recognition of characteristic structural features and the ability to draw Lewis structures and condensed structural formulas. Chapter 2 contains a new section on hydroboration of alkenes and the synthesis of alcohols. The importance of organoboranes is further illustrated in the new mini-essay "Footsteps on the Borane Trail" which traces the career in chemistry of H. C. Brown.

Most chapters in the first part of the book stress conversion problems which have been included to provide one more way for students to practice using the important reactions discussed. Most problems require no more than two or three steps. In my experience, the investment of time and effort in doing simple conversion problems leads to better understanding of organic chemistry.

Discussion of glycosides in Chapter 11 is expanded to include the structure of N-glycosides in anticipation of the chemistry of nucleic acids. A section on blood group substances and the ABO blood-type classification system has also been added. In the next chapter on amino acids and proteins there is expanded treatment of the acid-base properties of amino acids. Material on the relationship between primary structure and the three-dimensional shape of proteins has been added. There is a new section on enzyme cofactors including metal ions, coenzymes and vitamins, and prosthetic groups.

Chapter 12 contains a more detailed description of the structure of DNA with particular emphasis on base pairing and complementarity, the labelling of 3′ and 5′ ends of DNA and RNA strands, and the process and direction of transcription and translation. Protein synthesis is presented in greater detail and a section on inhibition of protein synthesis by streptomycin, tetracycline, and other antibiotics has been added.

The final chapter introduces infrared, ultraviolet-visible, and nuclear magnetic resonance spectroscopy. It may be taken up any time after Chapter 2 has been completed. All material involving spectroscopy has been planned so that it may be included or omitted as the instructor chooses. Independent sections about major spectral characteristics of each functional group appear at the ends of appropriate chapters. All discussions of spectroscopy are qualitative in nature and require no more background than what students receive in general chemistry. Problems containing spectral information are marked by the symbol so that they can be identified easily.

All mini-essays retained from the second edition have been revised and updated. Two new ones, "Footsteps on the Borane Trail" and "Acetic Acid: From What and How," have been added. The purpose of these short, optional articles is to reveal to students the role of organic chemistry in areas of concern to them. Furthermore, the mini-essays offer a glimpse of the human involvement in research and discovery while demonstrating that organic chemistry is an exciting field, full of challenge for the creative mind.

A new edition of the Study Guide has also been prepared. Its purpose is to guide students in their approach to solving organic chemistry problems and to provide complete and detailed solutions to all problems within the text.

For instructors, I have written a guide. It contains a summary of each chapter pointing out the degree of flexibility both in terms of material and sequence. Also presented is a suggested course outline and schedule for covering the material in approximately 40 lectures.

Many suggestions for changes in this edition were made by users of the previous edition, and I want to thank those who took the time to send me their recommendations. Furthermore, I want to express particular appreciation to reviewers of the various stages of the manuscript whose comments greatly improved the final version: Edwin C. Friedrich, University of California at Davis; Yadviga Halsey, University of Washington School of Medicine; Weston Sanford, Boston State College; Martin A. Schwartz, Florida State University; Jack W. Timberlake, University of New Orleans; Leroy G. Wade, Colorado State University; and Boris Weinstein, University of Washington.

I am much indebted to Brent Wurfel, a chemistry major at Beloit College, for his careful reading of many chapters and all the mini-essays. John Copp, another chemistry major, also read much of the manuscript. They saw the material as through the eyes of a student, and for this reason their many suggestions were especially valuable to me.

Finally, my appreciation to John Servideo of Willard Grant Press with whom it is always a pleasure to work.

William H. Brown

CONTENTS

CHAPTER 1

The Covalent Bond and the Geometry of Molecules

According to the simplest definition, organic chemistry is the study of the compounds of carbon. Perhaps the most remarkable feature of organic chemistry is that it is the chemistry of carbon and only a few other elements, chiefly hydrogen, nitrogen, and oxygen. In this chapter, we will describe how atoms of carbon, hydrogen, nitrogen, and oxygen combine to form molecules by sharing electron pairs. Then we will examine the three-dimensional shapes of some simple organic molecules. Finally, we will develop the concept of structural isomerism and functional groups. In subsequent chapters we will examine the reactions organic molecules undergo and the ways to convert one molecule into another.

1.1 ELECTRONIC STRUCTURE OF ATOMS

From a previous course in chemistry you are already familiar with the fundamentals of the electronic structure of atoms. Briefly, an atom contains a small, dense nucleus containing neutrons and positively charged protons. The nucleus is surrounded by a much larger, extranuclear space containing negatively charged electrons. Electrons are concentrated about the nucleus in regions of space called principal energy levels that are identified by the principal quantum numbers 1, 2, 3, and so on. Each principal energy level can contain up to $2n^2$ electrons, where n is the number of the principal energy level. Thus, the first energy level can contain 2 electrons; the second, 8 electrons; the third, 18 electrons; the fourth, 32 electrons; etc.

Each principal energy level is subdivided into regions of space called orbitals. The first principal energy level contains a single orbital called the 1*s* orbital. The second principal energy level contains one *s* orbital and three *p* orbitals. These orbitals are designated 2*s*, $2p_x$, $2p_y$, and $2p_z$. The third principal energy level contains one 3*s* orbital, three 3*p* orbitals, and five 3*d* orbitals. Each orbital can hold two electrons.

If we build an atom by surrounding the nucleus with enough electrons to neutralize its positive charge, the first orbital to fill is the 1*s*, that is, the orbital of lowest energy (the one closest to the nucleus). Next to fill is the 2*s* orbital, then the 2*p*, etc. Table 1.1 shows electron configurations of the first 18 elements of the periodic table.

When discussing the chemical properties of an element, we often focus on the outermost orbitals of the element, for it is the electrons in these orbitals that

Table 1.1 Electron configuration of the first 18 elements.

Element	Atomic number	Orbital								
		1*s*	2*s*	2p_x	2p_y	2p_z	3*s*	3p_x	3p_y	3p_z
H	1	1								
He	2	2								
Li	3	2	1							
Be	4	2	2							
B	5	2	2	1						
C	6	2	2	1	1					
N	7	2	2	1	1	1				
O	8	2	2	2	1	1				
F	9	2	2	2	2	1				
Ne	10	2	2	2	2	2				
Na	11	2	2	2	2	2	1			
Mg	12	2	2	2	2	2	2			
Al	13	2	2	2	2	2	2	1		
Si	14	2	2	2	2	2	2	1	1	
P	15	2	2	2	2	2	2	1	1	1
S	16	2	2	2	2	2	2	2	1	1
Cl	17	2	2	2	2	2	2	2	2	1
Ar	18	2	2	2	2	2	2	2	2	2

are involved in the formation of chemical bonds and in chemical reactions. To show the outermost electrons of an atom, we use a representation called a Lewis structure. A Lewis structure shows the symbol of the element surrounded by a number of dots equal to the number of electrons in the outer shell of that element. In Lewis structures, the atomic symbol represents the "core," that is, the nucleus and all completely filled inner shells. Outershell electrons are called valence electrons, and the energy level in which they are found is called the valence shell. Table 1.2 shows Lewis structures for the first eighteen elements of the periodic table.

The noble gases—helium, neon, and argon—have completely filled valence shells. The valence shell of helium is filled with two electrons; those of neon and argon are filled with eight electrons. The valence shells of all other elements in Table 1.2 are only partially filled.

You should compare these valence electron representations with the electron configurations given in Table 1.1. For example, beryllium is shown in

Table 1.2 Lewis structures for the first 18 elements.

IA	IIA	IIIA	IVA	VA	VIA	VIIA	VIIIA
H·							He:
Li·	Be:	Ḃ:	·Ċ:	·Ṇ̇:	:Ọ̇:	:F̤̈:	:N̤̈e:
Na·	Mg:	Ȧl:	·Ṡi:	·Ṗ̣:	:Ṩ:	:C̤̈l:	:Ä̤r:

Table 1.2 with two paired valence electrons; these are the two 2*s* electrons listed in Table 1.1. Carbon is shown with four valence electrons, two of which are paired and two of which are unpaired; these represent the two paired 2*s* electrons and the single $2p_x$ and $2p_y$ electrons listed in Table 1.1.

Notice also that carbon and silicon each have four valence electrons. Nitrogen and phosphorus each have five valence electrons. Oxygen and sulfur each have six electrons. Fluorine and chlorine each have seven valence electrons. Although the number of valence electrons for these pairs of atoms is the same, the shells in which their valence electrons are found are different. For C, N, O, and F, the valence electrons belong to the principal quantum number 2 shell. With eight electrons this shell is completely filled. For Si, P, S, and Cl, the valence electrons belong to the principal quantum number 3 shell. This shell is only partially filled with eight electrons; the 3*s* and 3*p* orbitals are fully occupied but the five 3*d* orbitals can accommodate an additional ten valence electrons. Because of this difference between the number and kind of orbitals in principal energy levels 2 and 3, we should expect differences in the covalent bonding of oxygen and sulfur, and of nitrogen and phosphorus. Such differences do exist, as we shall see.

1.2 THE FORMATION OF CHEMICAL BONDS

In 1916, Gilbert N. Lewis devised a beautifully simple hypothesis that unified many of the observations about chemical reactions of the elements. Lewis pointed out that the chemical inertness of the noble gases indicates a high degree of stability of the electron configurations of these elements: helium with an outer shell of two electrons, neon with shells of two and eight electrons, and argon with shells of two, eight, and eight electrons. According to the Lewis model of bonding, atoms bond together in such a way that each atom participating in a chemical bond acquires a completed outer shell electron configuration resembling that of the noble gas nearest it in the periodic table.

There are two ways by which atoms acquire completed outer shells:

(1) An atom may share electrons with another atom or atoms so that with the shared electrons it has a complete outer shell. The chemical bond formed by the sharing of electrons is called a covalent bond.

(2) An atom may lose or gain enough electrons to acquire a completely filled outer shell. An atom that gains electrons becomes a negatively charged ion, and an atom that loses electrons becomes a positively charged ion. The chemical bond between positively charged and negatively charged ions is called an ionic bond.

How can we determine if a given pair of atoms will bond together through the formation of a covalent bond or an ionic bond? We can do this by looking at the difference in electronegativity between the two atoms in question.

Electronegativity is a measure of the tendency for an atom in a molecule to attract a pair of electrons that it shares with another atom in a chemical bond.

Table 1.3 Electronegativities of common elements (Pauling scale).

(increasing → left to right; increasing ↑ bottom to top)

H 2.1						
Li 1.0	Be 1.5	B 2.0	C 2.5	N 3.0	O 3.5	F 4.0
Na 0.9	Mg 1.2	Al 1.5	Si 1.8	P 2.1	S 2.5	Cl 3.0
K 0.8	Ca 1.0					Br 2.8
Rb 0.8	Sr 1.0					I 2.5

A scale of electronegativities has been developed by Linus Pauling. On this scale, fluorine, the most electronegative element, is assigned an electronegativity of 4.0. Table 1.3 shows the electronegativities of the elements with which we shall deal most often. On this scale, oxygen has an electronegativity of 3.5, carbon an electronegativity of 2.5, and hydrogen an electronegativity of 2.1.

The electronegativity of most metals is close to 1.0. The electronegativity of a nonmetal depends on its position in the periodic table, but is always greater than 1.0. Electronegativity is a periodic property of the elements: it increases going up a column and moving from left to right across the periodic table.

An ionic bond is formed if the difference in electronegativity between two atoms is greater than about 1.7. The more electronegative atom gains electrons and becomes a negatively charged anion. The less electronegative atom loses electrons and becomes a positively charged cation.

A covalent bond is generally formed between two atoms when the difference in electronegativity between them is less than about 1.7. The simplest example of the formation of a covlaent bond is the hydrogen molecule. When two hydrogen atoms combine, the single electrons from each combine to form an electron pair.

$$H\cdot + \cdot H \longrightarrow H{:}H \qquad \Delta H = -104 \text{ kcal/mole}$$

According to the Lewis model, a pair of electrons in a covalent bond functions in two ways simultaneously: it is shared by two atoms and it fills the outer shell of each. In other words, for the purposes of acquiring a noble gas electron configuration, we consider each atom to "own" completely all electrons it shares in covalent bonds with other atoms.

What is the reason for the stability of covalent bonds? In the Lewis model, an electron pair forming a covalent bond occupies the region between the two nuclei and serves to shield one positively charged nucleus from the repulsive force of the other nucleus. At the same time, the electron pair attracts both nuclei. In other words, putting an electron pair in the space between two nuclei

bonds them together and fixes the distance between atoms to within very narrow limits. We call this distance the bond length.

Although all covalent bonds involve sharing of electrons, they differ widely in the degree of sharing. For example, consider the covalent bond between carbon and sulfur. Because these atoms have identical electronegativities, the electrons in the C—S covalent bond are shared equally. Covalent bonds in which the sharing of electrons is equal or almost equal are called nonpolar covalent bonds.

A different situation arises when two atoms joined by a covalent bond have different electronegativies. For example, the difference in electronegativity between hydrogen and chlorine is 0.9. The H—Cl bond is covalent, but the sharing of electrons is not equal. Electrons of the bond are attracted to chlorine much more strongly than to hydrogen. Chlorine has a higher concentration of electrons around it and, hence, has a partial negative charge. The partial negative charge is indicated by the symbol δ^- (partial negative charge). Hydrogen has a partial positive charge, indicated by the symbol δ^+ (partial positive charge). Covalent bonds in which sharing of electrons is unequal are called polar covalent bonds.

Chemical bonds are classified as nonpolar covalent, polar covalent, or ionic according to the following guidelines:

- **nonpolar covalent bond**: The electronegativity difference between the atoms bonded together is between 0.0 and 0.4 units.
- **polar covalent bond**: The electronegativity difference between the atoms bonded together is between 0.5 and 1.6 units. In a polar covalent bond, the more electronegative atom has a partial negative charge; the less electronegative atom has a partial positive charge.
- **ionic bond**: The electronegativity difference between two atoms is 1.7 units or greater.

The distribution of electrons in a nonpolar covalent bond, a polar covalent bond, and ionic bond are illustrated in Figure 1.1.

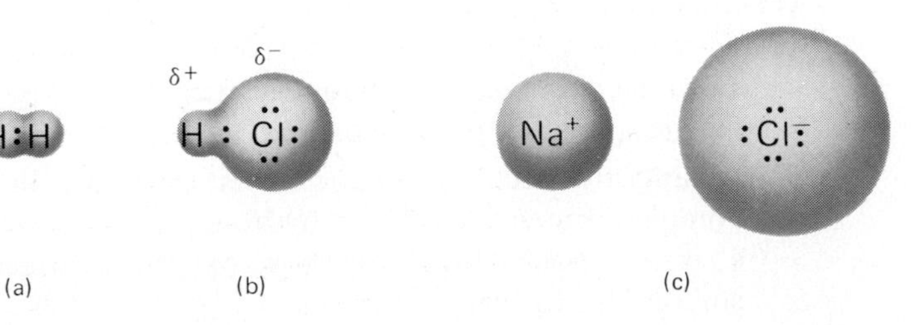

Figure 1.1 *Sharing of electrons in (a) nonpolar and (b) polar covalent bonds. (c) Transfer of electrons in an ionic bond.*

Example 1.1

Classify the bonds between the following pairs of atoms as nonpolar covalent, polar covalent, or ionic. For each polar covalent bond, show which atom bears the partial positive charge and which the partial negative charge.

(a) C—H **(b)** C—O **(c)** Na—O

Solution

(a) The electronegativity of carbon is 2.5 and that of hydrogen is 2.1. The difference is 0.4 unit, therefore, the C—H bond is nonpolar covalent.
(b) The electronegativity difference between carbon and oxygen is 1.0 unit (3.5 − 2.5), therefore, the C—O bond is polar covalent. Carbon is partially positive and oxygen is partially negative.

$$\overset{\delta+}{C}—\overset{\delta-}{O}$$

(c) The electronegativity difference between sodium and oxygen is 2.6 units (3.5 − 0.9), therefore, the Na—O bond is ionic.

PROBLEM 1.1 Classify the following bonds as nonpolar covalent, polar covalent, or ionic. For each polar covalent bond, show which atom bears the partial positive charge and which the partial negative charge.

(a) N—H **(b)** B—H **(c)** C—Mg

1.3 THE COVALENCE OF CARBON, HYDROGEN, NITROGEN, AND OXYGEN

Discussing the covalent bonding on carbon, hydrogen, nitrogen, and oxygen, we can focus our attention on just the electrons of the valence shell. Lewis structures for these four elements are shown in Table 1.2.

The single valence electron of hydrogen belongs to the first principal energy level. This shell is completely filled with two electrons. Thus, hydrogen can form only one covalent bond; hydrogen has a valence of one. Oxygen has six valence electrons, and in forming covalent bonds, acquires two more electrons; oxygen has a valence of two. By similar reasoning, nitrogen with five electrons in its outer shell has a valence of three and carbon with four electrons in its outer shell has a valence of four. The halogens (F, Cl, Br, I) have valences of one.

The following guidelines will help you write correct Lewis structures for covalent molecules and ions.

(1) Determine the number of valence electrons in the covalent molecule or ion. To do this, add the number of valence electrons contributed by each atom. For ions, add one electron for each negative charge on the ion; subtract one electron for each positive charge on the ion.

(2) Arrange the electrons in pairs so that each atom in the molecule or ion has a complete outer shell. Each hydrogen atom must be surrounded by two electrons. Each atom of carbon, oxygen, nitrogen, and halogen must be surrounded by eight electrons.

Table 1.4 Lewis structures for several small molecules.

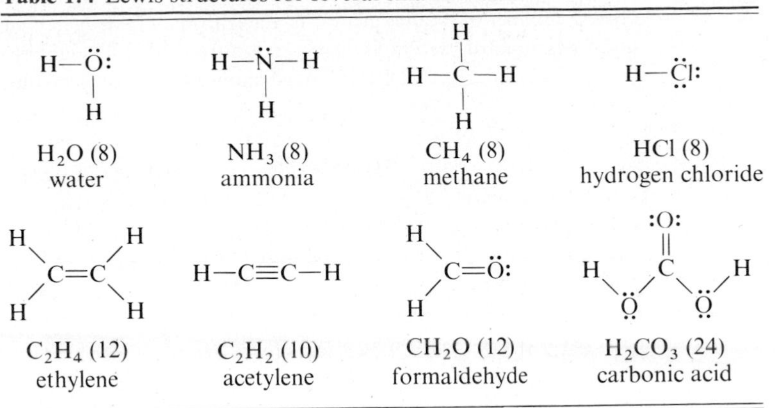

(3) A pair of electrons involved in a covalent bond (bonding electrons) is shown as a dash; an unshared pair of electrons is shown as a pair of dots.

(4) In a single bond, two atoms share one pair of electrons. In a double bond they share two pairs of electrons, and in a triple bond they share three pairs of electrons.

Shown in Table 1.4 are Lewis structures, molecular formulas, and names for several molecules. After the molecular formula of each is given the number of valence electrons it contains.

Notice that in these neutral, uncharged molecules, each hydrogen atom is surrounded by two valence electrons and each atom of carbon, nitrogen, and chlorine is surrounded by eight valence electrons. Furthermore, each carbon atom has four bonds; each nitrogen atom has three bonds and one unshared pair of electrons; each oxygen atom has two bonds and two unshared pairs of electrons. Each chlorine (and other halogens as well) has one bond and three unshared pairs of electrons.

Example 1.2

Draw Lewis structures, showing all valence electrons, for the following covalent molecules.

(a) CO_2 **(b)** CH_4O **(c)** CH_3Cl

Solution

(a) A Lewis structure for carbon dioxide must show 16 valence electrons; 12 from the two oxygens and 4 from carbon.

Ö=C=Ö (16 valence electrons)

With four shared and four unshared pairs of electrons, the Lewis structure shows the required 16 valence electrons. Furthermore, each atom of carbon and oxygen has a complete octet.

(b) A Lewis structure for CH_4O must show 14 valence electrons; 4 from carbon, 4 from the four hydrogens, and 6 from oxygen.

```
    H
    |     ..
H — C — O — H     (14 valence electrons)
    |     ..
    H
```

This structure shows five single bonds and two unshared pairs of electrons, and thus has the correct number of valence electrons. Each atom of hydrogen is surrounded by two electrons and each atom of carbon and oxygen is surrounded by eight electrons.

(c) A Lewis structure of CH_3Cl must show 14 valence electrons.

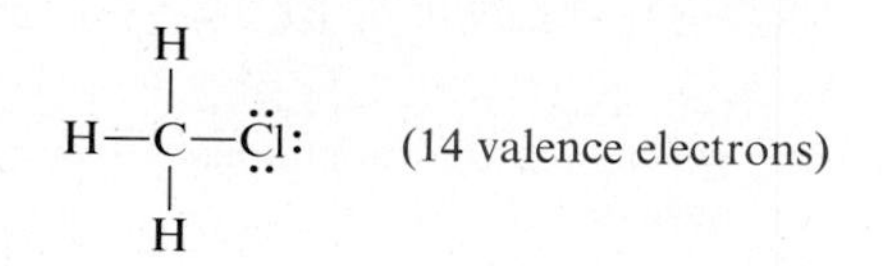

This structure shows four single bonds and three unshared pairs of electrons and thus has the correct number of valence electrons. Each atom has a complete valence shell.

PROBLEM 1.2 Draw Lewis structures, showing all valence electrons for the following covalent molecules.

(a) C_2H_6 **(b)** CS_2 **(c)** HCN

1.4 FORMAL CHARGE

Throughout this course we will deal not only with molecules but also with positively charged and negatively charged ions. Examples of positively charged ions are the hydronium ion, H_3O^+, and the ammonium ion, NH_4^+. An example of a negatively charged ion is the bicarbonate ion, HCO_3^-. It is important to be able to determine which atom (or atoms) in a polyatomic ion bears the positive or negative charge. The charge on an individual atom in an ion or molecule is called its formal charge. To derive formal positive or negative charges, first write a correct Lewis structure for the molecule or ion. Second, assign to each particular atom all of its unshared (nonbonding) electrons and half of its shared (bonding) electrons. Third, subtract this number from the number of valence electrons in the neutral, unbonded atom. The difference is the formal charge.

Example 1.3 Draw Lewis structures for the following ions and show which atom bears the formal charge.

(a) H_3O^+ **(b)** NH_4^+ **(c)** HCO_3^-

Solution **(a)** The Lewis structure of H_3O^+ shows 8 valence electrons; 3 from the three hydrogens, 6 from oxygen, minus 1 for the single positive charge. Hydrogens have no formal charge since each is assigned one valence electron, the same number as a hydrogen atom $(1 - 1 = 0)$. An oxygen atom has six valence electrons. The oxygen atom in H_3O^+ is assigned 5 electrons; two nonbonding electrons and one from each shared pair of electrons. Therefore, oxygen has the formal charge of $+1$ $(6 - 5 = 1)$.

assigned 5 valence electrons, formal charge +1

$$H-\ddot{O}^{+}-H \quad (\text{with a third H on O})$$

(b) A nitrogen atom has five valence electrons. In the ammonium ion, there are eight valence electrons. Nitrogen is assigned four valence electrons; one from each shared pair. Therefore, nitrogen has the formal charge of $+1$ $(5 - 4 = +1)$.

(c) In the bicarbonate ion, HCO_3^-, there are 24 valence electrons; 1 from hydrogen, 4 from carbon, 18 from the three oxygens, plus 1 additional electron for the single negative charge. Carbon is assigned four valence electrons and therefore has no formal charge (4 − 4 = 0). Two of the oxygens are assigned six valence electrons and have no formal charge (6 − 6 = 0). The third oxygen is assigned seven valence electrons and has a formal charge of −1 (6 − 7 = −1).

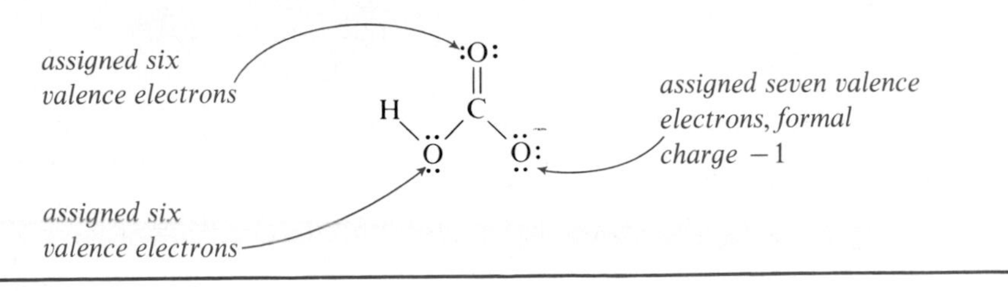

PROBLEM 1.3 Draw Lewis structures for the following ions and show which atoms bear the formal charges.

(a) CO_3^{2-} **(b)** CH_3^+ **(c)** CH_3^-

1.5 BOND ANGLES AND THE SHAPES OF MOLECULES

In the preceeding section, we used a shared pair of electrons as the fundamental unit of the covalent bond and we drew Lewis structures for several small molecules and ions containing various combinations of single, double, and triple bonds. Can we predict the shapes of these molecules? For example, if an atom is bonded to two other atoms, can we predict the angle they create about the central atom? In the water molecule, the central oxygen atom is bounded to two hydrogen atoms. Can we predict the H—O—H bond angle? In acetylene, H—C≡C—H, each carbon atom is bonded to two other atoms. Can we predict the H—C—C bond angle?

We can predict bond angles in these and other covalent molecules in a very straightforward way using the valence-shell electron-pair repulsion (VSEPR) model. According to the VSEPR model, an atom is surrounded by an outer shell of valence electrons. These valence electrons may be involved in the formation of single, double, or triple covalent bonds, or they may be unshared. Each of these combinations creates a negatively charged region of space. We know that like charges repel each other. Therefore, the various regions of electron density around an atom are spread out so that each is as far apart from the others as possible.

Let us use the valence-shell electron-pair repulsion model first to predict the shape of methane, CH_4. The Lewis structure of CH_4 shows a carbon atom surrounded by four separate regions of electron density. Each region of electron density consists of a pair of electrons forming a bond to a hydrogen atom. According to the VSEPR model, these regions of electron density radiate from carbon so that they are as far away from each other as possible. What is the H—C—H bond angle and what is the shape of the molecule?

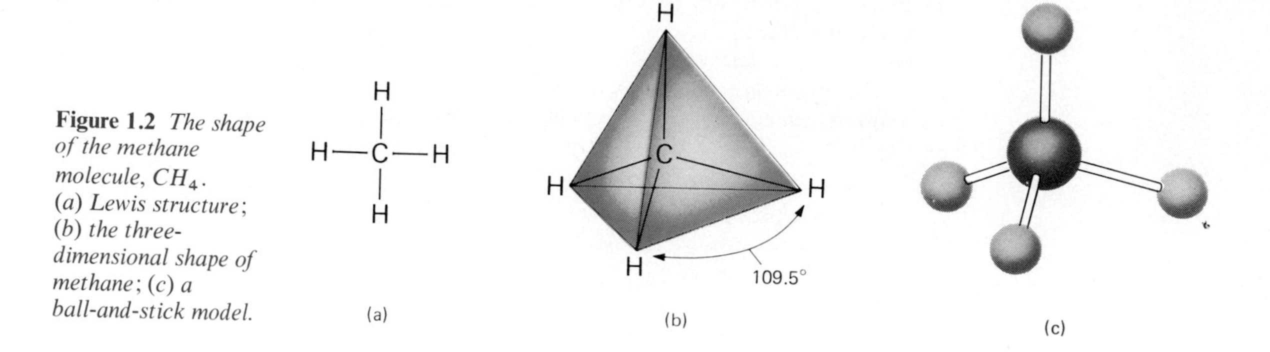

Figure 1.2 *The shape of the methane molecule, CH_4. (a) Lewis structure; (b) the three-dimensional shape of methane; (c) a ball-and-stick model.*

You can approximate the shape of CH_4 with a styrofoam ball and four toothpicks. Poke the toothpicks into the ball so that the free ends of the toothpicks are as far from one another as possible. If you have done this corectly, the angle between any two toothpicks is 109.5° and all angles are identical. If you cover this model with four triangular-shaped pieces of paper, you will have built a four-sided figure called a regular tetrahedron. Figure 1.2 shows a Lewis structure for methane, the tetrahedral arrangement of the four regions of electron density around carbon, and a ball-and-stick model.

According to the VSEPR model, the predicted H—C—H bond angle in methane is 109.5°. This angle has been measured experimentally and found to be 109.5°. Thus, the bond angle predicted by the VSEPR model is identical with that observed.

We can predict the shape of the ammonia molecule in exactly the same manner. The Lewis structure of NH_3 shows a central nitrogen atom surrounded by four regions of electron density. Three regions contain single pairs of electrons forming covalent bonds with hydrogen atoms. The fourth region contains an unshared pair of electrons. These four regions of electron density are arranged in a tetrahedral manner around the central nitrogen atom as shown in Figure 1.3.

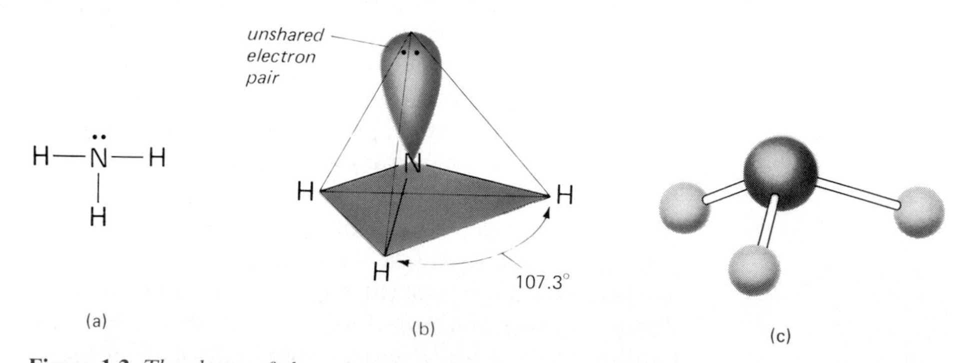

Figure 1.3 *The shape of the ammonia molecule, NH_3. (a) Lewis structure; (b) the three-dimensional shape of ammonia; (c) a ball-and-stick model (unshared pairs of electrons not shown).*

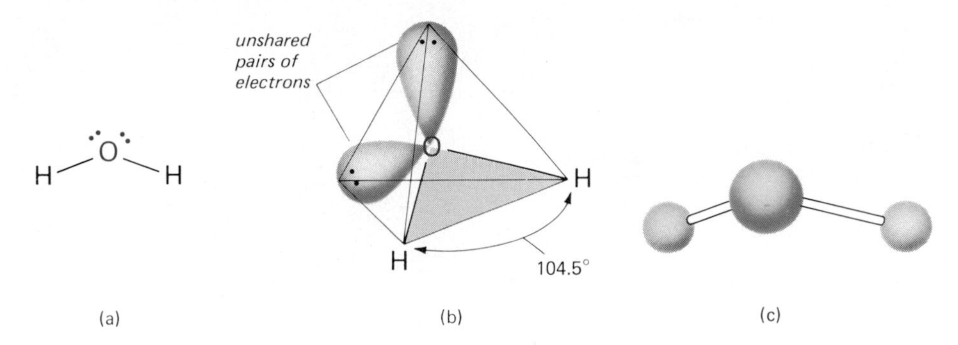

Figure 1.4 *The shape of the water molecule, H_2O. (a) Lewis structure; (b) the three-dimensional shape of water; (c) a ball-and-stick model (unshared pairs of electrons not shown).*

According to the VSEPR model, the four regions of electron density around nitrogen are arranged in a tetrahedral manner, therefore, we predict that each H—N—H bond angle should be 109.5°. The observed bond angle is 107.3°. This small difference between the predicted angle and the observed angle can be explained by proposing that the unshared pair of electrons on nitrogen repels adjacent electron pairs more strongly than do the bonding pairs of electrons.

Figure 1.4 shows a Lewis structure and a ball-and-stick model of the water molecule. In H_2O, oxygen is surrounded by four separate regions of electron density. Two of these regions contain pairs of electrons used to form covalent bonds with hydrogens; the other two contain unshared electron pairs.

The four regions of electron density are arranged in a tetrahedral manner around oxygen and, according to the VSEPR model, we would predict an H—O—H bond angle of 109.5°. Experimental measurements show that the actual bond angle is 104.5°, a value smaller than our prediction. This difference between the predicted and observed bond angle can be explained by proposing, as we did for NH_3, that unshared pairs of electrons repel adjacent pairs more strongly than do bonding pairs. Note that the distortion from 109.5° is greatest in H_2O which has two unshared pairs of electrons; it is smaller in NH_3 which has one unshared pair. There is no distortion in CH_4.

A general prediction emerges from our discussions of the shapes of CH_4, NH_3, and H_2O. Any time there are four separate regions of electron density around a central atom, we can predict a tetrahedral distribution of electron density and bond angles of approximately 109.5°.

In many of the molecules we will encounter, an atom is surrounded by three regions of electron density. Shown in Figure 1.5 are the Lewis structures of formaldehyde, CH_2O, and ethylene, C_2H_4.

According to the VSEPR model, a double bond is a single region of electron density. In formaldehyde, carbon is surrounded by three regions of electron density. Two regions contain single pairs of electrons forming single bonds to hydrogen atoms; the third contains a double pair of electrons forming the double bond to oxygen. In ethylene, each carbon atom is also surrounded by

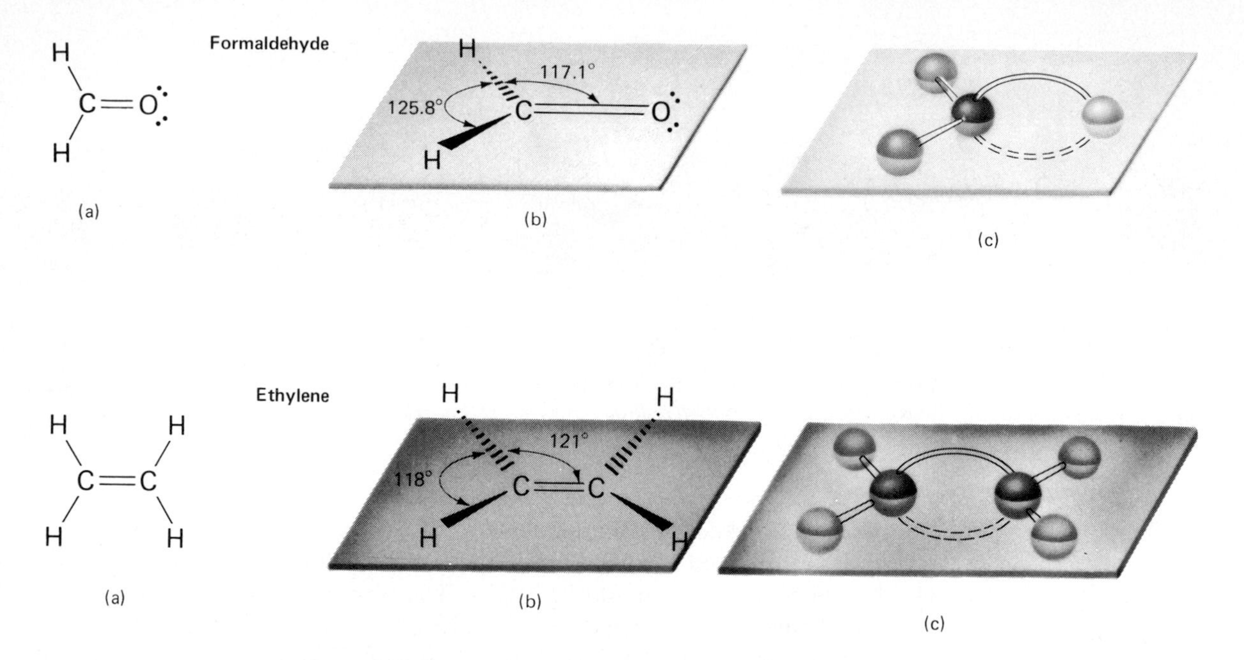

Figure 1.5 *Shapes of formaldehyde and ethylene molecules. (a) Lewis structures; (b) planar arrangements of three regions of electron density around carbon atoms; (c) ball-and-stick models. In these figures,* ◄ *represents a bond projecting in front of the plane of the paper and* ┉ *represents a bond projecting behind the plane of the paper.*

three regions of electron density: two contain single pairs of electrons and one contains a double pair of electrons. If you again experiment with a styrofoam ball and this time only three toothpicks, you will find that the free ends of the toothpicks are farthest apart if they are all in the same plane and make angles of 120° with each other. Thus, the predicted H—C—H and H—C—O bond angles in formaldehyde are 120°. The predicted H—C—H and H—C—C bond angles in ethylene are also 120°. We describe such arrangements of atoms as trigonal planar.

In still other types of molecules, a central atom is surrounded by only two regions of electron density. Shown in Figure 1.6 are Lewis structures and ball-and-stick models of carbon dioxide, CO_2, and acetylene, C_2H_2.

In carbon dioxide, carbon is surrounded by two regions of electron density; each contains a double pair of electrons and forms a double bond to an oxygen atom. In acetylene, each carbon is surround by two regions of electron density; one contains a single pair of electrons and forms a single bond to a hydrogen atom while the other contains a triple pair of electrons and forms a triple bond to a carbon atom. In each case, the two regions of electron density are farthest apart if they form a straight line through the central atom and a bond angle of 180°. Both carbon dioxide and acetylene are linear molecules.

Predictions of the valence-shell electron-pair repulsion model are summarized in Table 1.5.

:Ö=C=Ö: H—C≡C—H

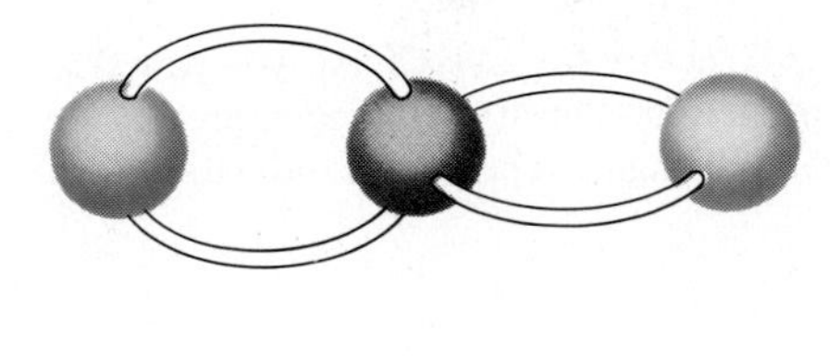

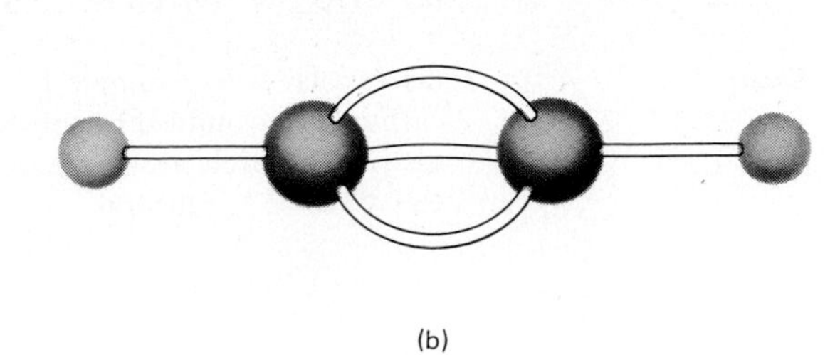

(a) (b)

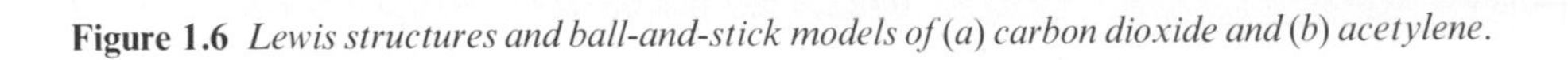

Figure 1.6 *Lewis structures and ball-and-stick models of (a) carbon dioxide and (b) acetylene.*

Table 1.5 Molecular shapes.

Number of regions of electron density around central atom	*Arrangement of regions of electron density in space*	*Predicted bond angle*	*Example*
4	tetrahedral	109.5°	H | H—C—H | H H—N̈—H | H Ö / \\ H H
3	trigonal planar	120°	H \\ C=Ö: / H H H \\ / C=C / \\ H H
2	linear	180°	:Ö=C=Ö: H—C≡C—H

Example 1.4 Predict all bond angles in the following molecules and ions.

(a) CH_3Cl **(b)** HCN **(c)** H_2CO_3 **(d)** HCO_3^-

Solution **(a)** In answer to Example 1.2, you drew a Lewis structure for CH_3Cl and you saw that carbon is surrounded by four separate regions of electron density. Therefore, predict the distribution of electron pairs to be tetrahedral, all bond angles to be 109.5°, and the shape of CH_3Cl to be tetrahedral.

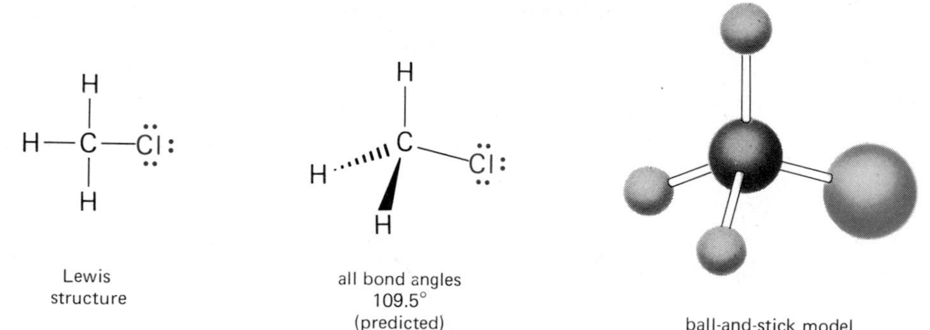

(b) In the Lewis structure of HCN, carbon is surrounded by two regions of electron density. Therefore, predict 180° for the H—C—N bond angle. The shape of HCN is linear.

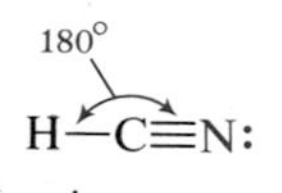

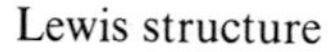
Lewis structure

(c) The Lewis structure of carbonic acid, H_2CO_3, is given in Table 1.4. Carbon is surrounded by three regions of electron density. Therefore, predict O—C—O bond angles of 120°. Each oxygen bonded to hydrogen is surrounded by four regions of electron density. Therefore predict H—O—C bond angles of 109.5°.

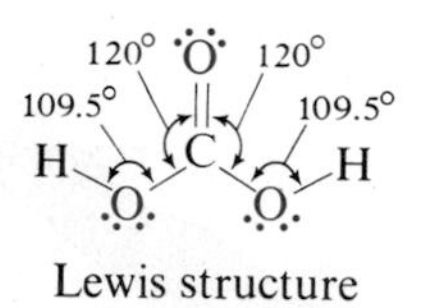

Lewis structure

(d) The bicarbonate ion, HCO_3^-, contains the same number of valence electrons as H_2CO_3. Therefore, the H—O—C and O—C—O bond angles are the same as those predicted for H_2CO_3.

PROBLEM 1.4 Predict all bond angles for the following molecules and ions.

(a) CH_4O **(b)** CH_5N **(c)** CH_2Cl_2 **(d)** NH_4^+

1.6 THE NEED FOR ANOTHER MODEL OF COVALENT BONDING

As much as the Lewis and valence-shell electron-pair repulsion models have helped us to formulate a clear picture of chemical bonding, they leave important questions unanswered. The most important of these is the relationship between molecular structure and chemical reactivity. For example, a carbon-carbon double bond is quite different in chemical reactivity from a carbon-carbon single bond. Most carbon-carbon single bonds are quite unreactive at room temperature, but carbon-carbon double bonds react with a variety of reagents under a variety of experimental conditions. The Lewis model gives us no way to account for these differences. Yet to discuss modern organic chemistry we must have a clear understanding of how the chemist accounts for them. Therefore, let us now approach the question of covalent bonding on a different and more sophisticated level—in terms of atomic orbitals and covalent bond formation by the overlap of atomic orbitals.

1.7 COVALENT BOND FORMATION BY THE OVERLAP OF ATOMIC ORBITALS

Modern bonding theory describes the formation of covalent bonds in terms of the overlap of atomic orbitals. The formation of a covalent bond between two atoms amounts to bringing the atoms up to each other in such a way that an atomic orbital of one atom overlaps with an atomic orbital of the other atom to form two new orbitals, called molecular orbitals, that encompass both nuclei. In the bonding molecular orbital, the electrons are concentrated in the region between the two nuclei and hold the nuclei together. In the other, the antibonding molecular orbital, the electrons are not concentrated between the two nuclei and no bonding results. In this text we shall be concerned only with bonding molecular orbitals. Like the atomic orbital, a molecular orbital can accommodate two electrons.

As an example, in forming the covalent bond in the hydrogen molecule, H_2, two hydrogen atoms approach each other so that their atomic 1*s* orbitals overlap (Figure 1.7). The molecular orbital resulting from the overlap of two 1*s* atomic orbitals is cylindrically symmetrical about the axis joining the two nuclei. Molecular orbitals that are cylindrically symmetrical about the bond axis are called sigma (σ) orbitals, and the bond is called a sigma bond.

Carbon, nitrogen, and oxygen form covalent bonds using atomic orbitals of the second principal energy level. Shapes and orientations in space of the four atomic orbitals of the second principal energy level are shown in Figure 1.8.

The three 2*p* orbitals are at angles of 90° to each other. If atoms of carbon, nitrogen, and oxygen were to use 2*p* orbitals for the formation of covalent bonds, we would expect to find bond angles of approximately 90° around each of these atoms. However, as we have seen in Section 1.5, bond angles of 90° are not observed. What we find instead are angles of approximately 109.5°, 120°, or 180°. How can we reconcile the difference between these observed bond

Figure 1.7 *Overlap of atomic 1s orbitals of two hydrogen atoms to form a molecular orbital. Each 1s atomic orbital contains one electron and the molecular orbital formed by their overlap contains two electrons.*

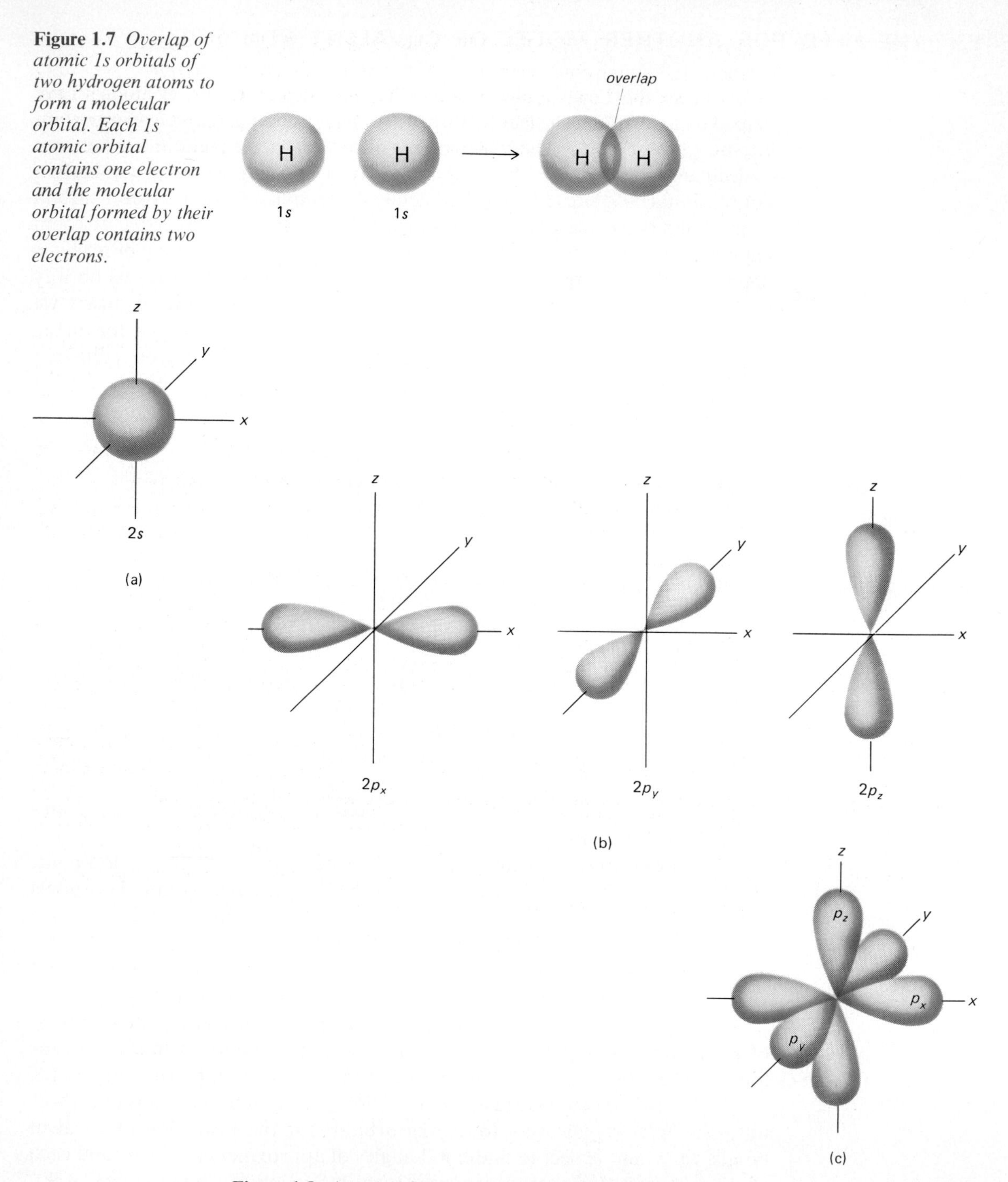

Figure 1.8 *Atomic orbitals of the second principal energy level; (a) shape of the 2s orbital; (b) shapes of the three 2p orbitals; (c) orientation in space of the three 2p orbitals relative to each other.*

angles and the angles predicted by using separate $2s$, $2p_x$, $2p_y$, and $2p_z$ orbitals for the formation of covalent bonds? Modern bonding theory tells us that atomic orbitals combine to form new atomic orbitals which then form bonds with the angles we do observe. The combination of atomic orbitals is called hybridization, and the new atomic orbitals formed are called hybrid atomic orbitals or, more simply, hybrid orbitals.

1.8 sp^3 HYBRID ORBITALS

Combination of one $2s$ orbital and three $2p$ orbitals produces four new orbitals called sp^3 hybrid orbitals (Figure 1.9). The four sp^3 orbitals are directed toward the corners of a regular tetrahedron. Thus, sp^3 hybridization always results in bond angles of approximately 109.5°.

Let us consider first the bonding in CH_4 in terms of the overlap of atomic orbitals. Carbon has four valence electrons. One electron is placed in each sp^3 orbital. Each partially filled sp^3 orbital forms a molecular orbital by overlap with a partially filled $1s$ orbital of hydrogen, and the hydrogen atoms occupy the four corners of a regular tetrahedron.

Second, let us consider the bonding in NH_3. Nitrogen has five valence electrons. One sp^3 orbital is filled with a pair of electrons while the other three sp^3 orbitals have one electron apiece. Overlapping these partially filled sp^3 orbitals with $1s$ orbitals of hydrogen gives the NH_3 molecule. Hydrogen atoms occupy three corners of a regular tetrahedron and the unshared pair of electrons occupies the fourth corner.

Finally, let us consider the bonding in H_2O. Oxygen has six valence electrons. Filling two sp^3 hybrid orbitals accounts for four of the six valence electrons, and placing one electron in each of the other two sp^3 orbitals accounts for the remaining valence electrons. Each partially filled sp^3 orbital forms a molecular orbital with a $1s$ orbital of hydrogen and hydrogen atoms occupy two corners of a regular tetrahedron. The remaining two corners of the tetrahedron are occupied by unshared pairs of electrons. Figure 1.10 shows orbital overlap and ball-and-stick models for methane, ammonia, and water.

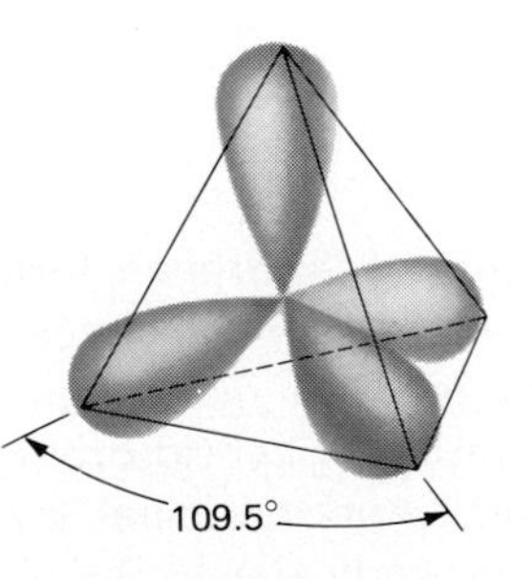

Figure 1.9 *sp^3 hybrid orbitals.*

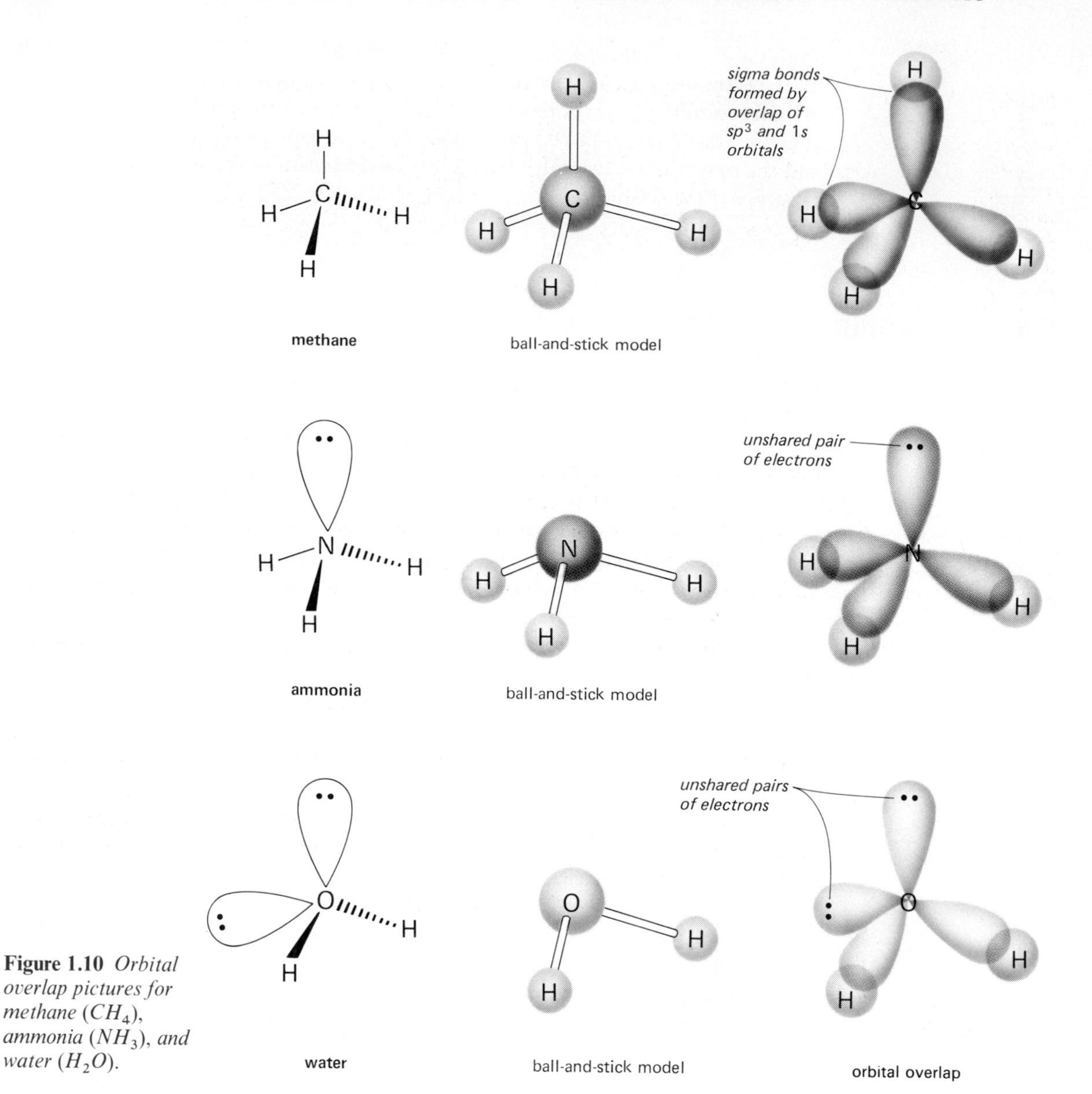

Figure 1.10 *Orbital overlap pictures for methane (CH_4), ammonia (NH_3), and water (H_2O).*

1.9 *sp*² HYBRID ORBITALS

Combination of one 2*s* orbital and two 2*p* orbitals forms three equivalent sp^2 hybrid orbitals. sp^2 Orbitals lie in a plane and are directed toward the corners of an equilateral triangle; the angle between sp^2 orbitals is 120°. This trigonal arrangement maximizes the separation of hybrid orbitals and minimizes their electrostatic interaction. The third 2*p* orbital is not involved in hybridization and consists of two lobes lying perpendicular to the plane of the sp^2 orbitals.

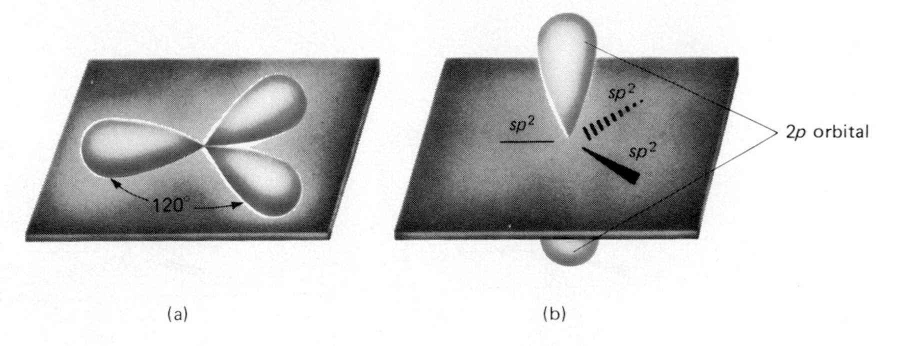

Figure 1.11 *An atom in the sp^2 hybridized state: (a) three sp^2 orbitals in a plane with 120° angles between them; (b) the remaining 2p orbital at a right angle to the plane of the sp^2 orbitals.*

Figure 1.11 shows the three equivalent sp^2 orbitals along with the remaining unhybridized $2p$ orbital.

sp^2 Hybrid orbitals are used to form double bonds. Consider ethylene, C_2H_4. The two carbon atoms form a sigma bond by the overlap of sp^2 orbitals (Figure 1.12). Each carbon also forms sigma bonds to two hydrogens by the overlap of sp^2 orbitals of carbon and $1s$ orbitals of hydrogen. The remaining $2p$ orbitals on adjacent carbon atoms overlap to form a pi bond. The pi bond consists of two sausage-shaped regions of electron density, one on either side of the plane formed by the carbon and hydrogen atoms. The pi bond joining the two carbon atoms is weaker than the sigma bond, because of the lesser degree of overlap of the $2p$ orbitals.

We can describe carbon-oxygen and carbon-nitrogen double bonds in the same manner as we have already described the carbon-carbon double bond. In formaldehyde, the simplest organic molecule containing a carbon-oxygen double bond, carbon forms sigma bonds to two hydrogens by overlap of sp^2 orbitals of carbon and $1s$ orbitals of hydrogen. Carbon and oxygen are joined by a sigma bond formed by the overlap of sp^2 orbitals and a pi bond formed by

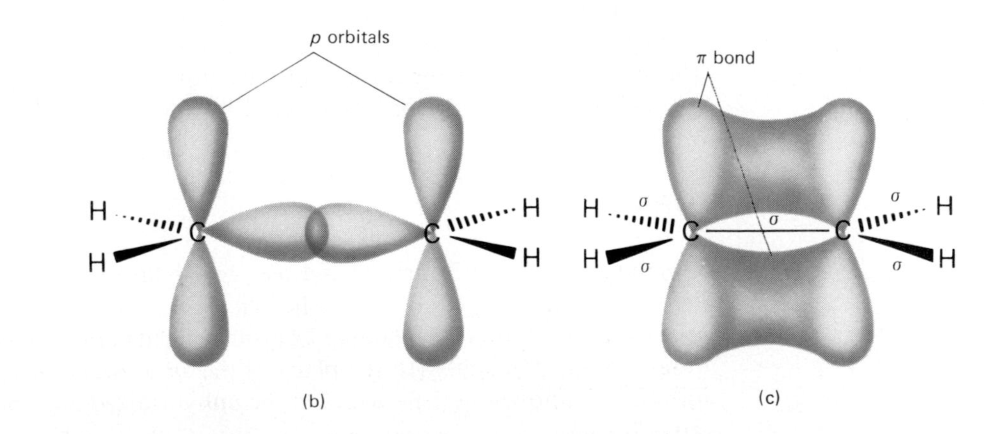

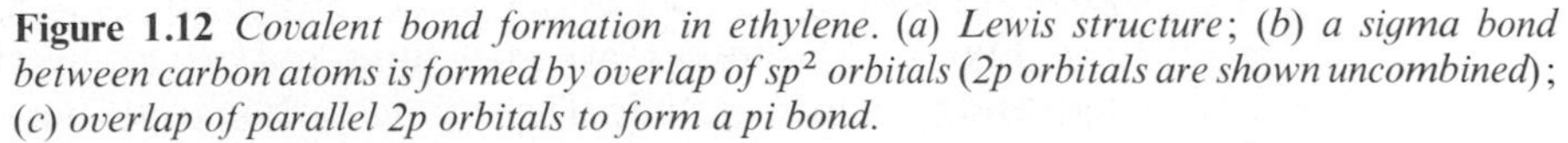
Figure 1.12 *Covalent bond formation in ethylene. (a) Lewis structure; (b) a sigma bond between carbon atoms is formed by overlap of sp^2 orbitals (2p orbitals are shown uncombined); (c) overlap of parallel 2p orbitals to form a pi bond.*

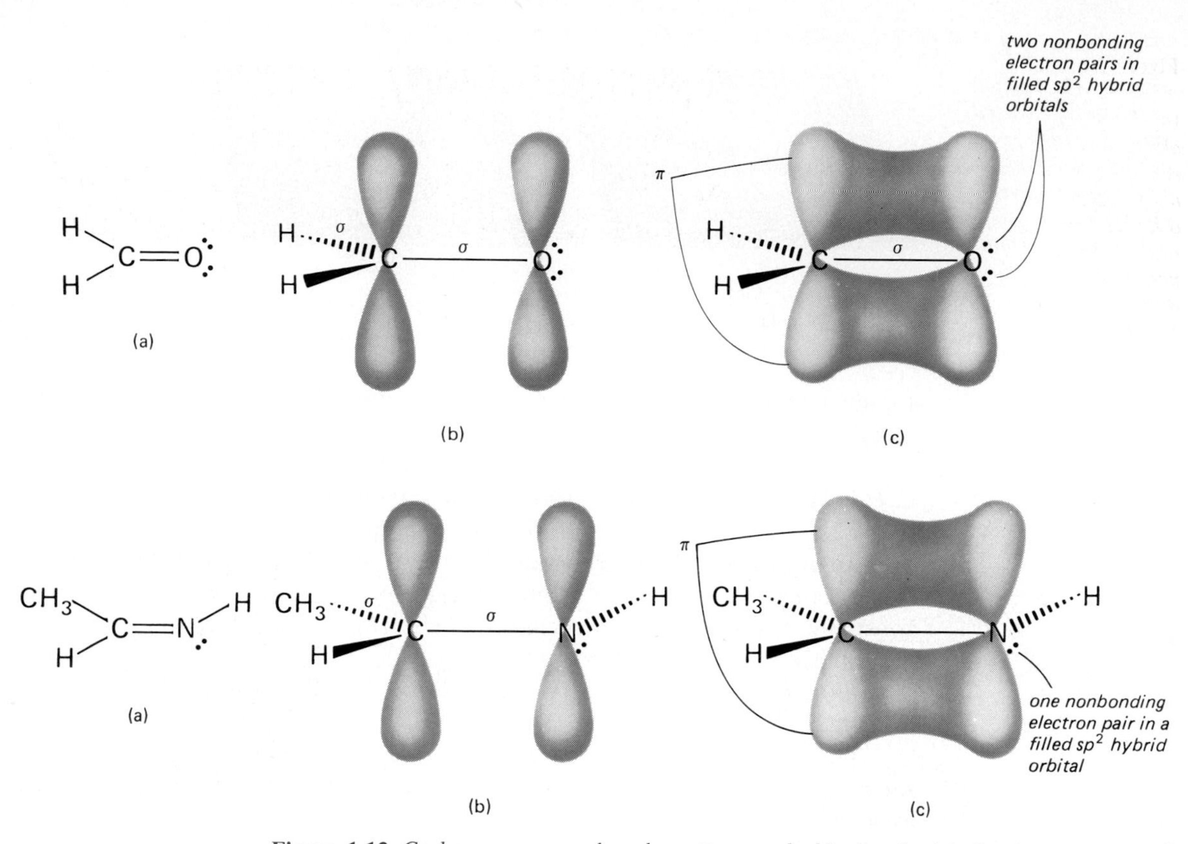

Figure 1.13 *Carbon-oxygen and carbon-nitrogen double bonds. (a) Lewis structures of formaldehyde (CH_2O) and ethylene imine (CH_3CHNH); (b) the sigma bond framework and nonoverlapping 2p orbitals; (c) overlap of parallel 2p orbitals to form a pi bond.*

the overlap of unhybridized 2*p* orbitals. Figure 1.13 shows the Lewis structure of formaldehyde, the sigma bond framework, and the overlap of 2*p* orbitals to form the pi bond. Similarly, the carbon-nitrogen double bond is a combination of one sigma bond and one pi bond (Figure 1.13).

1.10 *sp* HYBRID ORBITALS

Combination of one 2*s* orbital and one 2*p* orbital produces two equivalent *sp* hybrid orbitals. *sp* Hybrid orbitals lie in a plane at an angle of 180° with respect to the nucleus. The two remaining 2*p* orbitals lie in planes perpendicular to each other and perpendicular to the plane of the *sp* orbitals. In Figure 1.14, the *sp* orbitals are shown on the *x*-axis and the unhybridized 2*p* orbitals on the *y*-axis and the *z*-axis.

Figure 1.15 shows Lewis structures and orbital overlap pictures for acetylene and hydrogen cyanide. The carbon-carbon triple bond consists of one sigma

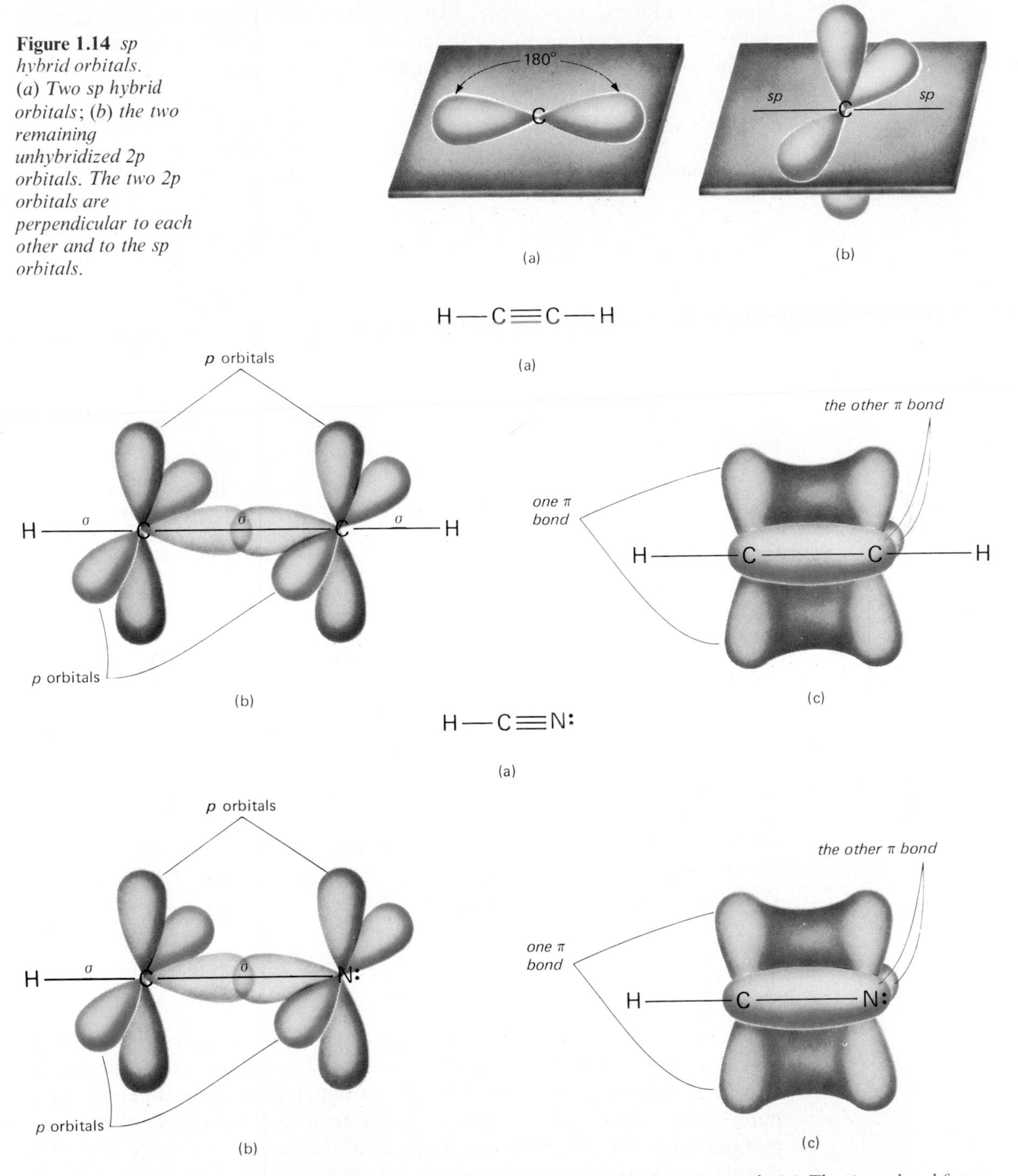

Figure 1.14 *sp hybrid orbitals. (a) Two sp hybrid orbitals; (b) the two remaining unhybridized 2p orbitals. The two 2p orbitals are perpendicular to each other and to the sp orbitals.*

Figure 1.15 *Covalent bonding in acetylene and hydrogen cyanide. (a) The sigma bond framework shown along with the 2p orbitals, and (b) formation of two pi bonds by the overlap of two sets of 2p orbitals.*

bond formed by the overlap of *sp* hybrid orbitals and two pi bonds. One pi bond is formed by the overlap of $2p_y$ orbitals and the second pi bond is formed by the overlap of $2p_z$ orbitals. The carbon-nitrogen triple bond also consists of one sigma bond and two pi bonds.

Example 1.5 Describe the bonding in the following molecules in terms of the orbitals involved and predict all bond angles.

```
        H                      H  :O:
        |                      |  ||
(a) H—C—Ö—H          (b) H—C—C—Ö—H
        |                      |
        H                      H
```

Solution The problem here is how to show clearly and concisely on a structural formula (1) the hybridization of each atom, (2) the orbitals involved in each covalent bond, and (3) all bond angles. This can be done easily in three separate diagrams. Labels on the first diagram point to atoms and show the hybridization of each atom. Labels on the second diagram point to bonds and show the type of bond, either sigma or pi. Labels on the third diagram point to atoms and show predicted bond angles about each atom.

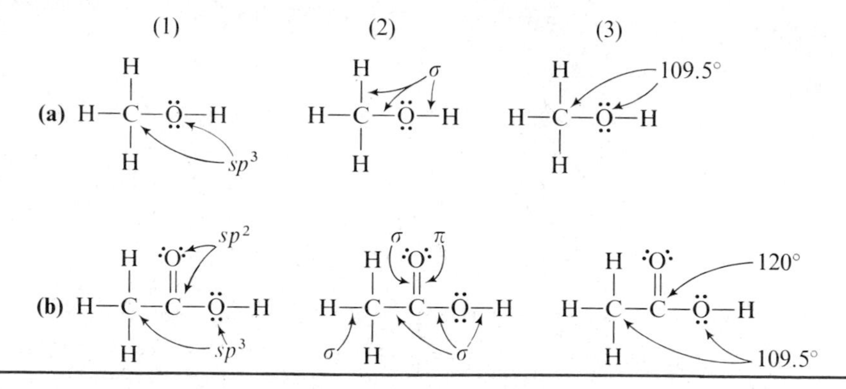

PROBLEM 1.5 Describe the bonding in the following molecules in terms of the orbitals involved and predict all bond angles.

```
        H     H              H                      H
        |     |              |                      |
(a) H—C—Ö—C—H      (b) H—C—C=C—H        (c) H—C—N̈—H
        |     |              |  |  |                |  |
        H     H              H  H  H                H  H
```

If you compare the valence-shell electron-pair repulsion model and orbital overlap model you will see that each gives good predictions of bond angles. However, the orbital overlap model gives a more useful understanding of the nature of double and triple bonds. First, the arrangement in space of electrons in a sigma bond is quite different from that of electrons in a pi bond. Second, a double bond is not just a combination of two identical bonds. Rather, the double bond is a combination of one sigma bond and one pi bond. Similarly, a triple bond is a combination of one sigma bond and two pi bonds. We will use the orbital overlap picture of covalent bonding in later chapters to help us understand why compounds containing double and triple bonds have quite different chemical properties from compounds containing only single bonds.

1.11 COVALENT BONDING—AN INTERIM SUMMARY

Following are the Lewis structures for several molecules we have discussed.

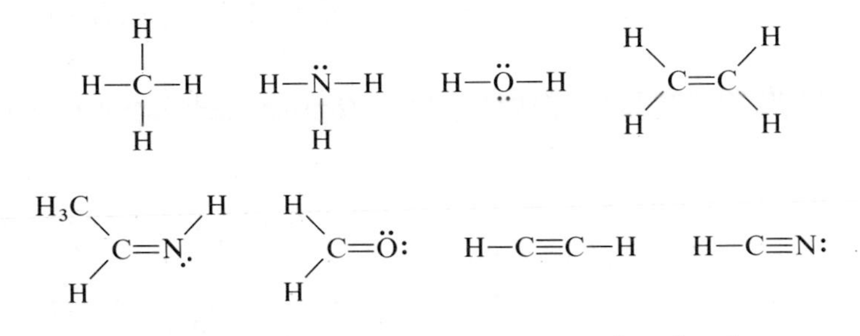

By examining the structures of these stable uncharged molecules, we can make certain generalizations:

(1) Carbon forms four covalent bonds. Bond angles on carbon are approximately 109.5° for four attached groups (sp^3), 120° for three attached groups (sp^2), and 180° for two attached groups (*sp*).

(2) Nitrogen forms three covalent bonds and has one unshared pair of electrons. Bond angles on nitrogen are approximately 109.5° for three attached groups (sp^3) and 120° for two attached groups (sp^2). We cannot specify a bond angle for an *sp* hybridized nitrogen since it requires three nuclei to create a bond angle. The unshared pair of electrons on nitrogen may lie in an sp^3 hybrid oribtal (three attached groups), in an sp^2 hybrid orbital (two attached groups), or in an *sp* hybrid orbital (one attached group).

(3) Oxygen forms two covalent bonds and has two unshared pairs of electrons. Bond angles about oxygen are approximately 109.5° for two attached groups (sp^3). The two unshared pairs of electrons on oxygen may lie in sp^3 hybrid orbitals (two attached groups) or in sp^2 hybrid orbitals (one attached group).

1.12 FUNCTIONAL GROUPS

In the preceding sections we examined different types of covalent bonds formed by carbon with hydrogen, nitrogen, and oxygen. These bonds combine in various ways to form certain characteristic structural features known as functional groups. The concept of the functional group is important to organic chemistry for two reasons. First, functional groups serve as a basis for naming of organic compounds. Second, and perhaps most important, they are the sites of chemical reaction. A particular functional group has very similar chemical properties whenever it is found in an organic molecule.

One way to begin the study of functional groups is to start with a water molecule, and then to replace first one and then both hydrogens by atoms of carbon. If one hydrogen is replaced by a carbon the substance formed is an alcohol. If both hydrogens are replaced by carbons, the substance formed is an ether.

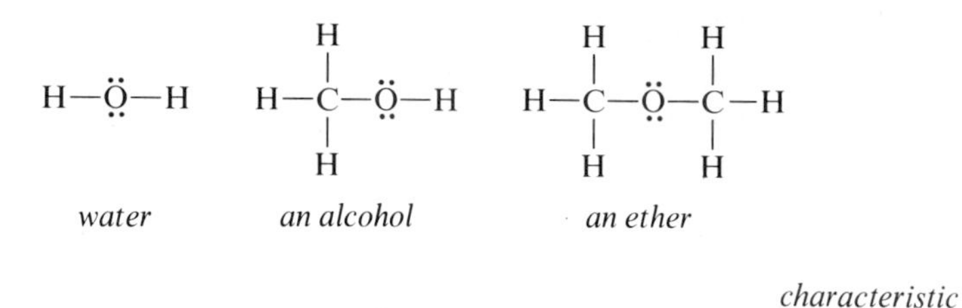

C—Ö—H   C—Ö—C   *characteristic structural feature*

The characteristic structural feature of an alcohol is an oxygen atom bonded to atoms of carbon and hydrogen. We say that an alcohol contains a hydroxyl, —OH, group. The characteristic structural feature of an ether is an atom of oxygen bonded to two carbon atoms, C—Ö—C. We can write formulas for the alcohol and the ether above in a more abbreviated form by using what are called condensed structural formulas. In a condensed structural formula H—C(H)(H)— is written CH_3— and indicates a carbon with three attached hydrogens. In the same manner —CH_2— indicates a carbon with two attached hydrogens and —CH(|)— indicates a carbon with one attached hydrogen. Following are Lewis structures and condensed structural formulas for the alcohol and ether of molecular formula C_2H_6O.

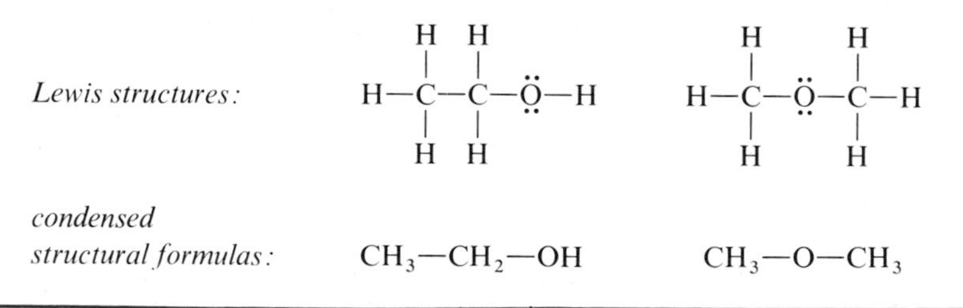

Example 1.6 There are two alcohols of molecular formula C_3H_8O. Draw Lewis structures and condensed structural formulas for each.

Solution The characteristic structural feature for an alcohol is an atom of oxygen bonded to one carbon and one hydrogen atom.

C—Ö—H

The molecular formula contains three carbon atoms. These can be bonded together in a chain with the —OH group attached to the first carbon of the chain or attached to the second carbon of the chain.

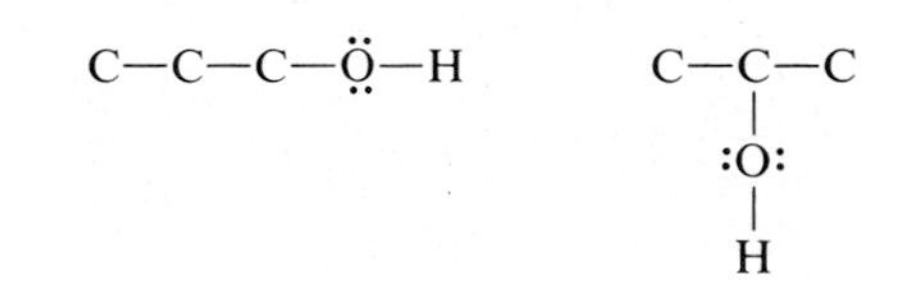

Finally, add seven hydrogens to satisfy the tetravalence of carbon and give the correct molecular formula.

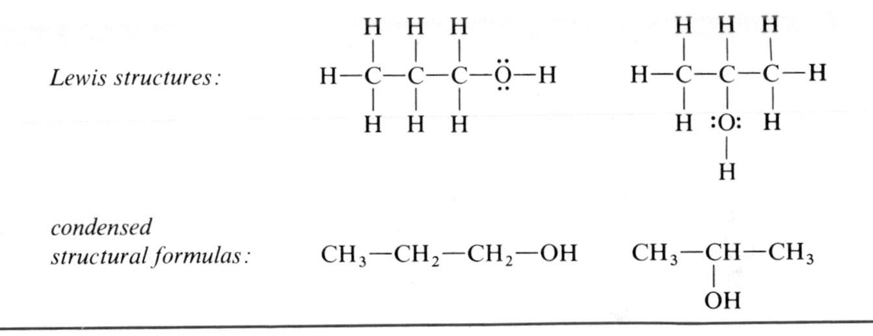

PROBLEM 1.6 There is one ether of molecular formula C_3H_8O. Draw a Lewis structure and a condensed structural formula for this substance.

Several times in this chapter, we have discussed the structure of formaldehyde. Formaldehyde belongs to a class of organic compounds known as aldehydes. The characteristic structural feature of an aldehyde is the presence of a $\overset{:O:}{\overset{\|}{-C-H}}$ functional group. Formaldehyde is the simplest molecule containing an aldehyde functional group, and it is the only aldehyde that contains two hydrogen atoms bonded to the C=O group. All other aldehydes have one carbon and one hydrogen bonded to the C=O. The characteristic structural feature of a ketone is the presence of a $\overset{:O:}{\overset{\|}{-C-}}$ functional group.

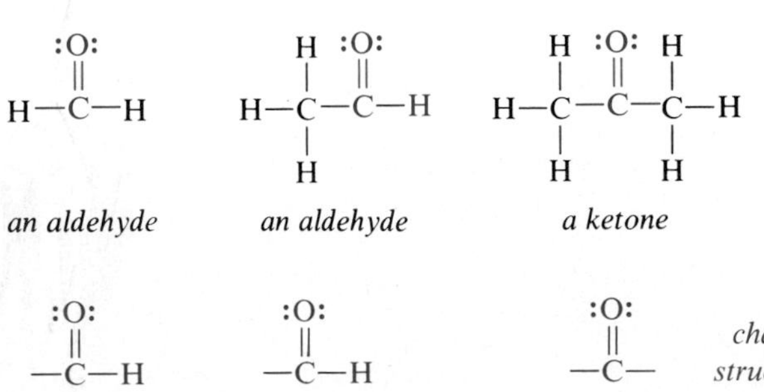

an aldehyde *an aldehyde* *a ketone*

characteristic structural feature

Example 1.7 Draw Lewis structures and condensed structural formulas for the two aldehydes of molecular formula C_4H_8O.

Solution The characteristic structural feature of an aldehyde is —C(=:O:)—H. First draw the characteristic structural feature of the aldehyde group and then add the remaining carbons. These may be attached in two different ways.

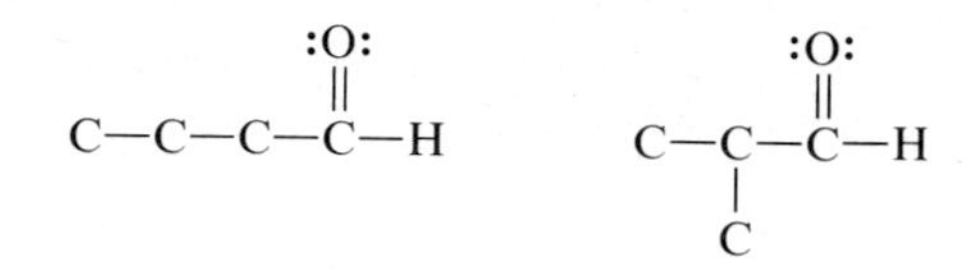

Finally, add seven hydrogens to complete the tetravalence of carbon and give the correct molecular formula.

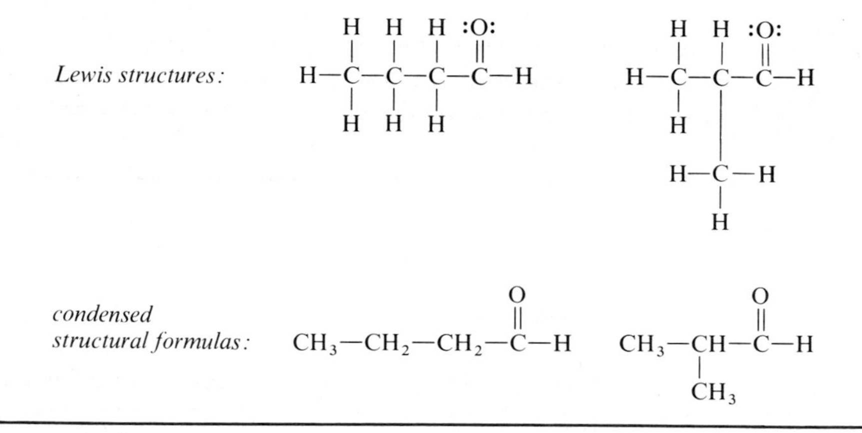

PROBLEM 1.7 Draw Lewis structures and condensed structural formulas for the three ketones of molecular formula $C_5H_{10}O$.

The characteristic structural feature of a carboxylic acid is the presence of a >C=O (a carbonyl group) bonded to an —OH (an hydroxyl) group—

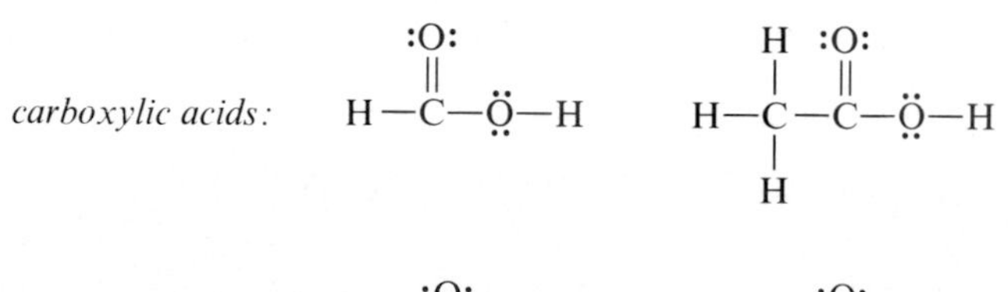

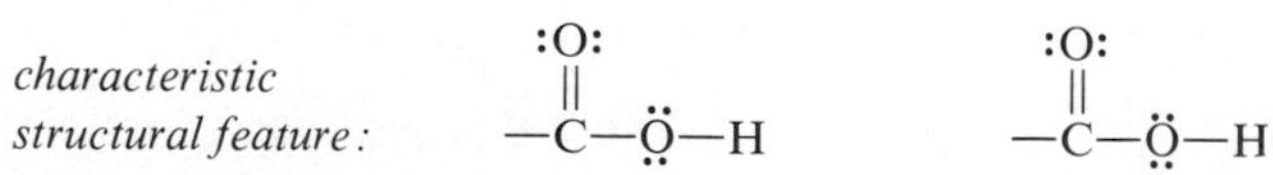

Example 1.8

Draw the Lewis structure and condensed structural formula for the single carboxylic acid of molecular formula $C_3H_6O_2$.

Solution

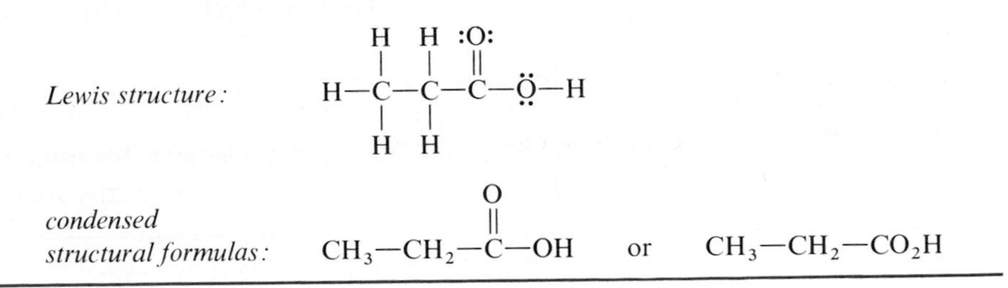

PROBLEM 1.8 Draw Lewis structures and condensed structural formulas for the two carboxylic acids of molecular formula $C_4H_8O_2$.

After studying this section, you should be able to do two things: (1) Given a Lewis structure or condensed structural formula, you should be able to recognize the few functional groups we have discussed so far; and (2) given a molecular formula and the characteristic structural feature of a particular functional group, you should be able to draw a Lewis structure and condensed structural formula for a molecule containing that functional group.

A list of the major functional groups discussed in this course is presented on the inside front cover of the text.

1.13 STRUCTURAL ISOMERISM

In Section 1.12 you saw that for most molecular formulas it is possible to draw more than one structural formula. For example, the following alcohol and the ether have the same molecular formula, C_2H_6O, but different structural formulas and different functional groups.

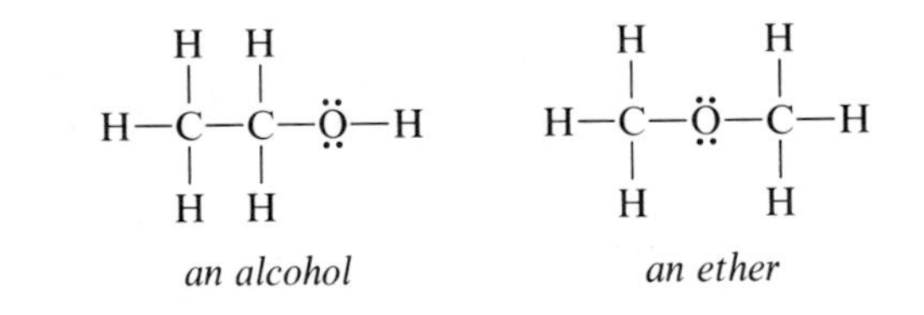

an alcohol *an ether*

Similarly, the following aldehyde and the ketone have the same molecular formula, C_3H_6O, but different structural formulas and different functional groups.

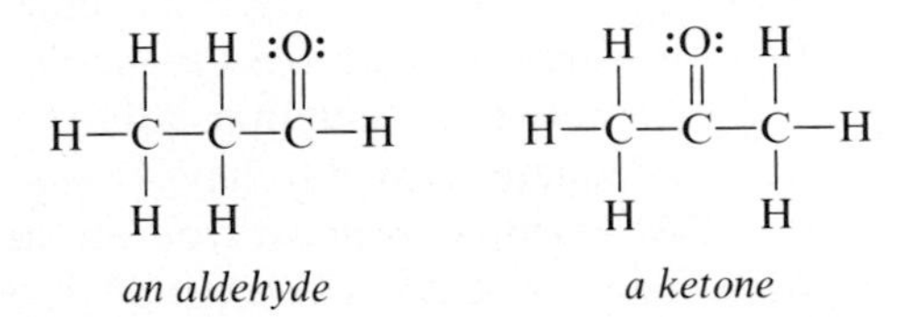

an aldehyde *a ketone*

The following alcohols have the same molecular formulas and the same functional group, but different structural formulas.

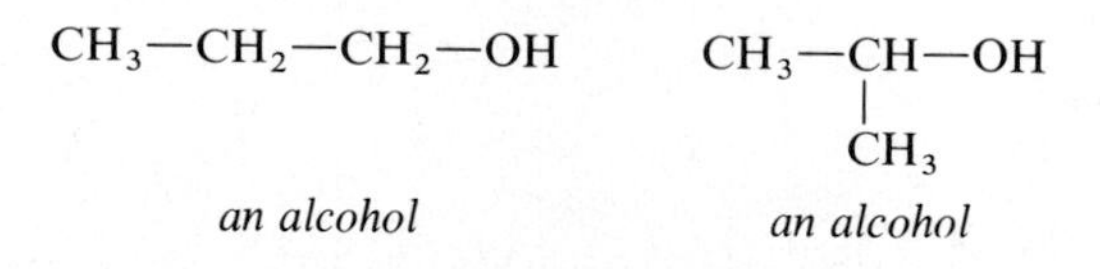

an alcohol *an alcohol*

Substances that have the same molecular formula but different structural formulas (different orders of attachment of atoms) are called structural isomers.

Example 1.9 Divide the following into groups of structural isomers.

(a) $CH_3—CH_2—C(=O)—O—H$ (b) $CH_3—O—CH_2—C(=O)—H$

(c) $CH_2=CH—CH_2—O—H$ (d) $CH_3—CH(CH_3)—C(=O)—H$

(e) $CH_3—C(=O)—O—CH_2—CH_2—CH_3$ (f) $CH_3—C(=O)—CH_2—CH_2—CH_2—OH$

Solution To determine which are structural isomers, first write the molecular formula of each substance and then compare them. All those that have the same molecular formula and different structural formulas are structural isomers. **(a)** and **(b)** have the molecular formula $C_3H_6O_2$ and are structural isomers. **(e)** and **(f)** have the molecular formula $C_5H_{10}O_2$ and are structural isomers. There are no structural isomers in this problem for **(c)** and **(d)**.

PROBLEM 1.9 Divide the following into groups of structural isomers.

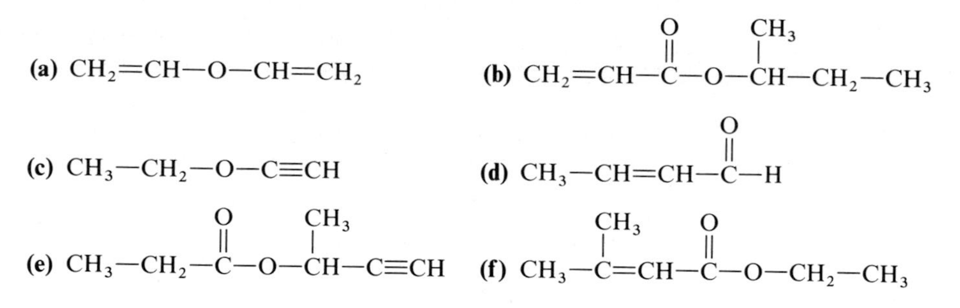

1.14 RESONANCE

Slowly, as the study of organic chemistry began to unfold, it became obvious that for a great many molecules and ions no single Lewis structure provided a truly accurate representation. All of the molecules for which structural theory was inadequate seemed to have two or more functional groups close together.

For example, suppose you are asked to draw a Lewis structure for the carbonate ion, CO_3^{2-}. You would probably begin by writing the carbon and

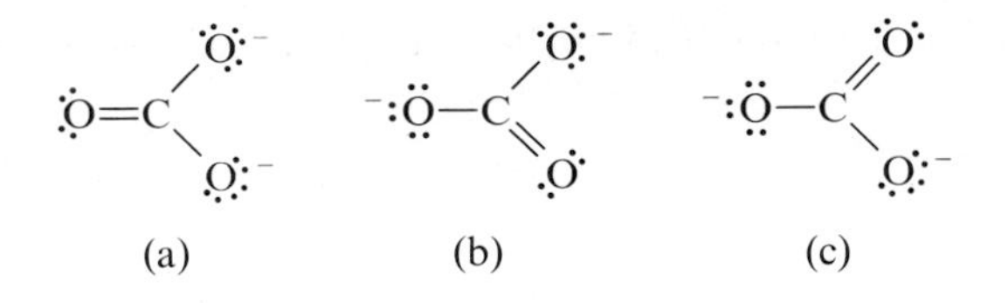

Figure 1.16 *Three Lewis structures for the carbonate ion.*

three oxygen atoms attached in the correct order, then determining the proper number of valence electrons, and finally building the ion by adding valence electrons to arrive at an acceptable Lewis structure. As you do this surely you would realize that any one of the three carbon-oxygen bonds could be written as a double bond with the other two carbon-oxygen bonds written as single bonds. These three possible Lewis structures for the carbonate ion are shown in Figure 1.16. Each structure implies that one carbon-oxygen bond is different from the other two. However, this is not so since it has been shown that all three bonds are identical.

To describe molecules and ions for which no single Lewis structure was adequate, it became necessary to refine still further the models for covalent bonding. Chemists use two such approaches. One is the molecular orbital approach, the second is the resonance or valence-bond approach. While most chemists will admit that the molecular orbital approach is more sophisticated and more useful for calculating molecular properties, the valence-bond approach is pictorially more useful for the organic chemists interested in understanding reactions of organic compounds. Therefore, it is this approach that we will use throughout the text.

The resonance or valence-bond method was developed by Linus Pauling. According to this method, many molecules and ions are best described by writing two or more valence-bond structures and considering the real molecule or ion as a hybrid of these structures. These valence-bond structures are known as contributing or resonance structures. We consider the real substance as a hybrid of the written structures by interconnecting them with double-headed arrows. Using this description, we would represent the resonance hybrid for the carbonate ion as shown in Figure 1.17.

It is important to remember that the carbonate ion has one and only one real structure. The problem is how to describe the real structure. The resonance method is an attempt to describe the real structure and at the same time retain the classical electron structures with electron pair bonds. Thus, while we fully realize that the carbonate ion is not accurately represented by contributing structure (a) or (b) or (c), we will continue to represent it as one of these contributing structures for convenience. Of course we will understand that what is intended is the resonance hybrid.

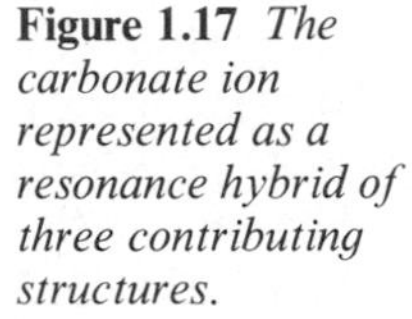

Figure 1.17 *The carbonate ion represented as a resonance hybrid of three contributing structures.*

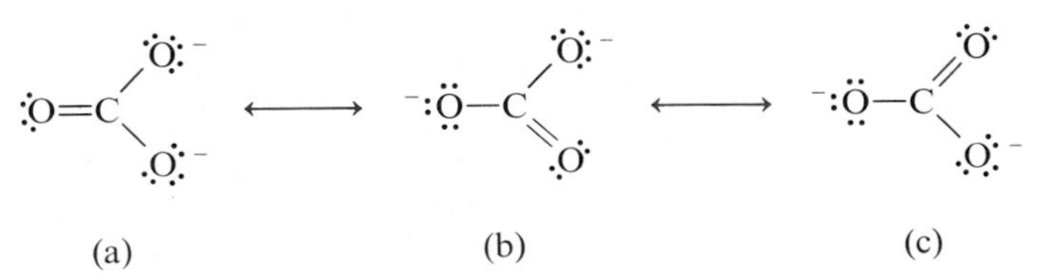

Following are resonance hybrids for the nitrite ion (NO_2^-) and the acetate ion ($CH_3CO_2^-$).

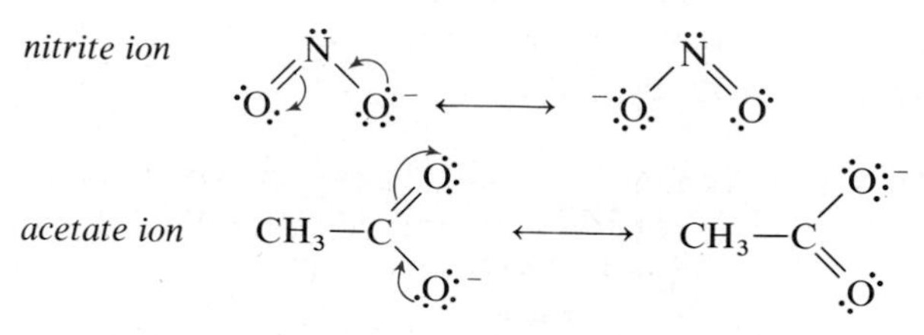

In these structures, curved arrows are used to show how one contributing structure is converted into another. A curved arrow indicates the movement of an electron pair, with the pair moving from the position indicated by the tail of the arrow to the position indicated by the head of the arrow. Following are contributing structures for the resonance hybrid of acetone. Note that in the second contributing structure, there is a formal positive charge on carbon and a formal negative charge on oxygen.

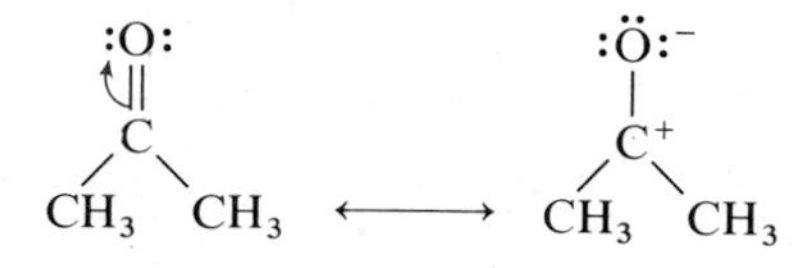

There are certain rules that must be followed in drawing acceptable contributing structures. First, all must show the correct number of valence electrons. Second, none may show more than 8 electrons in the valence shells of carbon, nitrogen, or oxygen. Third, the position of all nuclei must be the same in all contributing structures. This means that contributing structures differ only in the distribution of valence electrons within a molecule or ion. Fourth, all formal charges must be shown.

Example 1.10 Draw the contributing structure indicated by the curved arrow. Be certain to show all formal charges as appropriate.

Solution

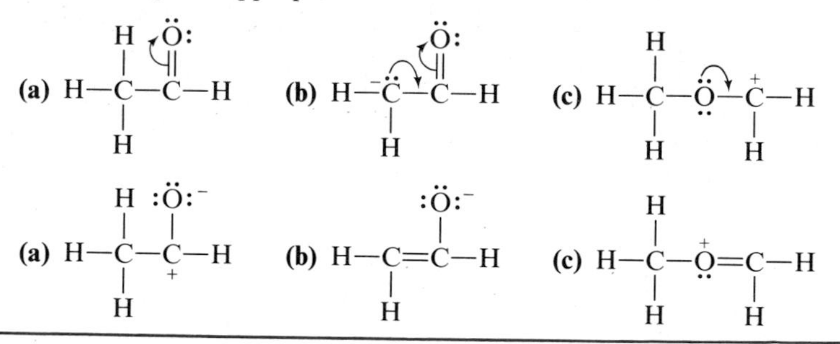

PROBLEM 1.10 Draw the contributing structure indicated by the curved arrow. Be certain to show all formal charges.

(a) H—C(=O:)—Ö:⁻ ⟷ (b) H—C(=O:)—Ö:⁻ ⟷ (c) H—CH₂—C(=O:)—NH—H ⟷

In this section we have seen that certain molecules and ions are best represented as resonance hybrids. The next question we must ask is how can we predict when resonance is or is not important. A few guidelines will help in making such predictions:

(1) Resonance is more important when there are two or more equivalent contributing structures as, for example, in the carbonate ion, the acetate ion, and the nitrite ion.

(2) Contributing structures that involve the creation of unlike charge are less important than those that do not involve the separation of unlike charge. For example, of the two contributing structures drawn for acetone, the second (separation of unlike charge) is less important in the hybrid than the first (no separation of unlike charge).

1.15 ORGANIC CHEMISTRY—THE UNIQUENESS OF CARBON

Why establish organic chemistry as a separate branch of chemistry? Or, what is so unique about the chemistry of carbon? The answer is due in part to the tendency of carbon atoms to bond together in long chains, the possibilities for structural and functional group isomerism, and the exceptionally large number of known organic compounds. But at a more fundamental level, the answer rests on the special stability of compounds containing carbon-carbon and carbon-hydrogen single bonds. C—C and C—H single bonds show little tendency to participate in chemical reactions. C—O and C—N single bonds, on the other hand, undergo a variety of chemical reactions. Why the marked difference in chemical reactivity between the two types of bonds? The explanation lies in the presence of unshared pairs of electrons on nitrogen and oxygen atoms. The unshared pairs of electrons on oxygen and nitrogen make these atoms susceptible to attack by electron deficient reagents such as H^+. However, when carbon is bonded by single bonds to four other carbon or hydrogen atoms, it has no unshared pairs of electrons and is not susceptible to attack by electron deficient reagents. Furthermore, a tetravalent carbon has a complete outer shell of electrons and is not susceptible to attack by electron rich reagents.

Herein lies the reason for the uniqueness of carbon compounds and of organic chemistry: the particular strength of C—C and C—H single bonds and their resistance to attack by either electron rich or electron deficient reagents.

PROBLEMS

1.11 Following is the electron configuration of O^{2-}.

$$O^{2-} \qquad 1s^2 2s^2 2p^6$$

Using the same notation, also write the electron configurations for F^-, Ne, Na^+, and Mg^{2+}. Compare these five electron configurations.

1.12 Following are names and structural formulas for several organo-metallic compounds (compounds in which there is a metal-to-carbon bond). Classify each carbon-metal bond as nonpolar covalent, polar covalent, or ionic. For each polar covalent bond, show which atom bears the partial positive charge and which the partial negative charge. The electronegativity for lead is 1.8; that for mercury is 1.9.

(a)
$CH_2—CH_3$
|
$CH_3—CH_2—Pb—CH_2—CH_3$
|
$CH_2—CH_3$
tetraethyl lead

(b) $CH_3—Hg—CH_3$
dimethyl mercury

(c)
$CH_2—CH$
|
$CH_3—CH_2—B—CH_2—CH_3$
triethyl borane

(d) $CH_3—Mg—Cl$
methylmagnesium chloride

(e) $CH_3—CH_2—CH_2—CH_2—Li$
butyl lithium

1.13 Write Lewis structures for the following molecules. Be certain to show all valence electrons.

(a) H_2O_2 **(b)** N_2H_4 **(c)** CH_3OH
(d) CH_3SH **(e)** CH_3NH_2 **(f)** CH_3Cl
(g) CH_3OCH_3 **(h)** C_2H_6 **(i)** C_2H_4
(j) C_2H_2 **(k)** CO_2 **(l)** H_2CO_3
(m) CH_2O **(n)** CH_3CHO **(o)** CH_3COCH_3
(p) HCO_2H **(q)** CH_3CO_2H **(r)** CH_3NH_2
(s) CH_3NNCH_3 **(t)** HCN **(u)** HNO_2
(v) NH_2OH **(w)** HNO_3 **(x)** H_2SO_4

1.14 Write a Lewis structure for each of the following ions. Be certain to show all valence electrons. Assign formal charges as appropriate.

(a) OH^- **(b)** H_3O^+ **(c)** NH_4^+ **(d)** NH_2^-
(e) HCO_3^- **(f)** CO_3^{2-} **(g)** Cl^- **(h)** Cl^+
(i) NO_2^- **(j)** NO_3^- **(k)** HCO_2^- **(l)** SO_4^{2-}

1.15 Following are Lewis structures showing all valence electrons. Assign formal charges to each structure as appropriate.

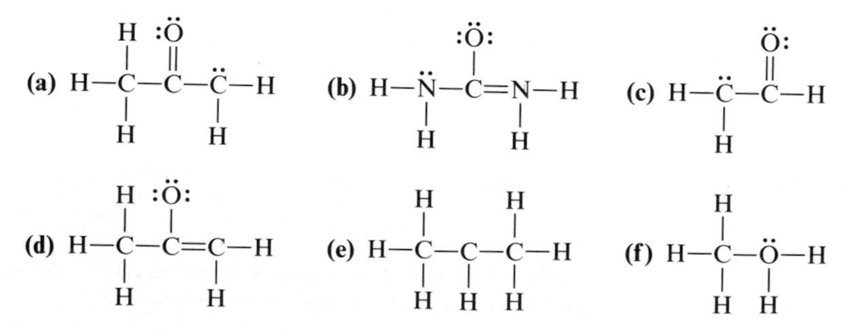

1.16 The following substances contain both ionic and covalent bonds. Draw a Lewis structure for each.

(a) $NaNH_2$ **(b)** CH_3ONa **(c)** $NaHCO_3$
(d) Na_2CO_3 **(e)** NH_4Cl **(f)** CH_3NH_3Cl

1.17 Following are Lewis structures for several molecules and ions. Using the valence-shell electron-pair repulsion model, predict bond angles about each circled atom.

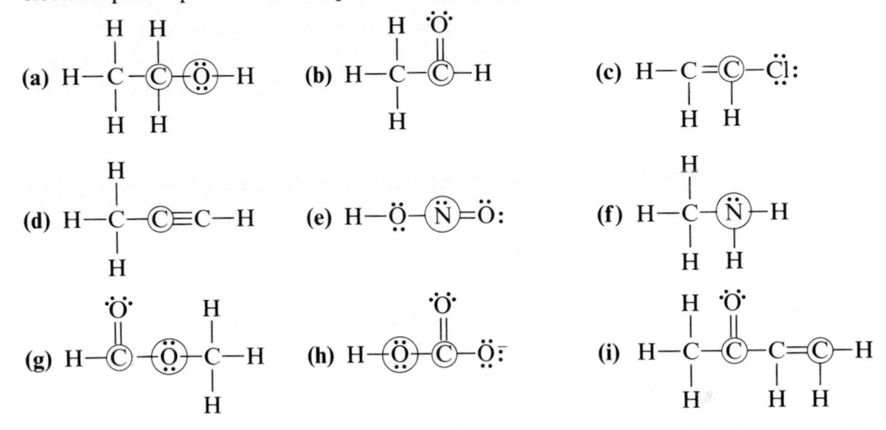

1.18 Add unshared pairs of electrons as necessary to complete valence shells of each atom in the following molecules or ions. Then assign formal positive or negative charges as appropriate.

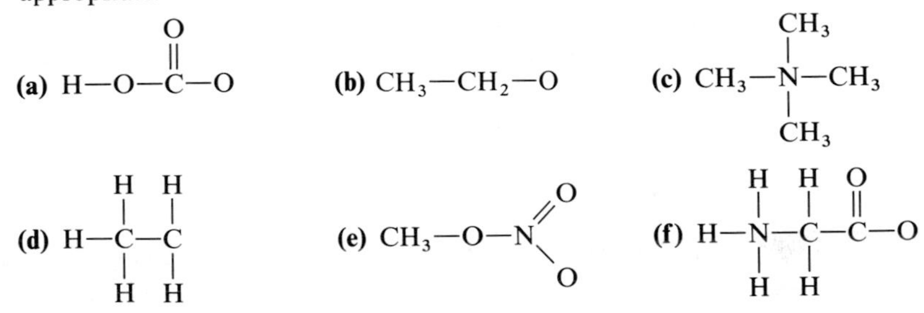

1.19 The covalence of carbon is 4. Draw the structural formula for an organic compound of carbon in which this covalence is satisfied by:

(a) four single bonds
(b) two single bonds and one double bond
(c) two double bonds
(d) one single and one triple bond

1.20 When discussing the reactions of organic compounds, we will encounter reactive intermediates in which the electron configuration around a carbon atom looks like one of the following:

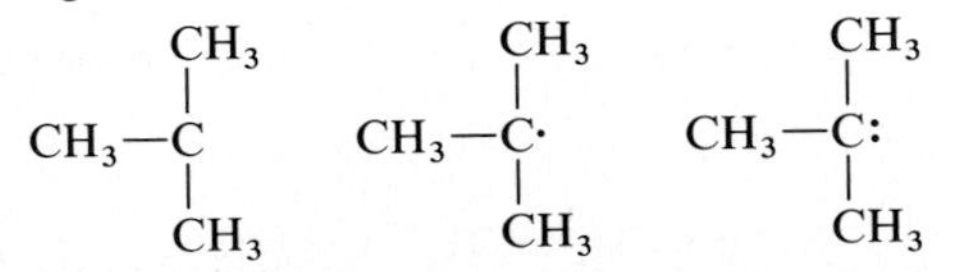

(a) For each example, determine the formal charge on the central carbon.
(b) Which structures are positively charged; negatively charged; uncharged; electron deficient?

1.21 In the beginning of the 19th century, organic chemistry was defined as a branch of chemistry dealing with compounds of carbon and hydrogen obtained from plant and animal sources. Since organic substances were derived from living organisms, they were thought to be different from inorganic substances in that they contained some kind of essential "vital force."

In 1828, the German scientist, Friedrich Wöhler, discovered that urea could be made by heating ammonium cyanate. Urea had previously been obtained only from urine, whereas ammonium cyanate was a typical inorganic or "nonorganic" salt. This experiment was significant in the history of organic chemistry for it demonstrated the interconversion of inorganic and organic substances.

Draw Lewis formulas for urea (a substance containing only covalent bonds), and ammonium cyanate (a substance containing both ionic and covalent bonds).

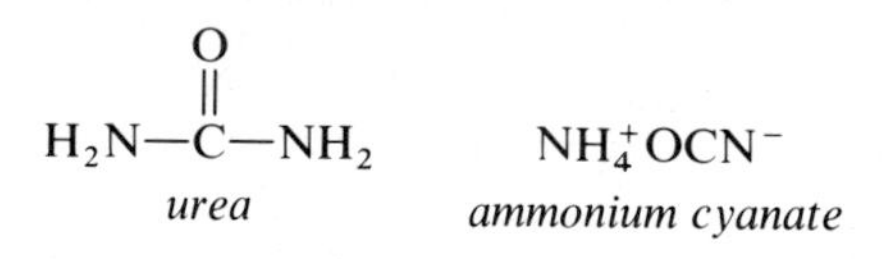

1.22 According to the molecular orbital theory of covalent bonding, sigma and pi bonds are similar in that each is formed by the overlap of atomic orbitals of adjacent atoms. In what ways do sigma and pi bonds differ?

1.23 What shape would you predict **(a)** for H_2O if covalent bond formation involved overlap of unhybridized $2p$ orbitals of oxygen with $1s$ orbitals of hydrogen; **(b)** for NH_3 if covalent bond formation involved overlap of unhybridized $2p$ orbitals of nitrogen with $1s$ orbitals of hydrogen? Compare your predictions with the experimentally determined bond angles in these two molecules.

1.24 Following are condensed structural formulas for a series of organic molecules. For each, write a Lewis structure and describe the bonding in terms of the orbitals involved. Based on your orbital description, predict all bond angles.

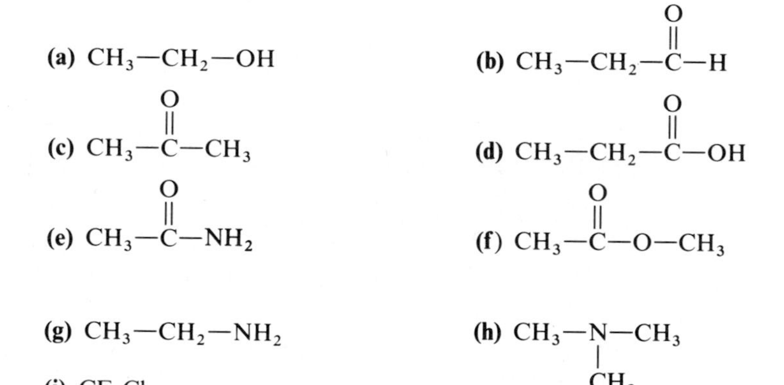

1.25 Name the functional groups in the following structural isomers:

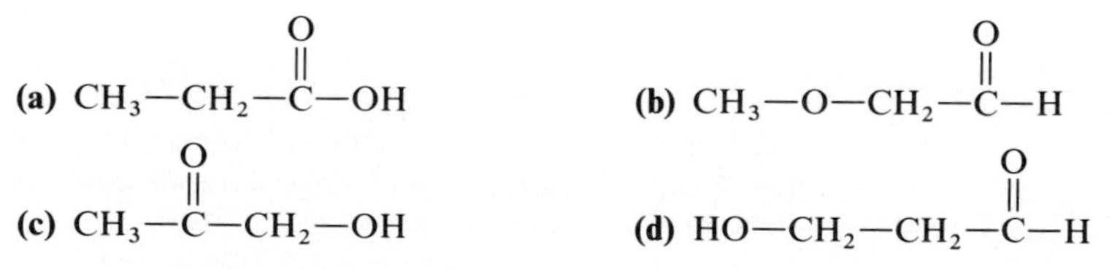
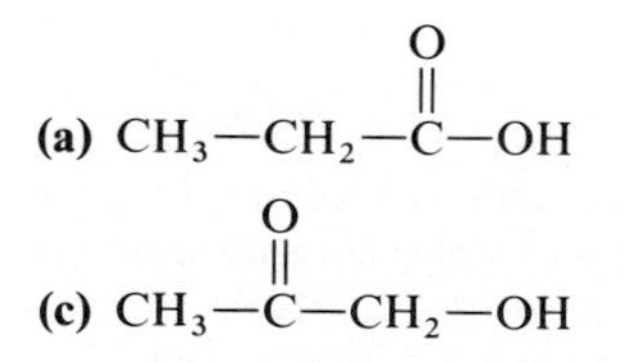

1.26 Write Lewis structures and condensed structural formulas for a compound of molecular formula C_4H_8O that contains the following functional groups:
(a) a carbon-carbon double bond and an alcohol
(b) a ketone
(c) an aldehyde
(d) a carbon-carbon double bond and an ether

1.27 Write Lewis structures and condensed structural formulas for:
(a) the eight alcohols of molecular formula $C_5H_{12}O$
(b) the six ethers of molecular formula $C_5H_{12}O$
(c) the eight aldehydes of molecular formula $C_6H_{12}O$
(d) the six ketones of molecular formula $C_6H_{12}O$
(e) the eight carboxylic acids of molecular formula $C_6H_{12}O_2$

1.28 Draw the contributing structures indicated by the curved arrows. Assign formal charges as appropriate.

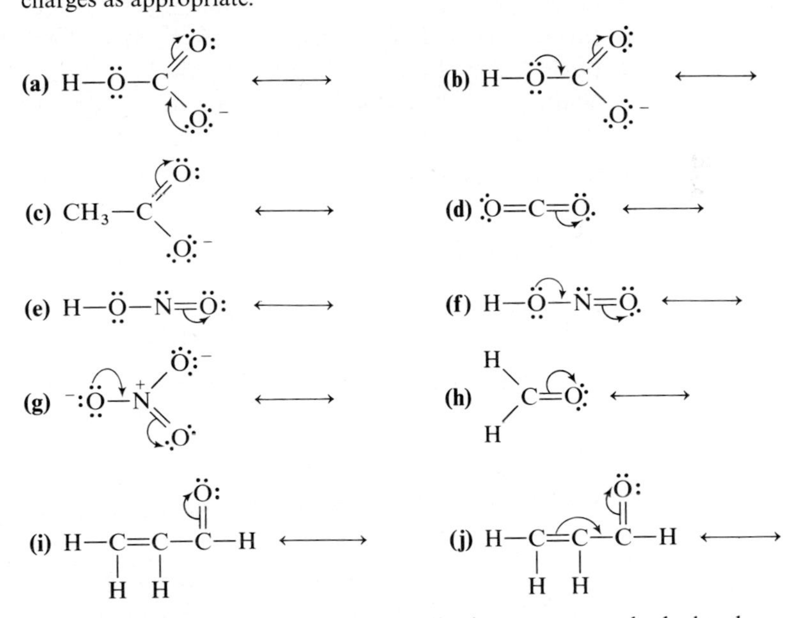

1.29 In Problem 1.28 you were given one contributing structure and asked to draw another. Label pairs of contributing structures that are equivalent. For pairs of contributing structures that are not equivalent, label the more important contributing structure, the less important contributing structure, and explain your reasoning.

1.30 Account for the fact that compounds containing C—O and C—N single bonds react with H^+, whereas those containing only C—C and C—H single bonds do not.

CHAPTER 2

Alkanes and Cycloalkanes

Compounds that consist solely of carbon and hydrogen are called hydrocarbons. If the carbon atoms in a hydrocarbon are joined together only by single covalent bonds, the hydrocarbon is called a saturated hydrocarbon or alkane. If any of the carbon atoms of the hydrocarbon are bonded together by one or more double or triple bonds, the hydrocarbon is called an unsaturated hydrocarbon. In this chapter we shall discuss the saturated hydrocarbons, the simplest organic substances from the structural point of view.

2.1 STRUCTURE OF ALKANES

Methane, CH_4, and ethane, C_2H_6, are the first members of the alkane family. Shown in Figure 2.1 are Lewis structures and ball-and-stick models of methane and ethane. The shape of methane is tetrahedral, and all H—C—H bond angles are 109.5°. Each of the carbon atoms in ethane is also tetrahedral, and all bond angles in ethane are 109.5°.

By increasing the number of carbon atoms in the chain, we can form the next members of the series: propane, C_3H_8; butane, C_4H_{10}; and pentane, C_5H_{12}.

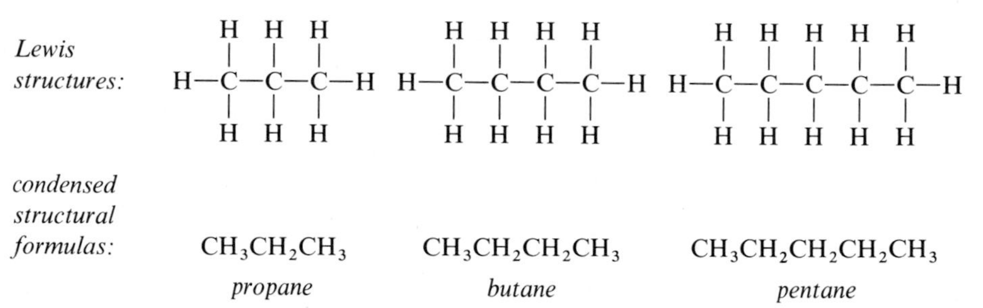

The condensed structural formulas for butane, pentane, and higher alkanes can be written in an even more abbreviated form. For example, the structural formula of pentane contains three $—CH_2—$ (methylene) groups in the middle of the chain. These can be grouped together and the structural formula written as $CH_3(CH_2)_3CH_3$. Table 2.1 shows the names, molecular formulas, and condensed structural formulas of the first ten alkanes.

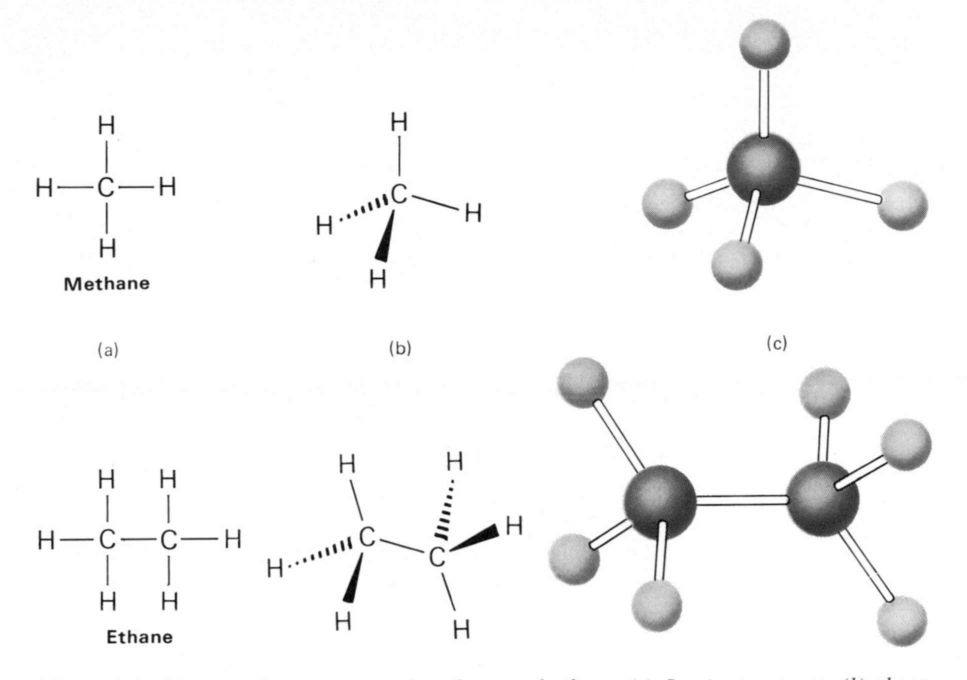

Figure 2.1 *Alkanes: the structures of methane and ethane. (a) Lewis structures; (b) three-dimensional structures; (c) ball-and-stick models.*

The names of the first four alkanes are methane, ethane, propane, and butane. The names of all higher alkanes consist of two parts: a prefix indicating the number of carbon atoms and the ending "ane." Pentane contains five carbon atoms; hexane contains six carbon atoms, heptane contains seven carbon atoms, etc.

Alkanes have the general formula C_nH_{2n+2}. Thus, if you know the number of carbon atoms in an alkane, it is an easy matter to determine the number of

Table 2.1 Names, molecular formulas, and condensed structural formulas of the first ten alkanes.

Name	*Molecular formula*	*Condensed structural formula*
methane	CH_4	CH_4
ethane	C_2H_6	CH_3CH_3
propane	C_3H_8	$CH_3CH_2CH_3$
butane	C_4H_{10}	$CH_3(CH_2)_2CH_3$
pentane	C_5H_{12}	$CH_3(CH_2)_3CH_3$
hexane	C_6H_{14}	$CH_3(CH_2)_4CH_3$
heptane	C_7H_{16}	$CH_3(CH_2)_5CH_3$
octane	C_8H_{18}	$CH_3(CH_2)_6CH_3$
nonane	C_9H_{20}	$CH_3(CH_2)_7CH_3$
decane	$C_{10}H_{22}$	$CH_3(CH_2)_8CH_3$

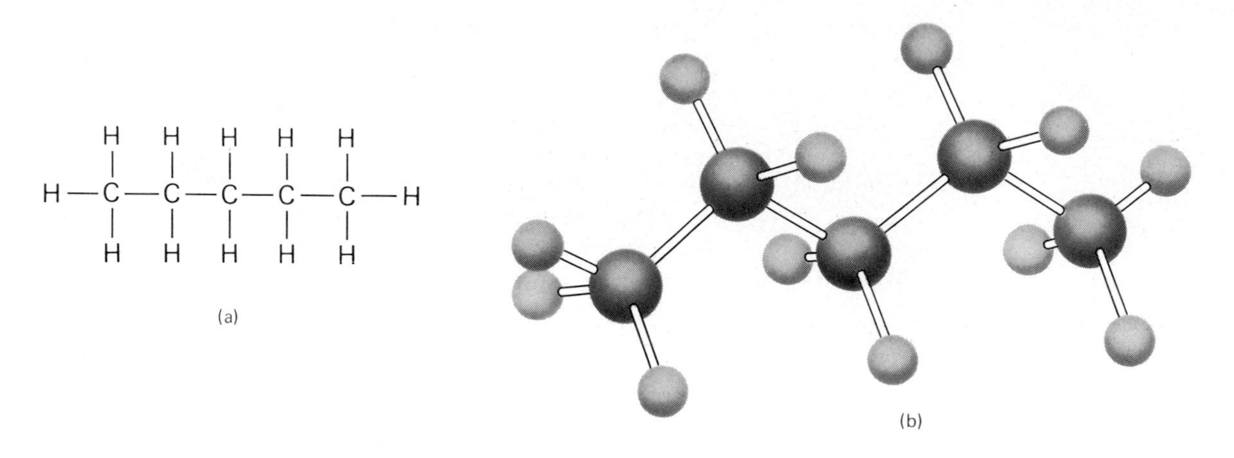

Figure 2.2 *The pentane molecule. (a) Lewis structure; (b) ball-and-stick model.*

hydrogen atoms. For example, decane with 10 carbon atoms must have 2 × 10 + 2 or 22 hydrogen atoms and a molecular formula of $C_{10}H_{22}$.

Although the three-dimensional shapes of propane and larger alkanes are more complex than those of methane and ethane, each carbon atom is still tetrahedral with bond angles of 109.5°. Figure 2.2 shows a Lewis structure and a ball-and-stick model of pentane. The three-dimensional shape of this molecule is best described as staggered or bent.

2.2 STRUCTURAL ISOMERISM IN ALKANES

Two or more compounds that have the same molecular formula but different orders of attachment of atoms are called structural isomers. We have already encountered several examples of structural isomerism in Section 1.13.

For the molecular formulas CH_4, C_2H_6, and C_3H_8, there is only one possible order of attachment for the carbon atoms. Therefore, methane, ethane, and propane have no structural isomers. For the molecular formula C_4H_{10}, there are two possible orders of attachment of the atoms. In one, the four carbon atoms are attached in a chain; in the other, they are attached three in a chain with the fourth carbon as a branch of the chain. These isomeric alkanes are named butane and 2-methyl-propane. (We will discuss how to name alkanes in the following section.)

$CH_3—CH_2—CH_2—CH_3$

butane

bp −0.5°C

$CH_3—CH(CH_3)—CH_3$

2-methylpropane

bp −10.2°C

Butane and 2-methylpropane are structural isomers. They have the same molecular formula, but a different order of attachment of the atoms. Structural isomers are different compounds and, therefore, have different physical and chemical properties. Notice that the boiling points of butane and 2-methylpropane differ by over 9°C.

There are three structural isomers of C_5H_{12}, five structural isomers of C_6H_{14}, eighteen of C_8H_{18}, and seventy-five of $C_{10}H_{22}$. It should be obvious that even for a rather small number of carbon and hydrogen atoms, a very large number of structural isomers is possible. In fact, the potential for structural and functional group individuality from just the basic building blocks of carbon, hydrogen, nitrogen, and oxygen is practically unlimited.

Example 2.1 Identify the following pairs as formulas of identical substances or formulas of structural isomers.

(a) $CH_3{-}CH_2{-}CH_2{-}CH_2{-}CH_2{-}CH_3$ and

$$\begin{array}{llll} CH_3{-}CH_2{-}CH_2 & & & \\ \quad\quad\quad\quad\quad | & & & \\ \quad\quad\quad\quad\quad CH_2{-}CH_2{-}CH_3 & & & \end{array}$$

(b)

$$\begin{array}{c} CH_3{-}CH_2{-}CH{-}CH_2{-}CH_2{-}CH_3 \\ \quad\quad\quad\quad | \quad\quad\quad\quad\quad\quad\quad \\ \quad\quad\quad\quad CH_3 \quad\quad\quad\quad\quad\quad \end{array} \quad \text{and} \quad \begin{array}{c} CH_3{-}CH_2{-}CH_2{-}CH{-}CH_3 \\ \quad\quad\quad\quad\quad\quad\quad | \quad\quad \\ \quad\quad\quad\quad\quad\quad\quad CH_2 \quad \\ \quad\quad\quad\quad\quad\quad\quad | \quad\quad \\ \quad\quad\quad\quad\quad\quad\quad CH_3 \quad \end{array}$$

(c)

$$\begin{array}{c} \quad\quad\quad\quad\quad\quad\quad\quad\quad CH_3 \\ \quad\quad\quad\quad\quad\quad\quad\quad\quad | \\ CH_3{-}CH{-}CH_2{-}CH \\ \quad\quad | \quad\quad\quad\quad\quad | \\ \quad\quad CH_3 \quad\quad\quad CH_3 \end{array} \quad \text{and} \quad \begin{array}{c} \quad\quad\quad\quad\quad\quad\quad CH_3 \quad\quad \\ \quad\quad\quad\quad\quad\quad\quad | \quad\quad\quad \\ CH_3{-}CH_2{-}CH{-}CH{-}CH_3 \\ \quad\quad\quad | \quad\quad\quad\quad\quad\quad \\ \quad\quad\quad CH_3 \quad\quad\quad\quad\quad \end{array}$$

Solution To determine if two formulas are identical or represent structural isomers, first find the longest chain of carbon atoms and number it from the end nearest the first branch. Then compare the lengths of each carbon chain and the size and location of any branches.

(a)

$$\overset{1}{C}H_3{-}\overset{2}{C}H_2{-}\overset{3}{C}H_2{-}\overset{4}{C}H_2{-}\overset{5}{C}H_2{-}\overset{6}{C}H_3 \quad\quad \begin{array}{l} \overset{1}{C}H_3{-}\overset{2}{C}H_2{-}\overset{3}{C}H_2 \\ \quad\quad\quad\quad\quad | \\ \quad\quad\quad\quad\quad \underset{4}{C}H_2{-}\underset{5}{C}H_2{-}\underset{6}{C}H_3 \end{array}$$

Each has an unbranched chain of six carbon atoms. Therefore, these structural formulas are identical and represent the same compound.

(b)

$$\begin{array}{c} \overset{1}{C}H_3{-}\overset{2}{C}H_2{-}\overset{3}{C}H{-}\overset{4}{C}H_2{-}\overset{5}{C}H_2{-}\overset{6}{C}H_3 \\ \quad\quad\quad\quad | \quad\quad\quad\quad\quad\quad\quad \\ \quad\quad\quad\quad CH_3 \quad\quad\quad\quad\quad\quad \end{array} \quad\quad \begin{array}{c} \overset{6}{C}H_3{-}\overset{5}{C}H_2{-}\overset{4}{C}H_2{-}\overset{3}{C}H{-}CH_3 \\ \quad\quad\quad\quad\quad\quad\quad | \quad\quad \\ \quad\quad\quad\quad\quad\quad {}_2\,CH_2 \quad \\ \quad\quad\quad\quad\quad\quad\quad | \quad\quad \\ \quad\quad\quad\quad\quad\quad {}_1\,CH_3 \quad \end{array}$$

Each has a chain of six carbon atoms with a CH_3 group on the third carbon atom of the chain. Therefore, the formulas are identical and represent the same compound.

(c)

$$\begin{array}{c} \quad\quad\quad\quad\quad\quad\quad\quad\quad {}^5CH_3 \\ \quad\quad\quad\quad\quad\quad\quad\quad\quad {}^4| \\ \overset{1}{C}H_3{-}\overset{2}{C}H{-}\overset{3}{C}H_2{-}CH \\ \quad\quad | \quad\quad\quad\quad\quad | \\ \quad\quad CH_3 \quad\quad\quad CH_3 \end{array} \quad\quad \begin{array}{c} \quad\quad\quad\quad\quad\quad\quad CH_3 \quad\quad \\ \quad\quad\quad\quad\quad\quad\quad {}^2| \quad\quad\quad \\ \overset{5}{C}H_3{-}\overset{4}{C}H_2{-}\overset{3}{C}H{-}CH{-}\overset{1}{C}H_3 \\ \quad\quad\quad | \quad\quad\quad\quad\quad\quad \\ \quad\quad\quad CH_3 \quad\quad\quad\quad\quad \end{array}$$

Both have chains of five carbon atoms with two CH_3 branches. Although the size of the branches is identical, they are on different locations on the chains. Therefore, these formulas represent structural isomers: they have the same molecular formula, C_7H_{16}, but different orders of attachment of the atoms.

PROBLEM 2.1 Identify the following pairs as formulas of identical substances or as formulas of structural isomers.

$$\text{(a)}\quad CH_3-\overset{\overset{\displaystyle CH_3}{|}}{\overset{\displaystyle CH_2}{\overset{|}{C}H}}-\underset{\underset{\displaystyle CH_2-CH_3}{|}}{CH}-CH_3 \quad\text{and}\quad CH_3-CH_2-\overset{\overset{\displaystyle CH_3}{|}}{CH}-CH_2-\overset{\overset{\displaystyle CH_3}{|}}{CH}-CH_3$$

$$\text{(b)}\quad CH_3-\overset{\overset{\displaystyle CH_3}{|}}{CH}-\underset{\underset{\displaystyle CH_2-CH_3}{|}}{CH}-CH_3 \quad\text{and}\quad CH_3-\overset{\overset{\displaystyle CH_3}{|}}{CH}-\overset{\overset{\displaystyle CH_3}{|}}{CH}-CH_2-CH_3$$

Example 2.2 Draw structural formulas for the five structural isomers of molecular formula C_6H_{14}.

Solution You should approach this type of problem (there will be more like it later) in a systematic manner. First, draw the structural isomer with all six carbon atoms in an unbranched chain. Then draw all structural isomers with five carbons in a chain and one carbon as a branch on the chain. Finally, draw all structural isomers with four carbons in a chain and two carbons as branches. By working in a systematic manner you will arrive at all possible structural isomers.

six carbons in an unbranched chain:

$$\overset{1}{CH_3}-\overset{2}{CH_2}-\overset{3}{CH_2}-\overset{4}{CH_2}-\overset{5}{CH_2}-\overset{6}{CH_3}$$

five carbons in a chain; one carbon as a branch:

$$\overset{1}{CH_3}-\underset{\underset{\displaystyle CH_3}{|}}{\overset{2}{CH}}-\overset{3}{CH_2}-\overset{4}{CH_2}-\overset{5}{CH_3} \qquad \overset{1}{CH_3}-\overset{2}{CH_2}-\underset{\underset{\displaystyle CH_3}{|}}{\overset{3}{CH}}-\overset{4}{CH_2}-\overset{5}{CH_3}$$

four carbons in a chain; two carbons as branches:

$$\overset{1}{CH_3}-\underset{\underset{\displaystyle CH_3}{|}}{\overset{\overset{\displaystyle CH_3}{2|}}{C}}-\overset{3}{CH_2}-\overset{4}{CH_3} \qquad \overset{1}{CH_3}-\overset{\overset{\displaystyle CH_3}{2|}}{CH}-\underset{\underset{\displaystyle CH_3}{|}}{\overset{3}{CH}}-\overset{4}{CH_3}$$

PROBLEM 2.2 Draw structural formulas for the three structural isomers of molecular formula C_5H_{12}.

2.3 NOMENCLATURE OF ALKANES

Ideally, every organic substance should have a name which clearly describes the structure of the compound and from which a structural formula can be drawn. For this purpose, chemists throughout the world have accepted a set of rules proposed by the International Union of Pure and Applied Chemistry (IUPAC). This system is known as the IUPAC system or, alternatively, as the Geneva system after the fact that the first meetings of the IUPAC were held in Geneva, Switzerland. There are two major components to the IUPAC name of an alkane: (1) a parent name which indicates the longest chain of carbon atoms in the formula and (2) substituent names which indicate groups attached to the parent chain.

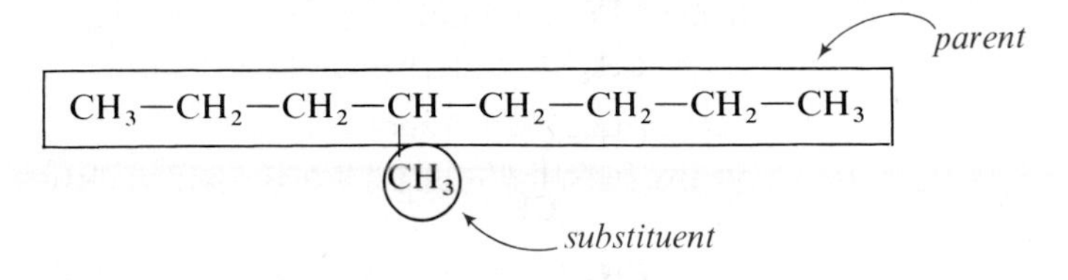

A substituent group derived from an alkane is called an alkyl group. Alkyl groups are named by dropping the -ane from the name of the parent alkane and adding the suffix -yl. For example, the alkyl substituent $CH_3CH_2—$ is named ethyl.

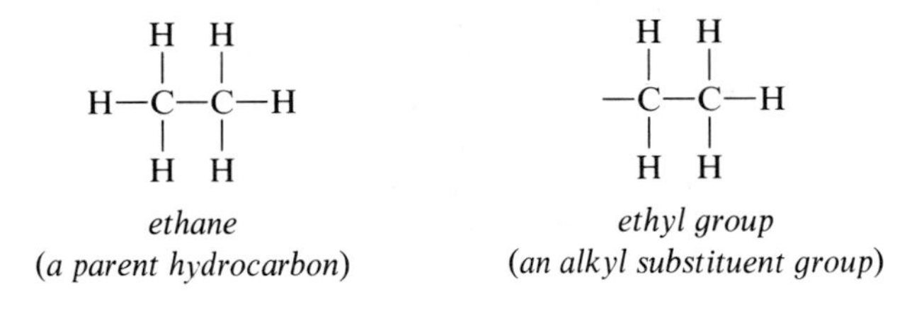

ethane
(*a parent hydrocarbon*)

ethyl group
(*an alkyl substituent group*)

Names and structural formulas for eight of the most common alkyl groups are given in Table 2.2

Table 2.2 Common alkyl substituent groups.

IUPAC Name	*Condensed structural formula*	*IUPAC Name*	*Condensed structural formula*
methyl	$—CH_3$	butyl	$—CH_2—CH_2—CH_2—CH_3$
ethyl	$—CH_2—CH_3$	isobutyl	$—CH_2—CH(CH_3)—CH_3$
propyl	$—CH_2—CH_2—CH_3$	*sec*-butyl	$—CH(CH_3)—CH_2—CH_3$
isopropyl	$—CH(CH_3)—CH_3$	*tert*-butyl	$—C(CH_3)_2—CH_3$

Following are the rules of the IUPAC system for naming alkanes.

(1) The general name of the saturated hydrocarbon is alkane.

(2) For branched-chain hydrocarbons, the hydrocarbon derived from the longest chain of carbon atoms is taken as the parent chain and the root or stem name is that of the parent chain.

(3) Groups attached to the parent chain are called substituents. Each substituent is given a name and a number. The number shows the carbon atom of the parent chain to which the substituent is attached.

(4) If the same substituent occurs more than once, the number of each carbon of the parent chain on which the substituent occurs is given. In addition, the number of times the substituent group occurs is indicated by a prefix di-, tri-, tetra-, penta-, or hexa-.

(5) If there is one substituent, number the parent chain from the end which gives the substituent the lowest number. If there are two or more substituents, number the parent from the end which gives the lowest number to the substituent encountered first.

(6) If there are two or more different alkyl substituents, list them in alphabetical order.

Example 2.3 Name the following compounds by the IUPAC system.

(a) 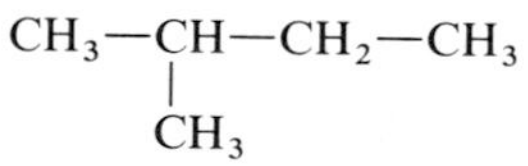 $CH_3-\underset{\displaystyle CH_3}{\underset{|}{CH}}-CH_2-CH_3$

(b) 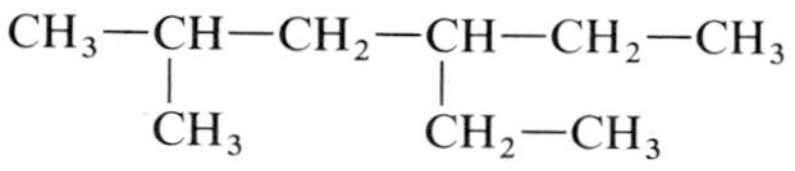 $CH_3-\underset{\displaystyle CH_3}{\underset{|}{CH}}-CH_2-\underset{\displaystyle CH_2-CH_3}{\underset{|}{CH}}-CH_2-CH_3$

(c) 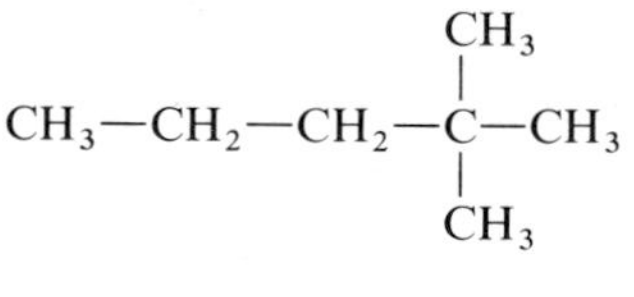 $CH_3-CH_2-CH_2-\overset{\displaystyle CH_3}{\overset{|}{\underset{\displaystyle CH_3}{\underset{|}{C}}}}-CH_3$

Solution **(a)** There are four carbon atoms in the longest chain. Therefore, the name of the parent chain is butane (rule 2). The butane chain must be numbered so that the single methyl group is on carbon 2 of the chain (rule 5). The correct name of this hydrocarbon is 2-methylbutane.

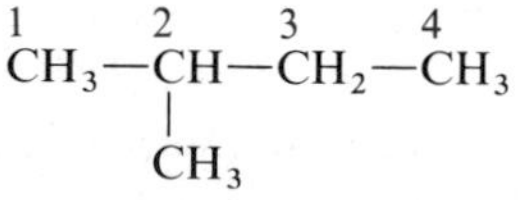

$$\overset{1}{CH_3}-\underset{\displaystyle CH_3}{\underset{|}{\overset{2}{CH}}}-\overset{3}{CH_2}-\overset{4}{CH_3} \qquad \text{2-methylbutane}$$

(b) The longest chain contains six carbon atoms. Therefore, the parent chain is a hexane (rule 2). There are two alkyl substituents on the hexane chain: a methyl group and an ethyl group. The hexane chain must be numbered to the right so that the substituent encountered first (the methyl group) is on carbon 2 of the chain (rule 5). The ethyl and methyl substituents are listed in alphabetical order (rule 6) to give the name 4-ethyl-2-methylhexane.

$$\overset{1}{CH_3}-\overset{2}{CH}(CH_3)-\overset{3}{CH_2}-\overset{4}{CH}(CH_2-CH_3)-\overset{5}{CH_2}-\overset{6}{CH_3}$$

4-ethyl-2-methylhexane

(c) The longest chain contains five carbon atoms and, therefore, the parent chain is a pentane (rule 2). The pentane chain must be numbered so that the substituents are on carbon 2 of the chain (rule 5). There are two substituents and each must have a name and a number (rule 4). Since the substituents are identical, they can be grouped together using the prefix "di-" (rule 4). The correct IUPAC name is 2,2-dimethylpentane.

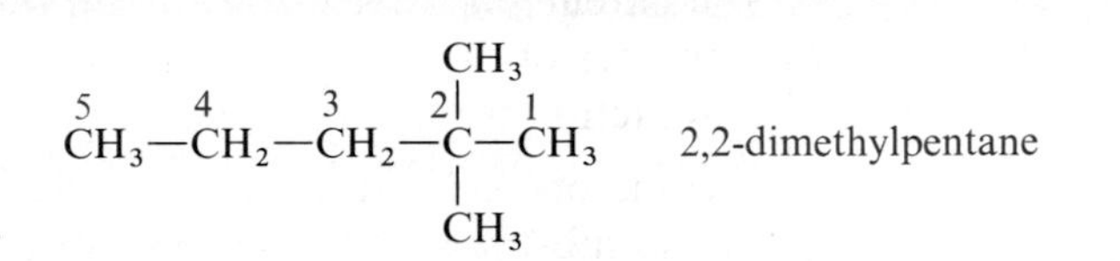

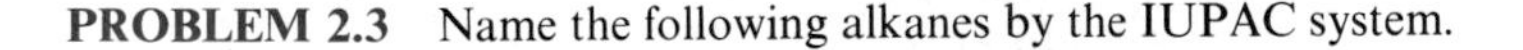

PROBLEM 2.3 Name the following alkanes by the IUPAC system.

(a) $CH_3-CH(CH_3)-CH_2-CH_2-CH(CH_2-CH_2-CH_3)-CH(CH_3)-CH_3$

(b) $CH_3-CH_2-CH_2-C(CH_2-CH_2-CH_3)(CH(CH_3)-CH_3)-CH_2-CH_2-CH_3$

2.4 COMMON NAMES OF ALKANES

In spite of the precision of the IUPAC system, routine communication in organic chemistry still relies on a combination of trivial, semisystematic, and systematic names. The reasons for this situation are rooted in both convenience and historical development.

In the older, semisystematic nomenclature, the total number of carbon atoms in the alkane, regardless of their arrangement, determines the name. The first three alkanes are methane, ethane, and propane. All alkanes of formula C_4H_{10} are called butanes, all alkanes of formula C_5H_{12} are called pentanes, and those of formula C_6H_{14} are called hexanes. For alkanes beyond propane, "normal" or the prefix "*n*-" is used to indicate that all carbons are joined in a continuous chain. The prefix "iso-" indicates that one end of an otherwise continuous chain terminates in the $(CH_3)_2CH—$ group. The prefix "neo-" indicates that one end of an otherwise continuous chain of carbon atoms terminates in the $(CH_3)_3C—$ group. Following are examples of common names.

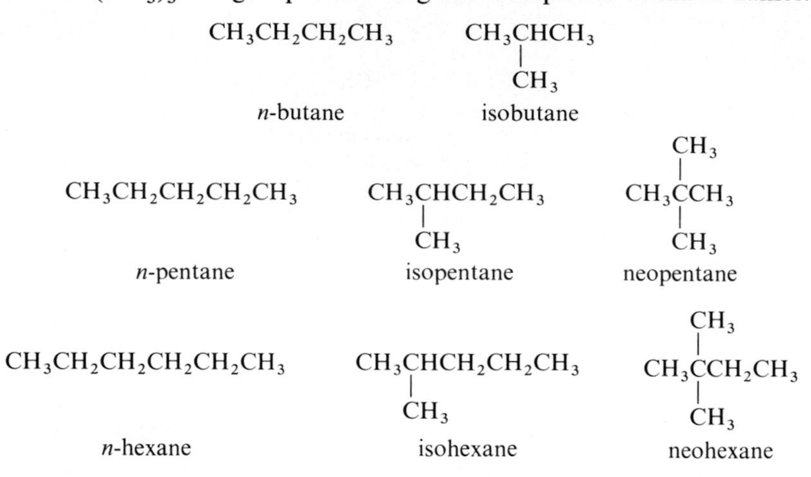

This system of common names has no good way of handling other branching patterns, and for more complex alkanes it is necessary to use the more flexible IUPAC system of nomenclature.

We shall strive, as far as possible, to use IUPAC names in this text, but unavoidably, there will be some use of other systems of nomenclature.

2.5 CYCLOALKANES

So far we have considered only chains of carbon atoms, often having one or more branches. However, the ends of these chains can be joined together to form rings of carbon atoms. Molecules that contain carbon atoms in the ring are called cyclic hydrocarbons. Furthermore, when all the carbons of the ring are saturated, the molecules are called cycloalkanes.

Cycloalkanes of ring size from three to over thirty are found in nature and, in principle, there is no limit to ring size. Five-membered rings (cyclopentanes) and six-membered rings (cyclohexanes) are especially abundant in nature and they have received special attention.

Figure 2.3 shows the structural formulas for cyclopropane, cyclobutane, cyclopentane, and cyclohexane. As a matter of convenience, the organic chemist does not usually write out the structural formulas for the cycloalkanes showing all carbons and hydrogens. Rather, the rings are represented by polygons of the same number of sides. For example, cyclopropane is represented by a triangle and cyclohexane by a regular hexagon.

To name cycloalkanes, prefix the name of the corresponding open-chain hydrocarbon by cyclo- and name each substituent on the ring. If there is only a single substituent on the cycloalkane ring, there is no need to give it a number. However, if there are two or more substituents, each substituent must have a number to indicate its position on the ring.

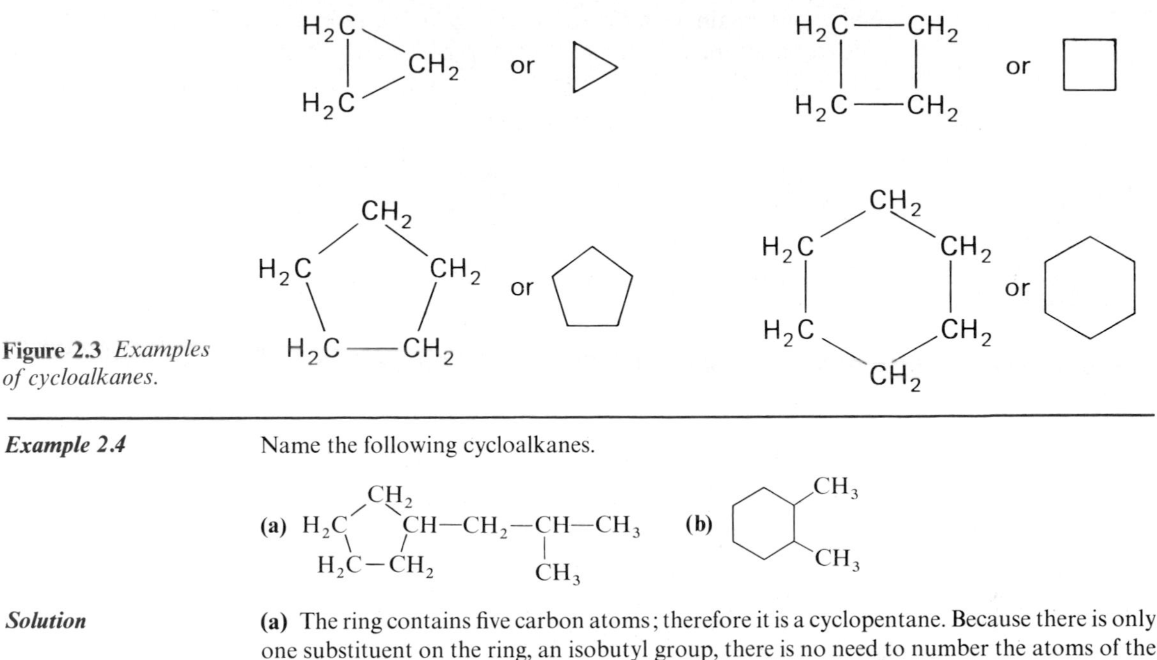

Figure 2.3 *Examples of cycloalkanes.*

Example 2.4 Name the following cycloalkanes.

Solution **(a)** The ring contains five carbon atoms; therefore it is a cyclopentane. Because there is only one substituent on the ring, an isobutyl group, there is no need to number the atoms of the ring. The IUPAC name is isobutylcyclopentane.

(b) The ring contains six carbon atoms; therefore it is a cyclohexane. Because there are two substituents on the ring, a numbering system must be used to locate the groups. The IUPAC name is 1,2-dimethylcyclohexane.

PROBLEM 2.4 Name the following cycloalkanes.

The use of carbon bonds to close a ring means that cycloalkanes contain two hydrogen atoms fewer than an alkane of the same number of carbon atoms. For example, compare the molecular formulas of cyclopropane (C_3H_6) and propane (C_3H_8) or the molecular formulas of cyclohexane (C_6H_{12}) and hexane (C_6H_{14}). The general formula of a cycloalkane is C_nH_{2n}.

2.6 CONFORMATIONS OF ALKANES AND CYCLOALKANES

Structural formulas are useful to represent a compound for they show us the order of attachment of the atoms. However, structural formulas usually make no attempt to show the actual three-dimensional shapes. As chemists try to understand more and more about the relationships between structure and the chemical and physical properties of molecules, it becomes increasingly important to understand more about the three-dimensional shapes of molecules.

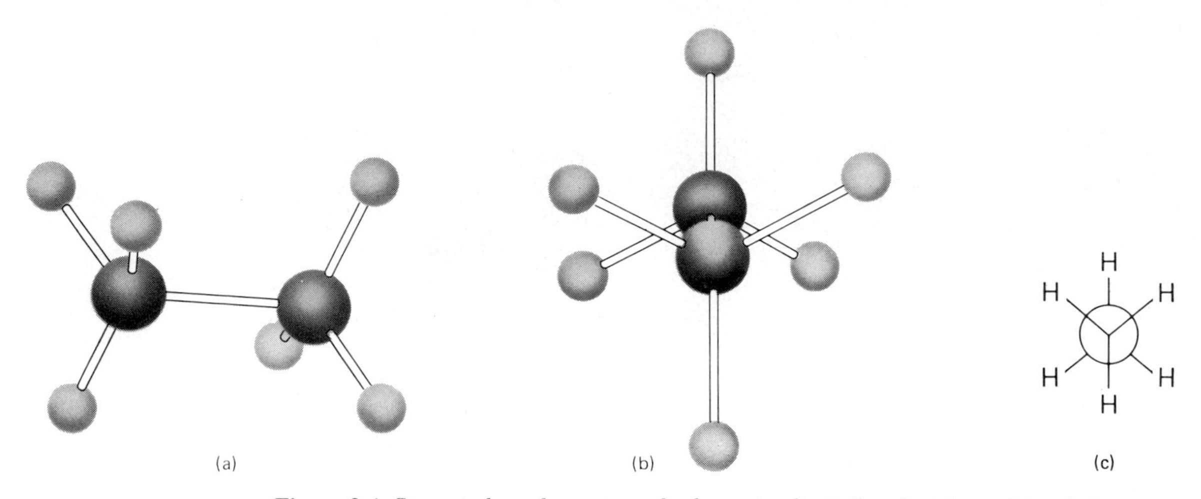

Figure 2.4 *Staggered conformation of ethane. (a, b) Ball-and-stick models of the same conformation; (c) a Newman projection.*

Most molecules can be twisted into a number of different three-dimensional shapes without breaking any bonds. These different shapes are called conformations. A simple molecule such as ethane has an infinite number of conformations. You can visualize all the conformations by holding one CH_3 group in your hand and twisting the other CH_3 group about the single bond joining the two carbon atoms.

Figure 2.4a shows a ball-and-stick model for one of the conformations of ethane. In Figure 2.4b, the same conformation is turned in space so that you are looking at it from another angle. Figure 2.4c is a Newman projection of this ethane conformation. It is important to realize that the three parts of Figure 2.4 depict the same conformation seen from different angles and drawn differently.

Newman projections are especially convenient for showing molecular conformations of alkanes. In the Newman projection of ethane (Figure 2.4c), you are looking at the C—C bond head-on. The three hydrogens nearer your eye are shown as lines from the center of the circle at angles of 120°. The three hydrogens of the carbon farther from your eye are shown by lines extending from the circumference of the circle. Remember, of course, that the bond angles about each of these carbon atoms are 109.5° and not 120° as the Newman projection formula might suggest at first.

Figures 2.5a and 2.5b show ball-and-stick models for another conformation of ethane viewed from two different perspectives. In this conformation, the hydrogen atoms on the back carbon are lined up or eclipsed with the hydrogen atoms on the front carbon. For this reason this conformation is called an eclipsed conformation.

The conformations of ethane are interconvertible by rotation about the carbon-carbon bond. At room temperature, ethane molecules undergo collision with sufficient energy that rotation about the carbon-carbon single bond from one conformation to another is easily possible. In fact, for a long time chemists believed there was completely free rotation about the carbon-carbon single

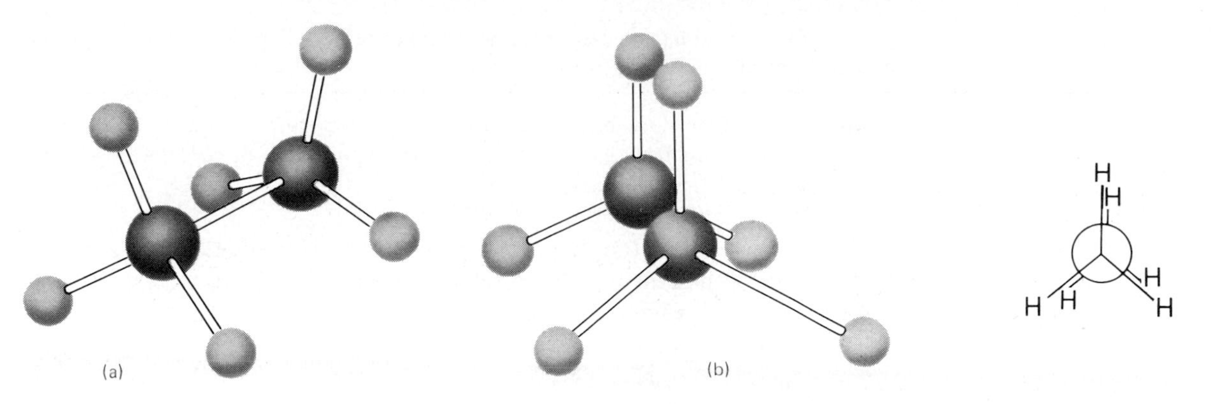

Figure 2.5 *Eclipsed conformation of ethane. (a, b) The same conformation viewed from two different perspectives; (c) a Newman projection of the eclipsed conformation.*

bond. However, more recent studies of ethane and other molecules have revealed that the rotation is not completely free, but rather is hindered by the size and the electrical character of the atoms joined to one or both of the connected carbon atoms. In the case of ethane, there is a preference for the staggered conformation over the eclipsed because the interaction between the hydrogen atoms is minimized.

Example 2.5 Draw Newman projections for the two staggered conformations of butane. Consider only conformations along the bond between carbon atoms 2 and 3 of the butane chain. Which of these is the preferred (more stable)?

Solution The condensed structural formula of butane is

$$\overset{1}{C}H_3-\overset{2}{C}H_2-\overset{3}{C}H_2-\overset{4}{C}H_3$$

First, view the molecule along the bond between carbon atoms 2 and 3. Then, to see the possible conformations asked for, hold carbon atom 2 in place and rotate carbon 3 about the single bond joining carbons 2 and 3.

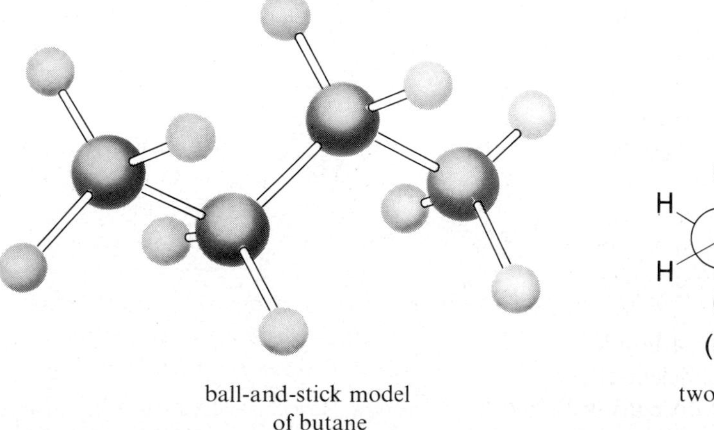

ball-and-stick model of butane

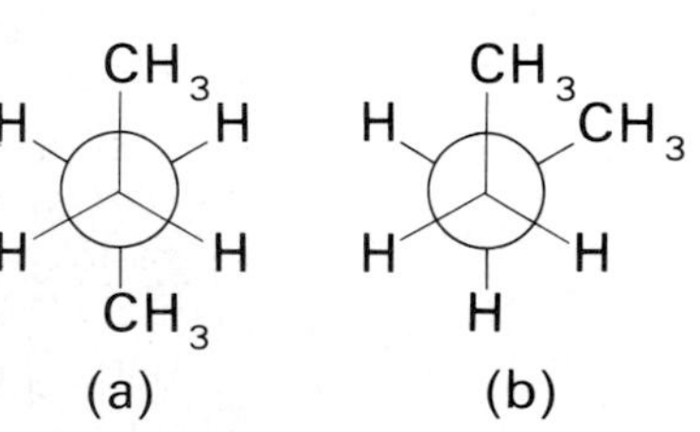

two staggered conformations of butane

Staggered conformation (a) is preferred (more stable) because in it the two $—CH_2$ groups are as far from each other as possible.

PROBLEM 2.5 Draw two eclipsed conformations of butane. Consider only conformations along the bond between carbon atoms 2 and 3. Which of these is the more stable conformation; the less stable conformation?

Next let us look at the shapes of cycloalkanes. The three carbon atoms of cyclopropane lie in a plane (three points determine a plane) with C—C—C bond angles of 60°. Figure 2.6 shows the Lewis structure and a ball-and-stick model of cyclopropane.

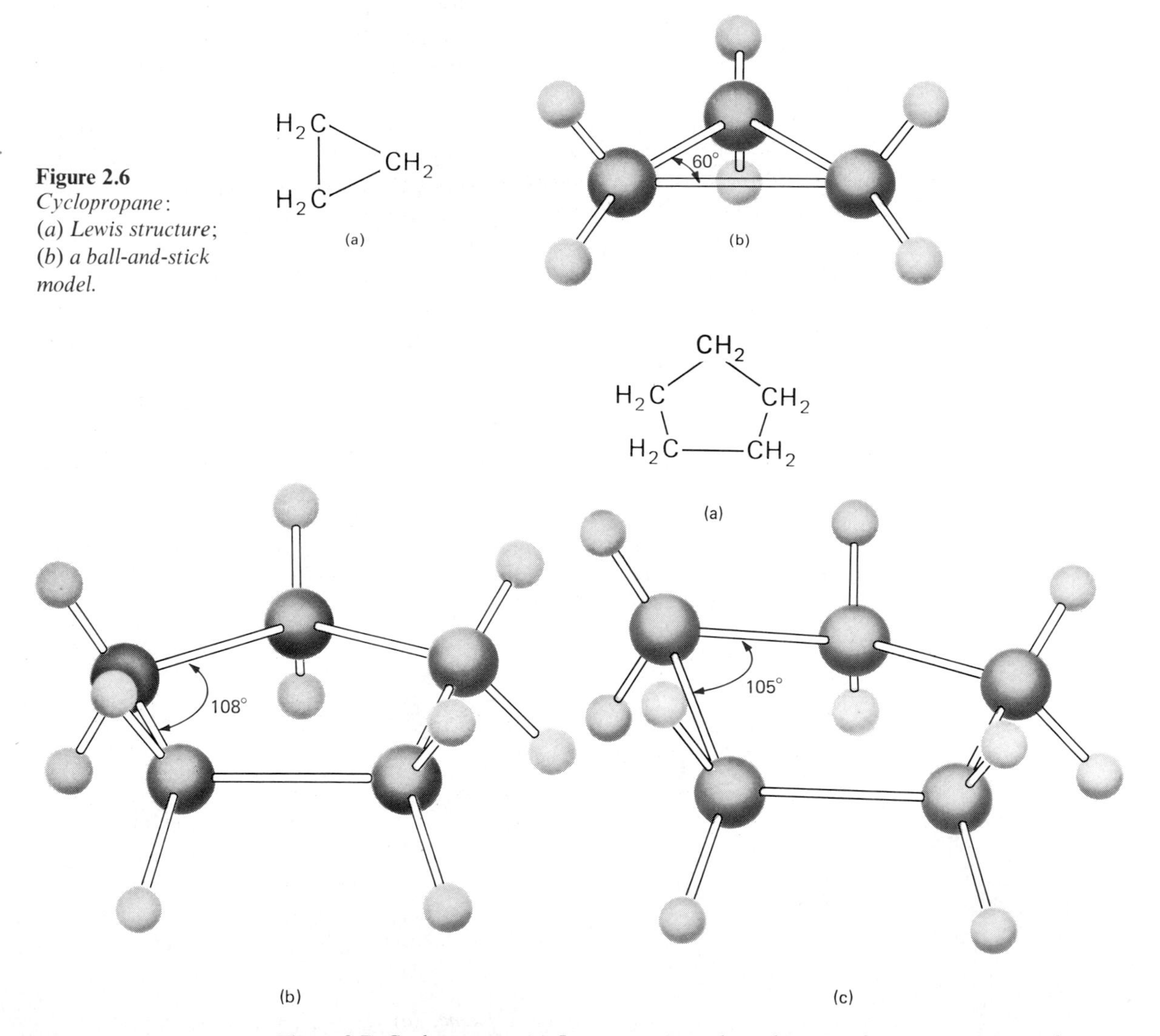

Figure 2.6 *Cyclopropane: (a) Lewis structure; (b) a ball-and-stick model.*

Figure 2.7 *Cyclopentane: (a) Lewis structure; (b) a planar conformation; (c) a nonplanar or puckered conformation.*

Table 2.3 Shapes of four cycloalkanes.

Name	*Planar representation*	C—C—C *Bond angle in planar molecule*	*Actual* C—C—C *bond angle*	*Ring shape*
cyclopropane	△	60°	60°	planar
cyclobutane	□	90°	88°	slightly puckered
cyclopentane	⬠	108°	105°	slightly puckered
cyclohexane	⬡	120°	109.5°	puckered, chair conformation most stable

All cycloalkanes larger than cyclopropane exist in dynamic equilibria between puckered conformations. Puckered conformations can be illustrated using cyclopentane as an example. Shown in Figure 2.7 are ball-and-stick models for two conformations of cyclopentane.

In Figure 2.7b, cyclopentane is shown as a planar conformation with all C—C—C bond angles of 108°. In this planar conformation there are ten fully eclipsed C—H interactions. The strain due to interactions of eclipsed hydrogens can be partially relieved by a slight puckering of the ring. In the puckered conformation of cyclopentane shown in Figure 2.7c, four carbon atoms are in a plane and the fifth is bent slightly out of the plane. In this nonplanar conformation, C—C—C bond angles are compressed to about 105°. Observed bond angles and shapes of cyclopropane, cyclobutane, cyclopentane, and cyclohexane are shown in Table 2.3.

Cyclohexane and all larger cycloalkanes exist in a number of different nonplanar conformations. For cyclohexane, the most stable of these is the chair conformation (Figure 2.8). In the chair conformation, all C—C—C bond

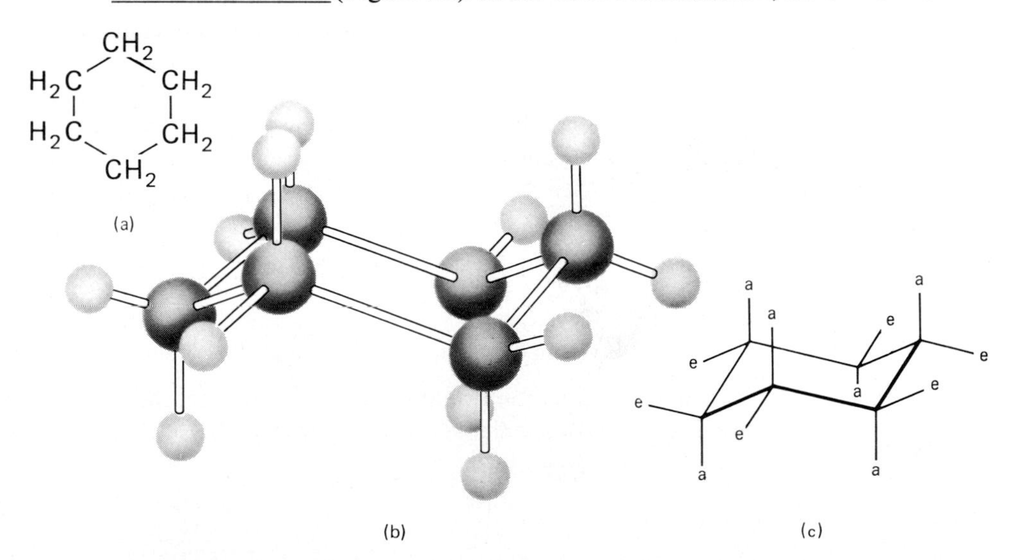

Figure 2.8 *Chair conformation of cyclohexane. (a) Lewis structure; (b) ball-and-stick model; (c) three-dimensional drawing. In the chair conformation, six hydrogens are equatorial, e, and six hydrogens are axial, a.*

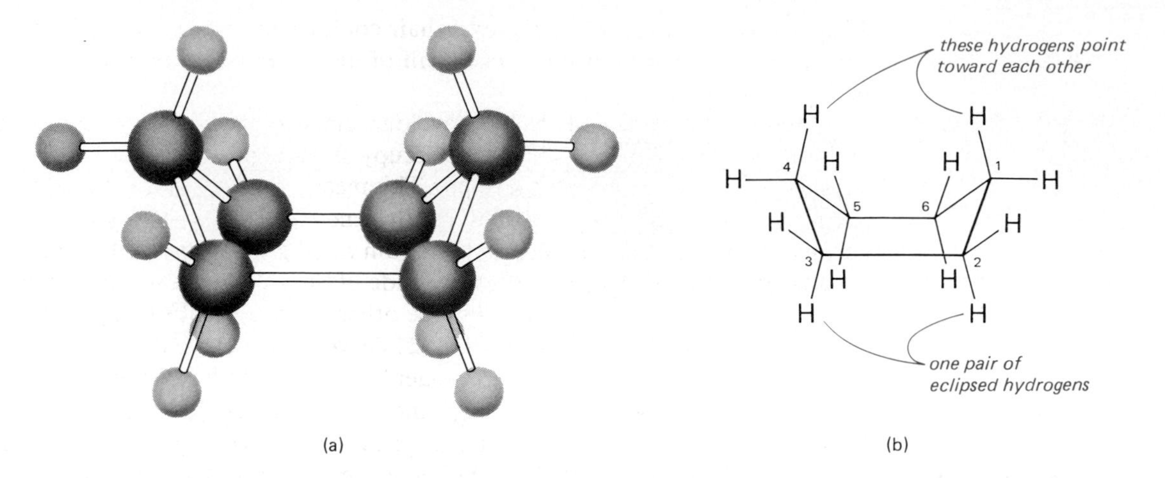

Figure 2.9 *Interactions in the boat conformation of cyclohexane. (a) Ball-and-stick model; (b) three-dimensional drawing.*

angles are 109.5° (the normal tetrahedral angle) and all C—H bonds are staggered. It is estimated that at any given time about 99.9% of cyclohexane molecules are in the chair conformation. When the arrangement of hydrogens on the chair conformation is considered, it can be seen that there are hydrogen atoms in two different geometrical situations. Six, called equatorial hydrogens, project in the plane of the ring and six, called axial hydrogens, are perpendicular to the plane of the ring. Each carbon of cyclohexane has two hydrogens; one hydrogen is axial, the other hydrogen is equatorial (Figure 2.8c).

There are many other nonplanar conformations of cyclohexane, one of which is the boat conformation. In the boat conformation (Figure 2.9), the hydrogens of carbons 2, 3 and 5, 6 are eclipsed (as in ethane) and the hydrogens of carbons 1 and 4 jut forward toward each other. Because of these interactions the boat form of cylohexane is of higher energy (less stable) than the chair form.

The chair and boat conformations can be interconverted by twisting about carbon-carbon bonds.

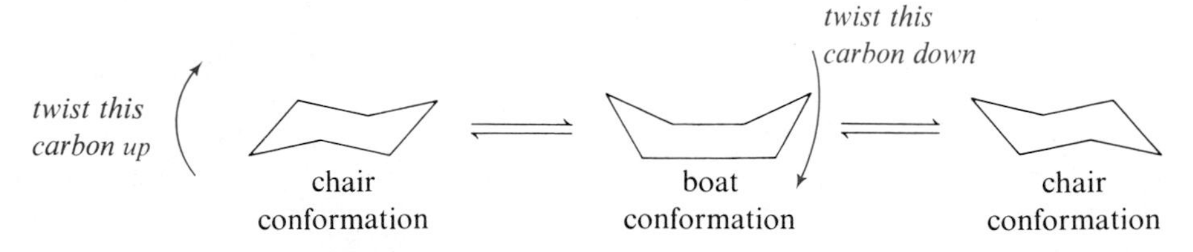

When one chair conformation is converted to the other chair, equatorial hydrogens become axial and axial hydrogens become equatorial.

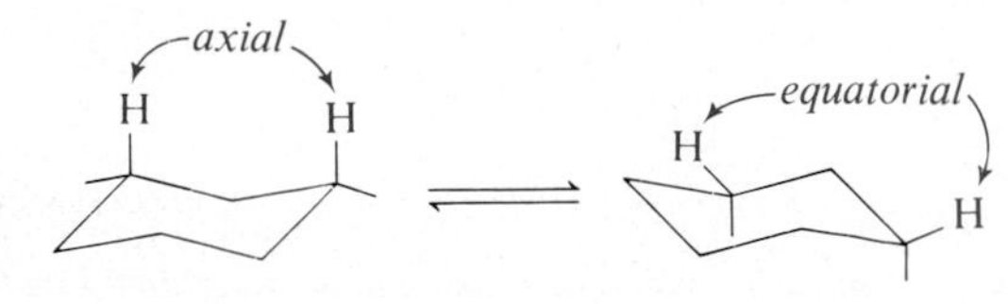

Thus, in cyclohexane, where the two chair conformations are readily interconverted, each hydrogen atom is axial half of the time and equatorial the other half of the time.

If one of the hydrogen atoms of cyclohexane is replaced by a methyl group or another substituent, the group will occupy an axial position in one chair and an equatorial position in the other chair. This means that the two chair conformations are no longer of equal stability. In general, bulky substituents do not occupy axial positions because of the strain arising from the interaction with the two other axial groups on the same side of the ring. Such is the case of the hydrogens in methylcyclohexane. On the other hand, a substituent group in an equatorial position is as far away as possible from the other atoms on the ring, causing the chair form with the substituent equatorial to be favored at equilibrium. When the substituent is methyl, most of the molecules exist in the chair conformation with the $—CH_3$ in the less crowded equatorial position, and only a small percentage (about 5%) of the molecules of methylcyclohexane have the methyl group in the axial position. The two chair conformations of methylcyclohexane are shown in Figure 2.10.

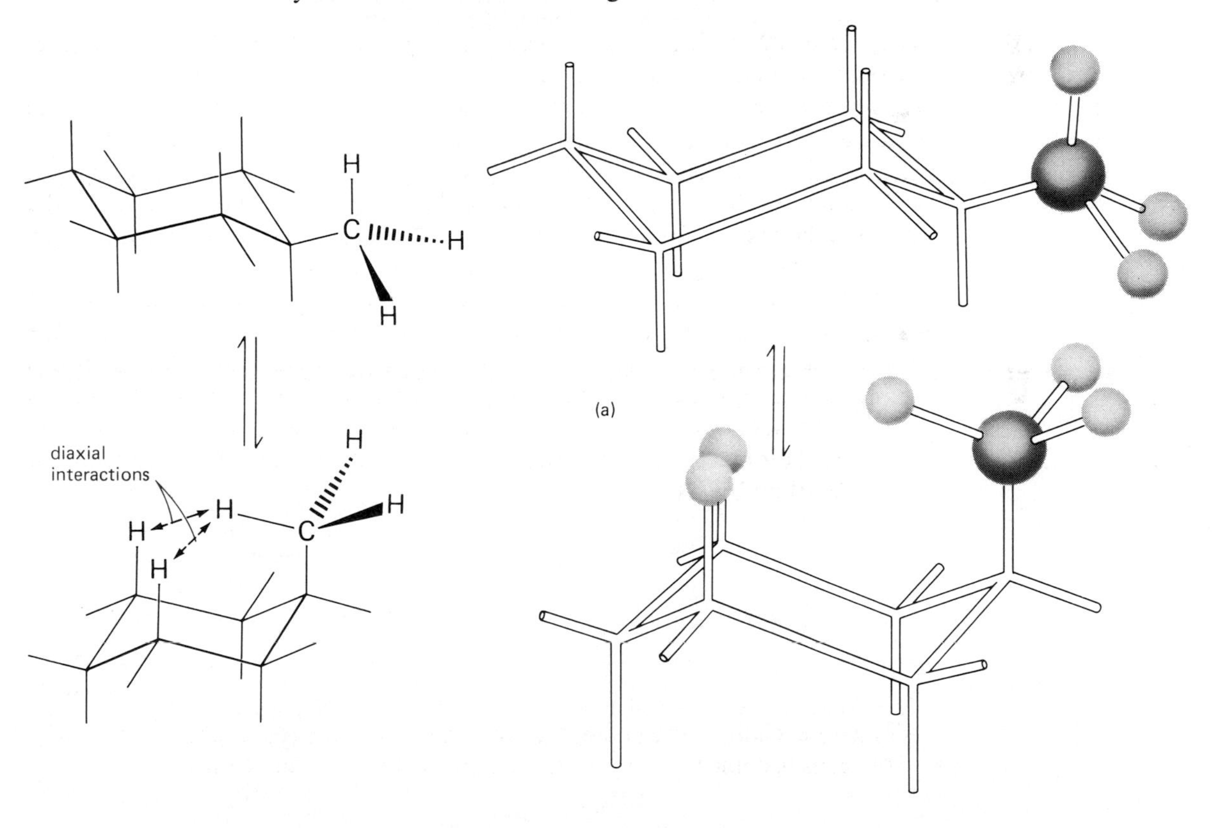

Figure 2.10 *Two chair conformations of methylcyclohexane. The two axial-axial interactions make conformation (b) considerably less likely than conformation (a).*

As the size of the substituent is increased, the preference for the conformations with the group equatorial is increased. When the group is as large as *tert*-butyl, the equatorial conformation is 10,000 times more abundant at room temperature than the axial and, in effect, the ring is "locked" into this chair conformation.

2.7 STEREOISOMERISM IN CYCLOALKANES

Because of the restricted rotation about the carbon-carbon single bonds imposed by the ring structure, cycloalkanes show a type of isomerism called stereoisomerism. To put this type of isomerism in perspective, recall our discussion of isomerism. Structural isomers have the same molecular formula but different orders of attachment of the atoms. With stereoisomerism we deal with compounds of the same molecular formula, the same order of attachment of the atoms, but different arrangements of the atoms in space.

The term *cis-trans* isomerism is applied to the type of stereoisomerism that depends on the arrangement of substituent groups, either on a cyclic structure, as we shall discuss here, or on a double bond, as will be shown in the following chapter.

The principle of *cis-trans* isomerism in cyclic structures can be illustrated by looking at models of 1,2-dimethylcyclopentane. To see these isomers, draw the cyclopentane ring as a planar pentagon and as though you are looking at it through the plane of the ring. In the following drawings, the carbon-carbon bonds of the ring projecting toward you are shown as heavy lines. Substituents attached to the ring project above and below the plane of the ring. In one of the isomers of 1,2-dimethylcyclopentane, both methyl groups are on the same side of the ring; in the other, they are on opposite sides of the ring. The prefix *cis*- indicates that the substituents are on the same side of the ring; the prefix *trans*- indicates that they are on opposite sides of the ring.

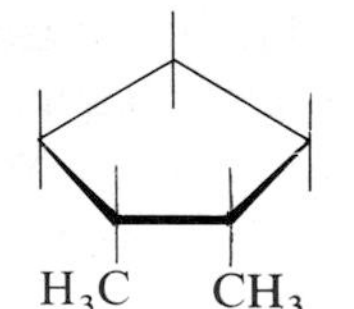

cis-1,2-dimethylcyclopentane

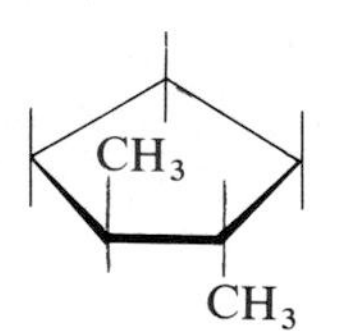

trans-1,2-dimethylcyclopentane

Example 2.6 Following are several cycloalkanes of molecular formula C_6H_{14}. State which show *cis-trans* isomerism and for each that does, draw both *cis*- and *trans*-isomers.

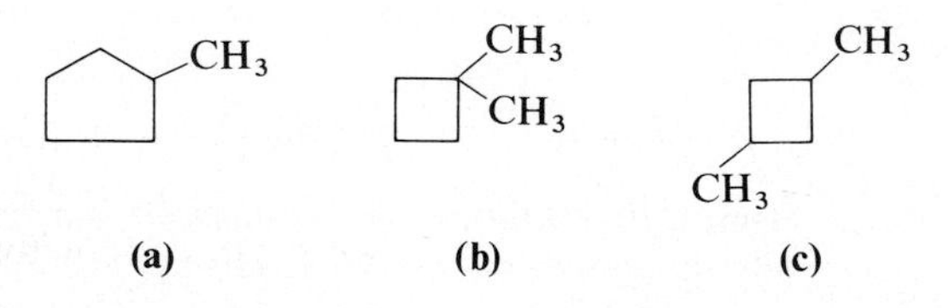

Solution

(a) Methylcyclopentane does not show *cis-trans* isomerism because it has only one substituent on the ring.
(b) 1,1-Dimethylcyclobutane does not show *cis-trans* isomerism because there is only one possible arrangement for the two methyl groups on the ring.
(c) 1,3-Dimethylcyclobutane does show *cis-trans* isomerism. To show this isomerism, draw the cyclobutane ring as though you are looking at the molecule through the plane of the ring. The two methyl groups are on the same side of the ring in the *cis*-isomer and on opposite sides in the *trans*-isomer.

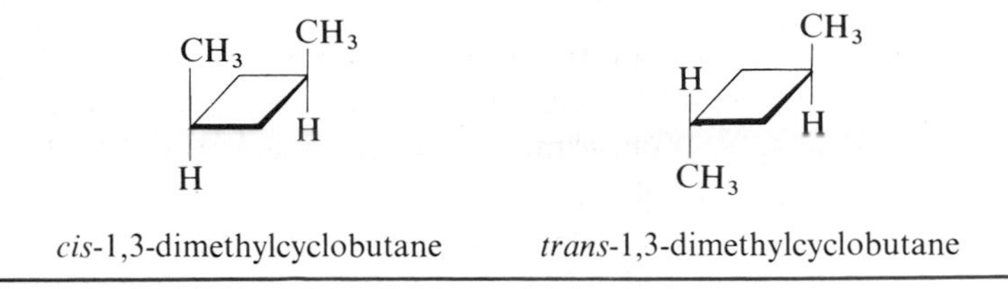

cis-1,3-dimethylcyclobutane *trans*-1,3-dimethylcyclobutane

PROBLEM 2.6 Following are several cycloalkanes of molecular formula C_7H_{16}. State which cycloalkanes show *cis-trans* isomerism and for each that does, draw both *cis-* and *trans*-isomers.

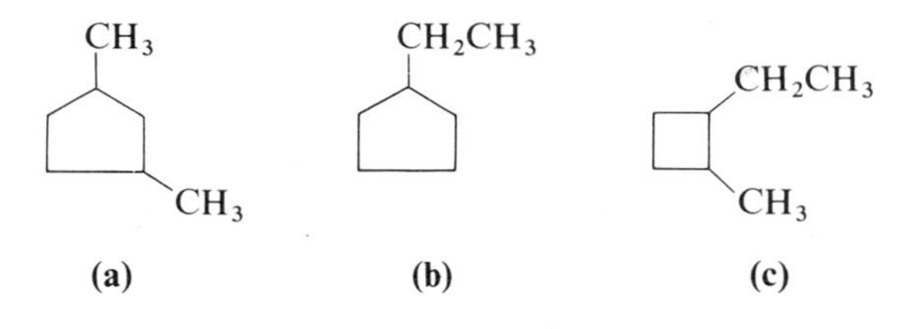

For cyclohexanes and larger rings, the situation is somewhat more complicated by the existence of various ring conformations. We will consider only the case of cyclohexane. There are two *cis-trans* isomers of 1,2-dimethylcyclohexane. How should we draw these isomers? For the purpose of deciding the number of *cis-trans* isomers in substituted cyclohexanes, it is adequate to draw the cyclohexane ring as a planar hexagon. Heavy lines show the carbon-carbon bonds of the cyclohexane ring nearest you.

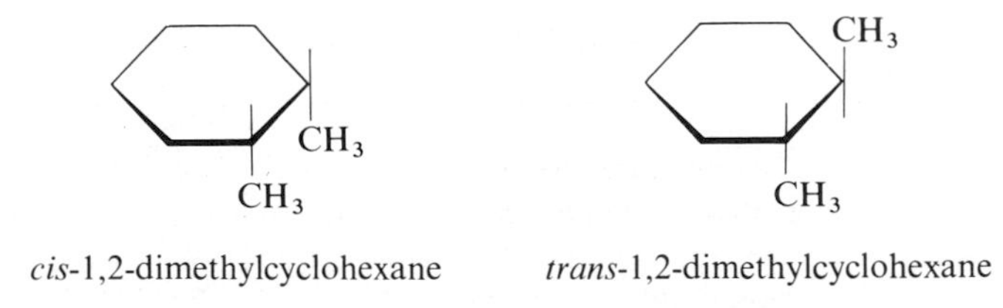

cis-1,2-dimethylcyclohexane *trans*-1,2-dimethylcyclohexane

Cis- and *trans*-isomers can also be drawn as nonplanar chair conformations. In one chair conformation of *trans*-1,2-dimethylcyclohexane, the two methyl groups are axial and in the other chair they are both equatorial. Of these two chair conformations, the one with the methyl groups equatorial is by far the more stable.

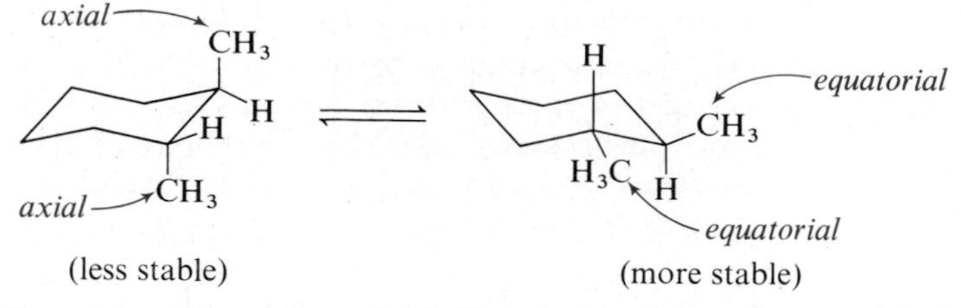

In the *cis* configuration of 1,2-dimethylcyclohexane, one of the $-CH_3$ groups occupies an equatorial position on the ring, and the other occupies an axial position.

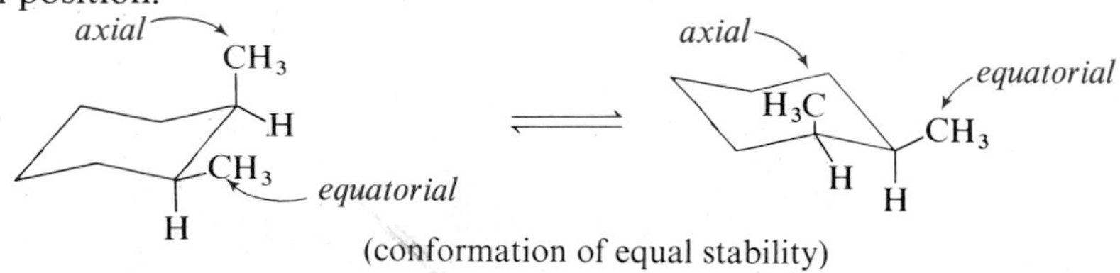

(conformation of equal stability)

These two chair conformations are of equal stability for in each, one methyl group is equatorial and one is axial.

Example 2.7

Following is a chair conformation of 1,3-dimethylcyclohexane.

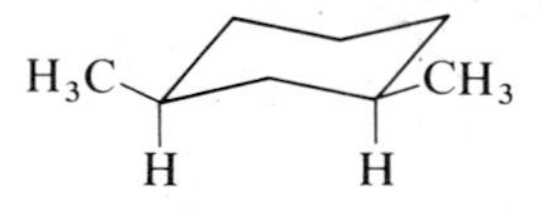

(a) Is this a chair conformation of *cis*-1,3-dimethylcyclohexane or of *trans*-1,3-dimethylcyclohexane?
(b) Draw a planar hexagon representation of this isomer.
(c) Draw the other chair conformation of this isomer. Of the two chair conformations, which is the more stable?

Solution

(a) The isomer is *cis*-1,3-dimethylcyclohexane because the two methyl groups are on the same side of the ring.

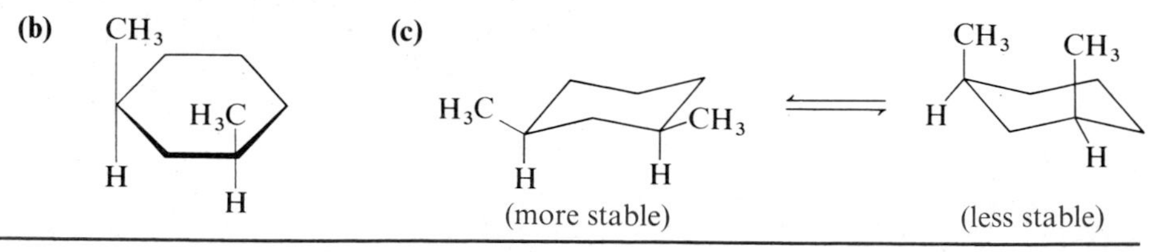

PROBLEM 2.7 Following is a planar hexagon representation of one isomer of 1,2,4-trimethylcyclohexane. Draw two chair conformations of this isomer and state which is the more stable.

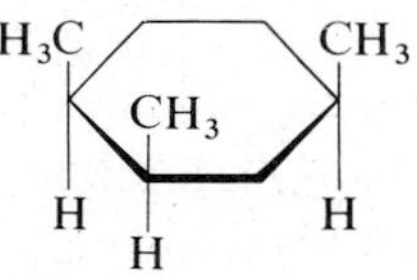

2.8 PHYSICAL PROPERTIES OF ALKANES

You are already familiar with the physical properties of some alkanes and cycloalkanes from your everyday experiences. The low molecular weight alkanes such as methane (marsh gas), ethane, propane, and butane are gases at room temperature and atmospheric pressure. Higher molecular weight alkanes such as those in gasoline and kerosine are liquids. The very high molecular weight alkanes such as those found in paraffin wax are solids. Melting points, boiling points, and densities of the first ten alkanes are listed in Table 2.4.

How can we account for these trends in physical properties in terms of the molecular structure of alkanes? Alkanes are nonpolar molecules, and the only forces of attraction between alkane molecules are dispersion forces. Dispersion forces occur because atoms and molecules, even those with no permanent dipole moments, behave as if their electron clouds are constantly moving in relation to the nucleus. This movement of electron clouds creates temporary dipole moments, i.e., temporary positive and negative charges. The attractive forces between the temporary dipole moments of adjacent atoms and molecules are called dispersion forces. Because alkanes interact only by dispersion forces, the boiling points of alkanes are lower than those of almost any other type of substances of the same molecular weights. As the number of atoms and the molecular weight of an alkane increase, the number of dispersion forces per molecule also increases. Therefore, the boiling points of alkanes increase as the molecular weight increases. This accounts for the fact that the boiling point of octane (C_8H_{18}) is considerably higher than that of butane (C_4H_{10}).

The melting points of alkanes also increase with increasing molecular weight. However, the increase is not as regular as that observed for boiling points.

The density of the alkanes listed in Table 2.4 is about 0.7g/mL. That of the higher molecular weight hydrocarbons found in crude oil and paraffin wax is about 0.8g/mL. Thus all alkanes are less dense than water (1.0g/mL).

We all know that oil and water do not mix. For example, gasoline and crude oil do not dissolve in water. We refer to these water-insoluble substances as

Table 2.4 Physical properties of alkanes.

Name	*Structural formula*	*mp* (°C)	*bp* (°C)	*density* (g/mL)
methane	CH_4	−182	−164	
ethane	CH_3CH_3	−183	−88	
propane	$CH_3CH_2CH_3$	−190	−42	
butane	$CH_3(CH_2)_2CH_3$	−138	0	
pentane	$CH_3(CH_2)_3CH_3$	−130	36	0.626
hexane	$CH_3(CH_2)_4CH_3$	−95	69	0.659
heptane	$CH_3(CH_2)_5CH_3$	−90	98	0.684
octane	$CH_3(CH_2)_6CH_3$	−57	126	0.703
nonane	$CH_3(CH_2)_7CH_3$	−51	151	0.718
decane	$CH_3(CH_2)_8CH_3$	−30	174	0.730

hydrophobic (water-hating). Alkanes do dissolve in nonpolar liquids such as carbon tetrachloride (CCl_4), carbon disulfide (CS_2), and of course other hydrocarbons.

Alkanes that are structural isomers of each other are different substances and therefore have different physical and chemical properties. Table 2.5 lists the boiling points, melting points, and densities of the five structural isomers of molecular formula C_6H_{14}. Notice that the boiling point of each of the branched-chain isomers is lower than hexane itself, and that the more branching there is, the lower the boiling point. Why does branching lower the boiling point? Remember that the only forces of attraction between alkane molecules are very weak dispersion forces and that boiling point depends on the extent of contact between molecules and the total number of intermolecular attractive forces. As branching increases, the shape of an alkane molecule becomes more compact, the contact between molecules decreases, and the number of dispersion forces decreases. Figure 2.11 shows structural formulas and ball-and-stick models of hexane and its most highly branched isomer, 2,2-dimethylbutane.

For any group of alkane structural isomers, it is usually observed that the least branched isomer has the highest boiling point and the most branched isomer has the lowest boiling point.

Example 2.8

Arrange the following in order of increasing boiling points.

(a) $CH_3CH_2CH_2CH_3$ $CH_3CH_2CH_2CH_2CH_2CH_3$ $CH_3(CH_2)_8CH_3$

(b) $CH_3(CH_2)_6CH_3$ $CH_3C(CH_3)_2CH_2CH(CH_3)CH_3$ $CH_3CH(CH_3)CH_2CH_2CH_2CH_2CH_3$

Solution

(a) All are unbranched alkanes. Therefore, as the number of carbon atoms in the chain increases, the dispersion forces between molecules increase and boiling points also increase.

$CH_3CH_2CH_2CH_3$	$CH_3CH_2CH_2CH_2CH_2CH_3$	$CH_3(CH_2)_8CH_3$
butane	hexane	decane
(bp −0.5°C)	(bp 69°C)	(bp 174°C)

(b) These three alkanes have the same molecular formula, C_8H_{18}, and are structural isomers of each other. Therefore, relative boiling points depend on the degree of branching. The most highly branched is 2,2,4-trimethylpentane; the least highly branched is octane. Dispersion forces are least between molecules of 2,2,4-trimethylpentane. Therefore, it has the lowest boiling point. Dispersion forces are greatest between molecules of octane and it has the highest boiling point.

$CH_3C(CH_3)_2CH_2CH(CH_3)CH_3$	$CH_3CH(CH_3)CH_2CH_2CH_2CH_2CH_3$	$CH_3(CH_2)_6CH_3$
2,2,4-trimethylpentane	2-methylheptane	octane
(bp 99°C)	(bp 118°C)	(bp 126°C)

Table 2.5 Physical properties of the isomeric alkanes of molecular formula C_6H_{14}.

Name	*bp* (°C)	*mp* (°C)	*density* (g/mL)
hexane	68.7	−95	0.659
2-methylpentane	60.3	−154	0.653
3-methylpentane	63.3	−118	0.664
2,3-dimethylbutane	58.0	−129	0.661
2,2-dimethylbutane	49.7	−98	0.649

$CH_3—CH_2—CH_2—CH_2—CH_2—CH_3$

hexane
bp 68.7°C

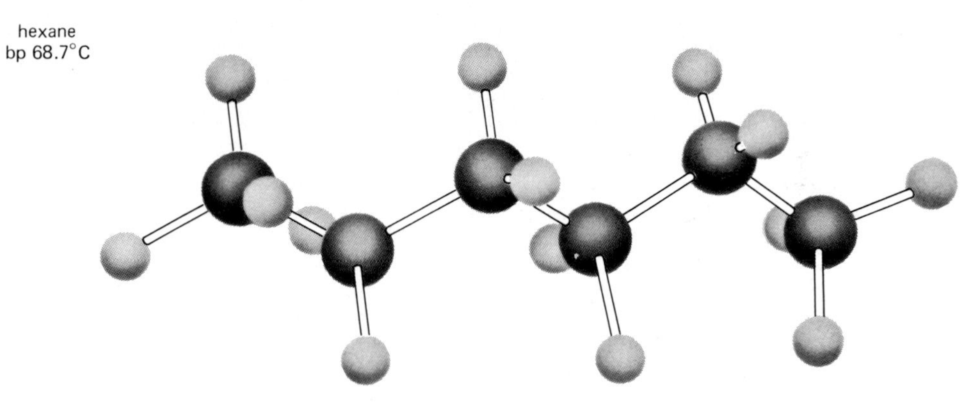

(a)

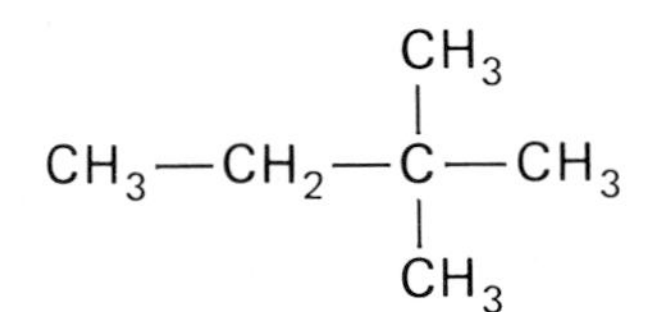

2, 2-dimethylbutane
bp 49.7°C

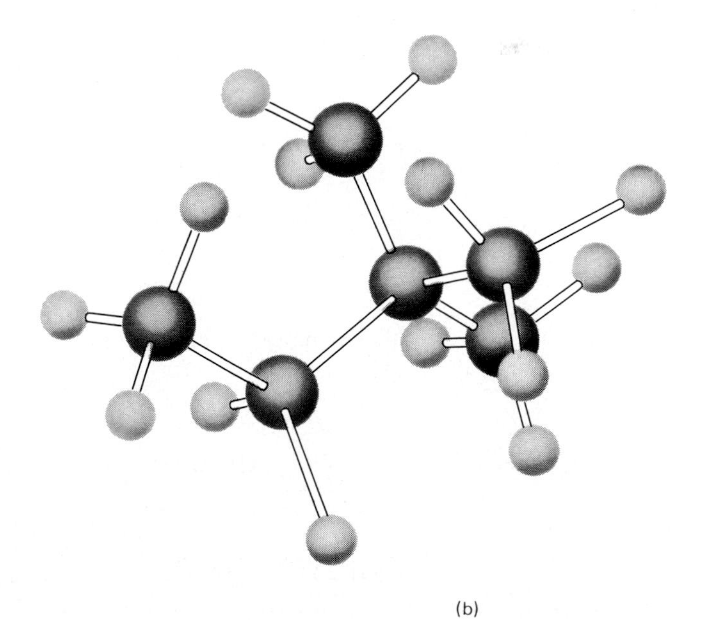

(b)

Figure 2.11 *Ball-and-stick models of (a) hexane and (b) 2,2-dimethylbutane.*

PROBLEM 2.8 Arrange the following in order of increasing boiling point.

(a) 2-methylbutane 2,2-dimethylpropane pentane
(b) 3,3-dimethylheptane nonane 2,2,4-trimethylpentane

2.9 REACTIONS OF ALKANES

As we have already seen in Section 1.15, saturated hydrocarbons are quite inert because a carbon atom bonded to four other carbon or hydrogen atoms has a completely filled valence shell and no unshared pairs of electrons. Therefore, such a carbon atom is not susceptible to attack by either electron-deficient or electron-rich reagents. However, saturated hydrocarbons do react with halogens and with oxygen. We will consider both of these reactions, the first because it is useful in the preparation of substituted alkanes, the second because it is the basis for the use of saturated hydrocarbons (and unsaturated hydrocarbons as well) as fuel.

A. HALOGENATION OF ALKANES

If a mixture of methane and chlorine gas is kept in the dark at room temperature, no detectable change occurs. However, if the mixture is heated or exposed to light, a reaction begins almost at once with the evolution of heat. The products are chloromethane (methyl chloride) and hydrogen chloride. What occurs is a substitution reaction—the replacement of a hydrogen atom on methane by chlorine and the production of an equivalent amount of hydrogen chloride.

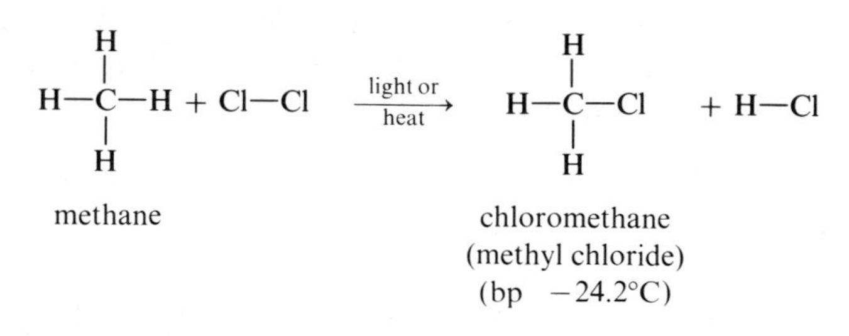

This reaction is shown in more detail in Figure 2.12

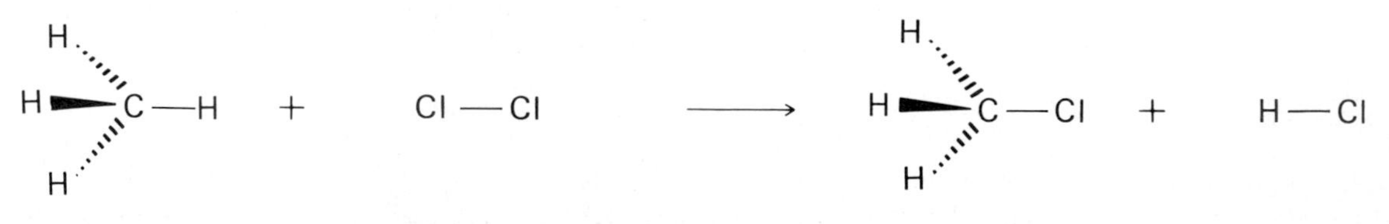

Figure 2.12 *The chlorination of methane.*

Reaction of ethane with chlorine gives chloroethane.

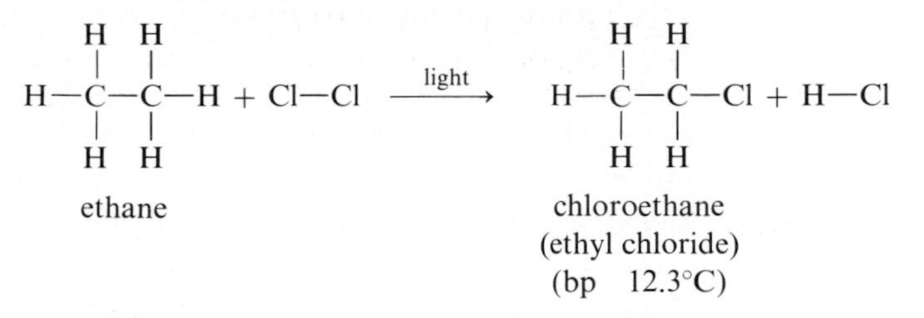

Reaction of propane with chlorine gives two isomeric chloropropanes.

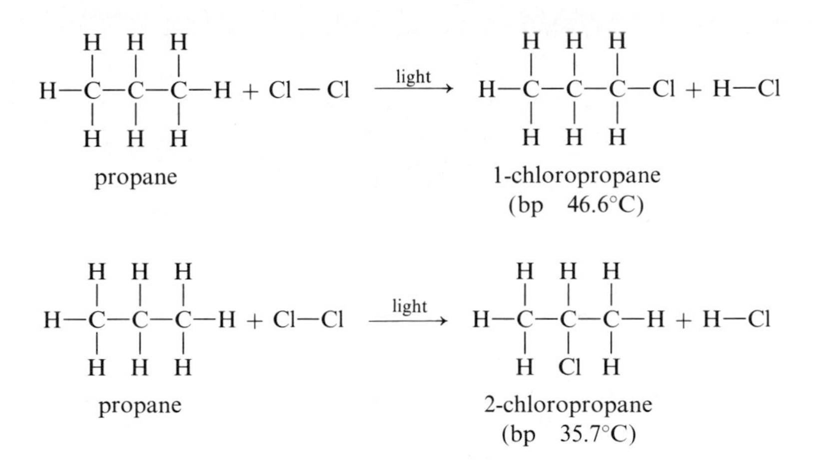

These are two different reactions and we have written them separately. However, it is much more common to show all of the different products that can be formed in one equation. The single equation below means that reaction of one mole of propane with one mole of chlorine produces one mole of HCl and a mixture totalling one mole of 1-chloropropane and 2-chloropropane combined.

$$CH_3{-}CH_2{-}CH_3 + Cl_2 \xrightarrow{\text{light}} CH_3{-}CH_2{-}CH_2{-}Cl + CH_3{-}\underset{\displaystyle Cl}{\underset{|}{CH}}{-}CH_3 + HCl$$

If chloromethane is allowed to react with more chlorine, further chlorination produces a mixture of dichloromethane (methylene chloride), trichloromethane (chloroform), and tetrachloromethane (carbon tetrachloride). These various chlorination products of methane have different boiling points and can be separated from one another by distillation.

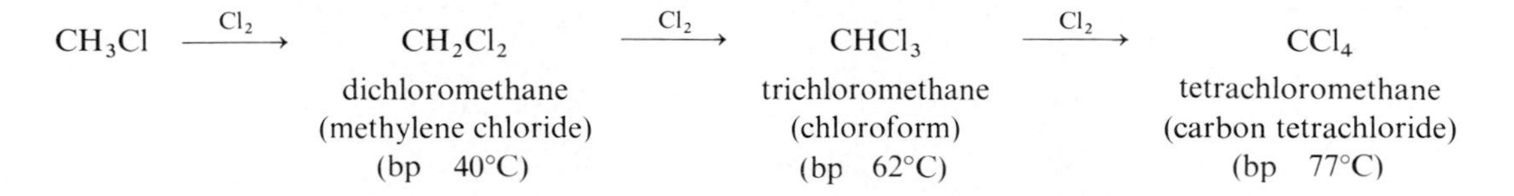

As you might expect, higher alkanes and those with more branching yield much more complex mixtures of chlorinated hydrocarbons.

Haloalkanes are named according to the rules we have already used for naming alkanes. The parent chain is numbered from the direction which gives the alkyl substituent encountered first the lowest number. Halogen substituents are not taken into consideration in numbering the parent alkane. Halogen substituents are indicated by the prefixes fluoro-, chloro-, bromo-, and iodo-.

Example 2.9 Name and draw structural formulas for all monochlorination products of the following reactions.

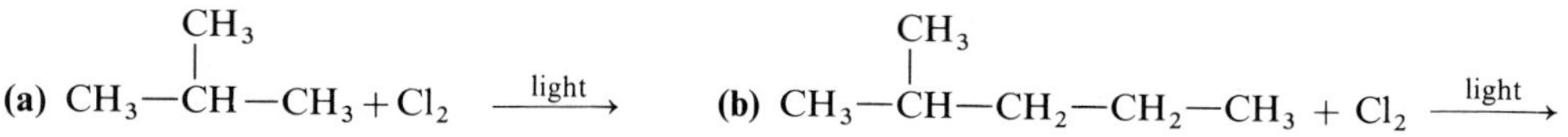

$$\textbf{(a)}\ CH_3-\overset{\displaystyle CH_3 \atop |}{C}H-CH_3 + Cl_2 \xrightarrow{\text{light}} \qquad \textbf{(b)}\ CH_3-\overset{\displaystyle CH_3 \atop |}{C}H-CH_2-CH_2-CH_3 + Cl_2 \xrightarrow{\text{light}}$$

Solution As in drawing all possible structural isomers of a given molecular formula, first devise a system and then follow it. The most direct way is to start at one end of the carbon chain and substitute —Cl for —H. Then do the same thing on each carbon until you come to the other end of the chain. There are only two monochlorination derivatives of 2-methylpropane.

(a)

$$Cl-CH_2-\overset{\displaystyle CH_3 \atop |}{C}H-CH_3 \qquad CH_3-\underset{| \atop \displaystyle Cl}{\overset{\displaystyle CH_3 \atop |}{C}}-CH_3$$

1-chloro-2-methylpropane   2-chloro-2-methylpropane

(b) There are five monochlorination derivatives of 2-methylpentane.

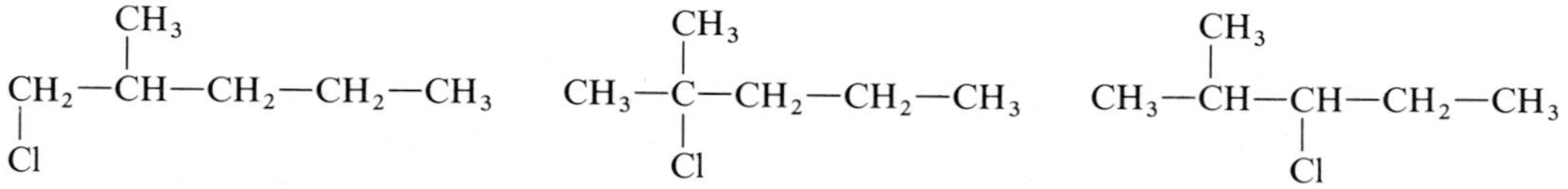

$$\underset{| \atop \displaystyle Cl}{CH_2}-\overset{\displaystyle CH_3 \atop |}{C}H-CH_2-CH_2-CH_3 \qquad CH_3-\underset{| \atop \displaystyle Cl}{\overset{\displaystyle CH_3 \atop |}{C}}-CH_2-CH_2-CH_3 \qquad CH_3-\overset{\displaystyle CH_3 \atop |}{C}H-\underset{| \atop \displaystyle Cl}{C}H-CH_2-CH_3$$

1-chloro-2-methylpentane   2-chloro-2-methylpentane   3-chloro-2-methylpentane

$$CH_3-\overset{\displaystyle CH_3 \atop |}{C}H-CH_2-\underset{| \atop \displaystyle Cl}{C}H-CH_3 \qquad CH_3-\overset{\displaystyle CH_3 \atop |}{C}H-CH_2-CH_2-\underset{| \atop \displaystyle Cl}{CH_2}$$

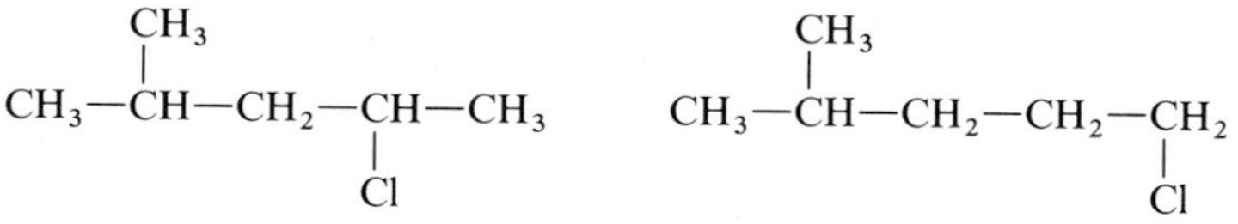

4-chloro-2-methylpentane   5-chloro-2-methylpentane

PROBLEM 2.9 Name and draw structural formulas for all monohalogenation derivatives of:

$$\textbf{(a)}\ CH_3-CH_2-CH_2-CH_2-CH_2-CH_3 + Cl_2 \xrightarrow{\text{light}}$$

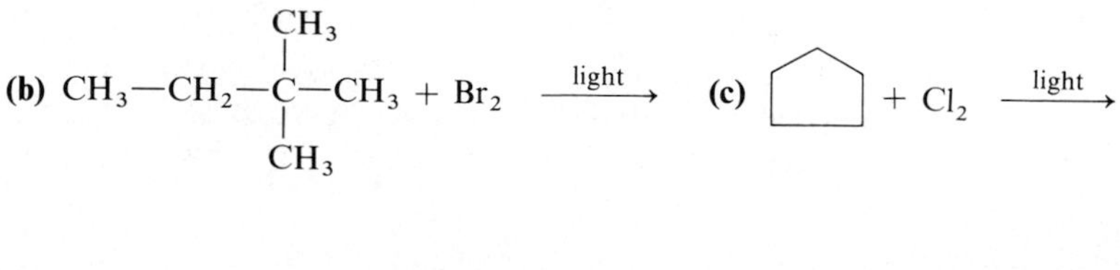

$$\textbf{(b)}\ CH_3-CH_2-\underset{| \atop \displaystyle CH_3}{\overset{\displaystyle CH_3 \atop |}{C}}-CH_3 + Br_2 \xrightarrow{\text{light}} \qquad \textbf{(c)}\ \text{(cyclopentane)} + Cl_2 \xrightarrow{\text{light}}$$

B. OXIDATION OF ALKANES

The reaction of alkanes with oxygen to form carbon dioxide and water is the basis for the use of hydrocarbons as a source of heat (natural gas and liquefied petroleum gas) and power (fuel for the internal combustion engine).

$$CH_4 + 2O_2 \longrightarrow CO_2 + 2H_2O \qquad \Delta H = -212 \text{ kcal/mole}$$

$$CH_3C(CH_3)_2CH_2CH(CH_3)CH_3 + \tfrac{25}{2}O_2 \longrightarrow 8CO_2 + 9H_2O \qquad \Delta H = -1304 \text{ kcal/mole}$$

2,2,4-trimethylpentane
"isooctane"

The preceding combustion reactions illustrate the burning of methane (the major component of natural gas) and 2,2,4-trimethylpentane (a hydrocarbon with an octane rating of 100) to form carbon dioxide, water, and the generation of heat energy.

2.10 SOURCES OF ALKANES

The two major sources of alkanes throughout the world are natural gas and petroleum. Natural gas consists of approximately 80% methane, 10% ethane, and a mixture of other relatively low-boiling alkanes chiefly propane, butane, and 2-methylpropane. Almost all of the ethane in natural gas is separated and used for the production of ethylene, and for this reason the natural gas supplied for home heating is greater than 90% methane.

Petroleum is a liquid mixture of literally thousands of substances, most of them hydrocarbons, formed from the decomposition of marine plants and animals. Petroleum and petroleum-derived products fuel automobiles, aircraft, and trains. They provide heat for buildings and fuel for electric generating plants. They provide most of the greases and lubricants required for the machinery of our highly industrialized society. Furthermore, petroleum, along with natural gas, provides close to 90% of the organic raw materials for the synthesis and manufacture of synthetic fibers, plastics, detergents, and a multitude of other products.

It is the task of the petroleum refinery to produce usable products, with a minimum of waste, from the thousands of different hydrocarbons in this liquid mixture. The various physical and chemical processes for this purpose fall into two broad categories: separation processes which simply separate the complex mixture into various fractions and conversion processes which alter the molecular structure of the hydrocarbon components themselves. An example of how these two processes have been used over the years to respond to changing economic and market demand can be seen in the yield of gasoline per barrel of crude oil. In 1920 the average yield of gasoline that could be obtained by separation processes was 11 gallons per 42-gallon barrel of crude, or 26%.

With the introduction of new and more sophisticated conversion processes, the yield has been increased to approximately 50%.

The fundamental process in refining petroleum is distillation. Practically all crude oil that enters a refinery today goes to distillation units where it is heated to temperatures as high as 370°C to 425°C and separated into fractions or "cuts." Each fraction contains a mixture of hydrocarbons that boils within a particular range. Following are the common names associated with several of these fractions along with the major uses of each.

(1) Gases boiling below 20°C are taken off at the top of the distillation column. This fraction is a mixture of low-molecular-weight hydrocarbons, predominantly propane, butane, and 2-methylpropane. These three hydrocarbons can be liquefied under pressure at room temperature. The liquefied mixture, known as liquefied petroleum gas (LPG) can be stored easily and shipped in metal tanks and is therefore a convenient source of gaseous fuel.

(2) Naphthas, bp 20° to 200°C, are a mixture of C_4 to C_{10} alkanes and cycloalkanes. The naphthas may also contain some aromatic hydrocarbons such as benzene, toluene, and xylene. The light naphtha fraction, bp 20 to 150°C, is the source of what is known as straight run gasoline and averages approximately 25% of crude petroleum. In a sense, the naphthas are the most valuable distillation fractions for they are useful not only as fuel but also as a source of raw materials for the organic chemical industry.

(3) Kerosene, bp 175 to 275°C, is a mixture of C_9 to C_{15} hydrocarbons and is used for fuel and heat.

(4) Gas oil, bp 200 to 400°C, is a mixture of C_{15} to C_{25} hydrocarbons. It is from this fraction that diesel fuel is obtained.

(5) Lubricating oil and heavy fuel oil distill from the column at temperatures over 350°C.

(6) Asphalt is the name given to the black, tarry residue remaining after the removal of the other volatile fractions.

Gasoline is a complex mixture of C_4 to C_{10} hydrocarbons. The quality of gasoline as a fuel for the internal combustion engine is expressed in terms of octane number or antiknock index. When an engine is running normally, the air-fuel mixture is ignited by the spark plug and burns smoothly as the flame moves outward from the plug, building up pressure that forces the piston down during the compression stroke. Engine knocking occurs when a portion of the air-fuel mixture explodes prematurely (usually as a result of heat developed during compression), and independently of ignition by the spark plug. The basic procedure for measuring the antiknock quality of gasoline was established in 1929. Two compounds were selected as reference fuels. One of these, 2,2,4-trimethylpentane, "isooctane", has very good antiknock properties and was assigned an octane number of 100. The other, heptane, has poor antiknock properties and was assigned an octane number of 0. The octane rating of a particular gasoline is that percent isooctane in a mixture of isooctane and heptane that has equivalent knock properties. For example, the knock properties of 2-methylhexane are the same as those of a mixture of 42% isooctane and 58% heptane. Therefore, the octane rating of 2-methylhexane is 42. Octane itself has an octane rating of −20, which means that it produces even more engine knocking than heptane.

Early in the 20th century it became obvious that conventional separation techniques could not meet the needs of the radically changing market for petroleum products. The automobile had created an enormous demand for gasoline. Yet refiners could only produce as much gasoline as crude oil contained naturally. They could meet the growing demand either by processing uneconomically large volumes of crude oil or by devising some means to convert other hydrocarbon fractions into hydrocarbons in the gasoline range. The need to adapt processing of crude oil stocks was made more imperative by the new demands for aviation fuels in the 1930s and 1940s and the demand for jet fuel beginning in the 1950s.

What was needed by the petroleum refining industry was a way to increase sharply the yield of gasoline from crude oil and, at the same time, significantly improve octane ratings. Both of these goals were achieved by a catalytic cracking process developed by the French engineer, Eugene Houdry. The first commercial plant to use the Houdry process was built in 1937 by Sun Oil. By 1976, catalytic cracking capacity in the United States reached 150 million gallons of feedstock a day. Catalytic cracking processes also produce substantial amounts of hydrogen (H_2) and low-molecular-weight hydrocarbons such as ethylene, propene, and butenes.

As demand for gasoline grew steadily, petroleum chemists began to seek ways to convert these low-molecular-weight by-product hydrocarbons into larger hydrocarbons in the gasoline range. Of the processes developed, alkylation is the most significant in the United States. The alkylation process can be illustrated by the reaction of 2-methylpropene and 2-methylpropane to form a branched-chain C_8 hydrocarbon.

$$\underset{\text{2-methylpropene}}{CH_3-\overset{\overset{CH_3}{|}}{C}=CH_2} + \underset{\text{2-methylpropane}}{CH_3-\overset{\overset{CH_3}{|}}{CH}-CH_3} \xrightarrow[\text{or HF}]{H_2SO_4} \underset{\text{2,2,4-trimethylpentane (isooctane)}}{CH_3-\underset{\underset{CH_3}{|}}{\overset{\overset{CH_3}{|}}{C}}-CH_2-\overset{\overset{CH_3}{|}}{CH}-CH_3}$$

The first commercial alkylation plant was built in 1938 by Humble Oil. By the end of World War II, refiners were producing more than 3 million gallons of alkylate daily. Current production of alkylate exceeds 30 million gallons per day. Alkylate is a prime ingredient in no-lead high-octane motor gasoline.

Catalytic reforming supplements catalytic cracking as a means of converting low-octane hydrocarbons into high-octane components. Catalytic reforming, for example, converts hexane into cyclohexane and then into benzene, a stable high-octane aromatic.

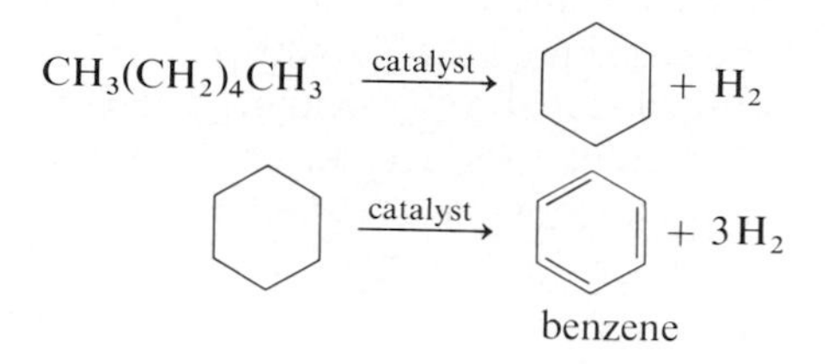

The first catalytic reforming process came into use in 1940. It used a silica-molybdena catalyst in the presence of hydrogen gas. In 1949 Universal Oil Products introduced a platinum catalyst. This process, called Platforming, is extremely effective and remains the dominant catalytic reforming process used today. The petroleum industry now uses catalytic reforming to treat more than 120 million gallons per day of feedstock, or close to 25% of the crude oil that enters refineries.

2.11 COMMERCIALLY IMPORTANT HALOGENATED HYDROCARBONS

Because of their physical and chemical properties, several of the halogenated hydrocarbons have found wide commercial use. These include applications as commerical solvents, refrigerants, dry cleaning agents, local and general anesthetics, and insecticides.

Carbon tetrachloride, CCl_4, is a dense, nonflammable liquid. With bp 77°C it is remarkably inert to most common reagents and laboratory conditions. It is immiscible with water, but it is a good solvent for oils and greases, and at one time found wide use in the dry cleaning industry. It is somewhat toxic, readily absorbed through the skin, and like all organic solvents should be used only with adequate ventilation. Prolonged exposure to carbon tetrachloride vapors results in liver and renal damage.

Chloroform, $CHCl_3$, is a colorless, rather sweet-smelling liquid, bp 61°C. It too is a widely used solvent for organic substances. In the past chloroform was used extensively as a general anesthetic for surgery, but it is rarely used for this purpose now because it is known to cause extensive liver damage.

Ethyl chloride, CH_3CH_2Cl, is used as a fast-acting, topically applied local anesthetic. It owes its anesthetic property more to its physical than chemical characteristics. Ethyl chloride boils at 13°C, and unless under pressure, it is a gas at room temperature. When sprayed on the skin, it evaporates rapidly and cools the skin surface and underlying nerve endings. Underlying nerve endings become anesthesized when skin temperature drops to about 30°C.

Halothane is a recently discovered and now widely used inhalation anesthetic. It has distinct advantages over other general inhalation anesthetics (as for example diethyl ether and cyclopropane) in that it is nonflammable, nonexplosive, and causes minimum discomfort to the patient. Although some cases of liver damage have been reported, halothane's record as a safe anesthetic is impressive.

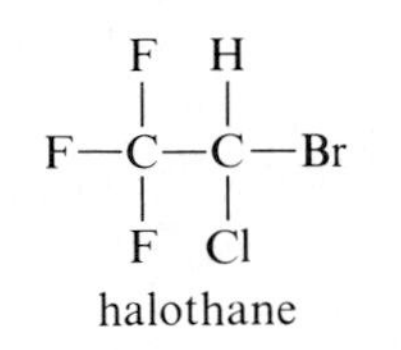

halothane

Some of the more complex chlorinated hydrocarbons are poisons and, as you well know, have found wide use as insecticides. Undoubtedly the best known, the cheapest, and the most astonishingly effective is DDT.

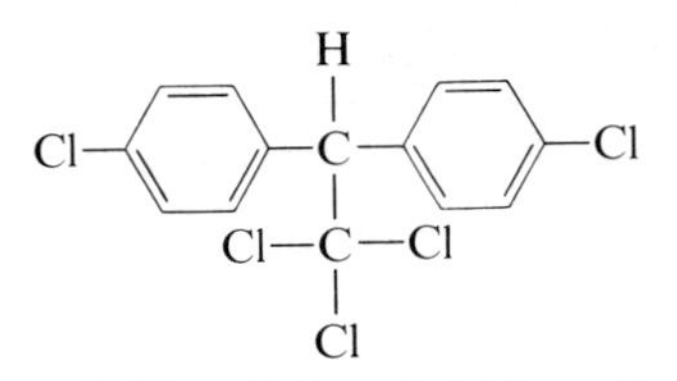

DDT (dichlorodiphenyltrichloroethane)

Other widely known and used polychlorinated hydrocarbons are Dieldrin, Aldrin, Chlordane, and Lindane (Problem 2.24).

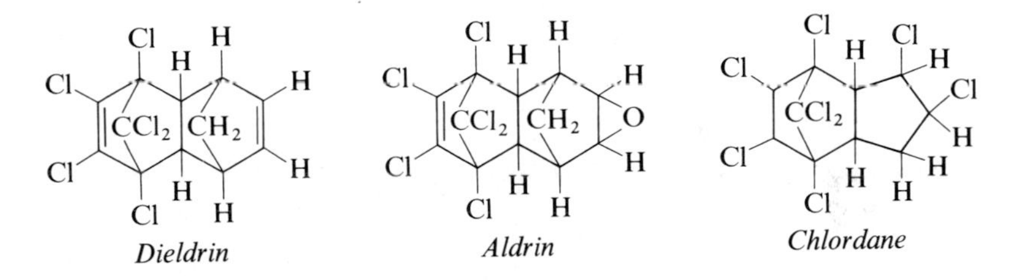

Dieldrin *Aldrin* *Chlordane*

The use of polychlorinated hydrocarbon insecticides is now carefully regulated and in some instances prohibited. In the search for more ecologically and environmentally sound methods of insect control, recent interest has turned to such means as insect attractants (see the mini-essay *Pheromones*) and insect growth regulators (see the mini-essay *Insect Juvenile Hormones*).

Of all the fluorinated hydrocarbons, those manufactured under the trade name Freon have had the most dramatic impact. The first of the Freons was developed in a search for new refrigerants—substances that would be nontoxic, nonflammable, odorless, and noncorrosive. In 1930, General Motors announced the discovery of just such a compound, dichlorodifluoromethane, which was marketed under the trade name Freon-12.

The Freons are manufactured by reacting a chlorinated hydrocarbon with hydrofluoric acid in the presence of an antimony pentafluoride or antimony chlorofluoride catalyst. Both Freon-11 and Freon-12 can be prepared from carbon tetrachloride. Freon-22, monochlorodifluromethane, is made from chloroform.

$$CCl_4 + HF \xrightarrow{SbF_5} \underset{\text{Freon-11}}{CCl_3F} + HCl$$

$$CCl_3F + HF \xrightarrow{SbF_5} \underset{\text{Freon-12}}{CCl_2F_2} + HCl$$

A major, new use of the Freons came during World War II with the development of aerosol insecticides for which they served as propellants. By 1974, U.S. production of Freons had grown to more than 1.1 billion pounds annually, almost one-half the world production.

Concern about the environmental impact of chlorofluorocarbons like Freon-11 and Freon-12 arose in 1974 when Drs. Sherwood Rowland and Mario Molina of the University of California, Irvine, announced their theory of ozone destruction by these substances. When used as aerosol propellants and refrigerants, chlorofluorocarbons escape to the lower atmosphere, but because of their general inertness do not decompose there. Slowly, they find their way to the stratosphere where they absorb ultraviolet radiation from the sun and then decompose. As they decompose, they set up a chemical reaction that may also lead to the destruction of the stratospheric ozone layer. What makes this a serious problem is that the stratospheric ozone layer acts as a shield for the earth against excess ultraviolet radiation. An increase in ultraviolet radiation reaching the earth may lead to the destruction of certain crops and agricultural species, climate modification, and even increased incidence of skin cancer in sensitive individuals. Controversy continues over the potential for ozone depletion and its impact on the environment. In the meantime, both government and the chemical industry have taken steps to sharply limit the production and use of chlorofluorocarbons.

IMPORTANT REACTIONS

(1) Halogenation (Section 2.9A)

$$CH_4 + Cl_2 \xrightarrow{\text{light}} CH_3Cl + HCl$$

(2) Oxidation (Section 2.9B)

$$CH_4 + 2O_2 \xrightarrow{\text{spark}} CO_2 + 2H_2O \qquad \Delta H = -212 \text{ kcal/mole}$$

PROBLEMS

2.10 Write names for the following structural formulas.

(a) $CH_3\underset{\displaystyle CH_3}{\underset{|}{C}}HCH_2CH_2CH_3$

(b) $CH_3\underset{\displaystyle CH_3}{\underset{|}{C}}HCH_2CH_2\underset{\displaystyle CH_3}{\underset{|}{C}}HCH_3$

(c) $CH_3CH_2\underset{\displaystyle CH_3}{\underset{|}{C}}HCH_2\underset{\displaystyle CH_2CH_3}{\underset{|}{C}}HCH_3$

(d) $(CH_3)_3CH$

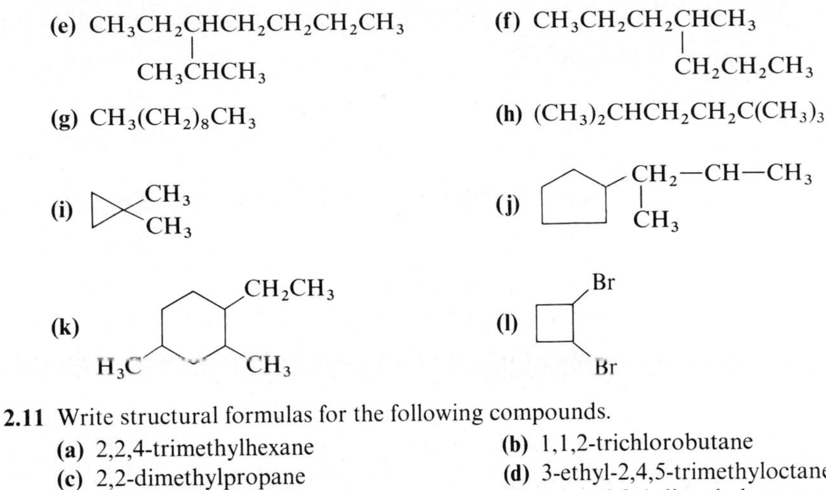

2.11 Write structural formulas for the following compounds.

(a) 2,2,4-trimethylhexane
(b) 1,1,2-trichlorobutane
(c) 2,2-dimethylpropane
(d) 3-ethyl-2,4,5-trimethyloctane
(e) 2-bromo-2,4,6-trimethyloctane
(f) 5-butyl-2,4-dimethylnonane
(g) 4-isopropyloctane
(h) 3,3-dimethylpentane
(i) 1,1,1-trichloroethane
(j) *trans*-1,3-dimethylcyclopentane
(k) *cis*-1,2-diethylcyclobutane
(l) 1,1-dichlorocycloheptane

2.12 **(a)** Name and draw structural formulas for all isomeric alkanes of molecular formula C_6H_{14}.
(b) Predict which has the lowest boiling point; the highest boiling point.

2.13 **(a)** Name and draw structural formulas for all isomeric alkanes of molecular formula C_7H_{16}.
(b) Predict which has the lowest boiling point; the highest boiling point.

2.14 There are 35 structural isomers of molecular formula C_9H_{20}. Name and draw structural formulas for the eight that have five carbon atoms in the longest chain.

2.15 Name and draw structural formulas for all cycloalkanes of molecular formula C_5H_{10}. Be certain to include *cis-trans* as well as structural isomers.

2.16 Explain why each of the following names is incorrect. Write a correct name.

(a) 1,3-dimethylbutane
(b) 4-methylpentane
(c) 2,2-diethylbutane
(d) 2-ethyl-3-methylpentane
(e) 4,4-dimethylhexane
(f) 2-propylpentane
(g) 2,2-diethylheptane
(h) 5-butyloctane
(i) 2-dimethylpropane
(j) 2-*sec*-butyloctane
(k) 4-isopentylheptane
(l) 1,3-dimethyl-6-ethylcyclohexane

2.17 Which of the following are identical compounds and which are structural isomers?

(a) $CH_3—CH_2—CH(Cl)—CH_3$

(b) $CH_3—CH(Cl)—CH_3$

(c) $CH_3—CH(Cl)—CH_2—CH_3$

(d) $CH_2(Cl)—CH_2—CH_2—CH_3$

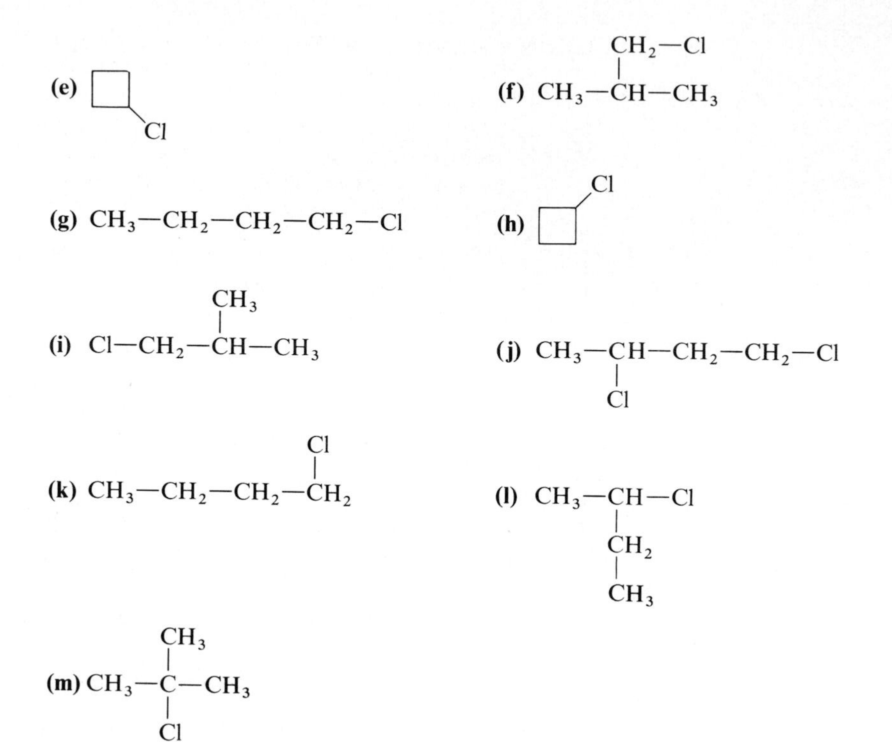

2.18 Draw a Newman projection for 1,2-dichloroethane:
 (a) where the Cl atoms are eclipsed by each other,
 (b) where the Cl atoms are eclipsed by hydrogen atoms,
 (c) in two different staggered conformations.

2.19 Draw and name the *cis*- and *trans*-isomers of dimethylcyclopropane.

2.20 Using a planar pentagon representation for the cyclopentane ring, draw structural formulas for the *cis*- and *trans*-isomers of 1,2-dimethylcyclopentane and for 1,3-dimethylcyclopentane.

2.21 Draw the alternative chair conformations for the *cis*- and *trans*-isomers of 1,2-dimethylcyclohexane; of 1,3-dimethylcyclohexane; and of 1,4-dimethylcyclohexane. Label axial and equatorial substituents.
 (a) For which isomers are the chair conformations of equal stability?
 (b) For which isomers is one chair conformation more stable than the other chair?

2.22 There are four *cis-trans* isomers of 2-isopropyl-5-methylcyclohexanol.

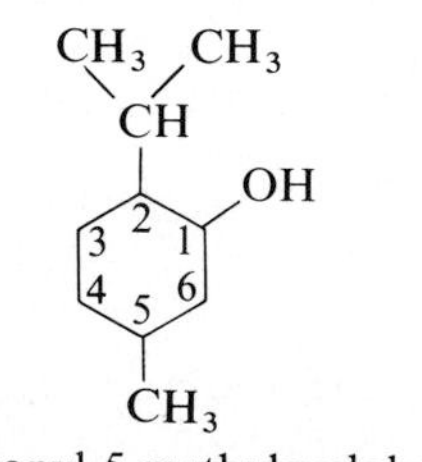

2-isopropyl-5-methylcyclohexanol

(a) Using a planar hexagon representation for the cyclohexane ring, draw structural formulas for the four *cis-trans* isomers. Label your answers (1), (2), (3), and (4).
(b) Draw the most stable chair conformation for each of your answers in part (a).
(c) Of the four *cis-trans* isomers, which would you predict to be the most stable, that is the one with the least interactions between atoms in the molecule? (If you have answered this part correctly, you have picked the isomer found in nature and named menthol.)

2.23 Following are planar hexagon representations for several substituted cyclohexanes. For each, draw the two chair conformations and state which of the two chair conformations is the more stable.

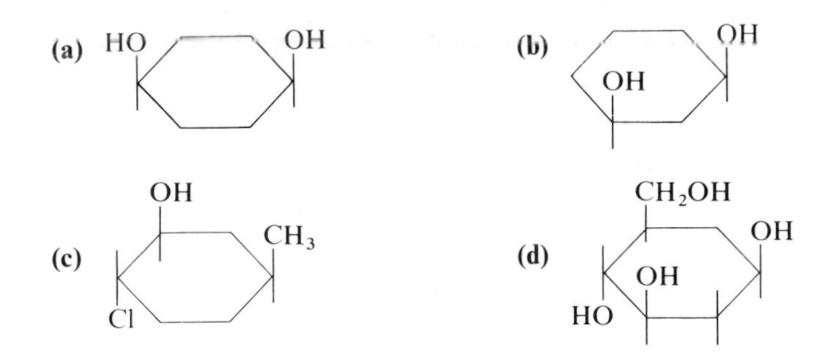

2.24 "Benzene hexachloride," more properly named 1,2,3,4,5,6,-hexachlorocyclohexane, is a mixture of various *cis-trans* isomers. The crude mixture is sold as the insecticide benzene hexachloride (BHC). The insecticidal properties of the mixture arise from one isomer known as the γ-isomer (gamma isomer), which is marketed under the trade names Lindane and Gammexane. The γ-isomer is *cis*-1,2,4,5-*trans*-3,6-hexachlorocyclohexane.

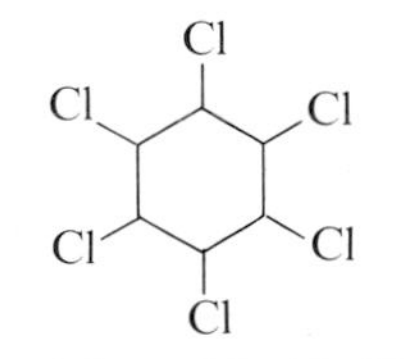

benzene hexachloride (BHC)

(a) Using a planar hexagon representation for the cyclohexane ring, draw a structural formula for the γ-isomer.
(b) Draw a chair conformation of the γ-isomer and label the chlorine substituents either axial or equatorial.
(c) Draw the other chair conformation of the γ-isomer and again label chlorine substituents axial or equatorial.
(d) Which of these two chair conformations of the γ-isomer would you predict to be the more stable? Why?

2.25 What is the major component of natural gas? of bottled or liquefied petroleum (LP) gas?

2.26 What generalization can you make about the densities of alkanes relative to that of water?

2.27 In a handbook of chemistry or any other suitable reference, find the densities of methylene chloride, chloroform, and carbon tetrachloride. Which of these substances are more dense than water; which are less dense?

2.28 What straight-chain alkane has about the same boiling point as water? (Refer to Table 2.4 for data on the physical properties of alkanes). Calculate the molecular weight of this alkane and compare it with water.

2.29 Draw structural formulas for Freon-11, Freon-12, and Freon-22. Explain why Freons such as these have become so widely used as refrigerants and aerosol propellants.

2.30 In 1974, Drs. Sherwood Rowland and Mario Molina proposed that the Freons used as aerosol propellants and refrigerants may have a very harmful effect on the environment. Explain the basis for this concern.

2.31 Account for the fact that saturated hydrocarbons are quite inert, that is, they are resistant to attack by most strong acids and bases as well as most oxidizing and reducing agents.

2.32 Name and draw structural formulas for all monochlorination products formed in the following reactions:

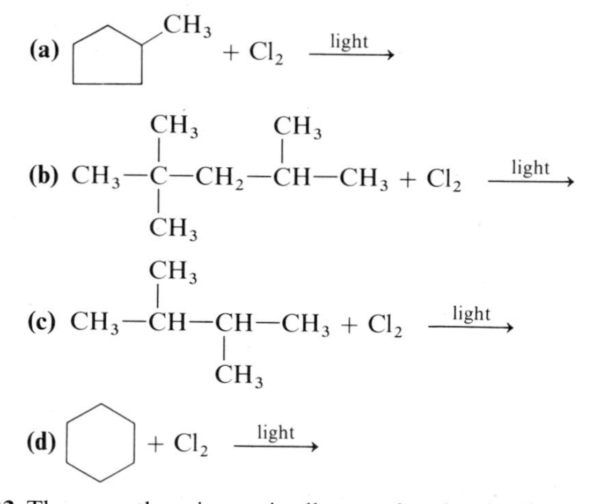

2.33 There are three isomeric alkanes of molecular formula C_5H_{12}. Isomer *A* gives a mixture of four monochlorination products when reacted with chlorine gas at 300°C. Under the same conditions, isomer *B* gives a mixture of three monochlorination products while isomer *C* gives only one monochlorination product. From this information assign structural formulas to isomers *A*, *B*, and *C*.

2.34 If the chlorination of propane were completely random, that is, equally probable at any one of the eight hydrogens, predict the relative percentage yields of 2-chloropropane and 1-chloropropane. Compare your prediction with the observed percentages of 48% 1-chloropropane and 52% 2-chloropropane.

2.35 Complete and balance the following combustion reactions. Assume that each hydrocarbon is converted completely to carbon dioxide and water.

(a) propane + O_2 $\longrightarrow$
(b) octane + O_2 $\longrightarrow$
(c) cyclohexane + O_2 $\longrightarrow$
(d) 3-methylpentane + O_2 $\longrightarrow$
(e) methylcyclopentane + O_2 $\longrightarrow$
(f) cyclobutane + O_2 $\longrightarrow$

2.36 Following are the heats of combustion per mole of methane, propane, and 2,2,4-trimethylpentane. Each is a major source of energy. On a gram-for-gram basis, which of these hydrocarbons is the best source of heat energy?

Hydrocarbon	*A major component of*	
CH_4	natural gas	212 kcal/mole
$CH_3CH_2CH_3$	LPG	531 kcal/mole
$CH_3C(CH_3)_2CH_2CH(CH_3)CH_3$	gasoline	1304 kcal/mole

CHAPTER 3

Alkenes and Alkynes

Unsaturated hydrocarbons are hydrocarbons that contain one or more carbon-carbon double or triple bonds. There are three classes of unsaturated hydrocarbons: alkenes, alkynes, and aromatic hydrocarbons. Alkenes contain one or more carbon-carbon double bonds and alkynes contain one or more carbon-carbon triple bonds. Following are structural formulas for acetylene, the simplest alkyne, and ethylene, the simplest alkene.

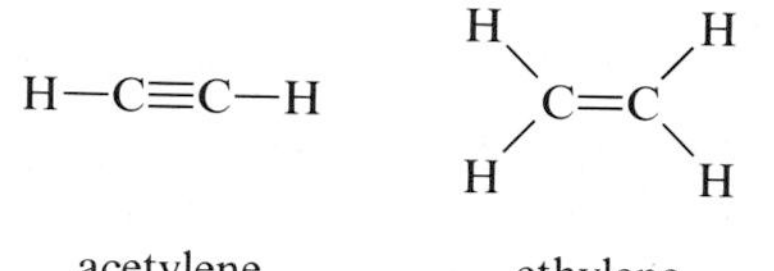

acetylene ethylene

The third class of unsaturated hydrocarbons are the aromatic hydrocarbons. The simplest aromatic hydrocarbon, benzene, has three carbon-carbon double bonds in a six-membered ring.

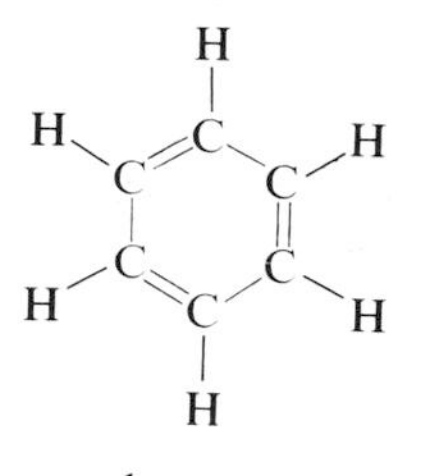

benzene

Benzene and other aromatic hydrocarbons have chemical properties quite different from those of alkenes and alkynes; for this reason, aromatic hydrocarbons are discussed separately in Chapter 6.

3.1 STRUCTURE OF ALKENES AND ALKYNES

In Chapter 1 we described the formation of double and triple bonds in terms of the overlap of atomic orbitals. Recall that to form bonds with three other atoms, carbon uses sp^2 hybrid orbitals formed by the combination of one $2s$ orbital and two $2p$ orbitals. The three sp^2 orbitals lie in a plane at angles of 120° to each other. The remaining $2p$ orbital of carbon is not hybridized and lies perpendicular to the plane created by the three sp^2 orbitals. The carbon-carbon double

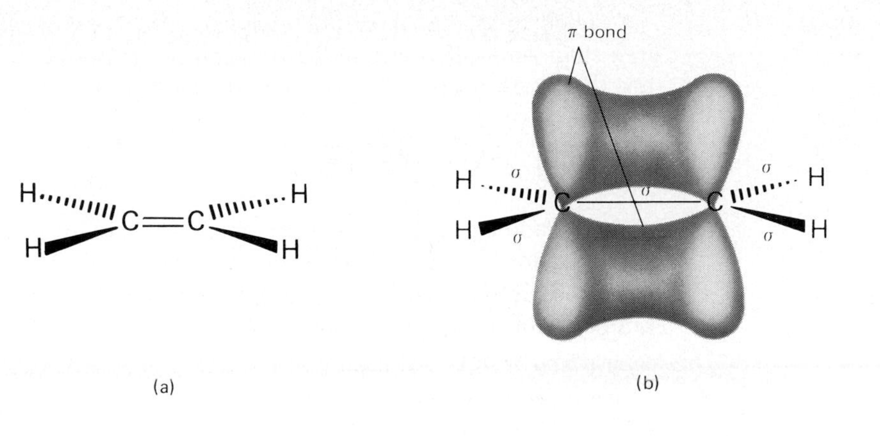

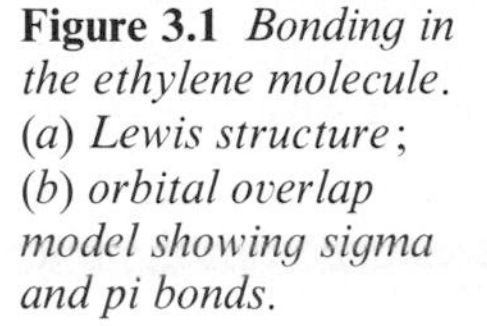

Figure 3.1 *Bonding in the ethylene molecule. (a) Lewis structure; (b) orbital overlap model showing sigma and pi bonds.*

bond consists of one sigma bond formed by the overlap of sp^2 hybrid orbitals of adjacent carbon atoms and one pi bond formed by the overlap of unhybridized $2p$ orbitals (Figure 3.1).

Because the overlap of parallel $2p$ orbitals is less than that of sp^2 orbitals, the pi bond of an alkene is not as strong as a sigma bond. One of the characteristic chemical reactions of alkenes is rupture of the pi bond and use of the two electrons from it to make two stronger sigma bonds with other atoms.

To bond with two other atoms in the formation of a triple bond, carbon uses sp hybrid orbitals formed by combination of one $2s$ orbital and one $2p$ orbital. These two sp hybrid orbitals lie in a straight line at an angle of 180°. The carbon-carbon triple bond consists of one sigma bond formed by the overlap of sp hybrid orbitals and two pi bonds. One pi bond is formed by the overlap of $2p_y$ atomic orbitals, the other by the overlap of $2p_z$ atomic orbitals (Figure 3.2).

Ethylene is the first member of the alkene family. The second member is propene, C_3H_6.

$$CH_3{-}CH{=}CH_2$$

propene

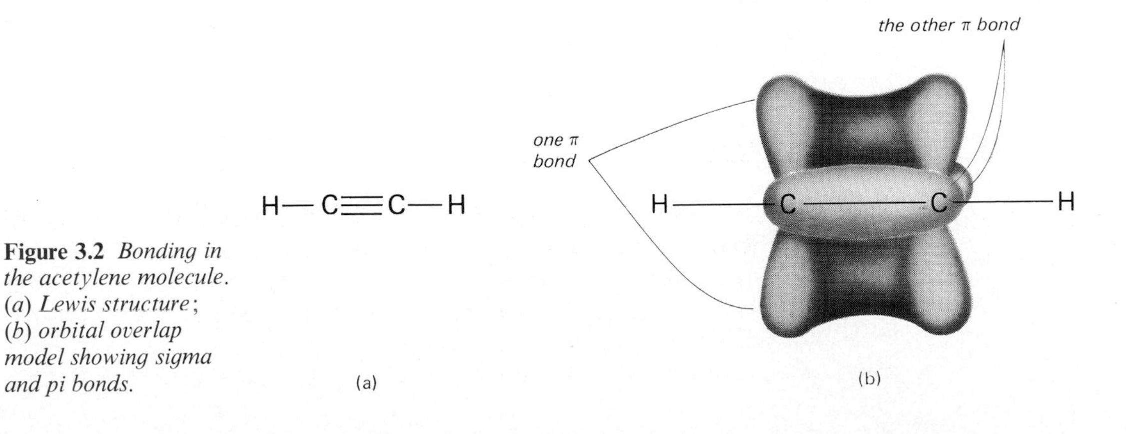

Figure 3.2 *Bonding in the acetylene molecule. (a) Lewis structure; (b) orbital overlap model showing sigma and pi bonds.*

There are three alkenes of formula C_4H_8. Two of these have four carbons in a chain and differ only in the position of the double bond. The third isomer has three carbons in a chain with a one-carbon substituent.

$$CH_3-CH_2-CH=CH_2 \qquad CH_3-CH=CH-CH_3 \qquad CH_3-\underset{\substack{|\\ CH_3}}{C}=CH_2$$

The number of alkene isomers increases rapidly as the number of carbons increases, for in addition to variations in chain length and branching, there are variations in the position of the double bond. Alkenes form a series of compounds with the general formula C_nH_{2n}.

Example 3.1 Draw structural formulas for all alkenes of molecular formula C_5H_{10}.

Solution Approach this type of problem systematically. First draw the carbon skeletons possible for molecules of five carbon atoms. There are three such skeletons.

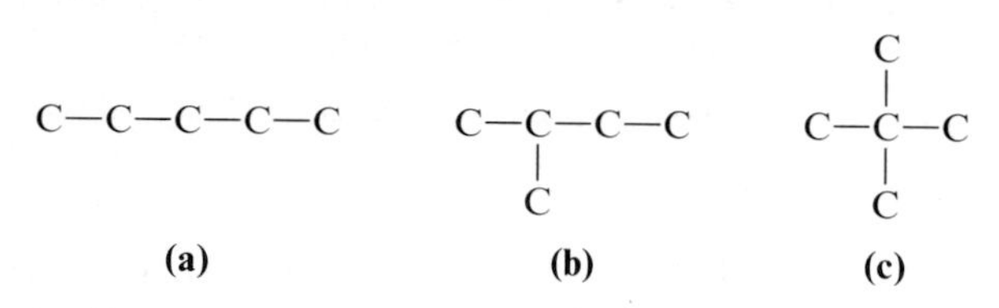

Note that skeleton (c) cannot contain a carbon-carbon double bond for the central carbon atom already has four bonds to it. Therefore, we need consider only skeletons (a) and (b) when drawing structural isomers for alkenes of molecular formula C_5H_{10}. Next, locate the double bond between carbon atoms along the chain. For carbon skeleton (a) there are two possible locations for the carbon-carbon double bond; for carbon skeleton (b) there are three possible locations for the carbon-carbon double bond.

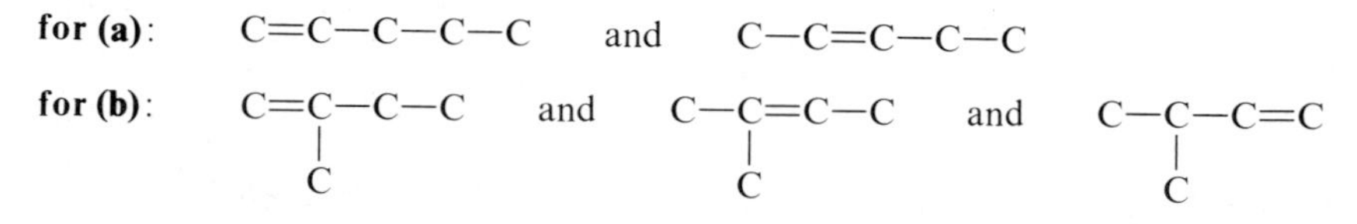

Finally, add hydrogen atoms to complete the tetravalence of carbon and give the correct molecular formula.

$$CH_2=CH-CH_2-CH_2-CH_3 \qquad CH_3-CH=CH-CH_2-CH_3$$

$$CH_2=\underset{\substack{|\\ CH_3}}{C}-CH_2-CH_3 \qquad CH_3-\underset{\substack{|\\ CH_3}}{C}=CH-CH_3 \qquad CH_3-\underset{\substack{|\\ CH_3}}{CH}-CH=CH_2$$

PROBLEM 3.1 Draw structural isomers for all alkenes of formula C_6H_{12} that have the following carbon skeletons.

$$\textbf{(a)}\ C-\overset{\substack{C\\ |}}{C}-C-C-C \qquad \textbf{(b)}\ C-\overset{\substack{C\\ |}}{C}-\overset{\substack{C\\ |}}{C}-C \qquad \textbf{(c)}\ C-\overset{\substack{C\\ |}}{\underset{\substack{|\\ C}}{C}}-C-C$$

Acetylene (ethyne) is the first member of the alkyne family. Propyne is the second.

$$\underset{\text{acetylene}}{H{-}C{\equiv}C{-}H} \qquad \underset{\text{propyne}}{CH_3{-}C{\equiv}C{-}H}$$

Alkynes of four or more carbons can have different locations for the triple bond and therefore show structural isomerism. There are two alkynes of molecular formula C_4H_6.

$$CH_3{-}CH_2{-}C{\equiv}CH \qquad CH_3{-}C{\equiv}C{-}CH_3$$

Alkynes form a series of compounds with the general formula C_nH_{2n-2}.

Example 3.2 There are three alkynes of molecular formula C_5H_8. Draw a structural formula for each.

Solution First draw all possible carbon skeletons that can contain a carbon-carbon triple bond. Next locate the triple bond and finally add hydrogens to complete the tetravalence of carbon. There are only two carbon skeletons for C_5H_8 that can contain a triple bond:

$$C{-}C{-}C{-}C{-}C \qquad C{-}\overset{\overset{\displaystyle C}{|}}{C}{-}C{-}C$$

There are two possible locations for the triple bond in the first carbon skeleton and only one in the second. Structural formulas for the three possible alkynes are

$$CH_3{-}CH_2{-}CH_2{-}C{\equiv}CH \qquad CH_3{-}CH_2{-}C{\equiv}C{-}CH_3 \qquad CH_3{-}\overset{\overset{\displaystyle CH_3}{|}}{C}H{-}C{\equiv}CH$$

PROBLEM 3.2 Draw structural formulas for the seven alkynes of molecular formula C_6H_{10}.

3.2 NOMENCLATURE OF ALKENES AND ALKYNES

The IUPAC names of alkenes are formed by changing the -ane ending of the parent alkane to -ene. Hence, $CH_2{=}CH_2$ is named ethene and

$$CH_3{-}CH{=}CH_2$$

is named propene. The IUPAC system retains the common name ethylene; thus there are two acceptable IUPAC names for C_2H_4, ethene and ethylene.

There is no chance for ambiguity in naming the first two members of the alkene family because ethylene and propene can contain a double bond in only one position. In butene and all higher alkenes there are isomers that differ in the location of the double bond. Therefore a numbering system must be used to locate the double bond. According to the IUPAC system, the longest carbon chain that contains the double bond is numbered in such a manner as to give the double-bonded carbons the lowest possible numbers. The position of the double bond is then indicated by the number of the first carbon of the double bond. Branched or substituted alkenes are named in a manner similar to alkanes.

$$\overset{6}{C}H_3\overset{5}{C}H_2\overset{4}{C}H_2\overset{3}{C}H{=}\overset{2}{C}H\overset{1}{C}H_3$$

2-hexene

$$\overset{6}{C}H_3\overset{5}{C}H_2\underset{\displaystyle CH_3}{\overset{4}{C}H}\overset{3}{C}H{=}\overset{2}{C}H\overset{1}{C}H_3$$

4-methyl-2-hexene

In the following alkene, there is a chain of five carbon atoms, but the longest chain that contains the double bond is a four-carbon chain. Therefore, the parent compound is a butene. Next the substituent groups are named, alphabetized, and numbered. Putting this information together gives the IUPAC name 2-ethyl-3-methyl-1-butene.

$$\overset{4}{C}H_3{-}\overset{\displaystyle CH_3}{\overset{3|}{C}H}{-}\underset{\displaystyle \underset{\displaystyle CH_3}{|}\atop CH_2}{\overset{2}{C}}{=}\overset{1}{C}H_2$$

2-ethyl-3-methyl-1-butene

Example 3.3 In Example 3.1, you drew formulas for alkenes of molecular formula C_5H_{10}. Give each an IUPAC name.

Solution First, number the carbon chain from the direction which gives the double bond the lowest numbers. When writing the name be certain that each substituent has a name and a number and that the location of the double bond is given.

$$\overset{1}{C}H_2{=}\overset{2}{C}H{-}\overset{3}{C}H_2{-}\overset{4}{C}H_2{-}\overset{5}{C}H_3$$

1-pentene

$$\overset{1}{C}H_3{-}\overset{2}{C}H{=}\overset{3}{C}H{-}\overset{4}{C}H_2{-}\overset{5}{C}H_3$$

2-pentene

$$\overset{1}{C}H_2{=}\underset{\displaystyle CH_3}{\overset{2}{C}}{-}\overset{3}{C}H_2{-}\overset{4}{C}H_3$$

2-methyl-1-butene

$$\overset{1}{C}H_3{-}\underset{\displaystyle CH_3}{\overset{2}{C}}{=}\overset{3}{C}H{-}\overset{4}{C}H_3$$

2-methyl-2-butene

$$\overset{4}{C}H_3{-}\underset{\displaystyle CH_3}{\overset{3}{C}H}{-}\overset{2}{C}H{=}\overset{1}{C}H_2$$

3-methyl-1-butene

PROBLEM 3.3 In Problem 3.1, you drew formulas for alkenes of three different carbon skeletons. Give each alkene an IUPAC name.

In naming cycloalkenes, the carbon atoms of the ring multiple bond are numbered 1 and 2 in the direction that gives substituents the smallest numbers possible.

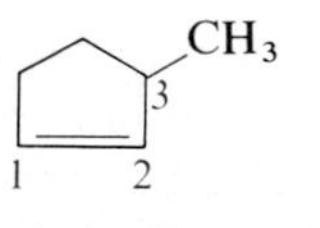

3-methylcyclopentene

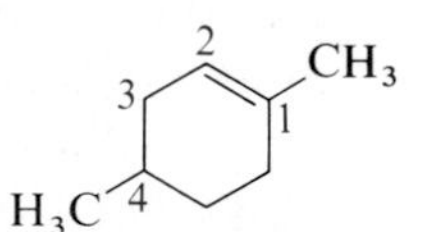

1,4-dimethylcyclohexene
(not 2,5-dimethylcyclohexene)

Alkenes that contain more than one double bond are called alkadienes, alkatrienes, or more simply dienes, trienes, etc.

$$\underset{\displaystyle CH_3}{CH_2{=}\overset{}{C}{-}CH{=}CH_2}$$
2-methyl-1,3-butadiene
(isoprene)

$$CH_2{=}CH{-}CH_2{-}CH{=}CH_2$$
1,4-pentadiene

Many alkenes, particularly the lower molecular weight ones, are known almost exclusively by their common names. For example, the common name of propene is propylene; that of 2-methylpropene is isobutylene. 1,3-Butadiene is called simply butadiene.

	$CH_3{-}CH{=}CH_2$	$CH_3{-}C(CH_3){=}CH_2$	$CH_2{=}CH{-}CH{=}CH_2$
IUPAC name:	propene	2-methylpropene	1,3-butadiene
common name:	propylene	isobutylene	butadiene

The use of common names is generally avoided for alkenes of more than four carbons because of the large number of structural isomers possible and the lack of suitable prefixes to distinguish one structural isomer from another.

In the IUPAC nomenclature of alkynes, the ending -yne indicates the presence of a triple bond. Hence, $HC{\equiv}CH$ is named ethyne, and $CH_3{-}C{\equiv}CH$ is named propyne. The IUPAC system retains the name acetylene. Therefore there are two acceptable IUPAC names for C_2H_2, acetylene and ethyne. For longer chain alkynes, the longest carbon chain that contains the triple bond is numbered in such a manner as to give triple-bonded carbons the lowest possible numbers. Then the location of the triple bond is indicated by the number of the first carbon of the triple bond.

$$CH_3{-}CH_2{-}C{\equiv}CH$$
1-butyne

$$CH_3{-}C{\equiv}C{-}CH_3$$
2-butyne

Example 3.4 Give IUPAC names for the following alkynes.

(a) $(CH_3)_2CH{-}C{\equiv}C{-}CH_3$ **(b)** $CH_3{-}CH_2{-}C{\equiv}C{-}CH_2{-}CH(CH_3)_2$

Solution **(a)** Number the carbon chain so that the triple bond is between carbons 2 and 3 of the chain. The correct IUPAC name is 4-methyl-2-pentyne.

$$\overset{5}{C}H_3{-}\underset{\displaystyle CH_3}{\overset{4}{C}H}{-}\overset{3}{C}{\equiv}\overset{2}{C}{-}\overset{1}{C}H_3$$
4-methyl-2-pentyne

(b) The correct IUPAC name is 6-methyl-3-heptyne.

$$\underset{1}{CH_3}{-}\underset{2}{CH_2}{-}\underset{3}{C}{\equiv}\underset{4}{C}{-}\underset{5}{CH_2}{-}\underset{6}{\overset{\displaystyle CH_3}{CH}}{-}\underset{7}{CH_3}$$

PROBLEM 3.4 Write structural formulas for the following alkynes.
(a) 1-hexyne **(b)** 2-hexyne **(c)** 3-ethyl-4-methyl-1-pentyne

3.3 STEREOISOMERISM IN ALKENES

Recall from Section 2.7 that stereoisomers have the same structural formula but different arrangements of atoms in space. *Cis-trans* isomerism is one type of stereoisomerism. In cycloalkanes, *cis-trans* isomerism depends on the arrangement of substituent groups in a ring. In alkenes, *cis-trans* isomerism depends on the arrangement of substituent groups on a double bond.

There are two important structural features of carbon-carbon double bonds which make it possible for the existence of *cis-trans* isomerism. (1) The two carbons of the double bond and the four atoms attached directly to them all lie in the same plane. (2) There is restricted rotation about the two carbons of the double bond. The double-bonded carbon atoms cannot be twisted to different conformations as we saw for ethane; to do so would require breaking the pi bond (Figure 3.3).

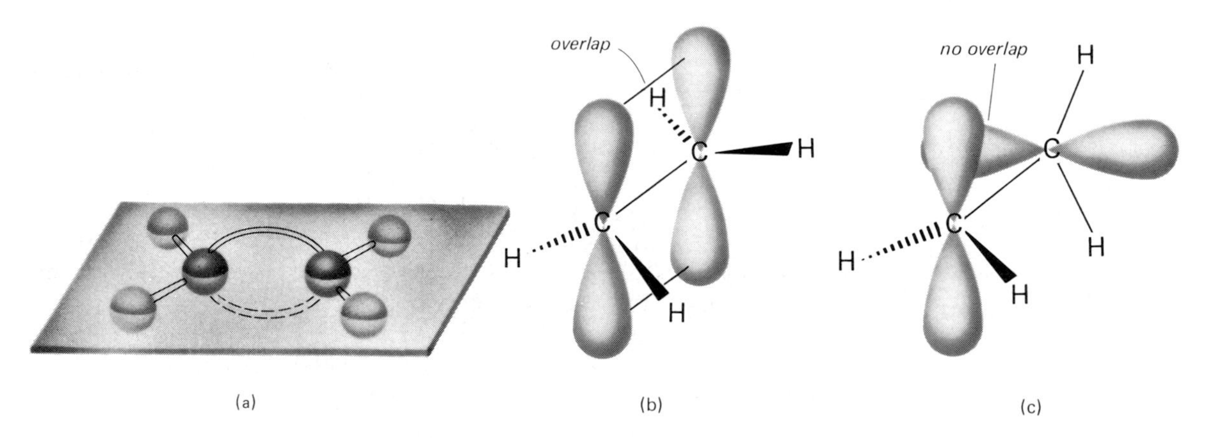

Figure 3.3 *Restricted rotation about the carbon-carbon double bond. (a) Ball-and-stick model; (b) orbital overlap model showing the pi bond; (c) the pi bond must be broken for the orbitals to rotate.*

Ethylene, propene, 1-butene, and 2-methylpropene have only one possible arrangement of the atoms. In 2-butene there are two possible arrangements. In one of them, the two methyl groups are on the same side of the double bond; this isomer is called *cis*-2-butene. In the other arrangement, the two methyl groups are on opposite sides of the double bond; this isomer is called *trans*-2-butene. Shown in Figure 3.4 are Lewis structures of *cis* and *trans*-2-butene. *Cis*- and *trans*-isomers are different substances and have different physical and chemical properties. The *cis*- and *trans*-isomers of 2-butene differ in melting points by 33°C and in boiling points by 3°C.

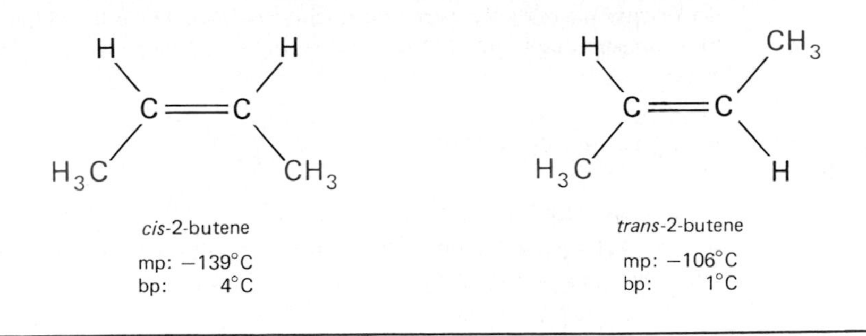

Figure 3.4 *The cis- and trans-isomers of 2-butene.*

Example 3.5

Which of the following alkenes show *cis-trans* isomerism?

(a) $CH_2{=}CHCH_2CH_2CH_3$ **(b)** $CH_3CH{=}CHCH_2CH_3$ **(c)** $CH_2{=}\underset{\displaystyle CH_3}{\underset{|}{C}}CH_2CH_3$

Solution

(a) 1-pentene. Begin by drawing the carbon-carbon double bond as accurately as possible to show the bond angles of 120° about the carbon-carbon double bond.

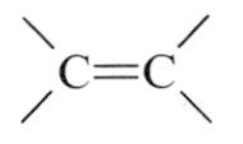

Next, complete the structural formula showing all four groups attached to the double bond.

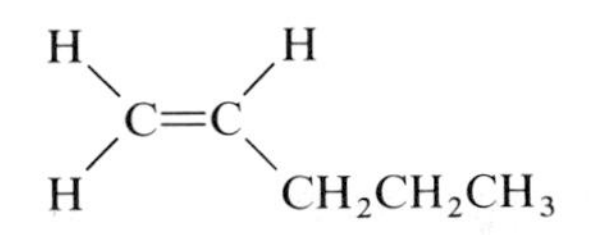

To determine if this molecule shows *cis-trans* isomerism, switch positions of the —H and $—CH_2CH_2CH_3$ groups and compare the two molecules you have drawn.

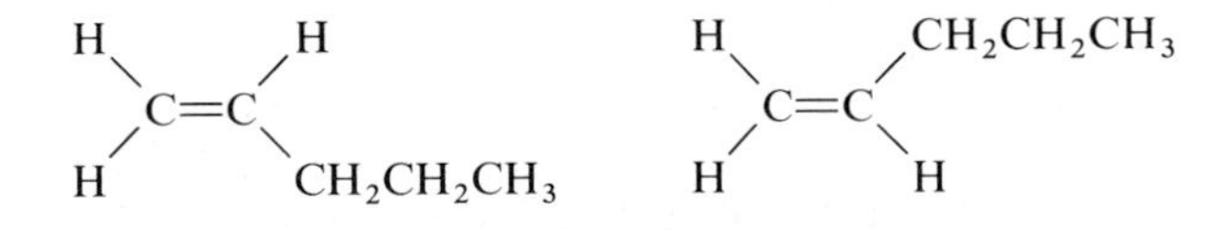

Are they the same or are they different? In this case they are the same molecule. To see this, imagine that you pick up either one, turn it over as you would turn your hand from palm down to palm up. If you have done this correctly, you will see that one structural formula fits exactly on top of the other (i.e., is superimposable on the other). Therefore, they are identical and 1-pentene does not show *cis-trans* isomerism.

(b) 2-pentene. Draw structural formulas for possible cis-trans isomers as you did in part (a) and then examine them to see if your drawings represent the same molecule or *cis-trans* isomers.

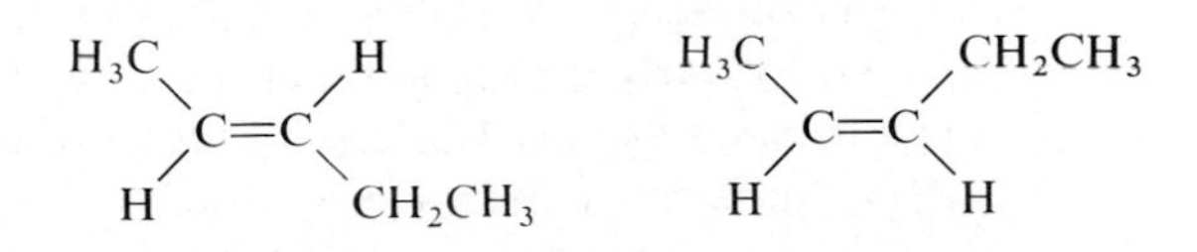

The orientation of the four groups attached to the double bond of the molecule on the left is different from that of the molecule on the right. Therefore, the drawings represent *cis-trans* isomers.

(c) 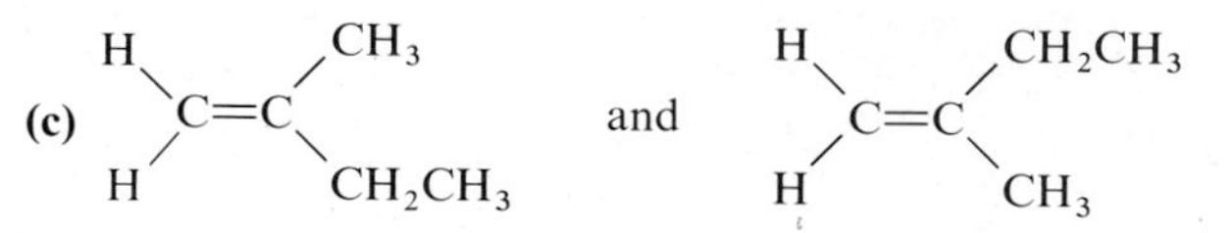

These two structural formulas are superimposable and therefore 2-methyl-1-butene does not show *cis-trans* isomerism.

From this example you should realize that:

(1) An alkene shows *cis-trans* isomerism only if each of the carbon atoms of the double bond has two different groups attached to it.

(2) An alkene does not show *cis-trans* isomerism if either one of the carbon atoms of the double bond has two identical groups attached to it.

PROBLEM 3.5 Which of the following alkenes show *cis-trans* isomerism? For each that does, draw structural formulas for the isomers.

(a) $CH_2{=}C(CH_3){-}CH_2{-}CH_2{-}CH_3$ **(b)** $CH_3{-}C(CH_3){=}CH{-}CH_2{-}CH_3$

(c) $CH_3{-}CH(CH_3){-}CH{=}CH{-}CH_3$ **(d)** $CH_3{-}CH(CH_3){-}CH_2{-}CH{=}CH_2$

An example of a biologically important molecule that can exist in a number of *cis-trans* isomers is Vitamin A.

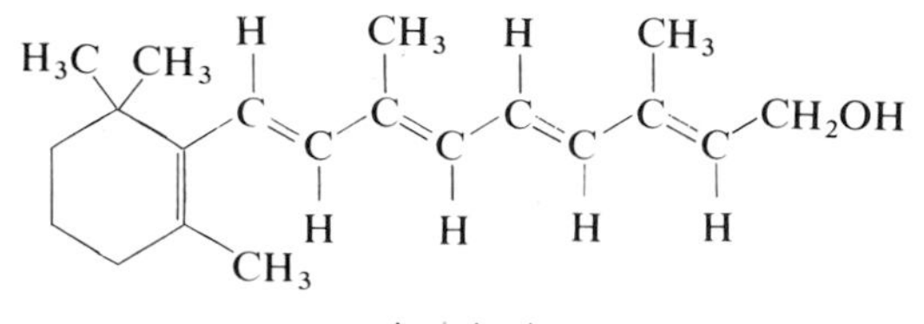

vitamin A

There are four carbon-carbon double bonds in the chain of atoms attached to the cyclohexene ring and each of the four can show *cis-trans* isomerism. Thus, there are 2 × 2 × 2 × 2 or 16 possible *cis-trans* isomers.

3.4 PHYSICAL PROPERTIES OF ALKENES AND ALKYNES

Alkenes and alkynes, like alkanes, are nonpolar compounds and their physical properties are much the same as those of the corresponding alkanes of the same number of carbon atoms and similar carbon skeletons. Alkenes and alkynes of 2, 3, and 4 carbon atoms are all gases at room temperature. Those of five or more carbons are all colorless liquids less dense than water. Alkenes and alkynes are insoluble in water but they are soluble in each other, in ethanol, and in other

Table 3.1 Physical properties of some alkenes and alkynes.

IUPAC name	*Structural formula*	*bp* (°C)	*mp* (°C)
ethene or ethylene	$CH_2{=}CH_2$	−104	−169
propene	$CH_3CH{=}CH_2$	−47	−185
1-butene	$CH_3CH_2CH{=}CH_2$	−6	−185
1-pentene	$CH_3CH_2CH_2CH{=}CH_2$	30	−138
cis-2-pentene	$CH_3CH_2(H)C{=}C(H)CH_3$ (cis: CH_3CH_2 and CH_3 on same side)	37	−151
trans-2-pentene	$CH_3CH_2(H)C{=}C(CH_3)H$ (trans: CH_3CH_2 and CH_3 on opposite sides)	36	−156
2-methyl-2-butene	$CH_3C(CH_3){=}CHCH_3$	39	−134
ethyne or acetylene	$HC{\equiv}CH$	−84	−81
propyne	$CH_3C{\equiv}CH$	−23	−101
1-butyne	$CH_3CH_2C{\equiv}CH$	8	−126
1-pentyne	$CH_3CH_2CH_2C{\equiv}CH$	40	−90

nonpolar organic solvents. Table 3.1 lists the physical properties of some alkenes and alkynes.

3.5 PREPARATION OF ALKENES

Alkenes are most often prepared in the laboratory by a process called an elimination reaction. In an elimination reaction, some small molecule (most commonly HOH, HCl, or HBr) is split away or eliminated from adjacent carbon atoms of a larger molecule. Removal of an acid compound such as HCl, HBr, or HI is called dehydrohalogenation and requires a strong base such as NaOH. For example, the dehydrohalogenation of 2-bromopropane (isopropyl bromide) in the presence of NaOH gives propene.

$$\underset{\text{2-bromopropane (isopropyl bromide)}}{CH_3{-}\underset{\substack{|\\Br}}{CH}{-}CH_3} + NaOH \longrightarrow \underset{\text{propene (propylene)}}{CH_3{-}CH{=}CH_2} + NaBr + HOH$$

In this process, two sigma bonds are broken (C—H and C—Br), and one new pi bond (C=C) and one new sigma bond (H—OH) are formed.

Dehydrohalogenation of chlorocyclohexane in the presence of a strong base gives cyclohexene.

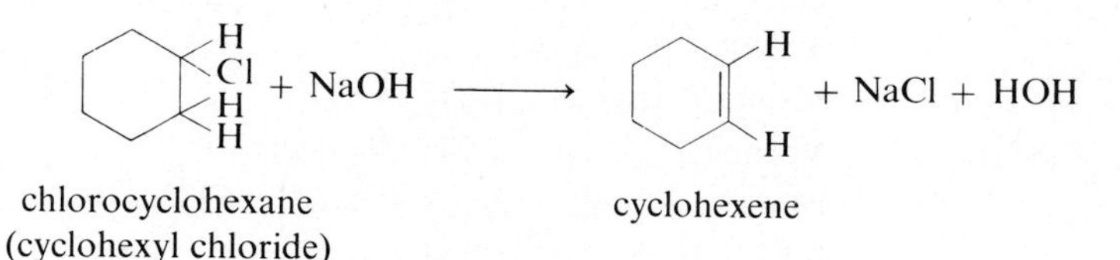

chlorocyclohexane (cyclohexyl chloride) — cyclohexene

Often, dehydrohalogenation of an alkyl halide leads to the formation of isomeric alkenes. In the dehydrohalogenation of 2-chlorobutane (*sec*-butyl chloride), —H and —Cl may be removed from carbon 1 and 2 to form 1-butene, or they may be removed from carbons 2 and 3 to form 2-butene. In practice, dehydrohalogenation of 2-chlorobutane gives 2-butene as the major product and 1-butene as the minor product.

$$\overset{1}{CH_3}-\underset{\underset{Cl}{|}}{\overset{2}{CH}}-\overset{3}{CH_2}-\overset{4}{CH_3} + NaOH \longrightarrow$$

2-chlorobutane

$$CH_3-CH{=}CH-CH_3 + CH_2{=}CH-CH_2-CH_3 + H_2O + NaCl$$

2-butene (major product) — 1-butene (minor product)

Where it is possible to obtain isomeric alkenes from the dehydrohalogenation of an alkyl halide, the alkene having the more substituents on the carbon-carbon double bond generally predominates.

Example 3.6

Draw structural formulas for the alkenes formed on dehydrohalogenation of the following alkyl halides. Predict which alkene is the major product and which is the minor product.

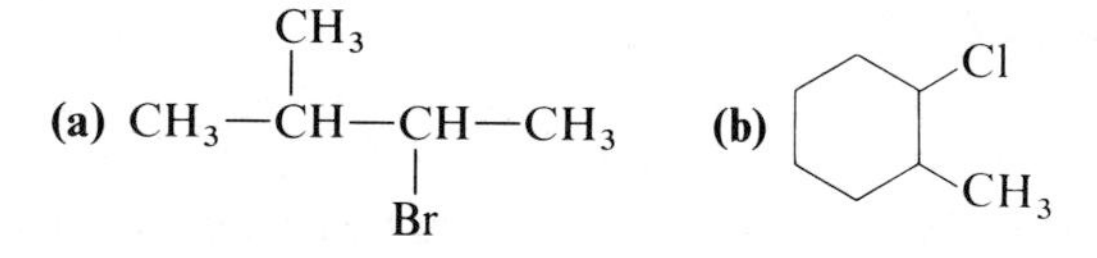

Solution

(a) Dehydrohalogenation of 3-bromo-2-methylbutane can result in loss of —H and —Br from either carbons 3 and 4 to give 3-methyl-1-butene, or from carbons 2 and 3 to give 2-methyl-2-butene.

$$\overset{1}{CH_3}-\overset{\overset{CH_3}{|}}{\overset{2}{CH}}-\underset{\underset{Br}{|}}{\overset{3}{CH}}-\overset{4}{CH_3} + NaOH \longrightarrow$$

$$CH_3-\overset{\overset{CH_3}{|}}{CH}-CH{=}CH_2 + CH_3-\overset{\overset{CH_3}{|}}{C}{=}CH-CH_3 + H_2O + NaBr$$

3-methyl-1-butene (minor product) — 2-methyl-2-butene (major product)

3-Methyl-1-butene has one alkyl substituent (an isopropyl group) on the double bond. 2-Methyl-2-butene has three alkyl substituents (three methyl groups) on the double bond. Therefore, predict that 2-methyl-2-butene is the major product and that 3-methyl-1-butene is the minor product.

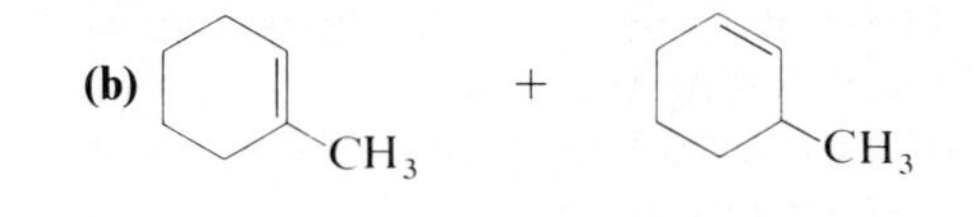

1-methylcyclohexene (major product) 3-methylcyclohexene (minor product)

The major product, 1-methylcyclohexene, has three alkyl substituents on the carbon-carbon double bond. The minor product, 3-methylcyclohexene, has only two.

PROBLEM 3.6 Draw structural formulas for the alkenes formed on dehydrohalogenation of the following alkyl halides. Where two alkenes are possible, predict which is the major product and which is the minor product.

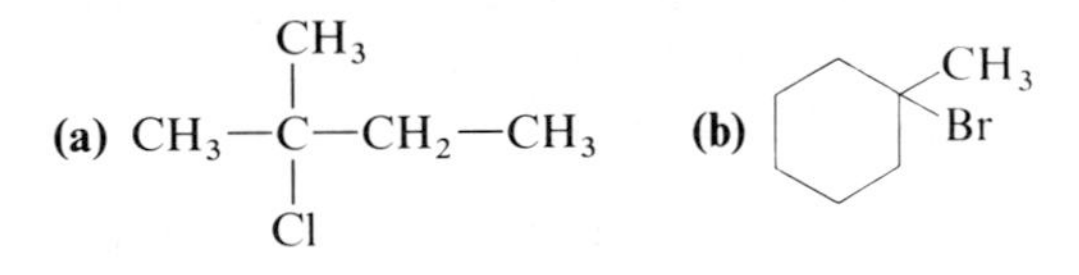

3.6 REACTIONS OF ALKENES

In contrast to alkanes, alkenes react readily with halogens, certain strong acids, a variety of oxidizing and reducing agents, and water in the presence of concentrated sulfuric acid. Reaction with these and most other reagents is characterized by addition to the double bond.

$$>C=C< + A{-}B \longrightarrow {-}\underset{A}{\underset{|}{C}}{-}\underset{B}{\underset{|}{C}}{-}$$

In addition reactions, there is rupture of one sigma bond (A—B) and one pi bond, and the formation of two sigma bonds. Consequently, addition reactions to the double bond are almost always energetically favorable because there is net conversion of one pi bond into one sigma bond.

In this section, we shall look at typical alkene reactions. As we examine these reactions and those of other functional groups in later chapters, we will be asking constantly, "How does each reaction occur?" or in the more common terminology of the chemist, "What is the mechanism of each reaction?"

Just what is a reaction mechanism? Literally, a reaction mechanism is an attempt to describe how and why a reaction occurs as it does. It is a description of how covalent bonds are broken and reformed, the role of catalysts and solvents, and the rates at which the various changes take place during the course of the reaction. Further, it is a description of the role of reactive intermediates, species formed during the reaction by breaking a covalent bond and having an atom in an abnormal valence state. In this chapter we will encounter two different types of reactive intermediates, carbocations, and free radicals. In Chapter 7 we will encounter a third type of reactive intermediate, the carbanion.

In our description of a reaction mechanism, we will try to specify (1) the steps in the reaction including the formation of any reactive intermediates, (2) the relative rates of the various steps, and (3) the structure of transition states.

How do we begin to describe a reaction mechanism? First, we assemble all of the variable experimental observations and facts about the particular chemical reaction under consideration. Next, through a combination of chemical sophistication, creative insight, and guesswork, we propose several sets of steps, or mechanisms, each of which will account for the overall chemical transformation. Then, we test each mechanism against the experimental observations, excluding those mechanisms that are not consistent with the facts. A mechanism becomes generally established by excluding reasonable alternatives and by showing that the mechanism is consistent with every test that the scientist can devise. This of course does not mean that a generally accepted mechanism is in fact a completely true and accurate description of the chemical events, but only that it is consistent with the mass of experimental evidence.

Before we go on to consider reactions and reaction mechanisms, we might also ask, why it is worth the trouble of chemists to establish them and your time to learn about them? Certainly one answer lies in the understanding and intellectual satisfaction derived from constructing models that accurately reflect the behavior of chemical systems. Problems of this kind present a particular fascination and exciting challenge to the creative scientist. But there is another more practical reason. Mechanisms provide a theoretical framework within which to organize a great deal of descriptive chemistry. For example, with some insight into how reagents add to particular alkenes, we can make generalizations and then use these generalizations to predict how the same reagents might add to other alkenes.

A. ADDITION OF HYDROGEN—REDUCTION

The addition of hydrogen to a carbon-carbon double bond reduces an alkene to an alkane. The process requires the presence of a metal catalyst and is called catalytic reduction. Because the reaction involves addition of hydrogen to a carbon-carbon double bond, it is also referred to as catalytic hydrogenation.

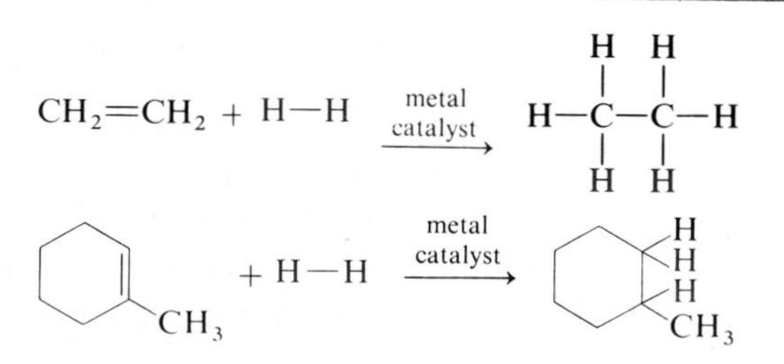

In hydrogenation, both hydrogen atoms are added from the same side of the alkene molecule. As an example, addition of hydrogen to 1,2-dimethylcyclopentene yields *cis*-1,2-dimethylcyclopentane.

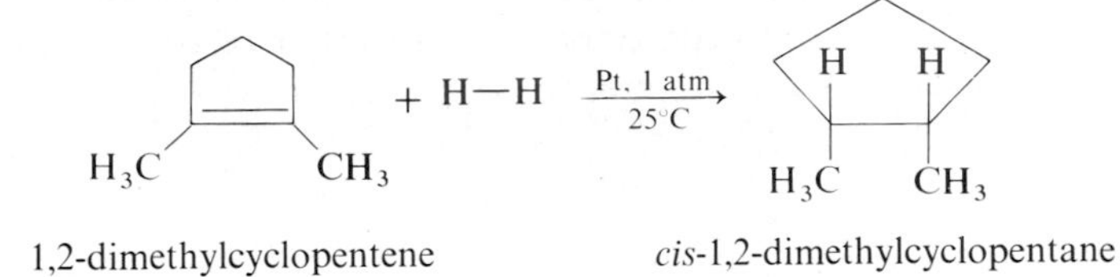

Although the addition of hydrogen to an alkene is exothermic, the reactions are immeasurably slow at moderate temperatures in the gas phase. However, they occur readily in the presence of finely divided platinum, palladium, or nickel. Separate experiments have shown that these and other transition metals near the center of the periodic table are able to adsorb large quantities of hydrogen gas. As a result the bond between the hydrogen atoms is weakened (Figure 3.5a). Alkenes also interact with metal surfaces forming reactive intermediates in which the 2*p* orbitals of carbon are used for bonding with the metal surface (Figure 3.5b). If hydrogen and the adsorbed alkene are positioned properly on the metal surface, hydrogen adds to the organic molecule to give a saturated hydrocarbon (Figure 3.5c) which is then desorbed.

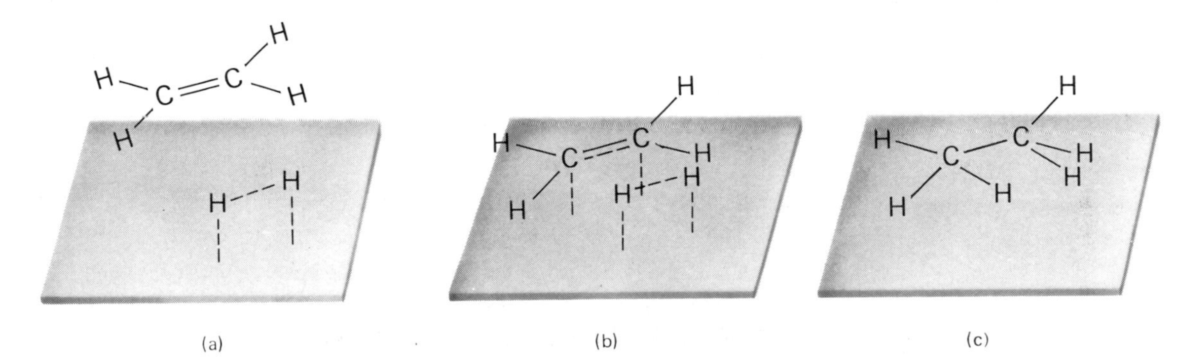

Figure 3.5 *Addition of hydrogen to ethylene involving a metal catalyst.*

B. ADDITION OF HYDROGEN HALIDES

Dry HCl, HBr, or HI add to alkenes to give alkyl halides. These additions may be carried out either with the pure reagents or in the presence of a polar solvent such as acetic acid.

$$CH_2{=}CH_2 + H{-}Cl \longrightarrow \underset{H}{CH_2}{-}\underset{Cl}{CH_2}$$

ethylene → chloroethane

Addition of HCl to ethylene gives chloroethane (ethyl chloride) as the only product. Addition of HBr to propene gives two products, 2-bromopropane (isopropyl bromide) and 1-bromopropane (*n*-propyl bromide) depending on the orientation of the addition. 2-Bromopropane is the major product of the addition and only minor amounts of 1-bromopropane are formed.

$$CH_3{-}CH{=}CH_2 + H{-}Br \longrightarrow CH_3{-}\underset{Br}{\underset{|}{C}H}{-}\underset{H}{\underset{|}{C}H_2} + CH_3{-}\underset{H}{\underset{|}{C}H}{-}\underset{Br}{\underset{|}{C}H_2}$$

propene — 2-bromopropane (major product) — 1-bromopropane (minor product)

After studying a large number of reactions involving the addition of HCl, HBr, and HI to alkenes, the Russian chemist, Vladimir Markovnikov, made the following empirical generalization: in the addition of HX to an alkene, the hydrogen atom adds to the carbon of the double bond with the greater number of hydrogen atoms.

Example 3.7

Name and draw structural formulas for the products of the following alkene addition reactions. Use Markovnikov's rule to predict which is the major product and which is the minor product.

(a) $CH_3{-}\overset{CH_3}{\overset{|}{C}}{=}CH_2 + HI \longrightarrow$ **(b)** 1-methylcyclopentene (CH_3 on the ring double bond) $+ HCl \longrightarrow$

Solution

(a) HI adds to 2-methylpropene (isobutylene) to form two products:

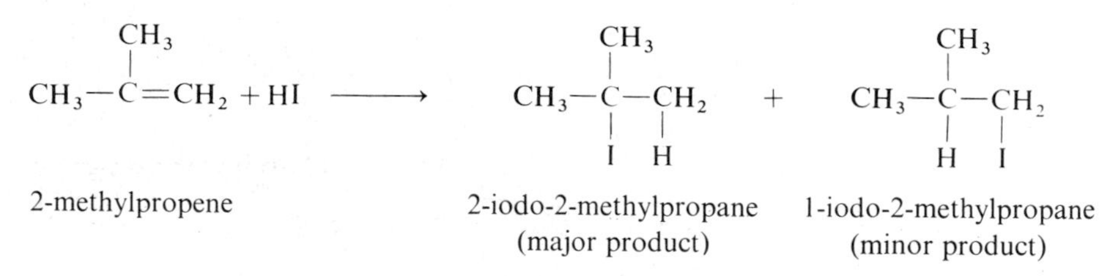

In forming 2-iodo-2-methylpropane (*tert*-butyl iodide), hydrogen adds to the carbon of the double bond bearing two hydrogens. In forming 1-iodo-2-methylpropane (isobutyl iodide), hydrogen adds to the carbon of the double bond bearing no hydrogens. Therefore, Markovnikov's rule predicts that 2-iodo-2-methylpropane is the major product and 1-iodo-2-methylpropane is the minor product.

(b) Addition of H—Cl to 1-methylcyclopentene forms two products:

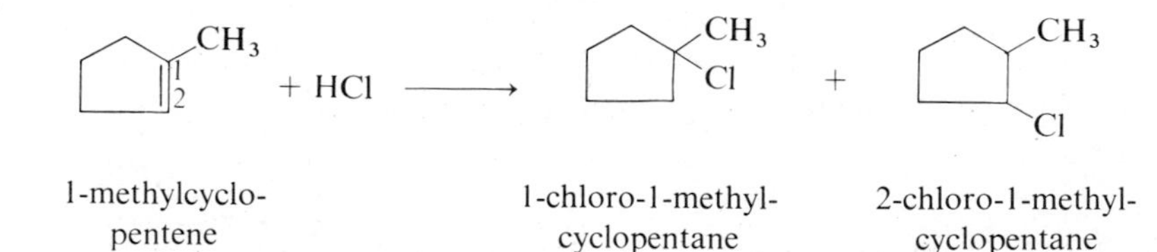

1-methylcyclo-pentene — 1-chloro-1-methyl-cyclopentane (major product) — 2-chloro-1-methyl-cyclopentane (minor product)

Carbon-1 of this cycloalkene contains no hydrogen atoms and carbon-2 contains one hydrogen. Therefore, according to Markovnikov's rule, hydrogen adds to carbon-2 and 1-chloro-1-methylcyclopentane is the major product. 2-Chloro-1-methylcyclopentane is the minor product.

PROBLEM 3.7 Name and draw structural formulas for the products of the following alkene addition reactions. Use Markovnikov's rule to predict which is the major product and which is the minor product.

$$\textbf{(a)}\ CH_3{-}CH{=}\underset{}{\overset{\overset{\large CH_3}{|}}{C}}{-}CH_3 + HI \longrightarrow$$

$$\textbf{(b)}\ CH_3{-}CH{=}CH{-}CH_2{-}CH_3 + HI \longrightarrow$$

It is important to remember that while Markovnikov's rule helps us to predict the major and minor products of alkene addition reactions, it does not explain why one product predominates over the other.

C. ADDITION OF WATER—HYDRATION

In the presence of an acid-catalyst, often 60% aqueous sulfuric acid, water adds to alkenes to produce alcohols. The addition of water to an alkene is called hydration. Hydrogen adds to the carbon of the double bond with the greater number of hydrogens; OH adds to the carbon of the double bond with the fewer hydrogens. Thus H—OH adds to alkenes in accordance with Markovnikov's rule.

$$\underset{\text{propene}}{CH_3{-}CH{=}CH_2} + H{-}OH \xrightarrow{H^+} \underset{\substack{\text{2-propanol}\\\text{(major product)}}}{CH_3{-}\underset{OH}{\underset{|}{CH}}{-}\underset{H}{\underset{|}{CH_2}}} + \underset{\substack{\text{1-propanol}\\\text{(minor product)}}}{CH_3{-}\underset{H}{\underset{|}{CH}}{-}\underset{OH}{\underset{|}{CH_2}}}$$

$$\underset{\text{2-methylpropene}}{CH_3{-}\overset{\overset{CH_3}{|}}{C}{=}CH_2} + H{-}OH \xrightarrow{H^+} \underset{\substack{\text{2-methyl-2-}\\\text{propanol}\\\text{(major product)}}}{CH_3{-}\underset{OH}{\underset{|}{\overset{\overset{CH_3}{|}}{C}}}{-}\underset{H}{\underset{|}{CH_2}}} + \underset{\substack{\text{2-methyl-1-}\\\text{propanol}\\\text{(minor product)}}}{CH_3{-}\underset{H}{\underset{|}{\overset{\overset{CH_3}{|}}{C}}}{-}\underset{OH}{\underset{|}{CH_2}}}$$

Example 3.8 Draw structural formulas for the products of the following hydration reactions. Use Markovnikov's rule to predict which is the major product and which is the minor product.

$$\textbf{(a)}\ CH_3{-}\overset{\overset{CH_3}{|}}{CH}{-}CH{=}CH_2 + H_2O \xrightarrow{H_2SO_4} \qquad \textbf{(b)}\ \text{1-methylcyclohexene (ring with } CH_3\text{)} + H_2O \xrightarrow{H_2SO_4}$$

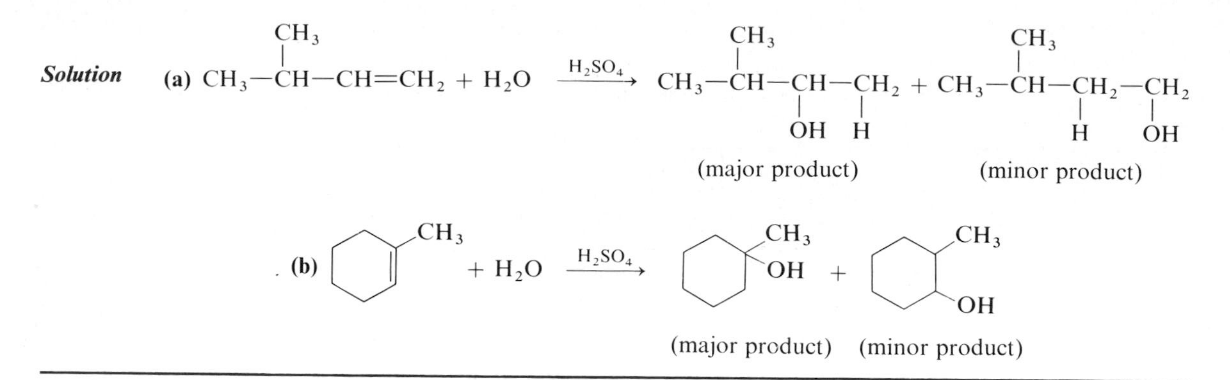

PROBLEM 3.8 Draw structural formulas for the products of the following alkene addition reactions. Use Markovnikov's rule to predict major and minor products.

$$\text{(a)}\ CH_3{-}\overset{\overset{\displaystyle CH_3}{|}}{C}{=}CH{-}CH_3 + H_2O \xrightarrow{H_2SO_4}$$

$$\text{(b)}\ CH_2{=}\overset{\overset{\displaystyle CH_3}{|}}{C}{-}CH_2{-}CH_3 + H_2O \xrightarrow{H_2SO_4}$$

D. ELECTROPHILIC ATTACK ON THE CARBON-CARBON DOUBLE BOND

It became apparent to chemists studying addition of HX and H_2O to alkenes that the initial attack on the carbon-carbon double bond is by H^+, or, in more general terms, the initial attack on the double bond is by an electrophilic (electron-loving or -seeking) reagent. This susceptibility of the double bond to attack by an electrophile is certainly consistent with our electronic formulation, which pictures the double bond as a center of high electron (negative) density.

To account for these reactions of alkenes, organic chemists have proposed a three-step ionic mechanism which correlates much that is known about these electrophilic additions. Let us consider first the addition of HBr to ethylene.

$$CH_2{=}CH_2 + HBr \longrightarrow CH_3{-}CH_2Br$$

In the addition of HBr to an ethylene, the first step is ionization of HBr to H^+ and Br^-. H^+ needs two electrons to complete its valence shell and therefore is electron-deficient. In the second step of this addition reaction, H^+ reacts with the alkene and the two electrons of the pi bond are used to form a new sigma bond between hydrogen and carbon.

Step 1: $H{-}Br \longrightarrow H^+ + Br^-$

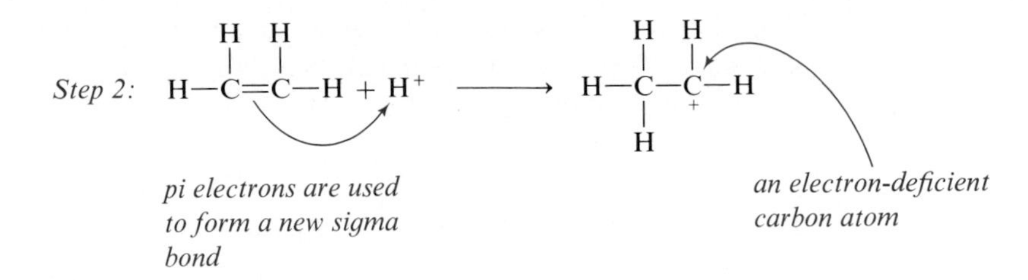

This step in turn leaves a carbon atom of the alkene with only six electrons in its valence shell. A carbon atom with six electrons in its valence shell bears a positive charge and is called a carbocation (carbon-containing cation). These reactive intermediates are also called carbonium ions by analogy with ammonium ion (NH_4^+) and hydronium ion (H_3O^+). We will use the term *carbocation* throughout the text. Nevertheless, you should be aware of the term carbonium ion for you will undoubtedly encounter it in the literature of chemistry and in your discussions with other chemists. In the final step of the reaction, the carbocation reacts with one of the pairs of electrons of the bromide ion to form a new sigma bond between carbon and bromine.

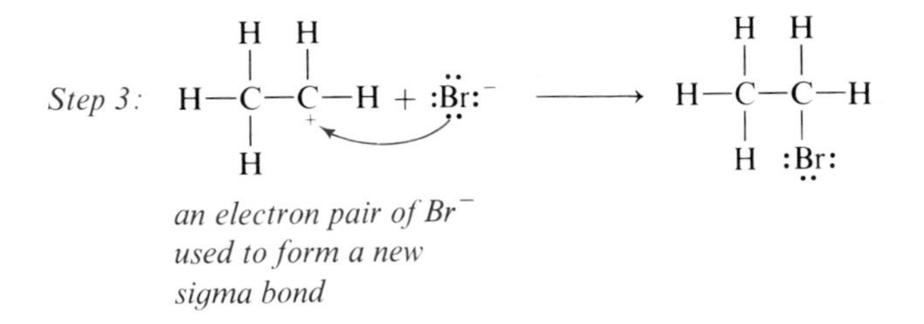

Note that in this three-step mechanism for the addition of HBr to ethylene, two bonds have been broken (one sigma bond between H—Br and one pi bond of the alkene) and two new sigma bonds have been formed.

This reaction mechanism is an example of electrophilic addition to an alkene. Literally, electrophilic means electron-loving or electron-seeking. Both H^+ and the carbocation intermediate, $CH_3—CH_2^+$, are electrophiles because they seek electrons to complete their valence shells.

The carbocation mechanism accounts for the observation that HBr adds to ethylene to give $CH_3—CH_2—Br$. But how does it account for the fact that addition of HBr to propene gives mainly 2-bromopropane and very little 1-bromopropane? In other words, how does it account for the observations generalized in Markovnikov's rule? The answer is that the mechanism as developed thus far does not account for them. At this point, we have two alternatives: either we discard the idea of the carbocation intermediate and start again, or we can refine the mechanism by introducing a new concept. In this case it is more convenient and fruitful to refine the mechanism and introduce a new concept, namely, the relative ease of formation of carbocations.

The carbocation mechanism proposes that two different carbocations can be formed depending on how H^+ adds to the double bond. The major carbocation formed is the more stable one. A tertiary carbocation (positive charge on a

3° carbon) is more stable than a secondary carbocation (positive charge on a 2° carbon), and a secondary carbocation is more stable than a primary carbocation (positive charge on a primary carbon).

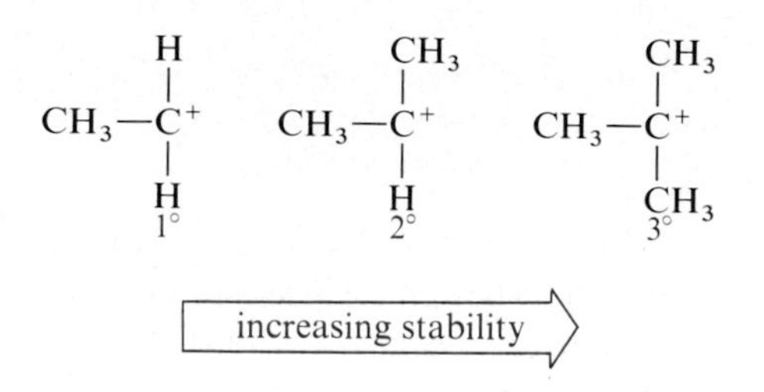

Addition of a proton to propene can give the propyl carbocation (a primary carbocation) or the isopropyl carbocation (a secondary carbocation).

$$CH_3{-}CH{=}CH_2 + H^+ \rightarrow CH_3{-}\overset{+}{C}H{-}CH_3$$

isopropyl carbocation (*a secondary carbocation*) — *more stable*

$$CH_3{-}CH{=}CH_2 + H^+ \rightarrow CH_3{-}CH_2{-}CH_2^+$$

propyl carbocation (*a primary carbocation*) — *less stable*

The secondary carbocation is more stable than the primary carbocation and therefore it is formed more easily. The secondary isopropyl carbocation reacts with halide ion to give the major product of the reaction. The primary propyl carbocation reacts with halide ion to give the minor product of the reaction.

$$CH_3{-}CH{=}CH_2 + H{-}Cl \longrightarrow \underset{\text{major product}}{CH_3{-}\underset{\displaystyle Cl}{\underset{|}{C}H}{-}CH_3} + \underset{\text{minor product}}{CH_3{-}CH_2{-}CH_2{-}Cl}$$

As another example, reaction of 2-methylpropene with a proton gives the isobutyl carbocation (a primary carbocation) and the *tert*-butyl carbocation (a tertiary carbocation). The tertiary cabocation is more stable than the primary carbocation and therefore is formed more readily.

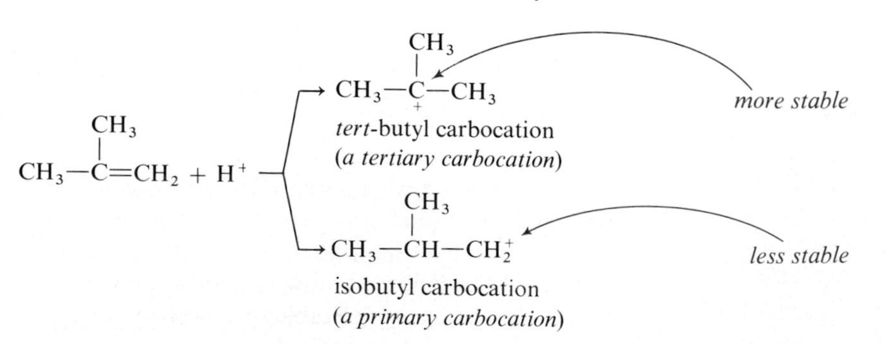

The tertiary carbocation reacts with halide ion to form the major product and the primary carbocation reacts with halide ion to form the minor product.

$$CH_3-\overset{\overset{CH_3}{|}}{C}=CH_2 + H-I \longrightarrow \underset{\text{major product}}{CH_3-\underset{\underset{I}{|}}{\overset{\overset{CH_3}{|}}{C}}-CH_3} + \underset{\text{minor product}}{CH_3-\overset{\overset{CH_3}{|}}{CH}-CH_2-I}$$

Acid-catalyzed hydration of alkenes follows much the same carbocation mechanism as the addition of hydrogen halides. For example in the presence of an acid catalyst, ethylene adds a molecule of water to form ethanol. The first step involves the reaction of the pi bond of the alkene with a proton to form a carbocation. In step 2, this reactive intermediate completes its valence octet by forming a new covalent bond with an unshared electron pair on the oxygen of a water molecule. Finally, loss of a proton in Step 3 results in the formation of an alcohol and regeneration of a proton. Adding Steps 1–3 gives the observed reaction.

Step 1: $H^+ + CH_2=CH_2 \longrightarrow CH_3-CH_2^+$

Step 2: $CH_3-CH_2^+ + :\ddot{O}-H \;(\text{with } H \text{ on O}) \longrightarrow CH_3-CH_2-\ddot{O}^+-H \;(\text{with } H \text{ on O})$

Step 3: $CH_3-CH_2-\ddot{O}^+-H \;(\text{with } H \text{ on O}) \longrightarrow CH_3-CH_2-\ddot{\underset{..}{O}}-H + H^+$

$$CH_2=CH_2 + H_2O \longrightarrow CH_3-CH_2-OH$$

Given the concept of the relative ease of formation of primary, secondary, and tertiary carbocations, this mechanism explains why hydration of propene gives mostly 2-propanol and only very little 1-propanol.

$$\underset{\text{propene}}{CH_3-CH=CH_2} + H_2O \xrightarrow{H_2SO_4} \underset{\substack{\text{2-propanol}\\ \text{(major product)}}}{CH_3-\underset{\underset{OH}{|}}{CH}-CH_3} + \underset{\substack{\text{1-propanol}\\ \text{(minor product)}}}{CH_3-CH_2-CH_2-OH}$$

Example 3.9 Write a mechanism for the formation of the major product of the following reaction. Identify all electrophiles in your mechanism.

$$CH_3-\overset{\overset{CH_3}{|}}{C}=CH-CH_3 + H-Br \longrightarrow \underset{\text{major product}}{CH_3-\underset{\underset{Br}{|}}{\overset{\overset{CH_3}{|}}{C}}-CH_2-CH_3} + \underset{\text{minor product}}{CH_3-\overset{\overset{CH_3}{|}}{CH}-\underset{\underset{Br}{|}}{CH}-CH_3}$$

Solution Propose a three-step mechanism involving a carbocation intermediate.

Step 1: $$H{-}Br \longrightarrow H^+ + Br^-$$

Step 2: $$CH_3{-}\overset{\overset{\textstyle CH_3}{|}}{C}{=}CH{-}CH_3 + H^+ \longrightarrow CH_3{-}\overset{\overset{\textstyle CH_3}{|}}{\underset{+}{C}}{-}CH_2{-}CH_3$$

(a tertiary carbocation)

Step 3: $$CH_3{-}\overset{\overset{\textstyle CH_3}{|}}{\underset{+}{C}}{-}CH_2{-}CH_3 + Br^- \longrightarrow CH_3{-}\overset{\overset{\textstyle CH_3}{|}}{\underset{\underset{\textstyle Br}{|}}{C}}{-}CH_2{-}CH_3$$

In this mechanism, both H^+ and the tertiary carbocation are electrophiles because each is electron-deficient and seeking electrons.

PROBLEM 3.9 Write a mechanism for the formation of the major product of the following reaction. Identify all electrophiles in your mechanism.

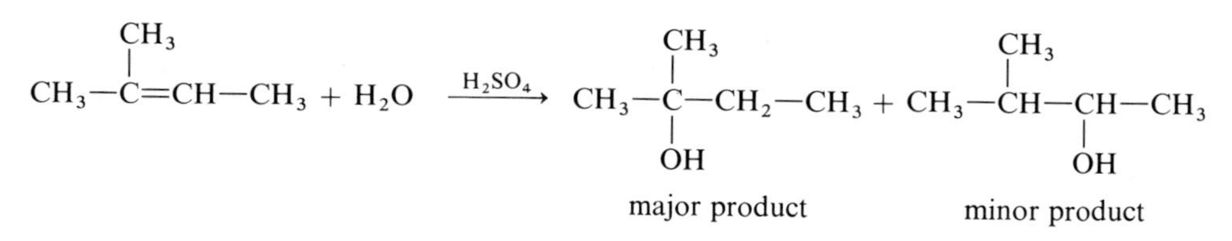

$$CH_3{-}\overset{\overset{\textstyle CH_3}{|}}{C}{=}CH{-}CH_3 + H_2O \xrightarrow{H_2SO_4} CH_3{-}\overset{\overset{\textstyle CH_3}{|}}{\underset{\underset{\textstyle OH}{|}}{C}}{-}CH_2{-}CH_3 + CH_3{-}\overset{\overset{\textstyle CH_3}{|}}{CH}{-}\underset{\underset{\textstyle OH}{|}}{CH}{-}CH_3$$

major product; minor product

E. HYDROBORATION

Hydroboration is the addition of borane, BH_3, to an alkene to form a trialkylborane. The overall reaction is the result of three separate reactions. BH_3 adds first to one molecule of alkene, then to a second, and finally to a third molecule of alkene until all three hydrogens of borane have been replaced by alkyl groups.

Step 1: $$CH_2{=}CH_2 + H{-}B\begin{smallmatrix} H \\ H \end{smallmatrix} \longrightarrow CH_3{-}CH_2{-}B\begin{smallmatrix} H \\ H \end{smallmatrix}$$

Step 2: $$CH_3{-}CH_2{-}B\begin{smallmatrix} H \\ H \end{smallmatrix} + CH_2{=}CH_2 \longrightarrow CH_3{-}CH_2{-}B\begin{smallmatrix} CH_2{-}CH_3 \\ H \end{smallmatrix}$$

Step 3: $$CH_3{-}CH_2{-}B\begin{smallmatrix} CH_2{-}CH_3 \\ H \end{smallmatrix} + CH_2{=}CH_2 \longrightarrow CH_3{-}CH_2{-}B\begin{smallmatrix} CH_2{-}CH_3 \\ CH_2{-}CH_3 \end{smallmatrix}$$

triethylborane
(a trialkylborane)

The addition of H and B to an alkene occurs in one step and does not involve formation of an intermediate carbocation. When borane adds to a substituted alkene, boron and hydrogen add simultaneously in such a way that boron becomes bonded to the less substituted carbon of the double bond.

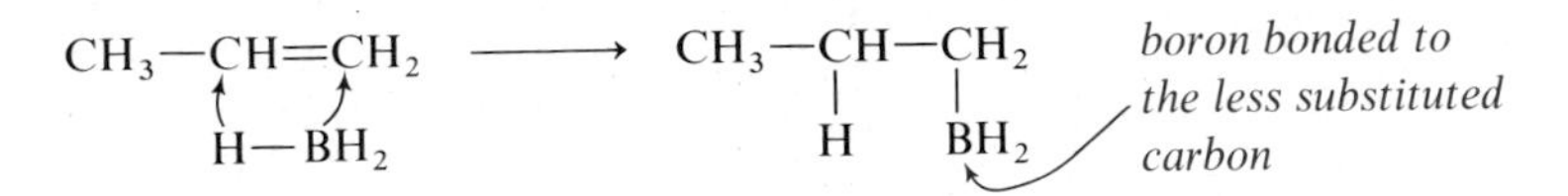

Trialkylboranes are readily oxidized by hydrogen peroxide in the presence of NaOH to yield an alcohol and sodium borate.

$$\underset{\text{tripropylborane}}{(CH_3CH_2CH_2)_3B} + \underset{\text{hydrogen peroxide}}{H_2O_2} + 3\,NaOH \longrightarrow \underset{\text{1-propanol}}{3\,CH_3CH_2CH_2OH} + \underset{\text{sodium borate}}{3\,Na_3BO_3}$$

The net reaction from hydroboration and subsequent oxidation is hydration of the double bond. To appreciate the value of this sequence, we might compare the product of acid-catalyzed hydration of 2-methylpropene with that obtained from hydroboration and hydrogen peroxide oxidation.

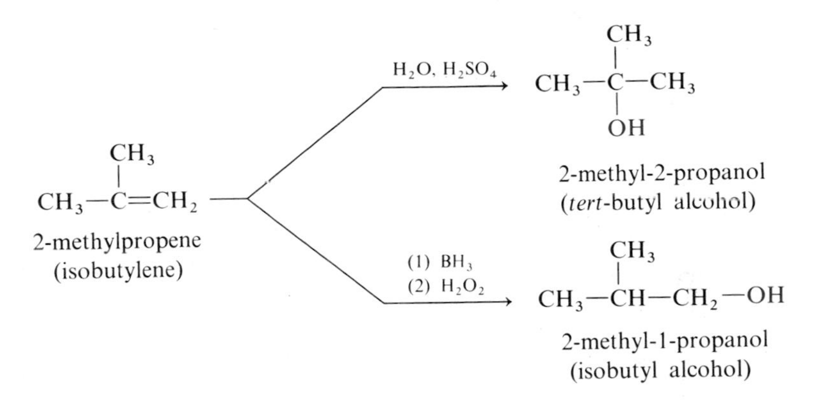

Thus, hydroboration-oxidation is so useful because it gives alcohols not obtainable by other methods from alkenes.

Example 3.10 Draw structural formulas for the trialkylborane and alcohol formed in the following reaction sequences.

(a) $CH_3-\overset{\overset{\large CH_3}{|}}{C}=CH-CH_3 \xrightarrow{BH_3}$ trialkylborane $\xrightarrow[NaOH]{H_2O_2}$ alcohol

(b) 1-methylcyclopentene (cyclopentene ring bearing CH_3 on the double-bond carbon) $\xrightarrow{BH_3}$ trialkylborane $\xrightarrow[NaOH]{H_2O_2}$ alcohol

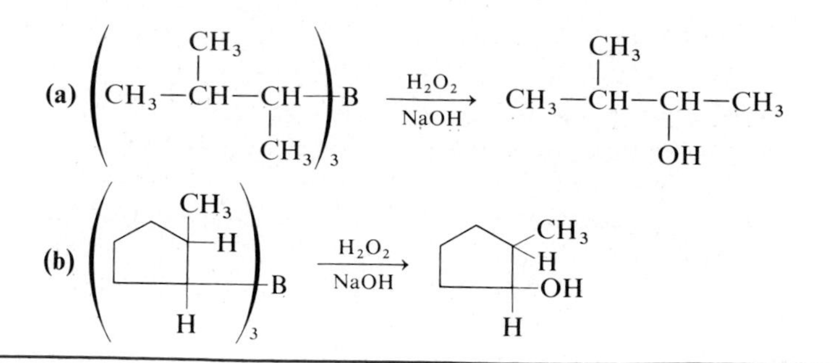

PROBLEM 3.10 Draw structural formulas for the trialkylborane and alkene that give the following alcohols.

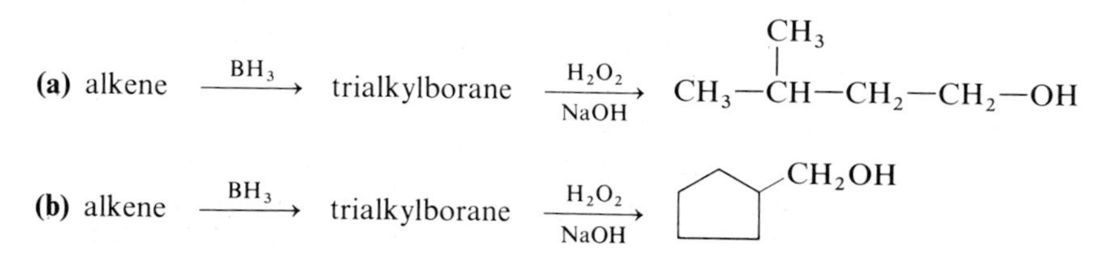

F. HALOGENATION OF ALKENES

Bromine and chlorine add readily to alkenes to form single covalent bonds on adjacent carbons. While iodine is generally too unreactive to add, the more reactive iodine monochloride (ICl) and iodine monobromide (IBr) do add readily. (See problem 3.28 for a quantitative application of iodine monobromide addition.) Halogenation with bromine or chlorine is carried out either with the pure reagents or by mixing the reagents in CCl_4 or some other inert solvent.

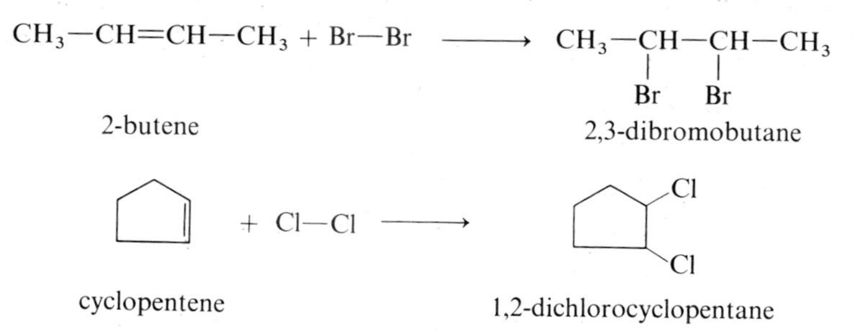

Addition of bromine is a particularly useful qualitative test for the detection of alkenes. A solution of bromine in carbon tetrachloride is red. Alkenes and dibromoalkanes are usually colorless. The rapid discharge of red color of a solution of bromine in carbon tetrachloride is a characteristic property of alkenes.

We might pause a moment to compare and contrast the halogenation of alkenes with that of alkanes (Section 2.9A). Recall that chlorine and bromine do not react with alkanes unless the halogen-alkane mixture is exposed to strong light or heated to temperatures of 250–400°C. The reaction which then occurs is one of substitution of halogen for hydrogen and the formation of an equivalent amount of HCl or HBr. The halogenation of most higher alkanes inevitably gives a complex mixture of monosubstitution products. In contrast, chlorine and bromine react readily with alkenes at room temperature in the dark. Furthermore, the reaction is addition of halogen to the two carbon atoms of the double bond with the formation of two new carbon-halogen bonds. If we know the position of the double bond, we can predict with certainty the location of the halogen atoms in the product.

The addition of Br_2 or Cl_2 to an alkene occurs by a two-step mechanism and a carbocation intermediate as illustrated by the reaction of bromine and propene to form 1,2-dibromopropane. The nonpolar bromine molecule becomes partially polarized as it approaches the carbon-carbon double bond, a region of high electron density. The polarized bromine molecule is then capable of acting as an electrophile and initiating attack on the double bond. In Step 1, the bromine atom which first bonds with carbon leaves behind the pair of electrons which formerly bonded it to the other bromine. The other bromine atom acquires a negative charge and becomes a bromide ion. The carbocation formed in Step 1 reacts with bromide ion in Step 2 to give the product, 1,2-dibromopropane.

Step 1: $:\ddot{Br}-\ddot{Br}: + CH_2=CH-CH_3 \longrightarrow :\ddot{Br}:^- + Br-CH_2-\overset{+}{C}H-CH_3$

Step 2: $:\ddot{Br}:^- + Br-CH_2-\overset{+}{C}H-CH_3 \longrightarrow Br-CH_2-CH(Br)-CH_3$

G. OXIDATION OF ALKENES TO GLYCOLS

Treatment of an alkene with a dilute, neutral solution of potassium permanganate, $KMnO_4$, oxidizes the alkene to a dialcohol with hydroxyl groups on adjacent carbons. Dialcohols with —OH groups on adjacent carbons are called glycols.

$$3CH_3CH=CH_2 + 2KMnO_4 + 4H_2O \longrightarrow 3CH_3CH(OH)-CH_2(OH) + 2MnO_2 + 2KOH$$

1,2-propanediol (a glycol) — manganese dioxide

Oxidation of cyclopentene produces a *cis* glycol.

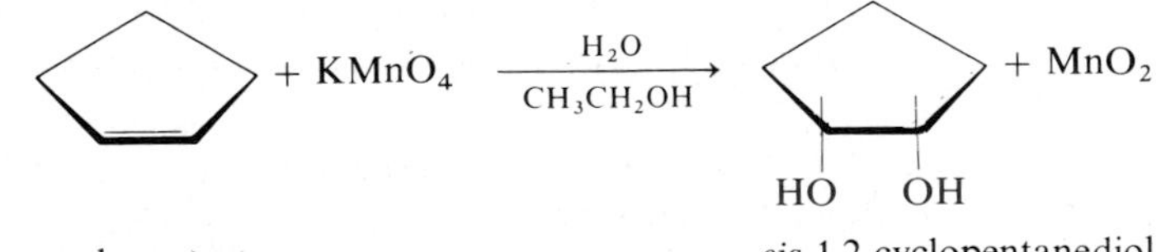

cyclopentene — *cis*-1,2-cyclopentanediol

The *cis* geometry of the product is accounted for by the formation of cyclic manganese ester as an intermediate in the oxidation.

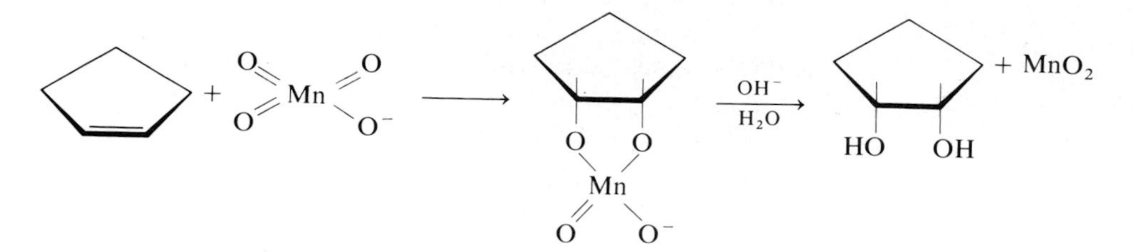

Reaction with permanganate is the basis for a qualitative test for alkenes. An aqueous solution of potassium permanganate is deep purple in color. When permanganate solution is added to an alkene, the purple color of permanganate disappears and a brown precipitate of MnO_2 appears. Therefore, the disappearance of the purple color and appearance of a brown precipitate is good evidence for the presence of an alkene. Unfortunately, this test is not completely specific for alkenes since alkynes and certain other easily oxidized functional groups also discharge the permanganate color.

H. OXIDATION BY OZONE-OZONOLYSIS

The reaction of an alkene with ozone, O_3, is a very important method for cleavage of carbon-carbon double bonds. In practice the alkene is dissolved in an inert solvent such as CCl_4 and a stream of ozone gas is bubbled through the solution. Ozone reacts with the alkene to form an ozonide as illustrated by the reaction of ozone with 2-methyl-2-pentene:

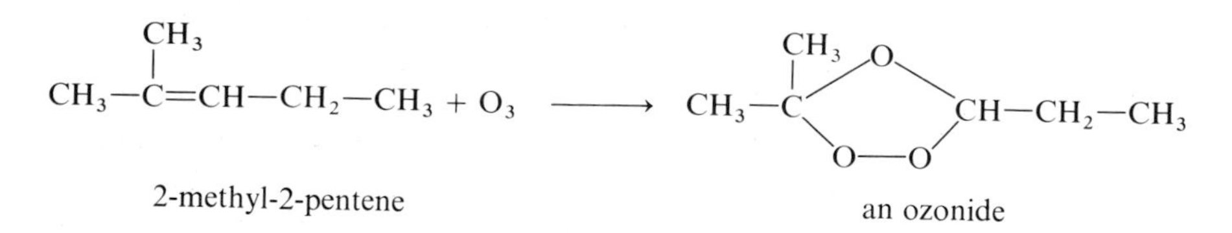

Ozonide intermediates are unstable and tend to explode. Therefore, ozonides are usually not isolated but instead are treated directly with water in the presence of powdered zinc. Under these conditions, carbon-carbon double bonds are cleaved and the carbon atoms of the alkene functional group are converted into aldehydes or ketones depending on the nature of the substituents on the original carbon-carbon double bond.

$$CH_3-\overset{CH_3}{\overset{|}{C}}\overset{O}{\diagup\diagdown}CH-CH_2-CH_3 \ (\text{ring closed by } O-O) \xrightarrow{H_2O,\ Zn} \underset{\text{(a ketone)}}{CH_3-\overset{CH_3}{\overset{|}{C}}=O} + \underset{\text{(an aldehyde)}}{H-\overset{O}{\overset{\|}{C}}-CH_2-CH_3}$$

Example 3.11 Draw structural formulas for the products of the following ozonolysis reactions. Name the new functional groups formed in each oxidation.

Solution

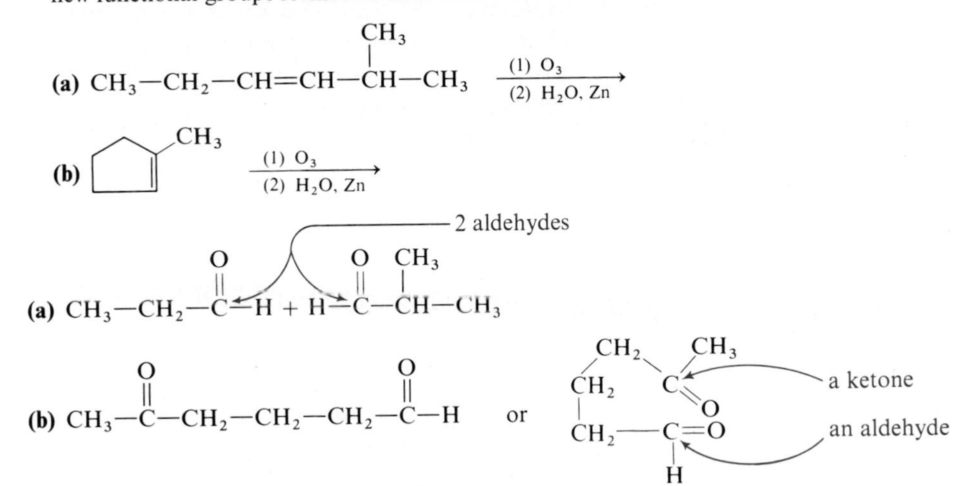

PROBLEM 3.11 Draw structural formulas for the products of ozonolysis of the following alkenes. Name the new functional groups formed in each oxidation.

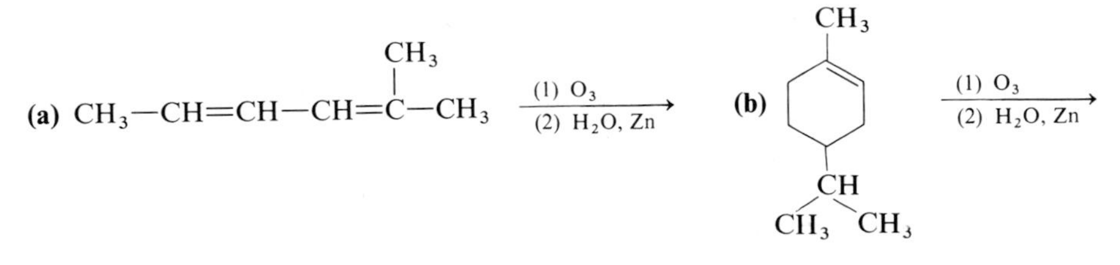

I. POLYMERIZATION OF SUBSTITUTED ETHYLENES

From the perspective of the chemical industry, the single most important reaction of alkenes is that of polymerization: the building together of many small units known as monomers (Greek, *mono* + *meros*, "single part") into very large, high-molecular weight polymers (Greek, *poly* + *meros*, "many parts"). In addition polymerization, monomer units are joined together without loss of atoms or groups of atoms. An example of addition polymerization is the formation of polyethylene from ethylene.

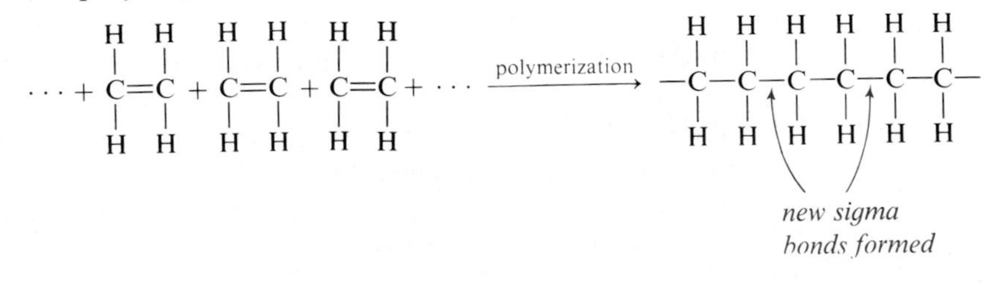

In the polymerization of ethylene, pi bonds are broken and the electrons are used to form new sigma bonds between the monomer units. The preceding example shows the addition polymerization of three ethylene units. In practice, hundreds of monomer units polymerize and molecular weights of polyethylene range from 50,000 to over 1,500,000. Polymerization reactions are more generally written in the following way.

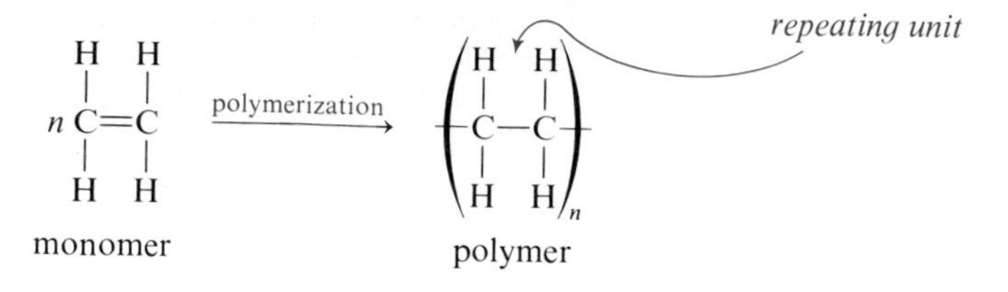

The subscript, *n*, in the formula of polyethylene indicates that the monomer-derived unit, $—CH_2CH_2—$, repeats *n* times in the polymer.

Polymerization of ethylene can be initiated by trace amounts of cations, anions, or free radicals. It is on the free radical polymerization of ethylene that we shall concentrate. A free radical is any atom or molecule that contains one or more unpaired electrons. Following are structural formulas for the chlorine radical (actually, a chlorine atom), the ethyl radical, and an alkoxide radical.

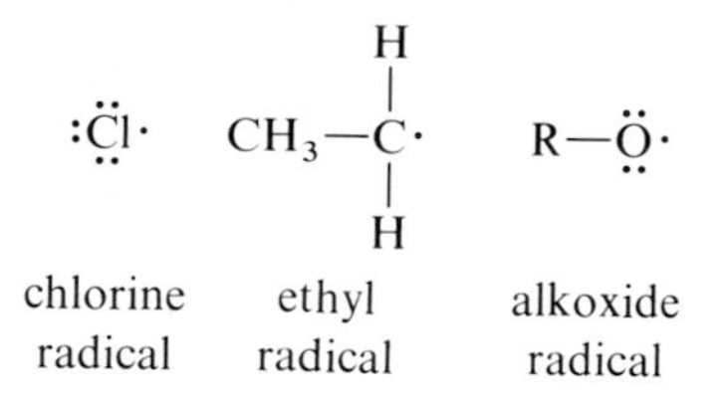

Thus far we have seen structural formulas for two classes of reactive intermediates, carbocations and free radicals. Let us compare the electronic structures of each. Consider for example the ethyl carbocation and the ethyl radical.

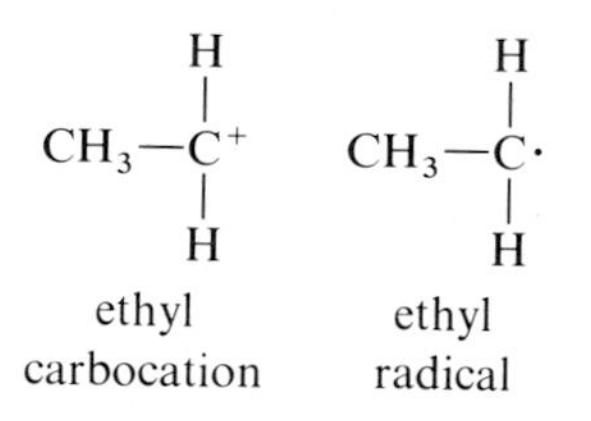

These reactive intermediates are similar in that each has a carbon atom in an abnormal valence state. In the carbocation, the carbon has only six valence electrons in its outer shell. In the free radical it has only seven. However, they differ in charge; the carbocation bears a positive charge while the free radical has no charge—it is neutral.

Free radicals from any number of sources will initiate the polymerization of ethylene. Organic peroxides are exceptionally good sources of radicals for this process because the weak —O—O— bond undergoes rupture at relatively low temperatures.

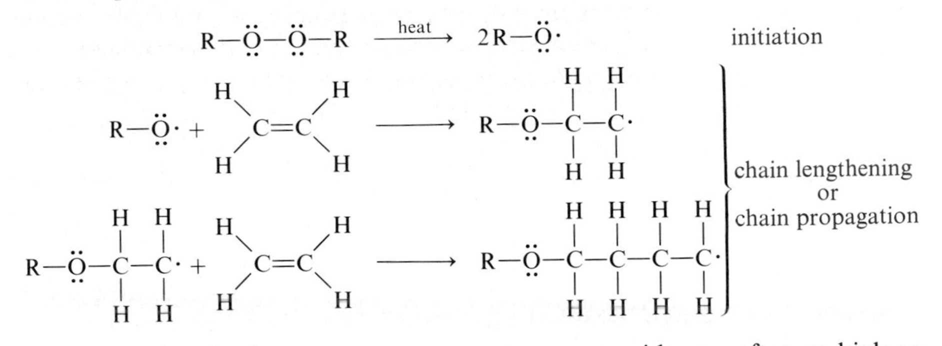

The chain-lengthening sequence occurs at a very rapid rate, often as high as thousands of additions per second, depending on the experimental conditions. The characteristic feature of this type of polymerization is a series of reactions each of which consumes a reactive particle (in this case, a free radical) and produces another reactive particle (another free radical). Such processes are called chain propagation.

In principle, a polymer chain might continue to grow until all monomer units are used up. In practice, however, the chain lengths generally do not exceed 10^4 or 10^5 monomer units (often they are far less). Chain growth or propagation is halted by what are called chain termination steps. One such chain termination process is radical combination.

$$\sim CH_2\cdot + \cdot CH_2\sim \longrightarrow \sim CH_2{-}CH_2\sim$$

growing chains → a radical combination product — *chain termination*

The other major chain termination process is disproportionation, in which a radical of one chain transfers a hydrogen atom (H·) to another chain to generate two nonradical containing molecules.

$$\sim CH_2{-}CH_2\cdot + \cdot CH_2{-}CH_2\sim \longrightarrow \sim CH_2CH_3 + CH_2{=}CH\sim$$

growing chains → chain disproportionation products — *chain termination*

Polymerization of propene can also be initiated by a free radical source. In principle, there are two different free radicals that might be formed by the reaction of R—Ö· and propene, a primary free radical and a secondary free radical.

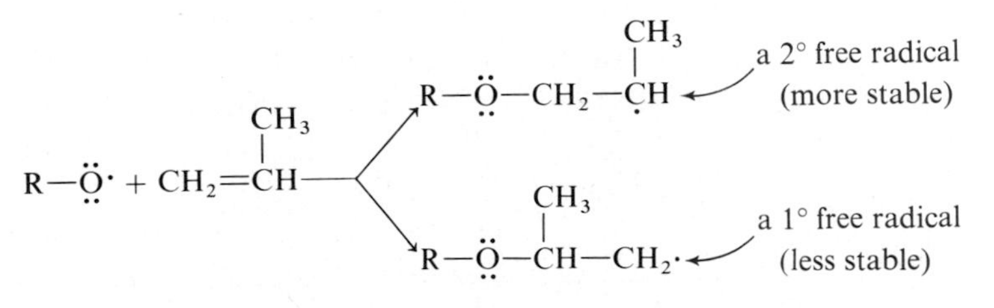

The order of stability of alkyl free radicals is the same as that of alkyl carbocations: a tertiary free radical is more stable than a secondary free radical, which is in turn more stable than a primary free radical. Therefore, polymerization of propene involves secondary free radical intermediates to give polypropylene with methyl groups repeating regularly on every other atom of the polymer chain.

$$nCH_3CH{=}CH_2 \xrightarrow{\text{polymerization}} \left(\begin{array}{c} CH_3 \\ | \\ CH{-}CH_2 \end{array} \right)_n$$

polypropylene
(fibers for carpeting
and clothing)

Table 3.2 lists several important polymers derived from substituted ethylenes along with their common names and uses.

The alkene polymers, mainly polyethylene and polypropylene, are the largest tonnage plastics in the world.

The tetrafluoroethylene polymers were discovered accidentally in 1938 by du Pont chemists during the search for new refrigerants (the Freons described in Section 2.11). One morning a cylinder of tetrafluoroethylene appeared to be empty (no gas escaped when the valve was opened) and yet the weight of the cylinder indicated it was full. The cylinder was opened and inside was found a waxy solid, the forerunner of Teflon. The solid proved to have very unusual properties: extraordinary chemical inertness, outstanding heat resistance, very high melting point and unusual frictional properties. In 1941 du Pont began limited production of Teflon. The small amount of polymer was preempted at once by the Manhattan Project where it was used in equipment to contain the highly corrosive UF_6 during the separation of the isotopes of uranium. In 1948

Table 3.2 Polymers derived from substituted ethylene monomers.

Monomer	*Monomer name*	*Polymer name or trade name*
$CH_2{=}CH_2$	ethylene	polyethylene, Polythene, for unbreakable containers and tubing
$CH_2{=}CHCH_3$	propylene	polypropylene, Herculon, fibers for carpeting and clothes
$CH_2{=}CHCl$	vinyl chloride	polyvinyl chloride, PVC, Koroseal
$CH_2{=}CCl_2$	1,1-dichloroethylene	Saran, food wrappings
$CH_2{=}CHCN$	acrylonitrile	polyacrylonitrile, Orlon, Acrylics
$CF_2{=}CF_2$	tetrafluoroethylene	polytetrafluoroethylene, Teflon
$CH_2{=}CHC_6H_5$	styrene	polystyrene, Styrofoam, for insulation
$CH_2{=}C(CH_3)CO_2CH_3$	methyl methacrylate	polymethyl methacrylate, Lucite, Plexiglas, for glass substitutes
$CH_2{=}CHCO_2CH_3$	methyl acrylate	polymethyl acrylate, Acrylics, for latex paints

du Pont built the first commercial Teflon plant and the product was used to make gaskets, bearings for automobiles, nonstick equipment for candy manufacturers and commercial bakers, seals for rotating equipment, and a number of other items. Teflon became a household word in 1961 with the introduction of nonstick frying pans in the U.S. market.

Much of the early interest in alkene polymers arose from the desire to make synthetic rubber and relieve the nearly total dependence on natural rubber. In the mid-1920s, several companies, most notably du Pont in the United States and I. G. Farbenindustrie in Germany, began research programs. By the mid-1930s, du Pont was producing the first synthetic rubber, Neoprene, on a commercial basis. Neoprene synthetic rubber is polychloroprene.

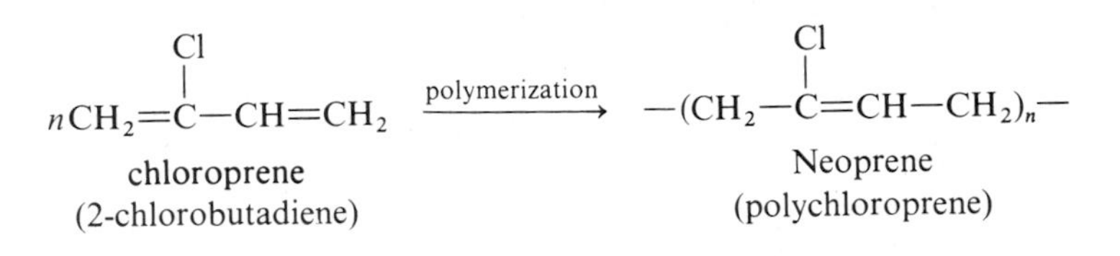

The years since the 1930s have seen extensive research and development in polymer chemistry and physics, and an almost explosive growth in plastics, coatings, and rubber technology has created a world-wide, multibillion dollar industry. A few basic characteristics account for this phenomenal growth. First, the raw materials for plastics are derived mainly from petroleum. With the development of efficient cracking processes (Section 2.10) the raw materials became generally cheap and plentiful. Second, within broad limits, scientists have learned how to tailor polymers to the requirements of the end use. Third, many plastics can be fabricated more cheaply than the competing materials. For example, plastic technology created the water-base (latex) paints that have revolutionized the coatings industry. Plastic films and foams have done the same for the packaging industry. The list could go on and on as we think of the manufactured items that surround us in our daily lives.

3.7 REACTIONS OF ALKYNES

Alkynes undergo addition reactions with hydrogen and the electrophilic addition reagents as do alkenes. A major difference, however, is that alkynes undergo addition much more slowly than do alkenes.

A. ADDITION OF HYDROGEN-REDUCTION

It is possible to add either one or two moles of hydrogen to an alkyne, depending on the conditions of the experiment. Addition of one mole of hydrogen in the presence of a metal catalyst leads to a *cis*-alkene. For this purpose it is necessary

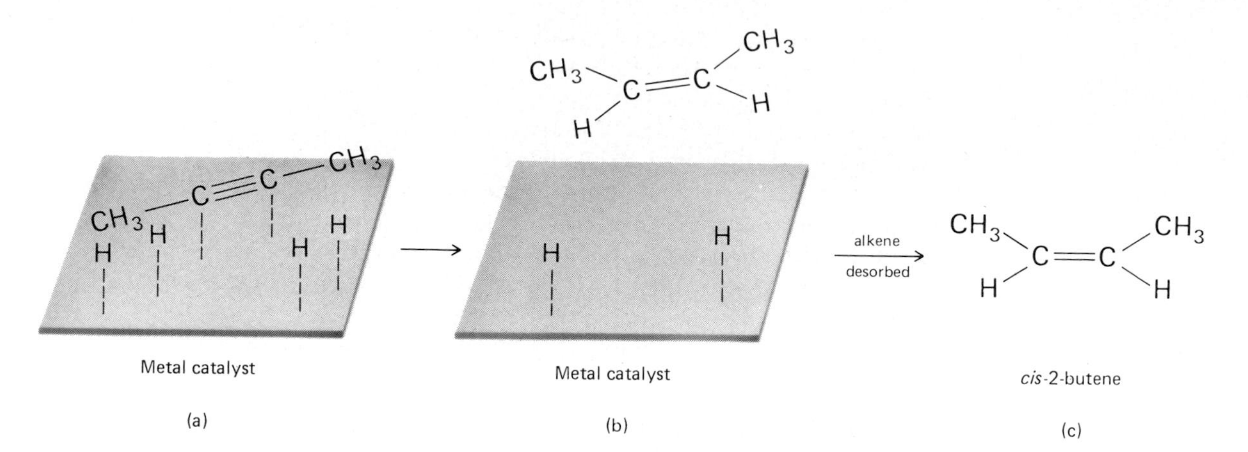

Figure 3.6 *Catalytic reduction of an alkyne to an alkene.*

to use a specially prepared catalyst, often palladium on $BaSO_4$, so that the addition stops after the addition of one mole of H_2.

$$CH_3{-}C{\equiv}C{-}CH_3 + H{-}H \xrightarrow{Pd/BaSO_4} \underset{H\ \ \ \ \ \ \ \ H}{\overset{H_3C\ \ \ \ \ \ \ CH_3}{C{=}C}}$$

The *cis* stereochemistry of the product is accounted for by assuming that both hydrogen and the alkyne are adsorbed on the catalytic surface and that both hydrogen atoms are delivered from the same side of the alkyne molecule (Figure 3.6). Addition of two moles of hydrogen to an alkyne gives an alkane.

$$\underset{\text{2-butyne}}{CH_3{-}C{\equiv}C{-}CH_3 + 2H_2} \xrightarrow{Pt} \underset{\text{butane}}{CH_3{-}CH_2{-}CH_2{-}CH_3}$$

B. ADDITION OF HYDROGEN HALIDES, BROMINE, AND CHLORINE

Acetylene reacts with one mole of HCl to form chloroethylene (vinyl chloride) and with two moles of HCl to form 1,1-dichloroethane.

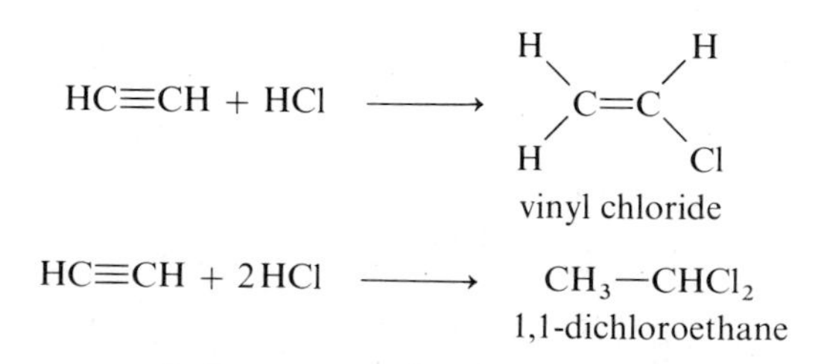

$$HC{\equiv}CH + HCl \longrightarrow \underset{H\ \ \ \ \ \ \ \ Cl}{\overset{H\ \ \ \ \ \ \ \ H}{C{=}C}}$$
vinyl chloride

$$HC{\equiv}CH + 2HCl \longrightarrow \underset{\text{1,1-dichloroethane}}{CH_3{-}CHCl_2}$$

Addition of HX to propyne and higher alkynes follows Markovnikov's rule. For example, propyne adds one mole of HCl to form 2-chloropropene. It adds two moles of HCl to form 2,2-dichloropropane:

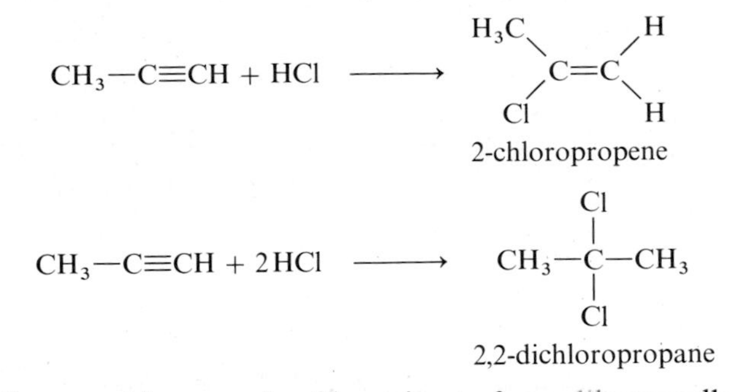

Similarly, alkynes add one mole of bromine to form dibromoalkenes and two moles of bromine to form tetrabromoalkanes:

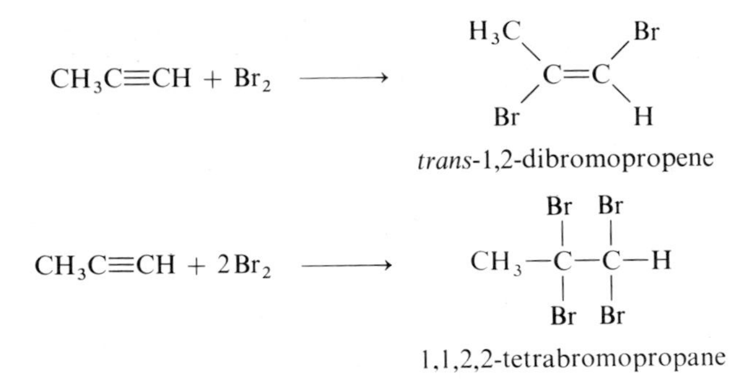

C. ADDITION OF WATER—HYDRATION

In the presence of sulfuric acid-mercuric sulfate catalyst, alkynes undergo addition of water to form substances called enols. Enols are so named because they contain both an alk<u>ene</u> and an alcoh<u>ol</u>. Enols are in equilibrium with an aldehyde or ketone and equilibrium greatly favors the formation of C=O rather than C=C. Therefore, the hydration of acetylene yields an aldehyde.

$$H{-}C{\equiv}C{-}H + H_2O \xrightarrow[HgSO_4]{H_2SO_4} \underset{\text{an enol}}{H{-}C(H){=}C(OH){-}H} \rightleftharpoons \underset{\text{an aldehyde}}{H{-}CH_2{-}C({=}O){-}H}$$

Addition of water to propyne and higher alkynes gives enols which are in equilibrium with ketones.

$$CH_3(CH_2)_4C{\equiv}C{-}H + H_2O \xrightarrow[HgSO_4]{H_2SO_4} \underset{\text{an enol}}{CH_3(CH_2)_4C(OH){=}C(H){-}H} \rightleftharpoons \underset{\text{a ketone}}{CH_3(CH_2)_4C({=}O){-}CH_3}$$

Thus, hydration of 1-alkynes is an important method for the synthesis of methyl ketones (substances containing the group $-\overset{\overset{O}{\|}}{C}-CH_3$). We will discuss the chemistry of aldehydes, ketones, and enols in much more detail in Chapter 7.

Acid catalyzed addition of acetic acid to acetylene gives vinyl acetate, an important monomer for the synthesis of substituted polyethylenes (Table 3.2).

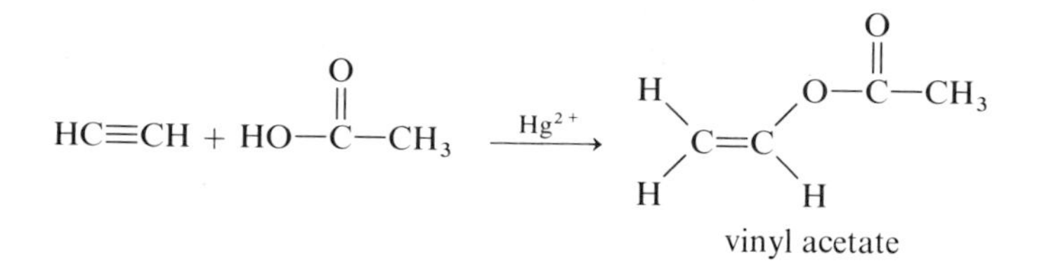

3.8 SPECTROSCOPIC PROPERTIES OF ALKENES

Both carbon-carbon double bonds and hydrogen atoms attached to carbon-carbon double bonds, often referred to as vinyl hydrogens, show absorption in the infrared region of the spectrum. The characteristic absorption frequencies for the stretching of these bonds are:

Type of bond	Frequency (cm^{-1})	Wavelength (μm)
=C—H (stretching)	3000 to 3200	3.13 to 3.33
C=C (stretching)	1620 to 1680	5.95 to 6.17

Each of these stretching frequencies can be seen in the infrared spectrum of 1-octene (Figure 3.7).

Notice the peak of medium intensity at 3080 cm^{-1} corresponding to stretching of the vinyl hydrogens and the peak of medium intensity at 1670 cm^{-1} corresponding to stretching of the carbon-carbon double bond. Also note the three closely spaced peaks around 2900 cm^{-1} corresponding to stretching of the alkane carbon-hydrogen bonds. You might compare the infrared spectrum of 1-octene, an alkene (Figure 3.7), with that of octane, an alkane (Figure 15.2), and note the similarities and differences.

Isolated carbon-carbon double bonds do not absorb in the near ultraviolet (200 to 400 nm) or visible (400 to 700 nm) region. However, conjugated double bond systems do absorb strongly in this region of the spectrum. Butadiene, the simplest conjugated diene, shows an absorption maximum in the ultraviolet spectrum at 217 nm. In general, as the number of double bonds in the conjugated system increases, the more the absorption maximum is shifted toward the visible region of the spectrum. 1,3,5-hexatriene with three conjugated carbon-carbon double bonds absorbs at 258 nm.

$$CH_2{=}CH{-}CH{=}CH_2 \qquad CH_2{=}CH{-}CH{=}CH{-}CH{=}CH_2$$

butadiene
217 nm

1,3,5-hexatriene
258 nm

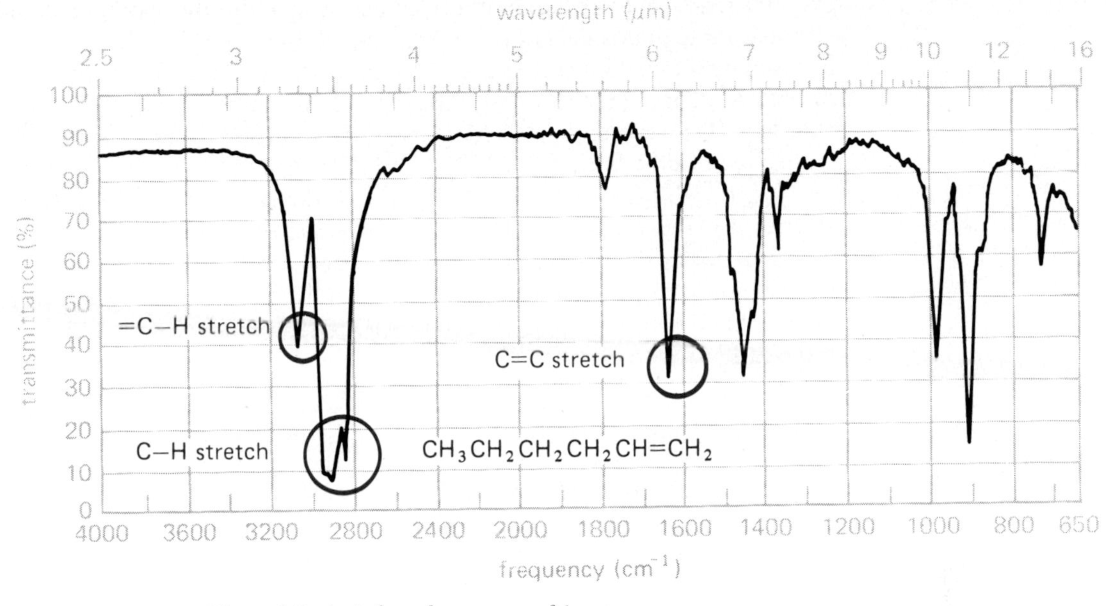

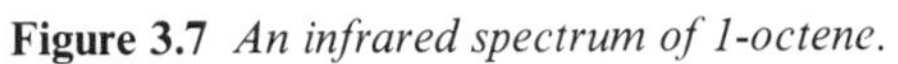

Figure 3.7 *An infrared spectrum of 1-octene.*

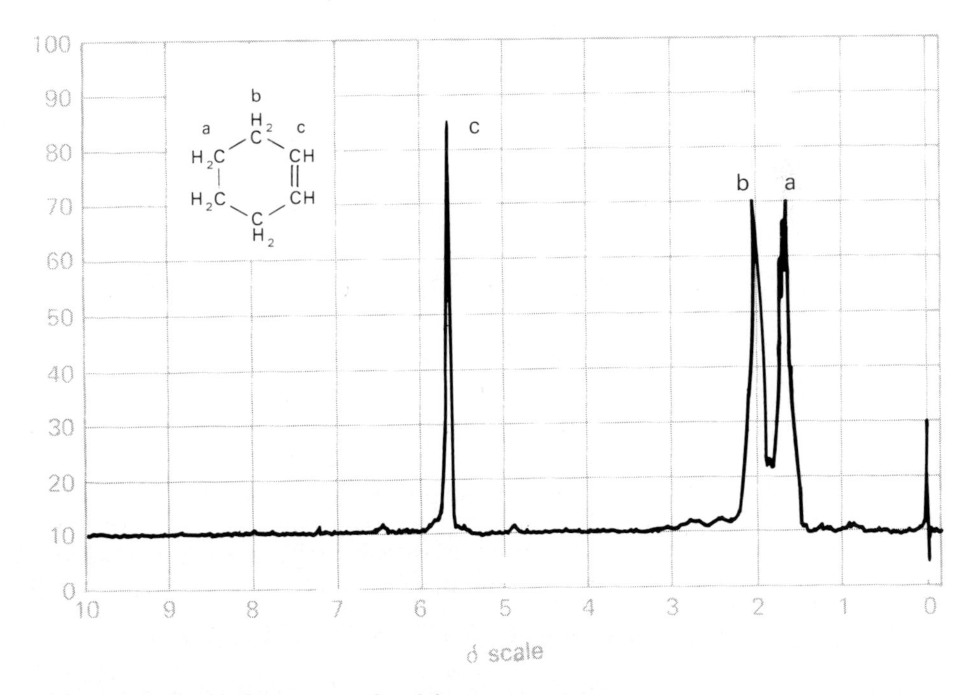

Figure 3.8 *An NMR spectrum of cyclohexene.*

The absorption maxima for molecules with five or more conjugated carbon-carbon double bonds generally occur in the visible region of the spectrum, and accordingly we say that these substances are colored. For example, the diterpene vitamin A alcohol (Section 13.4) which contains five conjugated double bonds is yellow. The tetraterpene β-carotene (Section 13.4) which contains eleven conjugated double bonds is red.

Vinyl hydrogens give rise to signals in the nuclear magnetic resonance spectrum in the range $\delta = 4.5\text{–}6.5$. In the case of cyclohexene (Figure 3.8), the signal due to the two vinyl hydrogens appears at $\delta = 5.7$. Notice that in this molecule there are three sets of chemically equivalent hydrogens and therefore you would predict three signals. The NMR spectrum indeed shows three signals in the ratio 4:4:2.

IMPORTANT REACTIONS

(1) Dehydrohalogenation of alkyl halides (Section 3.5):

$$CH_3{-}CH_2{-}\underset{\displaystyle Br}{\underset{|}{C}H}{-}CH_3 + NaOH \longrightarrow$$

$$CH_3{-}CH{=}CH{-}CH_3 + CH_3{-}CH_2{-}CH{=}CH_2 + HOH + NaBr$$

major product (first), *minor product* (second)

(2) Addition of hydrogen to alkenes (Section 3.6A):

$$CH_3{-}CH_2{-}CH{=}CH_2 + H_2 \xrightarrow{Pt} CH_3{-}CH_2{-}CH_2{-}CH_3$$

(3) Addition of HCl, HBr, and HI to alkenes (Section 3.6B):

$$CH_3{-}CH_2{-}CH{=}CH_2 + HCl \longrightarrow$$

$$CH_3{-}CH_2{-}\underset{\displaystyle Cl}{\underset{|}{C}H}{-}CH_3 + CH_3{-}CH_2{-}CH_2{-}\underset{\displaystyle Cl}{\underset{|}{C}H_2}$$

major product (first), *minor product* (second)

(4) Addition of water (hydration) of alkenes (Section 3.6C):

$$CH_3{-}CH{=}CH_2 + H_2O \xrightarrow{H_2SO_4} CH_3{-}\underset{\displaystyle OH}{\underset{|}{C}H}{-}CH_3 + CH_3{-}CH_2{-}\underset{\displaystyle OH}{\underset{|}{C}H_2}$$

major product (first), *minor product* (second)

(5) Hydroboration of alkenes (Section 3.6E):

$$3CH_3{-}CH{=}CH_2 + BH_3 \longrightarrow (CH_3{-}CH_2{-}CH_2)_3B$$

(a trialkylborane)

(6) Oxidation of trialkylboranes (Section 3.6E):

$$(CH_3{-}CH_2{-}CH_2)_3B \xrightarrow{H_2O_2,\ NaOH} 3CH_3{-}CH_2{-}CH_2{-}OH + Na_3BO_3$$

(7) Addition of Cl_2 and Br_2 to alkenes (Section 3.6F):

$$CH_3{-}CH{=}CH_2 + Br_2 \longrightarrow CH_3{-}\underset{\displaystyle Br}{\underset{|}{CH}}{-}\underset{\displaystyle Br}{\underset{|}{CH_2}}$$

(8) Oxidation of alkenes to glycols (Section 3.6G):

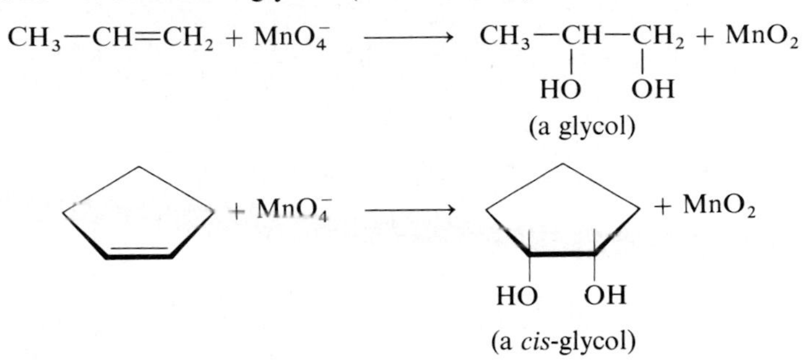

(a glycol)

(a *cis*-glycol)

(9) Ozonolysis of alkenes (Section 3.6H):

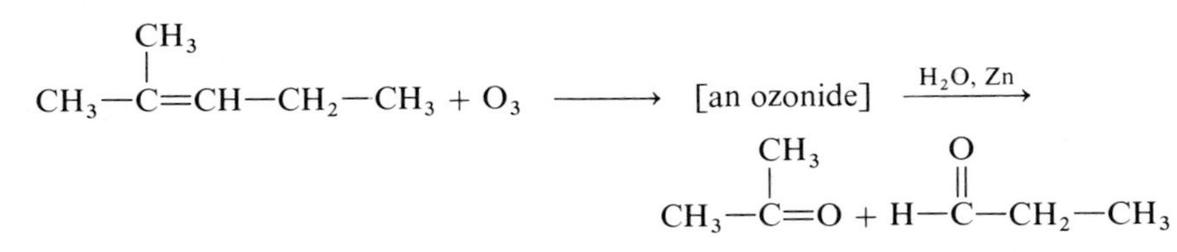

(10) Addition polymerization of ethylene and substituted ethylenes (Section 3.6I):

$$nCH_2{=}CHCl \longrightarrow {-}\left(CH_2{-}\overset{\displaystyle Cl}{\overset{|}{CH}}\right)_n{-}$$

vinyl chloride → polyvinylchloride (PVC)

(11) Addition of hydrogen to alkynes (Section 3.7A):

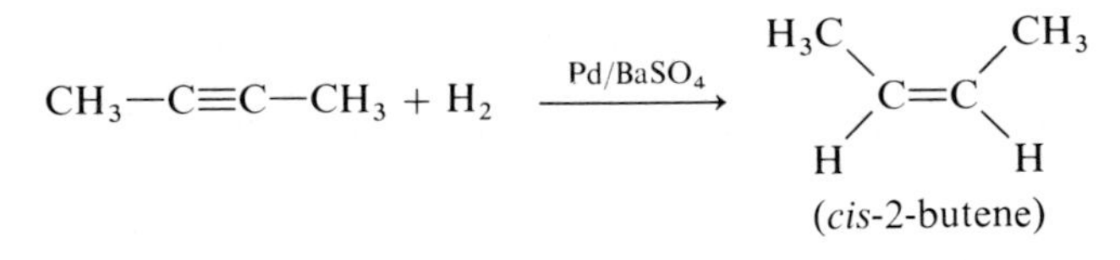

(*cis*-2-butene)

$$CH_3{-}C{\equiv}C{-}CH_3 + 2H_2 \xrightarrow{Pd} CH_3{-}CH_2{-}CH_2{-}CH_3$$

(12) Addition of HCl and HBr to alkynes (Section 3.7B):

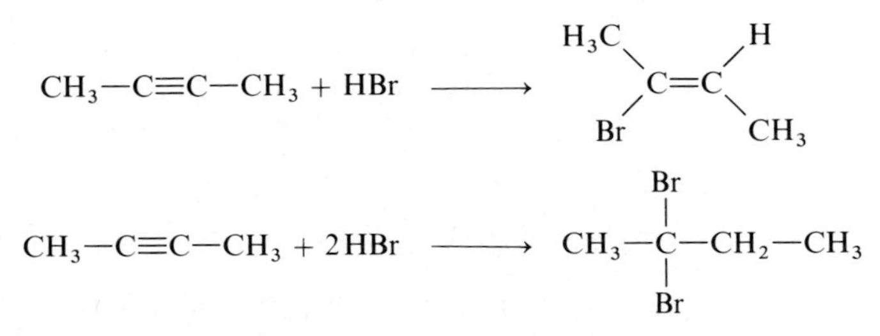

(13) Addition of Br_2 and Cl_2 to alkynes (Section 3.7B):

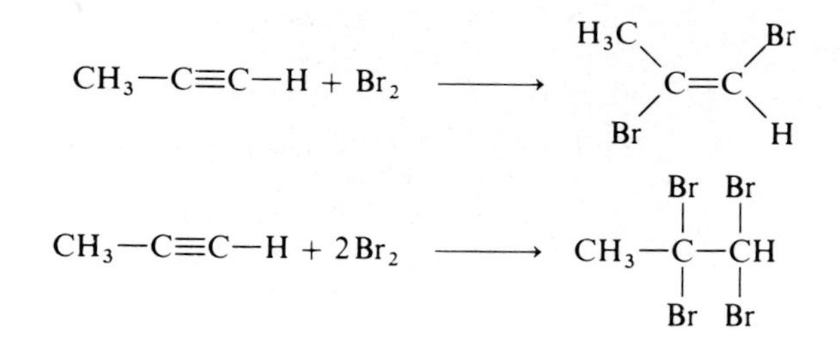

(14) Addition of water (hydration) of alkynes (Section 3.7C):

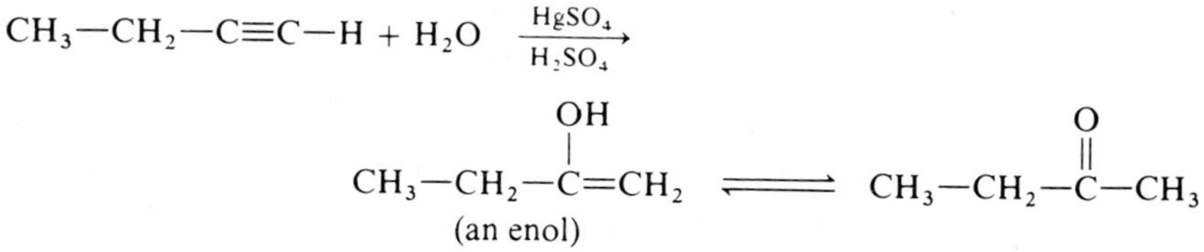

$$CH_3-CH_2-C\equiv C-H + H_2O \xrightarrow[H_2SO_4]{HgSO_4}$$

$$\underset{\text{(an enol)}}{CH_3-CH_2-\overset{\displaystyle OH}{\overset{|}{C}}=CH_2} \rightleftharpoons CH_3-CH_2-\overset{\displaystyle O}{\overset{\|}{C}}-CH_3$$

PROBLEMS

3.12 Name the following compounds.

(a) 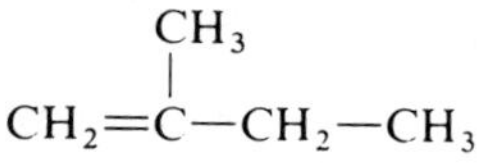 $CH_2=C(CH_3)-CH_2-CH_3$

(b) $CH_3-C(CH_3)=CH-CH_2-CH(CH_3)-CH_3$

(c) (Cl)(H)C=C(H)(Cl)

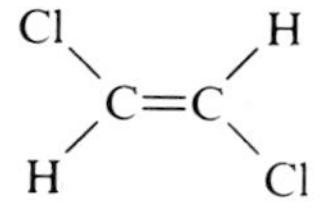

(d) 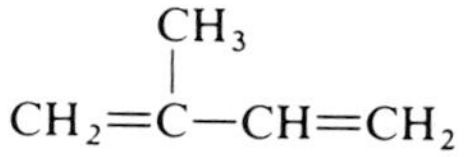 $CH_2=C(CH_3)-CH=CH_2$

(e) 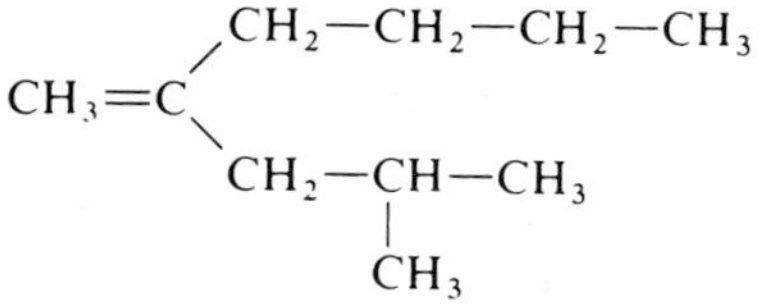 $CH_2=C(CH_2-CH_2-CH_2-CH_3)(CH_2-CH(CH_3)-CH_3)$

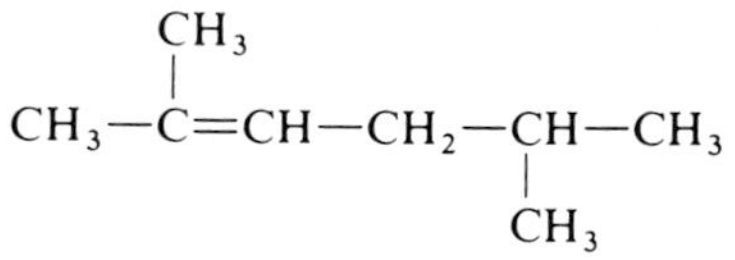

(f) 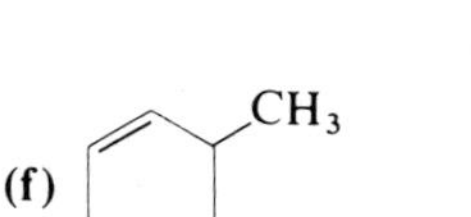

(g) 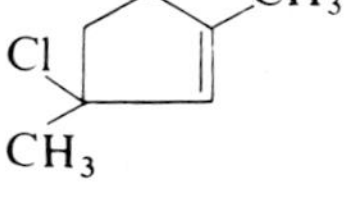 Cl, CH_3, CH_3 (substituted cyclopentene)

(h) 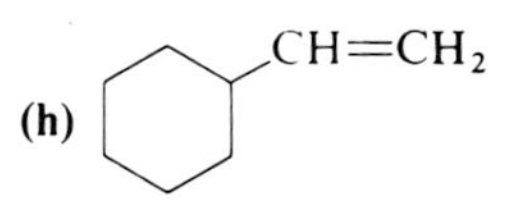

(i) $(CH_3)_3CCH_2CH=CH_2$

(j) $(CH_3)_2CHCH=C(CH_3)_2$

(k) $CH_2=CH-Cl$

(l) $CH_2=CH-CH=CH_2$

(m) $CH_3-CH_2-C\equiv CH$

(n) $CH_2=CH-CH_2Cl$

(o) 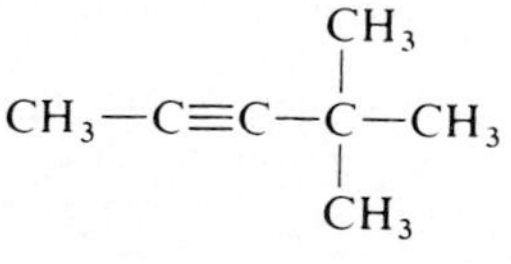 $CH_3-C\equiv C-C(CH_3)_2-CH_3$

3.13 Draw structural formulas for the following alkenes and alkynes.

(a) 2-methyl-3-hexene
(b) 2-methyl-2-hexene
(c) 2-methyl-1-butene
(d) 3-ethyl-3-methyl-1-pentene
(e) 2,3-dimethyl-2-butene
(f) 1-pentene
(g) 2-pentene
(h) 1-chloropropene
(i) 2-chloropropene
(j) 3-methylcyclohexene
(k) 1-isopropyl-4-methylcyclohexene
(l) 3-isopropyl-6-methylcyclohexene
(m) 1,5-hexadiene
(n) 1,4-hexadiene
(o) 2-chloro-1,3-butadiene
(p) 3-hexyne
(q) 5-isopropyl-3-octyne
(r) 3-chloro-1-butyne
(s) tetrachloroethylene
(t) tetrafluoroethylene

3.14 Which of the molecules in Problem 3.13 show *cis-trans* isomerism? For each that does, draw both *cis*- and *trans*-isomers.

3.15 Draw structural formulas for all compounds of molecular formula C_5H_{10} that are

(a) alkenes that do not show *cis-trans* isomerism.
(b) alkenes that do show *cis-trans* isomerism.
(c) cycloalkanes that do not show *cis-trans* isomerism.
(d) cycloalkanes that do show *cis-trans* isomerism.

3.16 Draw structural formulas for the four isomeric chloropropenes (C_3H_5Cl).

3.17 There are four *cis-trans* isomers of 2,4-heptadiene. Name and draw structural formulas for all four.

3.18 Draw structural formulas for the products of the following alkene addition reactions. Where two products are possible, state which is the major product and which is the minor product.

(a) $CH_3-\underset{}{\overset{CH_3}{\overset{|}{C}}}=CH-CH_3 + H_2O \xrightarrow{H_2SO_4}$

(b) $CH_3-\overset{CH_3}{\overset{|}{C}}=CH-CH_3 + HBr \longrightarrow$

(c) $CH_3-\overset{CH_3}{\overset{|}{C}}=CH-CH_3 + Br_2 \longrightarrow$

(d) $CH_3-\overset{CH_3}{\overset{|}{C}}=CH-CH_3 + H_2 \xrightarrow{Pt}$

(e) 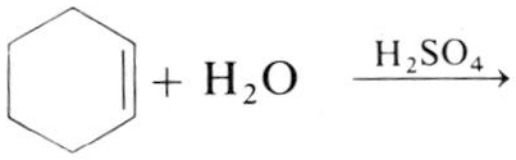 [cyclohexene] $+ H_2O \xrightarrow{H_2SO_4}$

(f) 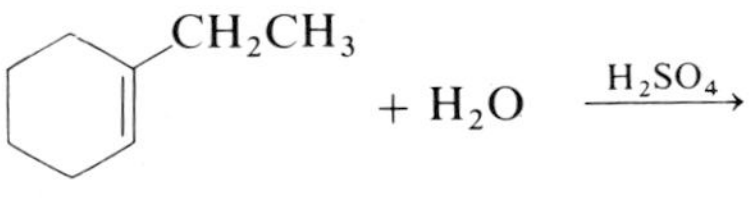 [1-ethylcyclohexene, CH_2CH_3] $+ H_2O \xrightarrow{H_2SO_4}$

(g) $CH_3-\overset{CH_3}{\overset{|}{C}}=CH-CH_3 + BH_3 \longrightarrow$

(h) 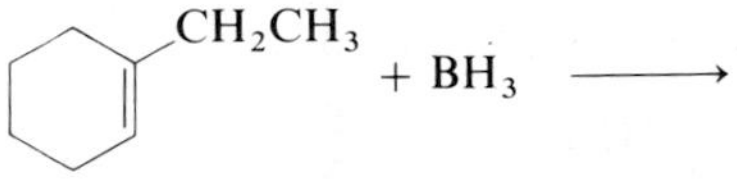 [1-ethylcyclohexene, CH_2CH_3] $+ BH_3 \longrightarrow$

(i) $CH_2{=}\overset{\overset{CH_3}{|}}{C}{-}CH_2{-}CH_3 + H_2O \xrightarrow{H_2SO_4}$

(j) 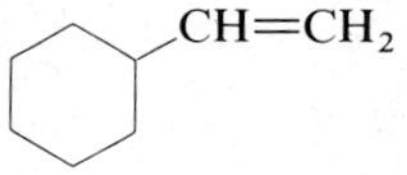 $+ Cl_2 \longrightarrow$ (k)  $+ HCl \longrightarrow$

(l) 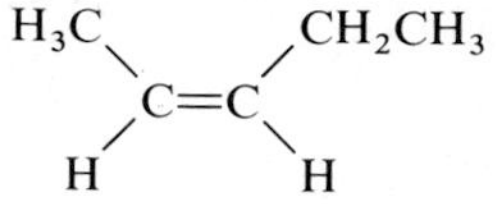 $+ H_2O \xrightarrow{H_2SO_4}$

(m) 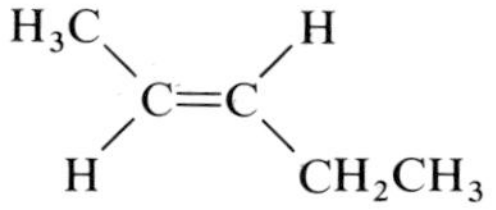 $+ H_2O \xrightarrow{H_2SO_4}$

(n) $CH_2{=}\overset{\overset{CH_3}{|}}{C}{-}CH{=}CH_2 + 2H_2 \xrightarrow{Pt}$

(o) $CH_3{-}\overset{\overset{H_3C}{|}}{C}{=}\overset{\overset{CH_3}{|}}{C}{-}CH_3 + Br_2 \longrightarrow$

(p) 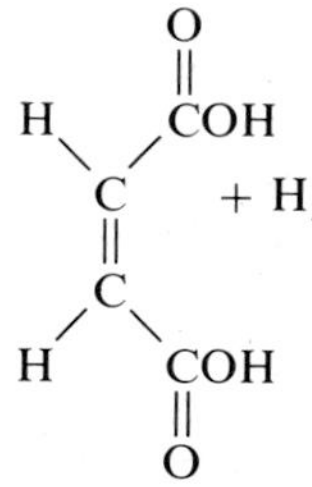 $+ H_2O \xrightarrow{H_2SO_4}$

(q) $\underset{\underset{Cl}{|}}{CH_2}{-}CH{=}CH_2 + MnO_4^- \longrightarrow$

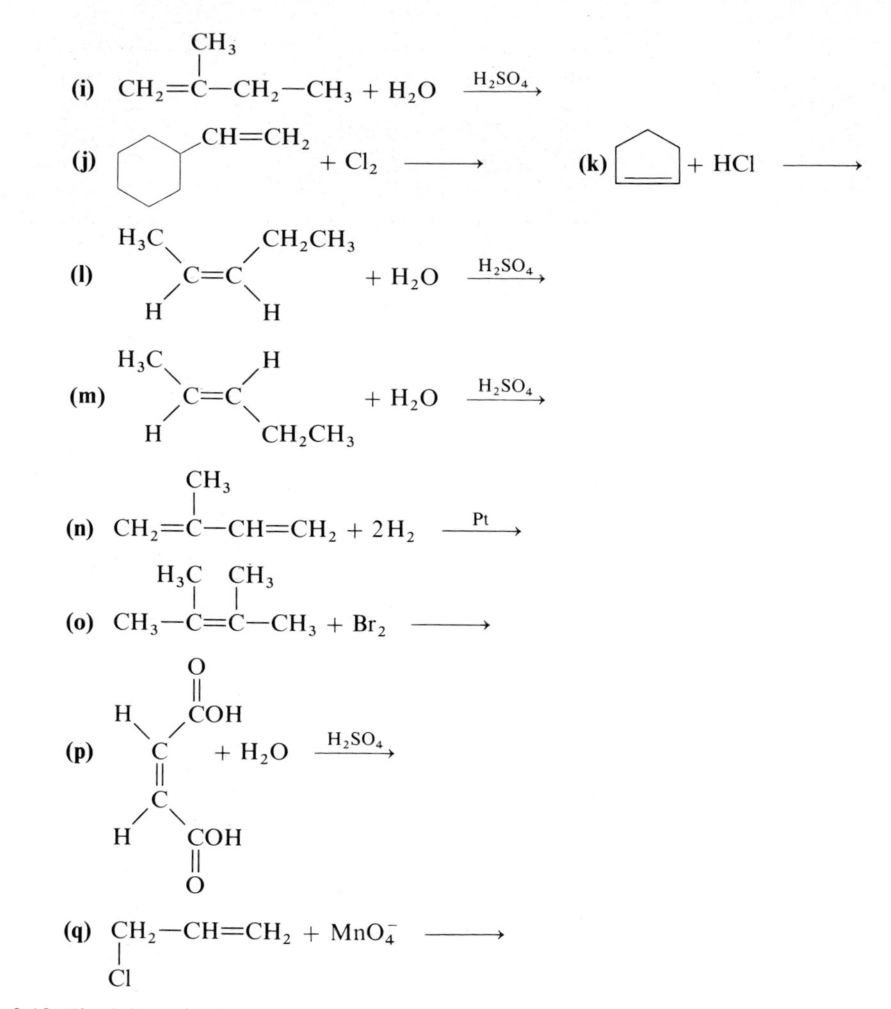

3.19 The following alkenes are reacted with BH_3 to form trialkylboranes, and the trialkylborane is then reacted with hydrogen peroxide in aqueous sodium hydroxide. Draw the structural formula of the alcohol produced in each case.

(a) $CH_3{-}\overset{\overset{CH_3}{|}}{C}{=}CH{-}CH_2{-}CH_3$

(b) 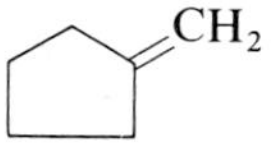

(c) 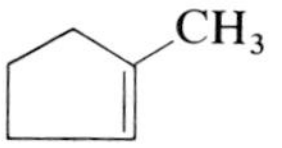

(d) $CH_2{=}CH(CH_2)_5CH_3$

(e) 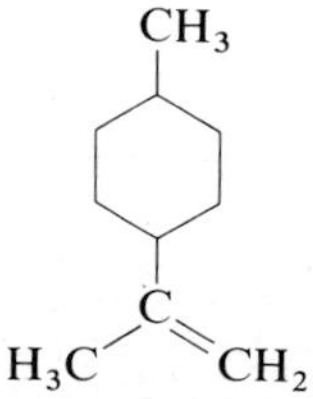

(f) $CH_2{=}\overset{\overset{CH_3}{|}}{C}{-}CH_2{-}CH_2{-}CH_3$

3.20 Draw structural formulas for the major alcohol formed on acid-catalyzed hydration of each alkene in Problem 3.19.

3.21 Each of the following is reacted with ozone to form an ozonide and the ozonide is treated with water in the presence of zinc. Draw structural formulas for the organic products formed from each alkene.

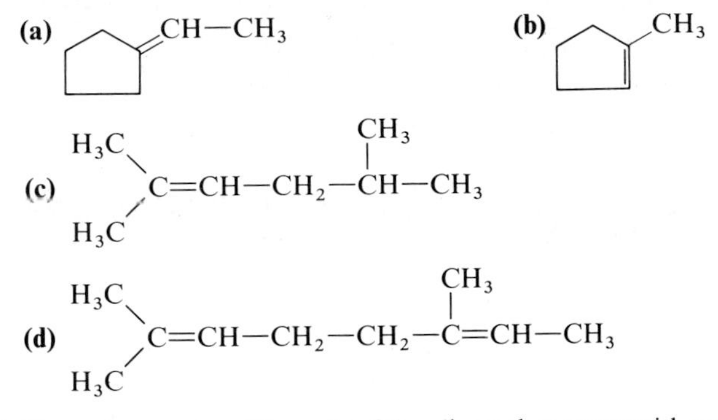

3.22 Draw the structural formula of the alkene that reacts with ozone and then water in the presence of powdered zinc to give the following products.

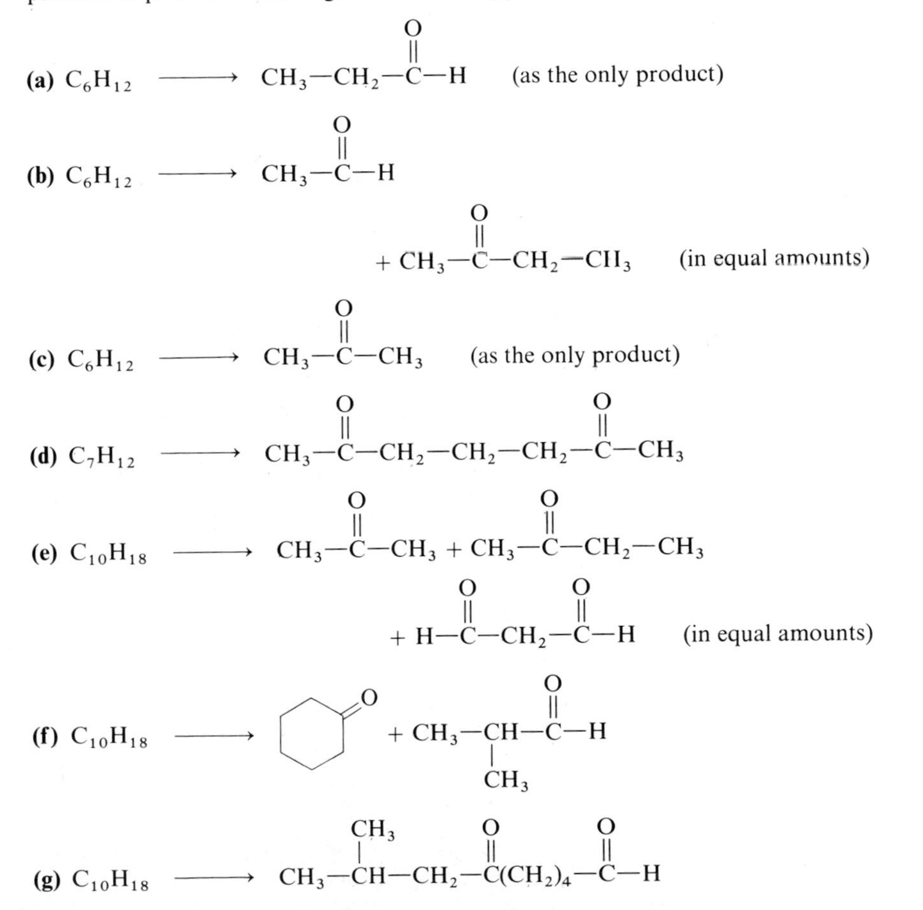

3.23 Draw structural formulas for the isomeric carbocations formed by the addition of H^+ to the following alkenes. Label each carbocation primary, secondary, or tertiary, and state which isomeric carbocation is formed more readily.

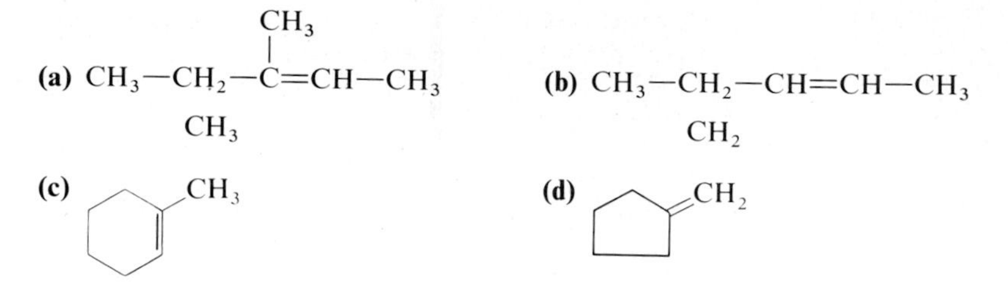

3.24 Write a reaction mechanism for the following alkene addition reactions. For each mechanism, identify all electrophiles.

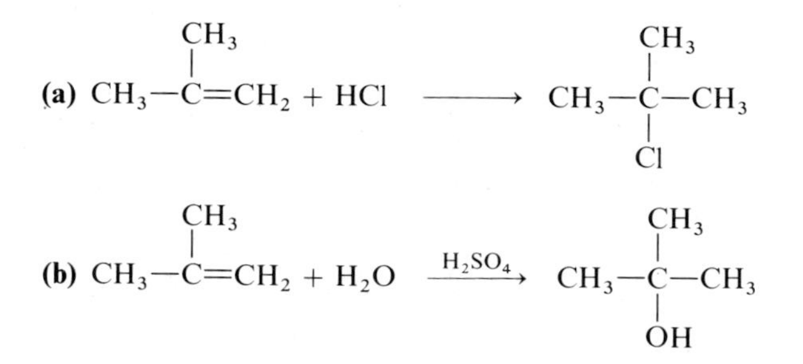

3.25 Terpin hydrate is prepared commercially by the addition of two moles of water to limonene in the presence of dilute sulfuric acid. Limonene is one of the main components of lemon, orange, caraway, dill, bergamot, and some other oils. Terpin hydrate is used medicinally as an expectorant for coughs. It may be given as terpin hydrate and codeine. Propose a structure for terpin hydrate and a reasonable mechanism to account for the formation of the product you have predicted.

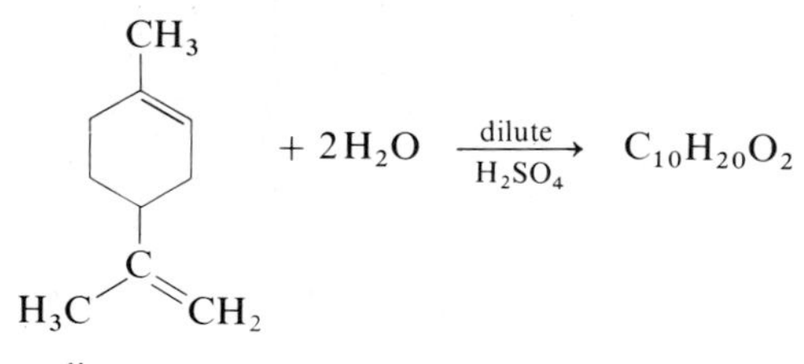

limonene terpin hydrate

3.26 The addition of hypochlorous acid, HOCl, to cyclohexene produces 2-chlorocyclohexanol.

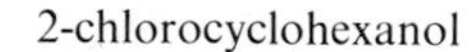

2-chlorocyclohexanol

Propose a mechanism for this reaction starting with the assumption that the addition is initiated by a chloronium ion, Cl^+.

3.27 Reaction of 2-methylpropene with methanol in the presence of H_2SO_4 yields a compound of formula $C_5H_{12}O$.

$$\begin{array}{c} CH_3 \\ | \\ CH_3{-}C{=}CH_2 \end{array} + CH_3OH \xrightarrow{H_2SO_4} C_5H_{12}O$$

Propose a structural formula for this compound and a mechanism to account for its formation.

3.28 Before the recent development of sensitive instrumental techniques, a number of methods were developed to measure the degree of unsaturation of fats and oils. One such method was to experimentally determine an "iodine number." In this procedure, equivalent amounts of I_2 and Br_2 are mixed in acetic acid to produce the highly reactive iodine monobromide, IBr. This reagent adds to alkenes as shown:

$$R{-}CH{=}CH{-}R + Br \longrightarrow R{-}\underset{\displaystyle I}{\underset{|}{CH}}{-}\underset{\displaystyle Br}{\underset{|}{CH}}{-}R$$

(a) Propose a mechanism to account for the addition of IBr to an alkene. Based on your mechanism would you expect the addition of IBr to propene to give

$$CH_3{-}\underset{\displaystyle I}{\underset{|}{CH}}{-}\underset{\displaystyle Br}{\underset{|}{CH_2}} \quad \text{or} \quad CH_3{-}\underset{\displaystyle Br}{\underset{|}{CH}}{-}\underset{\displaystyle I}{\underset{|}{CH_2}}?$$

(b) The iodine number is defined as the number of grams of iodine that adds to 100 grams of a fat or oil. (For definition of the terms *fat* and *oil* see Section 13.1.) The following table gives average molecular weights of three oils and one fat and their corresponding iodine numbers.

Fat or oil	*Average molecular weight (g/mole)*	*Iodine number*
corn oil	870–900	115–130
soybean oil	860–890	125–140
linseed oil	855–895	175–205
butter fat	700–720	25–40

Which of these substances is the most highly unsaturated? the least highly unsaturated? Explain your reasoning.

3.29 Which is the best means for the preparation of 1,2-dichloropentane? Explain your reasoning.

(a) $CH_3{-}CH_2{-}CH_2{-}CH_2{-}CH_3 + 2Cl_2 \xrightarrow{300^\circ}$

$$CH_3{-}CH_2{-}CH_2{-}\underset{\displaystyle Cl}{\underset{|}{CH}}{-}\underset{\displaystyle Cl}{\underset{|}{CH_2}} + 2HCl$$

(b) $CH_3{-}CH_2{-}CH_2{-}CH{=}CH_2 + Cl_2 \xrightarrow{CCl_4}$

$$CH_3{-}CH_2{-}CH_2{-}\underset{\displaystyle Cl}{\underset{|}{CH}}{-}\underset{\displaystyle Cl}{\underset{|}{CH_2}}$$

3.30 Draw the structural formula for an alkene or alkenes of molecular formula C_5H_{10} that gives the indicated compound as the major product. Note that in several parts of this problem (for example, in part a), there is more than one alkene that gives the same substance as the major product.

(a) $$C_5H_{10} + H_2O \xrightarrow{H_2SO_4} \begin{array}{c} \quad\; CH_3 \\ \quad\;\; | \\ CH_3{-}C{-}CH_2{-}CH_3 \\ \quad\;\; | \\ \quad\; OH \end{array} \quad \text{(2 alkenes)}$$

(b) $$C_5H_{10} + Br_2 \longrightarrow \begin{array}{c} CH_3 \\ | \\ CH_3{-}CH{-}CH{-}CH_2 \\ \qquad\;\;\; | \qquad | \\ \qquad\;\;\; Br \quad Br \end{array} \quad \text{(1 alkene)}$$

(c) $$C_5H_{10} + H_2 \xrightarrow{Pt} CH_3{-}CH_2{-}CH_2{-}CH_2{-}CH_3 \quad \text{(2 alkenes)}$$

(d) $$C_5H_{10} + H_2O \xrightarrow{H_2SO_4} \begin{array}{l} CH_3{-}CH{-}CH_2{-}CH_2{-}CH_3 \\ \qquad\;\; | \\ \qquad\; OH \end{array} \quad \text{(1 alkene)}$$

(e) $$C_5H_{10} + HCl \longrightarrow \begin{array}{c} \quad\; CH_3 \\ \quad\;\; | \\ CH_3{-}C{-}CH_2{-}CH_3 \\ \quad\;\; | \\ \quad\; Cl \end{array} \quad \text{(2 alkenes)}$$

(f) $$C_5H_{10} + BH_3 \longrightarrow [\text{trialkylborane}] \xrightarrow{H_2O_2/OH^-} \begin{array}{c} CH_3 \\ | \\ CH_3{-}CH_2{-}CH{-}CH_2 \\ \qquad\qquad\quad | \\ \qquad\qquad\quad OH \end{array} \quad \text{(1 alkene)}$$

(g) $$C_5H_{10} + MnO_4^- \longrightarrow \begin{array}{c} CH_3 \\ | \\ CH_3{-}CH{-}CH{-}CH_2 \\ \qquad\;\;\; | \qquad | \\ \qquad\;\;\; HO \quad OH \end{array} \quad \text{(1 alkene)}$$

(h) $$C_5H_{10} + BH_3 \longrightarrow [\text{trialkylborane}] \xrightarrow{H_2O_2/OH^-} HO{-}CH_2{-}CH_2{-}CH_2{-}CH_2{-}CH_3$$

3.31 Show the reagents and reaction conditions you might use to transform the given starting material into the desired product. Note that some transformations require only one step, while others require two steps.

(a) $$\begin{array}{c} CH_3 \\ | \\ CH_3{-}CH{-}CH_2{-}Cl \end{array} \xrightarrow{\text{(1 step)}} \begin{array}{c} CH_3 \\ | \\ CH_3{-}C{=}CH_2 \end{array}$$

(b) $$\begin{array}{c} CH_3 \\ | \\ CH_3{-}CH{-}CH_2{-}Cl \end{array} \xrightarrow{\text{(2 steps)}} \begin{array}{c} CH_3 \\ | \\ CH_3{-}C{-}CH_3 \\ | \\ Br \end{array}$$

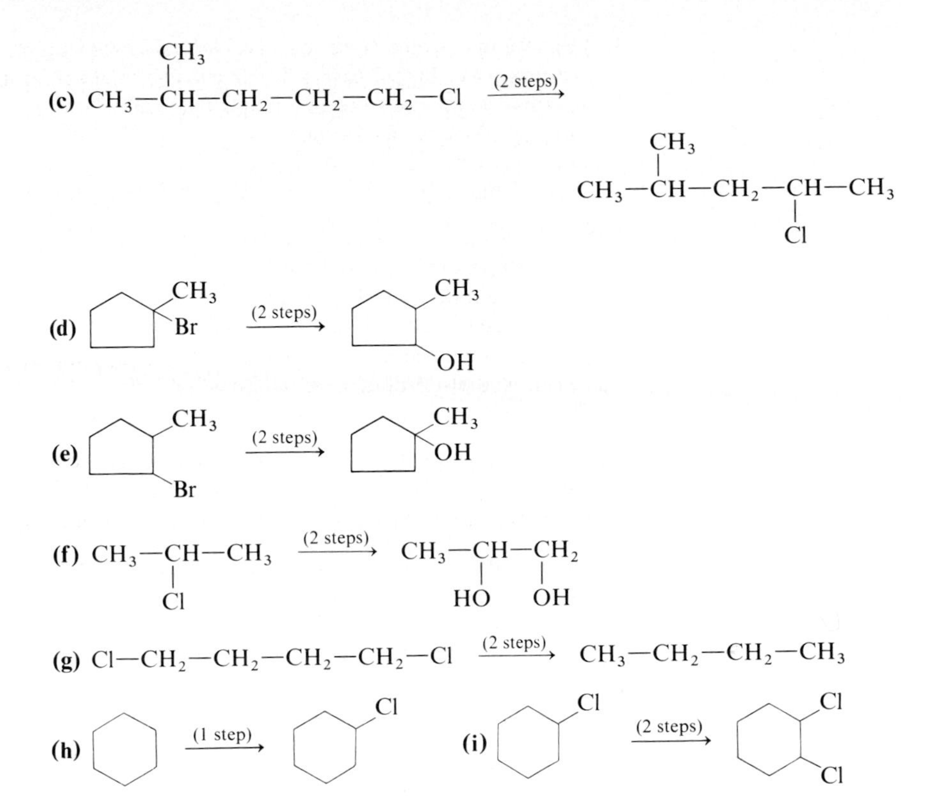

3.32 A structural formula for a section of polypropylene derived from three units of propylene monomer follows:

$$-CH_2-\underset{\displaystyle |}{\overset{\displaystyle CH_3}{CH}}-CH_2-\overset{\displaystyle CH_3}{CH}-CH_2-\overset{\displaystyle CH_3}{CH}-$$

Draw structural formulas for comparable sections of

(a) polyvinyl chloride
(b) Saran
(c) Teflon
(d) Orlon
(e) Styrofoam
(f) Plexiglas

3.33 Natural rubber is a polymer. The repeating unit in this polymer is

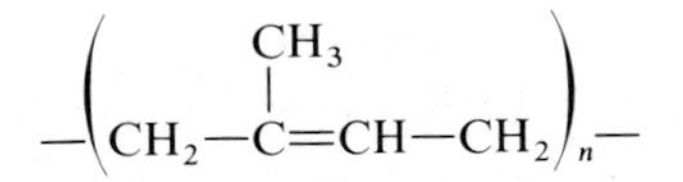

(a) Draw the structural formula for a section of natural rubber showing three repeating units.
(b) Draw the structural formula of the product of oxidation of natural rubber by ozone, and name the two new functional groups in the product.
(c) The smog prevalent in Los Angeles contains oxidizing agents. How might you account for the fact that this type of smog attacks natural rubber (automobile tires, etc.) but does not attack polyethylene or polyvinyl chloride?

3.34 Show how you might distinguish between the following pairs of compounds by a simple chemical test. In each case, tell what test you would perform, what you would expect to observe, and write an equation for each positive test.

(a) cyclohexane and 1-hexene
(b) 1-hexene and 2-chlorohexane
(c) 1,1-dimethylcyclopentane and 2,3-dimethyl-2-butene

3.35 List one major spectral characteristic that will enable you to distinguish between the following pairs of compounds.

(a) cyclohexane and 1-hexene (IR and NMR)
(b) 1-hexene and 2-chlorohexane (IR and NMR)
(c) 1,1-dimethylcyclopentane and 2,3-dimethyl-2-butene (NMR)
(d) 2-methyl-1,3-butadiene and 2-methyl-2-butene (UV)
(e) 2,3-dimethyl-1-butene and 2,3-dimethyl-2-butene (IR and NMR)

3.36 An alkyl bromide of molecular formula C_3H_7Br shows two signals in the NMR spectrum, a doublet at $\delta = 1.7$ and a septet at $\delta = 4.3$. The relative area of these two signals are in the ratio 6:1. Draw a structural formula for this alkyl bromide.

3.37 There are four structural isomers of molecular formula $C_3H_6Cl_2$.

(a) Draw structural formulas for each isomer.
(b) Show how you could distinguish between these four by the presence or absence of an NMR signal for $-CH_3$ hydrogens and the splitting pattern of any $-CH_3$ signals. (Hint: for one isomer the $-CH_3$ signal will be a triplet, for another it will be a doublet, and for a third it will be a singlet. For the fourth, there will be no signal due to $-CH_3$ hydrogens.)

Terpenes

A WIDE VARIETY of substances in the plant and animal world contain one or more carbon-carbon double bonds. In this essay, we will focus our attention on one group of naturally occurring alkenes—the terpene hydrocarbons. The characteristic structural feature of a terpene is a carbon skeleton that can be divided into two or more units that are identical with the carbon skeleton of isoprene. This generalization is known as the isoprene rule. In discussing terpenes and the isoprene rule, it is common to refer to the head and tail of an isoprene unit. The head of an isoprene unit is the carbon nearer the methyl branch. The tail is the carbon atom farther from the methyl branch.

$$CH_2{=}C(CH_3){-}CH{=}CH_2$$

isoprene

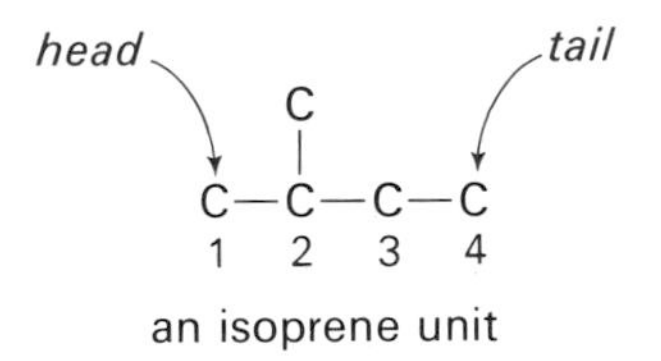

an isoprene unit

There are several important reasons for looking at this group of organic compounds. First, terpenes are among the most widely distributed compounds in nature. The number of terpenes found in bacteria, plants, and animals is staggering. Second, terpenes provide a glimpse at some of the wondrous diversity that nature generates from even a relatively simple carbon skeleton. Third, terpenes illustrate an important principle of the molecular logic of living systems, namely, that in building what might seem to be complex molecules, living systems piece together small subunits to produce complex but logically designed skeletal frameworks. In this mini-essay, we will show how to identify the skeletal framework of terpenes.

Probably the terpenes most familiar to you, at least by odor, are components of the so-called essential oils obtained by steam distillation or ether extraction of various parts of plants. Essential oils contain relatively low-molecular weight substances which are in large part responsible for characteristic plant fragrances. Many essential oils, particularly those from flowers, are used in perfumes.

An example of a terpene obtained from an essential oil is myrcene, $C_{10}H_{16}$, obtained from bayberry wax and from oils of bay and verbena. Its parent chain of eight carbon atoms contains three double bonds and two one-carbon branches. See Figure 1(a). In order to show structural features more clearly, organic chemists frequently represent structures such as that of myrcene using shorthand notation. Carbon-carbon bonds are shown as lines; carbon atoms are not shown, but are understood to be at junctions of lines and where lines end. Figures 1(a) and

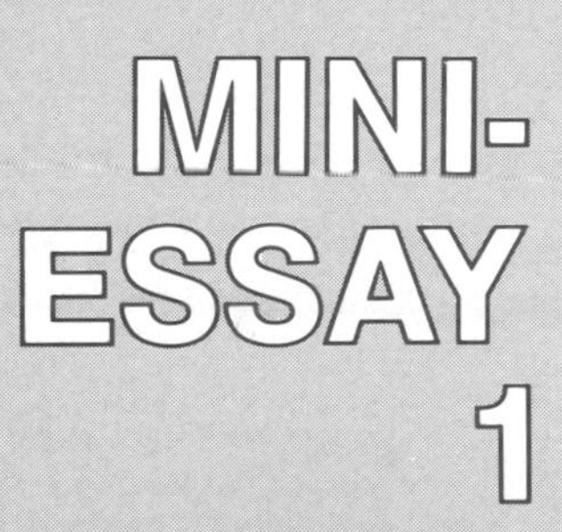

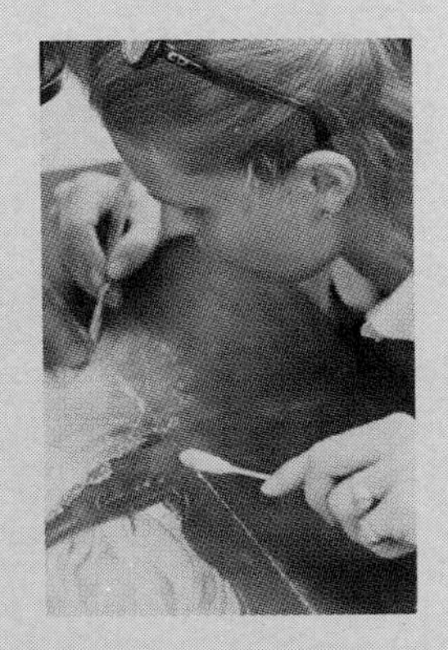

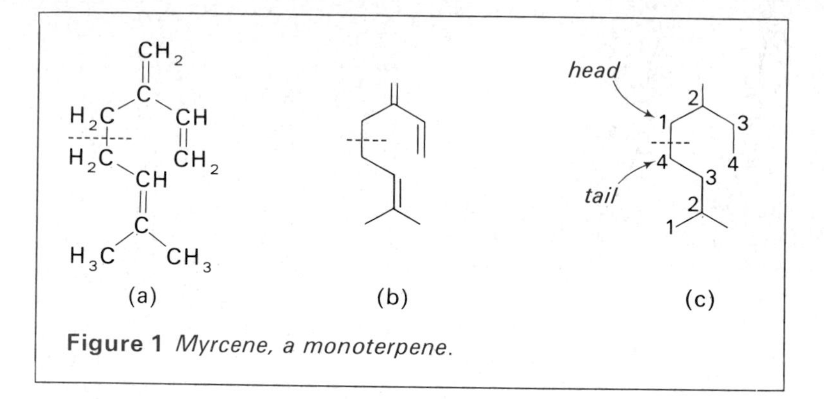

Figure 1 *Myrcene, a monoterpene.*

1(b) are equivalent representations of the structural formula of myrcene.

As you can see from the position of the dashed lines in Figures 1(a) and 1(b), or from the skeletal framework shown in Figure 1(c), myrcene can be divided into two isoprene units linked head-to-tail. Head-to-tail linkages of isoprene units are vastly more common in nature than are the alternative head-to-head or tail-to-tail patterns. Figure 2 shows structural formulas for six more terpenes. Geraniol and the aggregating pheromone of the bark beetle (see the mini-essay "Pheromones" for a discussion of the function of this substance) have the same carbon skeleton as myrcene but different locations of carbon-carbon double bonds. In addition, each has an —OH group. In the last four terpenes shown in Figure 2, the framework of carbon atoms present in myrcene, geraniol, and the bark beetle pheromone is cross-linked to form cyclic structures. To help you identify the points of cross-linkage and ring formation, the carbon atoms of the geraniol skeleton are numbered 1 through 8. Bond formation between carbon atoms 1 and 6 of the geraniol skeleton gives the carbon skeletons of limonene and menthol; between carbon atoms 1, 6, and 4, 7 gives the carbon skeleton of α-pinene: and between 1, 6 and 3, 7 gives camphor.

Terpenes are subdivided into categories depending on the number of isoprene units (see Table 1).

The terpenes we have seen thus far are all classified as monoterpenes. Shown in Figure 3 are structural formulas of several sesquiterpenes. For refer-

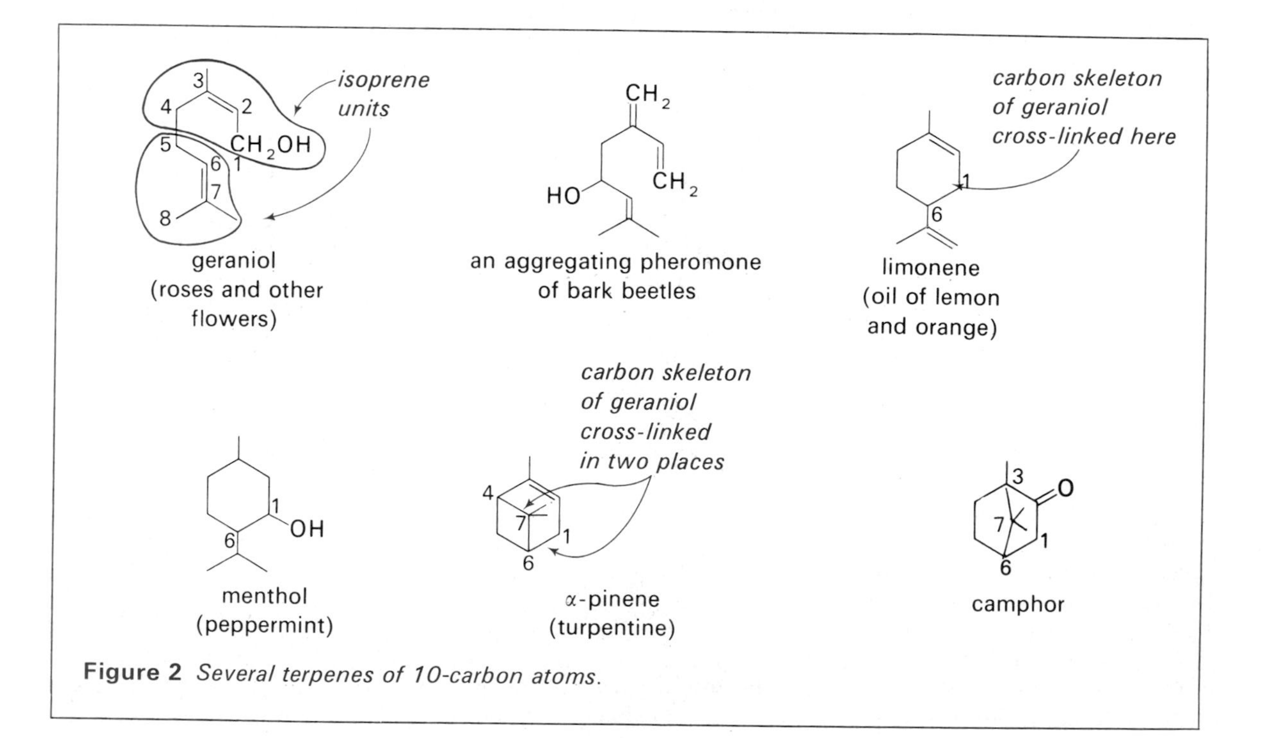

Figure 2 *Several terpenes of 10-carbon atoms.*

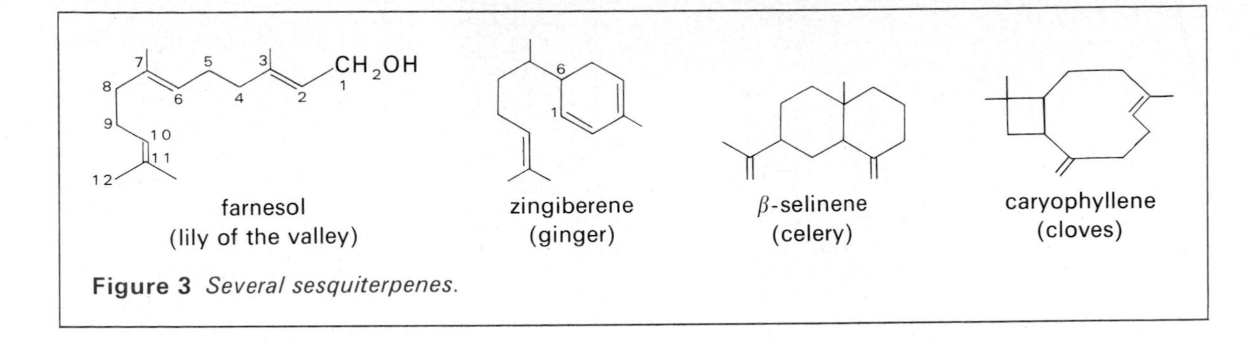

Figure 3 *Several sesquiterpenes.*

ence, the carbon atoms of the parent chain of farnesol are numbered 1 through 12. A bond between carbon atoms 1 and 6 of this skeleton gives the carbon skeleton of zingiberene. You should try to discover for yourself what patterns of cross-linking give the carbon skeletons of β-selinene and caryophyllene.

Structural formulas for vitamin A, a diterpene of molecular formula $C_{20}H_{30}O$, and β-carotene, a tetraterpene of molecular formula $C_{40}H_{56}$ are shown in Figure 4. Vitamin A consists of four isoprene units linked head-to-tail and cross-linked at one point to form a six-membered ring. The function of vitamin A is discussed in Section 13.4. β-Carotene can be divided into two diterpenes, each joined in a tail-to-tail manner. The function of β-carotene is also discussed in Section 13.4.

The synthesis of substances in nature is a fascinating area of research and one of the links between organic and biochemistry. How are terpenes synthesized in living systems? Are they, for example, constructed by joining together molecules of isoprene? However tempting this idea might be, it is not the way nature does it. A key intermediate in the biosynthesis of

Table 1 Categories of terpenes.

monoterpenes	2 isoprene units	10 carbon atoms
sesquiterpenes	3 isoprene units	15 carbon atoms
diterpenes	4 isoprene units	20 carbon atoms
triterpenes	6 isoprene units	30 carbon atoms
tetraterpenes	8 isoprene units	40 carbon atoms

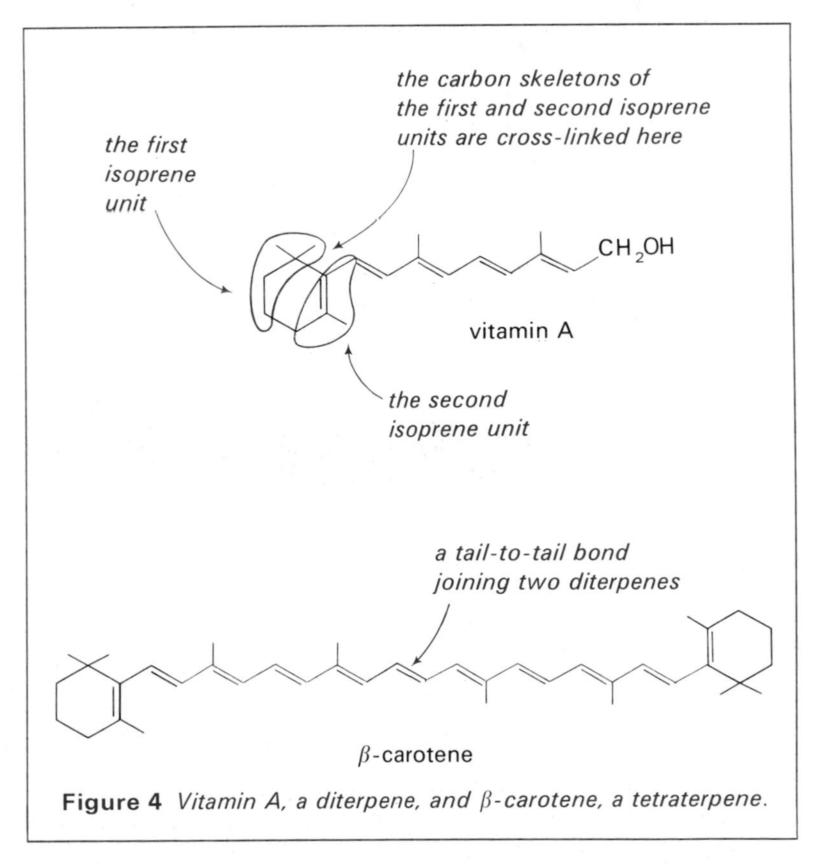

Figure 4 *Vitamin A, a diterpene, and β-carotene, a tetraterpene.*

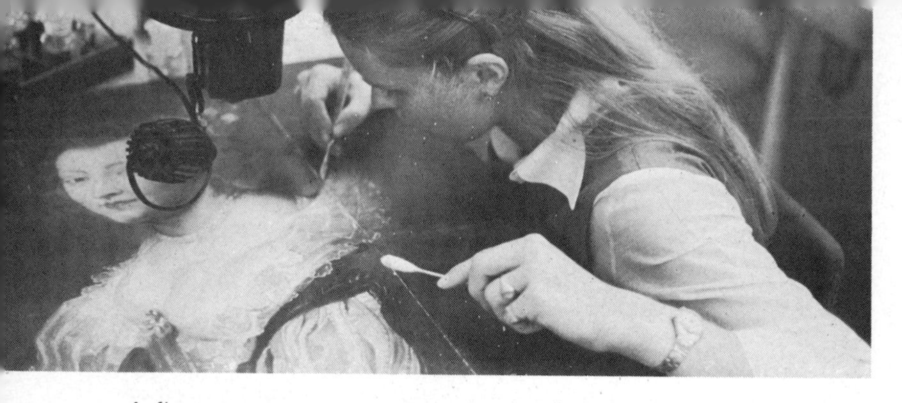

A fine arts conservator uses terpene-derived products for cleaning and restoring an oil painting. (*Fogg Art Museum, Harvard Univ., Barry Donahue*)

terpenes is the pyrophosphate ester of 3-methyl-3-buten-1-ol. This five carbon substance has the carbon skeleton of an isoprene unit and is itself derived in a series of enzyme-catalyzed reactions from three molecules of acetate. Reaction of two C_5 units gives geraniol pyrophosphate, a substance of 10 carbon atoms. This in turn reacts with a third C_5 unit to give farnesol pyrophosphate, a substance of 15 carbon atoms. Enzyme-catalyzed transformation of geraniol pyrophosphate gives monoterpenes; enzyme-catalyzed transformation of farnesol pyrophosphate gives sesquiterpenes. Combination of two C_{15} units by a tail-to-tail bond leads to triterpenes and combination of two C_{20} units by a tail-to-tail bond leads to tetraterpenes. The general scheme used by living systems for the biosynthesis of terpenes of 10, 15, 20, 30, and 40 carbon atoms is summarized in Figure 5. Note that most of these transformations involve several steps and that all are catalyzed by specific enzymes.

We have presented but a few of the terpenes that abound in nature but these examples should be enough to suggest to you their widespread distribution in living systems, the biological individuality that plants and animals achieve through their synthesis, and the structural pattern (the isoprene rule) that underlies this apparent diversity in structural formula. In the future, when you encounter molecules of 10, 15, 20, etc., carbon atoms derived from living systems, you might study their structural formulas to see if they are terpenes.

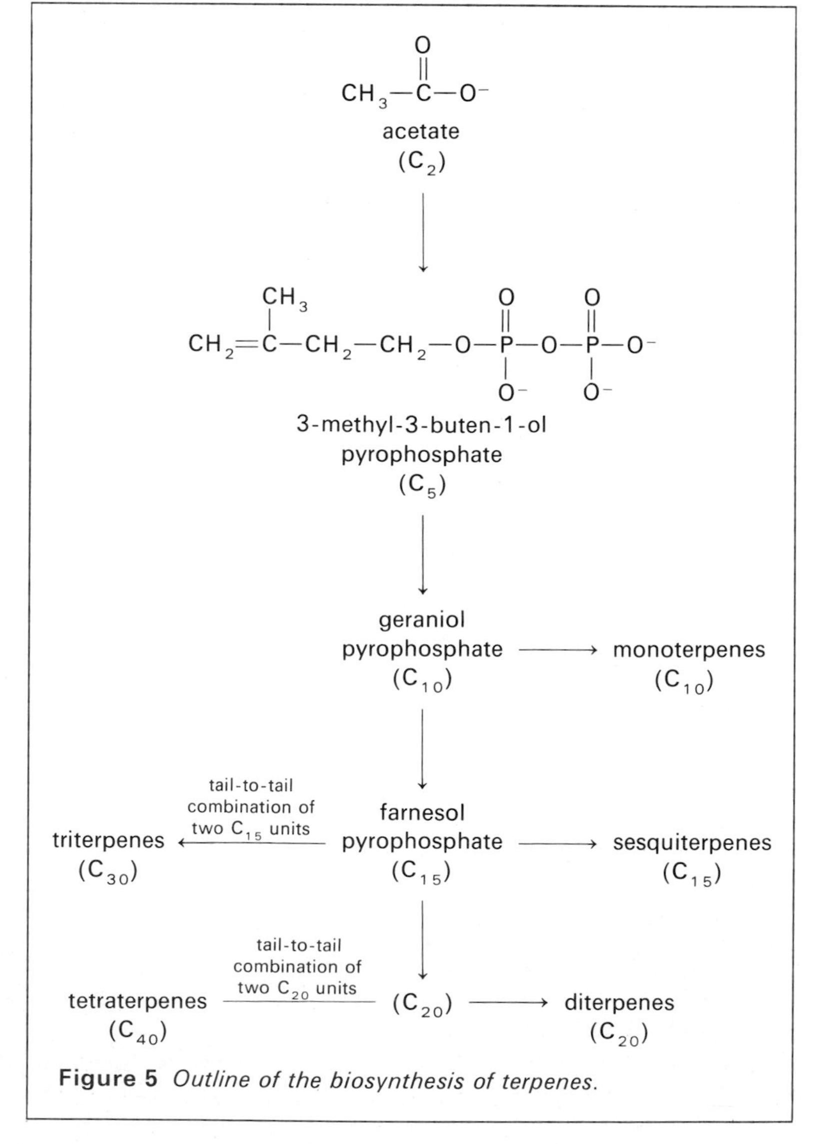

Figure 5 *Outline of the biosynthesis of terpenes.*

Ethylene: The Organic Chemical Industry's Number One Building Block

THE U.S. CHEMICAL INDUSTRY produces more ethylene, on a pound-per-pound basis, than any other organic chemical. Reports on this and other key chemicals can be found regularly in the weekly publication *Chemical and Engineering News* (C & EN). One such report is given in Figure 1. According to C & EN, ethylene production in 1980 totalled almost 28 billion pounds with a commercial value of approximately $6.25 billion.

Of the estimated 250 billion pounds of organic chemicals produced each year in the United States, approximately 100 billion pounds are derived from ethylene. Clearly in terms of its

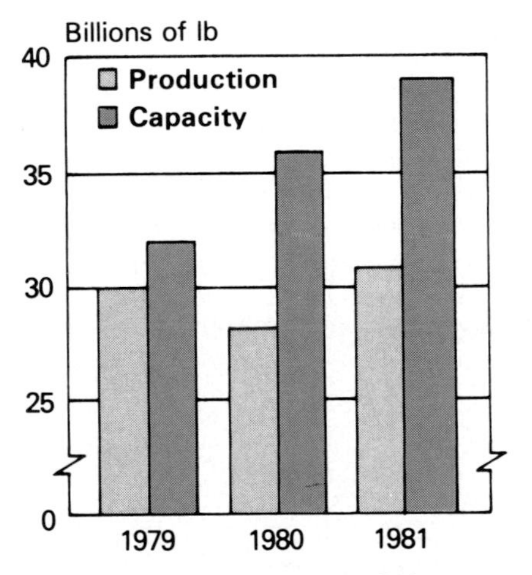

Figure 1 *Ethylene* $CH_2{=}CH_2$

HOW IT'S MADE: thermal (steam) or catalytic cracking of hydrocarbons ranging from natural gas-derived ethane to oil-derived gas oil (fuel oil).

MAJOR DERIVATIVES: polyethylenes 45%, *ethylene oxide* 20%, *vinyl chloride* 15%, *styrene* 10%.

MAJOR END USES: fabricated plastics 65%, *antifreeze* 10%, *fibers* 5%, *solvents* 5%.

COMMERICAL VALUE: $6.25 *billion for total production,* 1980.

MINI-ESSAY 2

volume and the volume of the chemicals derived from it, ethylene is the organic chemical industry's most important building block. This mini-essay will focus on how this vital starting material is produced, its major derivatives, its major end uses, and consumer products.

First, how do we obtain ethylene? More than 90% of all organic chemicals used by the chemical industry are derived from petroleum and natural gas. However, ethylene is not found in either of these resources. If we do not find ethylene in nature, then how do we make it from the raw materials available to us? The answer is not an easy one. In fact the answer differs from one part of the world to another depending on availability of raw materials and economic demand. As we shall see presently, how we make ethylene today may be quite different from how we will be forced to make it in the future.

Ethylene is produced by cracking of hydrocarbons. In the United States where there are vast reserves of natural gas, the major process for ethylene production has been thermal cracking, often in the presence of steam, of the small quantities of ethane, propane, and butane that can be recovered from natural gas. (Recall from Section 2.10 that natural gas is approximately 90% methane and 10% ethane.) For this reason ethylene generating plants constructed in the past have been concentrated near sites of natural gas reserves. Of the 34 ethylene plants in operation in the United States in 1979, 23 were located on the Gulf Coast of Texas and Louisiana and they accounted for over 80% of total U.S. ethylene production capacity.

When written as a balanced equation, thermal (steam) cracking of ethane seems simple enough

$$CH_3{-}CH_3 \xrightarrow{\text{thermal cracking}} CH_2{=}CH_2 + H_2$$

Actually the reaction is very complicated and several other substances are produced along with ethylene. For example, for every billion pounds of ethylene produced from ethane, there are also obtained 36 million pounds of propylene and 35 million pounds of butadiene as coproducts. These quantities of starting materials and products may seem enormous to you, but they reflect the scale on which the U.S. chemical industry operates. Although other low molecular weight alkanes can be cracked to give ethylene, thermal cracking of ethane gives the highest-percentage and highest-purity ethylene.

In the United States now, natural gas is the major source of the raw materials for the manufacture of ethylene. Approximately 10% of the natural gas consumed each year is used for this purpose. In Europe and Japan, however, supplies of natural gas are much more limited. As a result, these countries depend almost entirely on catalytic cracking of petroleum-derived naphtha for their ethylene. Thus, these countries produce not only aromatic hydrocarbons such as benzene, toluene, and xylene from naphtha (just as we do), but they also produce ethylene from it. The problem is that naphtha is in heavy demand as a source of straight run gasoline, as a feedstock for catalytic cracking and

Table 1 Major derivatives and end uses of ethylene.

Major derivatives of ethylene	*Structural formula*	*1979 production (billions of lbs)*	*Major end uses*
Polyethylene	$-(CH_2{-}CH_2)_n-$	12.8	Fabricated plastics
Ethylene oxide/ ethylene glycol	$CH_2{-}CH_2$ (bridged by O), $CH_2(OH){-}CH_2(OH)$	9.88	Antifreeze, polyester textile fibers, solvents
Vinyl chloride	$CH_2{=}CHCl$	7.54	Polyvinyl chloride fabricated plastics
Styrene	$C_6H_5{-}CH{=}CH_2$	7.48	Polystyrene, fabricated plastics and synthetic rubbers

Located along the Gulf Coast of Louisiana and Texas near the sites of natural gas reserves, catalytic cracking facilites, such as this refinery in Port Arthur, accounted for more than 80% of total ethylene production in the United States. (Gulf Oil Corp.)

reforming to produce higher-octane gasolines, and as a source of starting materials for the organic chemical industry.

The price of naphtha has increased steadily over the years in step with its increased world-wide demand. This upward spiral of demand and price has provided incentive in Europe and Japan, and more recently in the United States, to develop an ethylene-generating technology using higher-boiling petroleum fractions, particularly gas oil. Thus, while the world has depended in the past on natural gas and naphtha as raw materials for ethylene manufacture, it seems almost certain that in the future we will depend more and more on gas oil to meet this need. And we may also come to depend on coal as coal gasification and liquefaction technologies advance.

Now that we have seen how we make ethylene, let us turn to the second question: How do we use it? Each year ethylene is the starting material for the synthesis of almost 100 billion pounds of chemicals and polymers. As you can see from Table 1, its major derivatives are polyethylene (45%), ethylene oxide and ethylene glycol (20%), vinyl chloride (15%), and styrene (10%).

We will concentrate on just one major derivative of ethylene, namely, the fabricated polyethylene plastics that account for approximately 45% of all ethylene used in this country. The first commercial process for ethylene polymerization used peroxide catalysts at temperatures of 500°C and pressures of 1000 atmospheres, producing a polymer known as low density polyethylene (LDPE). Low density polyethylene is a soft, tough plastic. It has a density between 0.91 and 0.94 g/cm^3 and a melting point of about 115°C. Since LDPE's melting point is only slightly above 100°, it is not used for products that will be exposed to boiling water. LDPE is about 50–60% crystalline. Crystallinity of polymers is not quite the same as crystallinity of such common substances as sodium chloride and table sugar. The characteristic feature of the crystalline state is a high degree of order—the atoms are arranged regularly in a three-dimensional lattice built up of a simple unit that repeats over and over again. With crystals of low molecular weight substances like sodium chloride and table sugar, there is virtually complete order, i.e., 100% crystallinity. Although polymers do not crystallize in the conventional sense, they often have regions where their chains are precisely ordered with respect to each other and interact by noncovalent forces. Such regions are called crystallites. When we say that low density polyethylene is 50–60% crystalline, we mean that this percentage is composed of crystallites.

The major end use of low-density polyethylene is film for the packaging of such consumer

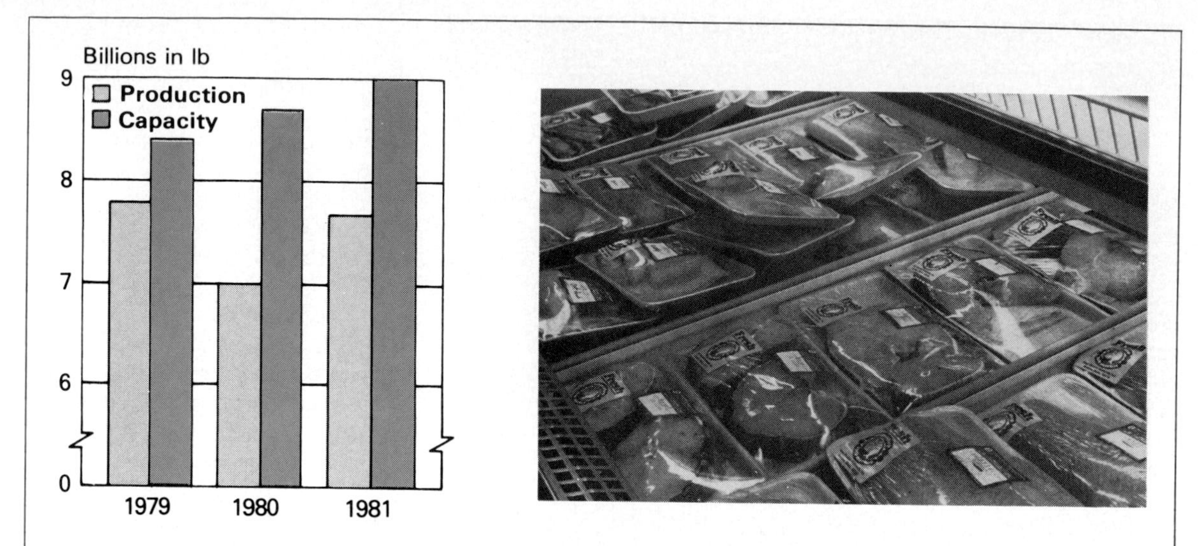

Figure 2(a) *Low-density polyethylene* $\sim\!\!\left[CH_2-CH_2\right]_n\!\!\sim$

HOW IT'S MADE: polymerization of ethylene at high and low pressures aided by initiators.

MAJOR FABRICATED FORMS: film, largely for packaging, 65%; *injection molding* 10%; *coatings* 10%; *extrusions* 5%.

COMMERCIAL VALUE: $2.8 *billion for total production,* 1980.

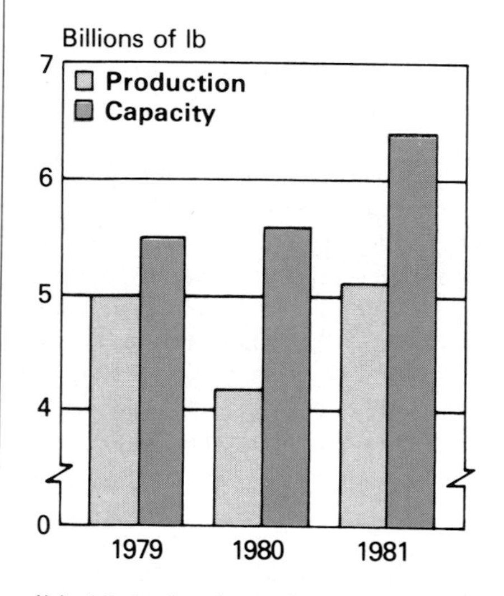

(b) *High-density polyethylene* $\sim\!\!\left[CH_2-CH_2\right]_n\!\!\sim$

HOW IT'S MADE: polymerization of ethylene under moderate pressure catalyzed by metal salts and alkyls.

MAJOR FABRICATED FORMS: Blow molding, mostly containers, 45%; *injection molding* 20%; *extrusions* 10%; *film* 5%.

COMMERCIAL VALUE: $1.75 *billion for total production,* 1980.

items as baked goods, vegetables and other produce; as coatings for cardboard, and paper; and perhaps most importantly, as trash bags. LDPE is cheap which makes it ideal for one-way packaging—from a LDPE film wrapped package to a LDPE trash bag and out!

An alternative method of ethylene polymerization uses catalysts composed of titanium chloride and organoaluminum compounds. With catalysts of this type, ethylene can be polymerized under conditions as low as 60°C and 20 atmospheres pressure. Polyethylene produced in this manner has a density of 0.96 g/cm^3 and is called high density polyethylene (HDPE). HDPE has a higher degree of crystallinity (90%) than LDPE and a higher melting point (135°C). It is best described as a hard, tough plastic. The physical properties and cost of HDPE relative to other materials make it ideal for the production of plastic bottles, lids, caps, etc. It is also molded into housewares such as mixing bowls and refrigerator and freezer containers. (Because of its higher melting point, HDPE can be used for such consumer products whereas LDPE cannot.)

Information from *Chemical and Engineering News* on high density polyethylene and low density polyethylene is summarized in Figure 2.

There are several techniques used in industry for the fabrication of polyethylene into consumer products. We will describe two of these, the first to show how HDPE is fabricated into bottles and the second to show how LDPE is fabricated into films.

Approximately 45% of all HDPE used in the United States is blow molded, mostly into containers. See Figure 3. Approximately 65% of all low-density polyethylene is used for the manufacture of films. Fabrication of LDPE into films is done by a variation of the blow-molding technique illustrated in Figure 3. A tube of LDPE, along with a jet of compressed air, is forced through an opening and blown into a giant, thin-walled bubble. The film is then cooled

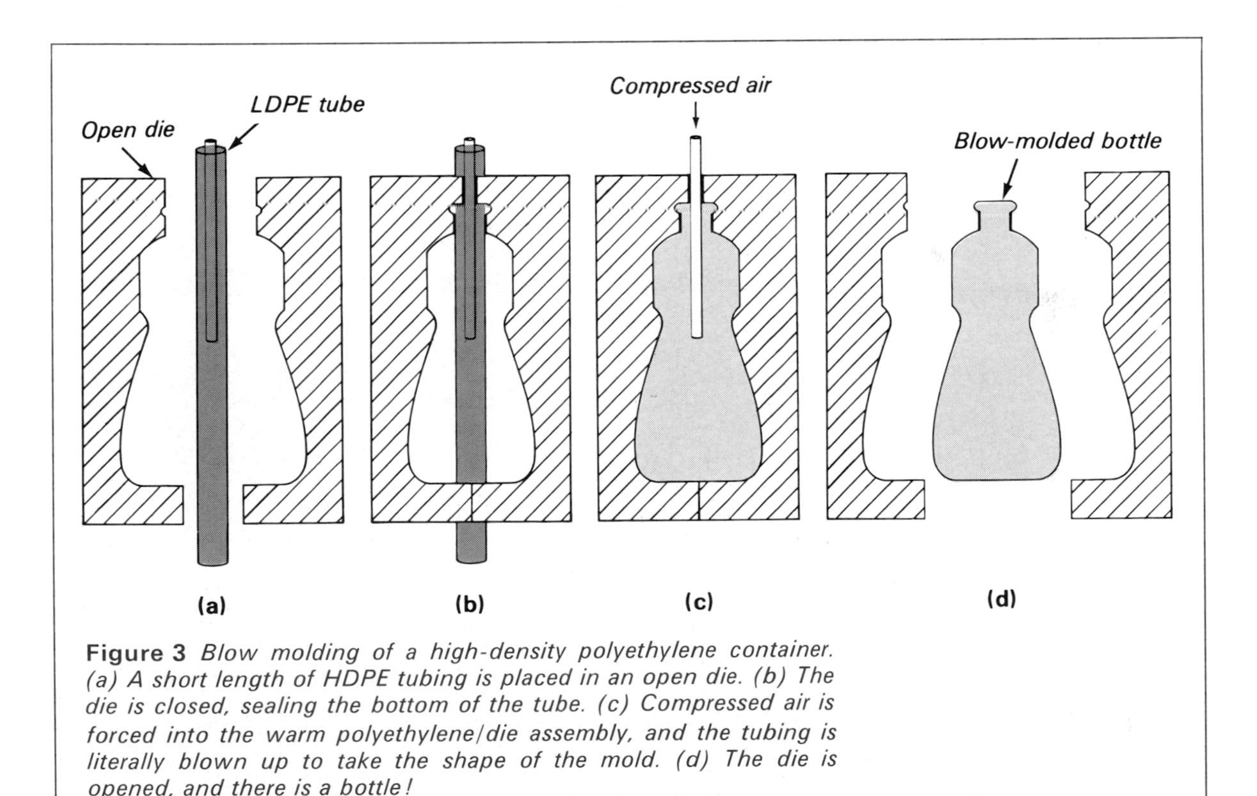

Figure 3 *Blow molding of a high-density polyethylene container. (a) A short length of HDPE tubing is placed in an open die. (b) The die is closed, sealing the bottom of the tube. (c) Compressed air is forced into the warm polyethylene/die assembly, and the tubing is literally blown up to take the shape of the mold. (d) The die is opened, and there is a bottle!*

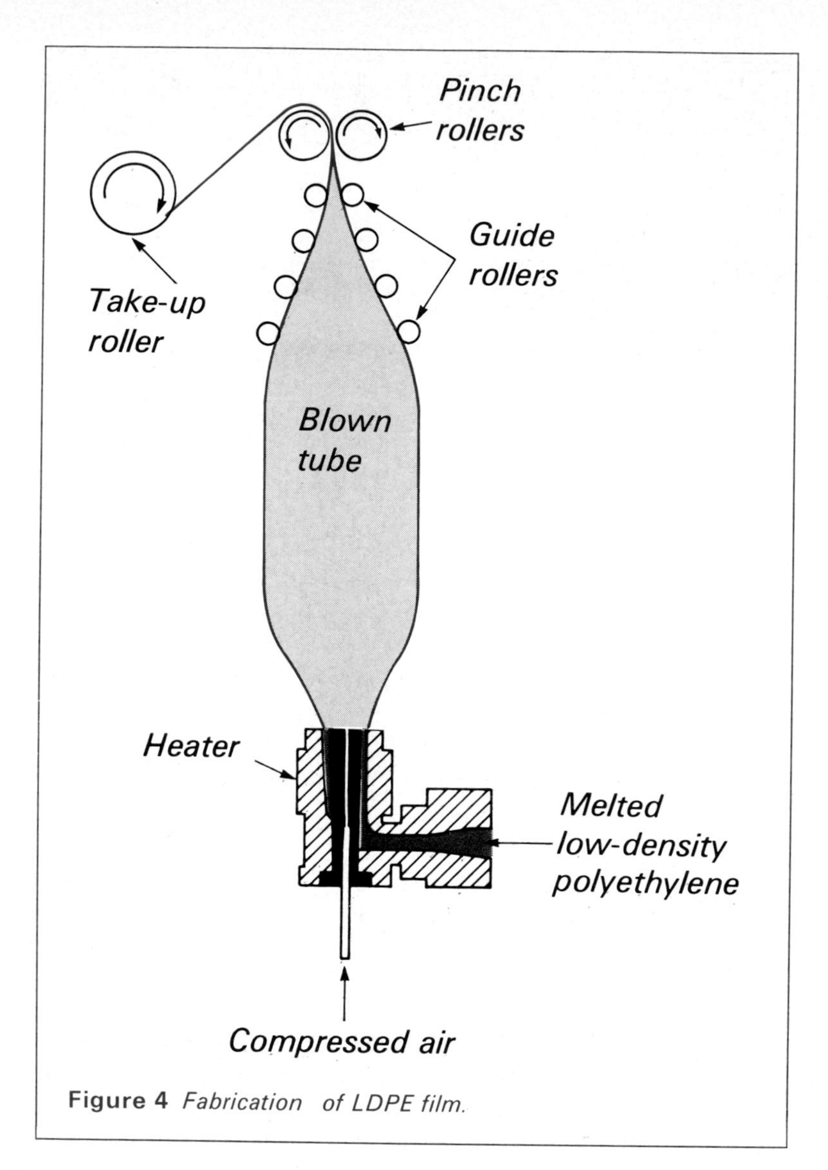

Figure 4 *Fabrication of LDPE film.*

and taken up onto a roller. This double-walled film can be slit down the side to give LDPE film or it can be sealed at points along its length to give LDPE trash bags (Figure 4).

Clearly our dependence on polyethylene is enormous, and given the nature of our industrialized society, it will continue to be enormous.

References

Chemistry in the Economy, (Washington, D. C.: American Chemical Society, 1973).

Bilmeyer, F. W., Jr., *Textbook of Polymer Science,* 2nd ed., (N.Y.: John Wiley & Sons, 1971).

"Facts and Figures," *Chemical and Engineering News* June 9, 1980.

Fernelius, W. C., Wittcoff, H. A., and Varnerid, R. E., eds., "Ethylene: The Organic Chemical Industry's Most Important Building Block," *Journal of Chemical Education* 56 (1979): 385–387.

Stinson, Stephen C., "Ethylene Technology Moves to Liquid Fuels," *Chemical and Engineering News* May 28, 1979.

Witcoff, H. A. and Reuben, B. G., *Industrial Organic Chemicals in Perspective,* (N.Y.: John Wiley & Sons, 1980).

CHAPTER 4

Stereoisomerism and Optical Activity

All compounds that have the same molecular formula and order of attachment of atoms but different orientations of their atoms in space are grouped collectively under the classification of stereoisomers. In this introduction to organic chemistry, we will deal with three categories of stereoisomers.

(1) Conformational isomers which differ only by rotation about single bonds (Section 2.6). Interconversion between conformational isomers is usually rapid at room temperature.
(2) *Cis-trans* isomers which differ in the orientation of substituent groups on a ring (Section 2.7) or on a carbon-carbon double bond (Section 3.3). Interconversion between *cis-* and *trans*-isomers does not occur readily even at elevated temperatures.
(3) Configurational isomers which differ in the orientation in space of atoms at one or more chiral centers. Configurational isomers are sometimes called optical isomers because they are often detected by their interaction with plane polarized light.

In this chapter we will deal with configurational isomers known as enantiomers, diastereoisomers, and *meso* compounds. We shall first examine how this type of stereoisomerism is detected in the laboratory, the structural features that give rise to it, the intimate relationship between configurational isomerism and *cis-trans* isomerism, and finally the significance of stereoisomerism in the biological world.

4.1 THE POLARIMETER

Ordinary light consists of waves vibrating in all possible planes perpendicular to its path (Figure 4.1). Certain substances, such as a Polaroid sheet (a plastic film containing properly oriented crystals of an organic substance embedded in it), selectively transmit light waves vibrating only in a specific plane. Light in which vibrations occur in only a specific plane is said to be plane polarized. No doubt you are familiar with the effect of polarizing sheets on light from experience with sunglasses or camera filters made of this material. If two polarizing

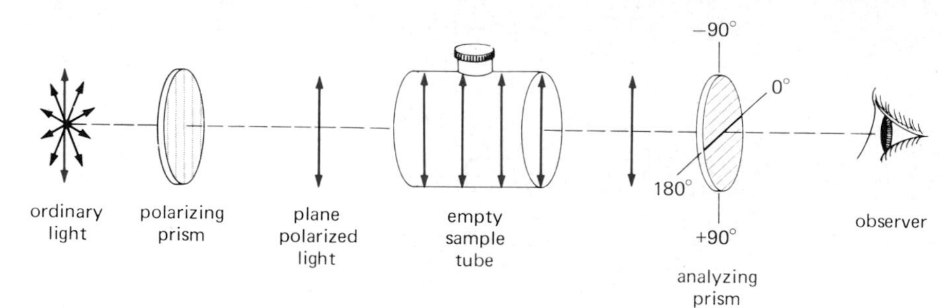

Figure 4.1 *Schematic diagram of a polarimeter with the sample tube empty.*

discs are placed in a light path in such a way that their polarizing axes are parallel to each other, then a maximum intensity of light passes through the second disc to you. However, if their polarizing axes are orientated perpendicular to each other (that is, at an angle of 90°) then no light passes through the second polarizing disc.

A polarimeter is an instrument used to measure quantitatively the effect of plane polarized light on samples of matter. A polarimeter consists of a light source, a polarizing prism, a sample tube, and an analyzing prism (Figure 4.1).

If the sample tube is empty, the intensity of light reaching you is a maximum when the polarizing axes of the two prisms are parallel. If the analyzer is turned either clockwise or counterclockwise, less light is transmitted. When the analyzer is at right angles to the polarizer, the field of view in the instrument is dark. We take this as the zero point, or 0° on the optical scale. If an optically active substance is placed in the sample tube, the plane of polarized light is rotated by the substance, and a certain amount of light passes through the analyzer to you. Turning the analyzer a few degrees clockwise or counterclockwise will restore the dark field of view. The number of degrees, α, that the analyzer is turned is equal to the number of degrees that the optically active substance has rotated the plane of the polarized light. If the analyzer must be turned to the right (clockwise) to restore the dark field, we say the substance is dextrorotatory. If the analyzer must be turned to the left (counterclockwise) to restore the dark field, we say the substance is levorotatory. In either case, the substance is optically active (see Figure 4.2).

The number of degrees, α, that the analyzing prism is turned is known as the

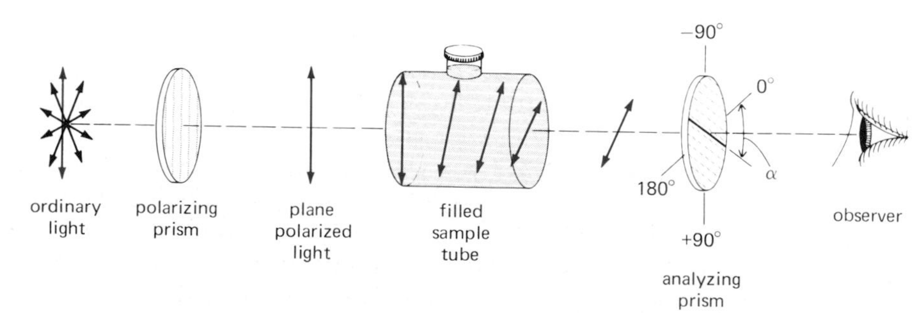

Figure 4.2 *Schematic diagram of a polarimeter with sample tube containing an optically active substance. To restore the dark field of view the analyzer has been rotated clockwise by α degrees.*

observed rotation, and depends on the structure of the compound, the length of the light path through the sample tube, the temperature, the wavelength of the light, and the solvent. The optical activity of a pure compound is reported as specific rotation. Specific rotation is defined as the rotation caused by a substance at a concentration of one gram per cubic centimeter in a sample tube ten centimeters long. Specific rotation depends on the temperature and the wavelength of the light, and these values must be reported. The most commonly used wavelength is the D line of sodium, the same line that is responsible for the yellow color of excited sodium vapor. In reporting either the observed or the specific rotation, it is common practice to indicate a dextrorotatory substance by a positive sign (+) and a levorotatory substance by a negative sign (−). Using these conventions, the specific rotation of sucrose (table sugar) in water at 25°C, using the D line of sodium as the light source, is reported as:

$$[\alpha]_{D}^{25^{\circ}C} = +66.5\ (H_2O)$$

4.2 STRUCTURE AND OPTICAL ISOMERISM

By the middle of the 19th century, a number of optically active compounds had been isolated from natural sources, and their structures had been determined. Among the compounds known at that time were

$CH_3-CH(OH)-C(=O)-OH$
lactic acid

$CH_3-CH_2-CH(CH_3)-CH_2-OH$
2-methyl-1-butanol
"active" amyl alcohol

$HO-C(=O)-CH(OH)-CH_2-C(=O)-OH$
malic acid

Lactic acid was one of the compounds most intensively investigated. It was originally isolated in 1780 from sour milk and found to be optically inactive. In 1807 lactic acid was also isolated from muscle tissue. Whereas the lactic acid from sour milk is optically inactive, that from muscle tissue is dextrorotatory and is designated (+)-lactic acid.

How can certain molecules with the same structural formulas have such different effects on the plane of polarized light? To answer this question we need to consider a type of stereoisomerism that deals with molecules, their mirror images, and a property of objects called chirality.

Figure 4.3a shows a perspective formula or stereorepresentation of a lactic acid molecule. The middle carbon of the lactic acid molecule is drawn at the center of a tetrahedron, and the four groups attached to it are directed toward the corners of the tetrahedron. In this stereorepresentation, one group ($-CO_2H$) is drawn in the plane of the paper, two groups ($-H$ and $-OH$) project behind the plane of the paper, and one group ($-CH_3$) projects in front of the plane of the paper. This molecule can be turned in space and redrawn in such a way (Figure 4.3b) that two groups are in the plane of the paper, a third projects to the rear, and the fourth projects forward. In Figure 4.3c, the same

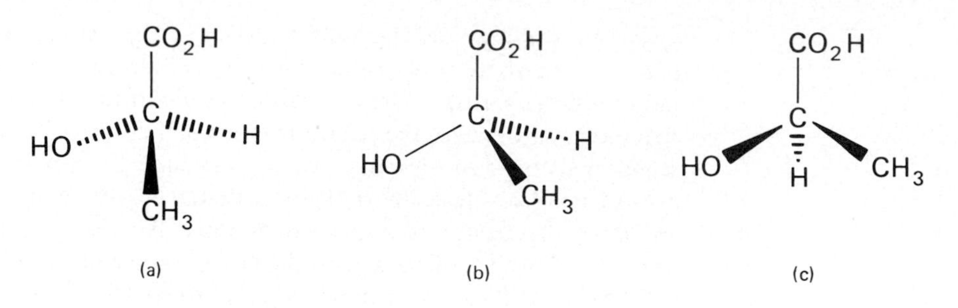

Figure 4.3 *Perspective formulas of lactic acid.* ▬ *represents a bond projecting in front of the plane of the paper.* — *represents a bond projecting in the plane of the paper.* ┆┆┆┆ *represents a bond projecting behind the plane of the paper.*

molecule is drawn again, this time with one group in the plane of the paper, two groups projecting forward and the remaining group to the back.

It is important to realize that all three drawings in Figure 4.3 are stereo-representations of the same molecule; they differ only in how the molecule is turned in space and the direction from which you look at it. Said another way, any one of these molecules can be picked up, turned in space, and placed directly on top of either of the other two in such a way that all four groups attached to the central carbon will coincide. In other words, each of the molecules drawn in Figure 4.3 is superimposable on the other two.

A molecule, like any other object in nature, has a mirror image. Shown in Figure 4.4 are ball-and-stick models of a lactic acid molecule and its mirror image.

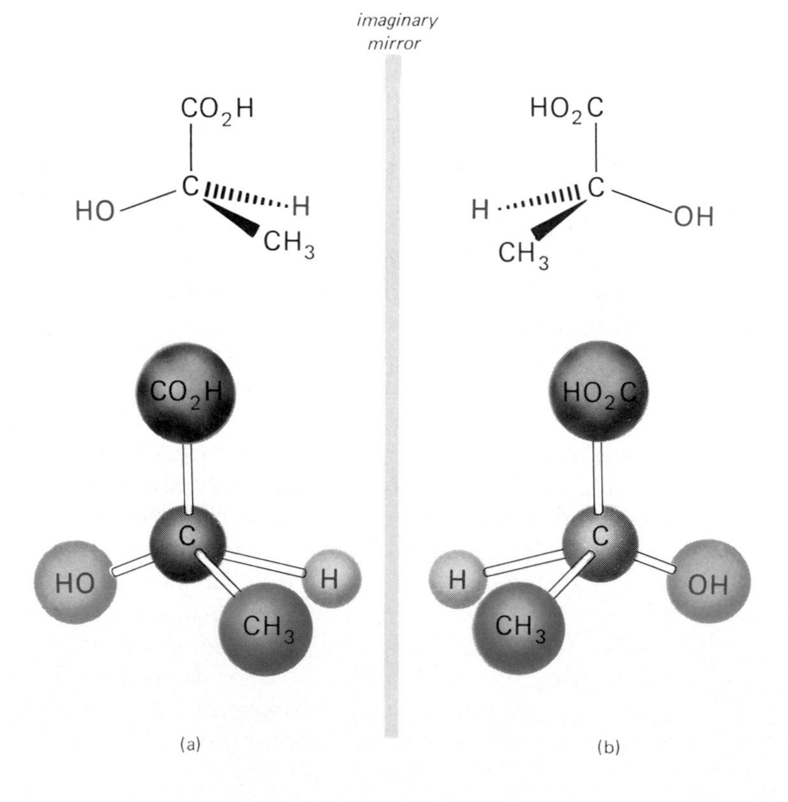

Figure 4.4 *Ball-and-stick models of (a) a lactic acid molecule and (b) its mirror image.*

In these stereorepresentations, the $—CO_2H$ groups are in the plane of the paper and parallel to the mirror; the —OH groups are also in the plane of the paper, but point away from the mirror. The —H groups are behind the plane of the paper and toward the mirror, and the $—CH_3$ groups are in front of the plane of the paper and toward the mirror.

The question is, what is the relationship between these two molecules. Are they the same or different substances? The answer is that they differ from each other just as a right hand differs from a left hand; they are related by reflection but they are not superimposable on each other. The ball-and-stick model of lactic acid shown in Figure 4.4b can be picked up and turned any direction in space, but as long as no bonds are broken, only two of the four groups attached to the central carbon in (b) can be made to coincide with those on (a). See Figure 4.5.

Objects that are not identical with their mirror images are said to be chiral (pronounced ki-ral, to rhyme with spiral; from Greek, *cheir*, "hand"). In a chiral molecule, the carbon that has four different groups attached to it is called a chiral carbon, or alternatively, an asymmetric carbon. It is important to realize that chirality is a property of an object, the property of being nonsuperimposable on its mirror image.

Molecules that are nonsuperimposable mirror images of each other are called enantiomers (Greek, *enantio*, "opposite" + *meros*, "part"). Many of the properties of enantiomers are identical: they have the same melting points, the same boiling points, the same solubilities in various solvents. Yet they are isomers and we can expect them to show some differences in their properties.

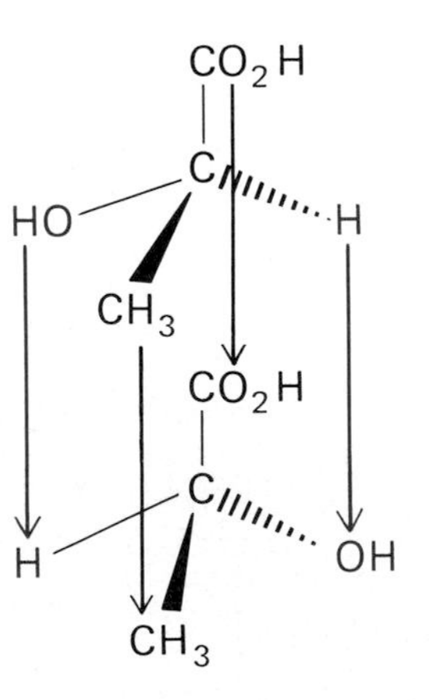

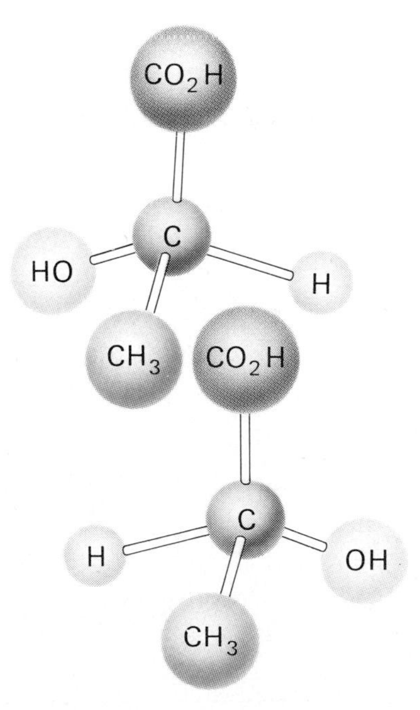

Figure 4.5 *Ball-and-stick models of a lactic acid molecule and its mirror image are not superimposable on each other. No matter how the mirror image is turned in space, only two of the four groups attached to the central carbon atom can be made to coincide.*

One important difference is optical activity. One member of a pair of enantiomers is dextrorotatory and the other member is levorotatory. For each, the absolute magnitude of the optical activity is the same, but rotations differ in sign as illustrated by the dextro and levorotatory isomers of 2-butanol.

$$CH_3-\underset{\underset{OH}{|}}{\overset{\overset{H}{|}}{C}}-CH_2-CH_3$$

$$(+)\text{-2-butanol} \quad [\alpha]_D^{20} = +13.9^\circ$$
$$(-)\text{-2-butanol} \quad [\alpha]_D^{20} = -13.9^\circ$$

We shall see other differences in enantiomeric pairs later. For example, one enantiomer may taste different from the other, or smell different. One enantiomer may be physiologically active whereas the other is physiologically inactive. We will suggest why this might be, later in the chapter.

Example 4.1 Draw stereorepresentations for the enantiomers of

$$\textbf{(a)}\ CH_3-\underset{\underset{OH}{|}}{CH}-CH_2-CH_3 \qquad \textbf{(b)}\ CH_3-CH_2-\overset{\overset{CH_3}{|}}{CH}-CH_2-OH$$

Solution In doing this problem, first locate the chiral (asymmetric) carbon atom and draw the four bonds from it arranged to show the tetrahedral geometry. Next, draw the four groups attached to the chiral carbon and finally draw the nonsuperimposable mirror image.

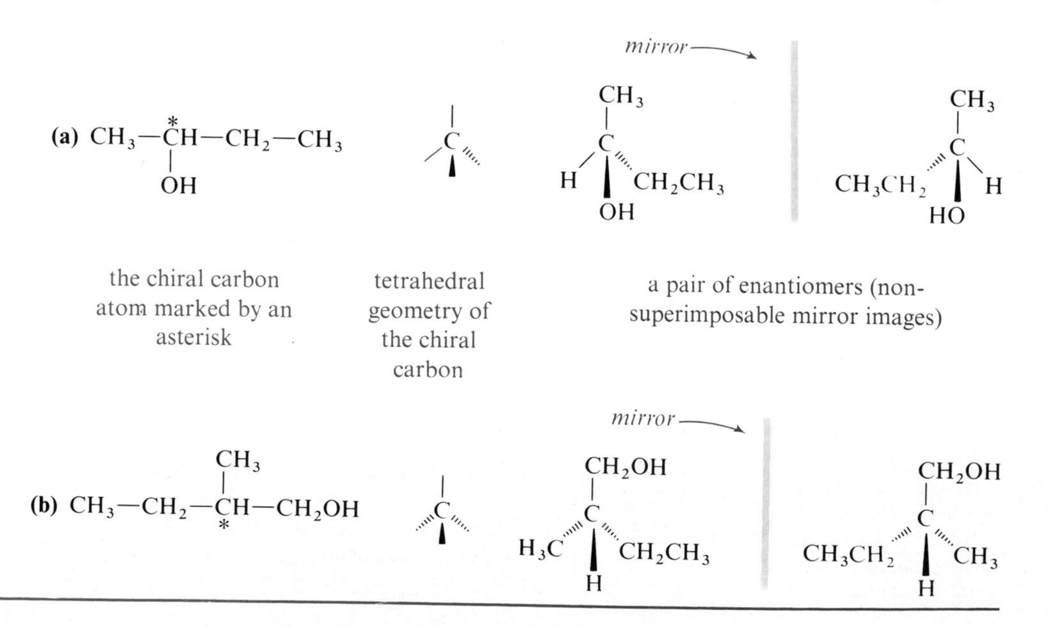

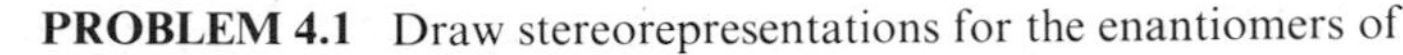

PROBLEM 4.1 Draw stereorepresentations for the enantiomers of

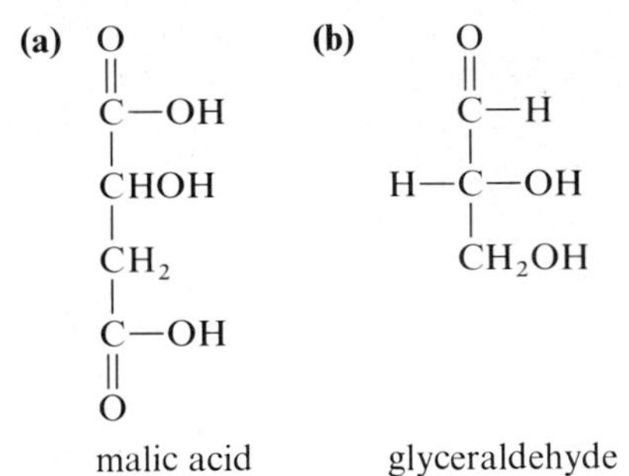

Every molecule has a mirror image. For lactic acid and all molecules that contain a single chiral carbon atom, the mirror images are nonsuperimposable and the compounds show optical isomerism. For many other compounds, the molecule and its mirror image are superimposable. Consider, for example, the amino acid glycine, a substance with no chiral carbon atoms.

$$H_2N{-}CH_2{-}\overset{\displaystyle O}{\overset{\|}{C}}OH$$

glycine

Figure 4.6 shows stereorepresentations of glycine and its mirror image. In these stereorepresentations, (a) and (b) are mirror images. But are they nonsuperimposable or are they superimposable? The answer is that they are superimposable mirror images. To see this, turn (b) by 180° about the C—CO_2H bond. If (b) is turned in space in this manner, it is possible to superimpose (b) directly on (a). Molecules that are superimposable on their mirror images are identical and therefore cannot show optical isomerism. They are achiral ("without handedness").

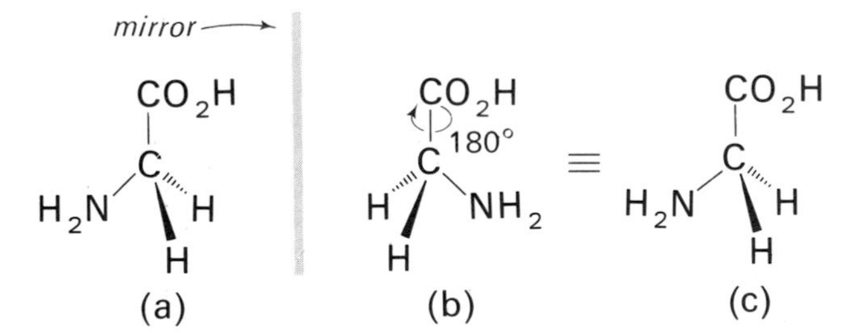

Figure 4.6
Stereorepresentations of glycine (a) and its mirror image (b). If (b) is rotated by 180° about the C—CO_2H bond, it can be seen that (b) is identical to (a).

4.3 MOLECULAR SYMMETRY

There are three recognized types of molecular symmetry, but for our purposes we shall consider only one, a plane of symmetry. A plane of symmetry is defined as a plane (often visualized as a mirror) cleaving the molecule in such a way that one side of the molecule is the mirror image of the other side. Any molecule that has a plane of symmetry is superimposable on its mirror image and therefore does not show enantiomerism.

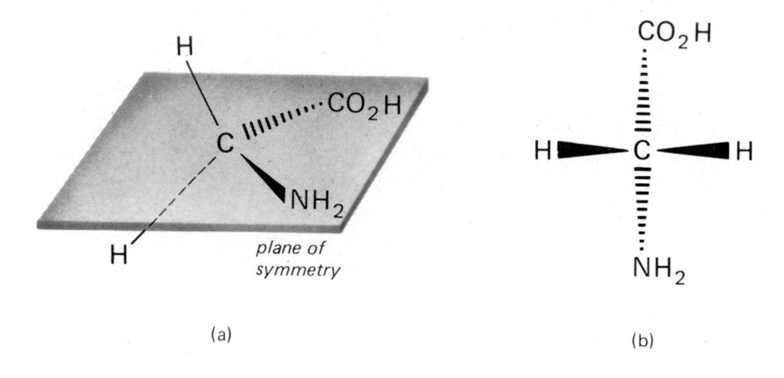

Figure 4.7 *Glycine. A plane of symmetry running through the axis of the C—C—N bond.*

Inspection of the projection formula of glycine (Figure 4.7a) shows that it possesses a plane of symmetry running through the molecule on the axis of the C—C—N bonds. Alternatively, we might rotate the molecule in space into a different projection formula (b), and see the plane of symmetry again running through the axis of the C—C—N bonds, this time oriented in a plane perpendicular to the plane of the paper.

Example 4.2 Which of the following compounds show optical isomerism? For each that does, draw stereorepresentations for both enantiomers. For each that does not, draw a stereorepresentation to show a plane of symmetry.

(a) $CH_2{=}CH{-}CH(OH){-}CH_2{-}CH_3$ **(b)** $CH_3{-}CH_2{-}CH(OH){-}CH_2{-}CH_3$

Solution Compound (a) has a chiral carbon atom, and therefore shows optical isomerism. Compound (b) has no chiral carbon and therefore does not show optical isomerism.

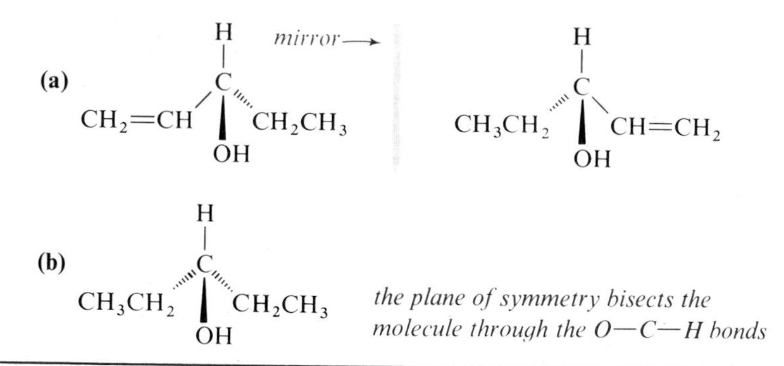

PROBLEM 4.2 Which of the following compounds show optical isomerism? For each that does, draw stereorepresentations for both enantiomers. For each that does not, draw a stereorepresentation to show a plane of symmetry.

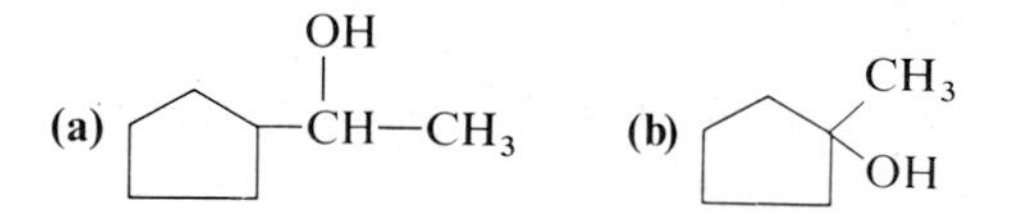

4.4 MULTIPLE CHIRAL CENTERS

Compounds that contain two or more chiral centers can exist in more than two configurational isomers. The maximum number of configurational isomers is 2^n, where n is the number of chiral centers.

As an example of a molecule with two different chiral centers, consider 2,3,4-trihydroxybutanal.

$$\underset{\text{OH}}{\underset{|}{CH_2}}-\underset{\text{OH}}{\underset{|}{\overset{*}{C}H}}-\underset{\text{OH}}{\underset{|}{\overset{*}{C}H}}-CHO$$

2,3,4-trihydroxybutanal

The two chiral carbons are marked by asterisks. There are four stereoisomers ($2^2 = 4$) of this substance. Each of these is drawn in Figure 4.8.

A and B are nonsuperimposable mirror images and therefore are a pair of enantiomers. C and D are also nonsuperimposable mirror images and therefore are a second pair of enantiomers. What is the relationship of A to C or A to D? The answer is that they are stereoisomers of each other but are not mirror images. Stereoisomers that are not mirror images of each other are called diastereomers. Of the optical isomers drawn in Figure 4.8, AB and CD are pairs of enantiomers; AC, AD, BC, and BD are pairs of diastereomers. Diastereomers have different chemical and physical properties, and sometimes they are even given different names. The diastereomers of trihydroxybutanal are given the common names of erythrose and threose.

The 2^n isomer number represents the maximum number of stereoisomers. Some molecules have special symmetry properties that reduce the isomer number. Tartaric acid is an example of a molecule possessing two chiral carbon atoms. The 2^n rule predicts four optical isomers, while in fact only three are known (Figure 4.9). Structures E and F are mirror images, and because they are not superimposable on each other, they constitute a pair of enantiomers. G and H are also mirror images, but they are superimposable. To see how, turn H by

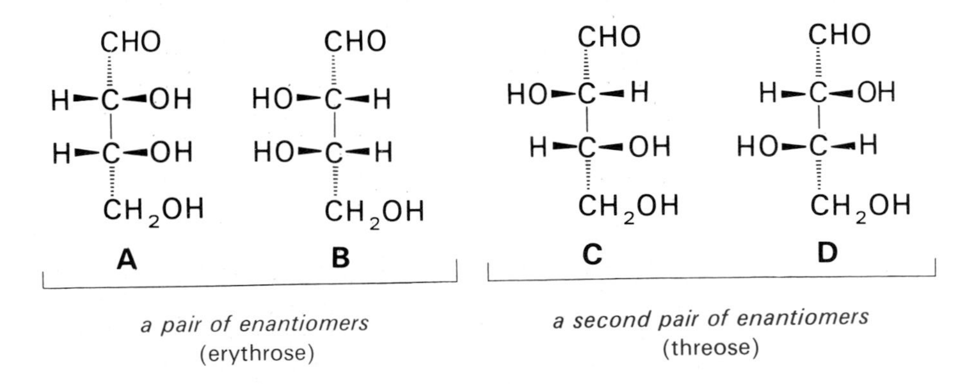

Figure 4.8 *Stereoisomers of 2,3,4-trihydroxybutanal, a substance with two dissimilar chiral carbon atoms.*

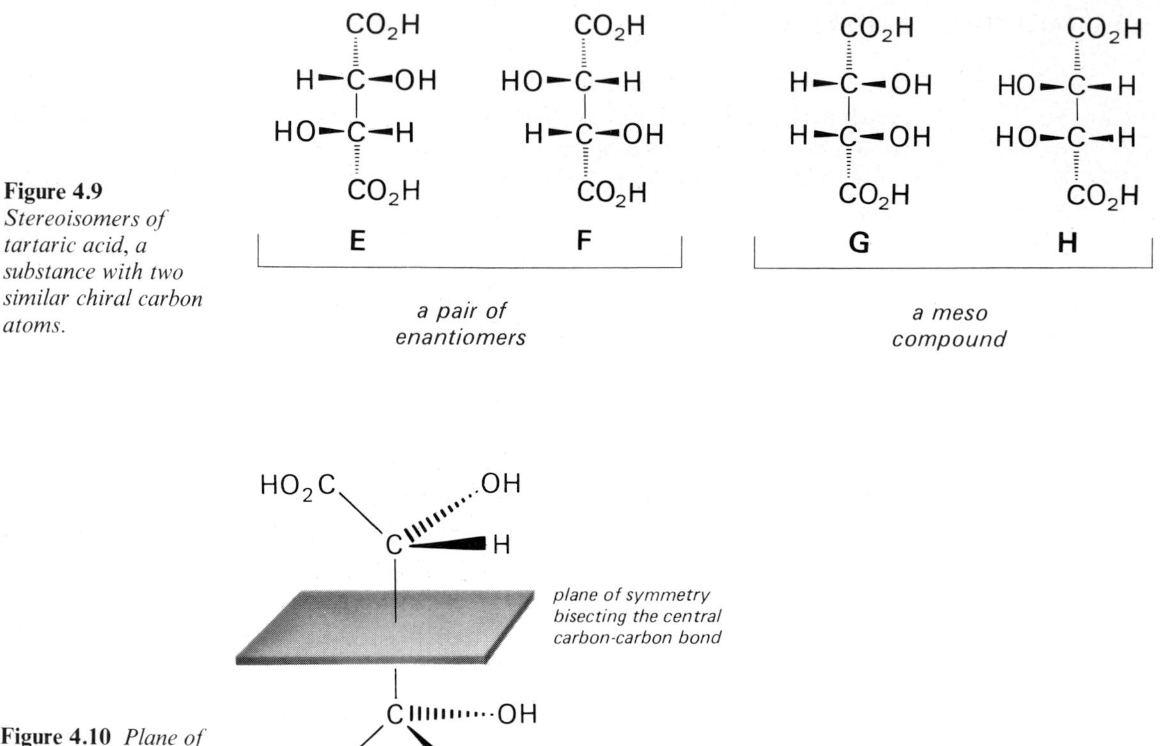

Figure 4.9 *Stereoisomers of tartaric acid, a substance with two similar chiral carbon atoms.*

Figure 4.10 *Plane of symmetry in meso-tartaric acid.*

180° in the plane of the paper and you will see that it fits exactly on G. In Section 4.3, we said that any molecule that has a plane of symmetry is superimposable on its mirror image. The plane of symmetry in G is shown in Figure 4.10.

Symmetrical substances that contain two or more chiral carbon atoms are called *meso* compounds. *Meso*-tartaric acid is a diastereomer of (−)-tartaric acid and of (+)-tartaric acid. Physical properties of the three optical isomers of tartaric acid are given in Table 4.1.

Table 4.1 Physical properties of the tartaric acids.

Acid	*mp* (°C)	$[\alpha]_D^{25°C}$
dextro	170	+12°
levo	170	−12°
meso	146	inactive

4.5 PREDICTING ENANTIOMERISM

There are three methods that you can use to determine whether a molecule will show enantiomerism. It is a necessary and sufficient condition for enantiomerism that a molecule and its mirror image be nonsuperimposable. Therefore, the first and most direct test is to build a model of the molecule and one of its mirror image. If the two are superimposable, then the substance is symmetric (or achiral) and does not show enantiomerism. If they are not superimposable, then the molecule does show enantiomerism.

A second method is to look for a plane of symmetry. If the molecule has a plane of symmetry, then the mirror images are identical, the molecule is achiral, and does not show enantiomerism.

Third, you can look for the presence or absence of a chiral center (an asymmetric carbon atom). For each chiral center there will be a maximum of 2^n enantiomers. Although looking for chiral centers is perhaps the easiest way to predict enantiomerism, you must remember that a molecule may have two or more chiral centers and still not show enantiomerism. Such molecules are said to be *meso* compounds.

4.6 THE RELATIONSHIP BETWEEN *CIS-TRANS* ISOMERISM AND ENANTIOMERISM

Thus far in our discussion of stereoisomerism we have clearly separated molecules into two classes—those showing *cis-trans* isomerism and those showing enantiomerism. Furthermore, most of the examples have dealt with molecules showing either one or the other type of stereoisomerism. This separation has been arbitrary for, in fact, vast numbers of molecules in the natural world have one or more chiral centers and contain one or more sites for *cis-trans* isomerism. The question we should be asking about these molecules is: "How many stereoisomers are possible?"

Following is the structural formula for 3-penten-2-ol.

$$CH_3{-}\underset{\displaystyle OH}{\underset{|}{C}H}{-}CH{=}CH{-}CH_3$$

3-penten-2-ol

How many stereoisomers are there of this molecule? The answer is there are four. To see these, first locate the chiral carbon atom and draw the four bonds from it arranged to show the tetrahedral geometry. Then attach the four groups to the chiral carbon atom, first with the *cis* configuration about the carbon-carbon double bond, then with the *trans* configuration. These are drawn in Figure 4.11. A and B are one pair of enantiomers, C and D are a second pair of enantiomers.

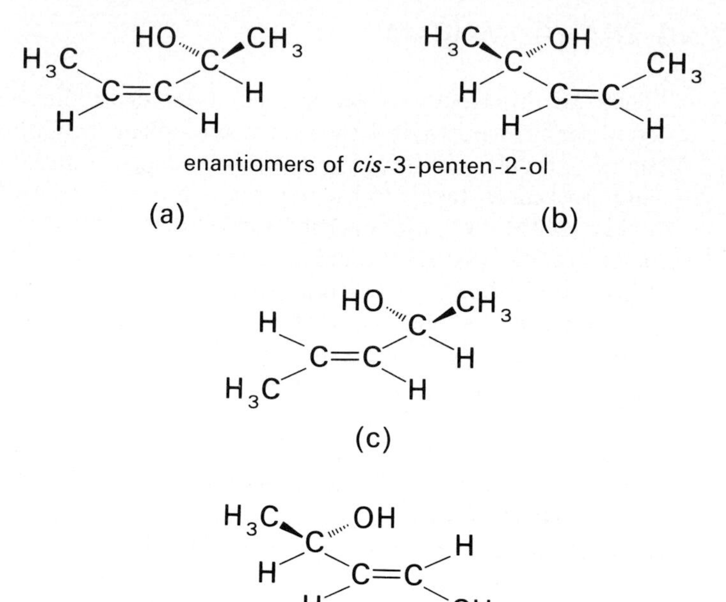

Figure 4.11 *The four stereoisomers of 3-penten-2-ol.*

Example 4.3

Following is the structural formula of 1,2-cyclopentanediol:

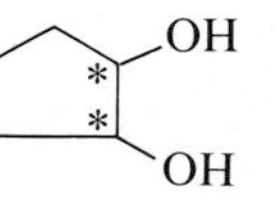

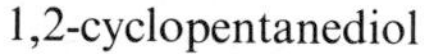

1,2-cyclopentanediol

This substance has two chiral carbon atoms, marked by asterisks. 1,2-Cyclopentanediol shows *cis-trans* isomerism.

(a) Draw a stereorepresentation for the *cis* isomer and its mirror image. Are they superimposable or nonsuperimposable?

(b) Draw a stereorepresentation for the *trans* isomer and its mirror image. Are they superimposable or nonsuperimposable?

(c) State the total number and kind of stereoisomers of 1,2-cyclopentanediol.

Solution

(a) To show *cis–trans* isomerism in derivatives of cyclopentane, draw the ring as a planar pentagon with the substituent groups above and below the plane of the ring. The *cis* isomer and its mirror image are superimposable on each other. Because *cis*-1,2-cyclopentanediol has two chiral carbon atoms yet it and its mirror image are superimposable, the *cis* isomer is a *meso* compound. There is a plane of symmetry in this molecule bisecting the —CH(OH)—CH(OH)— bond and cutting through the ring.

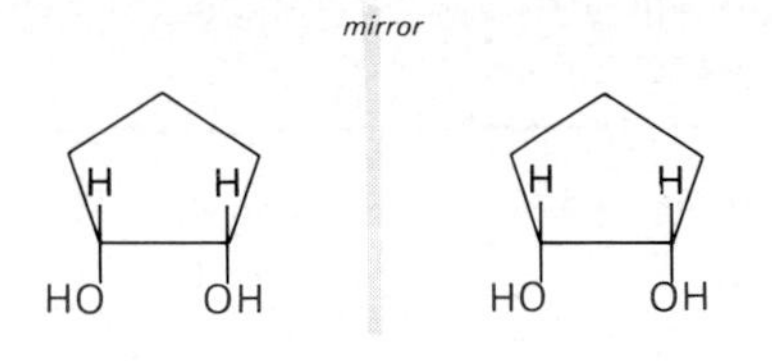

cis-1, 2-cyclopentanediol (a *meso* compound)

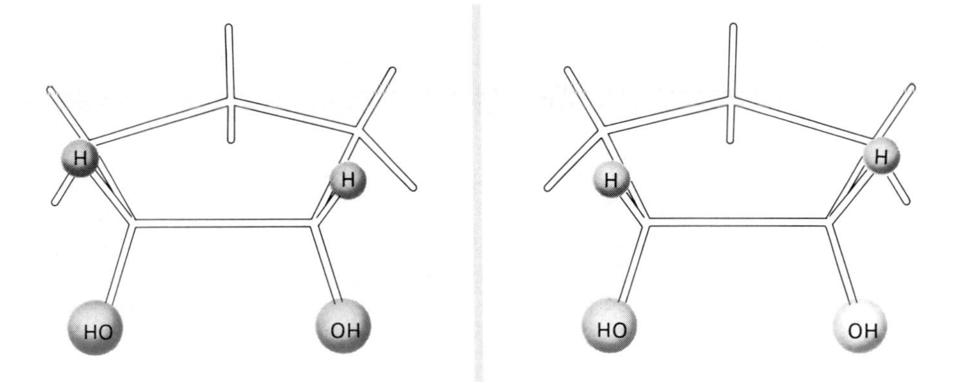

ball-and-stick models of *cis*-1, 2-cyclopentanediol

(b) The *trans* isomer and its mirror image are nonsuperimposable, and therefore, the *trans* isomer shows enantiomerism—it exists as a pair of enantiomers.

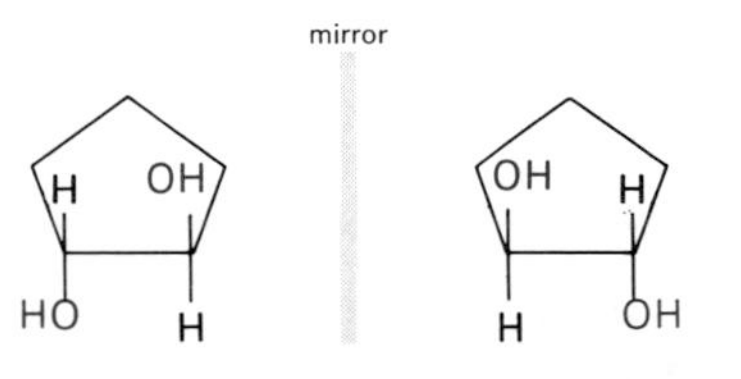

trans-1, 2-cyclopentanediol (a pair of enantiomers)

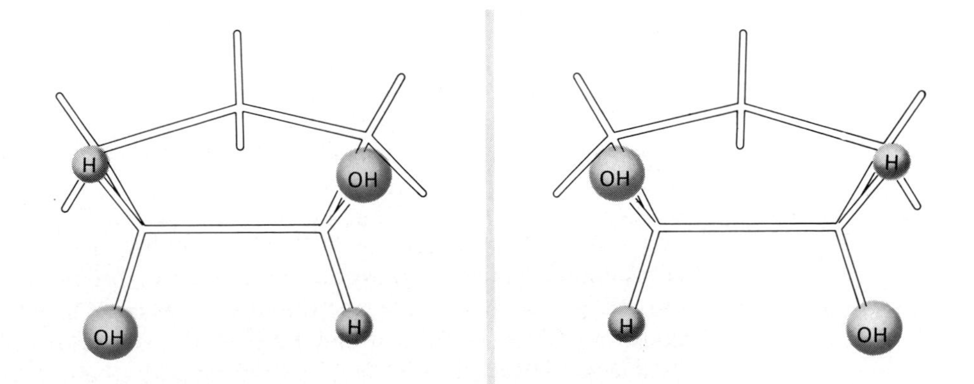

ball-and-stick models for the enantiomers of *trans*-1, 2-cyclopentanediol

(c) There are three stereoisomers of 1,2-cyclopentanediol: one pair of enantiomers and one *meso* compound.

PROBLEM 4.3 Following is the structural formula for 2-methylcyclopentanol.

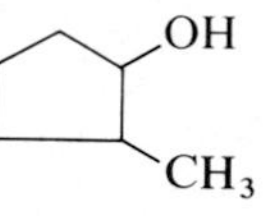

(a) Draw a stereorepresentation for the *cis* isomer and its mirror image. Are they superimposable or nonsuperimposable?
(b) Draw a stereorepresentation for the *trans* isomer and its mirror image. Are they superimposable or nonsuperimposable?
(c) State the total number and kind of stereoisomers of 2-methylcyclopentanol.

Earlier in this chapter we stated that the total number of enantiomers is given by the formula 2^n where n is the number of chiral centers. This is of course a maximum number and the actual number may be smaller due to the existence of *meso* compounds. Does this same rule apply to predicting the total number of stereoisomers? The answer is yes, if we redefine the rule slightly. For the molecule 2-methylcyclopentanol, there are two chiral centers and therefore the rule predicts four stereoisomers; in fact there are four. For 3-penten-2-ol, which contains only one chiral center, the 2^n rule can be used if we redefine n to be equal to the number of chiral centers plus the number of double bonds that are capable of *cis-trans* isomerism. The 2^n rule now predicts four stereoisomers, and in fact there are four.

4.7 CONVENTIONS FOR REPRESENTING ABSOLUTE CONFIGURATION

The most readily observed difference between enantiomers is the interaction with plane polarized light. For over a century after the discovery of optical activity and this type of stereoisomerism, the only method for distinguishing between enantiomers was by the sign of the rotation, hence the designations (+) and (−) in the naming of particular enantiomers. Furthermore, although it was possible to draw absolute configurations for a pair of enantiomers, there was no way to determine which enantiomer had which absolute configuration. Following are structural formulas for the lactic acid enantiomers.

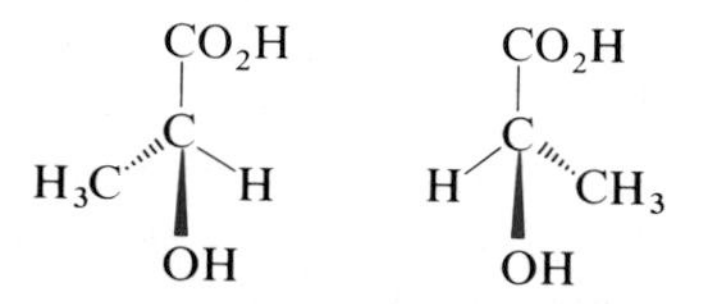

These are absolute configurations because each specifies how the four groups attached to the chiral center are oriented in space. Which is the absolute configuration of (+)-lactic acid and which is the absolute configuration of (−)-lactic acid? Until recently there seemed no way to answer this question. The problem was further compounded by the fact that there is no simple relationship between the sign of rotation and the absolute configuration of a molecule. For example, (+)-lactic acid can be converted into methyl lactate by a chemical procedure that in no way changes the absolute configuration of the chiral center.

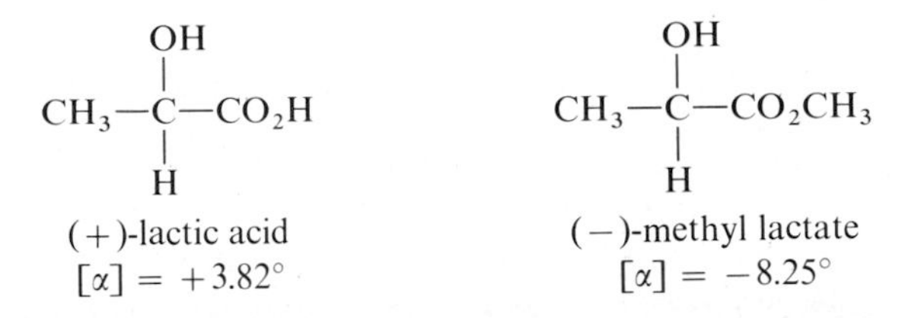

(+)-lactic acid
[α] = +3.82°

(−)-methyl lactate
[α] = −8.25°

The methyl ester of (+)-lactic acid is levorotatory and named (−)-methyl lactate. Whatever the absolute arrangement of the four groups on the chiral center is in (+)-lactic acid, it is the same in (−)-methyl lactate. In other words, these two substances have the same absolute configuration and yet both the direction and the magnitude of the specific rotations are different. While we do not know the absolute configuration of either (+)-lactic acid or (−)-methyl lactate, we do at least know that they both have the same configuration at the chiral carbon atom.

The question of the absolute configuration of lactic acid and a great many other organic substances has been solved, but only recently. Just how this has been done will be discussed in Chapter 11 along with the stereochemistry of carbohydrates. For now we shall simply say that the absolute configuration of the enantiomers of lactic acid and methyl lactate are

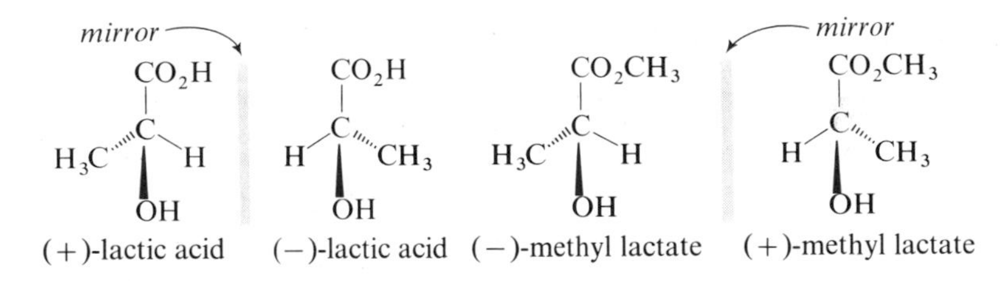

(+)-lactic acid (−)-lactic acid (−)-methyl lactate (+)-methyl lactate

There are two conventions currently used for specifying absolute configuration. One, involving the use of the letters D and L, was proposed by Emil Fischer near the turn of the century to deal with the absolute configuration of carbohydrates. We will discuss this convention in Chapter 11. More recently another set of rules has been proposed called the Cahn-Ingold-Prelog convention after the three chemists who proposed it. This system focuses on the chiral center and assigns an arbitrary order of priority to each of the atoms or groups bonded to it. Priority is assigned on the basis of the atomic number of the atom bonded directly to the chiral center; the higher the atomic number, the higher the priority. Thus typical substituents in order of decreasing priority are

$$I > Br > Cl > SH > F > OH > NH_2 > CH_3 > H$$

Obviously hydrogen has the lowest priority of any substituent. If two atoms attached to the chiral center are identical, then the priority is determined by referring to the next closest atoms.

$CH_3CH—$ (with CH_3 branch)	higher priority than $CH_3CH_2CH_2—$
$CH_3CH_2—$	higher priority than $CH_3—$
$CH_3O—$	higher priority than $HO—$

For the purposes of determining priority, a double-bonded atom is considered to have two single bonds to the same atom. Thus

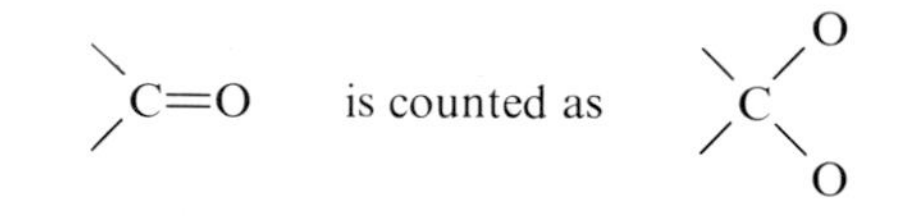

According to this last rule, an aldehyde group has a higher priority than a primary alcohol.

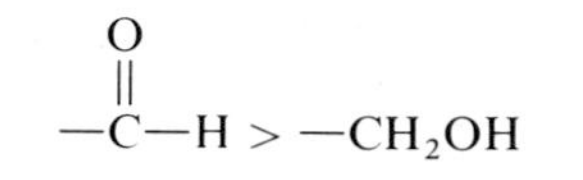

an aldehyde a primary alcohol

Using these rules, the order of priority of the four groups attached to the chiral carbon atom of lactic acid, ranked from highest (1) to lowest (4), is

$—OH$	$—CO_2H$	$—CH_3$	$—H$
(1)	(2)	(3)	(4)

To assign absolute configuration the molecule is turned so that it is viewed from the side directly opposite the group of lowest priority. For (+)-lactic acid, the molecule is oriented as

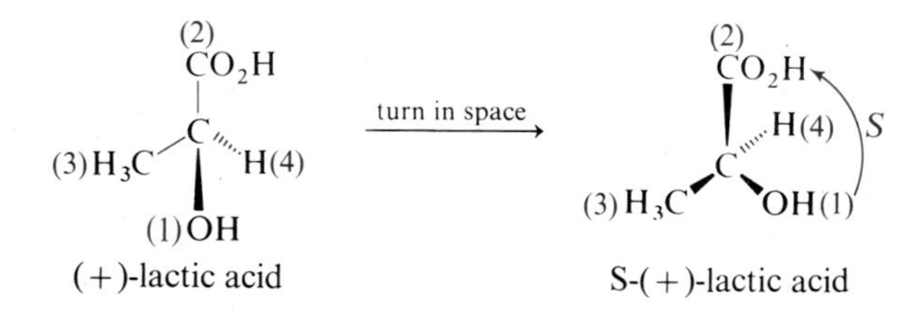

(+)-lactic acid S-(+)-lactic acid

Starting from the atom or group of highest priority (1), move to the group of priority (2), and then to priority (3). If this order of descending priority travels clockwise, then the absolute configuration of the chiral center is R (from Latin, *rectus*, "right"). If the order of descending priority travels counterclockwise, then the absolute configuration of the chiral center is S (from Latin, *sinister* "left").

Before leaving this discussion of absolute configuration and optical activity, let us repeat: to write "R-lactic acid" specifies the absolute configuration.

To write "(+)-lactic acid" specifies the dextrorotatory enantiomer. However, remember that there is no simple or necessary relationship between the absolute configuration (R or S) and the sign of the rotation (dextrorotatory or levorotatory). Only by isolating S-lactic acid and determining its interaction with polarized light can we determine that it is indeed dextrorotatory and therefore properly written S-(+)-lactic acid. It follows also that the enantiomer of S-(+)-lactic acid is R-(−)-lactic acid.

Example 4.4 Below is drawn a stereorepresentation of the absolute configuration of (+)-glyceraldehyde. Assign an (R) or (S) configuration to this enantiomer.

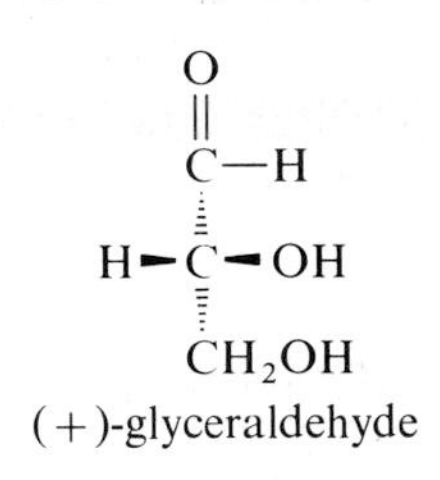

(+)-glyceraldehyde

Solution First, rank the groups attached to the chiral carbon in order of priority. For (+)-glyceraldehyde, the order is

OH	CH(=O)	CH_2OH	H
(1)	(2)	(3)	(4)

Second, turn the molecule in space so that the group of lowest priority (H) is away from you and the remaining three groups face you. Third, draw an arrow from the group of highest priority (OH) to the group of second-highest priority (CHO). For (+)-glyceraldehyde, the arrow turns clockwise and therefore the configuration is (R).

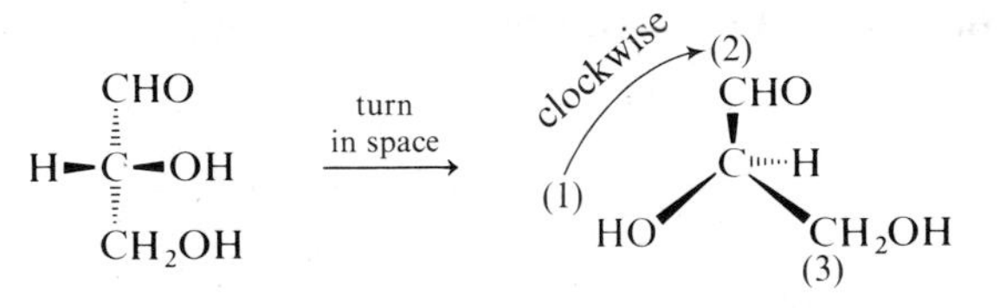

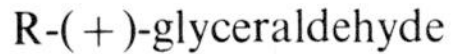
R-(+)-glyceraldehyde

The mirror image of R-(+)-glyceraldehyde is S-(−)-glyceraldehyde.

PROBLEM 4.4 Following is drawn a stereorepresentation for the absolute configuration of (+)-alanine. Assign an (R) or (S) configuration to this enantiomer.

O
‖
C—OH

H_2N—C—H

CH_3

(+)-alanine

4.8 RACEMIC MIXTURES

Let us return to the two lactic acids, (+)-lactic acid found in muscle tissue and optically inactive lactic acid found in sour milk. We have already demonstrated that lactic acid and its mirror image are not superimposable and that lactic acid shows optical isomerism. How is it that the lactic acid from fermentation is optically inactive while that from muscle is optically active? The answer is that the inactive form of lactic acid is a mixture of equal numbers of (+)-lactic acid molecules and (−)-lactic acid molecules. Because the mixture contains equal numbers of molecules that rotate the plane of polarized light to the right and to the left, the mixture does not rotate the plane of polarized light either to the right or left; it is optically inactive. Such a mixture, containing equal amounts of a pair of enantiomers, is called a racemic mixture.

4.9 RESOLUTION OF RACEMIC MIXTURES

The process of separating a pair of enantiomers into (+)- and (−)-isomers is called resolution. The first demonstration of this process was the historic resolution of racemic tartaric acid by Louis Pasteur in 1848. Below 25°C, the (+)-enantiomer of sodium ammonium tartrate crystallizes in one type of crystal, while the (−)-enantiomer crystallizes in a mirror-image crystal. By carefully hand-picking the crystals, Pasteur separated the mixture into two mirror image crystalline forms and examined their solutions separately in a polarimeter. He made the exciting discovery that a solution of one form rotated the plane of polarized light to the right and the mirror image form rotated it to the left. When equal weights of the two forms of crystals were dissolved in water, the solution of the mixture, like the starting material, had no effect on the plane of polarized light. Resolution of tartaric acid by Pasteur is particularly remarkable, for since that time very few additional examples have been found in which crystallization produces mirror image crystals that can be separated by hand. Fortunately, other methods of resolution are now known.

A second and more generally useful method of resolution is based on the fact that diastereomers have different physical properties. Pasteur observed that metal or ammonium salts of (+)-tartaric and of (−)-tartaric acid have identical solubilities in water and polar organic solvents. However, salts of (+)- and (−)-tartaric acid formed with certain naturally occurring amines such as strychnine or quinine do not have the same solubilities. (The amines named are chiral substances and are optically active.) Reaction of a racemic acid with an optically active amine (a base) forms a pair of diastereomeric salts.

$$\begin{matrix} (+)\text{-}RCO_2H \\ \\ (-)\text{-}RCO_2H \end{matrix} \quad + \quad (-)\text{-}R'NH_2 \longrightarrow \begin{matrix} (+)\text{-}RCO_2^- \ (-)\text{-}R'NH_3^+ \\ \\ (-)\text{-}RCO_2^- \ (-)\text{-}R'NH_3^+ \end{matrix}$$

racemic form of the acid — *a pure enantiomer of an amine* — *a pair of diastereomeric salts*

These salts can be separated by fractional crysallization. Treatment of the separated salts with mineral acid then liberates the original acid in optically pure (+) and (−) forms.

$$(+)\text{-RCO}_2^- \; (-)\text{-R'NH}_3^+$$
$$(-)\text{-RCO}_2^- \; (-)\text{-R'NH}_3^+$$

↓

separate by fractional crystallization due to different solubilities in water or a mixture of water and a polar organic solvent

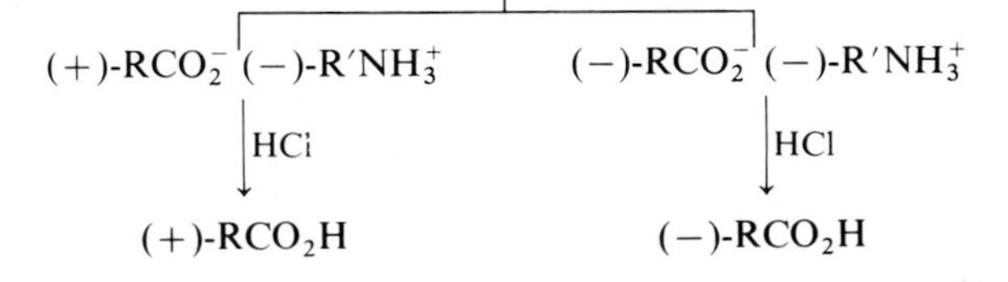

4.10 THE SIGNIFICANCE OF CHIRALITY IN THE BIOLOGICAL WORLD

Why is it so important to be able to describe stereoisomerism and recognize it in molecules? The reason is that we are interested in the reactions between molecules, particularly between organic molecules in the biological world. Except for inorganic salts and a relatively few low-molecular-weight organic substances, most of the compounds of living organisms, both plant and animal, are chiral. Although, in principle, these molecules can exist as a mixture of stereoisomers, almost invariably only one stereoisomer is found in nature. There are, of course, instances where both enantiomers can be found in nature, but they do not seem to exist together in the same biological system. We can make the further generalization that not only is just one enantiomer found in nature, but also only one enantiomer can be used or assimilated by an organism. Pasteur discovered in 1858–1860 that when the microorganism *Penicillium glaucum*, a green mold found in aging cheese and rotting fruit, is grown in a medium containing racemic tartaric acid, the solution slowly becomes levorotatory. The microorganism preferentially consumes or metabolizes (+)-tartaric acid. If the process is interrupted at the right time, (−)-tartaric acid crystallizes from solution in pure form. As another example, when racemic mevalonic acid is fed to rats, one enantiomer is excreted in the urine.

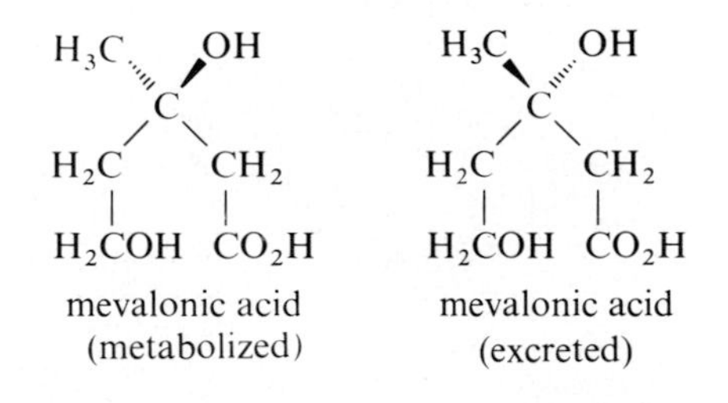

mevalonic acid (metabolized)   mevalonic acid (excreted)

Many other examples are known in which a mold or other microorganism uses one enantiomer of a racemic mixture.

The observations that only one enantiomer is found in a given biological system and that only one enantiomer is metabolized, should be enough to convince us that it is a chiral world in which we live. At least it is chiral at the molecular level. Essentially all chemical reactions in the biological world take place in a chiral environment. Perhaps the most conspicuous examples of chirality among biological molecules are the enzymes. To illustrate this point, consider the enzyme chymotrypsin which functions so efficiently in the intestine at pH 7–8 to catalyze the digestion of proteins. Chymotrypsin is made up of 241 amino acids and contains 251 chiral carbon atoms. The number of potential stereoisomers of chymotrypsin is 2^{251}, a number beyond comprehension! Fortunately, nature does not squander its precious resources and energies unnecessarily: only one of these stereoisomers is made in any given organism.

Enzymes catalyze biological reactions by first adsorbing on their surface the small molecule or molecules about to undergo reaction. The small molecule may be bound by an accumulation of noncovalent interactions such as hydrogen bonding or by actual covalent bond formation. Thus, whether the smaller molecule is chiral or not, it is now held in a chiral environment. Let us look in more detail at an example that illustrates how an enzyme might distinguish between a pair of enantiomers.

Glyceraldehyde, a key intermediate in the metabolism of carbohydrates, contains one chiral carbon and therefore shows optical isomerism. How might an enzyme discriminate between one enantiomer of glyceraldehyde and the other? It is generally agreed that an enzyme with specific binding sites for three of the four substituents on the chiral center can distinguish between two enantiomers. Assume, for example, that the enzyme involved in the catalysis of a glyceraldehyde reaction has three binding sites, one specific for —H, another for —OH, and a third for —CHO, and that the three sites are arranged on the enzyme surface as shown in Figure 4.12.

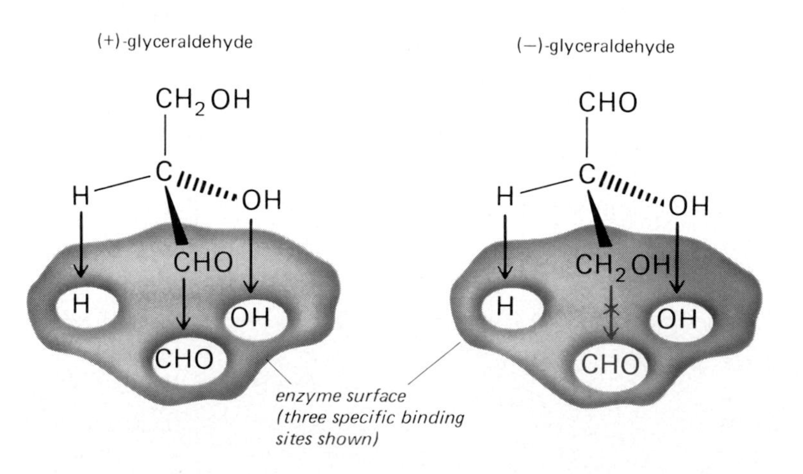

Figure 4.12 *Enzyme stereospecificity. A schematic diagram of an enzyme surface capable of interacting with R-(+)-glyceraldehyde at three binding sites but with S-(−)-glyceraldehyde at only two of the three potential binding sites.*

The enzyme can "recognize" R-(+)-glyceraldehyde (the natural or biologically active form) in the presence of S-(−)-glyceraldehyde because the correct enantiomer can be adsorbed with three groups attached to the appropriate binding sites; the other enantiomer can, at best, bind to only two of these sites.

With this insight into the interactions between molecules taking place in a chiral environment, such things as the selective metabolism of one enantiomer of mevalonic acid should be no surprise. Furthermore, it should be no surprise to discover that a molecule and its enantiomer have quite different psychological or physiological properties, since, after all, a molecule and its enantiomer are different substances. As an example (+)-leucine tastes sweet, while its enantiomer (−)-leucine is bitter.

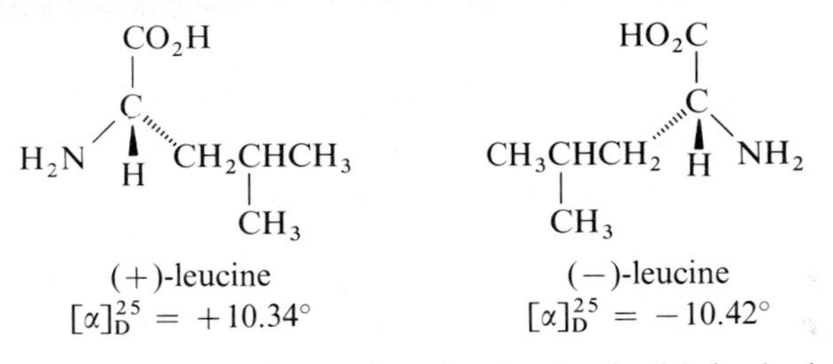

(+)-leucine $[\alpha]_D^{25} = +10.34°$ (−)-leucine $[\alpha]_D^{25} = -10.42°$

The fact that the interactions of molecules in the biological world are so very specific in geometry and chirality is not surprising. Just how these interactions are accomplished with such high precision and efficiency is one of the great challenges that modern science has only recently begun to *unravel.*

PROBLEMS

4.5 Which of the following compounds show optical isomerism? For each that does, draw stereorepresentations for both enantiomers.

(a) $CH_2{=}CH{-}CH(OH){-}CH_3$

(b) $CH_2{=}CH{-}CH(OH){-}CH_2{-}CH_3$

(c) $CH_3{-}CH_2{-}CH(OH){-}CH_2{-}CH_3$

(d) $CH_2(OH){-}CH_2{-}CH(OH){-}CH_2{-}CH_3$

(e) $CH_3{-}C(CH_3)(OH){-}CH{=}CH_2$

(f) $H{-}C(CO_2H)(CH_2OH){-}OH$

(g) $CH_3{-}CH(CH_3){-}CH(NH_2){-}CO_2H$

(h) $\bigcirc{-}OH$

(i) $CH_3{-}\overset{\overset{O}{\|}}{C}{-}CO_2H$

(j) $(H_3C)(H)C{=}C(CO_2H)(CH_3)$

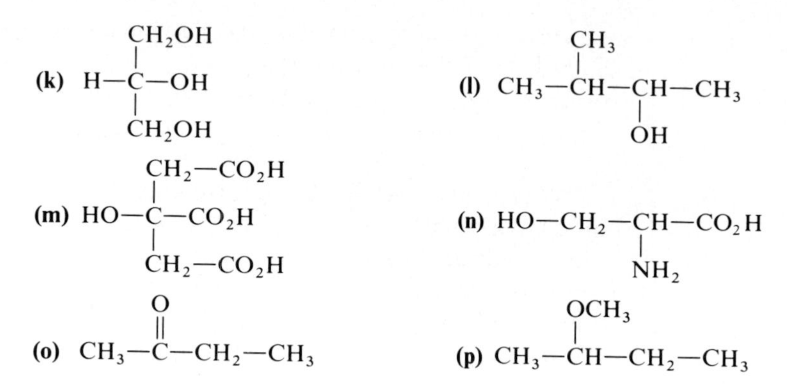

4.6 The following are absolute configurations for a series of small molecules each containing only one chiral center. Under each is written the name of the molecule. Determine whether the enantiomer drawn is R or S.

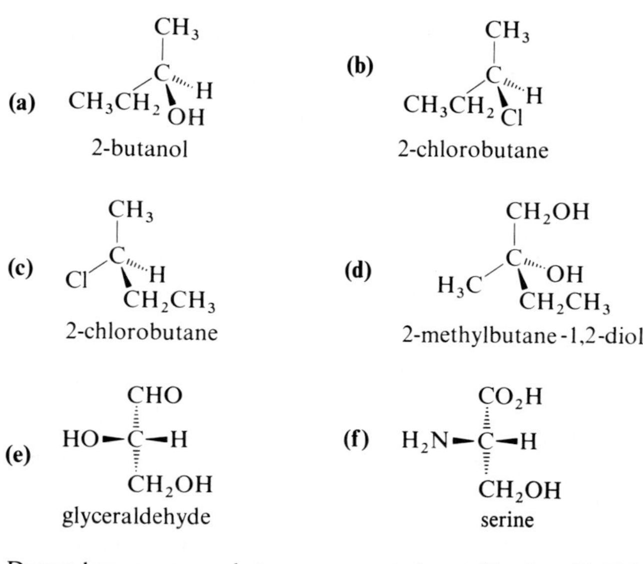

4.7 Drawn here are several stereorepresentations of lactic acid. Taking (a) as a reference structure, note that the other structures are stereorepresentations of lactic acid viewed from other perspectives. Which of the alternative representations are identical to (a) and which are mirror images of (a)?

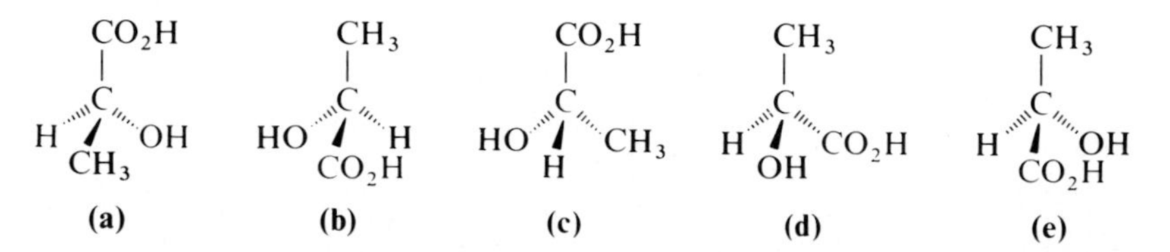

4.8 Explain the difference in molecular structure between *meso* tartaric acid and racemic tartaric acid.

4.9 Below the structural formula of each of the following molecules is given the number of stereoisomers, and in parentheses the number of these stereoisomers that are *meso* compounds. For each *meso* compound, draw an appropriate projection formula and indicate clearly the plane of symmetry.

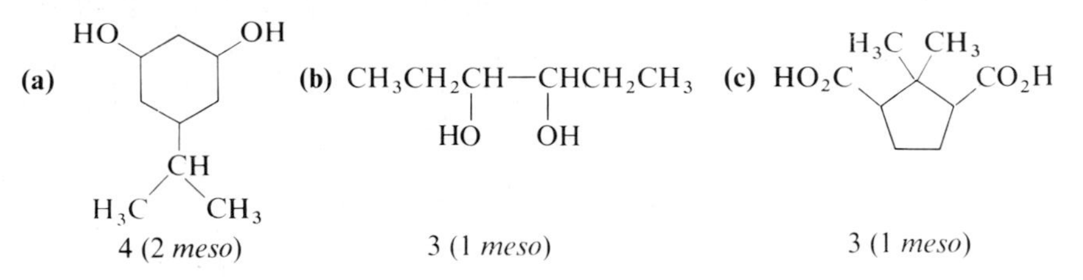

4.10 Inositol is widely distributed in plants and animals and is a growth factor for animals and microorganisms. It is used medically for treatment of cirrhosis of the liver, hepatitis, and fatty infiltration of the liver. There are nine stereoisomers of inositol; seven *meso* compounds and one pair of enantiomers. The most prevalent form of inositol in nature is *cis*-1,2,3,5-*trans*-4,6-cyclohexanehexol.

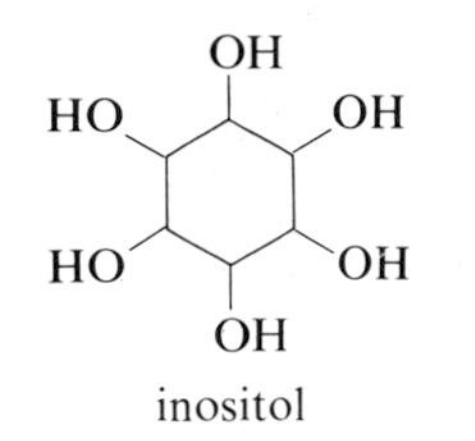

inositol

(a) Draw a chair conformation of the prevalent natural isomer and determine whether it is meso or shows enantiomerism.
(b) Draw chair conformations for the single pair of enantiomers.

4.11 Draw all stereoisomers for the following compounds. State which structures are enantiomers; which are diastereomers, which are *meso* compounds.

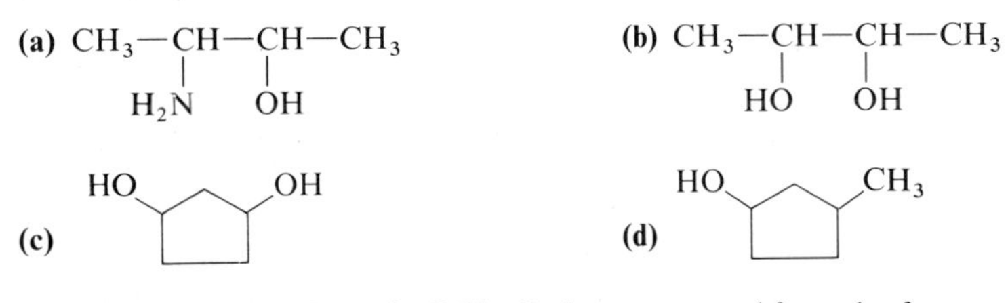

4.12 Using the molecular formula $C_6H_{12}O$, draw structural formulas for:
(a) four isomers with different functional groups
(b) a pair of acyclic *cis-trans* isomers
(c) a pair of cyclic *cis-trans* isomers
(d) a pair of conformational isomers
(e) a pair of enantiomers
(f) a pair of diastereomers

4.13 Draw the structural formula of at least one alkene of molecular formula C_5H_9Br that shows
(a) neither *cis-trans* isomerism nor enantiomerism
(b) *cis-trans* isomerism but not enantiomerism
(c) enantiomerism but not *cis-trans* isomerism
(d) both *cis-trans* isomerism and enantiomerism

4.14 Draw the stereoisomers of grandisol, a sex hormone secreted by the hind gut of the male boll weevil (*Anthonomus grandis*).

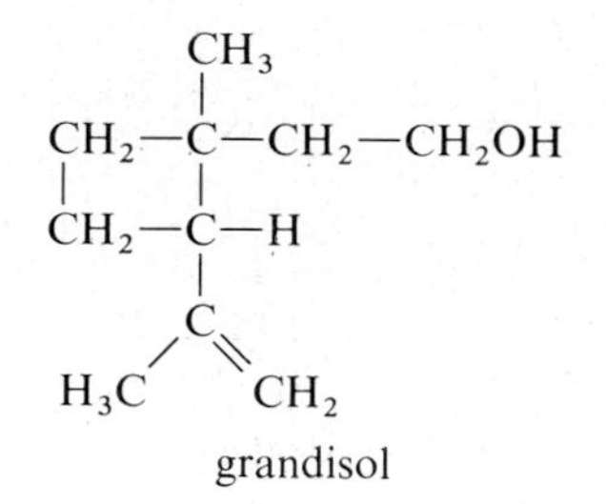

grandisol

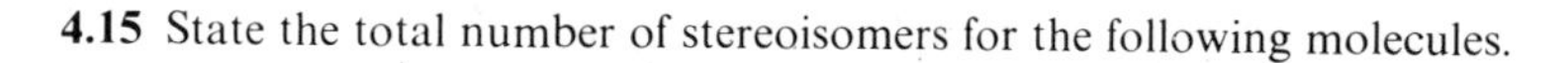

4.15 State the total number of stereoisomers for the following molecules.

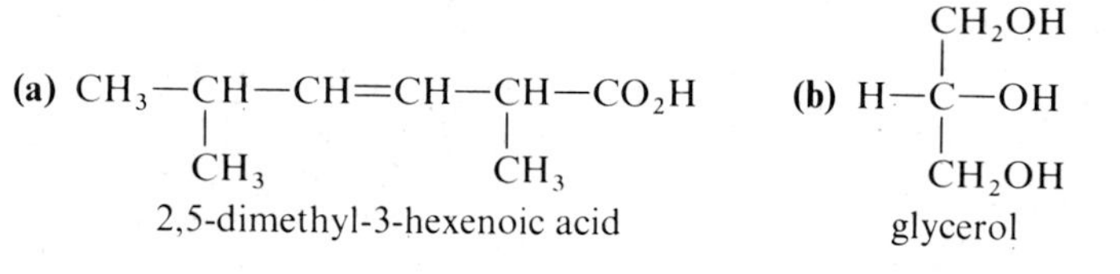

(a) 2,5-dimethyl-3-hexenoic acid

(b) glycerol

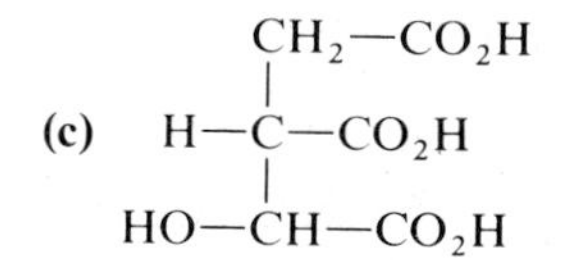

(c) isocitric acid

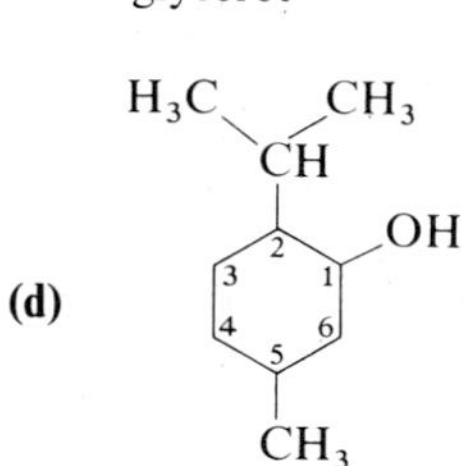

(d) 2-isopropyl-5-methylcyclohexanol (the all equatorial isomer is menthol)

(e) α-terpineol (camphor oil)

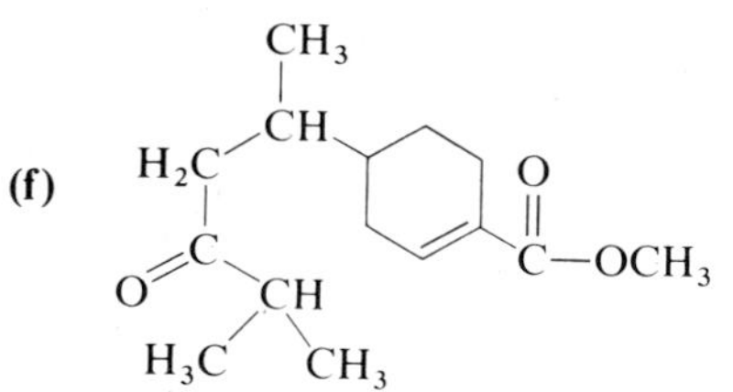

(f) juvabione (a juvenile hormone analog of balsam fir)

(g) $CH_3CH{=}CHCO_2H$

2-butenoic acid

(h)

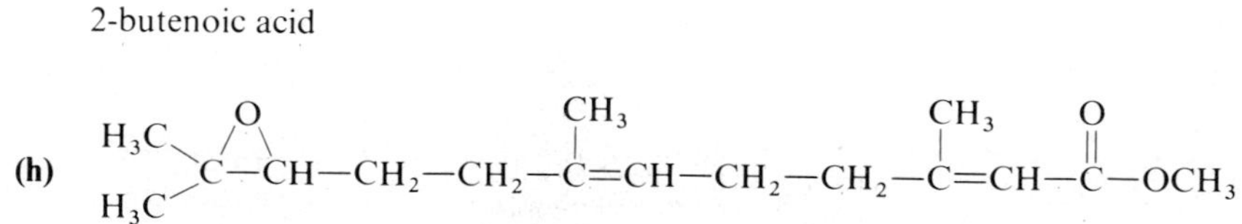

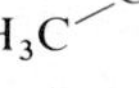

methyl 10,11-epoxyfarnesoate
(a synthetic insect juvenile hormone)

4.16 List three experimental methods for the resolution of racemic mixtures and explain each briefly. Which of these methods is the most widely used?

4.17 How might you explain the following:

(a) An enzyme is able to distinguish between a pair of enantiomers and catalyze a biochemical reaction of one enantiomer but not of its mirror image.

(b) The microorganism *Penicillium glaucum* preferentially metabolizes (+)-tartaric acid rather than (−)-tartaric acid.

(c) An enzyme catalyzes the conversion of a symmetrical, optically inactive molecule into one pure enantiomer uncontaminated by its enantiomer mirror image. For example, pyruvic acid is reduced in an enzyme-catalyzed reaction to S-lactic acid.

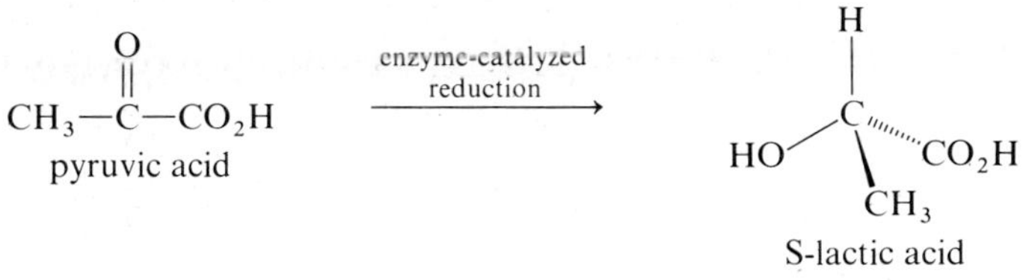

$$CH_3-\overset{\overset{\displaystyle O}{\|}}{C}-CO_2H \xrightarrow[\text{reduction}]{\text{enzyme-catalyzed}} HO-\overset{\overset{\displaystyle H}{|}}{\underset{\underset{\displaystyle CH_3}{|}}{C}}-CO_2H$$

pyruvic acid → S-lactic acid

CHAPTER 5

Alcohols, Ethers, and Thiols

Alcohols and ethers are derivatives of water in which one or both hydrogens are replaced by alkyl groups.

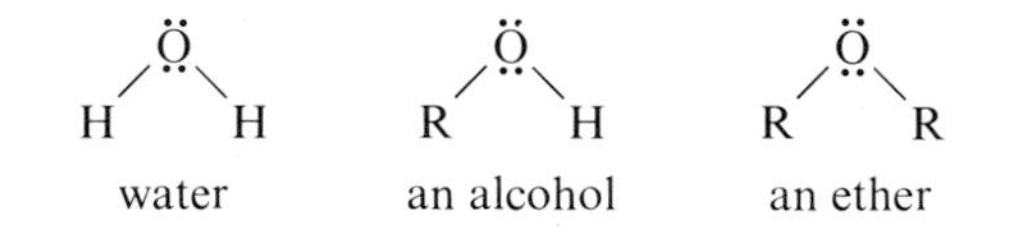

Closely related to alcohols are thiols, derivatives of hydrogen sulfide in which one hydrogen is replaced by an alkyl group:

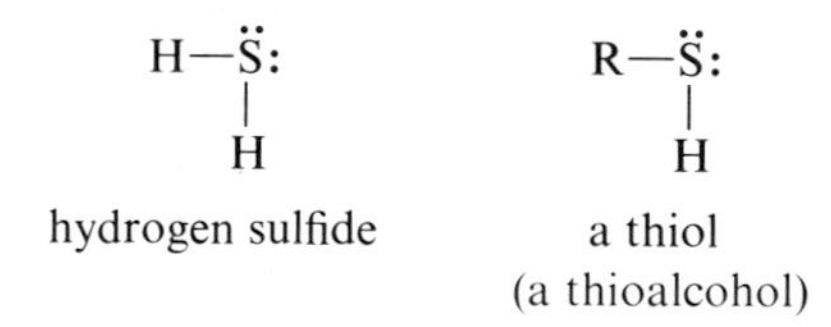

In this chapter, we will discuss some of the important physical and chemical properties of these three classes of compounds. In addition, we will introduce a fundamental type of organic reaction, namely, nucleophilic substitution at a saturated carbon atom.

5.1 STRUCTURE OF ALCOHOLS AND ETHERS

The characteristic structural feature of an alcohol is an —OH group bonded to a hydrocarbon chain. Figure 5.1 shows the Lewis structure of methanol, CH_3OH, the simplest alcohol. As we have already seen in Example 1.5, the oxygen atom of an alcohol is sp^3 hybridized. Two sp^3 hybrid orbitals are used to form sigma bonds to hydrogen and carbon, and the remaining sp^3 hybrid orbitals hold the two unshared pairs of electrons. Because oxygen is sp^3 hybridized, we predict 109.5° for the C—O—H bond angle. The measured bond angle for methanol is 108.9°, a value very close to the predicted value.

The characteristic structural feature of an ether is an oxygen atom bonded to two hydrocarbon chains. Figure 5.2 shows a Lewis structure of dimethyl ether, CH_3OCH_3, the simplest ether. In an ether, sp^3 hybrid orbitals of oxygen are

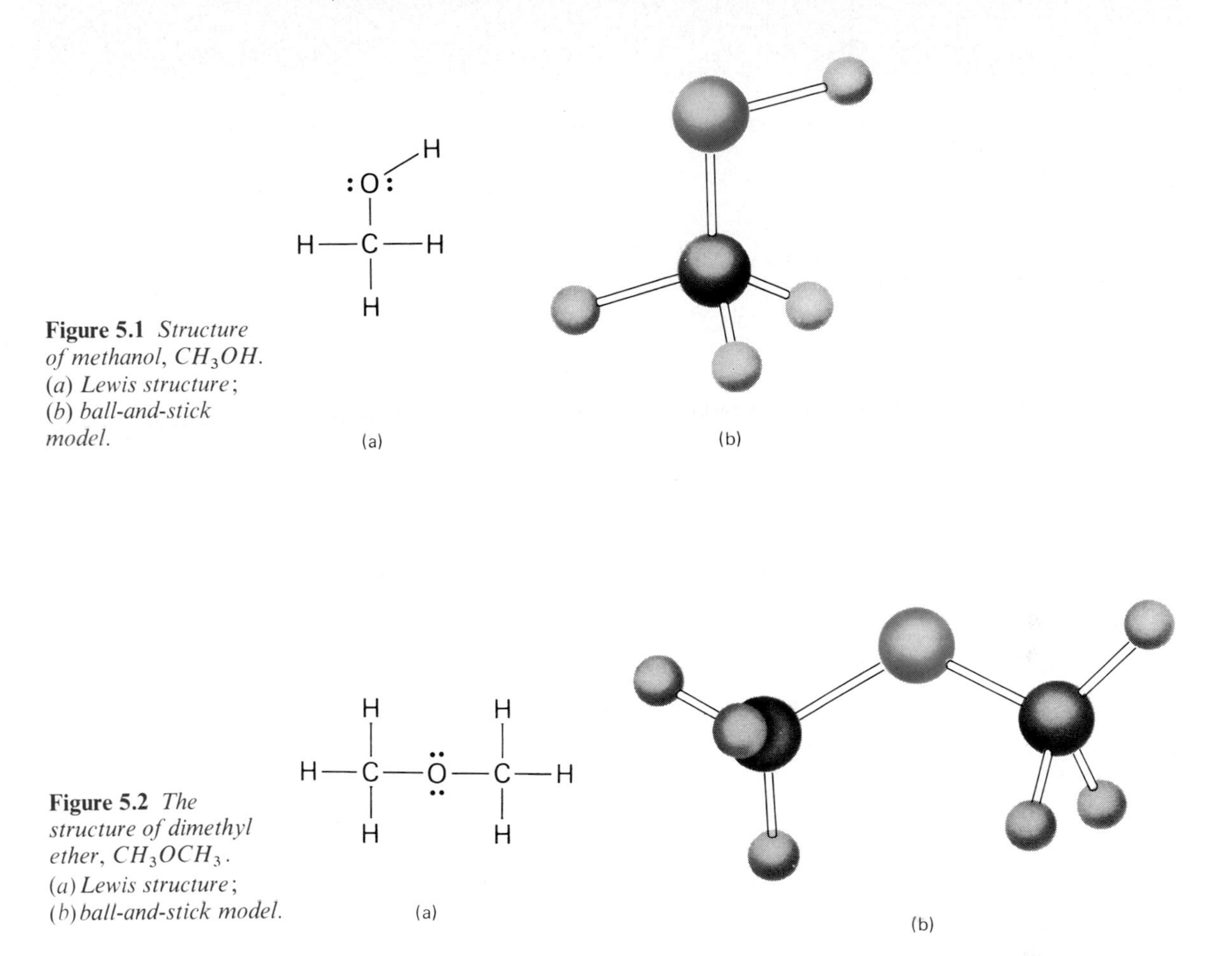

Figure 5.1 *Structure of methanol,* CH_3OH. *(a) Lewis structure; (b) ball-and-stick model.*

Figure 5.2 *The structure of dimethyl ether,* CH_3OCH_3. *(a) Lewis structure; (b) ball-and-stick model.*

used to form sigma bonds to the two hydrocarbon chains. The two unshared pairs of electrons are in the remaining sp^3 hybrid orbitals. The C—O—C bond angle in dimethyl ether is 110.3°, a value very close to the predicted angle of 109.5°. Figure 5.2 also shows a ball-and-stick model of dimethyl ether.

5.2 NOMENCLATURE OF ALCOHOLS AND ETHERS

The IUPAC system selects the longest continuous chain containing the —OH group as the parent compound. The alcohol is named by replacing the -e of the parent compound by -ol and adding a number to show the location of the —OH group. Common names for simple alcohols are derived by naming the alkyl group attached to —OH and then adding the word alcohol. Figure 5.3 lists IUPAC names and, in parentheses, common names for a variety of simple alcohols.

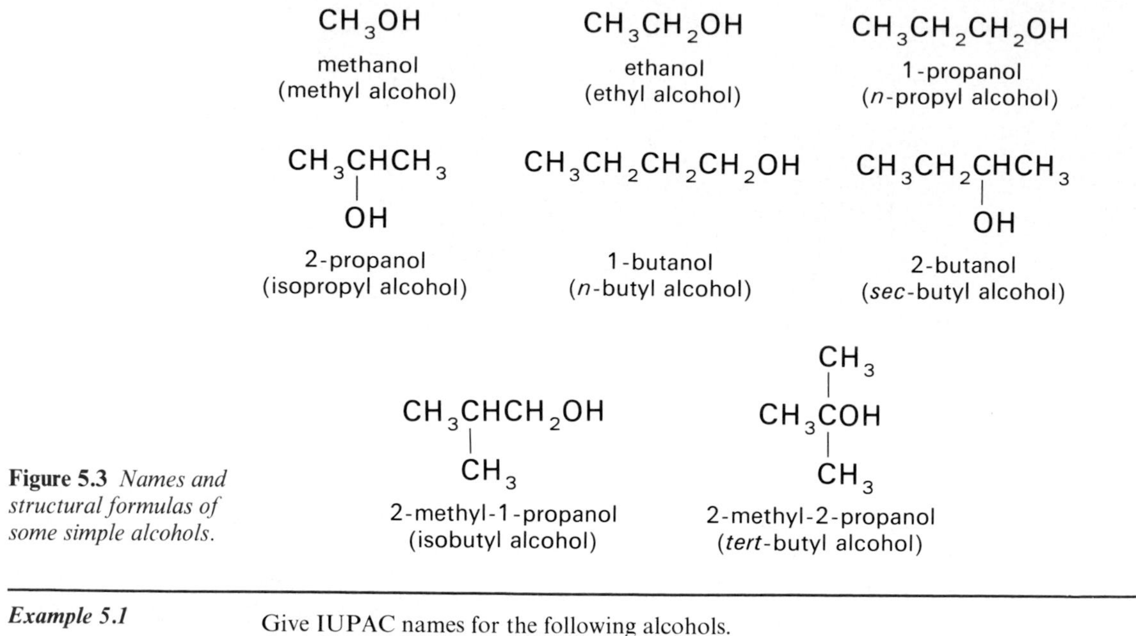

Figure 5.3 *Names and structural formulas of some simple alcohols.*

Example 5.1

Give IUPAC names for the following alcohols.

(a) $CH_3-\underset{|}{\overset{CH_3}{CH}}-CH_2-\underset{OH}{CH}-CH_3$ **(b)** (cyclohexane ring bearing OH and CH_3 on adjacent carbons)

Solution

(a) First select the longest chain of carbon atoms that contains the —OH group and number it to give —OH the lowest possible number. In this example, the longest chain is five carbon atoms and the —OH is on carbon 2. Therefore, this alcohol is a derivative of 2-pentanol. Finally, give the $—CH_3$ substituent a name and a number.

$$\overset{5}{CH_3}-\overset{CH_3}{\overset{4}{CH}}-\overset{3}{CH_2}-\underset{OH}{\overset{2}{CH}}-\overset{1}{CH_3} \qquad \text{4-methyl-2-pentanol}$$

(b) In cyclic alcohols, the carbon atoms of the ring are numbered starting with the carbon bearing the —OH group. Since the —OH is automatically on carbon 1, there is no need to give a number for its location. Therefore, the name of this substance is 2-methylcyclohexanol.

PROBLEM 5.1 Give IUPAC names for the following alcohols.

(a) $CH_3-CH_2-\underset{CH_2-CH_3}{CH}-CH_2-OH$ **(b)** (cyclopentane ring bearing OH and CH_3 on the same carbon)

We often refer to alcohols as primary (1°), secondary (2°), or tertiary (3°). This classification depends on whether the —OH group is on a primary carbon, a secondary carbon, or a tertiary carbon (Figure 5.4).

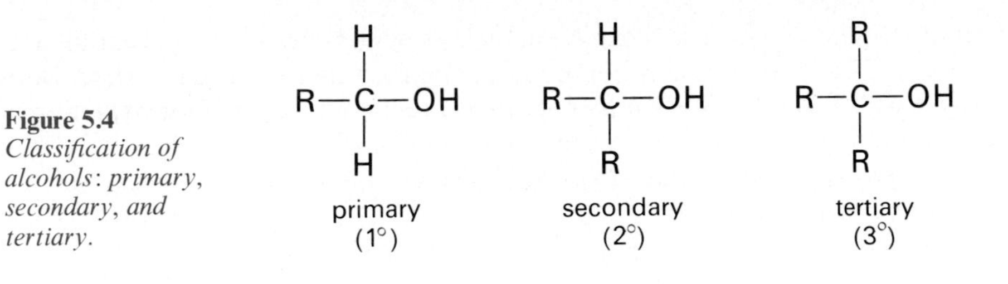

Figure 5.4 *Classification of alcohols: primary, secondary, and tertiary.*

Example 5.2 Classify the following alcohols as primary, secondary, or tertiary.

(a) $CH_3—CH(CH_3)—OH$ **(b)** cyclohexanol (ring with OH) **(c)** $CH_3—C(CH_3)_2—OH$ **(d)** cyclopentyl—CH_2OH

Solution

(a) A secondary alcohol. The carbon bearing the —OH group has two attached carbon groups and therefore is a secondary carbon.
(b) A secondary alcohol. The carbon bearing the –OH group is a secondary carbon.
(c) A tertiary alcohol. The carbon bearing the —OH group has three attached hydrocarbon groups and therefore is a tertiary carbon.
(d) A primary alcohol. The carbon bearing the —OH group has one attached carbon group and therefore is a primary carbon.

PROBLEM 5.2 Classify the following alcohols as primary, secondary, or tertiary.

(a) $CH_3—C(CH_3)_2—CH_2—OH$ **(b)** cyclopropyl—OH **(c)** $CH_2{=}CH—CH_2—OH$ **(d)** cyclopentane ring bearing OH and CH_3 on the same carbon

Compounds containing two hydroxyl groups are called diols, those containing three hydroxyl groups are called triols, and so on. Under each of the following examples is given the IUPAC name and, in parentheses, the common name.

$CH_2(OH)—CH_2(OH)$
1,2-ethanediol
(ethylene glycol)

$CH_2(OH)—CH(OH)—CH_2(OH)$
1,2,3-propanetriol
(glycerol, glycerine)

$CH_3—CH(OH)—CH_2(OH)$

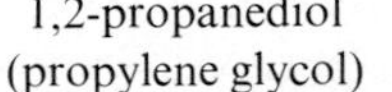

1,2-propanediol
(propylene glycol)

(cyclohexane ring with OH on adjacent carbons, one wedged, one hashed)
trans-1,2-cyclohexanediol

All of these diols and triols can be named as derivatives of the parent alkane. Yet as with so many organic compounds, common names have persisted. Both ethylene glycol and propylene glycol can be prepared by controlled oxidation of ethylene and propylene, hence their common names.

Molecules that contain two hydroxyl groups on the same carbon are almost never isolated

$$>C(OH)_2 \rightleftharpoons >C{=}O + H_2O$$

As we shall see in Section 7.5, compounds of this type are in equilibrium with aldehydes or ketones and equilibrium generally lies very far to the right.

For compounds containing two or more different functional groups, the IUPAC name must show the presence and location of each. For compounds containing a carbon-carbon double bond and an alcohol, the presence of the double bond is shown by changing the -ane of the parent compound to -ene and the presence of the —OH group is shown by changing the final -e to -ol. A number is used to show the location of each functional group. Compounds containing an alcohol and a carbon-carbon double bond are named as alcohols, and the carbon chain of the parent is numbered to give the —OH the lowest possible number.

Example 5.3 Name the following alcohols.

(a) $CH_2{=}CH{-}CH_2{-}OH$ (b) $CH_3{-}CH_2{-}CH{=}CH{-}CH(OH){-}CH_3$

Solution (a) The parent compound contains three carbons. Therefore this substance is named as a derivative of propane. The —OH group is on carbon 1 and the double bond is between carbons 2 and 3. The correct name is 2-propen-1-ol.

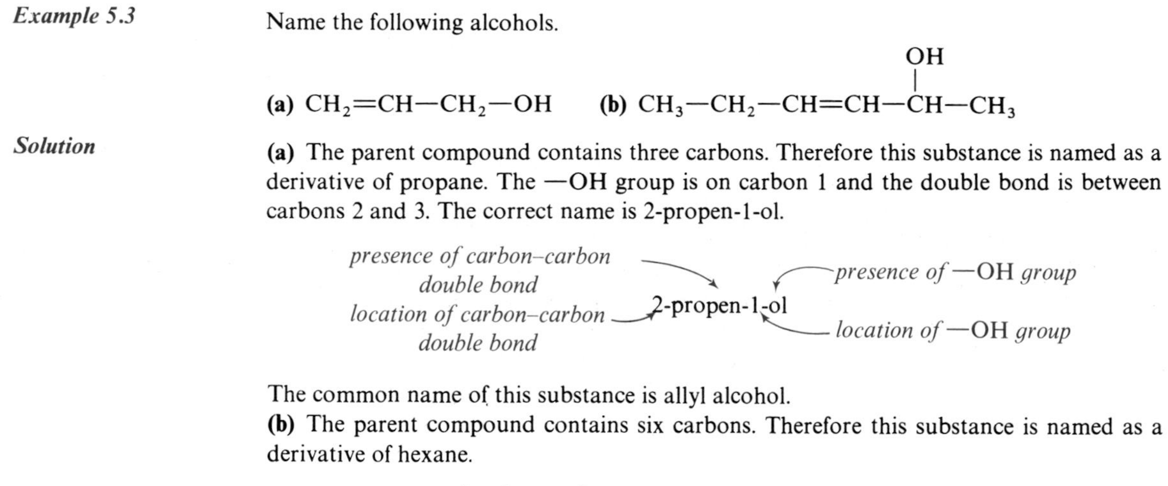

The common name of this substance is allyl alcohol.

(b) The parent compound contains six carbons. Therefore this substance is named as a derivative of hexane.

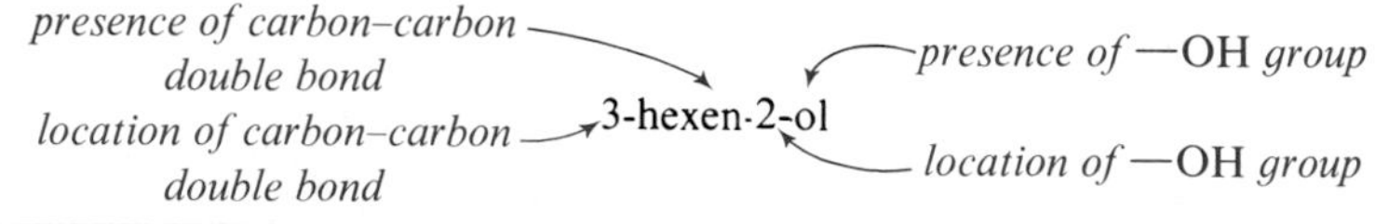

PROBLEM 5.3 Name the following alcohols.

(a) $CH_3{-}CH{=}CH{-}CH_2{-}OH$ (b) [cyclohexene ring with OH]

One last comment on the nomenclature of alcohols. The suffix -ol is generic to alcohols, and although names such as glycerol, menthol, and cholesterol contain no clues to their carbon skeleton, they do indicate that each compound contains one or more hydroxyl groups.

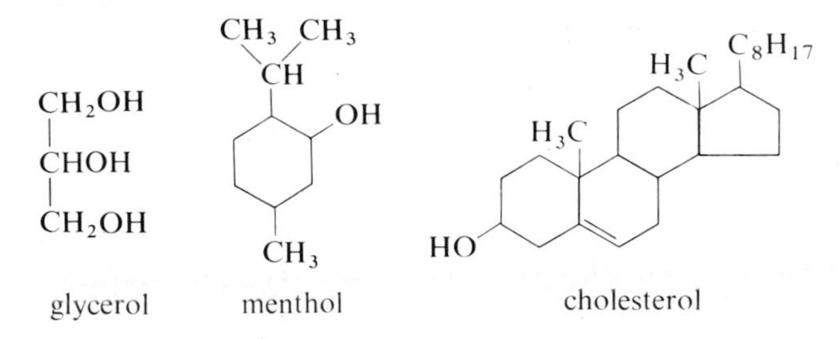

glycerol menthol cholesterol

To name ethers, list the names of the groups attached to oxygen in alphabetical order before the name "ether." In naming simple ethers, the prefix di- is sometimes not used.

$CH_3{-}O{-}CH_3$ $CH_3CH_2{-}O{-}CH_2CH_3$ $C_6H_{11}{-}O{-}CH_3$

dimethyl ether (methyl ether) diethyl ether (ethyl ether) cyclohexylmethyl ether

Heterocyclic ethers, that is, cyclic compounds in which the ether oxygen is one of the atoms in a ring, are given special names.

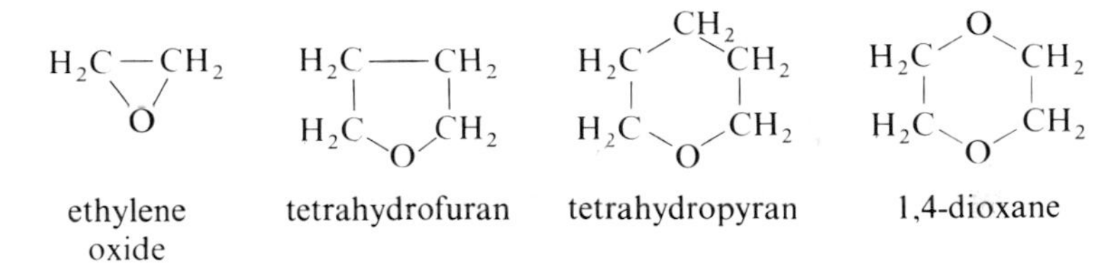

ethylene oxide tetrahydrofuran tetrahydropyran 1,4-dioxane

In ethers of more complex structure where there is no simple name for one of the attached alkyl groups or where the ether is only one of several functional groups, it may be necessary to indicate the $-OR$ group as an alkoxy group; methoxy- for $-OCH_3$, ethoxy- for $-OCH_2CH_3$, and so on.

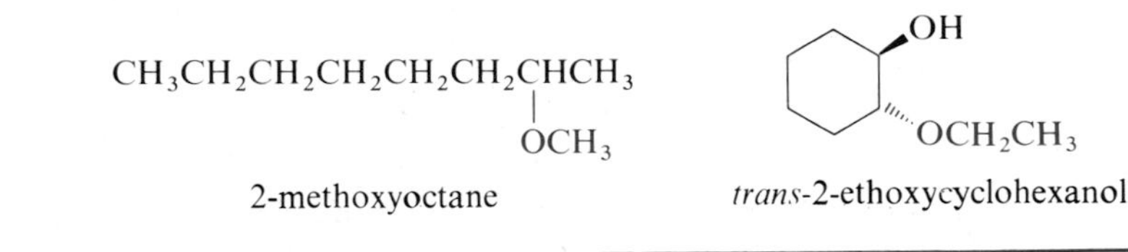

2-methoxyoctane *trans*-2-ethoxycyclohexanol

Example 5.4 Name the following ethers.

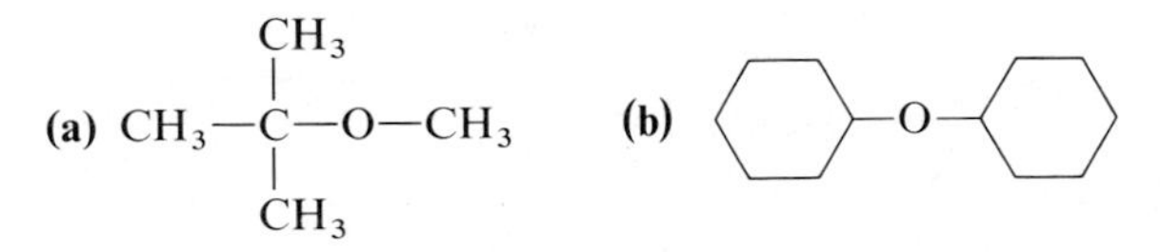

Solution

(a) Both alkyl groups have IUPAC names. Therefore this ether is named *tert*-butyl methyl ether.

(b) There are two cyclohexyl groups attached to oxygen. Therefore this ether is named dicyclohexyl ether.

PROBLEM 5.4 Name the following ethers.

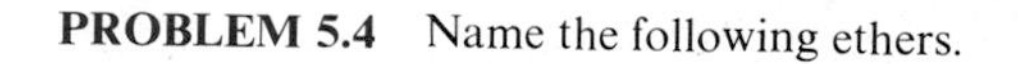

(a) $CH_3-\underset{\displaystyle CH_3}{\underset{|}{CH}}-CH_2-O-CH_2-CH_3$

(b) Cyclohexane ring bearing OCH_2CH_3

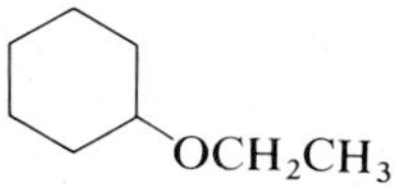

5.3 PHYSICAL PROPERTIES OF ALCOHOLS AND ETHERS

Because of the presence of the C—O—H group, alcohols are polar compounds. Oxygen is more electronegative than either carbon or hydrogen. Thus there are partial positive charges on carbon and hydrogen and a partial negative charge on oxygen (Figure 5.5).

In the pure state, there is extensive hydrogen bonding between the partially negative oxygen atoms and the partially positive hydrogen atoms. Figure 5.6 shows the association of ethanol molecules in the liquid state by hydrogen bonding.

Ethers are also polar molecules, with oxygen bearing a partial negative charge and each of the attached carbon atoms bearing a partial positive charge. Association by hydrogen bonding is not possible for ethers because there is no partially positive hydrogen atom on oxygen to participate in the formation of hydrogen bonds (Figure 5.7).

Table 5.1 lists the boiling points and solubilities in water for several groups of alcohols, ethers, and hydrocarbons of similar molecular weights.

Of the three classes of compounds compared in this table, alcohols have the highest boiling points because of hydrogen bonding between polar —OH groups. There is little interaction between ether molecules in the pure state. Therefore the boiling points of ethers are close to those of nonpolar hydrocarbons of comparable molecular weight. The effect of hydrogen bonding is dramatically illustrated by comparing the boiling points of ethanol (bp 78°C) and its structural isomer dimethyl ether (bp −24°C). Each compound has the same molecular formula, C_2H_6O, and hence the same molecular weight. The difference in boiling points between them is due to the presence of polar O—H groups in the alcohol. Alcohol molecules interact by hydrogen bonding; ether molecules do not. Compare also the boiling points of 1-propanol (97°C) and ethyl methyl ether (11°C); and 1-butanol (117°C) and diethyl ether (35°C).

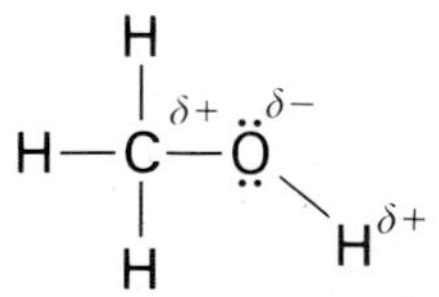

$$H-\overset{\displaystyle H}{\underset{\displaystyle H}{C}}{}^{\delta+}-\ddot{\underset{\cdot\cdot}{O}}{}^{\delta-}-H^{\delta+}$$

Figure 5.5 *Polarity of the C—O—H bonds in alcohols.*

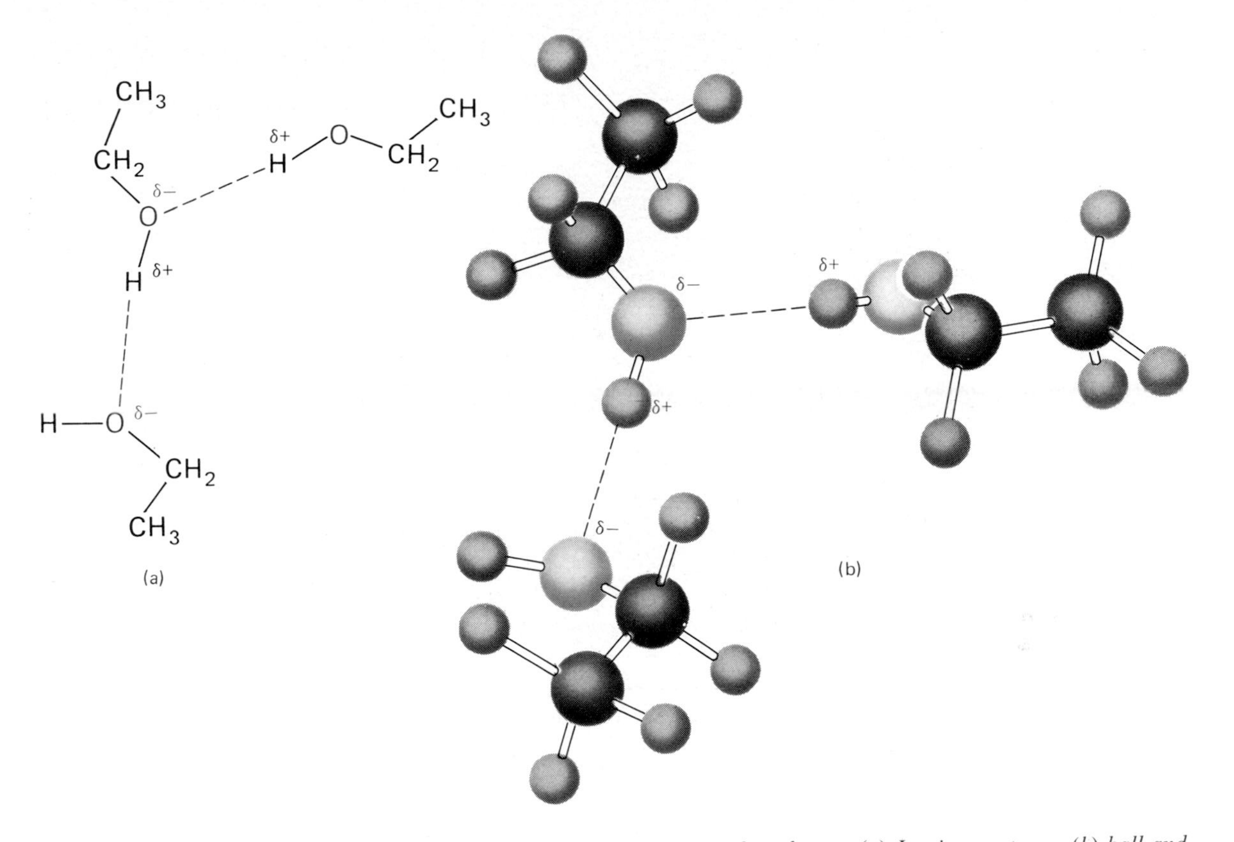

Figure 5.6 *The association of ethanol in the liquid state. (a) Lewis structures; (b) ball-and-stick models.*

The presence of additional hydroxyl groups in a molecule further increases the significance of hydrogen bonding, as you can see by comparing the boiling points of hexane (bp 69°C), 1-pentanol (bp 138°C), and 1,4-butanediol (bp 230°C). All three compounds have approximately the same molecular weight, but very different boiling points.

Boiling points of alcohols and ethers increase with increasing molecular weight due to the increased dispersion forces between hydrocarbon portions

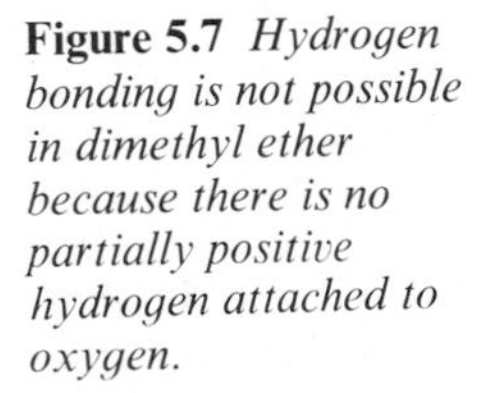

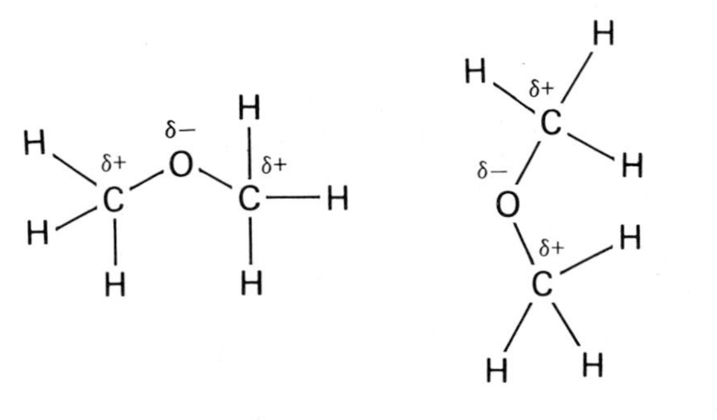

Figure 5.7 *Hydrogen bonding is not possible in dimethyl ether because there is no partially positive hydrogen attached to oxygen.*

Table 5.1 Boiling points and solubilities in water of several groups of alcohols, ethers, and hydrocarbons of similar molecular weight.

Structural formula	*Name*	*Molecular weight*	*bp* (C°)	*Solubility in water* ($g/100\ g\ H_2O$)
CH_3OH	methanol	32	65	soluble (∞)
CH_3CH_3	ethane	30	−89	insoluble
CH_3CH_2OH	ethanol	46	78	soluble (∞)
CH_3OCH_3	dimethyl ether	46	−24	soluble (7 g/100 g)
$CH_3CH_2CH_3$	propane	44	−42	insoluble
$CH_3CH_2CH_2OH$	1-propanol	60	97	soluble (∞)
$CH_3CH_2OCH_3$	ethyl methyl ether	60	11	soluble
$CH_3CH_2CH_2CH_3$	butane	58	0	insoluble
$CH_3CH_2CH_2CH_2OH$	1-butanol	74	117	soluble (8 g/100 g)
$CH_3CH_2OCH_2CH_3$	diethyl ether	74	35	soluble (8 g/100 g)
$CH_3CH_2CH_2CH_2CH_3$	pentane	72	36	insoluble
$CH_3CH_2CH_2CH_2CH_2OH$	1-pentanol	88	138	slightly soluble (2.3 g/100 g)
$CH_3CH_2CH_2CH_2OCH_3$	butyl methyl ether	88	71	slightly soluble
$HOCH_2CH_2CH_2CH_2OH$	1,4-butanediol	90	230	soluble (∞)
$CH_3OCH_2CH_2OCH_3$	ethylene glycol dimethyl ether	90	84	soluble (∞)
$CH_3CH_2CH_2CH_2CH_2CH_3$	hexane	88	69	insoluble

of the molecules. To see this, compare the boiling points of ethanol, 1-propanol, and 1-butanol.

Because they can interact by hydrogen bonding with water, alcohols and ethers are more soluble in water than are alkanes of comparable molecular weight (Figure 5.8). Methanol, ethanol, and 1-propanol are soluble in water in all proportions.

As molecular weight increases, the physical properties of alcohols and ethers become more like those of hydrocarbons of comparable molecular weight. Alcohols of higher molecular weight are much less soluble in water because of the increase in size of the hydrocarbon portion of the molecule. For example, 1-decanol is insoluble in water but soluble in ethanol and in nonpolar hydrocarbon solvents such as benzene and hexane.

Example 5.5

Arrange the following compounds in order of increasing boiling points.

$$CH_3{-}CH_2{-}OH \qquad CH_3{-}CH_2{-}Cl \qquad CH_3{-}CH_2{-}CH_3$$

Solution

Propane and ethanol have similar molecular weights. However, propane is a nonpolar hydrocarbon and the only interactions between molecules in the pure liquid are dispersion

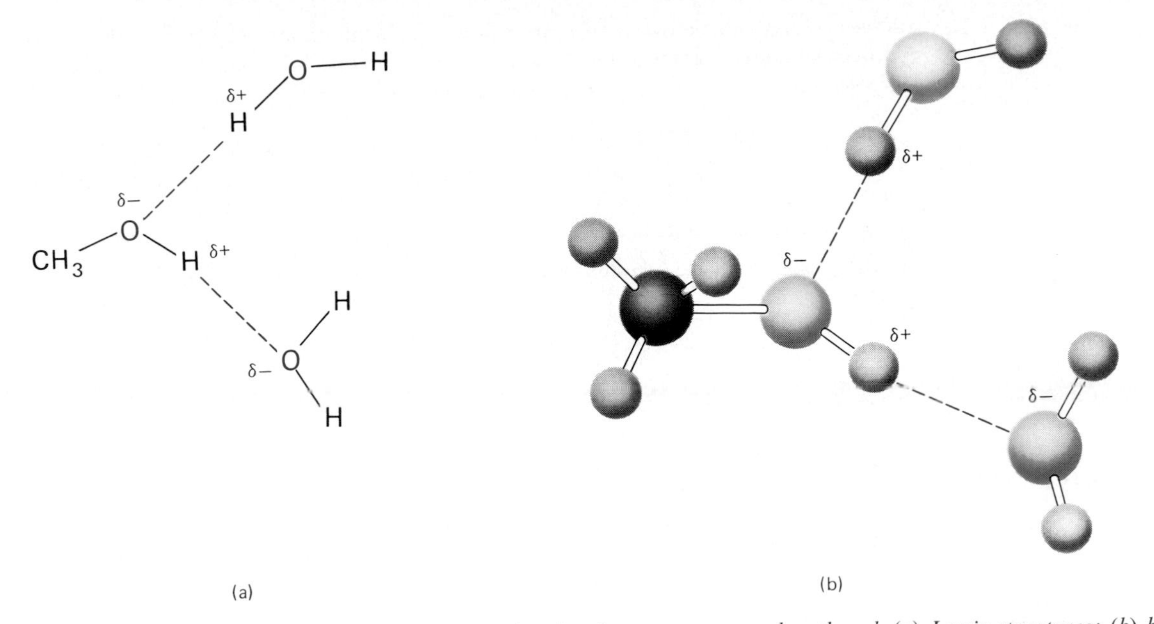

Figure 5.8 *Hydrogen bonding between water and methanol. (a) Lewis structures; (b) ball-and-stick models.*

forces (review Section 2.8). Ethanol is a polar compound and there is extensive hydrogen bonding between ethanol molecules in the pure liquid. Therefore, ethanol has a higher boiling point than propane. Chloroethane has a higher molecular weight than ethanol and is a polar molecule. However, it cannot associate by hydrogen bonding and therefore has a lower boiling point than ethanol. In order of increasing boiling points, they are

$CH_3CH_2CH_3$	CH_3CH_2Cl	CH_3CH_2OH
bp −87°C	bp 12°C	bp 78°C
(mw 44.1)	(mw 64.5)	(mw 46.1)

PROBLEM 5.5 Arrange the following compounds in order of increasing boiling points.

$$CH_3{-}O{-}CH_2{-}CH_2{-}O{-}CH_3 \qquad HO{-}CH_2{-}CH_2{-}OH$$
$$CH_3{-}O{-}CH_2{-}CH_2{-}OH$$

Example 5.6 Arrange the following compounds in order of increasing solubility in water.

$$CH_3CH_2CH_2CH_2CH_2CH_3 \qquad CH_3OCH_2CH_2OCH_3 \qquad CH_3CH_2OCH_2CH_3$$

Solution Water is a polar solvent. Hexane, C_6H_{14}, is a nonpolar hydrocarbon and therefore has the lowest solubility in water. Both diethyl ether and ethylene glycol dimethyl ether (1,2-dimethoxyethane) are polar compounds due to the presence of C—O—C bonds, and each interacts with water molecules by hydrogen bonding. Ethylene glycol dimethyl ether is more soluble in water than diethyl ether because the diether has more sites within the molecule for hydrogen bonding with water molecules. The water solubilities of these substances are given in Table 5.1.

$CH_3CH_2CH_2CH_2CH_2CH_3$	$CH_3CH_2OCH_2CH_3$	$CH_3OCH_2CH_2OCH_3$
insoluble	8 g/100 g water	soluble in all proportions

PROBLEM 5.6 Arrange the following compounds in order of increasing solubility in water.

$$Cl-CH_2-CH_2-Cl \qquad CH_3-CH_2-CH_2-OH \qquad CH_3-CH_2-O-CH_2-CH_3$$

5.4 PREPARATION OF ALCOHOLS

We have already seen two common laboratory methods for the preparation of alcohols, namely acid-catalyzed hydration of alkenes (Section 3.6C) and hydroboration of alkenes followed by reaction with hydrogen peroxide (Section 3.6E). These methods can be used to form isomeric alcohols as illustrated in the following example.

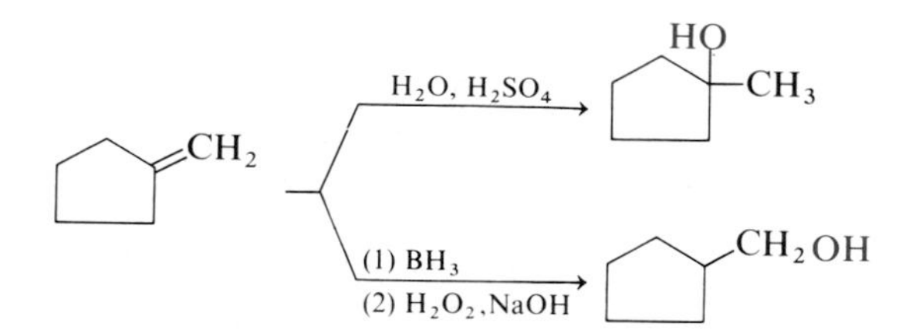

Methanol (methyl alcohol), commonly called wood alcohol, used to be prepared by the destructive distillation of wood, at least until 1923. When wood is heated to temperatures above 250°C without access to air, it decomposes to charcoal and a volatile fraction which partially condenses on cooling. This condensate contains methanol, acetic acid, and traces of acetone. At the present time, methanol is made on a large scale synthetically by reduction of carbon monoxide over a copper catalyst.

$$CO + 2H_2 \xrightarrow[260^\circ C,\ 100\ atm]{catalyst} CH_3OH$$

Ethanol (ethyl alcohol), or simply "alcohol" in nonscientific language, has been prepared since antiquity by the fermentation of sugars and starches by yeast, and is the basis for the preparation of alcoholic "spirits":

$$\underset{\text{glucose}}{C_6H_{12}O_6} \xrightarrow{enzymes} 2CH_3CH_2OH + 2CO_2$$

The sugars for the fermentation process come from a variety of sources: blackstrap molasses, a residue from the refining of cane sugar; various grains, hence the name "grain alcohol"; grape juice; or various vegetables. The immediate product of the fermentation process is an aqueous solution containing up to about 15% alcohol. This alcohol may be concentrated by distillation. Beverage alcohol may contain traces of flavor derived from the source (brandy from grapes, whiskeys from grains) or may be essentially flavor-free (vodka). The most important synthetic method for the preparation of ethanol is acid-catalyzed hydration of ethylene.

$$CH_2{=}CH_2 + H_2O \xrightarrow{H_3PO_4} CH_3{-}CH_2OH$$

Whatever the method of preparation of ethanol, it is first obtained as a mixture with water, which is then concentrated by distillation. Ordinary commercial ethanol is a mixture of 95% alcohol and 5% water which cannot be further purified by distillation. Absolute or 100% ethanol is prepared from 95% ethanol by techniques that selectively remove water from the mixture.

5.5 REACTIONS OF ALCOHOLS

Alcohols undergo a variety of important reactions, including conversion to alkyl halides, dehydration, and oxidation to aldehydes, ketones, and carboxylic acids. Thus, alcohols are valuable starting materials for the synthesis of other compounds.

A. REACTION WITH ACTIVE METALS

Like water, alcohols react with Li, Na, K, Mg, and other active metals to liberate hydrogen and form metal alkoxides.

$$2H_2O + 2Na \longrightarrow 2Na^+OH^- + H_2$$

$$2CH_3OH + 2Na \longrightarrow 2CH_3O^-Na^+ + H_2$$

$$2CH_3CH_2OH + Mg \longrightarrow (CH_3CH_2O^-)_2Mg^{2+} + H_2$$

Metal alkoxides are bases comparable in strength to sodium hydroxide. The following alkoxides are commonly used in organic reactions requiring a strong base in a nonaqueous solvent.

$CH_3O^-Na^+$	$CH_3CH_2O^-Na^+$	$CH_3C(CH_3)_2O^-K^+$
sodium methoxide	sodium ethoxide	potassium *tert*-butoxide

B. BASICITY OF ALCOHOLS

In the presence of strong acids, the oxygen atom of water behaves as a base and accepts a proton to form an oxonium ion. Similarly, alcohols behave as bases and react with protons to form oxonium ions.

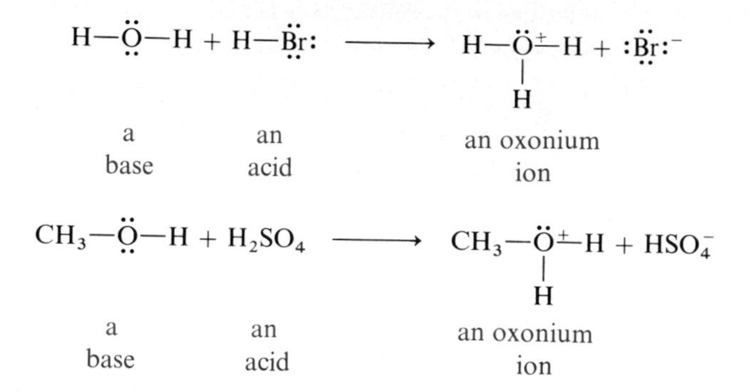

We can look at these as acid-base reactions with the oxygen atom of the ROH group acting as the base and HBr, H_3PO_4, or H_2SO_4 acting as acids. Such proton-transfer reactions are fundamental to all acid-catalyzed reactions of alcohols. Alternatively, we can look on them as reactions between a nucleophile and an electrophile. Oxygen is the nucleophile: it donates a pair of electrons to another atom to form a new covalent bond (in this case a new O—H bond). The proton is an electrophile: it is electron-deficient and accepts a pair of electrons from another atom in forming the new covalent bond.

Ethers also react with strong acids to form oxonium ions.

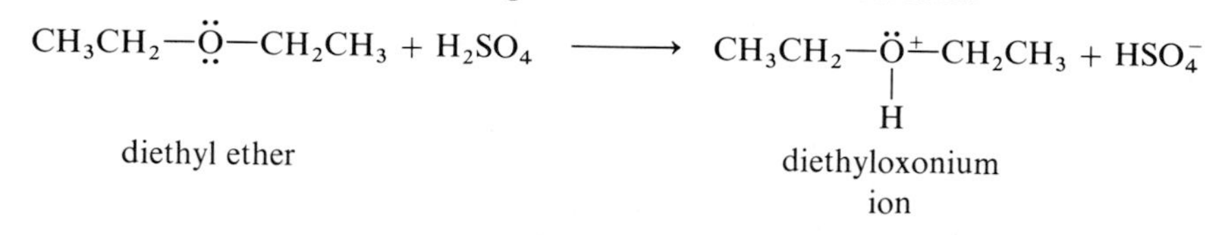

C. OXIDATION OF ALCOHOLS TO ALDEHYDES, KETONES, OR CARBOXYLIC ACIDS

Primary alcohols can be oxidized to aldehydes. The most commonly used oxidizing agents for this purpose are $K_2Cr_2O_7$ and CrO_3, each containing Cr(VI). Usually, $K_2Cr_2O_7$ dissolved in aqueous acid is added to a solution of the alcohol dissolved in acetone or a mixture of water and acetone.

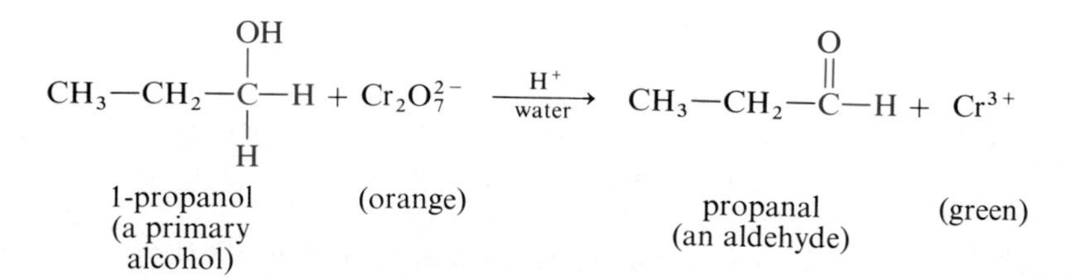

Aldehydes are very easily oxidized to carboxylic acids in a reaction that involves conversion of the aldehyde C—H bond to a C—OH bond.

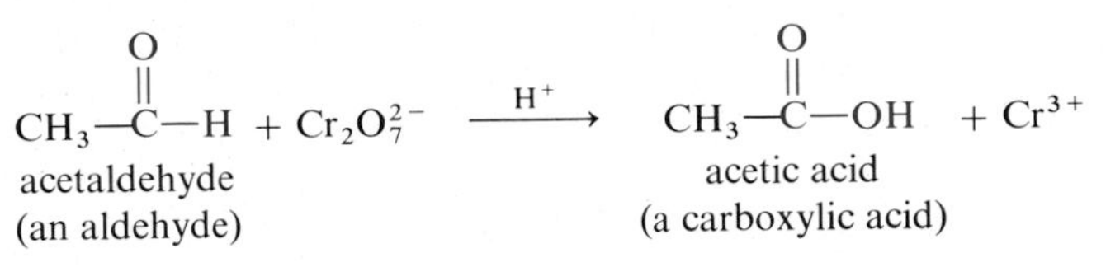

Under carefully controlled reaction conditions, it is possible to oxidize primary alcohols to aldehydes in good yield. However, unless special precautions are taken, primary alcohols are oxidized directly to carboxylic acids. In this oxidation, the aldehyde is an intermediate.

$$\underset{\text{ethanol}}{CH_3CH_2OH} + Cr_2O_7^{2-} \longrightarrow \left[CH_3-\overset{\overset{\displaystyle O}{\|}}{C}-H \right] \longrightarrow \underset{\text{acetic acid}}{CH_3-\overset{\overset{\displaystyle O}{\|}}{C}-OH} + Cr^{3+}$$

Oxidation of a secondary alcohol yields a ketone. This oxidation involves removal of hydrogen atoms from carbon and oxygen and the formation of a carbon-oxygen double bond.

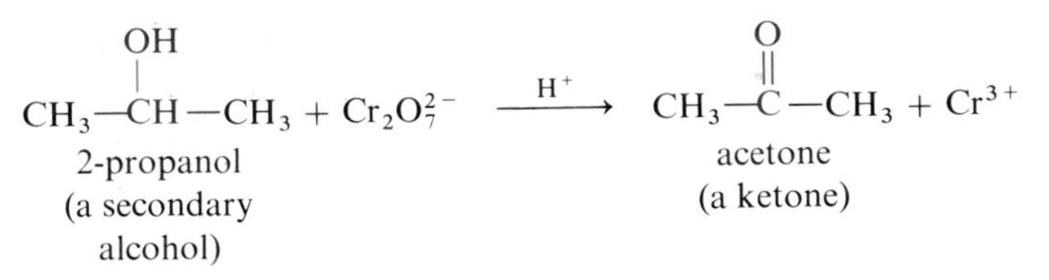

Oxidation of menthol yields menthone.

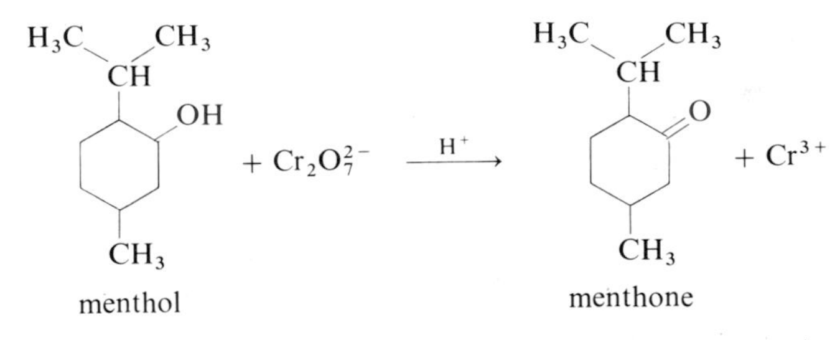

Tertiary alcohols are resistant to oxidation because the carbon bearing the —OH has no hydrogen atom on it. It is already bonded to three other carbon atoms and therefore cannot form a carbon-oxygen double bond.

$$\text{(cyclopentane ring)}\!<\!{}^{CH_3}_{OH} + Cr_2O_7^{2-} \xrightarrow{H^+} \text{no oxidation}$$

In the body, the substance used in the oxidation of most primary and secondary alcohols, as well as most aldehydes, is nicotinamide adenine dinucleotide, abbreviated NAD$^+$. The reactive group of NAD$^+$ is a substituted pyridine ring (Section 6.8). In the following structural formula, the remainder of this large organic molecule is abbreviated by the symbol R. The pyridine ring of NAD^+ accepts two electrons and one proton to form NADH.

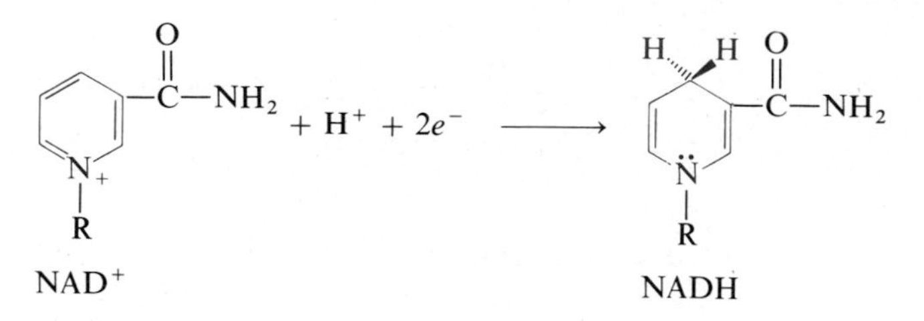

The following examples illustrate how NAD^+ is used in the oxidation of a primary alcohol, a secondary alcohol, and an aldehyde. Each of these oxidations is catalyzed by a specific enzyme.

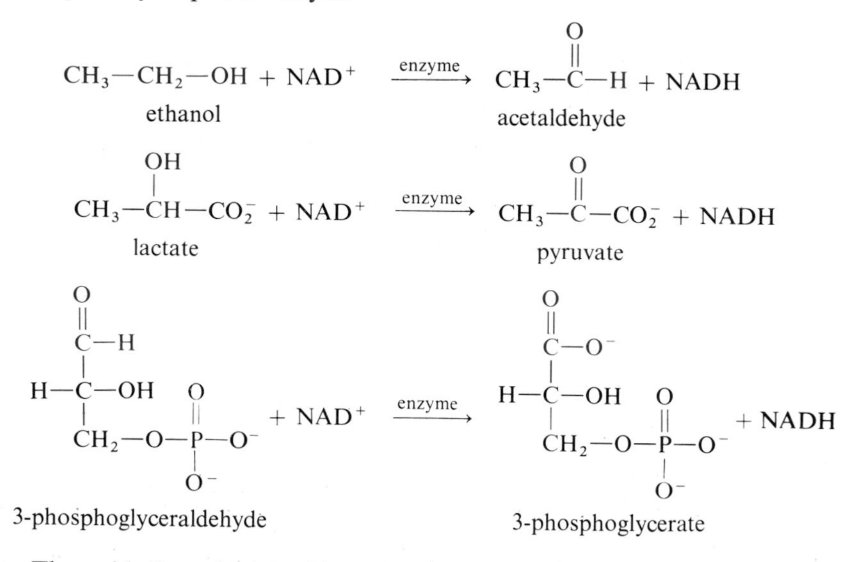

The oxidation of ethanol by NAD^+ occurs in the liver and is the first step in the detoxification of ingested alcohol. The oxidation of lactate to pyruvate also occurs to a large extent in the liver. During periods of strenuous activity, muscle cells metabolize glucose to produce energy, and in the process lactate accumulates. As lactate accumulates, muscle tissue becomes fatigued. To relieve this muscle fatigue, lactate is carried by the bloodstream to the liver, where it is oxidized to pyruvate which is in turn used to produce more energy. The final example, the oxidation of 3-phosphoglyceraldehyde to 3-phosphoglycerate, is a key step in the metabolism of glucose.

D. DEHYDRATION OF ALCOHOLS TO ALKENES

An alcohol can be converted to an alkene by elimination of a molecule of water from adjacent carbon atoms. Elimination of a molecule of water is called dehydration. In the laboratory, dehydration of an alcohol is most usually brought about by heating the alcohol with 85% phosphoric acid or concentrated sulfuric acid at temperatures from 100° to 200°C. For example, dehydration of ethanol yields ethylene.

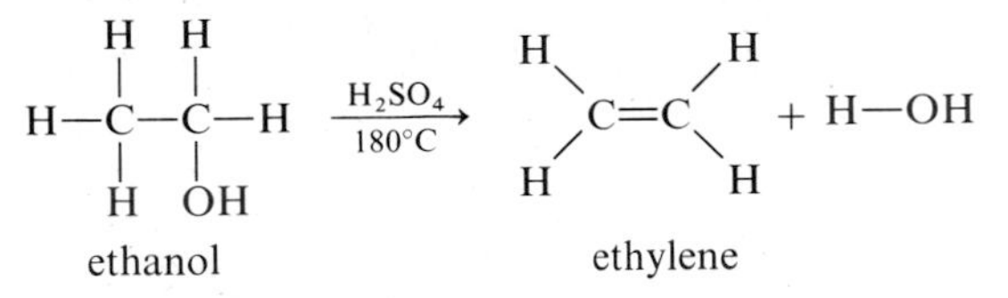

In this process, two sigma bonds (C—H and C—OH) are broken and one pi bond (C=C) and one sigma bond (H—OH) are formed. Dehydration of cyclohexanol in the presence of 85% phosphoric acid yields cyclohexene.

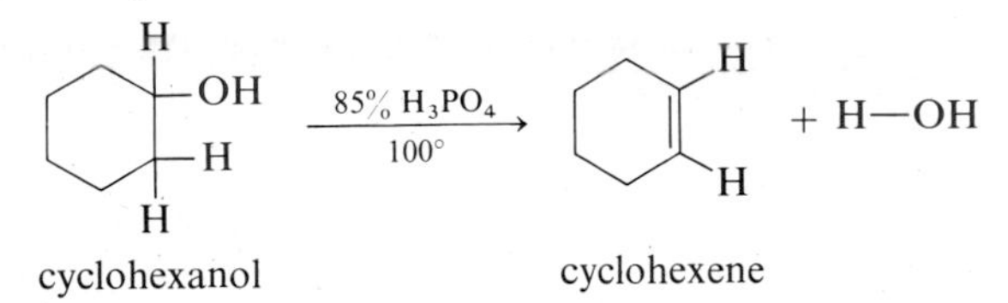

Often, acid-catalyzed dehydration of alcohols leads to the formation of isomeric alkenes. In the dehydration of 2-butanol, the —H and —OH may be removed from carbons 1 and 2 to form 1-butene, or they may be removed from carbons 2 and 3 to form 2-butene. In practice, acid-catalyzed dehydration of 2-butanol yields 80% 2-butene and 20% 1-butene.

$$\underset{\text{2-butanol}}{\overset{1}{C}H_3\overset{2}{C}H(OH)\overset{3}{C}H_2\overset{4}{C}H_3} \xrightarrow[\text{heat}]{85\%\ H_3PO_4} \underset{\substack{\text{2-butene} \\ 80\%}}{CH_3CH{=}CHCH_3} + \underset{\substack{\text{1-butene} \\ 20\%}}{CH_2{=}CHCH_2CH_3} + H_2O$$

Where it is possible to obtain isomeric alkenes from the acid-catalyzed dehydration of an alcohol, the alkene having more substituents on the double bond generally predominates.

Example 5.7

Draw structural formulas for the alkenes formed on acid-catalyzed dehydration of the following alcohols. Predict which alkene is the major product and which is the minor product.

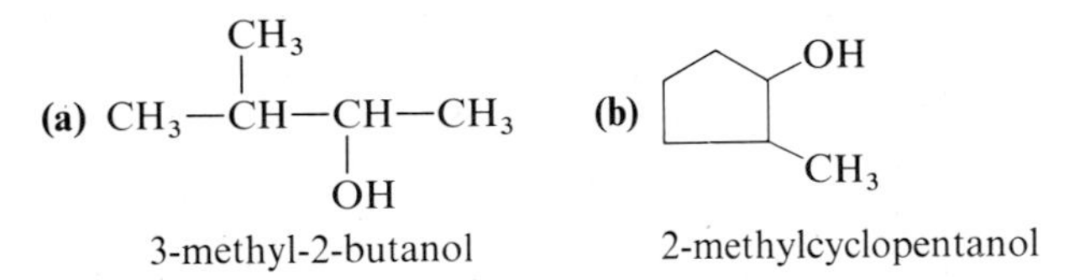

Solution

(a) Dehydration of 3-methyl-2-butanol can result in loss of —H and —OH from either carbons 1 and 2 to produce 3-methyl-1-butene, or from carbons 2 and 3 to produce 2-methyl-2-butene.

$$\underset{\text{3-methyl-2-butanol}}{\overset{4}{C}H_3\text{—}\overset{3}{C}H(CH_3)\text{—}\overset{2}{C}H(OH)\text{—}\overset{1}{C}H_3} \xrightarrow{85\%\ H_3PO_4} \underset{\substack{\text{3-methyl-1-butene} \\ \textit{(minor product)}}}{CH_3\text{—}CH(CH_3)\text{—}CH{=}CH_2} + \underset{\substack{\text{2-methyl-2-butene} \\ \textit{(major product)}}}{CH_3\text{—}C(CH_3){=}CH\text{—}CH_3} + H_2O$$

3-Methyl-1-butene has one alkyl substituent (an isopropyl group) on the double bond. 2-Methyl-2-butene has three alkyl substituents (three methyl groups) on the double bond. Therefore, predict that 2-methyl-2-butene is the major product and that 3-methyl-1-butene is the minor product.

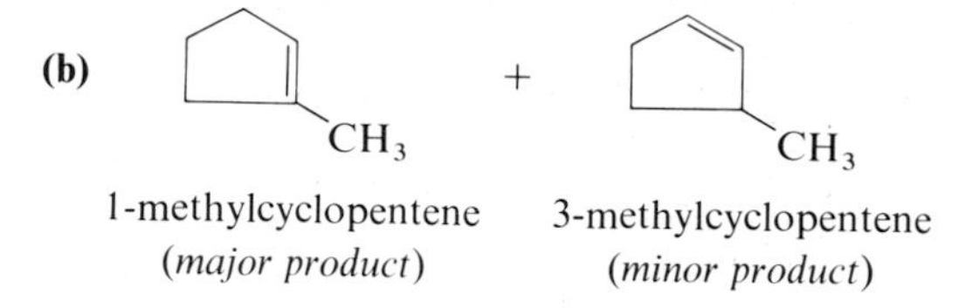

The major product, 1-methylcyclopentene, has three alkyl substituents on the carbon-carbon double bond. The minor product, 3-methylcyclopentene, has only two substituents on the double bond.

PROBLEM 5.7 Draw structural formulas for the alkenes formed on acid-catalyzed dehydration of the following alcohols. Where two alkenes are possible, predict which is the major product and which is the minor product.

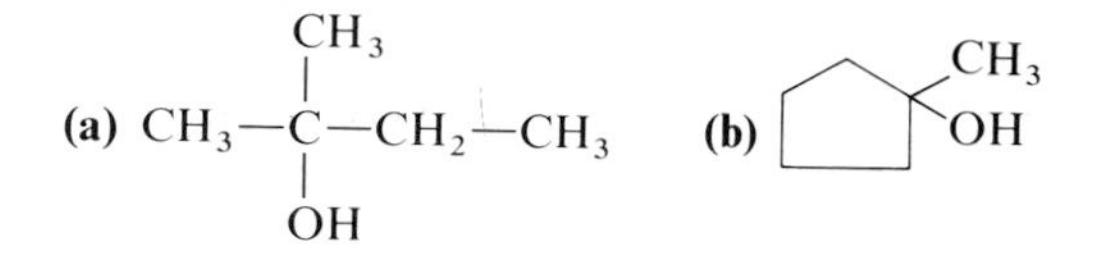

We have seen in this section that alcohols can be dehydrated in the presence of an acid catalyst to an alkene, and in Section 3.6C, we have seen that alkenes can be hydrated in the presence of an acid catalyst to an alcohol.

$$CH_3-\underset{HO}{CH}-\underset{H}{CH_2} \xrightarrow{H_2SO_4} CH_3-CH{=}CH_2 + HO-H \qquad \textit{a dehydration reaction}$$

$$CH_3-CH{=}CH_2 + HO-H \xrightarrow{H_2SO_4} CH_3-\underset{HO}{CH}-\underset{H}{CH_2} \qquad \textit{a hydration reaction}$$

In Section 3.6C we proposed a mechanism for acid-catalyzed hydration of alkenes to alcohols. This mechanism involved electrophilic attack on the pi bond of the alkene by H^+ (an electrophile) to generate a carbocation intermediate, and then reaction of the carbocation with a molecule of water to give the alcohol. The mechanism for acid-catalyzed dehydration of an alcohol to an alkene is just the reverse of this mechanism. In Step 1, the electrophile H^+ reacts with one of the unshared pairs of electrons on oxygen to form a new sigma bond between oxygen and hydrogen. The intermediate produced is called an oxonium ion.

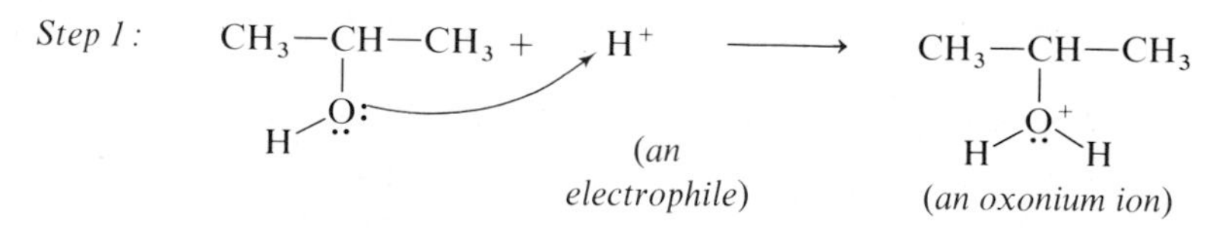

In Step 2, the C—O bond breaks to give a carbocation and a molecule of water.

Step 2:

$$CH_3-\underset{\underset{H\ \ \ \ H}{\overset{|}{\ddot{O}^+}}}{CH}-CH_3 \longrightarrow \underset{\text{a carbocation}}{CH_3-\overset{+}{C}H-CH_3} + H-\ddot{\underset{..}{O}}-H$$

Finally, in Step 3, a C—H bond on an adjacent carbon breaks to form H^+ and the pair of electrons of the C—H bond is used to form the pi bond of the alkene.

Step 3:

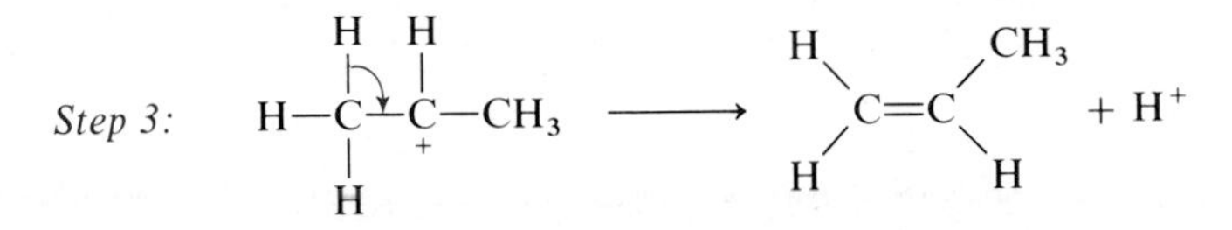

Since the reactive intermediate in this mechanism is a carbocation, we predict that the relative ease of dehydration of alcohols is the same as the ease of formation of carbocations. Tertiary alcohols dehydrate more readily than secondary alcohols because tertiary carbocations are formed more readily than secondary carbocations. For the same reason, secondary alcohols undergo acid-catalyzed dehydration more readily than primary alcohols. Thus, the ease of dehydration of alcohols is

$$3^\circ \text{ alcohols} > 2^\circ \text{ alcohols} > 1^\circ \text{ alcohols}$$

In Section 3.6C we discussed the acid-catalyzed hydration of alkenes to yield alcohols. In this section we discussed the dehydration of alcohols to yield alkenes. In fact, the hydration-dehydration reactions are reversible:

$$\gt C=C \lt + H_2O \rightleftharpoons -\underset{H}{\overset{|}{\underset{|}{C}}}-\underset{OH}{\overset{|}{\underset{|}{C}}}-$$

Large amounts of water favor alcohol formation, whereas operation under experimental conditions where water is removed favors formation of alkenes. Depending on the conditions, it is possible to use the hydration-dehydration equilibrium to prepare either alcohols or alkenes, each in high yields.

E. CONVERSION TO ALKYL HALIDES

An alcohol can be converted to an alkyl halide by reaction with HCl, HBr, or HI. These are called substitution reactions because one atom (a halogen) is substituted for another (—OH). We will discuss the mechanisms of these substitution reactions in the next section.

$$CH_3-\underset{H}{\overset{H}{\underset{|}{\overset{|}{C}}}}-OH + H-Br \longrightarrow CH_3-\underset{H}{\overset{H}{\underset{|}{\overset{|}{C}}}}-Br + H-OH$$

It has been found that tertiary alcohols react more rapidly than secondary alcohols, and that secondary alcohols react much more readily than primary alcohols. For example, mixing 2-methyl-2-propanol (a tertiary alcohol) with HCl for a few minutes at room temperature results in the conversion of the alcohol to 2-chloro-2-methylpropane. On the other hand, 1-butanol, a primary alcohol, must be heated with concentrated HCl to convert it to 1-chlorobutane.

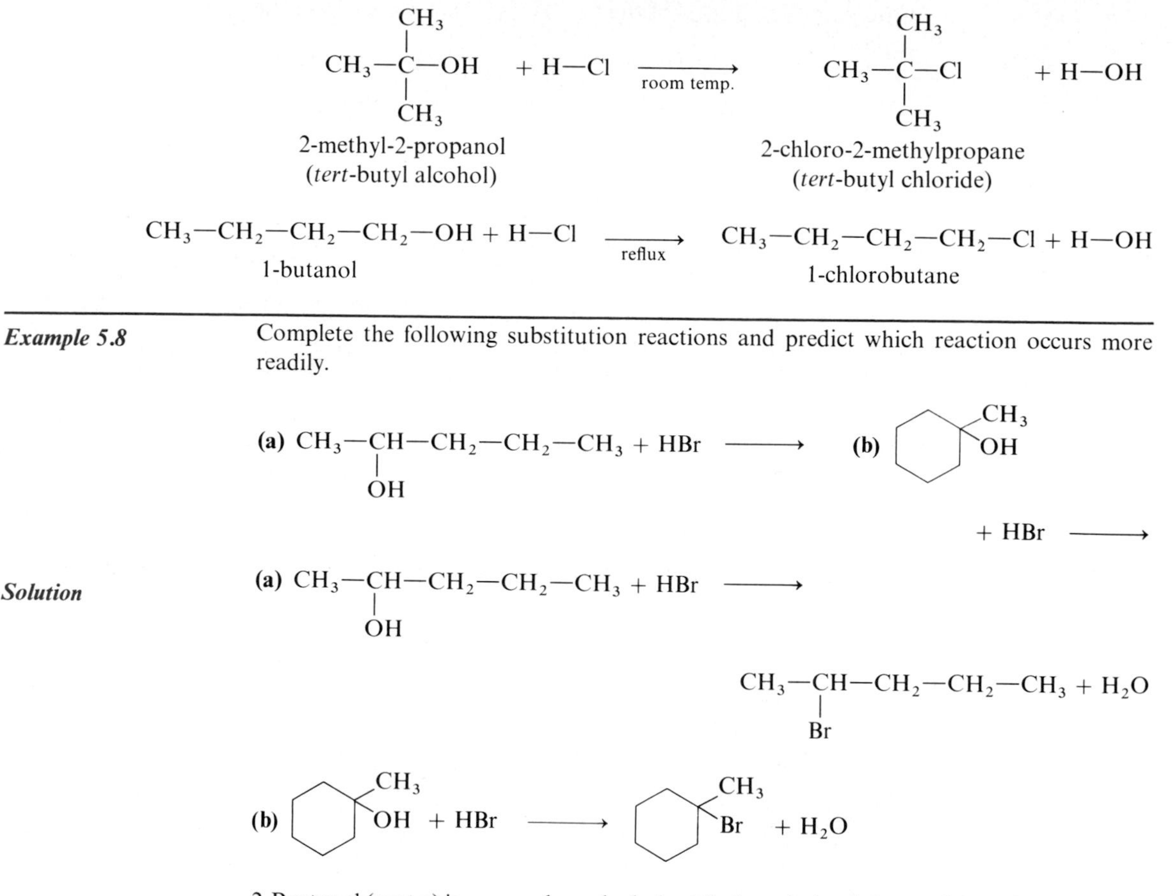

Example 5.8 Complete the following substitution reactions and predict which reaction occurs more readily.

(a) $CH_3-CH(OH)-CH_2-CH_2-CH_3 + HBr \longrightarrow$

(b) 1-methylcyclohexanol + HBr $\longrightarrow$

Solution

(a) $CH_3-CH(OH)-CH_2-CH_2-CH_3 + HBr \longrightarrow CH_3-CH(Br)-CH_2-CH_2-CH_3 + H_2O$

(b) 1-methylcyclohexanol + HBr $\longrightarrow$ 1-bromo-1-methylcyclohexane + H_2O

2-Pentanol (part a) is a secondary alcohol, while 1-methylcyclohexanol (part b) is a tertiary alcohol. Since tertiary alcohols react more readily with HBr than secondary alcohols, predict that reaction (b) occurs more readily than reaction (a).

PROBLEM 5.8 Complete the following substitution reactions and predict which reaction occurs more readily.

(a) $CH_3-C(CH_3)(OH)-CH_2-CH_3 + HI \longrightarrow$ (b) $CH_3-CH(CH_3)-CH(OH)-CH_3 + HI \longrightarrow$

A second method for the synthesis of alkyl chlorides is the use of thionyl chloride, $SOCl_2$, a particularly useful reagent because the by-products, HCl and SO_2, are given off as gases.

$$CH_3(CH_3)_5CH_2OH + SOCl_2 \xrightarrow{25^\circ} CH_3(CH_2)_5CH_2Cl + SO_2 + HCl$$

1-heptanol → 1-chloroheptane

The phosphorus halides, PCl_3, PBr_3, and PI_3, can also be used to convert alcohols to alkyl halides. The by-product of the reaction with phosphorus tribromide is the high-boiling phosphorus acid.

$$3CH_3CH(OH)CH_3 + PBr_3 \longrightarrow 3CH_3CH(Br)CH_3 + H_3PO_3$$

phosphorus tribromide; phosphorous acid

5.6 NUCLEOPHILIC SUBSTITUTION AT A SATURATED CARBON

In the previous section, we saw several examples of reactions in which one atom or group of atoms displaces another. In the following example, —Cl displaces —OH.

$$CH_3-C(CH_3)_2-OH + H^+ + Cl^- \longrightarrow CH_3-C(CH_3)_2-Cl + H-OH$$

In another displacement reaction, an alkyl halide can be converted to an alcohol by reaction in water or aqueous base.

$$CH_2{=}CH-CH_2-Cl + OH^- \xrightarrow{NaOH} CH_2{=}CH-CH_2-OH + Cl^-$$

3-chloropropene (allyl chloride) → 2-propenol (allyl alcohol)

These reactions are specific examples of a general reaction class known as nucleophilic substitution at a saturated carbon atom.

In the study of nucleophilic substitution at a saturated carbon atom, two general types of experiments have been particularly useful—kinetic studies and stereochemical studies. Kinetic studies measure reaction rates and can often tell us something about the timing of steps in the reaction sequence. Stereochemical studies tell us something about relationships between the configuration of the starting material and the products of a reaction. Let us examine experimental evidence from these studies and then formulate two general reaction mechanisms for nucleophilic substitution reactions.

First consider the reaction of 2-bromo-2-methylpropane (*tert*-butyl bromide) with hydroxide ion.

$$(CH_3)_3CBr + OH^- \xrightarrow{\text{aqueous ethanol}} (CH_3)_3COH + Br^-$$

2-bromo-2-methylpropane (*tert*-butyl bromide) → 2-methyl-2-propanol (*tert*-butyl alcohol)

The rate of conversion of *tert*-butyl bromide to *tert*-butyl alcohol depends only on the concentration of *tert*-butyl bromide; it is independent of the concentration of the hydroxide ion. The rate-determining step therefore involves only the *tert*-butyl bromide. In this reaction, hydroxide ion is a nucleophile, a substance that donates a pair of electrons to another atom. In the process of donating a pair of electrons, hydroxide ion displaces bromide ion. This reaction is also unimolecular (or first order) in that only one molecular species is involved in the rate-determining step. We refer to such a reaction as S_N1 (Substitution, Nucleophilic, Unimolecular). This designation of course does not tell us how the reaction takes place, only that it is a unimolecular nucleophilic substitution at a saturated carbon atom.

The rates of nucleophilic substitutions of methyl bromide and ethyl bromide by hydroxide ion are proportional to the concentration of both alkyl halide and hydroxide ion.

$$\underset{\text{methyl bromide}}{CH_3{-}Br} + OH^- \longrightarrow \underset{\text{methanol}}{CH_3{-}OH} + Br^-$$

$$\underset{\text{ethyl bromide}}{CH_3CH_2{-}Br} + OH^- \longrightarrow \underset{\text{ethanol}}{CH_3CH_2{-}OH} + Br^-$$

The rate-determining step in each of these reactions is bimolecular (or second order) because it involves two reacting species. We refer to such reactions as S_N2 (Substitution, Nucleophilic, Bimolecular). Again, the designation S_N2 does not tell us how the reaction takes place, only that it is a bimolecular nucleophilic substitution at saturated carbon.

The rate of reaction for isopropyl bromide is directly proportional to the concentration of isopropyl bromide and hydroxide ion (S_N2) when hydroxide ion was about 1 molar, but independent of hydroxide (S_N1) when hydroxide ion concentration is low. This reaction is either S_N1 or S_N2 depending on the experimental conditions.

Studies of the stereochemistry of S_N2 reactions have revealed that if the carbon undergoing substitution is a center of chirality, the product is opposite in configuration from the starting material. For example, reaction of S-2-bromobutane with sodium hydroxide in aqueous ethanol gives R-2-butanol.

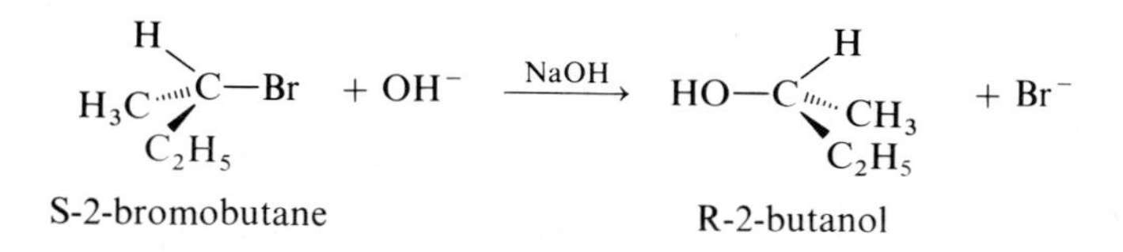

The stereochemical results of first-order substitution reactions are quite different. In an S_N1 reaction, if the carbon undergoing substitution is a center of chirality, the product is a racemic mixture. For example, reaction of (S)-3-

bromo-3-methylhexane with water in aqueous acetone gives racemic 3-methyl-3-hexanol.

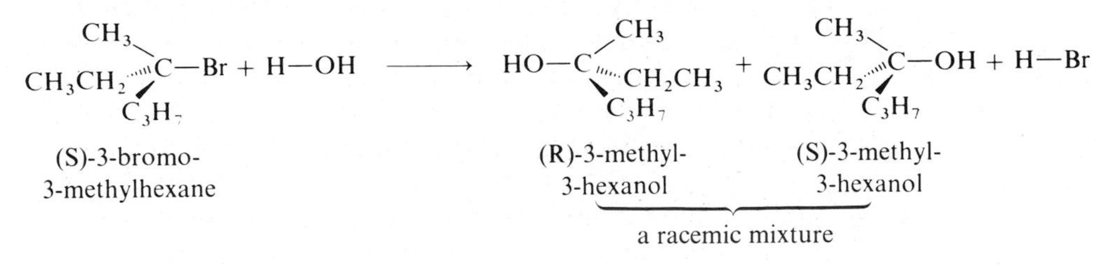

With the experimental observations from kinetic and stereochemical studies, we can now examine two different mechanisms that have been proposed for nucleophilic substitution reactions at saturated carbon. To account for S_N2 reactions, chemists propose a one-step mechanism in which the nucleophile attacks at the rear of the bond between carbon and the leaving group. We can illustrate this mechanism with the reaction of S-2-bromobutane and hydroxide ion. Hydroxide ion attacks carbon at the rear of the C—Br bond and begins to form a bond to carbon. At the same time, the carbon-bromine bond becomes progressively weaker and bromine is ejected as a bromide ion. We speak of this as a concerted or simultaneous process. During this change, the other three bonds to carbon flatten out and eventually end up on the other side of the chiral carbon (Figure 5.9).

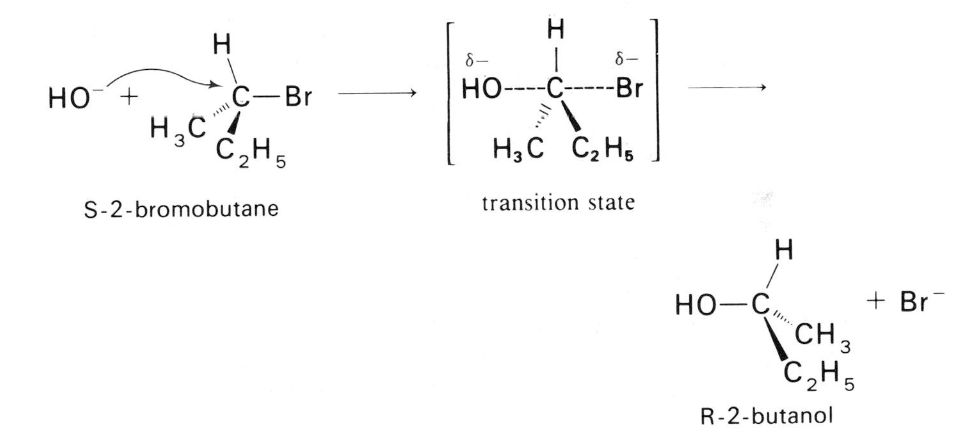

Figure 5.9 *Inversion of configuration during an S_N2 reaction.*

To account for S_N1 reactions, chemists propose a two-step mechanism. In step 1, the bond between carbon and the leaving group breaks to give a carbocation and an anion. In the second step, the carbocation reacts with a nucleophile to give the final product. Step 1 involves rupture of a covalent bond and is considerably slower than the second step which involves creation of a new covalent bond. The S_N1 mechanism is illustrated by the reaction of *tert*-butyl bromide with hydroxide ion to give *tert*-butyl alcohol and bromide ion.

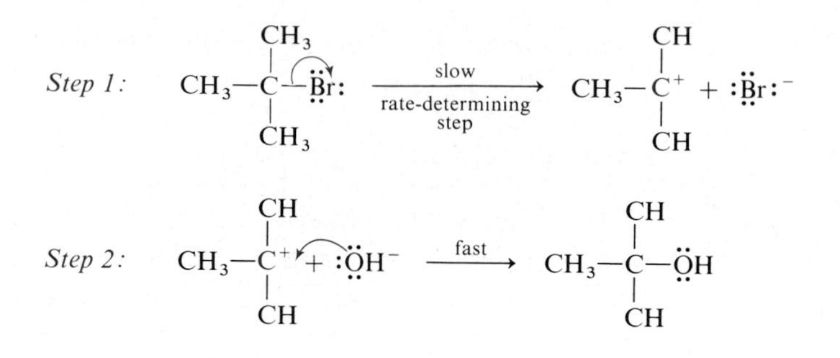

What about the stereochemistry of S_N1 reactions? Is this mechanism consistent with the observation that S_N1 reactions of optically active alkyl halides yield racemic products? The answer is yes. To see this, think about the geometry of carbocation. It is planar with bond angles of 120°, and therefore not chiral. The nucleophile can attack the carbocation with equal probability from either face of the plane of the molecule, and therefore produce a racemized product (Figure 5.10).

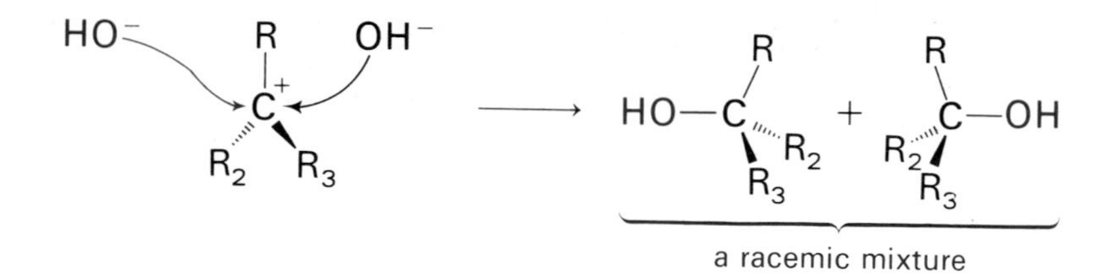

Figure 5.10 *Attack of a nucleophile from either side of a planar carbocation to produce equal amounts of two enantiomers.*

Table 5.2. Substances that can react as nucleophiles.

Nucleophile		*Name*
HS^-	strong nucleophile (increasing upward)	hydrosulfide ion
HO^-		hydroxide ion
$CH_3CH_2O^-$		ethoxide ion
CN^-		cyanide ion
Br^-		bromide ion
NH_3		ammonia
I^-		iodide ion
Cl^-		chloride ion
$CH_3-C(=O)-O^-$		acetate ion
H_2O		water
CH_3CH_2OH		ethanol

The importance of nucleophilic substitution lies in the large number of substances that can act as nucleophiles. Several of these are given in Table 5.2. Those near the top of the table react most readily in S_N2 reactions and are strong nucleophiles. Those near the bottom of the table react less rapidly in S_N2 reactions and are weak nucleophiles.

In practice, S_N2 reactions occur most readily when the leaving group is Cl^-, Br^-, or I^- and the carbon bearing the leaving group is primary.

Example 5.9 Draw Lewis structures for the following nucleophiles and write an equation to show the S_N2 reaction of each with 1-bromobutane.

(a) CN^- **(b)** NH_3 **(c)** $CH_3{-}\overset{\overset{\displaystyle O}{\|}}{C}{-}O^-$

Solution The Lewis structure of each nucleophile is shown in the following equations. A curved arrow shows a pair of electrons from the nucleophile displaicng bromine. Note that the structural formula of 1-bromobutane is drawn to emphasize the backside attack of the nucleophile on the C—Br bond.

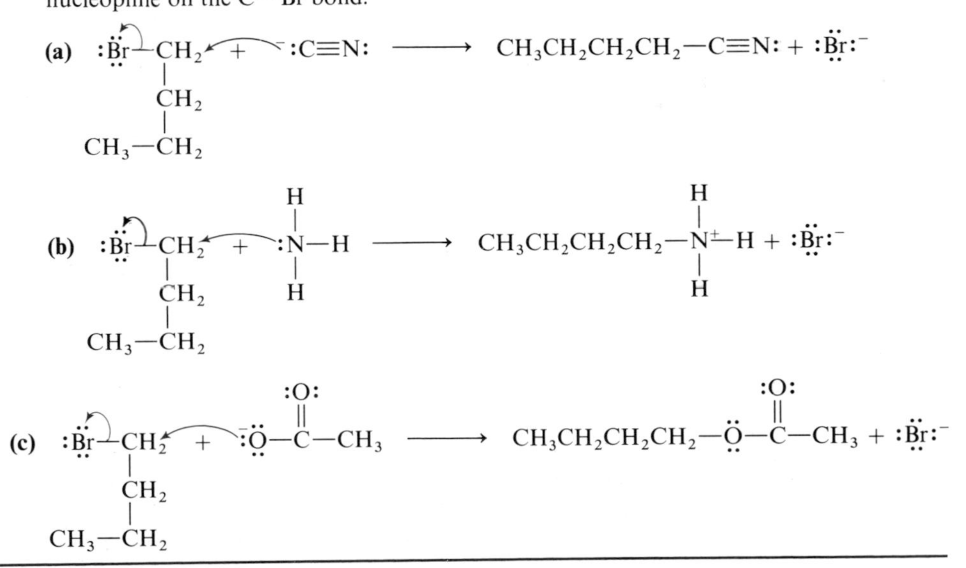

PROBLEM 5.9 Draw Lewis structures for the following nucleophiles and write an equation to show the S_N2 reaction of each with 1-chloropentane.

(a) HS^- **(b)** $CH_3CH_2O^-$ **(c)** I^-

The major side reaction during nucleophilic displacement reactions of Cl, Br, and I is dehydrohalogenation (Section 3.5).

$$\underset{\text{2-bromopropane}}{CH_3{-}\underset{\underset{\displaystyle Br}{|}}{CH}{-}CH_3} + Na^+ + OH^- \longrightarrow \underset{\text{propene}}{CH_3{-}CH{=}CH_2} + H{-}OH + Na^+ + Br^-$$

Dehydrohalogenation occurs most readily when the nucleophile is also a strong base, as for example OH^- or $CH_3CH_2O^-$, and the leaving group is on a secondary or tertiary carbon atom.

We can make the following generalizations about nucleophilic substitution versus elimination reactions.

(1) Primary halides react with good nucleophiles to give high yields of substitution products. The reactions proceed by an S_N2 mechanism. Elimination is generally only a minor side reaction.

(2) Secondary and tertiary alkyl halides react with nucleophiles to give mixtures of products corresponding to substitution and elimination. The greater the base strength of the nucleophile, the greater the percent elimination. Often, alkene mixtures formed by elimination are the major products.

Example 5.10

During the reaction of 2-iodobutane with sodium ethoxide in ethanol, substitution and elimination are competing reactions. Write structural formulas for the organic products of each reaction and state which reaction occurs more readily.

(a) $CH_3-\underset{\displaystyle I}{\underset{|}{CH}}-CH_2-CH_3 + CH_3CH_2O^-Na^+ \xrightarrow{\text{substitution}}$

(b) $CH_3-\underset{\displaystyle I}{\underset{|}{CH}}-CH_2-CH_3 + CH_3CH_2O^-Na^+ \xrightarrow{\text{elimination}}$

Solution

(a) Substitution gives ethyl *sec*-butyl ether, $CH_3-\underset{\underset{\displaystyle CH_2-CH_3}{|}}{\underset{\displaystyle O}{\underset{|}{CH}}}-CH_2-CH_3$

(b) Elimination gives 2-butene as the major product and 1-butene as the minor product:

$$CH_3-CH{=}CH-CH_3 + CH_2{=}CH-CH_2-CH_3$$

The reaction involves ethoxide ion, a good nucleophile and also a strong base. Iodine is on a secondary carbon. Therefore, elimination occurs more rapidly than substitution and the major product is the alkene mixture.

PROBLEM 5.10 During the reaction of iodoethane with potassium *sec*-butoxide, substitution and elimination are competing reactions. Write structural formulas for the organic products of each reaction and state which occurs more rapidly.

(a) $CH_3-CH_2-I + CH_3-\underset{\displaystyle O^-K^+}{\underset{|}{CH}}-CH_2-CH_3 \xrightarrow{\text{substitution}}$

(b) $CH_3-CH_2-I + CH_3-\underset{\displaystyle O^-K^+}{\underset{|}{CH}}-CH_2-CH_3 \xrightarrow{\text{elimination}}$

5.7 PREPARATION OF ETHERS

Most commercially available ethers are synthesized by acid-catalyzed dehydration of alcohols. Typical is the formation of diethyl ether from ethanol.

$$\underset{\text{ethanol}}{2CH_3CH_2OH} \xrightarrow[140°C]{H_2SO_4} \underset{\text{diethyl ether}}{CH_3CH_2{-}O{-}CH_2CH_3} + H_2O$$

Recall from Section 5.5D that acid-catalyzed dehydration of alcohols also yields alkenes. Generally, it is possible to maximize the formation of either ether or alkene by the choice of experimental conditions. For example, at 140°C the formation of ethylene from ethanol is minimized and diethyl ether is the major product. At 180°C ethylene is the major product. Since a single alcohol is used as a starting material, intermolecular dehydration of alcohols yields ethers having two identical alkyl groups. Such ethers are called symmetrical ethers. Another method for the synthesis of ethers is the Williamson ether synthesis. This scheme can be used for the synthesis of symmetrical as well as unsymmetrical ethers.

The Williamson synthesis involves nucleophilic displacement (S_N2) of an alkyl halide by a metal alkoxide. Recall from Section 5.5A that metal alkoxides are formed by reaction of an alcohol with sodium or potassium metal. An example of the Williamson ether synthesis is the preparation of methyl isopropyl ether from sodium isopropoxide and iodomethane.

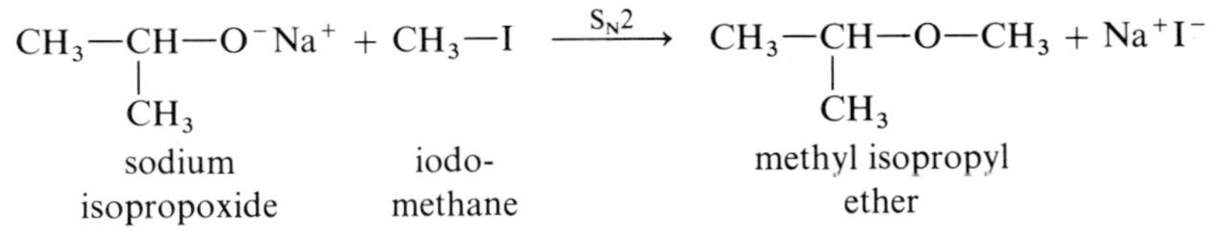

When planning a Williamson ether synthesis it is best to use a combination of alkyl halide and metal alkoxide that will maximize the S_N2 reaction and minimize the formation of alkene side-products. Recall from Section 5.6 that S_N2 reactions are most favorable when the halide is displaced from a primary rather than a secondary or tertiary carbon atom. With this in mind, let us plan a synthesis for ethyl *tert*-butyl ether. There are two possible choices of starting materials, an ethyl halide and a metal *tert*-butoxide, or a *tert*-butyl halide and a metal ethoxide. Each path is illustrated:

$$(CH_3)_3C{-}O^-K^+ + CH_3{-}CH_2{-}Br \xrightarrow[\text{substitution}]{S_N2} \underset{\text{(major product)}}{(CH_3)_3C{-}O{-}CH_2{-}CH_3} + KBr$$

$$(CH_3)_3C{-}Cl + CH_3{-}CH_2{-}O^-Na^+ \xrightarrow[\text{elimination}]{} \underset{\text{(major product)}}{CH_3{-}\overset{CH_3}{\overset{|}{C}}{=}CH_2} + CH_3{-}CH_2{-}OH + NaCl$$

The yield of the desired ethyl *tert*-butyl ether is greatest using a primary halide. Since yields of substitution product over elimination product are greatest with primary halides, the Williamson synthesis of ethers should always use primary halides whenever possible.

5.8 REACTIONS OF ETHERS

Ethers resemble hydrocarbons in their resistance to chemical reaction. They do not react with oxidizing agents such as potassium dichromate or potassium permanganate. They are not affected by most acids or strong bases at moderate temperatures. It is precisely this general inertness to chemical reaction on the one hand, and good solvent characteristics on the other, that make ethers excellent solvents in which to carry out many organic reactions.

Two hazards must be avoided when working with ethers. First, low molecular weight ethers are highly flammable. Consequently, sparks and open flames must be avoided in any area where low molecular weight ethers are being used. Ether fires cannot be extinguished with water because ethers float on water and continue to burn. Fortunately, carbon dioxide extinguishers are effective. Second, ethers react with oxygen to form hydroperoxides.

$$CH_3{-}CH_2{-}O{-}CH_2{-}CH_3 + O_2 \longrightarrow CH_3{-}CH_2{-}O{-}\underset{\text{}}{\overset{\overset{\displaystyle O{-}O{-}H}{|}}{C}}H{-}CH_3$$

a hydroperoxide

Hydroperoxides are dangerous because they are highly explosive. Commonly used ethers such as diethyl ether and tetrahydrofuran often become contaminated with hydroperoxides on prolonged storage and exposure to air and light. For this reason, purification of ethers by treatment with a reducing agent such as alkaline ferrous sulfate is frequently necessary before use.

5.9 ETHER AND ANESTHESIA

Prior to the middle of the 19th century, surgery was performed only when absolutely necessary because there were no truly effective general anesthetics. More often than not, patients were drugged, hypnotized, or simply tied down. In 1772 Joseph Priestley isolated nitrous oxide, and in 1799 Sir Humphrey Davy demonstrated its anesthetic effect and named it "laughing gas." In 1844, an American dentist, Horace Wells, administered nitrous oxide while extracting teeth. His first anesthetic trials were successful, and he introduced nitrous oxide into general dental practice. However, one patient awakened prematurely, screaming with pain, and another died. Wells was forced to withdraw from practice, became embittered, depressed, and finally insane. He committed suicide at the age of 33. In the same period, a Boston chemist, Charles Jackson, etherized himself with diethyl ether and persuaded a dentist, William Morton, to use it. Subsequently they persuaded a surgeon, John Warren, to give a public demonstration of surgery under anesthesia. The operation was completed painlessly, and soon general anesthesia by diethyl ether was widely adopted for surgical operations.

Diethyl ether is still a widely used inhalation anesthetic because it is easy to administer and causes excellent muscle relaxation. Blood pressure, pulse rate, and respiration are usually only slightly affected by diethyl ether. Diethyl ether's chief drawbacks are its irritating effect on the respiratory passages and its after-effect of nausea. It is also explosive.

Divinyl ether, $CH_2{=}CH{-}O{-}CH{=}CH_2$, is also used as an inhalation anesthetic. It does not have a nauseous after-effect and it is a rapid-acting, short-term anesthetic. It is favored by doctors for office use, but must be used with caution to prevent too deep a level of unconsciousness. Another widely used anesthetic is halothane, $C_2HBrClF_3$ (Section 2.11). It is nonflammable, nonexplosive, and causes minimum discomfort to the patient. Although there have been a few cases of liver damage caused by its use, its record as a safe anesthetic is impressive.

The mechanism of action of anesthetics is unknown. Correlations have been observed between anesthetic potency and certain physical properties such as the ratio of oil solubility/water solubility, vapor pressure, and absorbability. These generalizations have led to various theories about the mode of action of anesthetics, but correlations do not prove a theory, so that no one yet knows how these agents produce the state of unconsciousness.

5.10 EPOXIDES

Ethylene oxide and other three-membered ring ethers are classed together as epoxides and they show chemical reactivities quite different from those of other ethers.

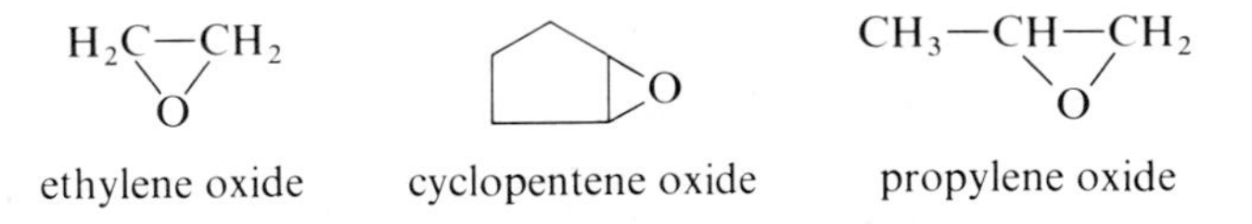

ethylene oxide cyclopentene oxide propylene oxide

The value of epoxides in laboratory and industrial chemistry lies in the fact that they can be prepared easily from alkenes and in turn undergo facile ring openings with a wide variety of nucleophiles. This versatility is well illustrated by ethylene oxide. With the abundant supplies of ethylene available from thermal and catalytic cracking processes in the petroleum industry and the development of the technology for the direct conversion of ethylene to ethylene oxide, this substance has grown from a laboratory curiosity to a 4 billion pound-per-year intermediate used in the production of thousands of consumer products. The largest use of ethylene oxide (50%) is for the synthesis of ethylene glycol whose major uses are in automobile antifreeze and polyester fibers and films. Ethylene oxide is also the starting point for the synthesis of many other materials including ethanolamines, glycol ethers, and polyethylene glycols. These derivatives are used in the production of synthetic rubbers, synthetic fibers, resins, paints, adhesives, molded articles, solvents, brake fluids, and cosmetics.

$$H_2C\overset{O}{—}CH_2 + H—OH \longrightarrow HO—CH_2—CH_2—OH$$
ethylene glycol

$$H_2C\overset{O}{—}CH_2 + NH_3 \longrightarrow H_2N—CH_2—CH_2—OH$$
ethanolamine

$$H_2C\overset{O}{—}CH_2 + CH_3—OH \longrightarrow CH_3—O—CH_2—CH_2—OH$$
2-methoxyethanol

$$H_2C\overset{O}{—}CH_2 + HO—CH_2—CH_2—OH \longrightarrow HO—CH_2—CH_2—O—CH_2—CH_2—OH$$
diethylene glycol

5.11 THIOLS

Sulfur analogs of alcohols are known as thioalcohols or thiols. In these names, the letters thi indicate the presence of a sulfur atom in place of an oxygen atom. Following are Lewis structures of methanethiol, the simplest thiol, and methanol.

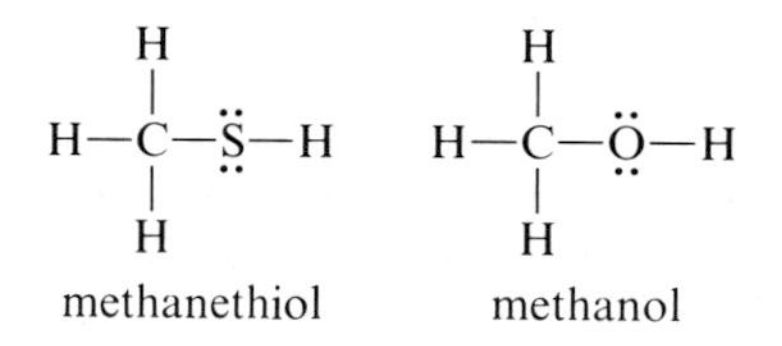

Oxygen and sulfur are both in Column VIA of the periodic table; therefore each has six electrons in its valence shell. For oxygen these valence electrons are in the second principal energy level; for sulfur, they are in the third principal energy level. Lewis structures of methanethiol and methanol show both sulfur and oxygen forming two covalent bonds; each atom also has two unshared pairs of electrons.

Following are IUPAC names and structural formulas for several low molecular weight thiols. In the common system of nomenclature, thiols are called mercaptans.

$$CH_3—CH_2—SH$$
ethanethiol
(ethyl mercaptan)

$$CH_3—CH_2—CH_2—CH_2—SH$$
1-butanethiol
(*n*-butyl mercaptan)

$$CH_3—CH_2—\underset{}{\overset{SH}{\overset{|}{C}}}H—CH_3$$
2-butanethiol
(*sec*-butyl mercaptan)

One of the most characteristic properties of lower molecular weight thiols is their odor! The odor of skunk's spray, for example, is due primarily to low molecular weight thiols.

The physical properties of thiols are quite different from those of alcohols, primarily because of the difference in electronegativity between sulfur and oxygen. The electronegativities of sulfur and hydrogen are almost identical; therefore the S—H bond is nonpolar covalent. In comparison, the electronegativity difference between oxygen and hydrogen is 0.9 unit (3.0 − 2.1); therefore the O—H bond is polar covalent.

$$CH_3{-}S{-}H \qquad CH_3{-}\overset{\delta-}{O}{-}\overset{\delta+}{H}$$

a nonpolar covalent bond *a polar covalent bond*

As a result of these differences in polarity of the O—H bond and the S—H bond, thiols do not associate by hydrogen bonding. Consequently, thiols have lower boiling points and are less soluble in water and other polar solvents than alcohols of comparable molecular weight.

Thiols are weakly acidic, and form insoluble salts with heavy metals such as lead(II) and mercury(II), as shown in the following reaction:

$$2CH_3{-}S{-}H + HgCl_2 \longrightarrow \underset{\text{an insoluble salt}}{CH_3{-}S{-}Hg{-}S{-}CH_3} + 2HCl$$

One important group of biomolecules containing —SH groups are the proteins. Reactions of protein —SH groups with heavy metal cations form salts, which change both the chemical and physical properties of these essential biomolecules. The reaction of protein —SH groups with heavy metal cations, for example, is one primary chemical basis for heavy metal poisoning.

A second important reaction of thiols is oxidation to disulfides by oxidizing agents such as oxygen or hydrogen peroxide. The characteristic structural feature of a disulfide is the presence of a —S—S— group.

$$CH_3{-}CH_2{-}S{-}H + H{-}S{-}CH_2{-}CH_3 + \tfrac{1}{2}O_2 \longrightarrow \underset{\text{a disulfide}}{CH_3{-}CH_2{-}S{-}S{-}CH_2{-}CH_3} + H_2O$$

Disulfide bonds are especially important in the structure of proteins (Chapter 12). Disulfides are reduced to thiols by hydrogen in the presence of a heavy metal catalyst or other reducing agent.

$$CH_3{-}CH_2{-}S{-}S{-}CH_2{-}CH_3 + H_2 \xrightarrow{Pt} 2CH_3{-}CH_2{-}SH$$

5.12 SPECTROSCOPIC PROPERTIES OF ALCOHOLS AND ETHERS

Alcohols show characteristic absorption of infrared radiation associated with stretching of the O—H and C—O bonds. The position of the O—H stretching frequency depends on whether the hydroxyl group is free or hydrogen-bonded.

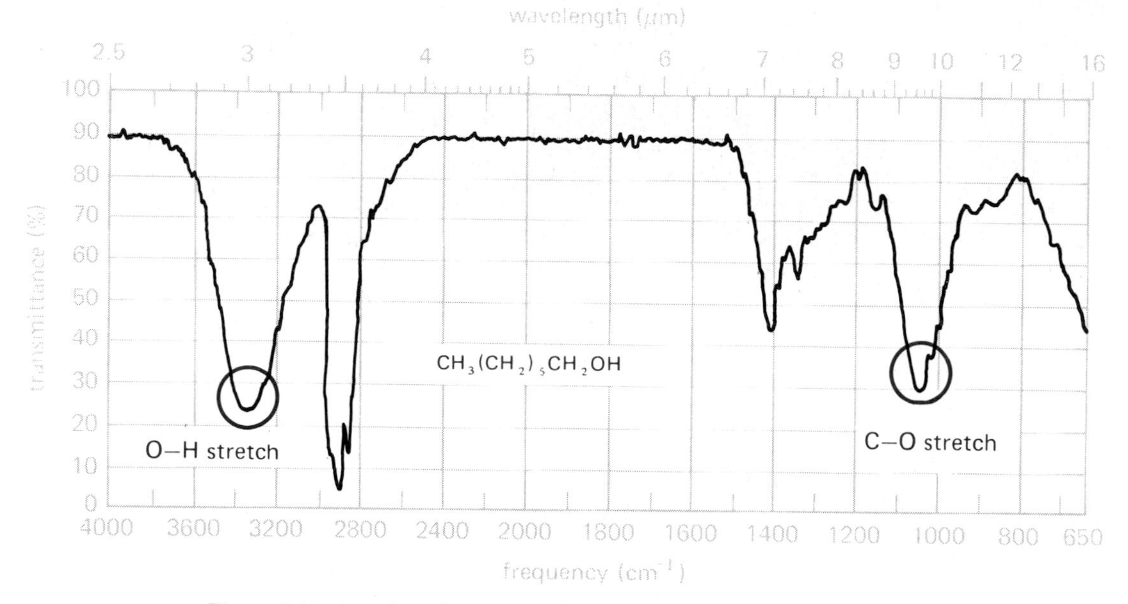

Figure 5.11 *An infrared spectrum of 1-hexanol.*

In dilute solutions the unassociated O—H bond absorbs at 3500 to 3650 cm^{-1}. However, in concentrated solutions in which the hydroxyl group is extensively hydrogen bonded, the position of the O—H stretching frequency is shifted and appears as a broad peak in the region 3200 to 3400 cm^{-1}. Stretching of the C—O bond appears between 1000 cm^{-1} and 1250 cm^{-1}. In the infrared spectrum of 1-hexanol (Figure 5.11) the broad peak at 3330 cm^{-1} is associated with stretching of the hydrogen-bonded O—H and the broad peak at 1050 cm^{-1} is associated with C—O stretching. The peaks between 2900 cm^{-1} and 3000 cm^{-1} are associated with stretching of the various C—H bonds within the molecule.

Ethers do not contain an O—H functional group and hence show no absorption between 3200 to 3600 cm^{-1}. The presence or absence of absorption in this region of the infrared spectrum can be used to distinguish alcohols from isomeric ethers. Ethers do, however, show absorption in the region 1000 to 1250 cm^{-1} due to C—O stretching.

Saturated alcohols and ethers do not absorb ultraviolet radiation. In fact, ethanol and diethyl ether are commonly used solvents for the recording of ultraviolet spectra.

In the NMR spectrum, the proton of the O—H group generally appears as a singlet within the range $\delta = 1$ to 5. Often the singlet is broad and difficult to see. The exact position and sharpness of the NMR signal depends on the extent of hydrogen bonding, temperature, and structure of the particular alcohol under examination. The NMR spectrum of 1-propanol (Figure 5.12) shows four signals corresponding to the four sets of chemically equivalent protons in the molecule. These signals are in the ratio of 3:2:1:2.

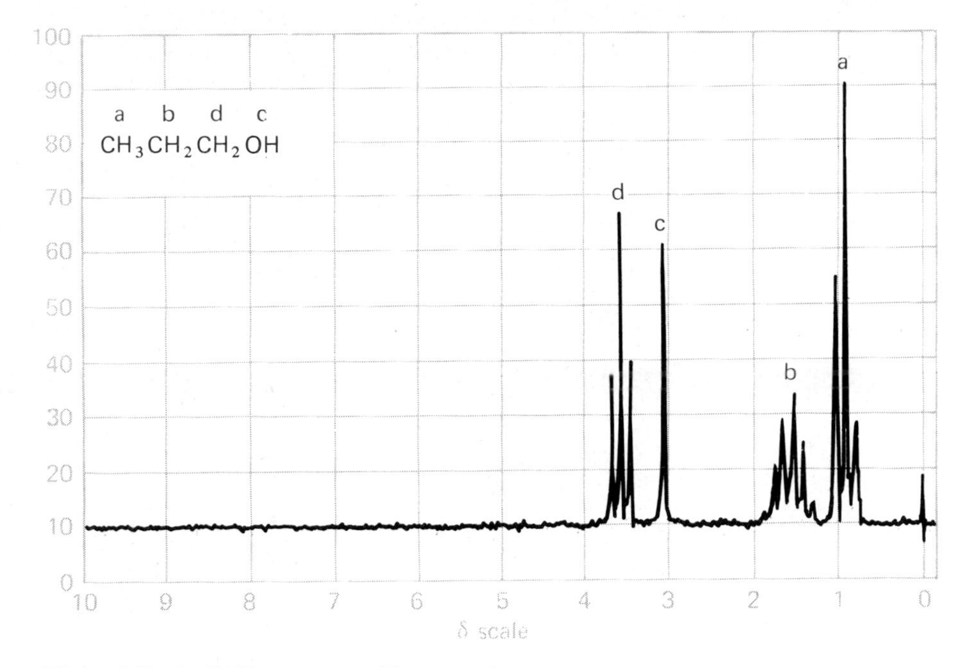

Figure 5.12 *An NMR spectrum of 1-propanol.*

Note that the signal for the hydroxyl proton at $\delta = 3.1$ appears as a sharp singlet even though there are two protons on the adjacent $—CH_2—$ group. Hydroxyl protons generally appear as singlets because of the rapid rate at which they exchange between alcohol molecules. Notice also that the signal at $\delta = 3.6$ for the protons of the $—CH_2—$ group adjacent to O—H appears as a triplet indicating that it is split only by the two protons of the adjacent methylene group and not also by the proton of the O—H group.

IMPORTANT REACTIONS

(1) Reaction of alcohols with metals to form metal alkoxides (Section 5.5A).

$$2CH_3CH_2OH + 2Na \longrightarrow 2CH_3CH_2O^-Na^+ + H_2$$

(2) Reaction of alcohols and ethers with strong acids to form oxonium ions (Section 5.5B).

$$CH_3CH(CH_3)—\ddot{O}H + H_3PO_4 \longrightarrow CH_3CH(CH_3)—\overset{H}{\underset{\cdot\cdot}{O^+}}—H + H_2PO_4^-$$

an oxonium ion

$$CH_3CH_2—\ddot{\underset{\cdot\cdot}{O}}—CH_2CH_3 + H_2SO_4 \longrightarrow CH_3CH_2—\overset{H}{\underset{\cdot\cdot}{O^+}}—CH_2CH_3 + HSO_4^-$$

an oxonium ion

(3) Oxidation of alcohols (Section 5.5C).

(a) Primary alcohols are oxidized to aldehydes or carboxylic acids depending on the experimental conditions.

$$CH_3{-}CH_2{-}OH + Cr_2O_7^{2-} \xrightarrow{H^+} CH_3{-}\overset{\overset{\displaystyle O}{\|}}{C}{-}H + Cr^{3+}$$

$$CH_3{-}CH_2{-}OH + Cr_2O_7^{2-} \xrightarrow{H^+} CH_3{-}\overset{\overset{\displaystyle O}{\|}}{C}{-}OH + Cr^{3+}$$

(b) Secondary alcohols are oxidized to ketones.

$$CH_3{-}\overset{\overset{\displaystyle OH}{|}}{C}H{-}CH_3 + Cr_2O_7^{2-} \xrightarrow{H^+} CH_3{-}\overset{\overset{\displaystyle O}{\|}}{C}{-}CH_3 + Cr^{3+}$$

(c) Tertiary alcohols have no hydrogen atom attached to the carbon bearing the —OH groups and are resistant to oxidation.

(4) Acid-catalyzed dehydration of alcohols (Section 5.5D).

$$CH_3{-}CH_2{-}\underset{\underset{\displaystyle OH}{|}}{C}H{-}CH_3 \xrightarrow{H_2SO_4} \underset{(major\ product)}{CH_3{-}CH{=}CH{-}CH_3} + \underset{(minor\ product)}{CH_3{-}CH_2{-}CH{=}CH_2} + H_2O$$

(5) Conversion to alkyl halides (Section 5.5E).

$$CH_3{-}CH_2{-}\underset{\underset{\displaystyle OH}{|}}{C}H{-}CH_3 + HBr \longrightarrow CH_3{-}CH_2{-}\underset{\underset{\displaystyle Br}{|}}{C}H{-}CH_3 + H_2O$$

$$CH_3(CH_2)_5CH_2OH + SOCl_2 \longrightarrow CH_3(CH_2)_5CH_2Cl + SO_2 + HCl$$

$$3CH_3CH_2CH_2OH + PI_3 \longrightarrow 3CH_3CH_2CH_2I + H_3PO_3$$

(6) Nucleophilic substitution at saturated carbon, S_N2 (Section 5.6).

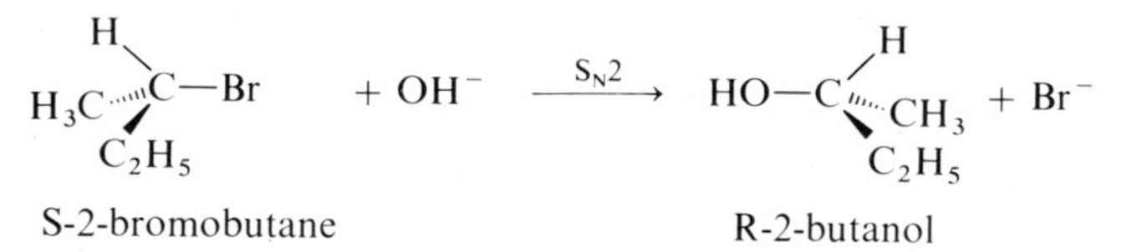

(7) Nucleophilic substitution at saturated carbon, S_N1 (Section 5.6).

$$\underset{\text{S-3-bromo-3-methylhexane}}{CH_3CH_2CH_2\underset{\underset{\displaystyle Br}{|}}{\overset{\overset{\displaystyle CH_3}{|}}{C}}CH_2CH_3} + H_2O \xrightarrow{S_N1} \underset{\text{RS-3-methyl-3-hexanol}}{CH_3CH_2CH_2\underset{\underset{\displaystyle OH}{|}}{\overset{\overset{\displaystyle CH_3}{|}}{C}}CH_2CH_3} + HBr$$

(8) Formation of hydroperoxides by ethers (Section 5.8).

$$CH_3{-}CH_2{-}O{-}CH_2{-}CH_3 + O_2 \longrightarrow CH_3{-}CH_2{-}O{-}\overset{\overset{\displaystyle O{-}O{-}H}{|}}{C}H{-}CH_3$$

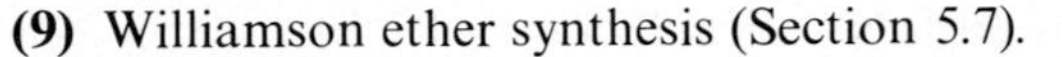

(9) Williamson ether synthesis (Section 5.7).

$$CH_3\underset{\displaystyle CH_3}{\underset{|}{C}}H{-}O^-Na^+ + CH_3CH_2{-}I \longrightarrow CH_3\underset{\displaystyle CH_3}{\underset{|}{C}}H{-}O{-}CH_2CH_3 + Na^+I^-$$

(10) Reaction of thiols with heavy metal cations to form salts (Section 5.11).

$$2CH_3CH_2SH + PbCl_2 \longrightarrow (CH_3CH_2S)_2Pb + 2HCl$$

(11) Oxidation of thiols to disulfides (Section 5.11).

$$2CH_3CH_2CH_2SH + H_2O_2 \longrightarrow CH_3CH_2CH_2{-}S{-}S{-}CH_2CH_2CH_3 + 2H_2O$$

(12) Reduction of disulfides to thiols (Section 5.11).

$$CH_3CH_2CH_2{-}S{-}S{-}CH_2CH_2CH_3 + H_2 \xrightarrow{Pt} 2CH_3CH_2CH_2{-}SH$$

PROBLEMS

5.11 Name each of the following:

(a) $CH_3CH_2CH_2CH_2OH$

(b) $HOCH_2CH_2CH_2CH_2OH$

(c) 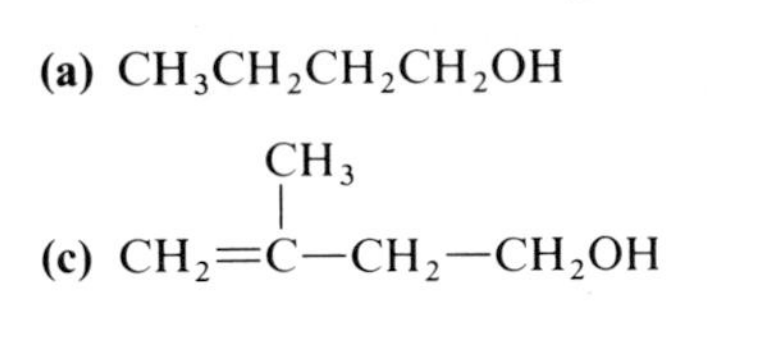
$CH_2{=}C(CH_3){-}CH_2{-}CH_2OH$

(d) $CH_3OCH_2CH_2OH$

(e) 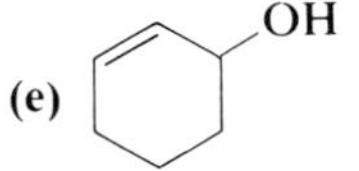

(f)

(g) $CH_3CH_2CH_2OCH(CH_3)CH_3$

(h) $HOCH_2CH_2OH$

(i) $CH_3{-}CH(HO){-}CH(OH){-}CH_3$

(j) 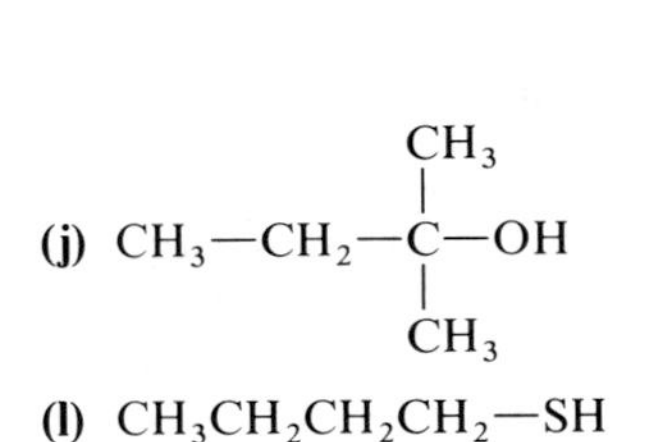
$CH_3{-}CH_2{-}C(CH_3)_2{-}OH$

(k) $CH_3{-}CH(HO){-}CH_2Cl$

(l) $CH_3CH_2CH_2CH_2{-}SH$

(m) 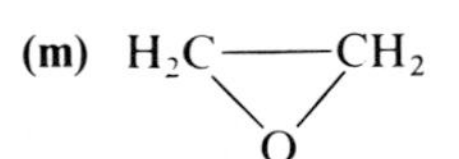

(n) 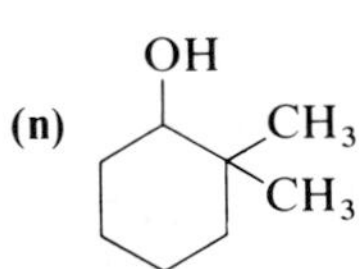

5.12 Write structural formulas for each of the following:

(a) isopropyl methyl ether
(b) propylene glycol
(c) 2-methyl-2-propylpropane-1,3-diol
(d) 1-chloro-2-hexanol
(e) 5-methyl-2-hexanol
(f) 2-isopropyl-5-methylcyclohexanol
(g) 2,2-dimethyl-1-propanol
(h) *tert*-butyl alcohol
(i) cyclopropyl methyl ether
(j) ethylene glycol

5.13 Name and draw structural formulas for the eight isomeric alcohols of molecular formula $C_5H_{12}O$. Classify each as primary, secondary, or tertiary. Which of the eight show enantiomerism?

5.14 Name and draw structural formulas for the six isomeric ethers of molecular formula $C_5H_{12}O$.

5.15 Arrange the following sets of compounds in order of decreasing boiling points (highest boiling point to lowest boiling point).

(a) $CH_3CH_2CH_3$ $CH_3CH_2CH_2CH_2CH_2CH_2CH_3$ $CH_3CH_2CH_2CH_2CH_3$

(b) N_2H_4 H_2O_2 CH_3CH_3

(c) CH_3CO_2H CH_3CH_2OH CH_3OCH_3

(d) $CH_3CH(OH)CH_3$ $CH_3{-}CH(OH){-}CH_2(OH)$ $CH_2(OH){-}CH(OH){-}CH_2(OH)$

5.16 Arrange the following compounds in order of decreasing solubility in water and explain the principles on which you have based your answers.
(a) ethanol; butane; diethyl ether
(b) 1-hexanol; 1,2-hexanediol; hexane

5.17 Diethyl ether has a lower boiling point than 1-butanol, yet each of these compounds shows about the same solubility in water (8 grams per 100 mL of water). How do you account for these observations?

$CH_3CH_2CH_2CH_2OH$	$CH_3CH_2OCH_2CH_3$
1-butanol	diethyl ether
(bp 117°C)	(bp 35°C)

5.18 How do you account for the fact that ethanol (molecular weight 46, bp 78°C) has a boiling point more than 43° higher than that of ethanethiol (molecular weight 62, bp 35°C)?

5.19 Both propanoic acid and methyl acetate have the molecular formula $C_3H_6O_2$. One of these compounds is a liquid, boiling point 141°C. The other is also a liquid, boiling point 57°C.

$CH_3{-}CH_2{-}C({=}O){-}OH$	$CH_3{-}C({=}O){-}O{-}CH_3$
propanoic acid	methyl acetate

(a) Which of these two compounds has the boiling point of 141°C; the boiling point of 57°C? Explain the basis for your prediction.
(b) Which compound is more soluble in water? Explain.

5.20 Compounds that contain the N—H bond also show considerable evidence of association. Would you expect this association to be stronger or weaker than that in compounds containing O—H groups? (Hint: remember the table of relative electronegatives.)

5.21 Complete the following reactions. Where two or more organic products are formed, predict which is the major product and which is the minor product.

(a) $CH_3{-}CH_2{-}CH_2{-}OH \xrightarrow[\text{heat}]{H_3PO_4}$ **(b)** $CH_3{-}CH_2{-}\underset{\large OH}{\underset{|}{CH}}{-}CH_3 \xrightarrow[\text{heat}]{H_2SO_4}$

(c) 2-methylcyclohexanol (cyclohexane ring with OH and CH_3 on adjacent carbons) $+ Cr_2O_7^{2-} \xrightarrow[\text{heat}]{H^+}$

(d) $HO{-}CH_2{-}CH_2{-}SH + PbCl_2 \longrightarrow$

(e) $HS{-}CH_2{-}CH_2{-}CH_2{-}SH + O_2 \longrightarrow$

(f) ring: CH_2–S–S–$CH(CH_2)_4CH_3$–CH_2–(back to CH_2) $+ H_2 \xrightarrow{Pt}$

(g) $HO{-}CH_2CH_2CH_2CH_2CH_2{-}OH + Cr_2O_7^{2-} \xrightarrow[\text{heat}]{H^+}$

(h) cyclopentane with CH_2OH on two adjacent carbons $+ 2HI \longrightarrow$

(i) $CH_2{=}CH{-}CH_2OH + H_2 \xrightarrow{Pt}$

(j) $HOCH_2CH_2CH_2CH_2CH_2CH_2OH + 2HCl \longrightarrow$

(k) $Cl{-}CH_2(CH_2)_4CH_2{-}Cl + 2Na^+CN^- \longrightarrow$

(l) $CH_2{=}CH{-}CH_2OH + Cl_2 \longrightarrow$

(m) $CH_3{-}\overset{\large CH_3}{\overset{|}{\underset{\large CH_3}{\underset{|}{C}}}}{-}OH + K \longrightarrow$

(n) $CH_3CH_2O^-Na^+ +$ cyclohexyl–Br $\longrightarrow$

(o) $CH_3CH_2{-}I +$ cyclohexyl–O^-K^+ $\longrightarrow$

(p) cyclopentane with OH on one carbon and two CH_3 on the adjacent carbon $+ SOCl_2 \longrightarrow$

(q) $CH_3CH_2OH + Na \longrightarrow$

(r) $CH_2{=}CH{-}CH_2Cl + CH_3CH_2O^-Na^+ \longrightarrow$

(s) cyclohexane with CH_3 and CH_2OH on the same carbon $+ Cr_2O_7^{2-} \xrightarrow[\text{heat}]{H^+}$

5.22 Predict the relative ease with which the following alcohols undergo acid-catalyzed dehydration. Draw a structural formula for the major product of each dehydration.

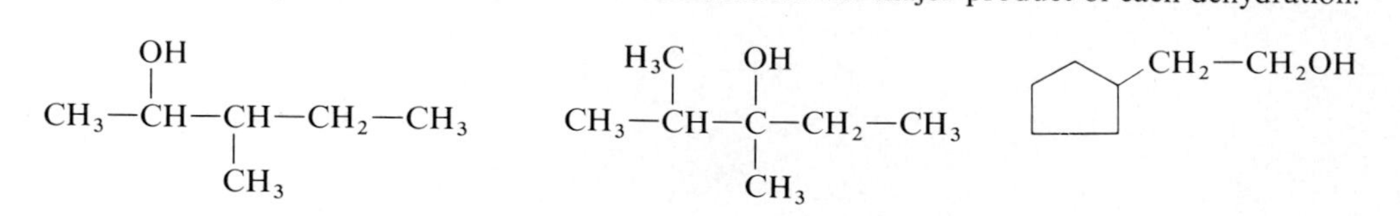

5.23 Propose a mechanism for the acid-catalyzed dehydration of cyclohexanol to give cyclohexene and water. One of the by-products of this reaction is dicyclohexyl ether. How might you account for the formation of this by-product?

5.24 What is meant by the terms S_N1 and S_N2? Compare and contrast S_N1 and S_N2 reactions in terms of (a) rate-determining step, (b) the stereochemistry of the product as related to that of the starting material.

5.25 Hydrolysis of 2-bromooctane yields 2-octanol:

$$\underset{\text{2-bromooctane}}{CH_3(CH_2)_5\underset{|\atop Br}{C}HCH_3} + H_2O \longrightarrow \underset{\text{2-octanol}}{CH_3(CH_2)_5\underset{|\atop OH}{C}HCH_3} + HBr$$

Assume that the starting material is R-2-bromooctane. Use suitable stereorepresentations to show the course of this reaction and the stereochemistry of the product if the reaction takes place by an S_N1 mechanism; by an S_N2 mechanism.

5.26 Arrange the following in order of increasing reactivity to S_N2 displacement by iodide ion, I^-.

(a) 2-chloro-2-methylbutane
(b) 1-chloropentane
(c) 3-chloro-2-methylbutane

5.27 Arrange the compounds in Problem 5.26 in order of increasing reactivity toward S_N1 displacement by H_2O.

5.28 The compounds in Problem 5.26 are each reacted with $CH_3CH_2O^-Na^+$.
(a) Which gives the highest ratio of substitution to elimination?
(b) Which gives the highest ratio of elimination to substitution?

5.29 Draw structural formulas for the major products of each reaction.

(a) $CH_3-\overset{CH_3\atop |}{\underset{|\atop CH_3}{C}}-Br + CH_3CH_2CH_2O^-Na^+ \longrightarrow$

(b) $CH_3-\overset{CH_3\atop |}{\underset{|\atop CH_3}{C}}-O^-Na^+ + CH_3CH_2CH_2Br \longrightarrow$

5.30 Write equations to show the best combination of reactants to prepare the following ethers by the Williamson ether synthesis.

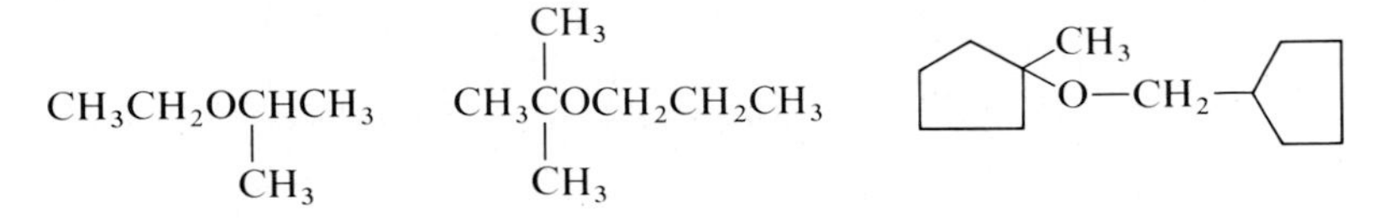

5.31 The following conversions can be carried out in either one or two steps. Show the reagents you would use and draw structural formulas for the intermediate formed in any conversion that requires two steps.

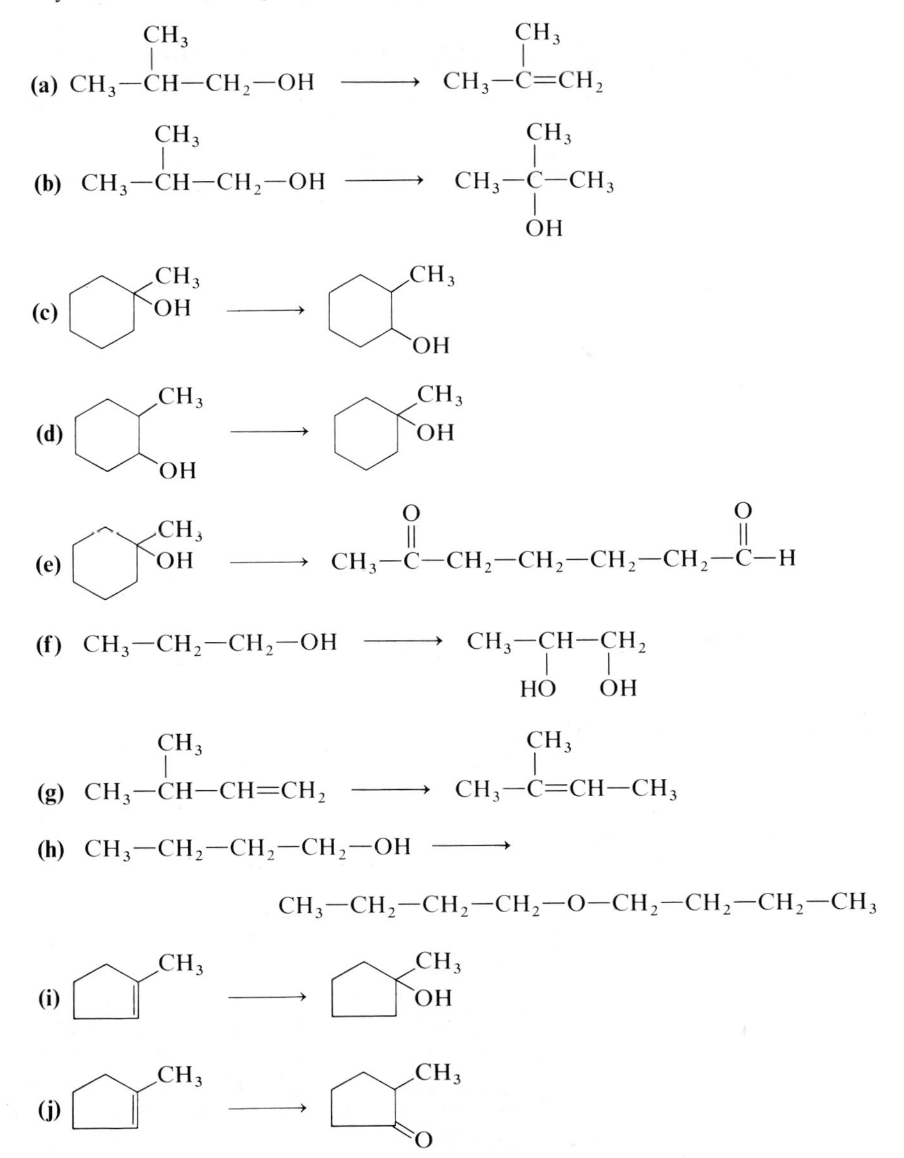

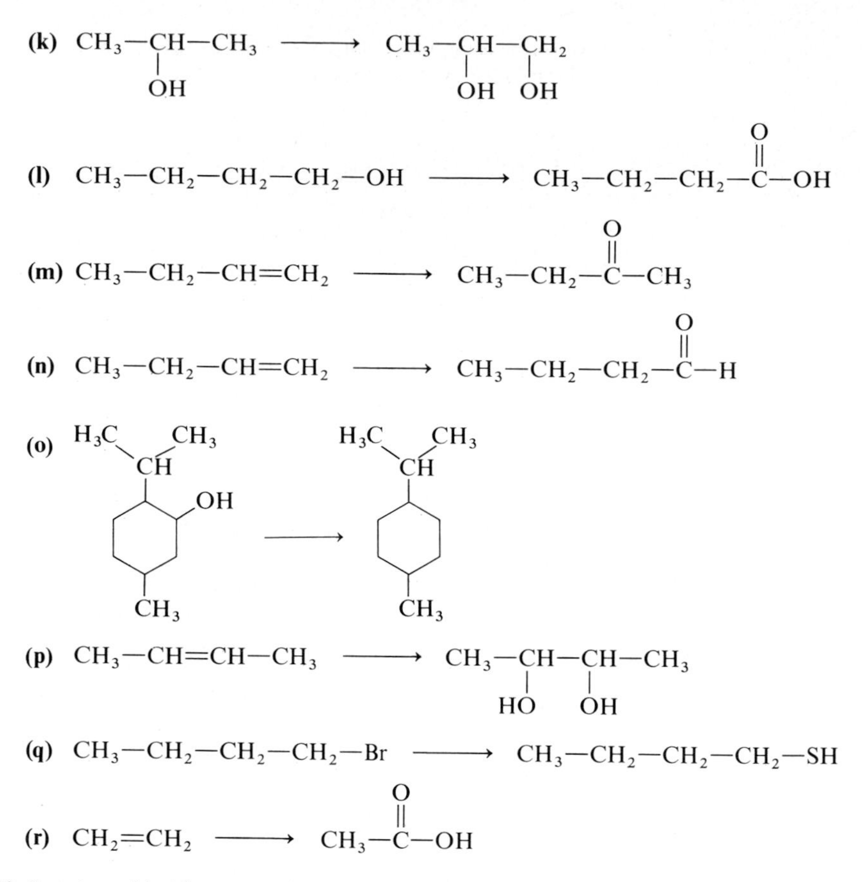

5.32 Starting with 1-butene and 2-methylpropene, show how you might synthesize each of the four isomeric alcohols of molecular formula $C_4H_{10}O$.

5.33 Hydration of fumaric acid, catalyzed by aqueous sulfuric acid, produces malic acid.

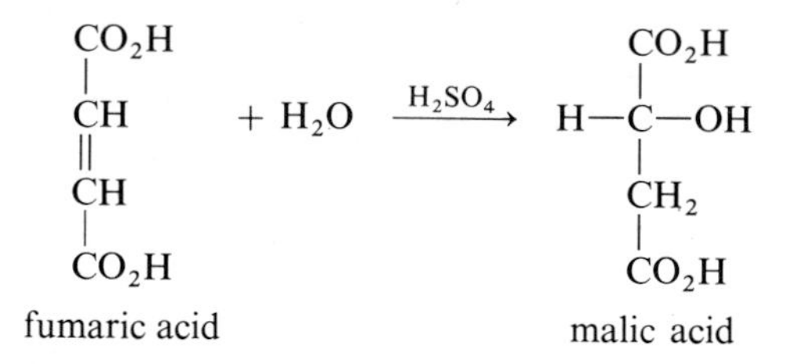

fumaric acid   malic acid

(a) Propose a mechanism for this hydration.

(b) Based on your mechanism, would you predict the malic acid so formed to be optically active or racemic? Explain your reasoning.

(c) The hydration of fumaric acid is one of the steps of the Krebs or tricarboxylic acid cycle. The biological hydration is catalyzed by the enzyme fumarase and produces optically active malic acid. How might you account for the fact that the hydration of fumaric acid catalyzed by aqueous sulfuric acid produces racemic malic acid while the same reaction, catalyzed by fumarase, produces only one enantiomer?

5.34 One of the reactions in the metabolism of glucose is the isomerization of citric acid to isocitric acid. The isomerization is catalyzed by the enzyme aconitase.

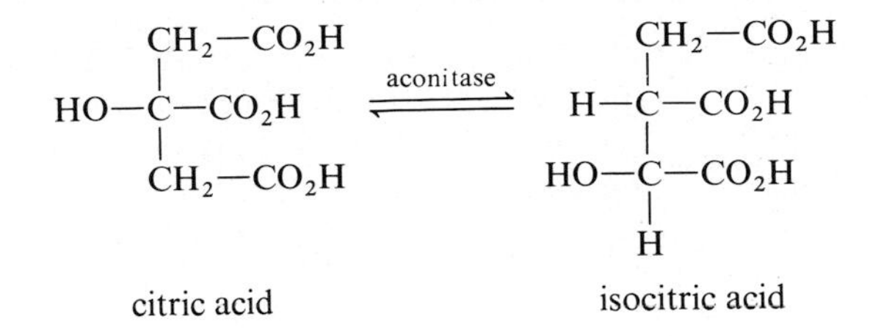

Propose a reasonable mechanism to account for this isomerization. (Hint: within its structure aconitase has groups that can function as acids.)

5.35 Compound *A* ($C_5H_{10}O$) is optically active, decolorizes a solution of bromine in carbon tetrachloride, and also decolorizes dilute potassium permanganate. Treatment of *A* with hydrogen gas over a platinum catalyst yields compound *B* ($C_5H_{12}O$). Treatment of *B* with warm phosphoric acid forms compound *C* (C_5H_{10}). Ozonolysis of *C* yields two compounds, CH_3CHO and CH_3CH_2CHO, in equal amounts. Only compound *A* is optically active. Propose structural formulas for compounds *A*, *B*, and *C* consistent with these observations.

5.36 Show how you might distinguish between the following pairs of compounds by a simple chemical test. In each case, tell what test you would perform, what you would expect to observe, and write an equation for each positive test.

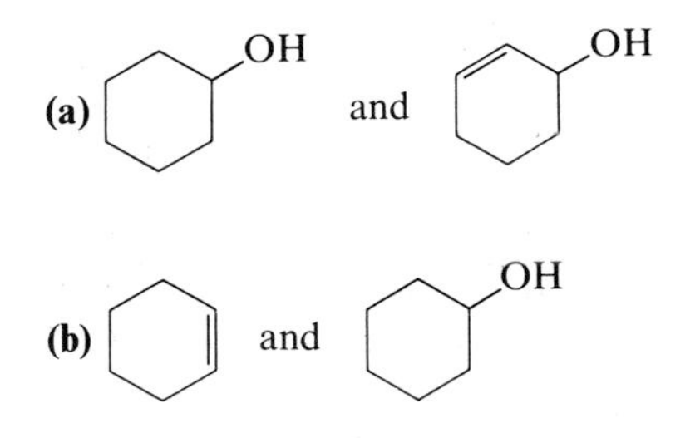

(c) CH_3CH_2OH and $CH_3OCH_2CH_2OCH_3$

5.37 List one major spectral characteristic that will enable you to distinguish between the following pairs of compounds.

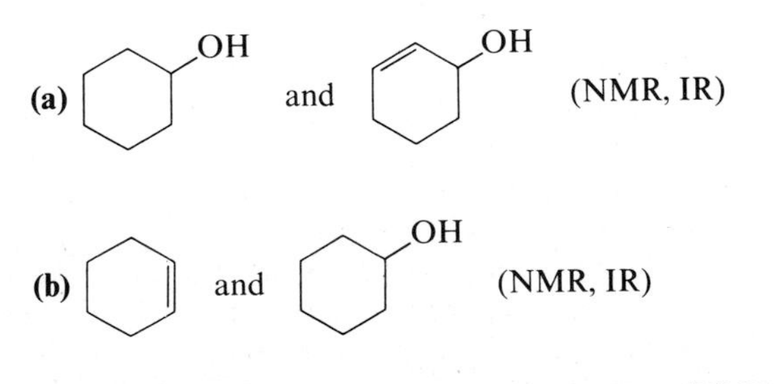

(c) CH_3CH_2OH and $CH_3OCH_2CH_2OCH_3$ (NMR, IR)

5.38 Predict the number of signals and the splitting pattern of each signal in the NMR spectrum of 2,2-dimethyl-1-propanol (neopentyl alcohol). Also predict the ratio of the areas under the signals.

5.39 An alcohol of molecular formula $C_5H_{12}O$ shows signals in the NMR spectrum at $\delta = 0.9$, 1.2, 1.7, and 5.0. The areas of these signals are in the ratio 3:6:2:1.

(a) Draw a structural formula for this alcohol.

(b) Predict the splitting pattern of each signal.

CHAPTER 6

Benzene and the Concept of Aromaticity

Benzene is a liquid with a boiling point of 80°C. It was first isolated by Michael Faraday in 1825 from the oily liquid that collected in the illuminating gas lines of London. Its molecular formula C_6H_6, suggests a high degree of unsaturation. (Remember that a saturated alkane of six carbons has the formula C_6H_{14} and that a saturated cycloalkane of six carbons has the formula C_6H_{12}.) Considering this high degree of unsaturation, you might expect benzene to be highly reactive and show reactions characteristic of alkenes and alkynes. Surprisingly, benzene does not undergo characteristic alkene reactions. For example, benzene does not react with bromine, hydrogen chloride, hydrogen bromide, or other reagents that usually add to double and triple bonds. When it does react, benzene typically does so by substitution in which a hydrogen atom is replaced by another atom or group of atoms. For example, benzene reacts with bromine in the presence of ferric bromide to form bromobenzene and hydrogen bromide.

$$\underset{\text{benzene}}{C_6H_6} + Br{-}Br \xrightarrow{FeBr_3} \underset{\text{bromobenzene}}{C_6H_5Br} + HBr$$

The terms aromatic and aromatic compounds have been used to classify benzene and a number of its derivatives because many of them have distinctive odors. However, after a time it became clear that a sounder classification for these compounds should be based not on aroma but on chemical reactivities. Currently, the term "aromatic" is used to refer to the unusual chemical properties of benzene and its derivatives. Aromatic hydrocarbons are resistant to typical alkene addition reactions. When they do react, they do so by substitution.

6.1 THE STRUCTURE OF BENZENE

Let us put ourselves in the mid-19th century to examine the evidence on which chemists attempted to build a model for the structure of benzene. First, the molecular formula of benzene is C_6H_6 and it seemed clear that the benzene molecule must be highly unsaturated. Yet, benzene does not show the chemical properties of the only unsaturated compounds known at that time, namely the alkenes. Benzene does undergo chemical reactions but its characteristic reaction is substitution rather than addition. When benzene, $C_6H_5{-}H$, is reacted with bromine in the presence of a ferric bromide catalyst, only one substance of molecular formula C_6H_5—Br is formed. Therefore, chemists concluded, all six hydrogens of benzene must be equivalent. Finally, when bromobenzene, C_6H_5Br is reacted further with bromine, three substances of molecular formula $C_6H_4Br_2$ are formed.

$$\underset{\text{bromobenzene}}{C_6H_5Br} + Br_2 \xrightarrow{FeBr_3} \underset{\text{dibromobenzene}}{C_6H_4Br_2} + HBr$$

For Kekulé and his contemporaries, the problem was to incorporate these observations, along with the accepted tetravalence of carbon, into a structural formulation of the benzene molecule. Before we examine the structural formula for benzene proposed by Kekulé, we should note that the problem of an adequate description of the structure of benzene and other aromatic hydrocarbons has occupied the efforts of chemists for over a century. Only since the 1930s has a general understanding of this problem been realized.

A. KEKULÉ'S MODEL OF BENZENE

In 1865, Kekulé proposed that the six carbons of benzene are arranged in a six-membered ring with one hydrogen attached to each carbon. To maintain the then-established tetravalence of carbon, he further proposed that the ring contains three double bonds which shift back and forth so rapidly that the two forms, Ia and Ib, cannot be separated.

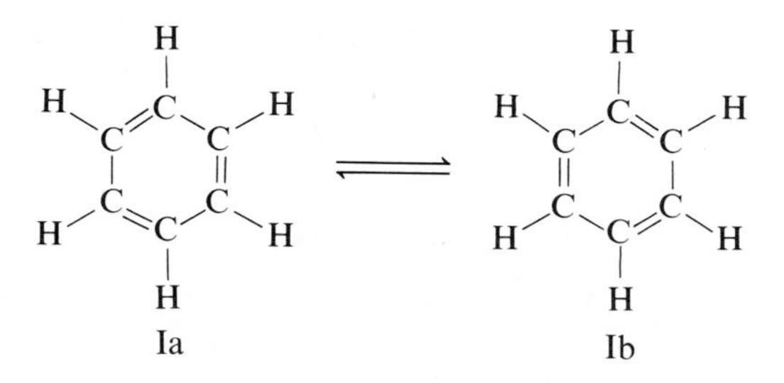

This proposal accounted nicely for the fact that bromination of benzene gives only one substance of molecular formula C_6H_5Br, but three substances of molecular formula $C_6H_4Br_2$.

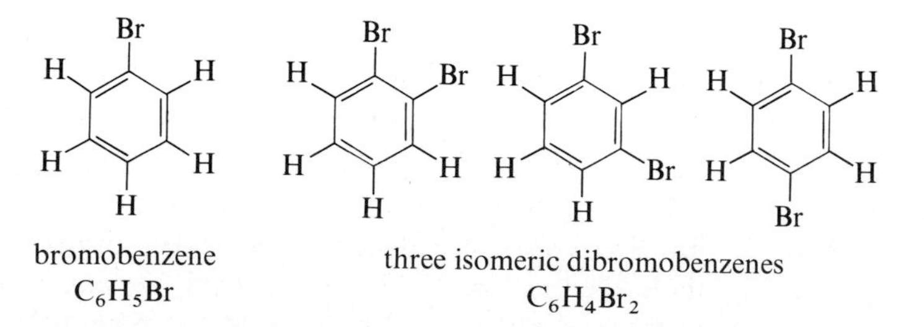

bromobenzene C_6H_5Br

three isomeric dibromobenzenes $C_6H_4Br_2$

Although Kekulé's proposal was consistent with many of the experimental observations, it did not totally solve the problem and was contested for years. The major objection was that it did not account for the unusual chemical behavior of benzene. For example, if benzene contained three double bonds, as Kekulé proposed, then, his critics argued, it should react with Br_2 just as alkenes do. Yet, it does not. Therefore, they argued, benzene cannot have the three double bonds as Kekulé had suggested. The controversy over the structural formula of benzene continued from Kekulé's time well into the 20th century.

B. RESONANCE MODEL

Linus Pauling's theory of resonance provided the first adequate description of the structure and unusual reactivities of benzene. Recall from Section 1.14 that according to the resonance theory, when a substance can be written as two or more contributing structures, the actual molecule does not conform to any one of the contributing structures but exists as a resonance hybrid of them all. The two major contributors to the benzene hybrid are

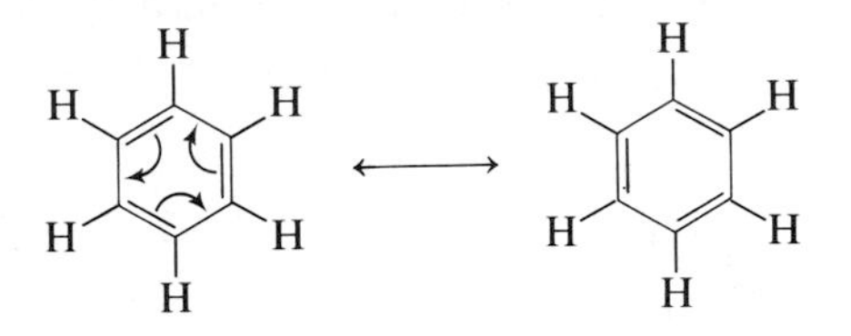

One of the consequences of resonance is a marked increase in stability of the hybrid over the stability of any one of the contributing structures. The extra energy stabilizing the hybrid is called resonance energy. Resonance stabilization is particularly large and important in benzene and other aromatic hydrocarbons. It is for this reason that benzene and other aromatic hydrocarbons do not undergo typical alkene addition reactions.

C. THE MOLECULAR ORBITAL MODEL

We can also describe the structure of benzene in terms of the overlap of atomic orbitals, for just as this approach gave us a clear understanding of the carbon-carbon double bond (as one sigma bond and one pi bond) so too will it give us a clear understanding of the bonding in benzene. Benzene is a regular hexagon with bond angles of 120°. For this type of bonding, each carbon uses sp^2 hybrid orbitals. The six carbon atoms are joined together in a regular hexagon by sigma bonds formed by the overlap of sp^2 hybrid orbitals. Each carbon is further joined to one hydrogen by a sigma bond formed from the overlap of $sp^2 - 1s$ orbitals. These twelve C—C and C—H sigma bonds form the skeletal framework of the ring (Figure 6.1a). Each carbon also has a single $2p$ orbital containing one electron. Overlap of these six parallel $2p$ orbitals forms three bonding pi-molecular orbitals. Because of the symmetry of the molecule, the pi electron density is completely symmetrical around the ring with half in the "upper lobe" and half in the "lower lobe" (Figure 6.1b).

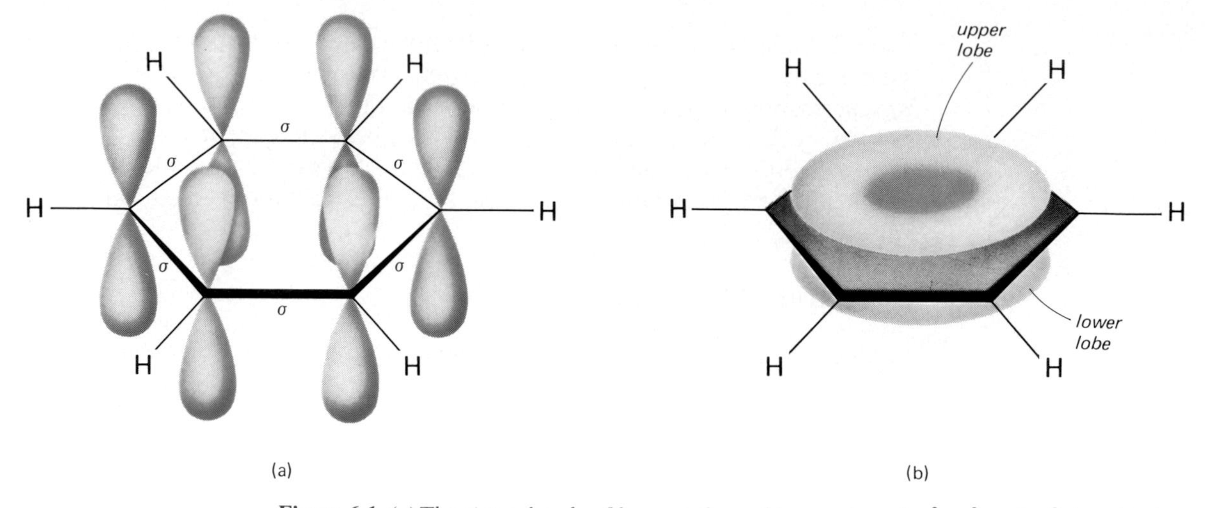

Figure 6.1 (*a*) *The sigma bonds of benzene formed by overlap of* sp^2-sp^2 *and* sp^2-$1s$ *orbitals. The six 2p orbitals are uncombined.* (*b*) *Overlap of six 2p orbitals to form the pi cloud.*

Calculations show that the energy released when six $2p$ orbitals combine to form the pi system of benzene is greater than the energy released when three pairs of $2p$ orbitals combine to form three ordinary pi bonds like those in ethylene. In the molecular orbital model, the extra energy stabilizing the molecule is called delocalization energy.

The completely symmetrical distribution of electrons in the pi cloud of benzene is shown by the symbol of a regular hexagon with a circle inside.

benzene

We have now seen two theories of the structure of benzene and each uses a particular symbol to represent this substance. The resonance approach uses two Lewis structures connected by a double-headed arrow. The molecular orbital approach uses a regular hexagon with an inscribed circle. Throughout the remainder of the book, we will use a single Lewis structure to represent benzene. We must realize, however, that this symbol is not an adequate representation of the benzene molecule, but to the extent that the symbol serves us in communication, it is useful. Furthermore, this symbol allows us to count all valence electrons and to draw contributing structures more easily.

6.2 THE RESONANCE ENERGY OF BENZENE

It has been possible to measure the resonance or delocalization energy of benzene. The most direct method is to compare the heats of hydrogenation of the actual benzene molecule with its hypothetical counterpart 1,3,5-cyclohexatriene, a six-membered ring with three separate, noninteracting pi bonds. Since cyclohexatriene does not exist, we can only estimate its heat of hydrogenation. Hydrogenation of an alkene is an exothermic reaction. The conversion of cyclohexene to cyclohexane liberates 28.6 kcal/mole.

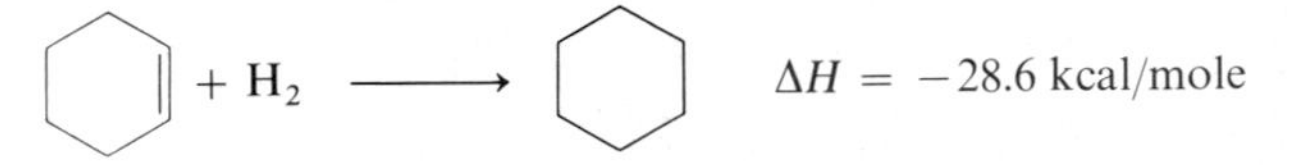

From this value we can estimate that hydrogenation of the hypothetical cyclohexatriene should release $3 \times 28.6 = 85.8$ kcal/mole. When benzene is reduced to cyclohexane, the heat released is 49.8 kcal/mole.

$$\text{C}_6\text{H}_6 + 3\text{H}_2 \longrightarrow \text{C}_6\text{H}_{12} \qquad \Delta H = -49.8 \text{ kcal/mole}$$

The difference in energy between the real molecule, benzene, and the hypothetical molecule, cyclohexatriene, is $(85.8 - 49.8) = 36$ kcal/mole. In other words the resonance or delocalization energy of benzene is approximately 36 kcal/mole.

6.3 NOMENCLATURE OF AROMATIC HYDROCARBONS

In the IUPAC system, monosubstituted alkyl benzenes are named as derivatives of benzene, as for example, ethylbenzene. For many of the simpler monosubstituted benzenes, the IUPAC system retains common names. Examples are toluene, styrene, and cumene.

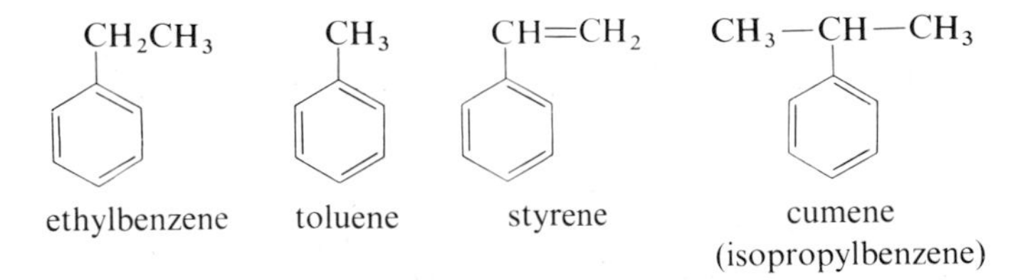

ethylbenzene toluene styrene cumene (isopropylbenzene)

When there are two substituents on the benzene ring, three structural isomers are possible. The prefixes *ortho* (*o*), *meta* (*m*), and *para* (*p*) are used to locate the substituents. *Ortho* substituents are on carbons 1, 2 of the benzene ring; *meta* substituents are on carbons 1, 3; and *para* substituents are on carbons 1, 4. The IUPAC system retains the name xylene for the dimethylbenzenes.

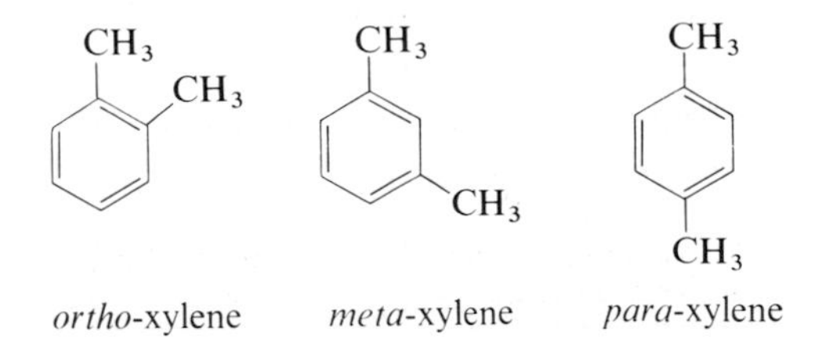

ortho-xylene *meta*-xylene *para*-xylene

With three or more substituents, a numbering system must be used.

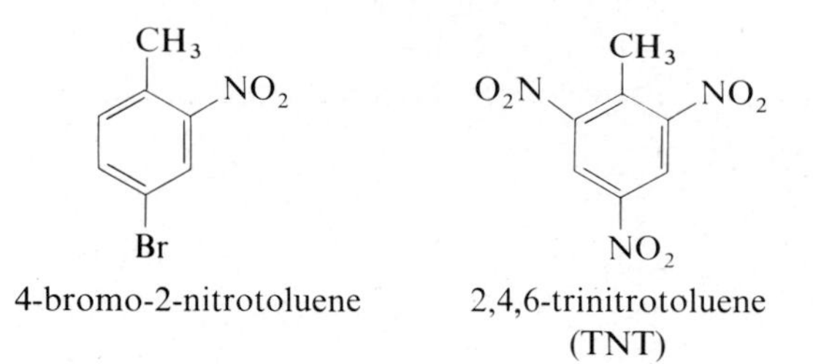

4-bromo-2-nitrotoluene

2,4,6-trinitrotoluene (TNT)

In naming more complex molecules, the benzene ring is often named as a substituent on a parent chain. In this case, the hydrocarbon group C_6H_5— is called a phenyl group.

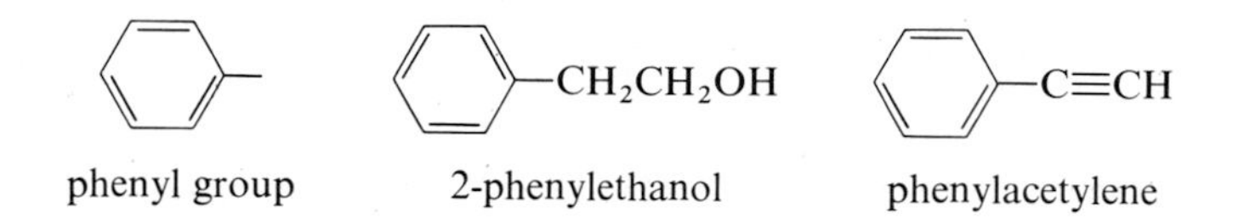

phenyl group

2-phenylethanol

phenylacetylene

Closely related to benzene are numerous polynuclear aromatic hydrocarbons having one or more six-membered rings fused together. For each an IUPAC numbering system is used to locate substituents.

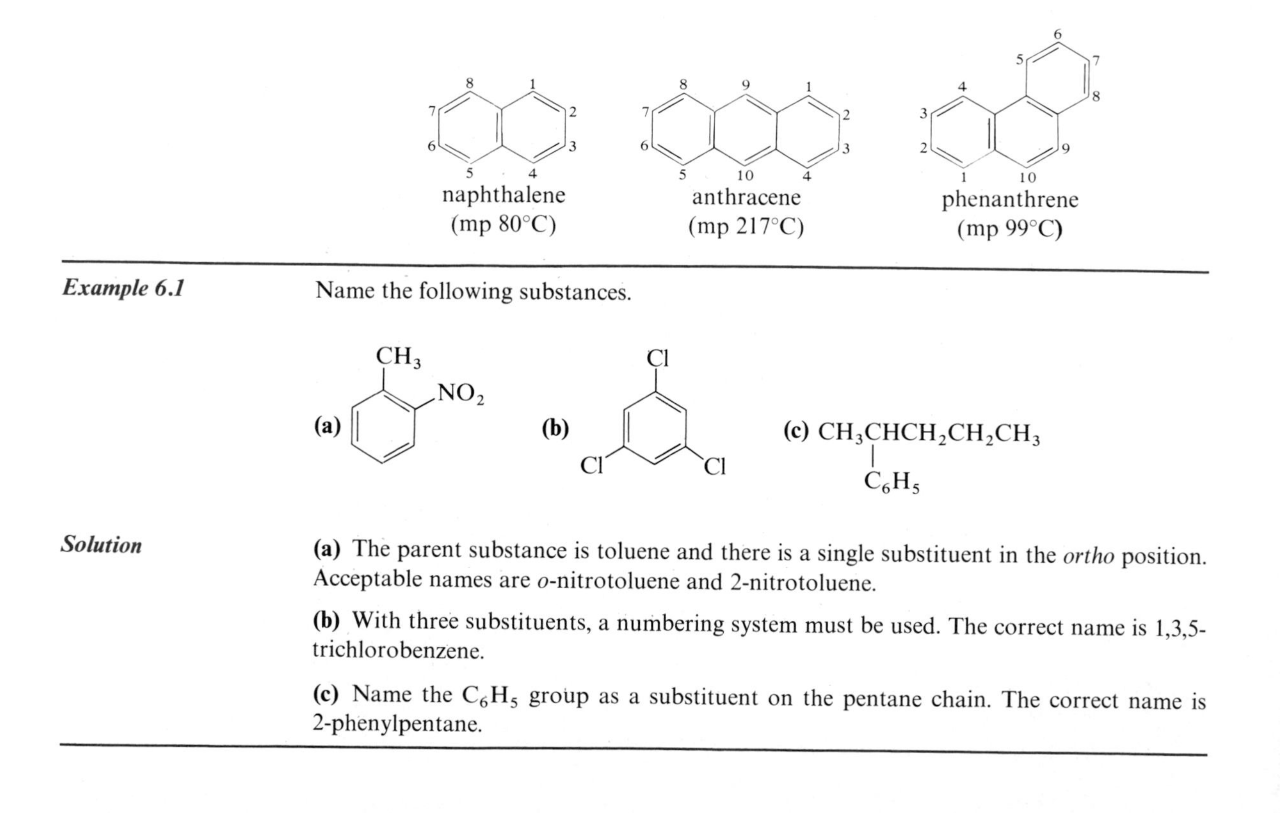

naphthalene (mp 80°C)

anthracene (mp 217°C)

phenanthrene (mp 99°C)

Example 6.1

Name the following substances.

Solution

(a) The parent substance is toluene and there is a single substituent in the *ortho* position. Acceptable names are *o*-nitrotoluene and 2-nitrotoluene.

(b) With three substituents, a numbering system must be used. The correct name is 1,3,5-trichlorobenzene.

(c) Name the C_6H_5 group as a substituent on the pentane chain. The correct name is 2-phenylpentane.

PROBLEM 6.1 Draw structural formulas for the following substances.

(a) *p*-chlorotoluene **(b)** 1-phenylcyclohexene **(c)** *m*-dinitrobenzene

6.4 PHENOLS

The characteristic structural feature of a phenol is the presence of a hydroxyl group bonded directly to a benzene ring. A structural formula of phenol, the simplest member of this class of compounds, is shown in Figure 6.2.

Other phenols are named either as derivatives of the parent hydrocarbon or by common names.

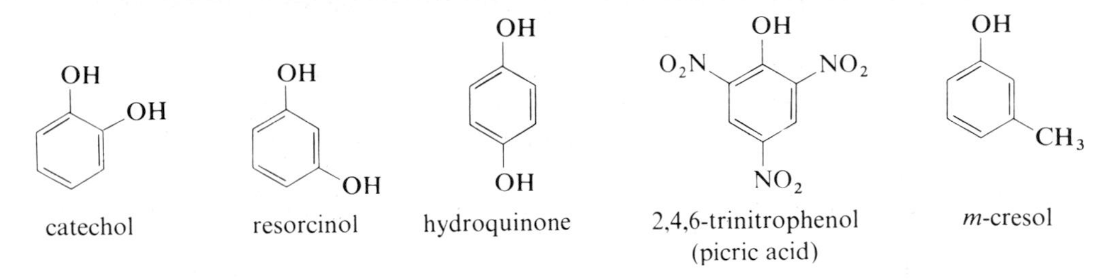

Phenols are widely distributed in nature. Phenol itself and the isomeric cresols (*ortho*-, *meta*-, and *para*-cresol) are found in coal tar and petroleum. Thymol and vanillin are important constituents of thyme and vanilla beans.

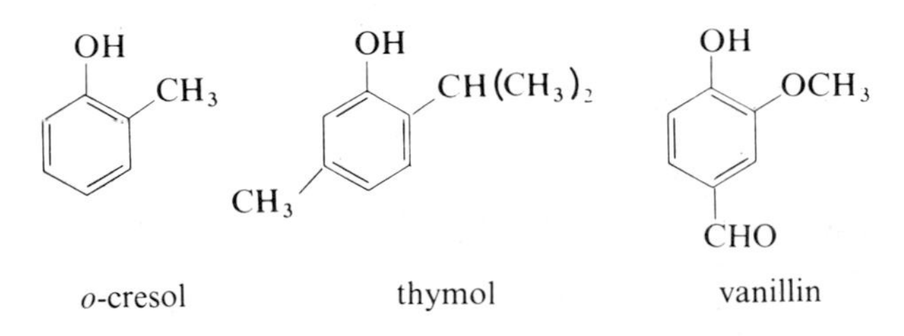

Phenol, or carbolic acid as it was once called, is a low-melting solid only slightly soluble in water. In sufficiently high concentrations, it is corrosive to all kinds of cells. In dilute solutions, it has some antiseptic properties, and was used for the first time in the 19th century by Joseph Lister for antiseptic surgery. Its medical use is now limited, as it has been replaced by antiseptics that are both

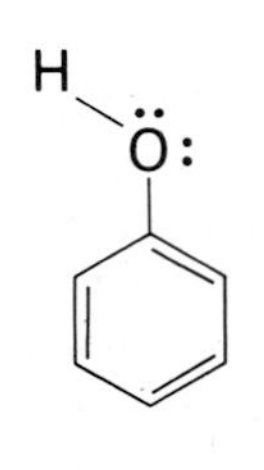

Figure 6.2 *The structure of phenol.*

more powerful and have fewer undersirable side effects. Among these is *n*-hexylresorcinol (Sucrets and mouthwashes), a substance widely used in household preparations as a mild antiseptic and disinfectant.

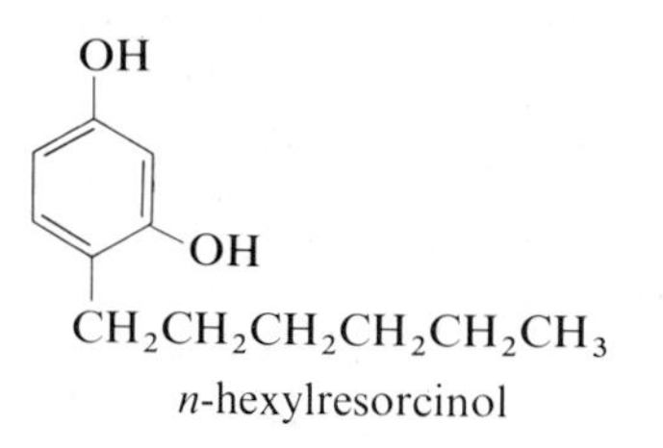

n-hexylresorcinol

Phenols and alcohols both contain the hydroxyl group, —OH. However, phenols are grouped together as a separate class of compounds because they have different chemical properties than alcohols. For us the most important of these differences is that phenols are significantly more acidic than alcohols. You can see just how large this difference in acidity is by comparing the acid dissociation constants for phenol and ethanol.

$$C_6H_5-\ddot{\underset{..}{O}}-H \rightleftharpoons C_6H_5-\ddot{\underset{..}{O}}\!:^- + H^+ \qquad K_a = \frac{[C_6H_5O^-][H^+]}{[C_6H_5OH]} = 1 \times 10^{-10}$$

$$CH_3CH_2-\ddot{\underset{..}{O}}-H \rightleftharpoons CH_3CH_2-\ddot{\underset{..}{O}}\!:^- + H^+ \qquad K_a = \frac{[CH_3CH_2O^-][H^+]}{[CH_3CH_2OH]} = 10^{-16}$$

The acid dissociation constant for phenol is approximately one million, or 10^6, times larger than that of ethanol. We can account for this enhanced acidity by using the resonance model and looking at the relative stabilities of the ethoxide ion and the phenoxide ion. There is no possibility for resonance stabilization in the ethoxide ion. However, for the phenoxide ion we can draw five contributing structures, two Kekulé structures with the negative charge on oxygen plus three additional ones delocalizing the negative charge to the *ortho* and *para* positions of the aromatic ring (Figure 6.3).

Since phenoxide ion is stabilized more by resonance than is ethoxide ion, the equilibrium for the ionization of phenol lies farther to the right than that for the ionization of ethanol. That is, phenol is a stronger acid than ethanol. Note, however, that we have used the resonance theory only in qualitative terms. While we can predict with confidence that phenol is a stronger acid than ethanol, at this point we have little or no way to predict just how much stronger it might be.

Another way to compare the relative acid strengths of alcohols and phenols is to look at the hydrogen ion concentration and pH of 0.1 molar water solutions of ethanol, phenol, and hydrochloric acid. These values are shown in Table 6.1.

Alcohols are neutral substances, and the hydrogen ion concentration of 0.1M ethanol in water is the same as that of pure water. A 0.1M solution of phenol is slightly acidic. Hydrochloric acid is a strong acid (completely dissociated in water) and the hydrogen ion concentration of 0.1M HCl is 0.1M. The pH of this solution is 1.0.

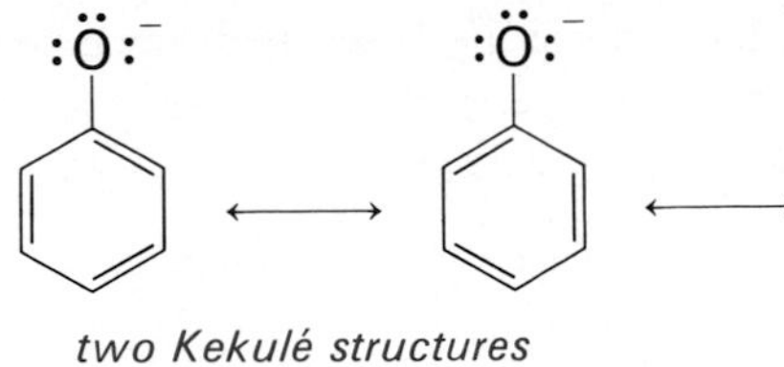

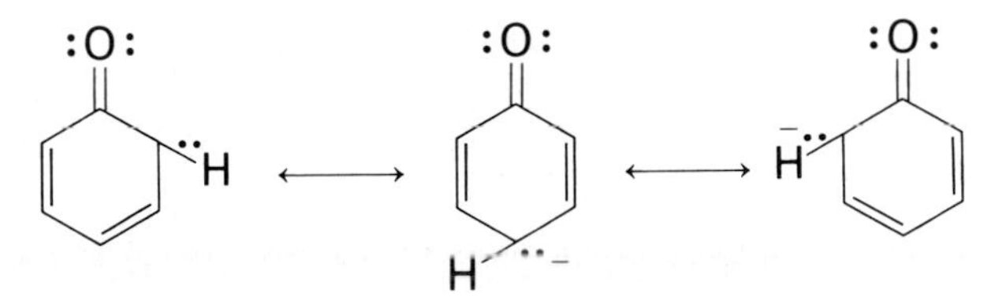

Figure 6.3 *Five contributing structures for the phenoxide ion.*

Because they are weak acids, phenols react with strong bases such as sodium hydroxide to form salts. However, phenols do not react with weaker bases such as sodium bicarbonate.

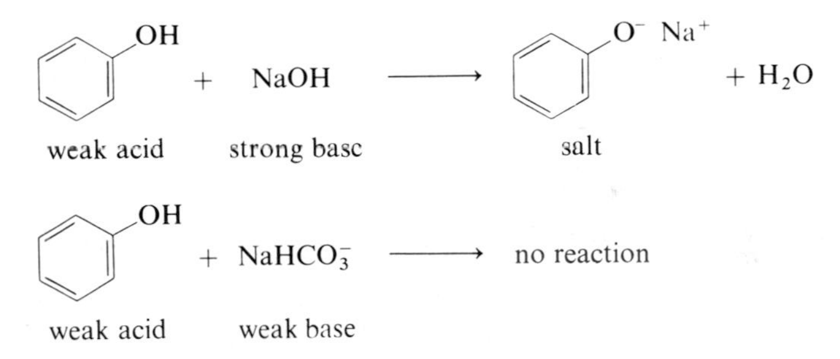

The fact that phenols are weakly acidic whereas alcohols are neutral provides a very convenient way to separate phenols from water-insoluble alcohols. Suppose we want to separate phenol from cyclohexanol. Each is only slightly soluble in water, so they cannot be separated on the basis of their water

Table 6.1 Relative acidities of 0.1M solutions of ethanol, phenol, and hydrochloric acid.

Dissociation equation	$[H^+]$	pH
$CH_3CH_2OH \rightleftharpoons CH_3CH_2O^- + H^+$	10^{-7}	7.0
$C_6H_5OH \rightleftharpoons C_6H_5O^- + H^+$	3.3×10^{-6}	5.4
$HCl \rightleftharpoons Cl^- + H^+$	0.1	1.0

solubilities. They can, however, be separated because of their differences in acidity.

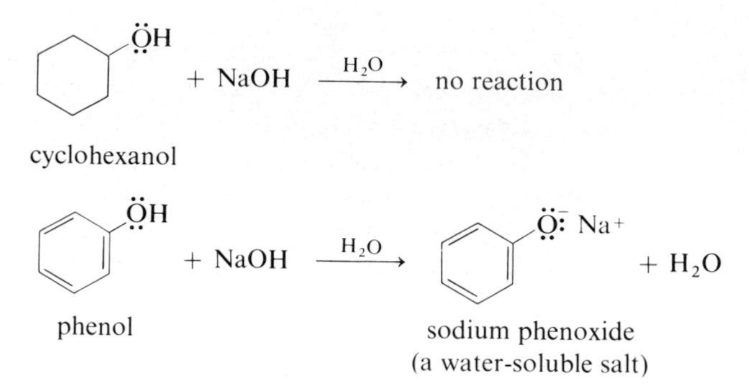

Phenol reacts with aqueous sodium hydroxide to form the water-soluble salt sodium phenoxide. Cyclohexanol remains as a water-insoluble layer. The water layer containing the sodium phenoxide can be separated from the water-insoluble cyclohexanol. Addition of a strong acid (for example, aqueous hydrochloric acid) converts sodium phenoxide to phenol.

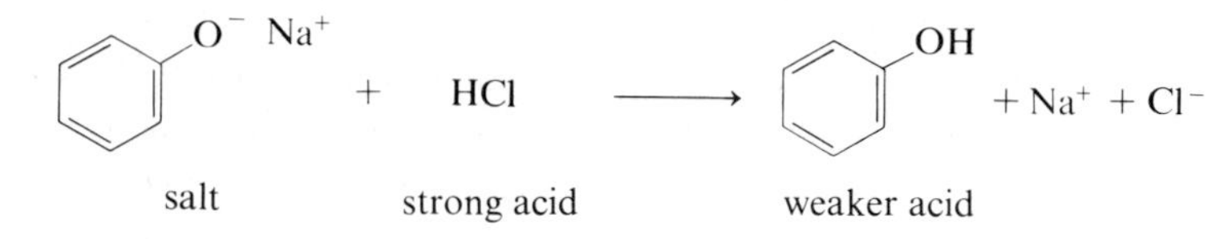

6.5 ELECTROPHILIC AROMATIC SUBSTITUTION

The most striking characteristic of aromatic hydrocarbons is their tendency to react by substitution rather than by addition. Following are examples of a few of the more common aromatic substitution reactions.

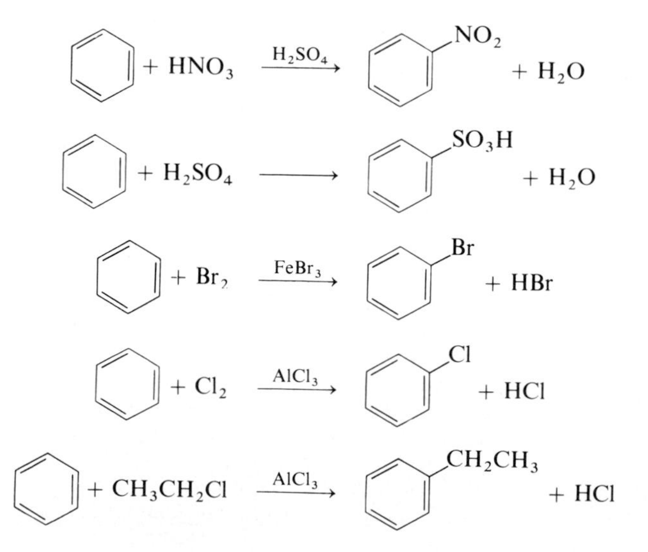

When studying these reactions we will deal with the questions of how they occur and why benzene, in contrast to alkenes, typically undergoes substitutions rather than additions.

The characteristic feature of nearly all aromatic substitution reactions is that the benzene ring is attacked by an electrophilic reagent. We can illustrate this principal by the bromination of benzene.

Benzene and most substituted benzenes react very slowly with bromine. However, in the presence of a Lewis acid catalyst such as $AlBr_3$ or $FeBr_3$, the reaction proceeds quite rapidly. The function of the catalyst is to polarize the Br—Br bond and develop positive character on one of the bromine atoms.

$$\mathrm{:\ddot{B}r-\ddot{B}r:} + \mathrm{Fe(Br)_3} \rightleftharpoons \mathrm{:\ddot{B}r^{\delta+}\cdots \ddot{B}r\cdots \overset{\delta-}{Fe}(Br)_3} \rightleftharpoons \mathrm{:\ddot{B}r^{+}\ :\ddot{B}r-\overset{-}{Fe}(Br)_3}$$

One of the bromine atoms now bears at least a partial positive charge and is therefore an electrophile. Reaction between Br^+ and the pi electrons of the benzene ring produces a resonance-stabilized carbocation (Figure 6.4). Loss of a proton then regenerates the aromatic ring. In the structural formulas in Figure 6.4, a curved arrow shows the reaction of a pair of electrons of the benzene ring and the bromonium ion, Br^+, to form a resonance-stabilized carbocation. Curved arrows also show how the first contributing structure is converted to the second and the second to the third. The curved arrow on the third contributing structure shows how it loses a proton, H^+, to give bromobenzene. Note that there are six hydrogens on each of the contributing structures for the resonance-stabilized carbocation. Two of these are shown to remind you that the carbon bearing the positive charge has a hydrogen bonded to it, and the carbon bearing the bromine also has a hydrogen bonded to it.

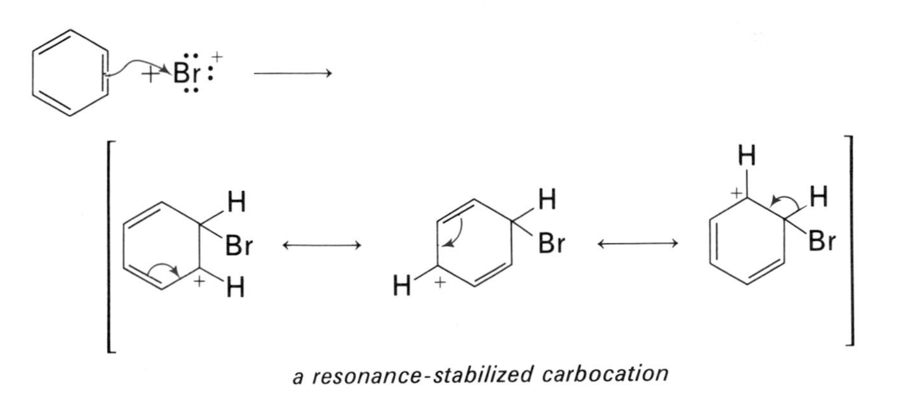

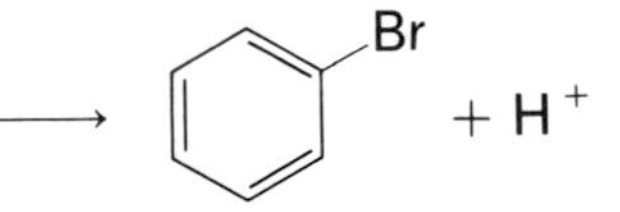

Figure 6.4 *A mechanism for the bromination of benzene. Electrophilic aromatic substitution.*

The major difference between halogen addition to an alkene and halogen substitution on an aromatic ring centers on the fate of the carbocation intermediate formed in the first step. Recall from Section 3.6F that addition of bromine to propene is a two-step process.

Step 1: $$CH_3{-}CH{=}CH_2 + :\ddot{Br}{-}\ddot{Br}: \longrightarrow CH_3{-}\overset{+}{C}H{-}CH_2{-}\ddot{Br}: + :\ddot{Br}:^-$$

Step 2: $$CH_3{-}\overset{+}{C}H{-}CH_2{-}\ddot{Br}: + :\ddot{Br}:^- \longrightarrow CH_3{-}\underset{:\ddot{Br}:}{\underset{|}{C}H}{-}CH_2{-}\ddot{Br}:$$

In the case of an alkene, the carbocation intermediate reacts with a nucleophile to complete the addition. In the case of an aromatic hydrocarbon, the resonance-stabilized carbocation intermediate loses a proton to regenerate the aromatic ring and regain the large resonance stabilization. There is no such resonance stabilization to be regained in the case of an alkene.

Chlorination, nitration, sulfonation, and alkylation of benzene can be formulated in much the same way. In nitration, sulfuric acid facilitates the formation of the nitronium ion, NO_2^+, which then reacts with the aromatic ring.

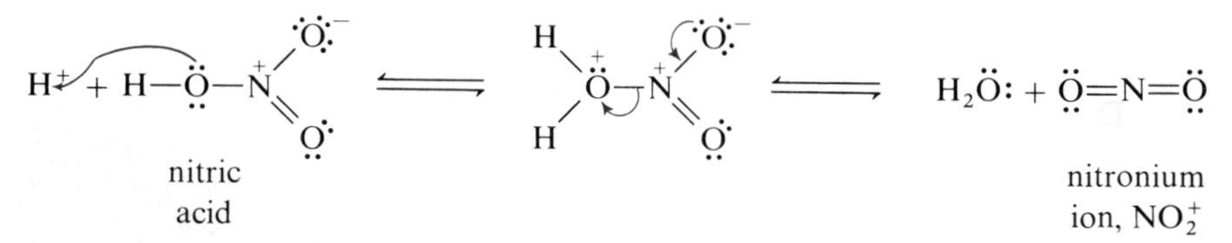

Sulfonation of aromatic compounds can be brought about with either concentrated or fuming sulfuric acid. The latter reagent contains sulfur trioxide, SO_3, dissolved in concentrated sulfuric acid and is very much more reactive. In either case, the electrophile appears to be sulfur trioxide itself.

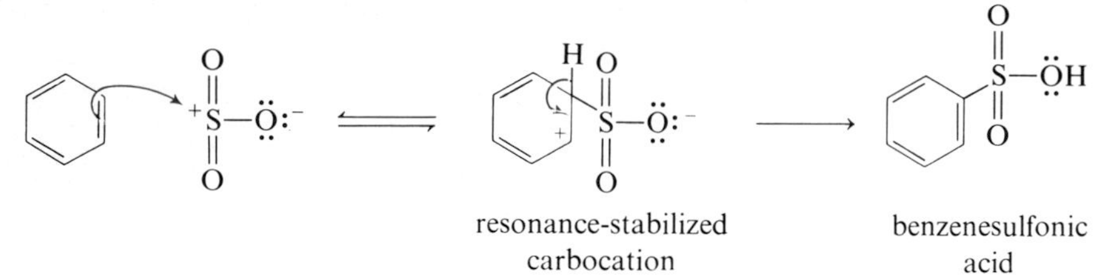

Benzenesulfonic acid is an acid comparable in strength to sulfuric acid.

$$C_6H_5{-}SO_3H + H_2O \rightleftharpoons C_6H_5{-}SO_3^- + H_3O^+ \qquad K_a = 2 \times 10^{-1}$$

Example 6.2 Write contributing structures for the resonance-stabilized carbocation intermediate formed by the reaction of benzene and the nitronium ion, NO_2^+. Use curved arrows to show how one contributing structure is converted to the next.

Solution

H, NO_2 $\longleftrightarrow$ H, NO_2 $\longleftrightarrow$ H, NO_2

PROBLEM 6.2 Write contributing structures for the resonance-stabilized carbocation intermediate formed by the reaction of benzene and SO_3. Use curved arrows to show how one contributing structure is converted to the next.

The reaction of an alkyl halide with benzene in the presence of a Lewis acid catalyst is known as the *Friedel-Crafts reaction.*

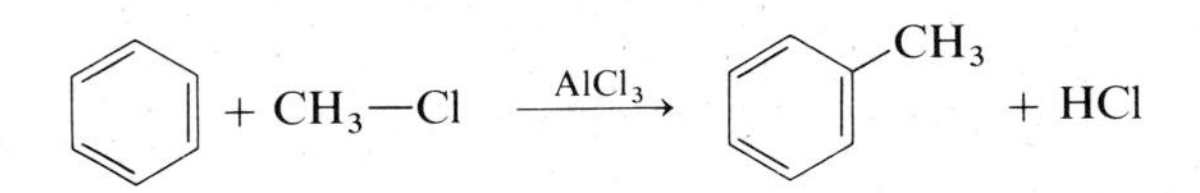

The function of the aluminum chloride is to polarize the C—Cl bond of the alkyl halide and generate carbocation character on the alkyl group.

$$CH_3{-}\ddot{\underset{..}{Cl}}{:} + Al(\ddot{Cl}{:})_3 \rightleftharpoons \overset{\delta+}{CH_3}\cdots\ddot{Cl}\cdots\overset{\delta-}{Al}(\ddot{Cl}{:})_3 \rightleftharpoons CH_3^+ \; {:}\ddot{Cl}{-}\overset{-}{Al}(\ddot{Cl}{:})_3$$

An alkyl carbocation, or something close to it in structure and polarity, is the electrophile.

6.6 DISUBSTITUTED DERIVATIVES OF BENZENE

There are relatively few substitution reactions of benzene itself, and the number of products that can be synthesized by direct substitution is limited. As we move to di-, tri-, and polysubstitution, the number of derivatives that can be synthesized becomes greater and at the same time the chemistry becomes more complex and more challenging. Our concern in this section is to examine the influence of a substituent already on the ring. As we do so, it will become clear that a substituent group on benzene directs the position at which further substitution takes place.

The directing effect can be seen by comparing the products of nitration of anisole and nitrobenzene. Nitration of anisole yields a mixture of *p*-nitroanisole and *o*-nitroanisole. Of these two products, the *para* isomer predominates. More important, there is virtually no *m*-nitroanisole formed.

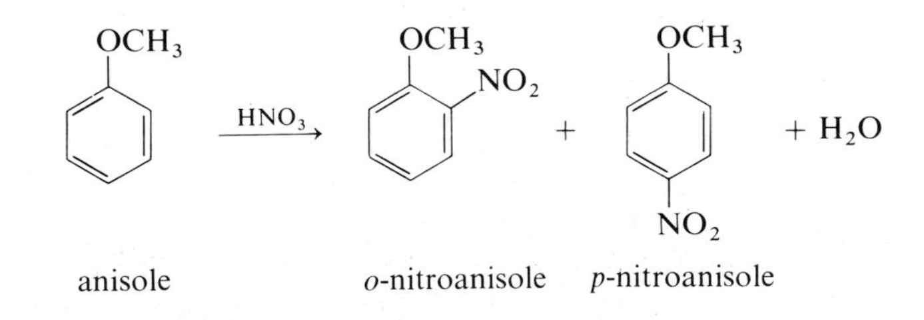

Because it directs the entering nitro group (or any other entering group for that matter) to the *ortho* and *para* positions, methoxyl is said to be an *ortho,para-directing group.*

Nitration of nitrobenzene yields a mixture consisting of approximately 93% of the *meta* isomer and less than 7% of the *ortho* and *para* isomers combined.

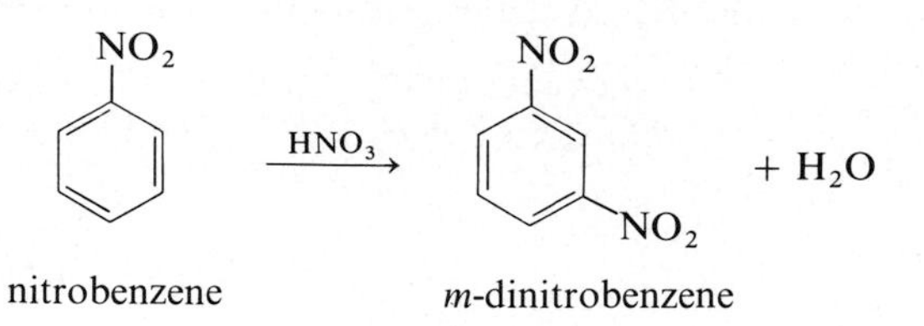

Because it directs the entering group to the *meta* position, nitro is said to be a *meta*-directing group.

Substitution reactions such as these have been done for a wide variety of monosubstituted derivatives of benzene and the directing influences of various functional groups determined. While these directing influences are not absolute, for most functional groups and in particular for the *ortho,para*-directing groups, there is a very high directional specificity; generally no more than 5 to 10% of the "unpredicted" isomer is formed. Listed in Table 6.2 are the directing influences for many of the important functional groups presented in this text.

If we compare these *ortho-*, *para-*, and *meta*-directing groups for structural similarities and differences, we can make the following generalization. If the atom attached directly to the aromatic ring bears a multiple bond, the group is a *meta*-director; otherwise it is an *ortho,para*-director. Of the groups listed in Table 6.2, the only two exceptions to this generalization are phenyl and the trimethylammonium group, $-N(CH_3)_3^+$.

We can illustrate the usefulness of this generalization by considering the

Table 6.2 Directing effects of some common functional groups.

ortho,para-directing	*meta-directing*
$-F, -Cl, -Br, -I$	$-\overset{+}{N}(=O)-O^-$, $-S(=O)_2-OH$
$-CH_3, -C_2H_5, -R$	
$-OH, -OCH_3, -OR$	
$-NH\overset{O}{\overset{\Vert}{C}}R$	$-\overset{O}{\overset{\Vert}{C}}OH, -\overset{O}{\overset{\Vert}{C}}OR, -\overset{O}{\overset{\Vert}{C}}NH_2$
$-NH_2, -NHR, -NR_2,$	$-\overset{O}{\overset{\Vert}{C}}H, -\overset{O}{\overset{\Vert}{C}}R$
$-O\overset{O}{\overset{\Vert}{C}}R$	$-\overset{+}{N}(CH_3)_3$
$-C_6H_5$	

synthesis of two different disubstituted derivatives of benzene. Suppose we wish to prepare *m*-bromonitrobenzene from benzene. Such a conversion can be done in two steps, nitration and bromination. If the steps are carried out in just this order, the product is indeed the desired *m*-bromonitrobenzene. The nitro group is a *meta*-director and therefore directs the incoming bromine atom to the *meta* position.

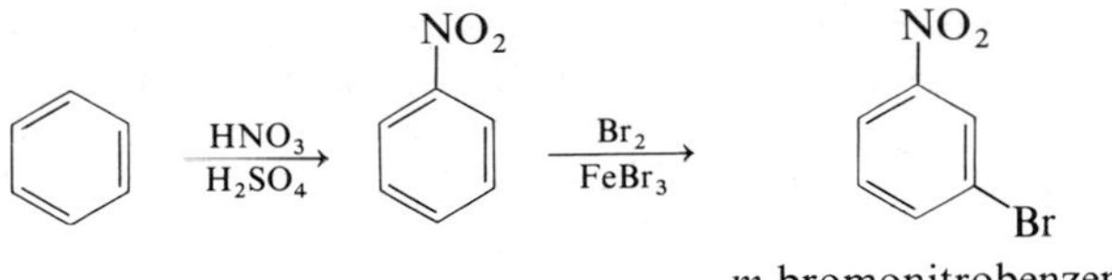

But what happens if we reverse the order of these two steps? Bromination produces bromobenzene. Bromine is an *ortho,para*-directing group and directs the incoming nitro group to the *ortho* and *para* positions. Bromination followed by nitration yields a mixture of the *ortho* and *para* isomers.

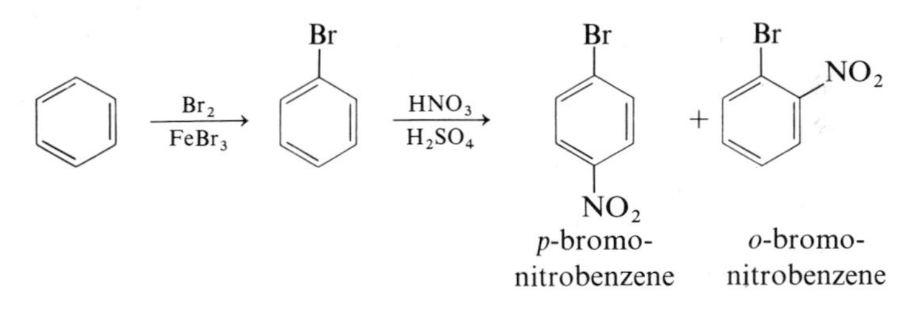

Clearly the order in which electrophilic aromatic substitution reactions are carried out is critical. As another example, consider the conversion of toluene into *p*-nitrobenzoic acid. The $—NO_2$ group can be introduced using a nitrating mixture of nitric and sulfuric acids. The —COOH group can be produced by oxidation of the $—CH_3$ group (Section 8.4).

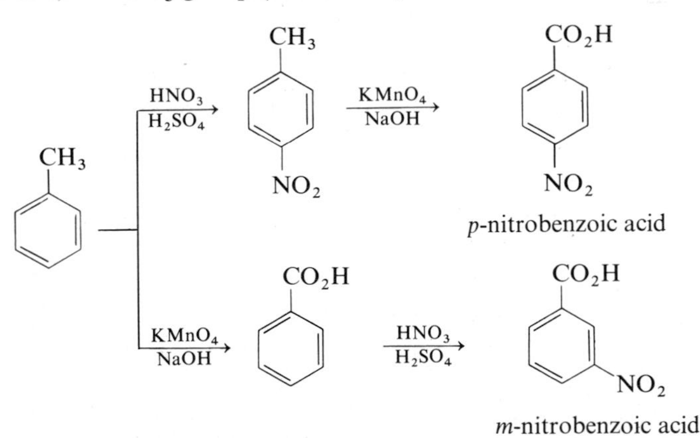

Nitration of toluene yields a product with the two substituents in the desired *para* relationship. Nitration of benzoic acid yields a product with the substituents *meta* to each other. Again we see that the order in which the steps are performed is critical. Note in this last example that we have shown the nitration of toluene producing only the *para* isomer. Since $—CH_3$ is an *ortho,para*-directing group, both *ortho* and *para* isomers are formed. However, in problems of this type where you are asked to prepare one or the other of these isomers, we will assume that there are chemical or physical methods that can be used to separate the desired isomer in pure form.

Example 6.3 Complete the following electrophilic aromatic substitution reactions. Where you predict *meta* substitution, show only the *meta* product. Where you predict *ortho-para* substitution, show both products.

Solution

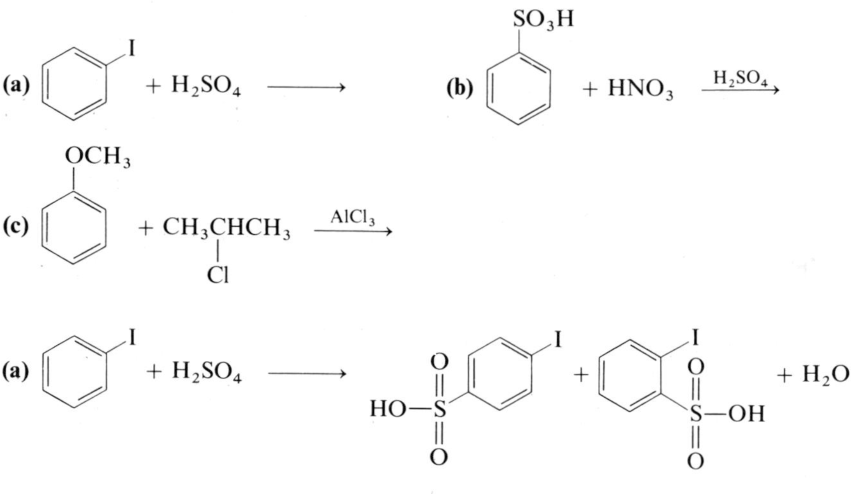

p-iodobenzene-sulfonic acid

o-iodobenzene-sulfonic acid

Iodine directs the incoming group *ortho-para.*

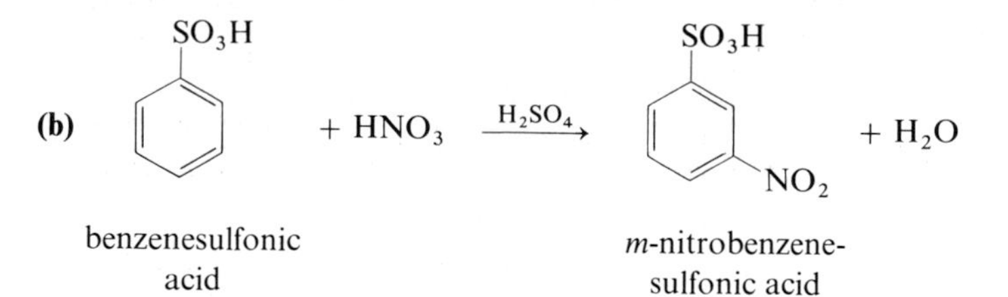

benzenesulfonic acid

m-nitrobenzene-sulfonic acid

The sulfonic acid group directs the incoming group meta.

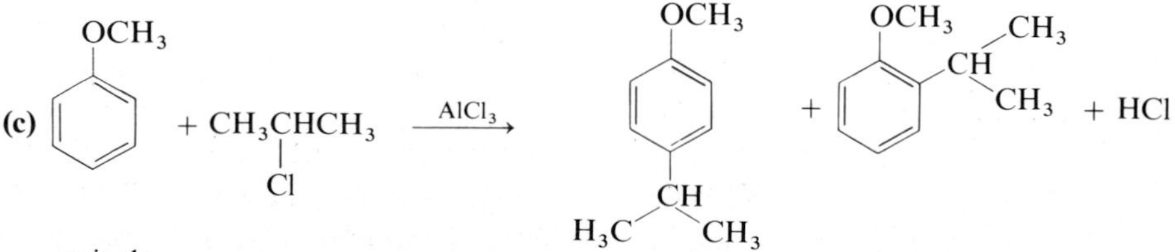

anisole

p-methoxycumene

o-methoxycumene

The $—OCH_3$ group directs the incoming group *ortho,para*. Note that the products may be named as derivatives of anisole (e.g., *p*-isopropylanisole) or as derivatives of cumene (e.g., *p*-methoxycumene).

PROBLEM 6.3 Complete the following electrophilic aromatic substitution reactions.

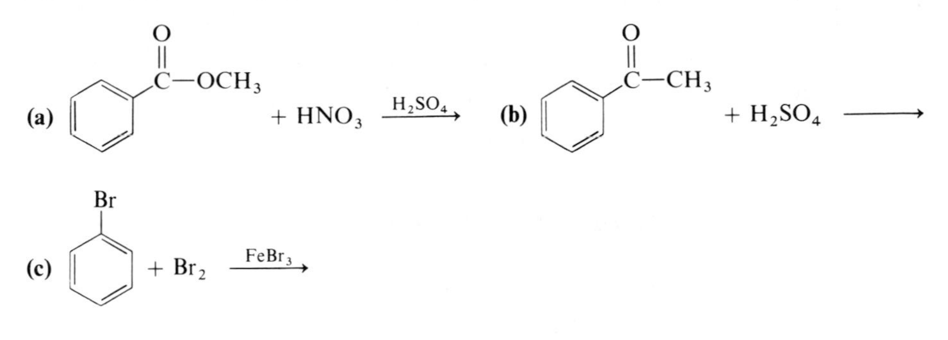

6.7 THEORY OF DIRECTING INFLUENCES

We have seen that substituents on a benzene ring exert a directing influence. Now let us propose a mechanism that accounts for this relationship between structure and reactivity. We will do this by proposing a mechanism for each particular reaction and then examining the relative stabilities of the various reaction intermediates.

First consider the nitration of anisole. The rate of electrophilic aromatic substitution is determined by the slowest step in the mechanism. For nitration of anisole (and in fact for every substitution reaction we shall consider) the slow step is the attack by the electrophile on the aromatic ring. Electrophilic attack by the nitronium ion, NO_2^+, produces a resonance-stabilized cation. Shown in Figure 6.5 is the resonance-stabilized cation formed by attack of the electrophile meta to the $—OCH_3$ group. Also shown in Figure 6.5 is the resonance-stabilized cation formed by attack para to the $—OCH_3$ group.

Attack of the nitronium ion in the position *meta* to the methoxyl group yields the resonance-stabilized cation (a) to (c). Attack in the *para* position produces the resonance-stabilized cation (d) to (g). (For the purposes of this discussion we will ignore for the moment the fact that attack can also be in the *ortho* position.) For each mode of attack we can draw three contributing structures that place the positive charge on carbon atoms of the ring. These three structures are the only important ones that can be drawn for *meta* attack. In the case of *para* attack, there is a fourth important contributing structure, one involving the methoxyl group and placing the positive charge on oxygen. Since the cation produced in *para* attack is stabilized more by resonance (a greater delocalization of the positive charge) than is the cation produced by *meta* attack, anisole undergoes nitration in the *para* position rather than in the *meta* position.

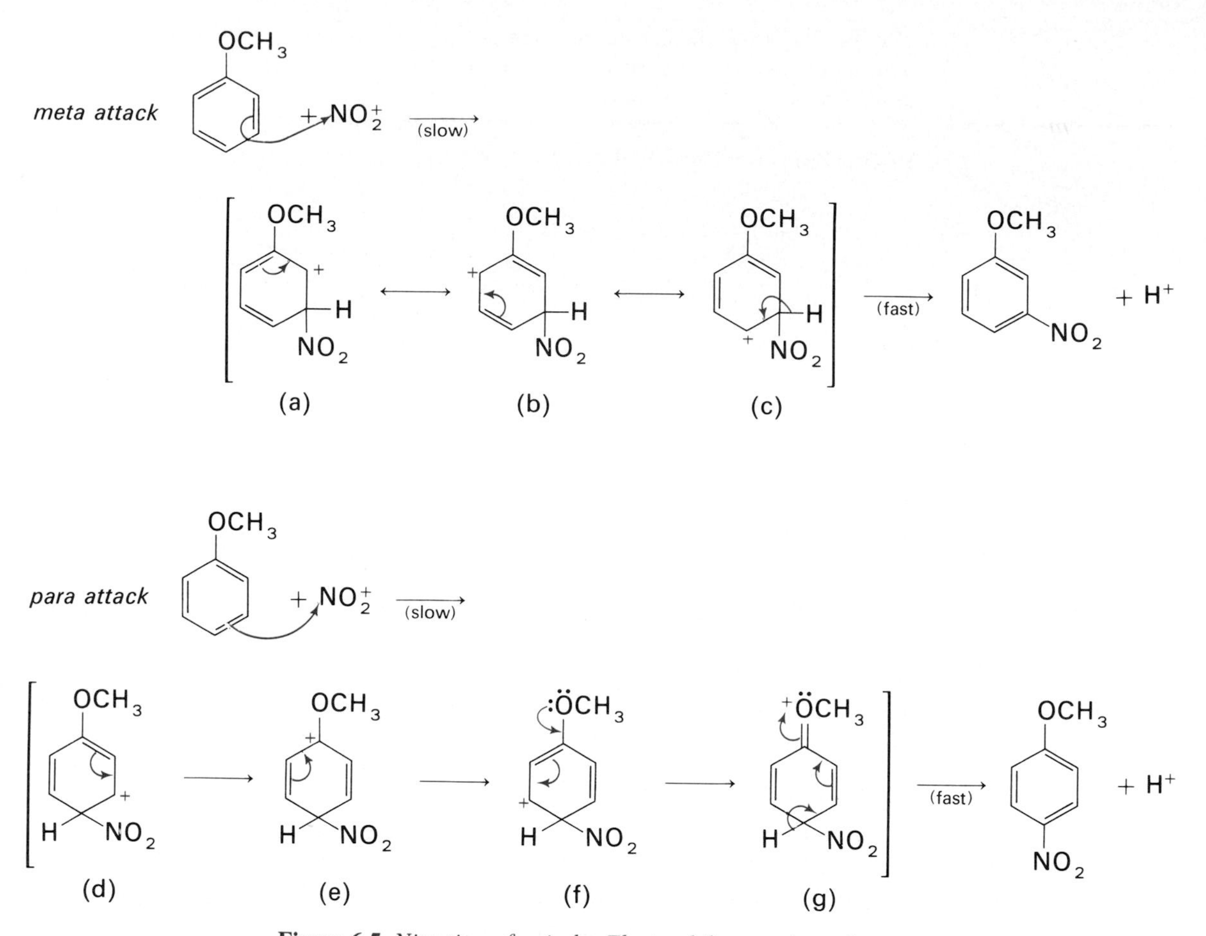

Figure 6.5 *Nitration of anisole. Electrophilic attack at the meta position and at the para position.*

Now let us examine the *meta*-directing influence of the nitro group in much the same manner as we have done for the *ortho,para*-directing influence of the methoxyl group. Shown in Figure 6.6 are the resonance-stabilizing cations formed by *meta* and *para* attack of the nitronium ion on nitrobenzene.

Each cation is a hybrid of three contributing structures and there are no additional important ones that can be drawn. There are two significant factors accounting for the markedly different influence of $—NO_2$ compared to $—OCH_3$. First, the nitro group cannot supply electrons to the ring as did the methoxyl group. Second, if we draw a Lewis structure for the nitro group, we realize that the nitrogen atom bears a formal positive charge.

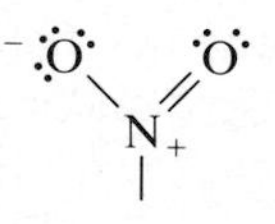

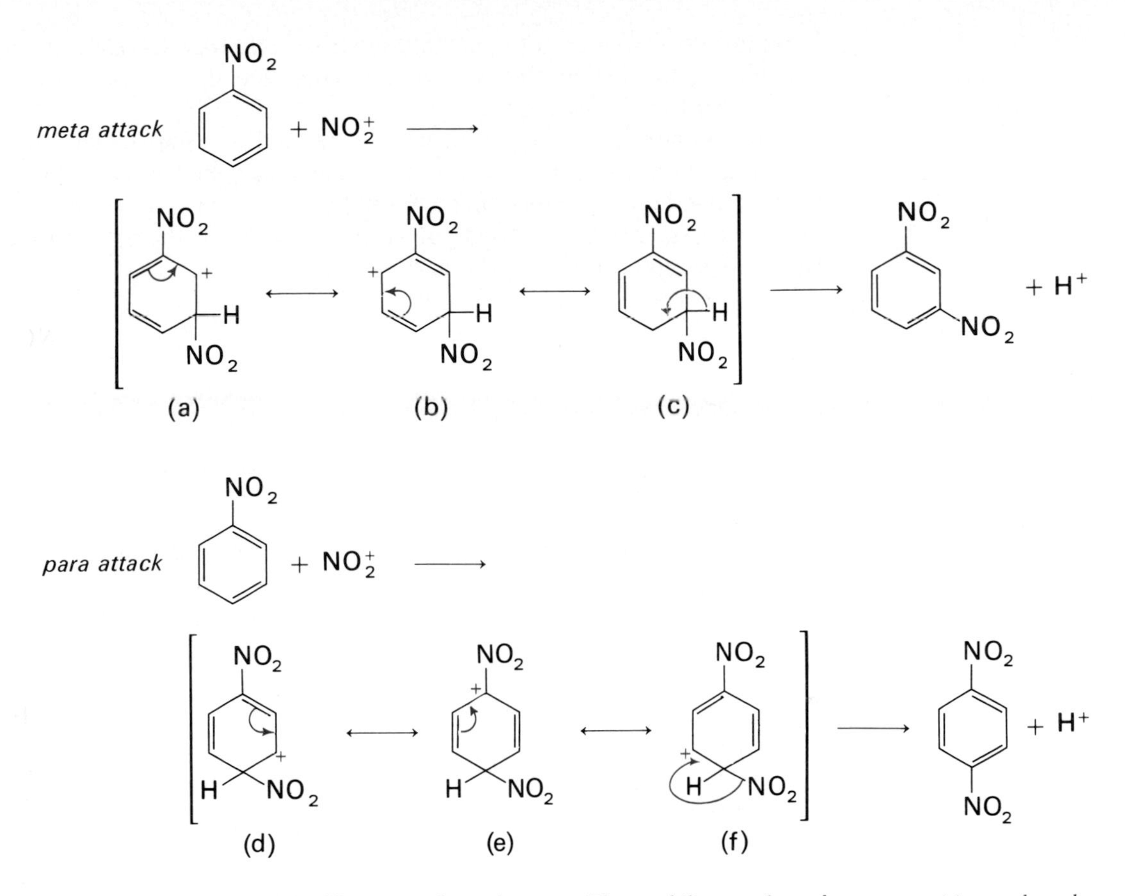

Figure 6.6 *Nitration of nitrobenzene. Electrophilic attack at the meta position and at the para position.*

If we now re-examine contributing structure (e) in Figure 6.6 we see that there are positive charges on adjacent atoms.

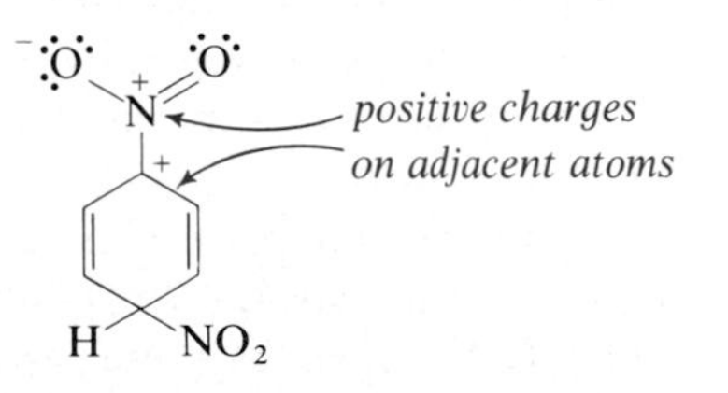

None of the other contributing structures for either *meta* or *para* attack places positive charges on adjacent atoms. This build-up of positive charge on adjacent atoms is unfavorable since like charges repel each other. Therefore contributing structure (e) is less important than either (d) or (f) in stabilizing the cation produced in *para* attack. As a consequence, resonance stabilization for *meta* attack is greater (three important contributing structures) than for *para* attack (only two important contributing structures). Therefore the reaction proceeds by *meta* rather than by *para* attack.

The same type of argument can be applied to each of the other *meta*-directing groups listed in Table 6.2. In each instance, the atom attached directly to the aromatic ring bears either a formal positive charge or at least a partial positive charge because of bond polarization. Typical is the polarization of the carbonyl group in benzaldehyde and in acetophenone.

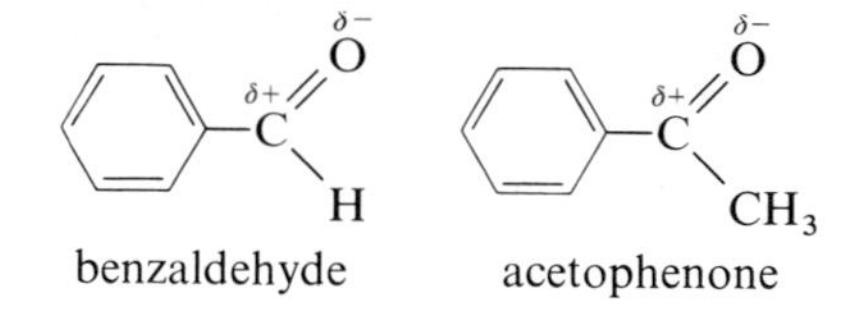

benzaldehyde acetophenone

We can summarize the various empirical generalizations presented in these two sections and the underlying theory as follows:

(1) The position of substitution is controlled by the group already present on the ring, not by the entering group.

(2) If the atom attached to the aromatic ring is part of a multiple-bond, the group is *meta*-directing; otherwise it is *ortho,para*-directing. While the directing influence of a particular group is not 100% effective, generally less than 5 to 10% of the "unpredicted" product is formed. Of the groups listed in Table 6.2, the only exceptions to these generalizations are phenyl which directs *ortho,para* and trimethylammonium which directs *meta*. All of these directing influences can be accounted for in terms of resonance stabilization of the cation formed on attack of the electrophile.

6.8 THE CONCEPT OF AROMATICITY

With the development of a structural theory for benzene and other aromatic hydrocarbons, it became clear that the unusual chemical and physical properties of benzene are due to the presence of six pi electrons in a cyclic, fully conjugated system. The question that followed was: are there other cyclic, fully conjugated molecules that also have aromatic properties? The answer is yes. Several classes of molecules have (1) a substantial resonance energy and (2) tend to undergo substitution reactions like benzene rather than addition reactions like the alkenes.

Pyridine and pyrimidine are heterocyclic analogs of benzene in which first one CH group and then two are replaced by nitrogen atoms. Following are structural formulas for these substances along with the numbering system used to specify the location of substituents.

pyridine pyrimidine

The nitrogen atoms in pyridine and pyrimidine are sp^2 hybridized and each contributes one of the six pi electrons in the ring. The unshared pair of electrons on each nitrogen atom lies in an sp^2 orbital in the plane of the ring and is not a part of either six pi electron system. We can represent each molecule as a hybrid of two benzene-like contributing structures.

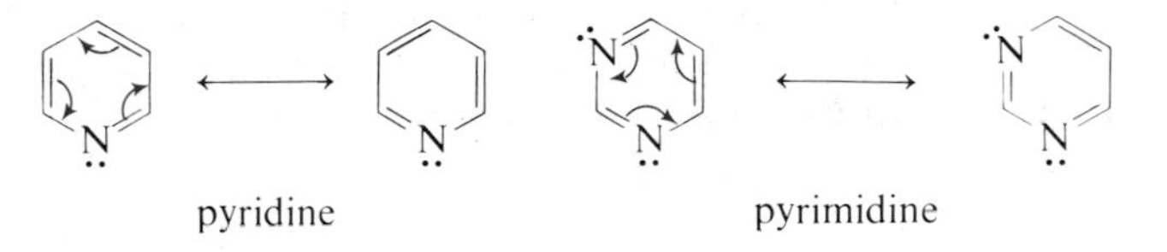

pyridine pyrimidine

The resonance energy of pyridine is about 32 kcal/mole, just slightly less than that of benzene itself. The resonance energy of pyrimidine is 26 kcal/mole.

The five-membered rings pyrrole, furan, and imidazole also show aromatic character.

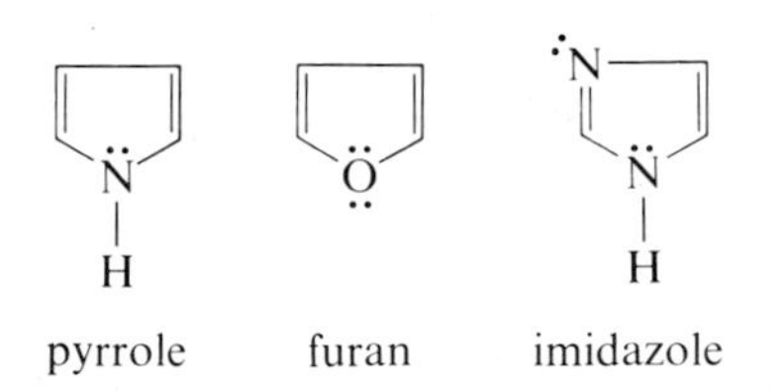

pyrrole furan imidazole

In each molecule, the heteroatom is sp^2 hybridized. In pyrrole and furan, the six pi electrons are derived from four 2p electrons of the ring carbon atoms and two 2p electrons of the heteroatom. If you compare the origin of the six pi electrons in pyrrole and pyridine, you will note an important difference. In pyridine, the unshared pair of electrons on nitrogen is not a part of the six pi electrons; in pyrrole the unshared pair on nitrogen is very much a part of the aromatic sextet. The resonance energy of pyrrole is about 21 kcal/mole; that of furan is about 16 kcal/mole.

Nature abounds in compounds having two or more heterocyclic aromatic rings fused together. Two such substances especially important in the biological world are purine and indole.

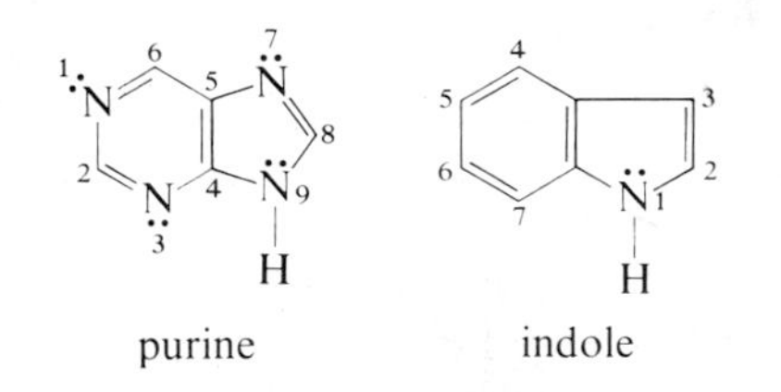

purine indole

Purine contains two fused rings, a six-membered pyrimidine ring, and a five-membered imidazole ring. Substances derived from purine and pyrimidine are building blocks of deoxyribonucleic acid (DNA) and ribonucleic acid (RNA). The structure of DNA and RNA are discussed in Chapter 14. Indole contains two fused rings, a six-membered benzene ring, and a five-membered pyrrole ring. Substances derived from indole include the essential amino acid

L-tryptophan (Chapter 12), the neutro-transmitter serotonin, and many other substances of plant and animal origin.

Thus, we can see that benzene is by no means unique in its aromatic properties. Rather, it is but one of a group of cyclic substances which have stabilities and chemical reactivities quite different from those of alkenes and alkynes.

6.9 SPECTROSCOPIC PROPERTIES

Benzene and other aromatic hydrocarbons show characteristic infrared absorption associated with both =C—H stretching and C=C stretching. The C—H stretching frequencies occur between 3000 cm^{-1} and 3100 cm^{-1} and generally consist of weak to medium intensity bands. Recall from Section 3.8 that this is the same region of the spectrum in which alkene =C—H stretching frequencies occur. In addition, aromatic hydrocarbons also show a number of bands between 1430 cm^{-1} and 1665 cm^{-1} corresponding to stretching of carbon-carbon bonds of the aromatic ring. These two sets of stretching frequencies can be seen in the infrared spectrum of *n*-propyl benzene (Figure 6.7).

The conjugated system of benzene is responsible for several absorption maxima in the ultraviolet-visible region of the spectrum. The most important of these absorptions occur at 200 nm, and at 257 nm. Toluene shows ultraviolet absorption at 265 nm, a value close to that of benzene itself. Groups attached to the benzene ring, particularly those which contain pi electrons and

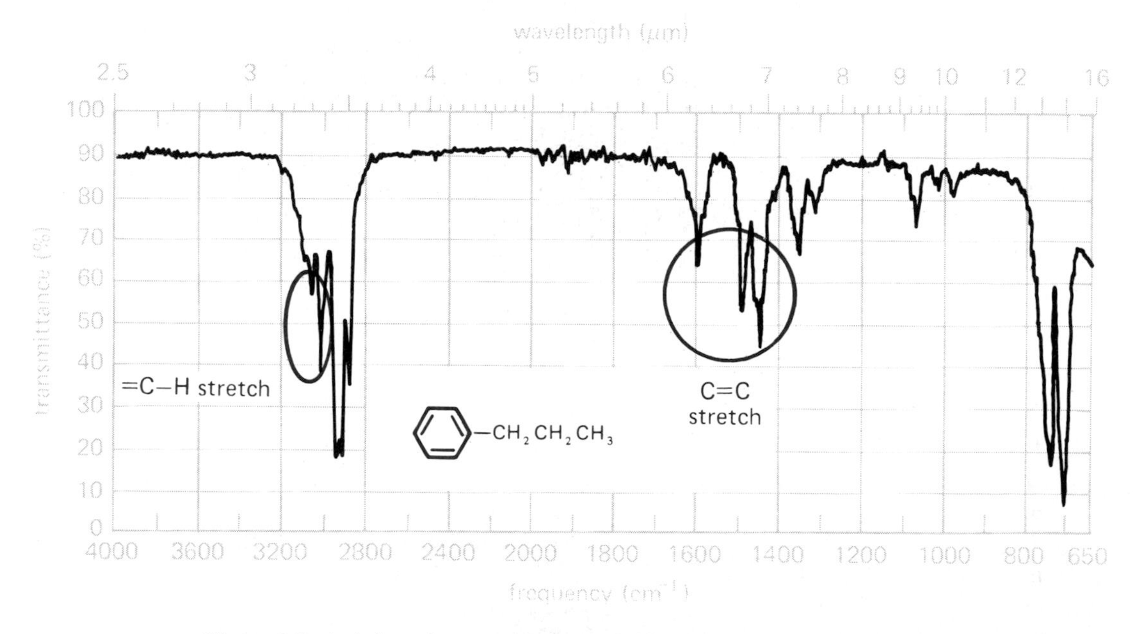

Figure 6.7 *An infrared spectrum of n-propyl benzene.*

can thereby extend the conjugated system, cause a shift of the absorption maxima toward the visible region of the spectrum. Styrene in which an additional carbon-carbon double bond is conjugated with the aromatic ring absorbs at 282 nm. Naphthalene, containing two fused aromatic rings, absorbs at 314 nm.

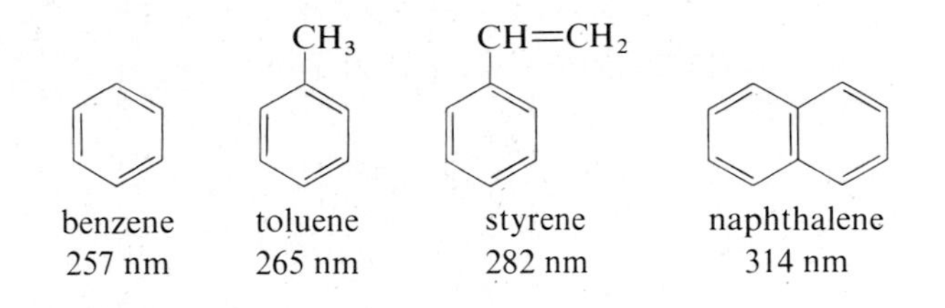

The NMR signals for protons attached to aromatic rings generally occur in the range $\delta = 5.6$ to 8.3. The NMR spectrum of benzene itself shows a single sharp peak at $\delta = 7.3$ due to the six chemically equivalent hydrogens of the aromatic ring. That of ethyl benzene (Figure 6.8) shows a singlet at $\delta = 7.3$ corresponding to the five protons of the $—C_6H_5$ group. Notice also from Figure 6.8 that the three protons of the $—CH_3$ group appear as a triplet and the two protons of the $—CH_2—$ group appear as a quartet.

Recall from Section 3.8 that NMR signals due to vinyl hydrogens (those attached to alkene carbon-carbon double bonds) typically occur in the range $\delta = 4.5$ to 5.5. Therefore, it is generally possible to distinguish between aromatic hydrogens and vinyl hydrogens by differences in chemical shifts.

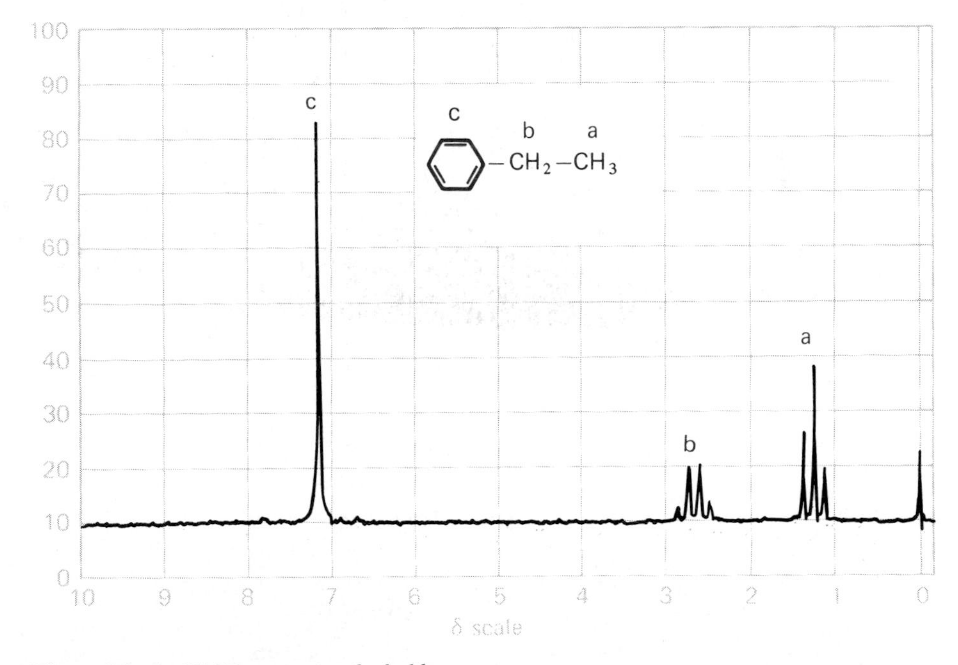

Figure 6.8 *An NMR spectrum of ethyl benzene.*

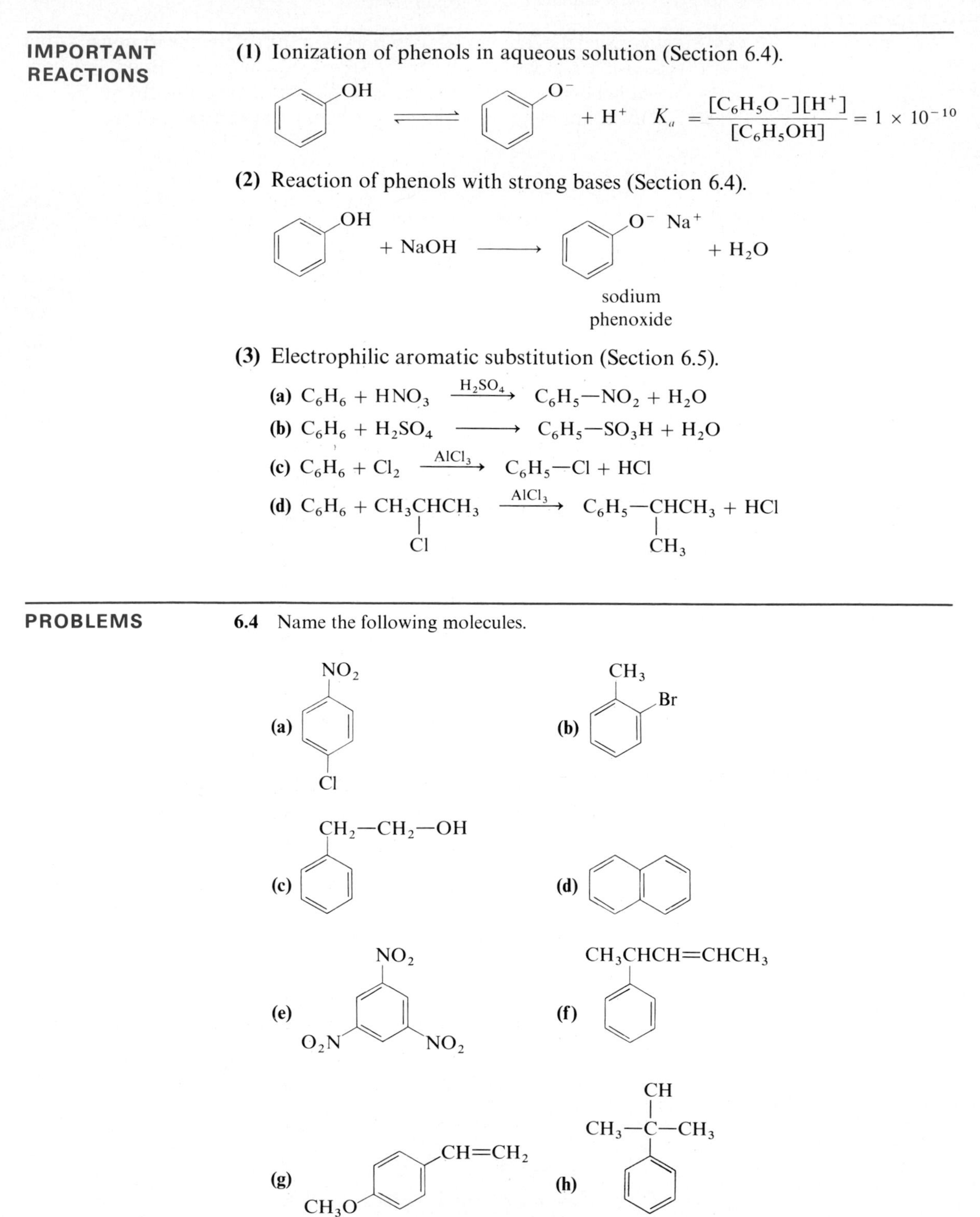

IMPORTANT REACTIONS

(1) Ionization of phenols in aqueous solution (Section 6.4).

$$C_6H_5OH \rightleftharpoons C_6H_5O^- + H^+ \qquad K_a = \frac{[C_6H_5O^-][H^+]}{[C_6H_5OH]} = 1 \times 10^{-10}$$

(2) Reaction of phenols with strong bases (Section 6.4).

$$C_6H_5OH + NaOH \longrightarrow C_6H_5O^-\ Na^+ + H_2O$$

sodium phenoxide

(3) Electrophilic aromatic substitution (Section 6.5).

(a) $C_6H_6 + HNO_3 \xrightarrow{H_2SO_4} C_6H_5{-}NO_2 + H_2O$

(b) $C_6H_6 + H_2SO_4 \longrightarrow C_6H_5{-}SO_3H + H_2O$

(c) $C_6H_6 + Cl_2 \xrightarrow{AlCl_3} C_6H_5{-}Cl + HCl$

(d) $C_6H_6 + CH_3CH(Cl)CH_3 \xrightarrow{AlCl_3} C_6H_5{-}CH(CH_3)CH_3 + HCl$

PROBLEMS

6.4 Name the following molecules.

(a) NO_2, Cl

(b) CH_3, Br

(c) $CH_2{-}CH_2{-}OH$

(d)

(e) NO_2, O_2N, NO_2

(f) $CH_3CHCH{=}CHCH_3$

(g) $CH{=}CH_2$, CH_3O

(h) CH, $CH_3{-}C{-}CH_3$

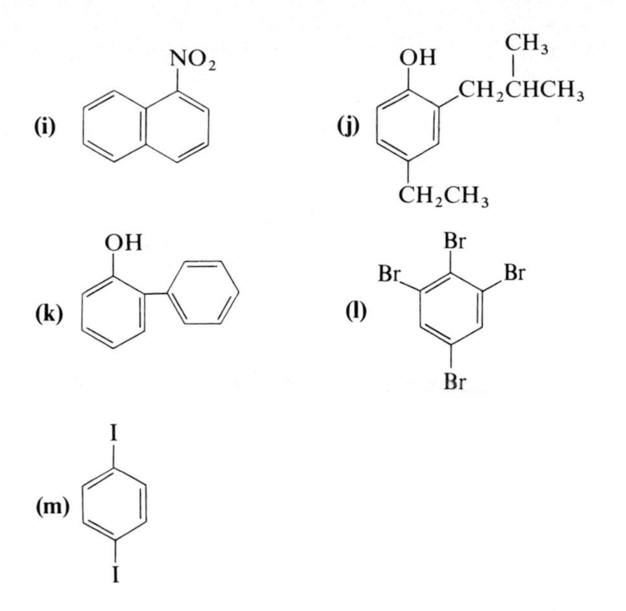

6.5 Draw structural formulas for the following molecules:

(a) *m*-dibromobenzene
(b) 2,4,6-trinitrophenol
(c) *p*-chloroiodobenzene
(d) 4-phenyl-2-pentanol
(e) *p*-cresol
(f) *p*-xylene
(g) anthracene
(h) pentachlorophenol
(i) 2-phenyl-2-pentanol
(j) diphenylmethane
(k) styrene
(l) 1-phenylcyclohexanol

6.6 Name and draw structural formulas for all derivatives of benzene having the following molecular formulas.

(a) $C_6H_3Br_3$ **(b)** C_8H_{10} **(c)** C_9H_{12}

6.7 Wilhelm Körner devised one of the earliest methods for determining the relative orientations of substituent groups on an aromatic ring. This method was first applied to determine the structural formulas of the three isomeric dibromobenzenes, $C_6H_4Br_2$. Let us call them isomer *A* (mp +87°C), isomer *B* (mp +7°C), and isomer *C* (mp −7°C). Each isomer can be converted into one or more mononitroderivatives, $C_6H_3Br_2NO_2$. Isomer *A* yields one mononitroderivative; isomer *B* yields two mononitroderivatives; and isomer *C* yields three mononitroderivatives. From this information assign structural formulas to *A*, *B*, and *C*.

6.8 There are three isomeric tribromobenzenes, melting points 44°, 88°, and 122°C. Show how the Körner method might be applied to determine the structural formulas of these isomers.

6.9 Draw five principal resonance contributing structures for the following:

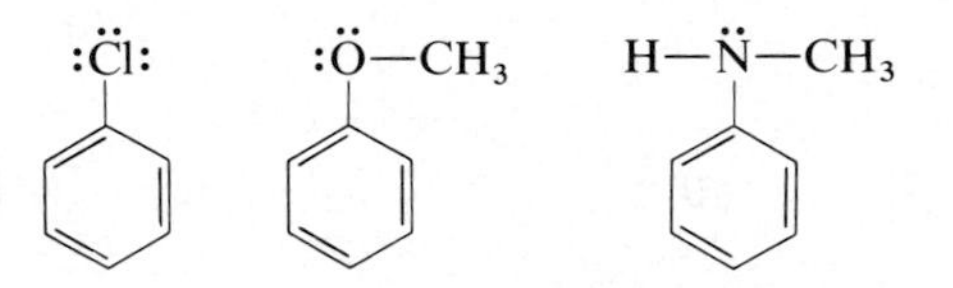

6.10 Draw the three Kekulé structures for naphthalene and the four Kekulé structures for anthracene.

6.11 Using the resonance theory, account for the fact that *p*-nitrophenol is a stronger acid than phenol itself.

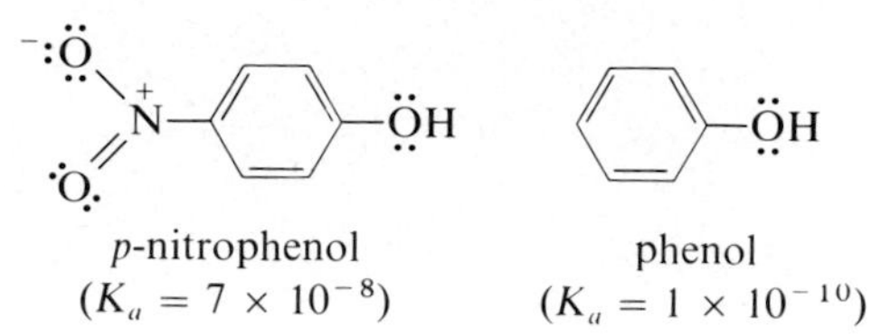

The substitution of three nitro groups on phenol leads to a dramatic increase in acidity. Picric acid (2,4,6-trinitrophenol) is comparable in strength to hydrochloric acid.

6.12 Following are structural formulas for benzyl alcohol and *ortho*-cresol.

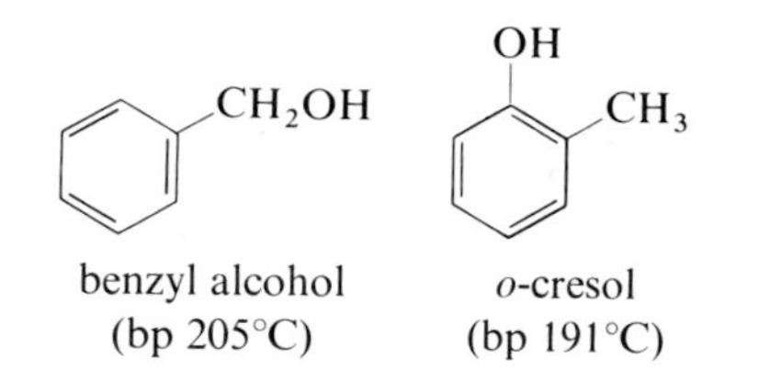

Describe a procedure you might use to separate a mixture of these two compounds and recover each in pure form.

6.13 Complete the following electrophilic aromatic substitution reactions.

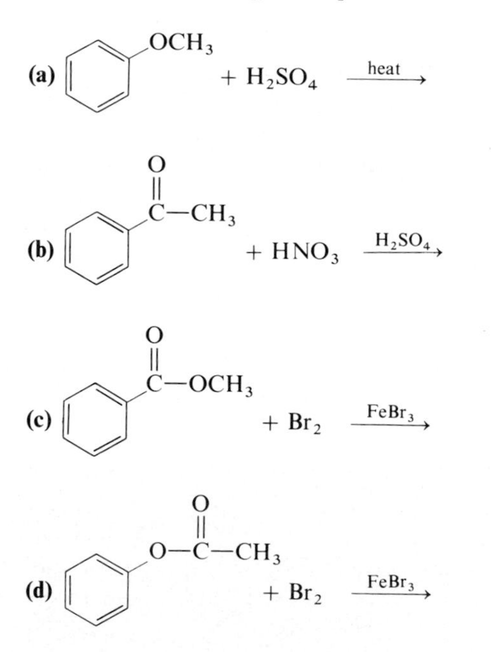

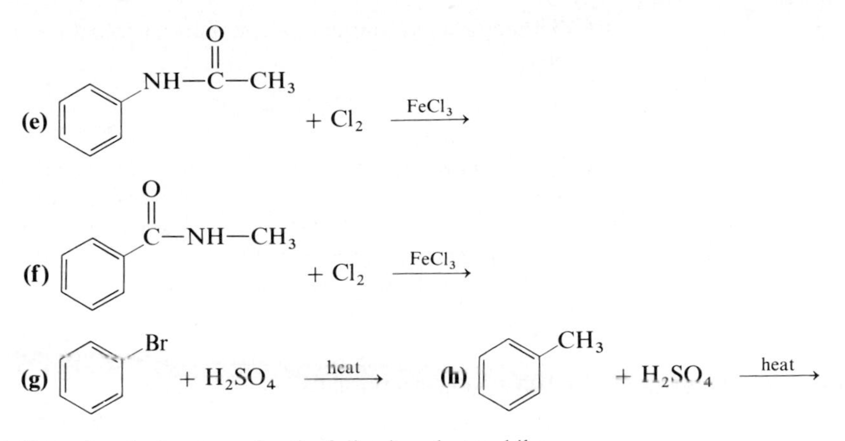

6.14 Draw Lewis structures for the following electrophiles.

(a) Br^+ **(b)** Cl^+ **(c)** NO_2^+

(d) $CH_3\underset{\displaystyle CH_3}{\underset{|}{C}}H^+$ **(e)** HSO_3^+ **(f)** $CH_2{=}CH{-}CH_2^+$

6.15 Draw contributing structures for the resonance-stabilized cations formed in the following electrophilic aromatic substitution reactions. Assume attack as indicated over each arrow.

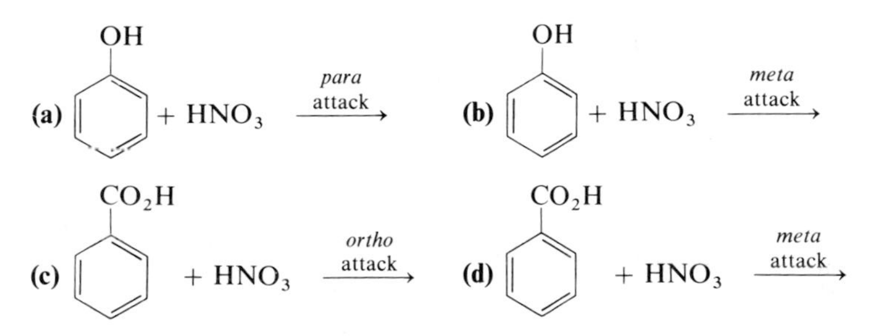

6.16 According to modern organic theory, the factor favoring *ortho,para* substitution is the ability of the atom attached directly to the ring to participate in stabilizing the intermediate cation. Participation by the oxygen of methoxyl is shown in contributing structure (g) in Figure 6.5. Draw comparable contributing structures showing resonance stabilization by the following *ortho,para*-directing groups.

(a) —OH **(b)** $-NHCH_3$ **(c)** —Cl

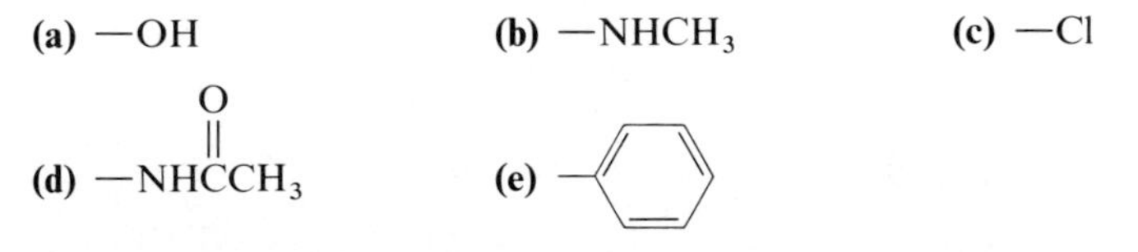

6.17 Show the reagents and conditions you would use to convert benzene into the following:

(a) toluene
(b) *p*-bromotoluene
(c) *o*-bromobenzoic acid
(d) *m*-bromobenzoic acid
(e) 2,4,6-trinitrotoluene (TNT)
(f) *p*-dichlorobenzene (a moth repellant)
(g) *p*-chlorobenzenesulfonic acid
(h) *o*-nitrobenzoic acid
(i) *m*-nitrobenzoic acid

6.18 Propose a mechanism to account for the following reaction.

$$2C_6H_6 + CH_2Cl_2 \xrightarrow{AlCl_3} C_6H_5{-}CH_2{-}C_6H_5 + 2HCl$$

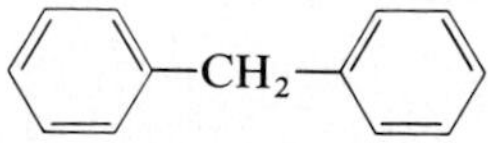

6.19 Show how you might distinguish between the following pairs of compounds by a simple chemical test. In each case, tell what test you would perform, what you would expect to observe, and write an equation for each positive test.

(a) benzene and cyclohexene
(b) ethylbenzene and styrene

6.20 List one major spectral characteristic that will enable you to distinguish between the following compounds.

(a) benzene and cyclohexene (UV, NMR)
(b) methylcyclopentane and benzene (IR, UV, NMR)
(c) toluene and styrene (UV, NMR)
(d) *p*-xylene and ethylbenzene (NMR)
(e) *p*-chlorotoluene and benzyl chloride (NMR)
(f) toluene and methylcylohexane (IR, UV, NMR)
(g) ethylbenzene and isopropylbenzene (NMR)
(h) *p*-diethylbenzene and 1,3,5-trimethylbenzene (NMR)

CHAPTER 7

Aldehydes and Ketones

In Chapter 3 we studied the physical and chemical properties of compounds containing carbon-carbon double bonds. In this and several following chapters, we will study the physical and chemical properties of compounds containing the carbonyl group, C=O. The carbonyl group is one of the most important functional groups in organic chemistry because it is the central structural feature of aldehydes, ketones, carboxylic acids, esters, amides, and anhydrides. The chemical properties of the carbonyl group are straightforward; understanding its few reaction themes leads very quickly to understanding a wide variety of reactions.

7.1 THE STRUCTURE OF ALDEHYDES AND KETONES

The characteristic structural feature of an aldehyde is the presence of a carbonyl group bonded to a hydrogen atom. In formaldehyde, the simplest aldehyde, a carbonyl group, is bonded to two hydrogen atoms. In all other aldehydes, the carbonyl group is bonded to one hydrogen and one carbon. Following are Lewis structures of formaldehyde and acetaldehyde.

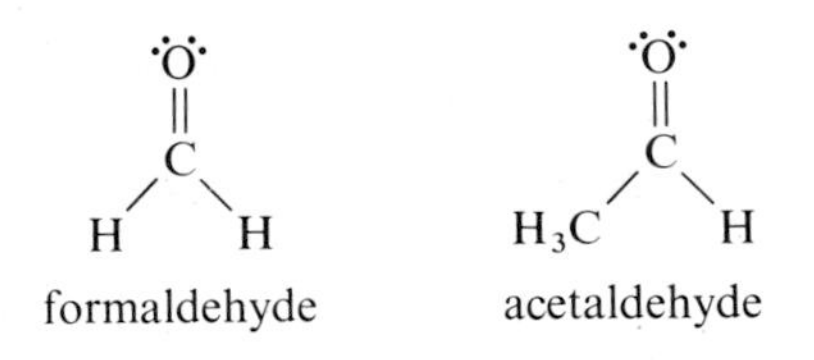

The characteristic structural feature of a ketone is a carbonyl group bonded to two carbon atoms. A Lewis structure of acetone, the simplest ketone, is

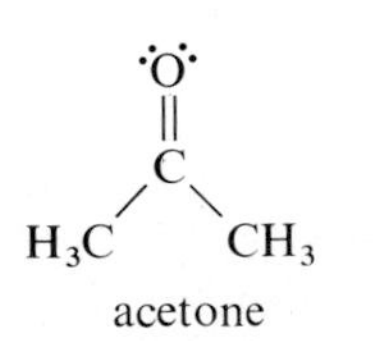

In forming covalent bonds with three other atoms, carbon uses sp^2 hybrid orbitals. The carbon-oxygen double bond consists of one sigma bond formed

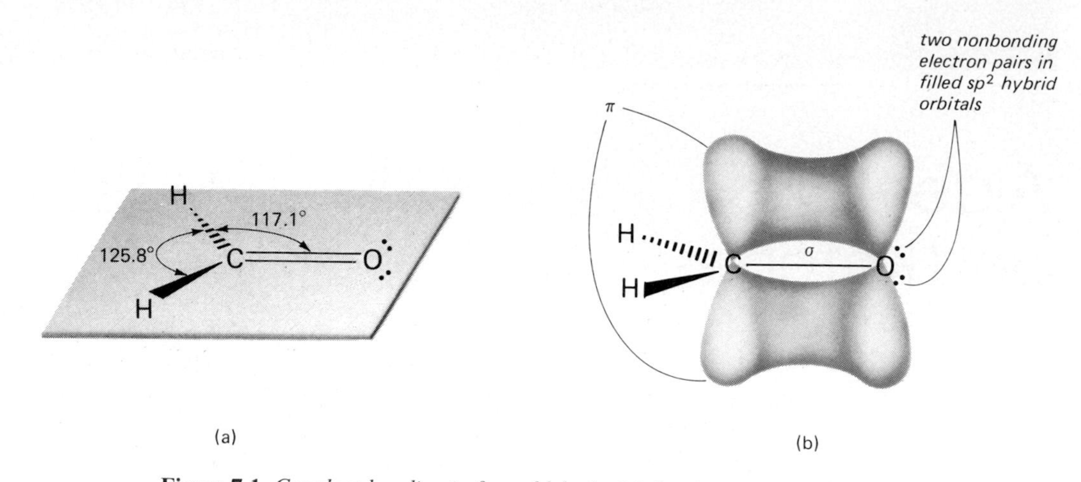

Figure 7.1 *Covalent bonding in formaldehyde. (a) Lewis structure; (b) covalent bond formation by the overlap of atomic orbitals.*

by the overlap of sp^2 orbitals and one pi bond formed by the overlap of parallel $2p$ orbitals. Figure 7.1 shows an orbital overlap diagram of the bonding in formaldehyde.

Formaldehyde is a trigonal planar molecule. The H—C—O bond angle is 117.1°, a value close to the predicted angle of 120° about an sp^2 hybridized carbon atom.

7.2 NOMENCLATURE OF ALDEHYDES AND KETONES

The IUPAC system of nomenclature for aldehydes follows the familiar pattern of selecting as the parent compound the longest chain of carbon atoms that contains the functional group. The aldehyde group is shown by changing the -e of the parent compound to -al. The aldehyde functional group can only occur on the end of the parent compound, so there is no need to use the number 1 to locate its position. Following are structural formulas and IUPAC names for methanal, ethanal, and 4-methylpentanal.

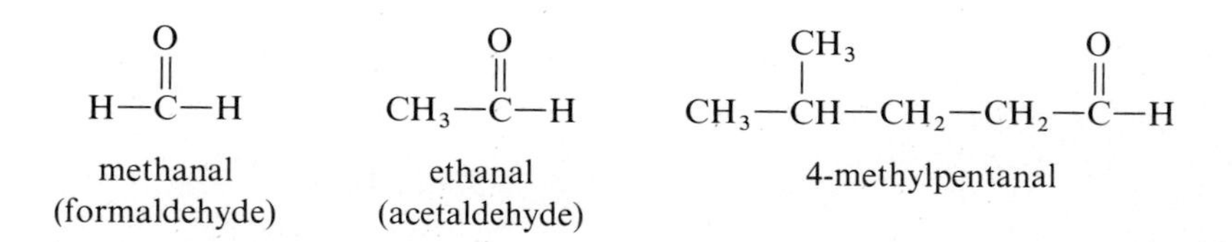

The IUPAC system retains certain common names. For example, formaldehyde and acetaldehyde are retained as IUPAC names for methanal and ethanal.

Other common names retained in the IUPAC system are benzaldehyde and cinnamaldehyde.

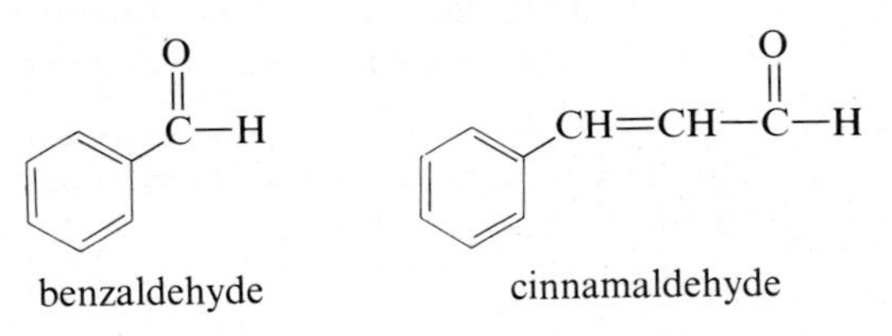

benzaldehyde cinnamaldehyde

The IUPAC system names ketones by selecting as the parent compound the longest chain that contains the carbonyl group, and then indicating the carbonyl group by changing the -e of the parent compound to -one. The parent chain is numbered from the direction that gives the carbonyl group the lowest number. The IUPAC system retains the common names acetone and acetophenone.

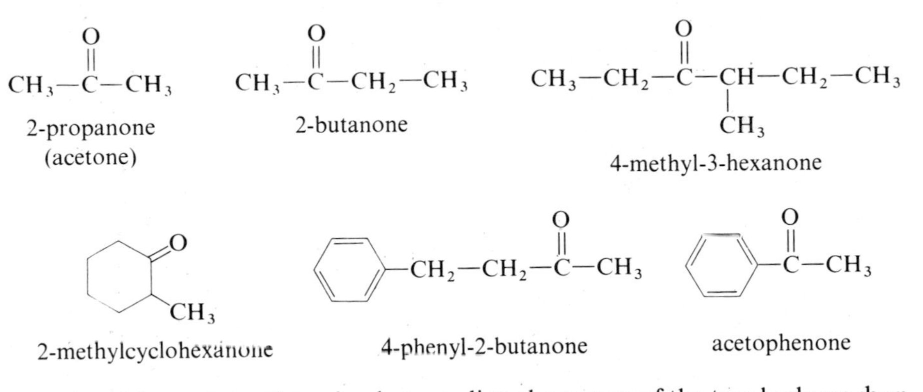

2-propanone (acetone) 2-butanone 4-methyl-3-hexanone

2-methylcyclohexanone 4-phenyl-2-butanone acetophenone

An older system of naming ketones lists the names of the two hydrocarbon groups attached to the carbonyl group, followed by the word *ketone*. In the following examples, the IUPAC name is given first, followed by the names using this older system.

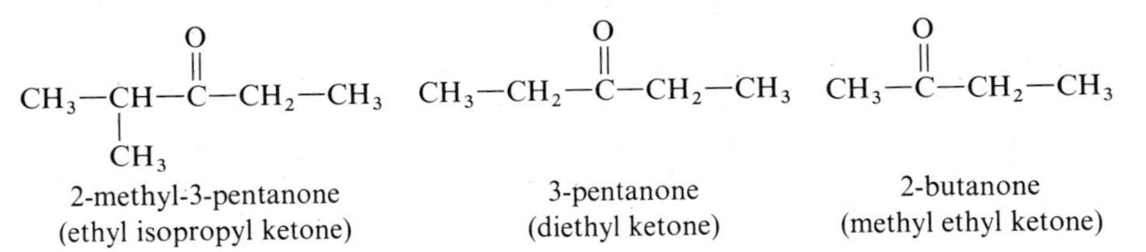

2-methyl-3-pentanone (ethyl isopropyl ketone) 3-pentanone (diethyl ketone) 2-butanone (methyl ethyl ketone)

In naming more complicated aldehydes and ketones, the carbonyl group takes precedence over alkene, hydroxyl, and most other functional groups.

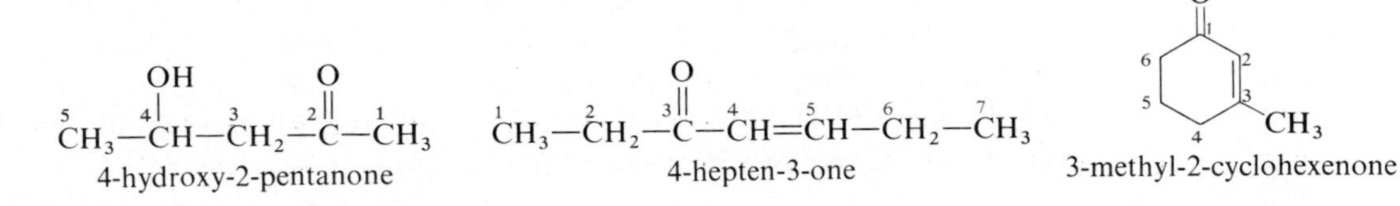

4-hydroxy-2-pentanone 4-hepten-3-one 3-methyl-2-cyclohexenone

In the first example the —OH group is indicated by 4-hydroxy. In the name 4-hepten-3-one, the heptane root is changed to 4-heptene to indicate the presence of an alkene between carbons 4 and 5 and the terminal -e is converted to -3-one to indicate the presence of a carbonyl group at carbon 3.

Example 7.1 Give IUPAC names for the following compounds.

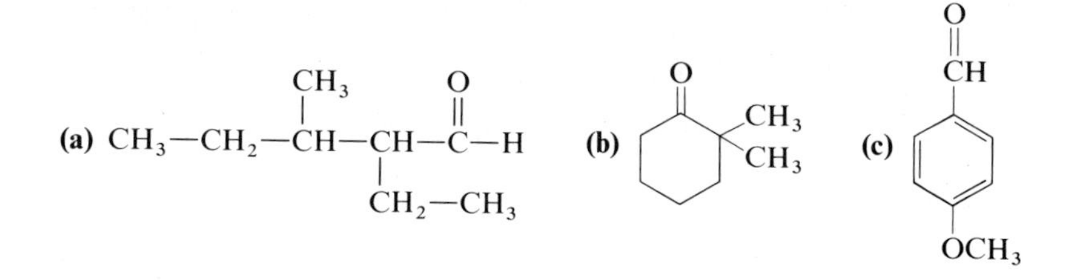

Solution **(a)** Number the longest chain that contains the aldehyde group, and then give each substituent a name and a number. In this molecule the longest carbon chain is 6 carbons, but the longest chain that contains the aldehyde is 5 carbons.

$$\overset{5}{CH_3}-\overset{4}{CH_2}-\overset{3}{CH}(CH_3)-\overset{2}{CH}(CH_2-CH_3)-\overset{1}{C}(=O)-H$$

2-ethyl-3-methylpentanal

(b) Number the six-membered ring beginning with the carbon bearing the carbonyl group. The name is 2,2-dimethylcyclohexanone.

(c) This molecule is a derivative of benzaldehyde. The methoxy substituent is located by numbering the carbons of the ring or by using the *ortho, meta, para* system. The name is either 4-methoxybenzaldehyde or *p*-methoxybenzaldehyde.

PROBLEM 7.1 Give an acceptable name for the following compounds.

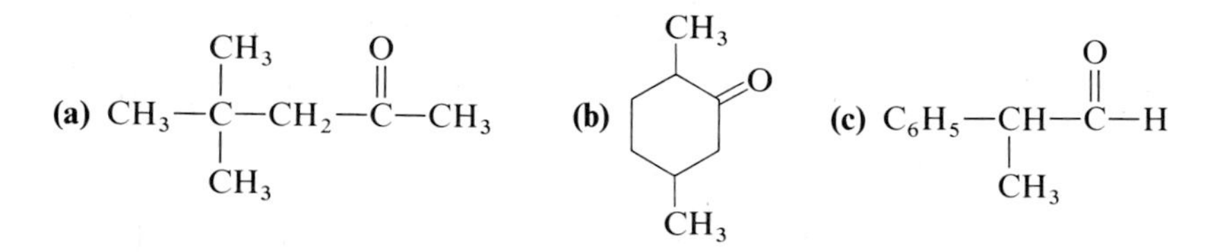

A great variety of aldehydes and ketones have been isolated from natural sources, and are best known by their common or trivial names. Such names are usually derived from a source from which the compound can be isolated, or they may refer to a characteristic property of the compound. Figure 7.2 shows structural formulas for several aldehydes and ketones of natural occurrence, together with their common names.

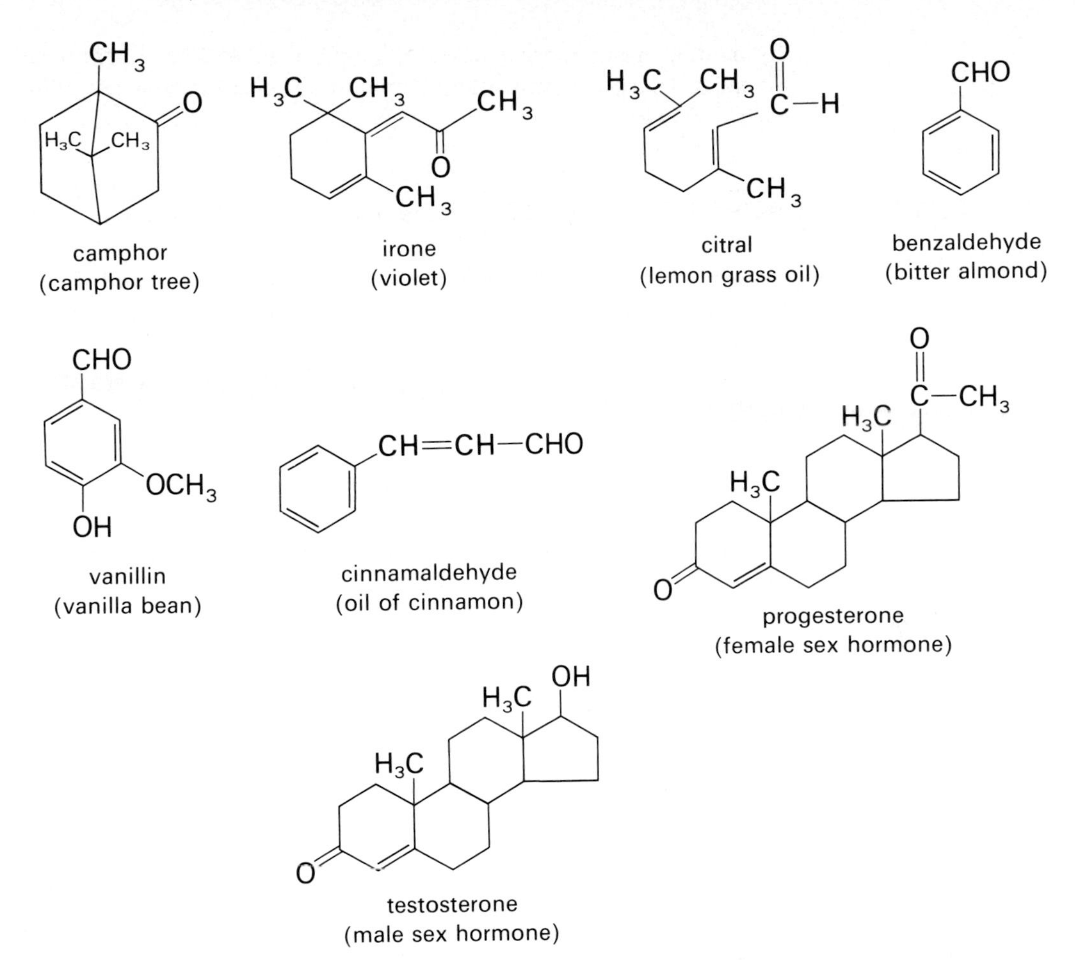

Figure 7.2 *Common names of some aldehydes and ketones of natural occurrence.*

Natural or dextrorotatory camphor is obtained from the bark of the camphor tree (*Cinnamomum camphora*) native to the island of Taiwan. Camphor has some medicinal uses as a weak antiseptic and as an analgesic; it is used in certain liniments. Aldehydes and ketones are in particular abundance in plant essential oils. Irone, the fragrant principal of the violet, is isolated from the rhizomes of iris. Citral is the major constituent (70%) of lemon grass oil. Because of its presence in almond seed, benzaldehyde has been known as oil of bitter almond. Benzaldehyde is also the chief constituent of oils extracted from kernels of the peach, cherry, laurel, and other fruits. Vanillin is the fragrant constituent of the vanilla bean. Cinnamaldehyde is the chief constituent of the oil of cinnamon.

Testosterone (Section 13.5), one of the most prominent members of the male sex hormone group, is produced in the testes and is involved in the development of accessory sex functions in the male. Progesterone (Section 13.5), secreted by ovarian tissue (*corpus luteum*), is one of several sex hormones involved in stimulating growth of the uterine mucosa in preparation for implantation of the fertilized ovum.

7.3 PHYSICAL PROPERTIES OF ALDEHYDES AND KETONES

Oxygen is more electronegative than carbon (3.5 compared to 2.5); therefore the carbon-oxygen double bond is polar covalent. The oxygen atom of a carbonyl group bears a partial negative charge and the carbon atom a partial positive charge, as illustrated for formaldehyde:

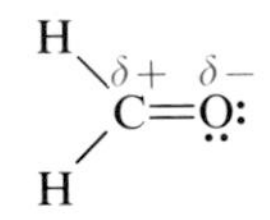

Alternatively, the carbonyl group may be pictured as a hybrid of two major contributing structures.

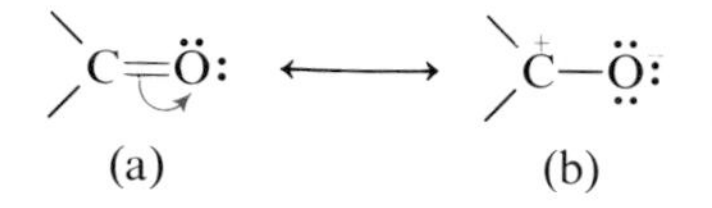

Structure (b) places the negative charge on the more electronegative oxygen atom and the positive charge on the less electronegative carbon atom.

Because they are polar substances and can interact in the pure state by dipole-dipole interaction, aldehydes and ketones have higher boiling points than nonpolar compounds of comparable molecular weight. Table 7.1 lists the

Table 7.1 Boiling points of substances of similar molecular weight.

Structural formula	*IUPAC name*	*Molecular weight*	*bp* (°C)
$CH_3CH_2CH_2CH_2CH_3$	pentane	72	36
$CH_3CH_2CH_2OCH_3$	methyl propyl ether	74	39
$CH_3CH_2CH_2\overset{O}{\overset{\|\|}{C}}—H$	butanal	72	76
$CH_3CH_2\overset{O}{\overset{\|\|}{C}}CH_3$	2-butanone	72	80
$CH_3CH_2CH_2CH_2OH$	1-butanol	74	117
$CH_3CH_2\overset{O}{\overset{\|\|}{C}}—OH$	propanoic acid	74	141

boiling points of six compounds of comparable molecular weight. Pentane, a nonpolar hydrocarbon, has the lowest boiling point. While methyl propyl ether is a polar compound, there is little association between molecules in the liquid state. Hence its boiling point is only slightly higher than that of pentane. Both butanal and 2-butanone are polar compounds, and because of the association between a partially positive carbon of one molecule and a partially negative oxygen of another molecule, their boiling points are higher than those of pentane and methyl propyl ether. Aldehydes and ketones have no partially positive hydrogen atom attached to oxygen and cannot associate by hydrogen bonding. Therefore, their boiling points are lower than those of alcohols and carboxylic acids, substances that can associate by hydrogen bonding.

Aldehydes and ketones interact with water molecules as hydrogen-bond acceptors, and therefore low-molecular-weight aldehydes and ketones are more soluble in water than nonpolar compounds of comparable molecular weight.

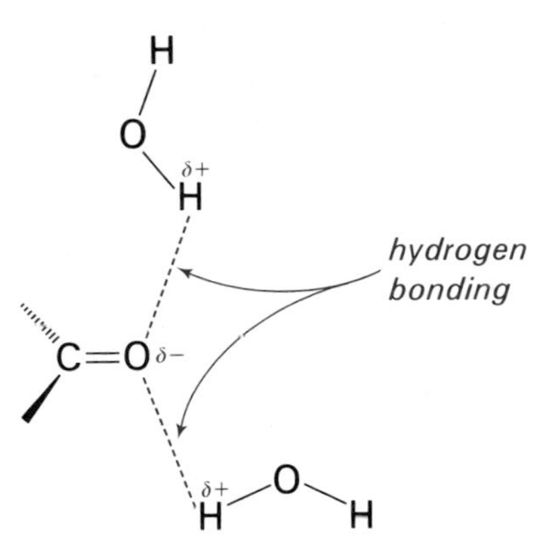

Table 7.2 lists boiling points and solubility in water for several aldehydes and ketones.

Table 7.2 Physical properties of aldehydes and ketones.

Structural formula	*IUPAC name*	*Common name*	*bp* (°C)	*Solubility* (g/100 g H_2O)
HCHO	methanal	formaldehyde	−21	very
CH_3CHO	ethanal	acetaldehyde	20	∞
CH_3CH_2CHO	propanal	propionaldehyde	49	16
$CH_3(CH_2)_2CHO$	butanal	butyraldehyde	76	7
$CH_3(CH_2)_3CHO$	pentanal	valeraldehyde	103	slightly
$CH_3(CH_2)_4CHO$	hexanal	caproaldehyde	129	slightly
CH_3COCH_3	2-propanone	acetone	56	∞
$CH_3COCH_2CH_3$	2-butanone	methyl ethyl ketone	80	26
$CH_3COCH_2CH_2CH_3$	2-pentanone	methyl propyl ketone	102	6
$CH_3CH_2COCH_2CH_3$	3-pentanone	diethyl ketone	101	5

7.4 PREPARATION OF ALDEHYDES AND KETONES

We have already encountered two general methods for the preparation of aldehydes and ketones, namely (1) oxidation of substituted alkenes and (2) oxidation of primary and secondary alcohols. Reaction of an alkene with ozone followed by treatment of the ozonide with water in the presence of powdered zinc results in cleavage of the carbon-carbon double bond and formation of two carbonyl groups (Section 3.6H).

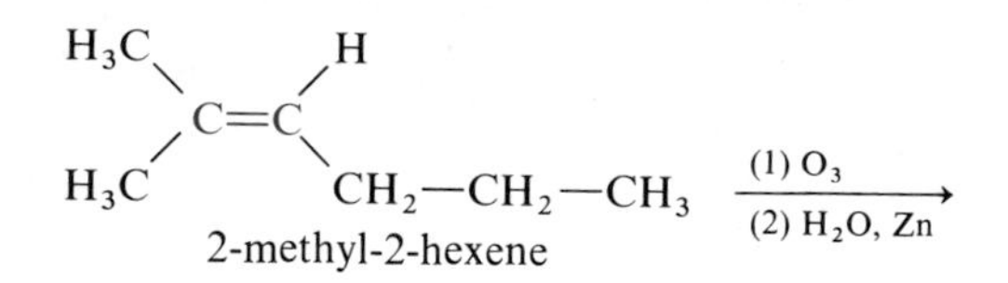

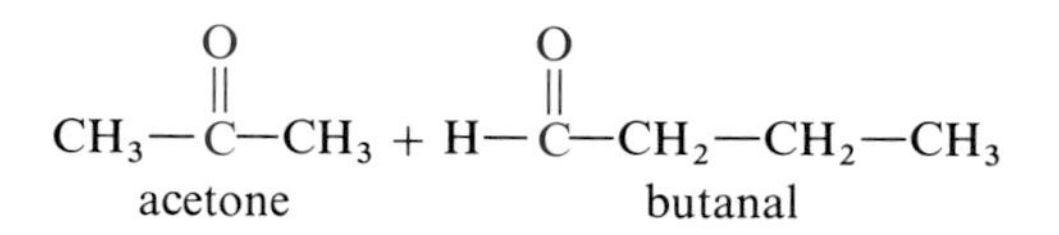

In Section 5.5C we saw that oxidation of primary alcohols by $K_2Cr_2O_7$ or CrO_3 gives aldehydes.

$$\underset{\text{1-hexanol}}{CH_3CH_2CH_2CH_2CH_2CH_2OH} + Cr_2O_7^{2-} \xrightarrow{H^+} \underset{\text{hexanal}}{CH_3CH_2CH_2CH_2CH_2\overset{\overset{\displaystyle O}{\|}}{C}H} + Cr^{3+}$$

Under carefully controlled conditions it is possible to obtain aldehydes in high yield. However, unless care is taken, primary alcohols are oxidized directly to carboxylic acids. In these oxidations, aldehydes are intermediates. Oxidation of secondary alcohols gives ketones.

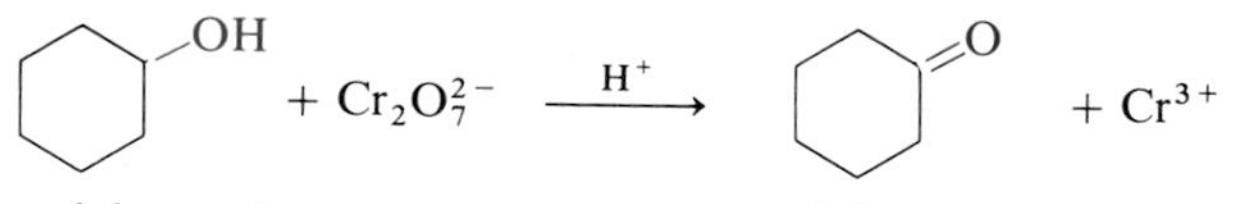

7.5 REACTIONS OF ALDEHYDES AND KETONES

The carbonyl group is polar with the carbonyl carbon bearing a partial positive charge and the carbonyl oxygen bearing a partial negative charge. Because of this polarity, carbonyl groups react with both electrophiles (as for example H^+) and with nucleophiles.

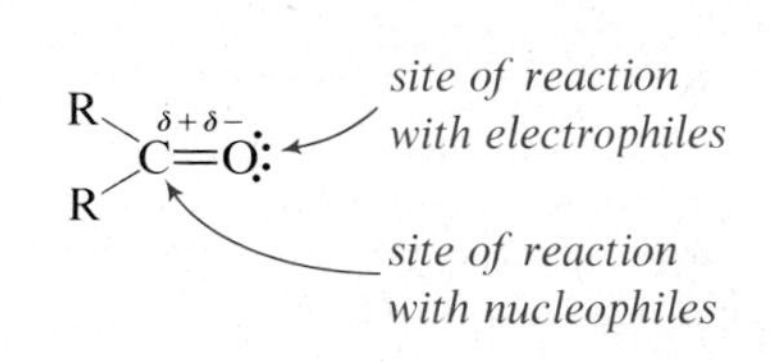

A nucleophile is any atom or group of atoms with an unshared pair of electrons that can be shared with another atom or group of atoms to form a new covalent bond. In nucleophilic additions to a carbonyl group, the nucleophile adds to the carbonyl carbon to produce a tetrahedral carbonyl addition compound. In the following general reaction, the nucleophilic reagent is written H—Nu: to emphasize the presence of the unshared pair of electrons in the nucleophile.

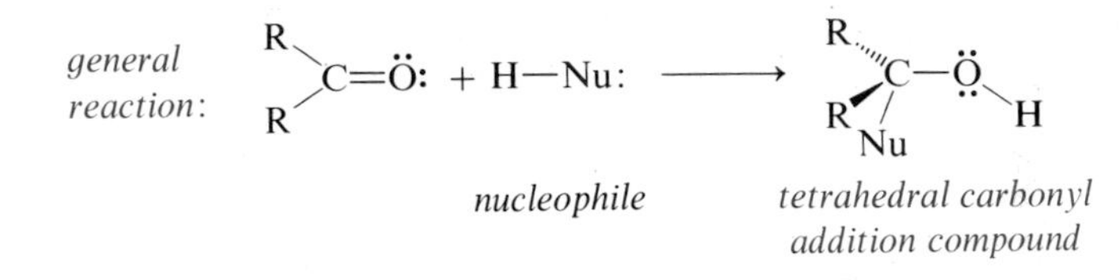

As an example of nucleophilic addition to a carbonyl group, water adds reversibly to aldehydes and ketones to form hydrated carbonyl compounds. In water at 20°C, formaldehyde is over 99% hydrated.

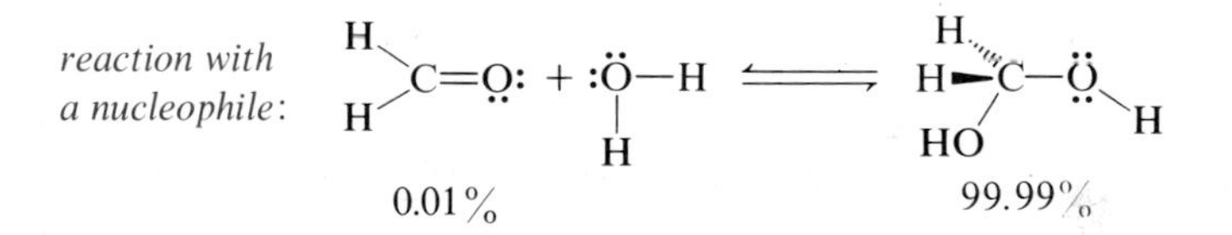

In this example, hydrogen adds to the oxygen of the carbonyl group and oxygen adds to the carbon of the carbonyl group. Under the same experimental conditions, acetaldehyde is about 58% hydrated. The extent of hydration of most ketones is very small.

An electrophile is any atom or group of atoms that can accept a pair of electrons to form a new covalent bond. The most common electrophilic reagent in carbonyl reactions is the proton, H^+. Reaction of a carbonyl group with a proton gives a resonance-stabilized cation.

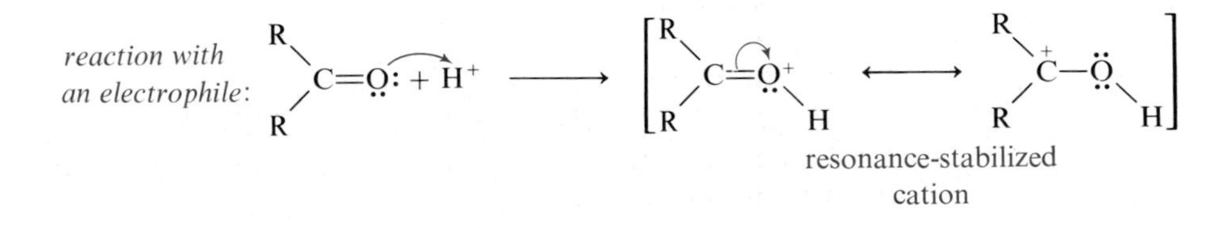

Protonation of a carbonyl group increases the electron deficiency of the carbonyl carbon and makes it even more reactive to nucleophiles.

A. ADDITION OF ALCOHOLS: FORMATION OF ACETALS AND KETALS

Alcohols add to aldehydes and ketones in the same manner as described for the addition of water. Addition of one molecule of alcohol to an aldehyde forms a hemiacetal. The comparable reaction with a ketone forms a hemiketal.

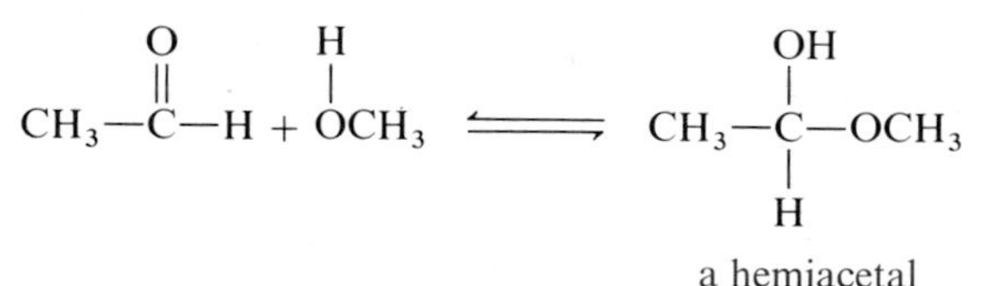

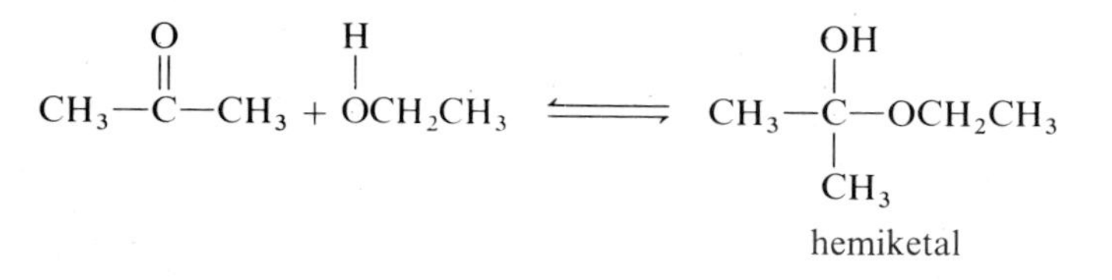

Following are the characteristic structural features of a hemiacetal and a hemiketal.

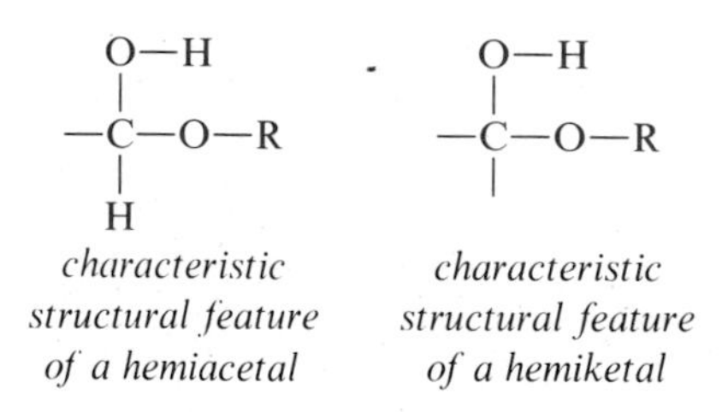

Hemiacetals and hemiketals are only minor components of an equilibrium mixture except in one very important case. When a hydroxyl group is part of the same molecule that contains the carbonyl group and a five- or six-membered ring can form, the substance exists almost entirely in the cyclic hemiacetal or cyclic hemiketal form.

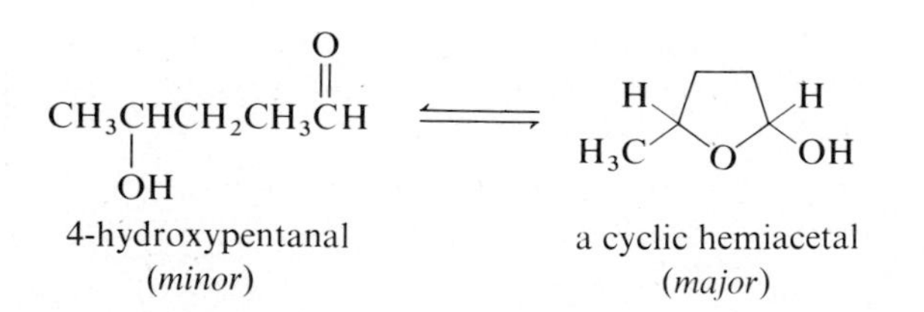

Hemiacetals and hemiketals react further with alcohols to form acetals and ketals. These reactions are catalyzed by acids.

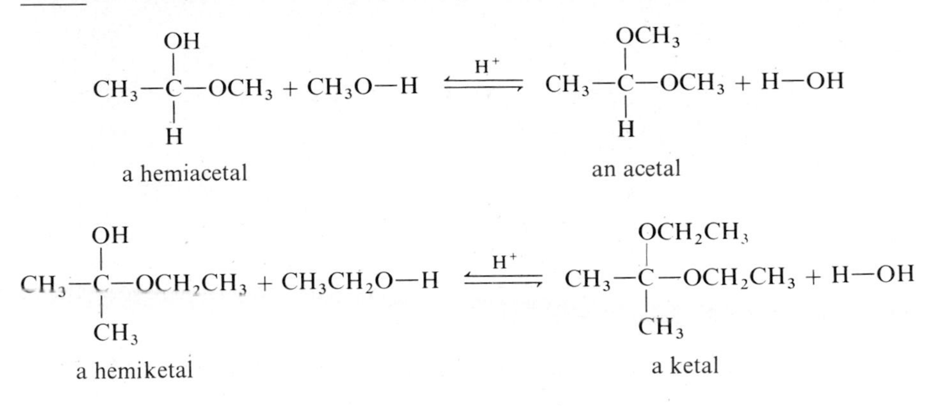

Following are the characteristic structural features of an acetal and a ketal.

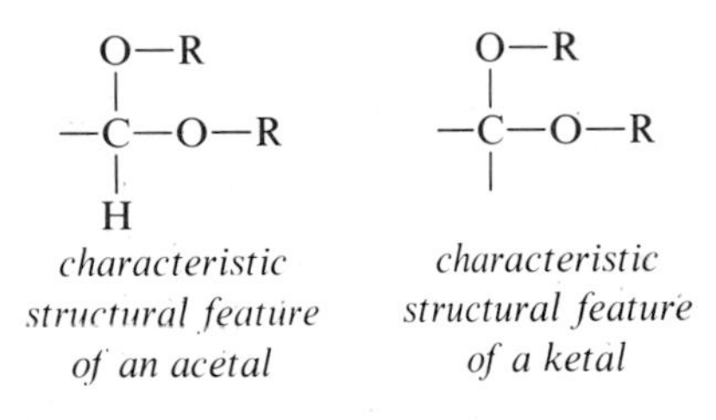

characteristic structural feature of an acetal

characteristic structural feature of a ketal

Acetal and ketal formation are equilibrium reactions, and in order to obtain high yields of an acetal, it is necessary to remove water from the reaction mixture and thus favor the formation of the product acetal.

The formation of acetals and ketals is catalyzed by acid. Their hydrolysis in water is also catalyzed by acid. However, acetals and ketals are stable and unreactive to aqueous base.

Example 7.2 For the following, show the reaction of each carbonyl group with one molecule of alcohol to form a hemiacetal or hemiketal and then with a second molecule of alcohol to form an acetal or ketal.

(a) $CH_3CH_2\overset{O}{\overset{\|}{C}}CH_3 + 2CH_3CH_2OH \xrightarrow{H^+}$

(b) $C_6H_5-\overset{O}{\overset{\|}{C}}-H + 2CH_3OH \xrightarrow{H^+}$

(c) cyclohexanone $+ HO-CH_2CH_2-OH \xrightarrow{H^+}$

Solution

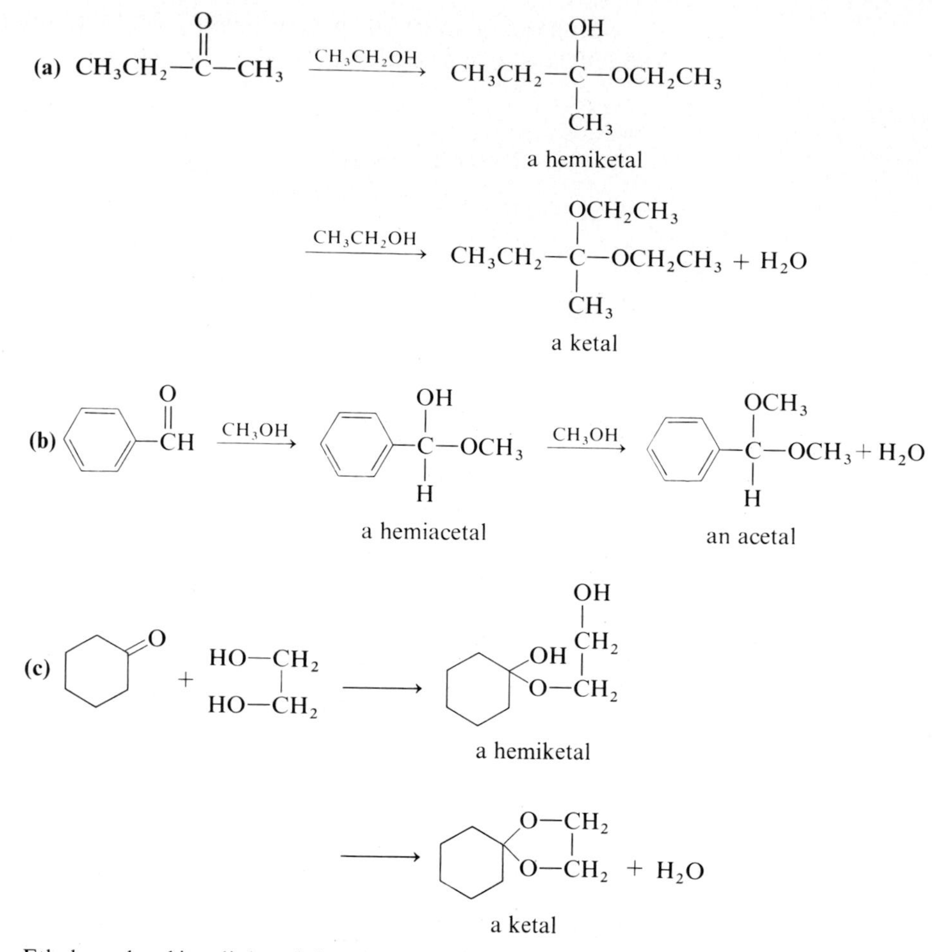

Ethylene glycol is a diol, and therefore one molecule of ethylene glycol provides both —OH groups needed for ketal formation.

PROBLEM 7.2 Following are structural formulas for one ketal and two acetals. Draw structural formulas for the alcohols and aldehyde or ketone from which each is formed. The reaction of an acetal (or ketal) with water to form an aldehyde (or ketone) and two alcohols is called hydrolysis.

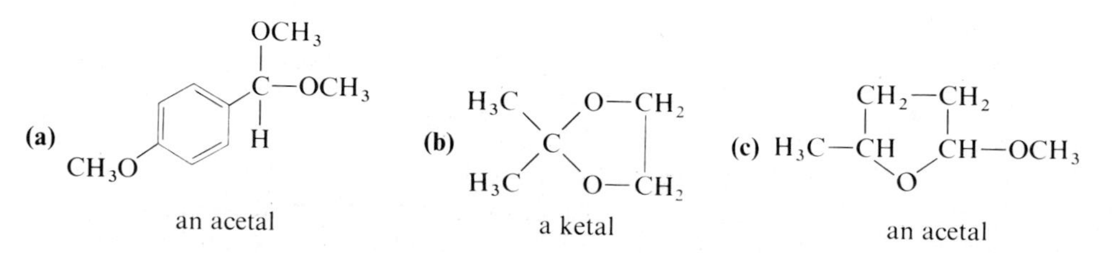

A mechanism for the acid-catalyzed conversion of acetaldehyde hemiacetal into acetaldehyde dimethylacetal follows. In Step 1, H^+, an electrophile, reacts with the —OH group of the hemiacetal to form an oxonium ion. In Step 2, the oxonium ion loses a molecule of water to form a resonance-stabilized cation.

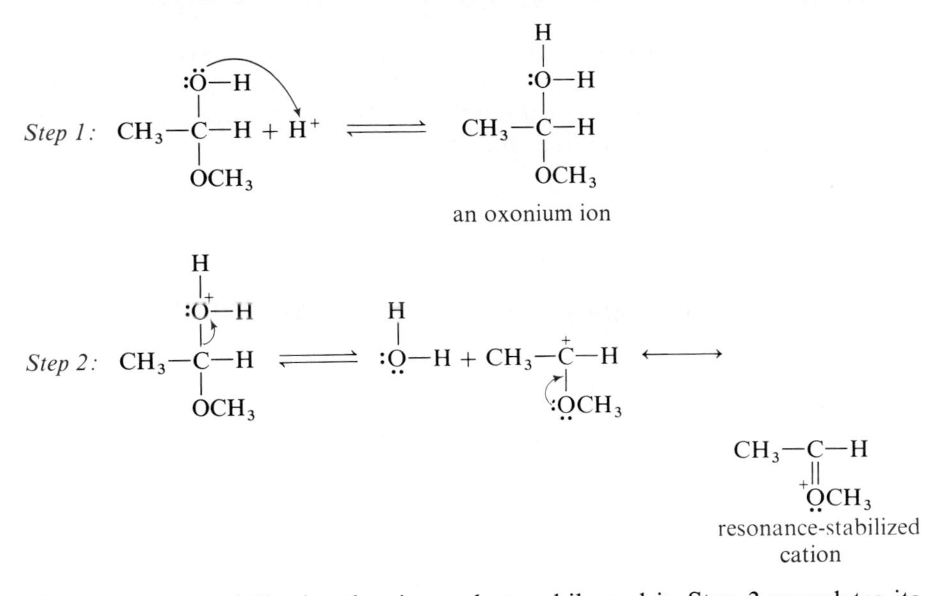

The resonance-stabilized cation is an electrophile and in Step 3 completes its valence shell by reacting with a nucleophile to form a new oxonium ion. Loss of H^+ in Step 4 gives the acetal. Note that H^+ is used in Step 1 of this mechanism but is regenerated in Step 4.

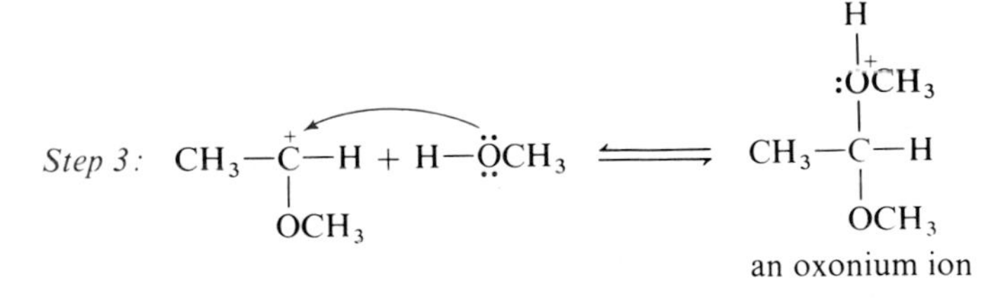

Step 4: $CH_3-C(H)(\overset{+}{O}HCH_3)(OCH_3) \rightleftharpoons CH_3-C(H)(OCH_3)(OCH_3) + H^+$

There are marked similarities between certain steps in this mechanism and steps in two other acid catalyzed reactions we have already studied. Steps 1 and 2 of the mechanism of acetal formation are similar to the first two steps in the acid-catalyzed dehydration of an alcohol to form an alkene (Section 5.5D): protonation of —OH and loss of a molecule of water to form a cation. Steps 3 and 4 are similar to the second and third steps in the acid-catalyzed hydration of an alkene to form an alcohol (Section 3.6C): reaction of a carbocation with an oxygen atom to form an oxonium ion followed by the loss of H^+.

B. ADDITION OF AMMONIA AND ITS DERIVATIVES: FORMATION OF SCHIFF BASES

Ammonia and amines of the type R—NH_2 add to the carbonyl group of aldehydes and ketones to form tetrahedral carbonyl addition compounds. In these additions, the amine nitrogen adds to the carbonyl carbon and an amine hydrogen adds to the carbonyl oxygen:

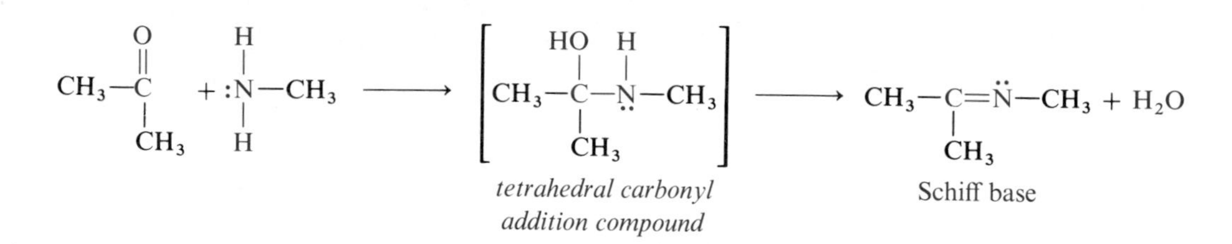

These tetrahedral carbonyl addition compounds readily lose a molecule of water to form compounds called Schiff bases. The characteristic structural feature of a Schiff base is the presence of a carbon-nitrogen double bond.

Aldehydes and ketones also react with hydrazine, $H_2N—NH_2$, and with certain of its derivatives to give products that contain carbon-nitrogen double bonds. The most widely used hydrazine derivative is 2,4-dinitrophenylhydrazine. The special value of this reagent is that it reacts with virtually all aldehydes and ketones to give water-insoluble solids. For this reason, 2,4-dinitrophenylhydrazine is especially useful as a qualitative test for the presence of aldehydes and ketones.

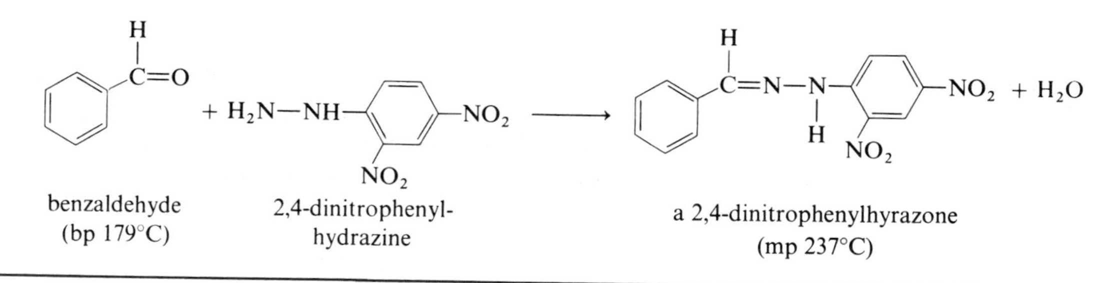

Example 7.3

Write structural formulas for the tetrahedral carbonyl addition compound and Schiff base formed in each of the following reactions.

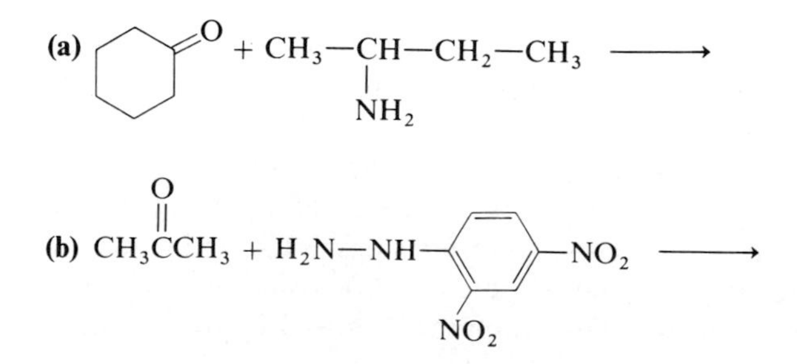

Solution

(a)

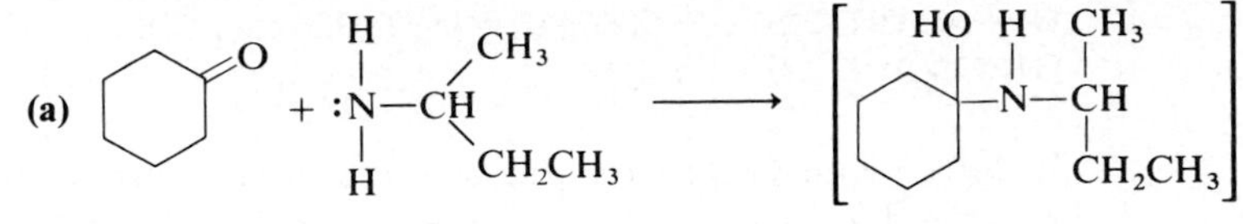

tetrahedral carbonyl addition compound

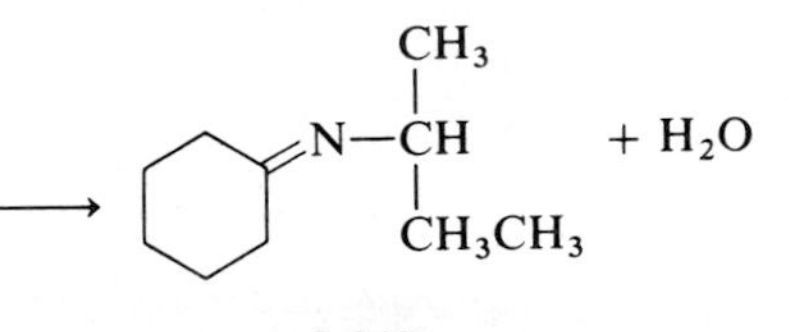

Schiff base

(b)

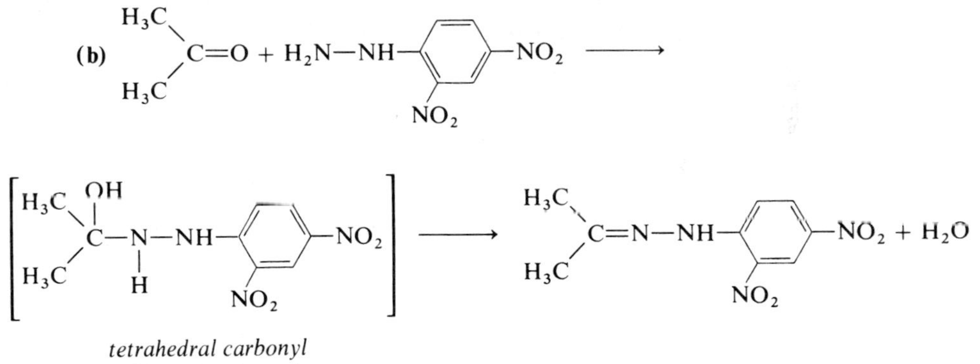

tetrahedral carbonyl addition compound

PROBLEM 7.3 Each of the following compounds is a Schiff base. Write structural formulas for the amine and aldehyde or ketone formed when each is reacted with water. The reaction of a Schiff base with water to form an amine and aldehyde or ketone is called hydrolysis.

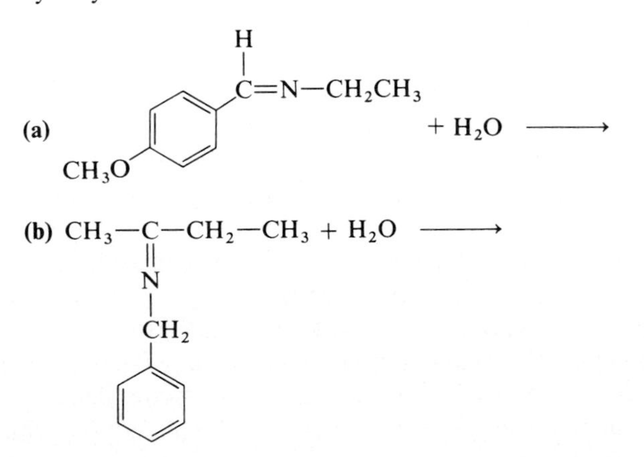

C. ADDITION OF ORGANOMETALLICS: THE GRIGNARD REACTION

In 1901 Victor Grignard discovered that magnesium metal reacts with alkyl or aryl halides to give a reagent containing a carbon-metal bond. For example, methyl iodide reacts with magnesium in diethyl ether to give an ether-soluble substance called methylmagnesium iodide.

$$CH_3I + Mg \xrightarrow{\text{ether}} CH_3MgI$$

methylmagnesium iodide (a Grignard reagent)

The electronegativity difference between carbon and magnesium is approximately 1.2 units. Therefore the C—Mg bond is polar covalent with carbon bearing a partial negative charge and magnesium bearing a partial positive charge. The carbon-magnesium bond can also be written as an ionic bond to emphasize its polarity.

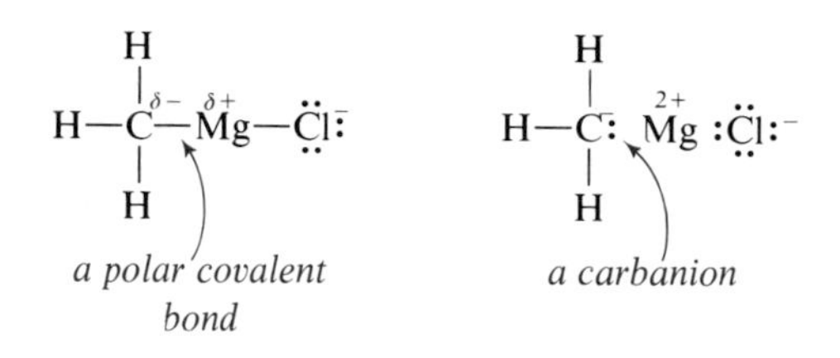

A Grignard reagent behaves as if it were a carbanion, a substance containing a carbon atom with an unshared pair of electrons and bearing a negative charge. Carbanions are strong bases and react with even such weak acids as water and alcohols to give hydrocarbons.

$$CH_3CH_2MgBr + H_2O \longrightarrow CH_3CH_3 + HOMgBr$$

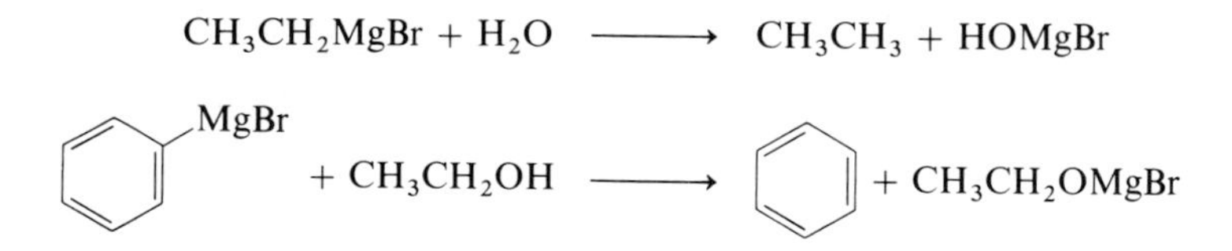

For this reason diethyl ether used as a solvent for Grignard reactions must be anhydrous (free of H_2O) and free from traces of ethanol.

Carbanions are also good nucleophiles and add to carbonyl groups to form tetrahedral carbonyl addition compounds. In the case of a Grignard reaction, the tetrahedral carbonyl addition compound is a magnesium alkoxide, the magnesium salt of an alcohol. The electronegativity difference between magnesium and oxygen is approximately 2.2 units. In the following examples, the magnesium oxygen bond is written $—O^-MgBr^+$ to emphasize its ionic nature. Hydrolysis of a magnesium alkoxide in aqueous acid gives an alcohol.

Reaction of a Grignard reagent with formaldehyde followed by hydrolysis in aqueous acid gives a primary alcohol; reaction with any other aldehyde followed by hydrolysis gives a secondary alcohol. Reaction of a Grignard reagent with a ketone followed by hydrolysis gives a tertiary alcohol.

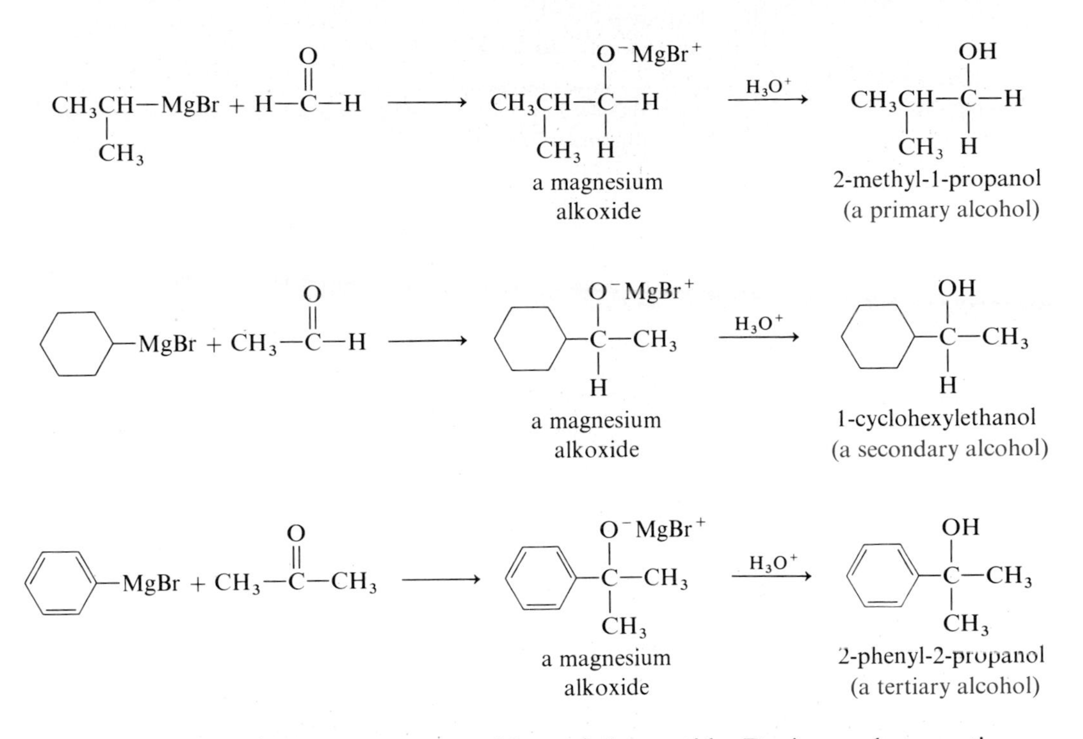

Grignard reagents also add to ethylene oxide. During such a reaction, a carbon-oxygen bond of the highly strained epoxide ring is broken and a new carbon-carbon bond is formed. Reaction of ethylene oxide and a Grignard reagent is a very convenient method for lengthening a carbon chain by two atoms.

$$C_6H_5{-}MgBr + H_2C{-}CH_2 \text{ (epoxide, O)} \longrightarrow C_6H_5{-}CH_2{-}CH_2{-}O^-MgBr^+ \xrightarrow{H_3O^+} C_6H_5{-}CH_2{-}CH_2{-}OH$$

Finally, Grignard reagents add to carbon dioxide to give magnesium salts of carboxylic acids. Treatment of these salts with dilute hydrochloric acid gives the free carboxylic acid. Carbonation of Grignard reagents is a very convenient method for converting alkyl or aryl halides to carboxylic acids as illustrated by the following conversion of bromobenzene to benzoic acid.

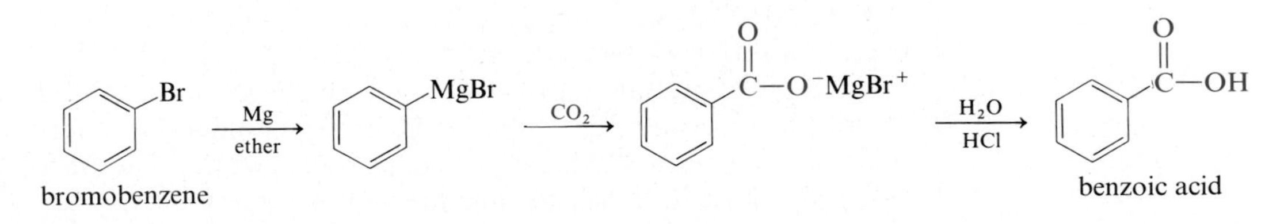

The special value of the Grignard reaction is that it is an excellent method for forming new carbon-carbon bonds. For the discovery and study of this class of organometallic compounds, Victor Grignard was awarded the Nobel Prize in Chemistry in 1912.

Example 7.4 Following is the structural formula of 2-phenyl-2-butanol, a tertiary alcohol that can be synthesized from three different combinations of a Grignard reagent and ketone. Show each combination.

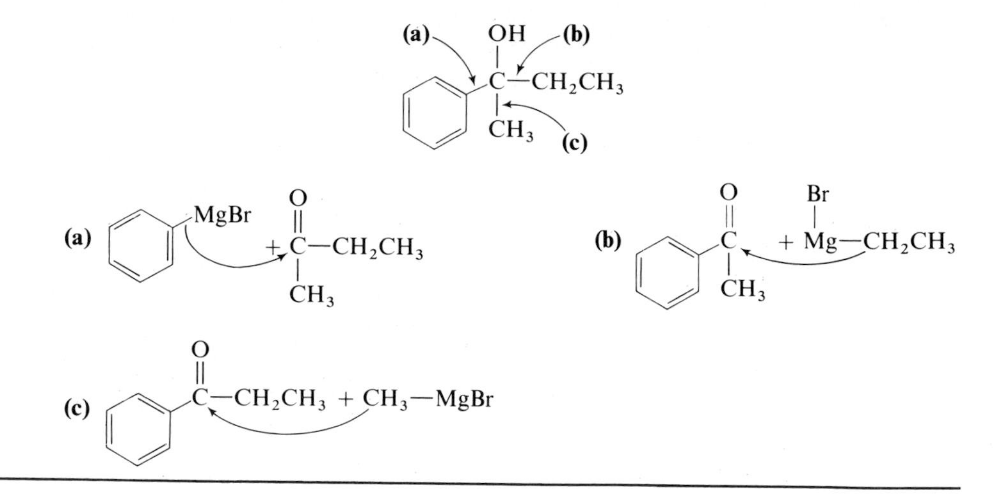

Solution

PROBLEM 7.4 What combinations of Grignard reagent and aldehyde can be used to synthesize the following secondary alcohols?

(a) $CH_3CH_2CH_2CH_2CHCH_3$ (with OH on the CH)

(b) cyclopentyl—$CHCH_2CH_3$ (with OH on the CH)

D. OXIDATION-REDUCTION OF ALDEHYDES AND KETONES

We saw in Section 5.5C that aldehydes are oxidized to carboxylic acids by oxidizing agents such as $K_2Cr_2O_7$. Ketones are not oxidized by these reagents. Aldehydes are also oxidized to carboxylic acids by silver ion in ammonium hydroxide. This reagent, known as Tollens' reagent, is selective for the oxidation of aldehydes—it will not oxidize alkenes, alkynes, alcohols, or ketones. The use of Ag^+ in NH_4OH is a convenient way to distinguish between aldehydes and ketones, both compounds that contain a carbonyl group.

In Tollens' test, a solution of silver nitrate in ammonium hydroxide is added to the substance suspected of being an aldehyde. Within a few minutes at room temperature, silver ion is reduced to metallic silver and the aldehyde is oxidized to a carboxylic acid. In the alkaline medium necessary for this test, the carboxylic acid reacts with ammonium hydroxide to form a water-soluble ammonium salt.

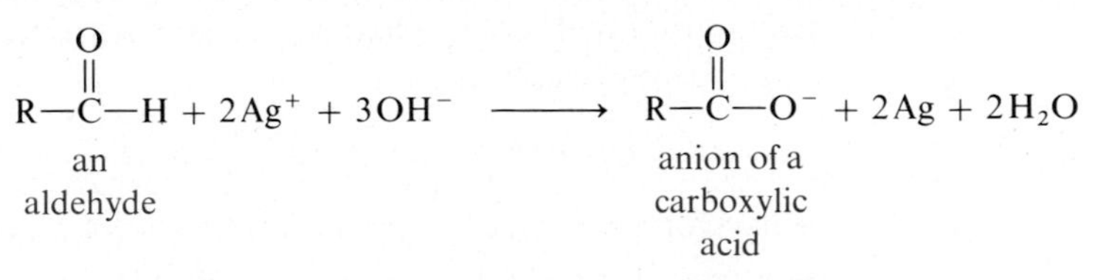

If the test is done properly, metallic silver deposits as a mirror on a glass surface. For this reason, the test is commonly called the silver mirror test.

Aldehydes are reduced to primary alcohols, and ketones to secondary alcohols by catalytic reduction. The most commonly used catalysts are palladium, platinum, or nickel.

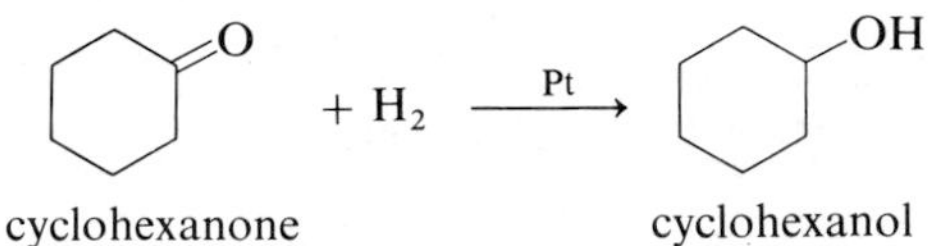

A second and more common laboratory method for the reduction of aldehydes and ketones uses the metal hydrides lithium aluminum hydride and sodium borohydride.

$$Li^+\ H{-}\overset{H}{\underset{H}{Al^-}}{-}H \qquad Na^+\ H{-}\overset{H}{\underset{H}{B^-}}{-}H$$

lithium aluminum hydride — sodium borohydride

Reaction of either with an aldehyde or ketone forms a metal alkoxide. Treatment of the metal alkoxide with water gives the alcohol.

$$4CH_3\overset{O}{\overset{\|}{C}}{-}H + NaBH_4 \longrightarrow \underset{\text{a metal alkoxide}}{(CH_3CH_2O)_4B^-Na^+} \xrightarrow{H_2O} 4CH_3CH_2OH$$

$$4\,(H_3C)_2C{=}O + LiAlH_4 \longrightarrow \underset{\text{a metal alkoxide}}{\left((H_3C)_2CH{-}O\right)_4Al^-Li^+} \xrightarrow{H_2O} 4CH_3\overset{OH}{\overset{|}{C}}HCH_3$$

These two metal hydrides have quite different chemical reactivities. $NaBH_4$ reduces only aldehydes and ketones. Reductions with $NaBH_4$ are carried out using water or water-alcohol mixtures as a solvent. $LiAlH_4$ is a much stronger reducing agent. It reduces not only aldehydes and ketones, but esters, amides, and anhydrides as well. Furthermore, $LiAlH_4$ reacts violently with even such weak acids (proton donors) as water or alcohols and consequently reductions using lithium aluminum hydride must be carried out in dry ether or some other nonhydroxylic solvent.

$$4CH_3CH_2OH + LiAlH_4 \longrightarrow (CH_3CH_2O)_4Al^-Li^+ + 4H_2$$

Both sodium borohydride and lithium aluminum hydride are selective in that they reduce aldehydes and ketones but do not reduce carbon-carbon double bonds or carbon-carbon triple bonds.

$$\underset{\text{3-pentenal}}{CH_3CH{=}CHCH_2\overset{O}{\overset{\|}{C}}{-}H} \xrightarrow[\text{(or } NaBH_4)]{LiAlH_4} \xrightarrow{H_2O} \underset{\text{3-penten-1-ol}}{CH_3CH{=}CHCH_2CH_2OH}$$

Example 7.5

Draw structural formulas for the products of the following reactions.

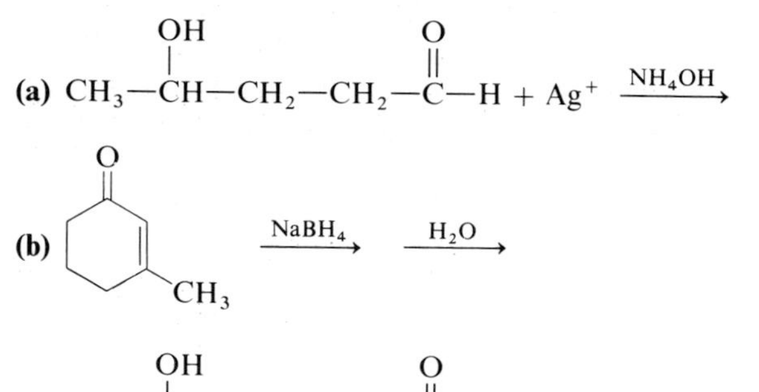

(a) $CH_3{-}\overset{OH}{\overset{|}{C}H}{-}CH_2{-}CH_2{-}\overset{O}{\overset{\|}{C}}{-}H + Ag^+ \xrightarrow{NH_4OH}$

(b) (3-methylcyclohex-2-enone, CH_3) $\xrightarrow{NaBH_4} \xrightarrow{H_2O}$

(c) $CH_3{-}\overset{OH}{\overset{|}{C}H}{-}CH_2{-}CH_2{-}\overset{O}{\overset{\|}{C}}{-}H + Cr_2O_7^{2-} \xrightarrow[\text{heat}]{H^+}$

Solution

(a) $CH_3{-}\overset{OH}{\overset{|}{C}H}{-}CH_2{-}CH_2{-}\overset{O}{\overset{\|}{C}}{-}OH + Ag$. Silver ion (as a silver-ammonia complex) is a weak oxidizing agent. It oxidizes aldehydes to carboxylic acids, but does not oxidize primary, secondary, or tertiary alcohols.

(b)

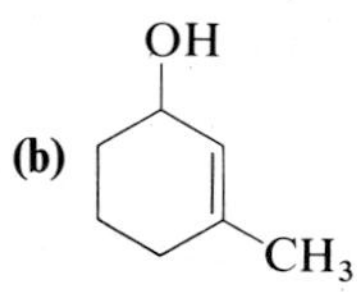

Sodium borohydride reduces aldehydes and ketones to alcohols but does not reduce carbon-carbon double bonds.

(c) $CH_3{-}\overset{O}{\overset{\|}{C}}{-}CH_2{-}CH_2{-}\overset{O}{\overset{\|}{C}}{-}OH + Cr^{3+}$. Potassium dichromate in acid solution is a strong oxidizing agent. Under these conditions, aldehydes are oxidized to carboxylic acids and secondary alcohols to ketones.

PROBLEM 7.5 Draw the structural formula for a substance with the given molecular formula that will undergo oxidation to give the product shown.

(a) $C_7H_{14}O + Cr_2O_7^{2-} \xrightarrow[\text{heat}]{H^+} C_6H_{11}-\overset{\overset{\displaystyle O}{\|}}{C}-OH + Cr^{3+}$ (cyclohexyl ring)

(b) $C_7H_{14}O + Cr_2O_7^{2-} \xrightarrow[\text{heat}]{H^+}$ (cyclopentanone ring with O, bearing CH_3 and CH_3 on the adjacent carbon) $+ Cr^{3+}$

(c) $C_7H_6O + Ag^+ \xrightarrow{NH_4OH} C_6H_5-\overset{\overset{\displaystyle O}{\|}}{C}-O^- + Ag$ (benzene ring)

(d) $C_6H_{14}O_2 + Cr_2O_7^{2-} \xrightarrow[\text{heat}]{H^+} CH_3-\overset{\overset{\displaystyle O}{\|}}{C}-CH_2-CH_2-\overset{\overset{\displaystyle O}{\|}}{C}-CH_3 + Cr^{3+}$

E. TAUTOMERISM

A carbon atom adjacent to a carbonyl group is defined as an α-carbon and hydrogen atoms attached to an α-carbon are defined as α-hydrogens.

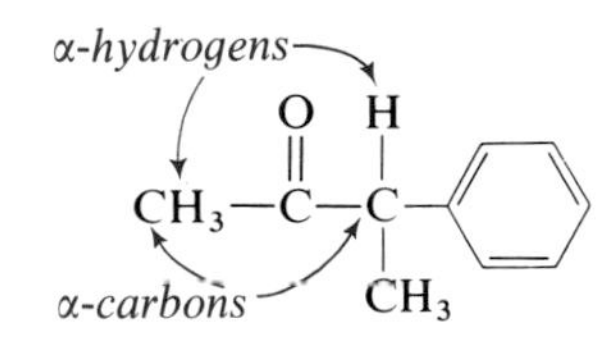

Aldehydes and ketones with an α-hydrogen are in equilibrium with an isomer formed by the migration of a proton from the α-carbon to oxygen. This new isomer is called an enol, a name derived from the IUPAC designation of it as both an alkene and an alcohol.

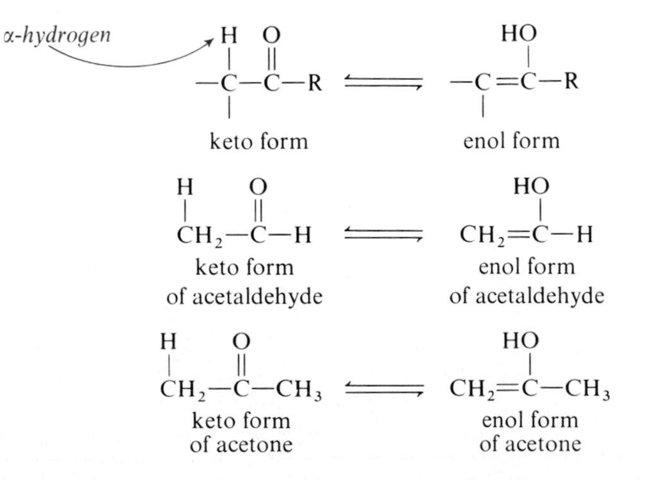

The rearrangement of a proton and a double bond is called tautomerism. In the case of simple aldehydes and ketones, the keto form predominates over the enol form at equilibrium by factors of well over 1000 to 1.

When the α-hydrogen is flanked by a second carbonyl group, as in 2,4-pentanedione, the equilibrium shifts toward the enol form. Liquid 2,4-pentanedione consists of an equilibrium mixture containing approximately 80% of the enol form.

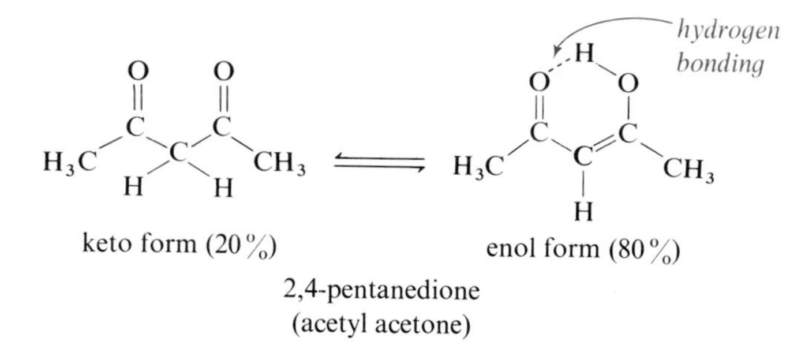

Note that the enol form of 2,4-pentanedione is stabilized by hydrogen bonding between the carbonyl oxygen and the O—H of the enol.

Example 7.6

Write the indicated number of enol structures for the following compounds.

Solution

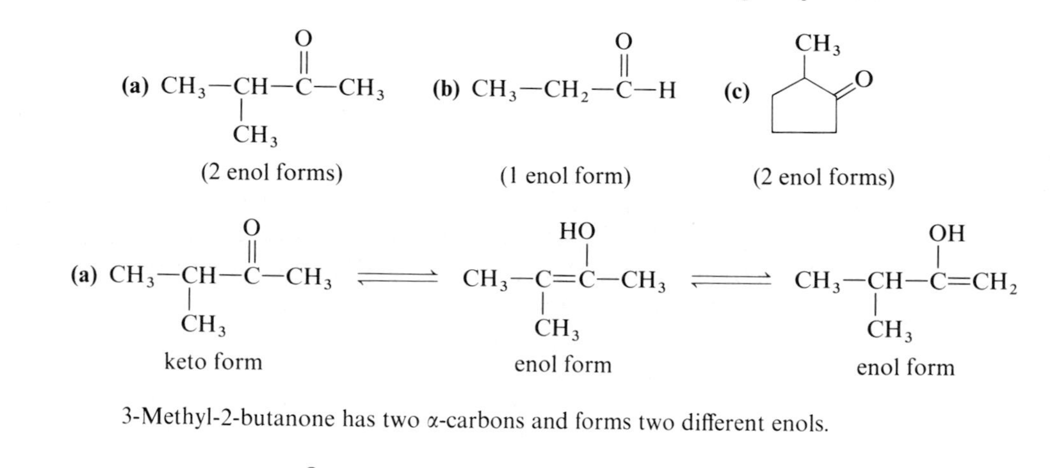

3-Methyl-2-butanone has two α-carbons and forms two different enols.

(b) $CH_3—CH_2—C(=O)—H \rightleftharpoons CH_3—CH=C(OH)—H$

Propanal has only one α-carbon and forms only one enol.

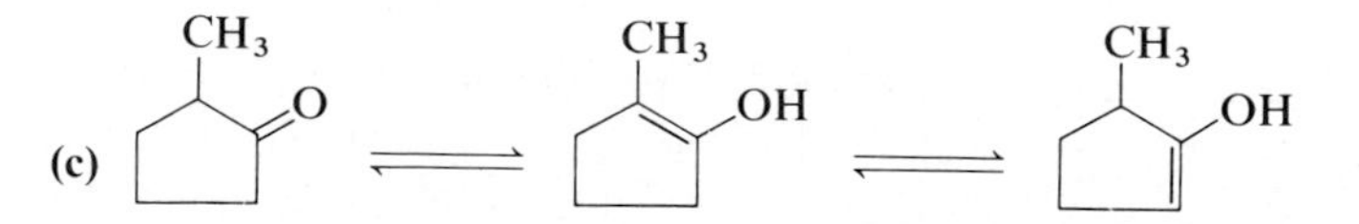

2-Methylcyclopentanone has two α-carbons and forms two different enols.

PROBLEM 7.6 Following are enol forms. Draw the structural formula of the keto form of each.

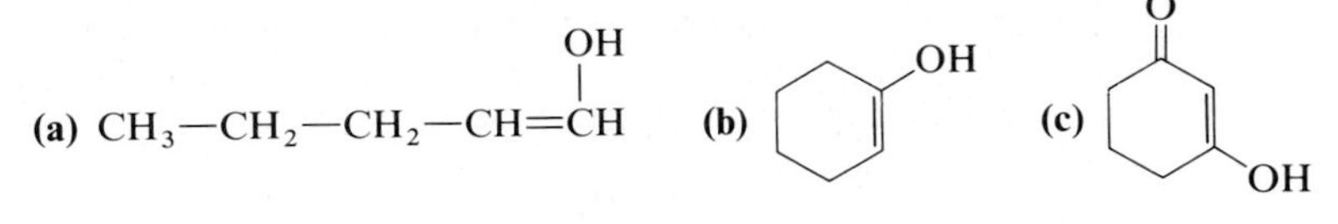

F. THE ALDOL CONDENSATION

Because hydrogen and carbon have similar electronegativities, there is normally no appreciable polarity to the C—H bond, and no tendency for the C—H bond to ionize or to show any acidic properties. However, in the case of hydrogens alpha to a carbonyl group, the situation is different. The reaction of an aldehyde or ketone with an α-hydrogen and a strong base forms a carbon-containing anion.

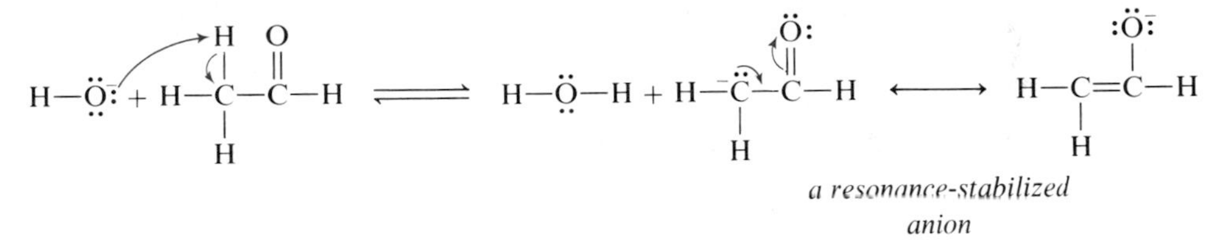

Two factors contribute to increase the acidity of an α-hydrogen compared to other C—H bonds. First, the presence of the adjacent polar covalent C=O bond polarizes the electron pair of the C—H bond so that the hydrogen may be removed as a proton by any strong base. The second and perhaps more important factor in the acidity of α-hydrogens is the fact that the resulting anion is a hybrid of two important contributing structures, one with a negative charge on carbon and the other with a negative charge on oxygen.

The most important reaction of an anion derived from an aldehyde or ketone is addition to the carbonyl group of another aldehyde or ketone. In this reaction, a new carbon-carbon single bond is formed between the α-carbon of one molecule and the carbonyl carbon of the other molecule. For example, reaction of two molecules of acetaldehyde in aqueous NaOH yields 3-hydroxybutanal, commonly known as aldol. The name aldol is derived from the structural features of 3-hydroxybutanal—it is both an aldehyde and an alcohol.

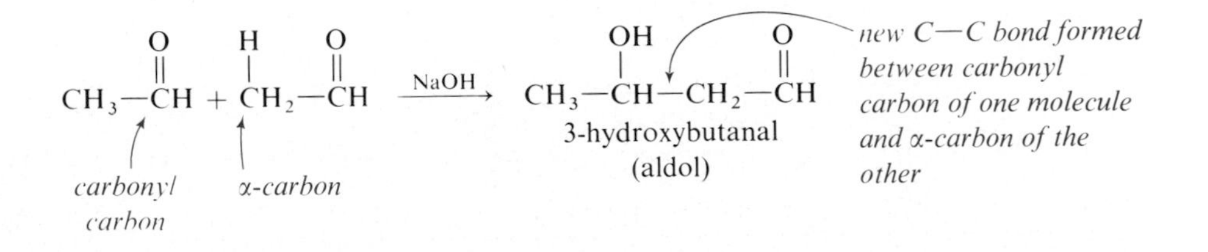

This type of reaction is known as an aldol condensation. Aldol condensation between two molecules of propanal yields 3-hydroxy-2-methylpentanal.

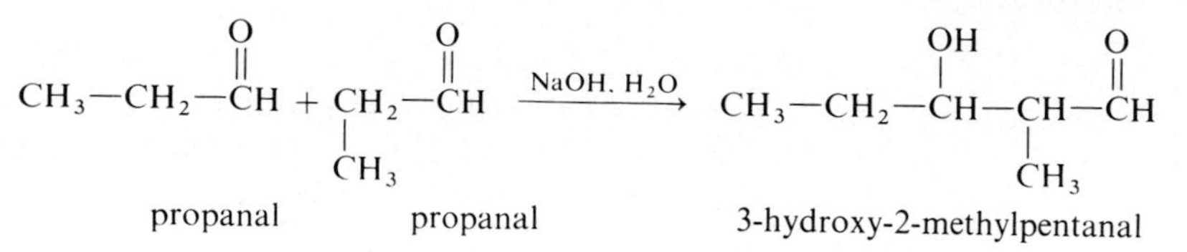

Ketones also undergo aldol condensation, as illustrated by the condensation of acetone in the presence of barium hydroxide.

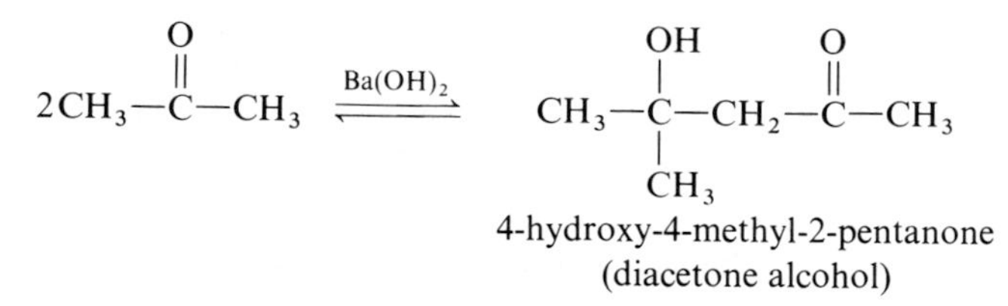

The characteristic structural feature of a product of an aldol condensation is a β-hydroxyaldehyde or a β-hydroxyketone.

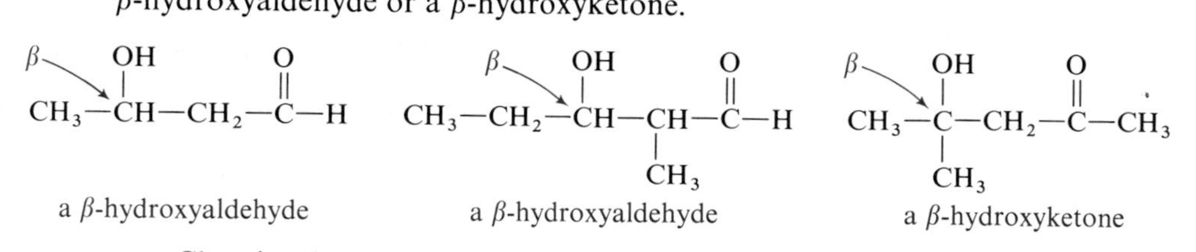

Chemists have proposed a three-step mechanism for the aldol condensation. This mechanism is illustrated for the aldol condensation of acetaldehyde to form 3-hydroxybutanal.

Step 1 uses a molecule of base to form the resonance-stabilized anion of acetaldehyde. Nucleophilic addition of this anion to another molecule of acetaldehyde in Step 2 gives a tetrahedral carbonyl addition compound, which reacts in step 3 with water to give the final product and to regenerate a molecule of base.

Step 1:

$$H{-}CH_2{-}\overset{O}{\overset{\|}{C}}{-}H + H{-}O^- \rightleftharpoons H{-}\overset{-}{\ddot{C}}H{-}\overset{O}{\overset{\|}{C}}{-}H + H_2O$$

Step 2:

$$CH_3{-}\overset{:\ddot{O}}{\overset{\|}{C}}H + H{-}\overset{-}{\ddot{C}}H{-}\overset{O}{\overset{\|}{C}}{-}H \rightleftharpoons CH_3{-}\overset{:\ddot{O}:^-}{\overset{|}{C}}H{-}CH_2{-}\overset{O}{\overset{\|}{C}}{-}H$$

Step 3:

$$CH_3{-}\overset{:\ddot{O}:^-}{\overset{|}{C}}H{-}CH_2{-}\overset{O}{\overset{\|}{C}}{-}H + H_2O \rightleftharpoons CH_3{-}\overset{OH}{\overset{|}{C}}H{-}CH_2{-}\overset{O}{\overset{\|}{C}}{-}H + H{-}O^-$$

The ingredients in the key step of the aldol condensation are an anion and a carbonyl group. In a self-condensation, both roles are played by one kind of molecule. Mixed aldol condensations are also possible as for example the mixed aldol condensation of acetone and formaldehyde. Formaldehyde cannot function as an anion because it contains no α-hydrogen, but it can function as a particularly good anion acceptor because its carbonyl group is bonded to two hydrogens. Acetone forms an anion easily, but its carbonyl group is bonded to two methyl groups (considerably larger than H) and it is therefore a relatively poor anion acceptor compared to the carbonyl group of formaldehyde. Consequently, the mixed aldol condensation of acetone and formaldehyde gives 4-hydroxy-2-butanone.

$$CH_3-\overset{\overset{O}{\|}}{C}-CH_3 + H-\overset{\overset{O}{\|}}{C}-H \xrightarrow{OH^-} \underset{\text{4-hydroxy-2-butanone}}{CH_3-\overset{\overset{O}{\|}}{C}-CH_2-\overset{\overset{OH}{|}}{CH_2}}$$

In the case of mixed aldol condensations where there is no appreciable difference in reactivity between the two compounds, mixtures of products result. For example, in the condensation of equimolar quantities of propanal and butanal, both α-carbons are alike and both carbonyls are also alike. As a consequence, reaction of these two aldehydes in base gives a mixture of all four possible aldol condensation products.

β-Hydroxyaldehydes and β-hydroxyketones are very easily dehydrated. Often the conditions necessary to bring about the condensation itself cause dehydration. Alternatively, warming the aldol product in dilute mineral acid leads to dehydration. The major product of the loss of water is one with the carbon-carbon double bond α-β to the carbonyl group.

$$\underset{\substack{\text{3-hydroxybutanal}\\\text{(aldol)}}}{CH_3-\overset{\overset{OH}{|}}{CH}-CH_2-\overset{\overset{O}{\|}}{C}-H} \xrightarrow[\text{warm}]{\text{dilute HCl}} \underset{\substack{\text{2-butenal}\\\text{(crotonaldehyde)}}}{CH_3-\overset{\beta}{C}H=\overset{\alpha}{C}H-\overset{\overset{O}{\|}}{C}-H} + H_2O$$

***

Example 7.7

Write structural formulas for the products of the following aldol condensations and for the unsaturated compounds produced by loss of water.

(a) $2CH_3-\overset{\overset{O}{\|}}{C}-CH_3 \xrightarrow{\text{base}}$ **(b)** $C_6H_5-\overset{\overset{O}{\|}}{C}-H + CH_3-\overset{\overset{O}{\|}}{C}-CH_3 \xrightarrow{\text{base}}$

Solution

(a) $2CH_3-\overset{\overset{O}{\|}}{C}-CH_3 \xrightarrow{\text{aldol condensation}}$

$$CH_3-\underset{\underset{CH_3}{|}}{\overset{\overset{OH}{|}}{C}}-CH_2-\overset{\overset{O}{\|}}{C}-CH_3 \longrightarrow CH_3-\underset{\underset{CH_3}{|}}{C}=CH-\overset{\overset{O}{\|}}{C}-CH_3 + H_2O$$

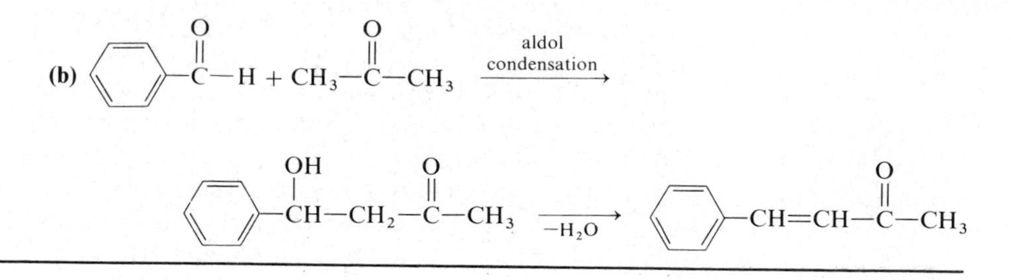

PROBLEM 7.7 Draw structural formulas for the products of the following aldol condensations and for the unsaturated compounds produced by loss of water.

(a) $C_6H_5-\overset{O}{\overset{\|}{C}}-CH_3 \xrightarrow{\text{base}}$ (b) $C_6H_5-\overset{O}{\overset{\|}{C}}-H + CH_3-\overset{O}{\overset{\|}{C}}-H \xrightarrow{\text{base}}$

The double bonds of alkenes, aldehydes, and ketones are readily reduced by catalytic hydrogenation. Hence, aldol condensation is often used for the preparation of saturated alcohols. For example, acetaldehyde may be converted into 1-butanol by first making 2-butenal.

$$2CH_3CHO \xrightarrow[\text{then dehydration}]{\text{aldol condensation}} \underset{\text{2-butenal}}{CH_3-CH=CH-CHO} \xrightarrow{2H_2,\ Pt} \underset{\text{1-butanol}}{CH_3CH_2CH_2CH_2OH}$$

Alternatively, if the β-hydroxyaldehyde is isolated, selective oxidation of the aldehyde group produces a β-hydroxycarboxylic acid.

$$\underset{\text{aldol}}{CH_3-\overset{OH}{\overset{|}{C}H}-CH_2-\overset{O}{\overset{\|}{C}}-H} + 2Ag^+ \xrightarrow{NH_4OH} \underset{\substack{\text{3-hydroxybutanoic acid} \\ (\beta\text{-hydroxybutyric acid})}}{CH_3-\overset{OH}{\overset{|}{C}H}-CH_2-\overset{O}{\overset{\|}{C}}-OH} + 2Ag$$

Example 7.8

Show how the following products can be synthesized from the indicated starting materials by way of aldol condensation reactions.

(a) $CH_3-\overset{O}{\overset{\|}{C}}-CH_3 \longrightarrow CH_3-\underset{\underset{CH_3}{|}}{CH}-CH_2-\overset{OH}{\overset{|}{C}H}-CH_3$

(b) $C_6H_5-\overset{O}{\overset{\|}{C}}-H + CH_3-\overset{O}{\overset{\|}{C}}-H \longrightarrow C_6H_5-CH=CH-CH_2OH$

Solution

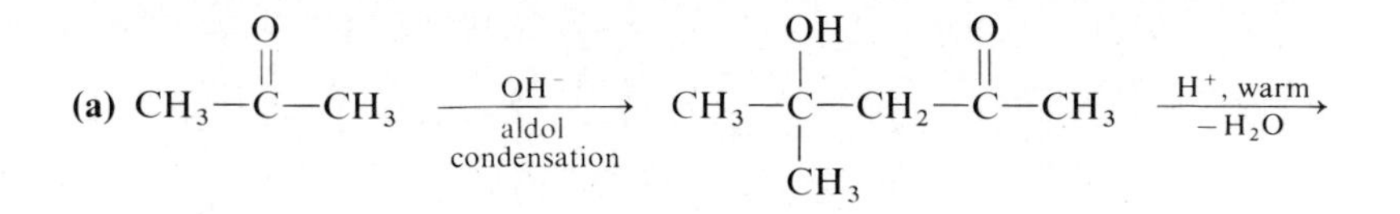

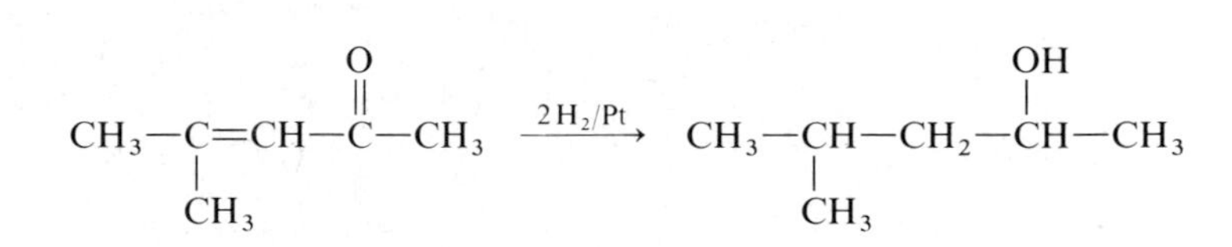

Aldol condensation of acetone yields 4-hydroxy-4-methyl-2-pentanone. Warming this β-hydroxyketone in acid leads to dehydration and formation of an α,β-unsaturated ketone. Catalytic reduction gives the desired alcohol.

(b) $C_6H_5-\overset{O}{\overset{\|}{C}}-H + CH_3-\overset{O}{\overset{\|}{C}}-H \xrightarrow[OH^-]{\text{aldol condensation}} C_6H_5-\overset{OH}{\overset{|}{C}H}-CH_2-\overset{O}{\overset{\|}{C}}-H \xrightarrow[-H_2O]{\text{warm}}$

$$C_6H_5-CH=CH-\overset{O}{\overset{\|}{C}H} \xrightarrow{LiAlH_4} \xrightarrow{H_2O} C_6H_5-CH=CH-CH_2OH$$

Mixed aldol condensation between benzaldehyde and acetaldehyde produces a β-hydroxyaldehyde that readily undergoes dehydration under the conditions used for the aldol condensation. Reduction of the aldehyde using $LiAlH_4$ or $NaBH_4$ forms the desired 3-phenyl-2-propenol.

PROBLEM 7.8 Show how the following products can be synthesized from the indicated starting materials by way of aldol condensations.

(a) $CH_3-CH_2-\overset{O}{\overset{\|}{C}}-H \longrightarrow CH_3-CH_2-CH_2-\underset{CH_3}{\underset{|}{C}H}-CH_2-OH$

(b) $CH_3-CH_2-CH_2-\overset{O}{\overset{\|}{C}}-H \longrightarrow CH_3-CH_2-CH_2-CH=\underset{CH_2-CH_3}{\underset{|}{C}}-\overset{O}{\overset{\|}{C}}-OH$

7.6 SPECTROSCOPIC PROPERTIES

Aldehydes and ketones show characteristic infrared absorption between 1710 and 1740 cm^{-1} associated with stretching of the carbonyl group. This absorption is generally very sharp and very intense as can be seen in the spectrum of menthone (Figure 7.3).

Because few other bond vibrations absorb energy in this region of the spectrum, absorption between 1680 and 1750 cm^{-1} is a reliable means for confirming

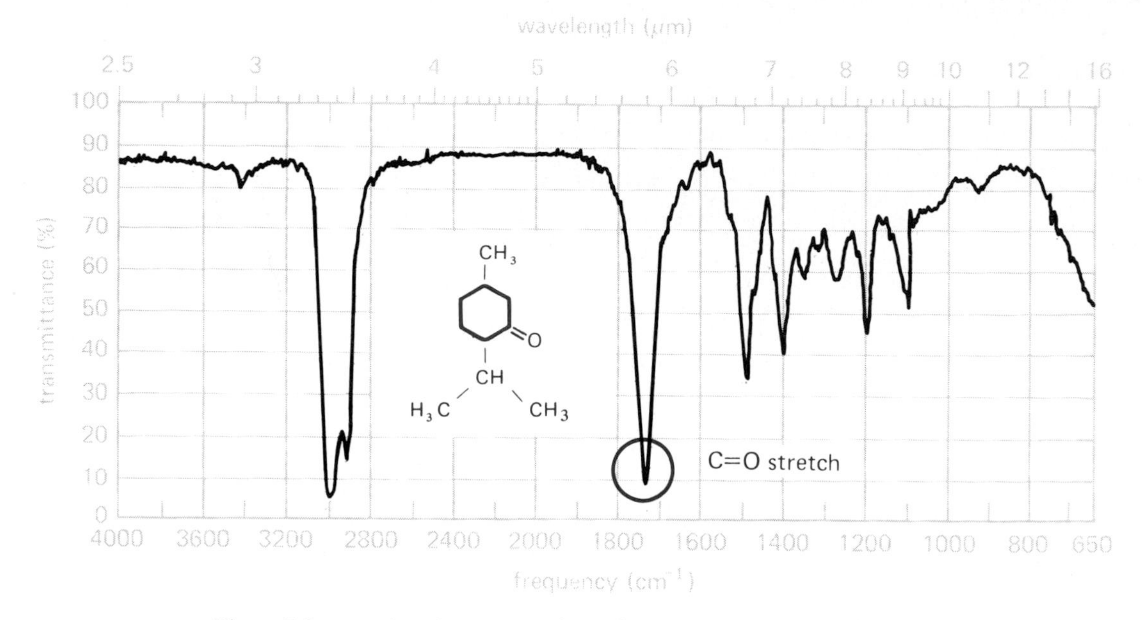

Figure 7.3 *An infrared spectrum of menthone.*

the presence of a carbonyl group. However, since several different functional groups contain the carbonyl group, it is not possible to tell from absorption in this region alone whether the carbonyl-containing substance is an aldehyde, ketone, or as we shall see in the next two chapters, a carboxylic acid, ester, or amide. Aldehydes show absorption (generally, two closely spaced peaks) between 2720 and 2830 cm^{-1} due to the stretching of the aldehyde C—H bond and can be distinguished from ketones by careful examination of this region of the infrared spectrum. For example, compare the spectrum of 3-methylbutanal (Figure 7.4) with that of menthone (Figure 7.3) and note the presence of the aldehyde C—H stretching frequencies at 2740 cm^{-1} and 2855 cm^{-1}.

Simple aldehydes and ketones show only weak absorption in the ultraviolet region of the spectrum. This absorption is due entirely to electronic excitation in the carbonyl group. If, however, the carbonyl group is conjugated with a carbon-carbon double bond as in the α,β-unsaturated ketone 3-pentene-2-one, the intensity of the absorption is sharply increased. 3-Pentene-2-one itself shows absorption at 224 nm.

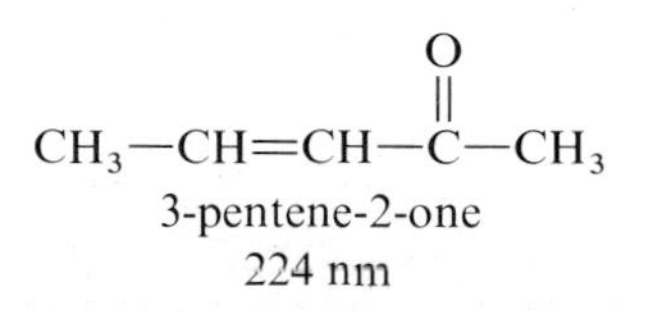

As with alkenes and aromatic hydrocarbons, the greater the extent of the system in conjugation with the carbonyl group, the more the absorption maximum is shifted toward the visible region of the spectrum.

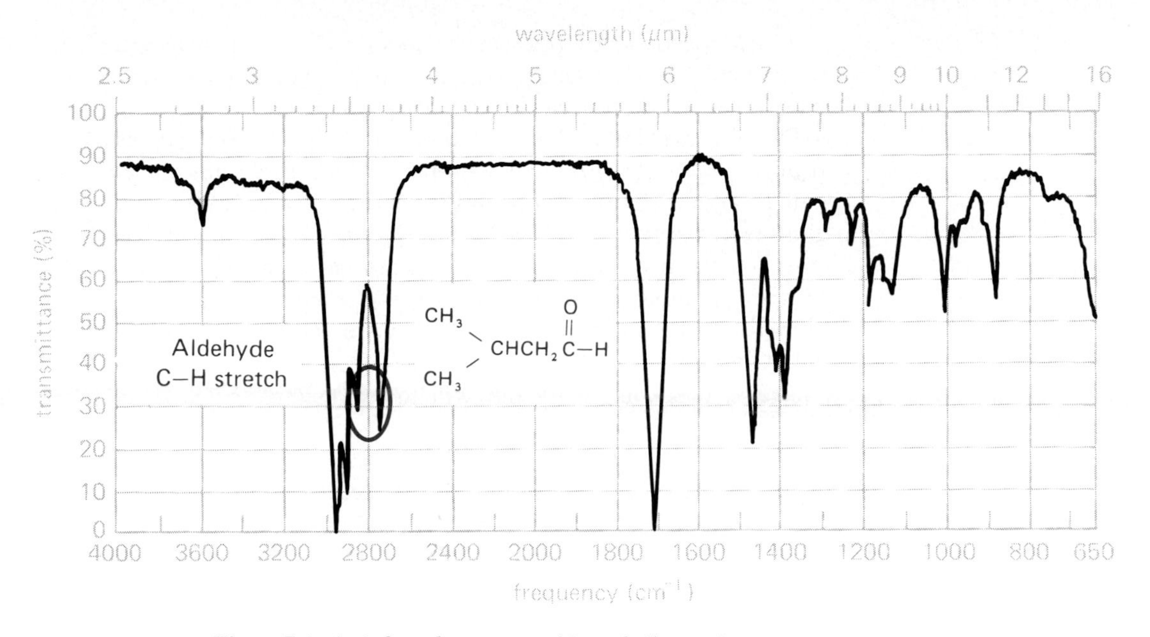

Figure 7.4 *An infrared spectrum of 3-methylbutanal.*

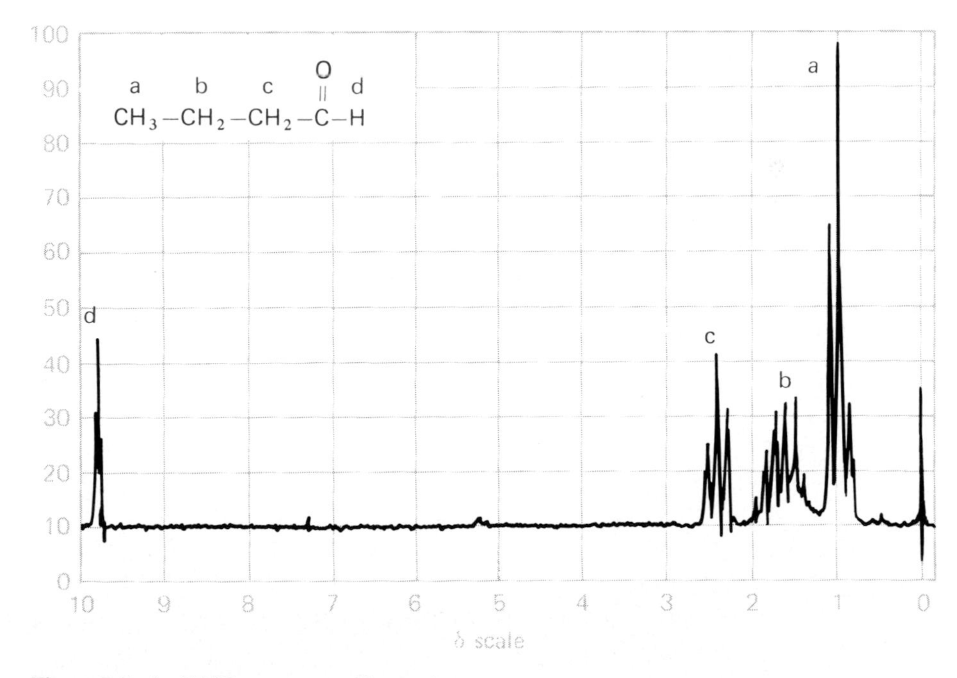

Figure 7.5 *An NMR spectrum of butanal.*

NMR spectroscopy is an important means for identifying aldehydes and for distinguishing between aldehydes and other carbonyl-containing compounds, that is, ketones, carboxylic acids, esters, and amides. For example, the NMR spectrum of butanal shows a very closely spaced triplet due to the aldehyde proton at $\delta = 9.5$ (Figure 7.5). Ketones, of course, do not contain a hydrogen bonded to the carbonyl group, and therefore do not give rise to a signal in this region. Notice also in the NMR spectrum of butanal that the signal at $\delta = 2.4$ due to the protons on the $-CH_2-$ group adjacent to the carbonyl group is split into a complex pattern. It is split into a triplet by the two protons of the adjacent $-CH_2-$ group, and each peak of the triplet is in turn split into a doublet by the single aldehyde proton. In general, we will not attempt to analyze splitting of this complexity but will instead simply refer to such signals as multiplets.

IMPORTANT REACTIONS

(1) Addition of water: hydration (Section 7.5).

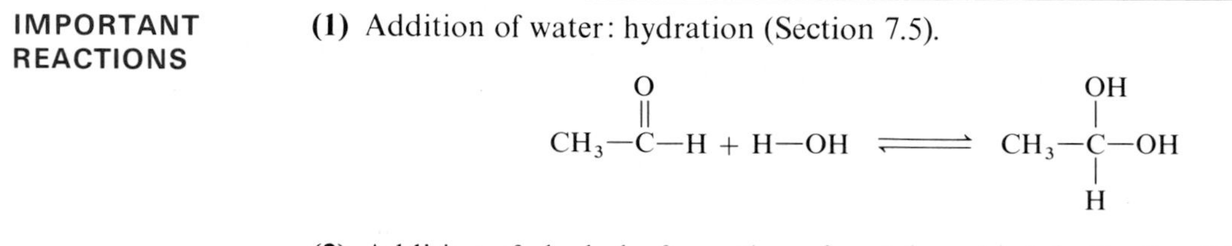

(2) Addition of alcohols: formation of acetals and ketals (Section 7.5A).

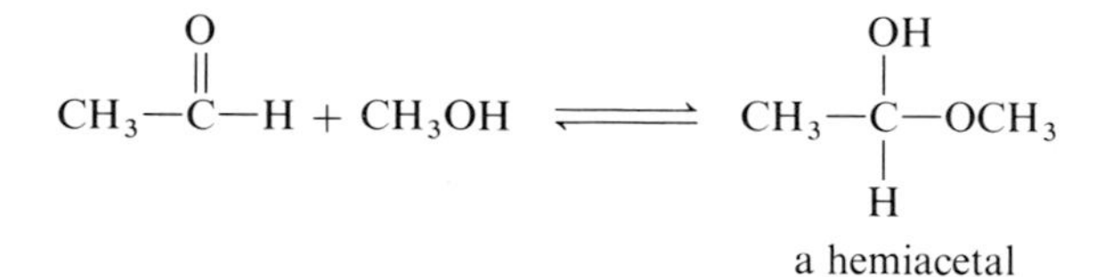

a hemiacetal

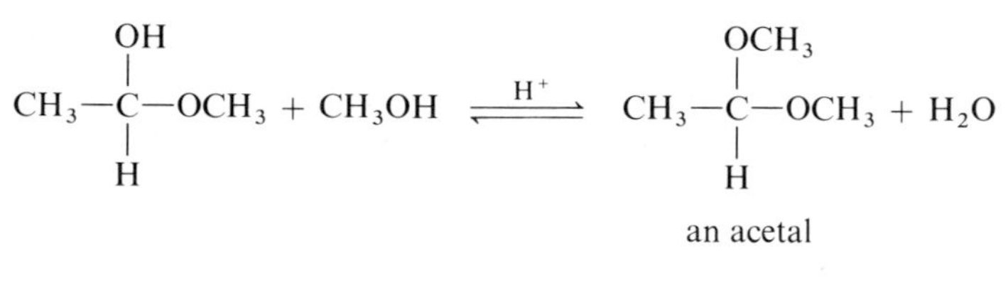

an acetal

a ketal

(3) Addition of ammonia and its derivatives: formation of Schiff bases (Section 7.5B).

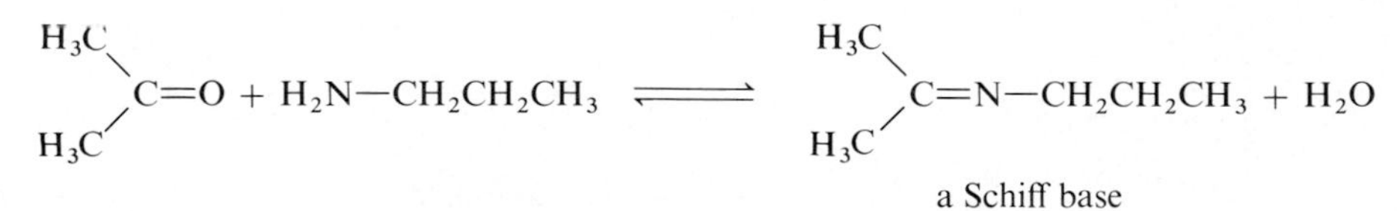

a Schiff base

(4) Addition of Grignard reagents followed by hydrolysis in aqueous acid (Section 7.5C).

(a) addition to formaldehyde gives primary alcohols:

$$C_6H_5MgBr + H{-}\overset{\overset{\displaystyle O}{\|}}{C}{-}H \longrightarrow C_6H_5{-}CH_2OH$$

(b) addition to aldehydes other than formaldehyde gives secondary alcohols:

$$C_6H_5MgBr + CH_3\overset{\overset{\displaystyle O}{\|}}{C}{-}H \longrightarrow C_6H_5{-}\overset{\overset{\displaystyle OH}{|}}{C}H{-}CH_3$$

(c) addition to ketones gives tertiary alcohols:

$$C_6H_5MgBr + CH_3\overset{\overset{\displaystyle O}{\|}}{C}CH_3 \longrightarrow C_6H_5{-}\underset{\underset{\displaystyle CH_3}{|}}{\overset{\overset{\displaystyle OH}{|}}{C}}{-}CH_3$$

(d) addition to ethylene oxide gives primary alcohols:

$$C_6H_5MgBr + \underbrace{H_2C{-}CH_2}_{O} \longrightarrow C_6H_5{-}CH_2{-}CH_2\ OH$$

(e) addition to carbon dioxide gives carboxylic acids:

$$C_6H_5MgBr + O{=}C{=}O \longrightarrow C_5H_5{-}\overset{\overset{\displaystyle O}{\|}}{C}{-}OH$$

(5) Oxidation of aldehydes to carboxylic acids (Section 7.5D).

$$CH_3{-}\overset{\overset{\displaystyle O}{\|}}{C}{-}H + Ag^+ \xrightarrow{NH_4OH} CH_3{-}\overset{\overset{\displaystyle O}{\|}}{C}{-}OH + Ag$$

$$CH_3{-}\overset{\overset{\displaystyle O}{\|}}{C}{-}H + Cr_2O_7^{2-} \xrightarrow{H^+} CH_3{-}\overset{\overset{\displaystyle O}{\|}}{C}{-}OH + Cr^{3+}$$

(6) Reduction of aldehydes and ketones to alcohols (Section 7.5D).

$$CH_3{-}CH_2{-}CH_2{-}\overset{\overset{\displaystyle O}{\|}}{C}{-}H + H_2 \xrightarrow{Pt} CH_3{-}CH_2{-}CH_2{-}CH_2{-}OH$$

$$CH_3{-}CH_2{-}\overset{\overset{\displaystyle O}{\|}}{C}{-}CH_3 + NaBH_4 \longrightarrow CH_3{-}CH_2{-}\overset{\overset{\displaystyle OH}{|}}{C}H{-}CH_3$$

(7) Tautomerism (Section 7.5E).

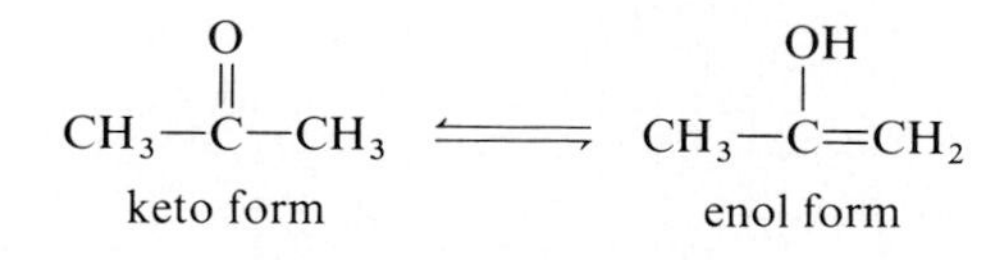

(8) The aldol condensation (Section 7.5F).

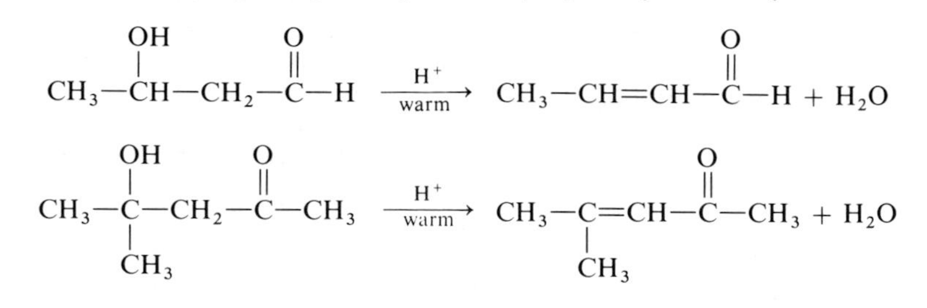

(9) Dehydration of β-hydroxyaldehydes and β-hydroxyketones (Section 7.5F).

$$CH_3-CH(OH)-CH_2-CHO \xrightarrow[\text{warm}]{H^+} CH_3-CH=CH-CHO + H_2O$$

$$CH_3-C(OH)(CH_3)-CH_2-CO-CH_3 \xrightarrow[\text{warm}]{H^+} CH_3-C(CH_3)=CH-CO-CH_3 + H_2O$$

PROBLEMS

7.9 Name the following compounds.

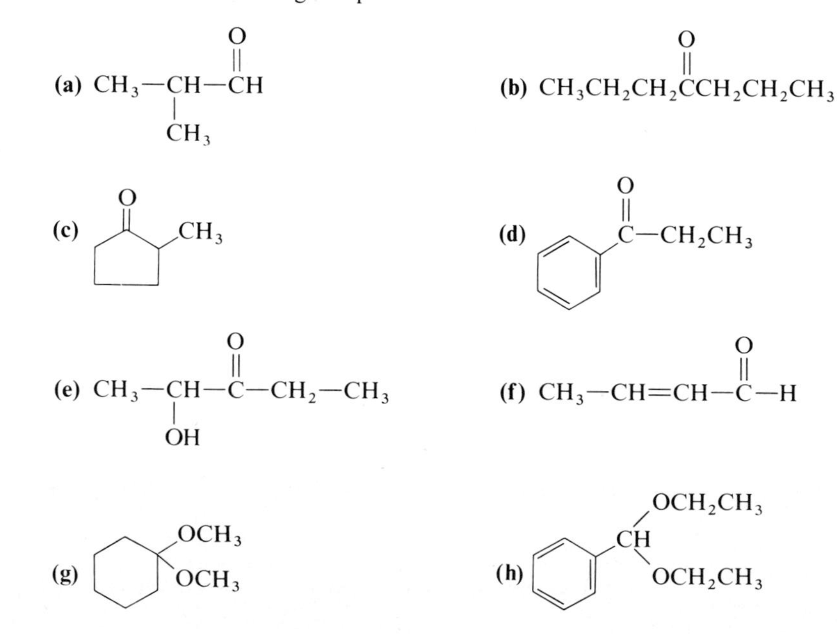

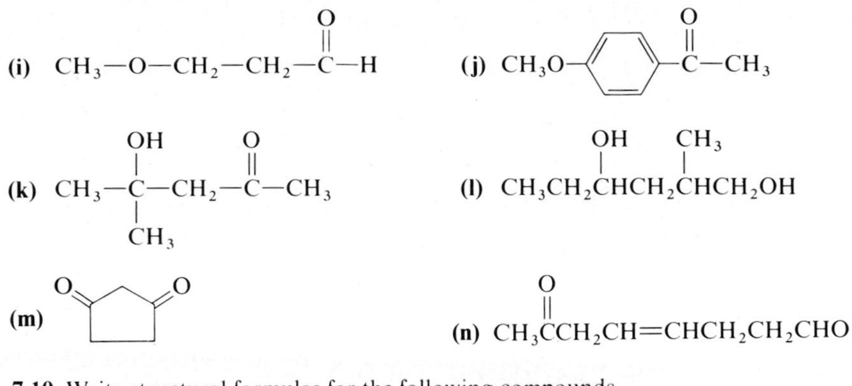

7.10 Write structural formulas for the following compounds.

(a) cycloheptanone
(b) propanal
(c) 2-methylpropanal
(d) benzaldehyde
(e) 3,3-dimethyl-2-butanone
(f) the diethylketal of acetone
(g) hexanal
(h) 2-decanone
(i) propenal
(j) *p*-bromocinnamaldehyde
(k) 2,5-hexanedione
(l) 3-methoxy- 4-hydroxybenzaldehyde (vanillin from the vanilla bean)
(m) 3-phenyl-2-propenal (from oil of cinnamon)
(n) 3,7-dimethyl-2,6-octadienal (from orange blossom oil)

7.11 Draw the structures for all aldehydes of molecular formula C_4H_8O; of molecular formula $C_5H_{10}O$. Give each an IUPAC name.

7.12 Draw the structures of all ketones of molecular formula C_4H_8O; of molecular formula $C_5H_{10}O$. Give each an IUPAC name.

7.13 Complete the following reactions. Where there is no reaction, write *N.R.*

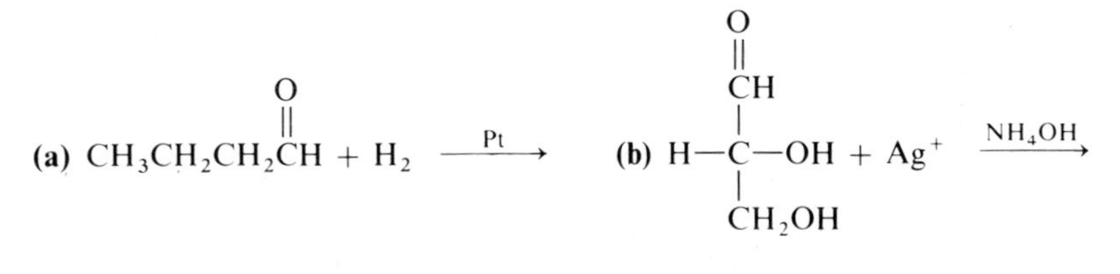

(c) $CH_3-CH(OH)-CH_2-C(=O)-CH_3 + Cr_2O_7^{2-} \xrightarrow{H^+}$

(d) $CH_3-CH(OH)-CH_2-C(=O)-CH_3 + H_2 \xrightarrow{Pt}$

(e) cyclopentanone $+ Ag^+ \xrightarrow{NH_4OH}$

(f) cyclopentanone $+ H_2 \xrightarrow{Pt}$

(g) $CH_3CH_2CH_2CH(=O) + 1\,CH_3CH_2OH \longrightarrow$

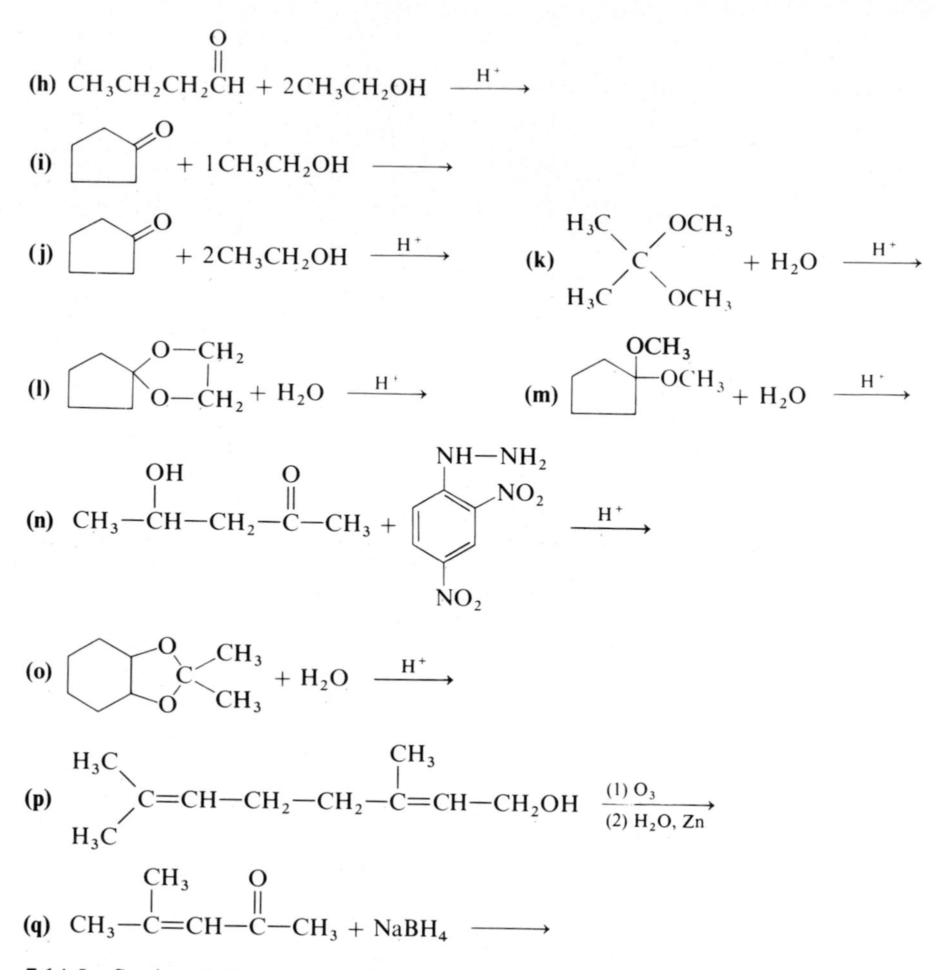

7.14 In Section 7.5A you saw that 4-hydroxypentanal forms a five-membered cyclic hemiacetal.

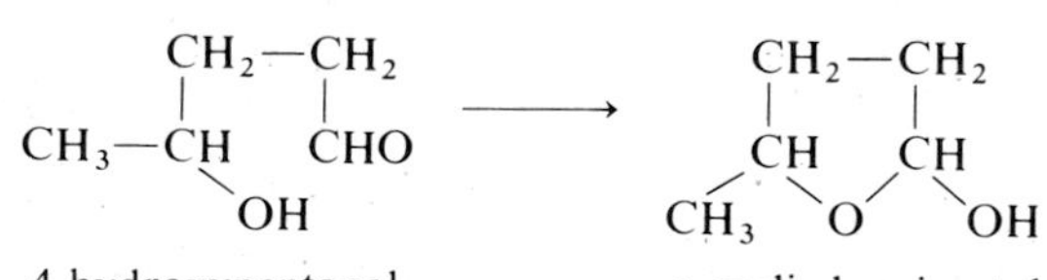

How many stereoisomers are possible for this cyclic hemiacetal? Draw stereorepresentations of each.

7.15 Propose a mechanism to account for the formation of a cyclic acetal from 4-hydroxypentanal and one molecule of methyl alcohol.

$$CH_3CH(OH)CH_2CH_2CHO + CH_3OH \xrightarrow{H^+} \text{(cyclic acetal: } H_3C, H, O, H, OCH_3) + H_2O$$

If the carbonyl oxygen of 4-hydroxypentanal were enriched with oxygen-18, would you predict that the oxygen-18 would appear in the cyclic acetal or in the water?

7.16 5-Hydroxyhexanal readily forms a six-membered cyclic hemiacetal.

$$\underset{\text{5-hydroxyhexanal}}{CH_3-\overset{\overset{\displaystyle OH}{|}}{C}H-CH_2-CH_2-CH_2-\overset{\overset{\displaystyle O}{\|}}{C}H} \longrightarrow \textit{a cyclic hemiacetal}$$

(a) Draw a structural formula for this cyclic hemiacetal.
(b) How many stereoisomers are possible for 5-hydroxyhexanal?
(c) How many stereoisomers are possible for this cyclic hemiacetal?
(d) Draw planar hexagon representations for each stereoisomer of the cyclic hemiacetal.
(e) Draw chair conformations for each stereoisomer of the cyclic hemiacetal and label substituent groups axial or equatorial.

7.17 Glucose, a polyhydroxyaldehyde, forms a six-membered cyclic hemiacetal in which the oxygen on carbon-5 of the chain reacts with the aldehyde on carbon-1.

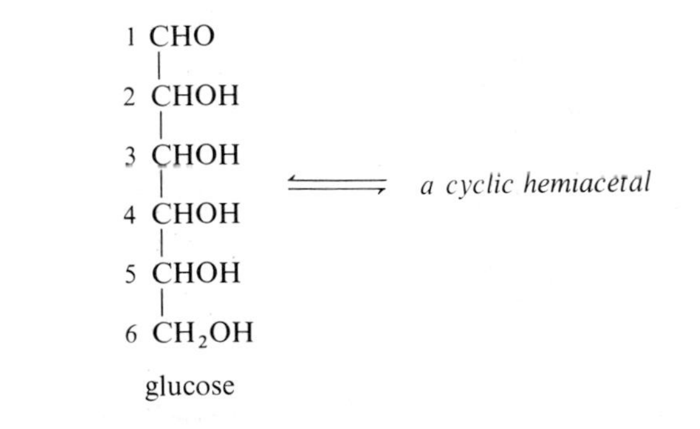

(a) How many chiral carbon atoms are there in glucose? How many stereoisomers are possible for a molecule of this structure?
(b) Draw a structural formula for the cyclic hemiacetal of glucose (do not worry about showing stereochemistry).
(c) How many chiral carbon atoms are there in the cyclic hemiacetal formed by glucose? How many stereoisomers are possible for the cyclic hemiacetal?

7.18 Acetaldehyde reacts with ethylene glycol in the presence of a trace of sulfuric acid to give a cyclic acetal of formula $C_4H_8O_2$. Draw the structural formula of this acetal and propose a mechanism for its formation.

7.19 Draw structural formulas for the products of hydrolysis of the following acetals and ketals in aqueous acid.

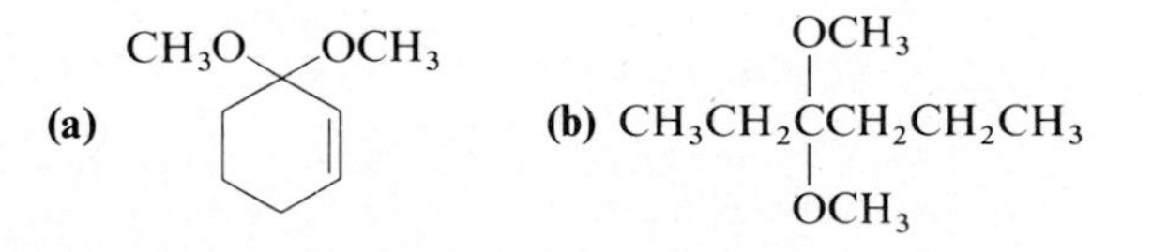

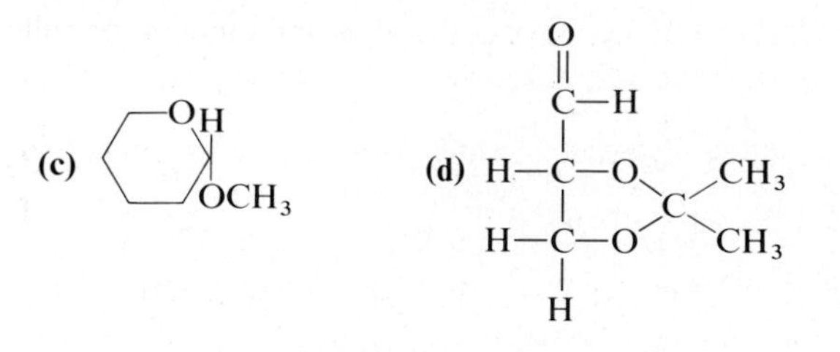

7.20 In Section 5.8 we stated that ethers such as diethyl ether and tetrahydrofuran are quite resistant to the action of strong aqueous acid. Acetals and ketals, however, in which there are two ether linkages to the same carbon, undergo hydrolysis in dilute aqueous acid (Problem 7.19). How might you account for this marked difference in chemical reactivity between ethers on the one hand and acetals and ketals on the other?

7.21 Draw structural formulas for the magnesium alkoxide formed by reaction of the following with propylmagnesium bromide. Also draw the structural formula of the alcohol formed by hydrolysis of the magnesium alkoxide in aqueous acid.

(a) CH_2O **(b)** CO_2 **(c)** $H_2C—CH_2$ (epoxide, O bridging)

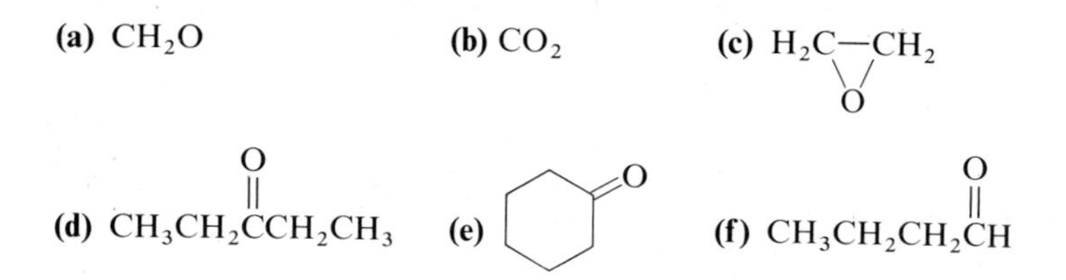

7.22 Suggest a method of synthesis for the following alcohols starting with any aldehyde, ketone, or epoxide and an appropriate Grignard reagent. In parentheses below each compound is shown the number of different combinations of Grignard reagents and aldehydes, ketones, or epoxides that might be used.

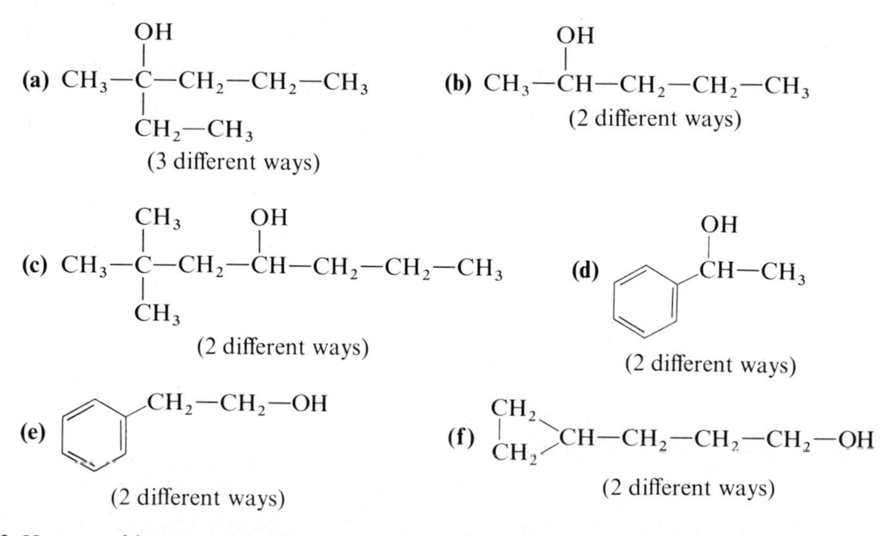

7.23 How would you account for the fact that Grignard reagents add readily to carbon-oxygen double bonds but do not add to carbon-carbon double bonds?

7.24 Draw the indicated number of enol forms for the following aldehydes and ketones.

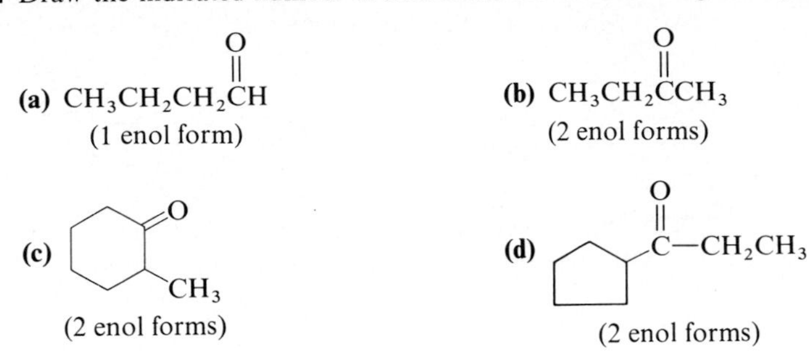

7.25 The following structures are enol forms. Draw structural formulas for the keto form of each.

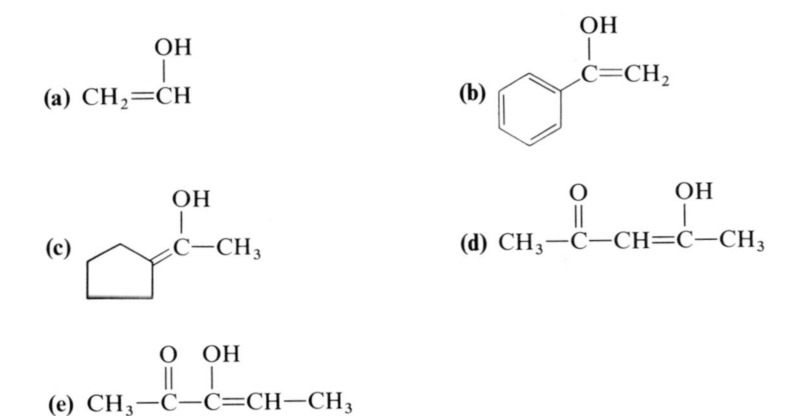

7.26 The following substance belongs to a class of compounds called enediols because it contains an alkene and two alcohols. An enol contains one hydroxyl group on a carbon-carbon double bond; an enediol contains two hydroxyl groups on a carbon-carbon double bond. Draw structural formulas for the two carbonyl-containing compounds with which this enediol is in equilibrium. *Hint*: One is a hydroxyaldehyde, the other a hydroxyketone.

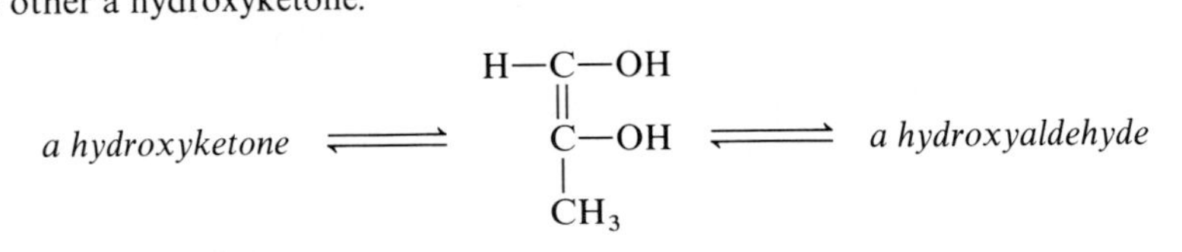

7.27 How would you account for the fact that in dilute aqueous alkali glyceraldehyde is converted into an equilibrium mixture of glyceraldehyde and dihydroxyacetone?

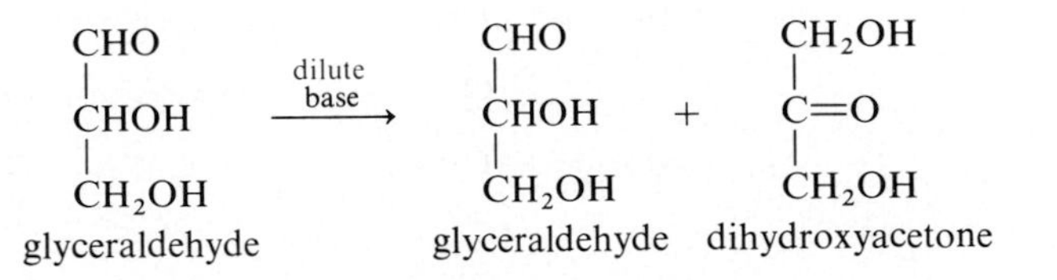

7.28 In dilute-aqueous base, the α-hydrogens of butanal show acidity, but neither the β- nor γ-hydrogens show any acidity whatsoever. Account for this difference in acidities.

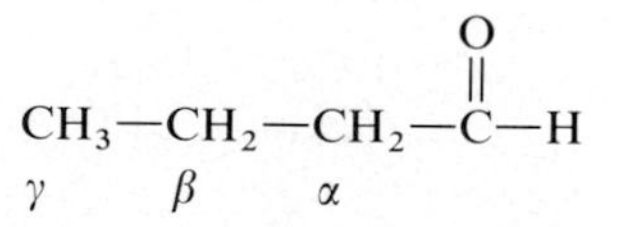

7.29 Draw structural formulas for the products of the following aldol condensations and for the unsaturated aldehyde or ketones produced by dehydration of the aldol product.

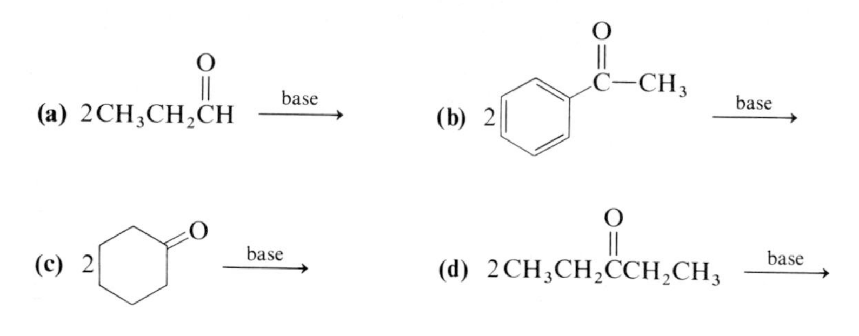

7.30 Draw structural formulas for the products of the following mixed aldol condensations and for the unsaturated aldehydes or ketones produced by dehydration of the aldol products.

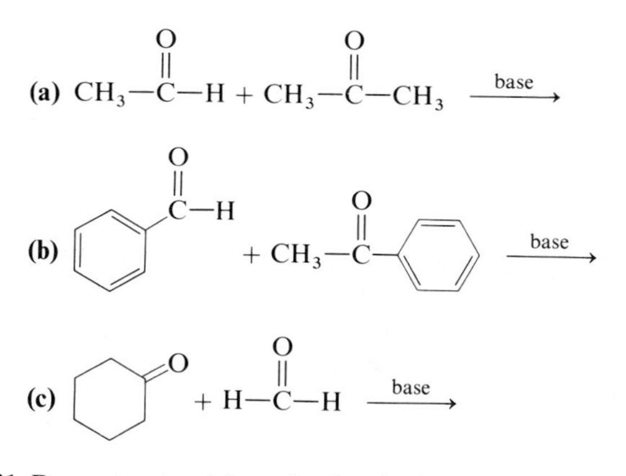

7.31 Draw structural formulas for the four possible aldol condensation products from a mixture of propanal and butanal.

7.32 Show reagents and conditions to illustrate how the following products can be synthesized from the indicated starting materials by way of aldol condensation reactions.

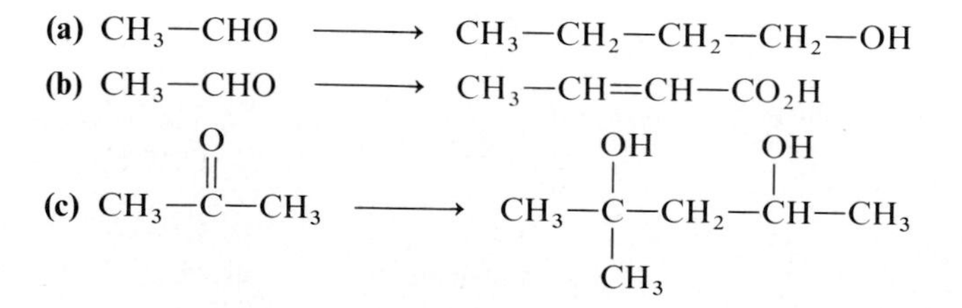

(a) $CH_3{-}CHO \longrightarrow CH_3{-}CH_2{-}CH_2{-}CH_2{-}OH$

(b) $CH_3{-}CHO \longrightarrow CH_3{-}CH{=}CH{-}CO_2H$

(c) $CH_3{-}C(=O){-}CH_3 \longrightarrow CH_3{-}C(OH)(CH_3){-}CH_2{-}CH(OH){-}CH_3$

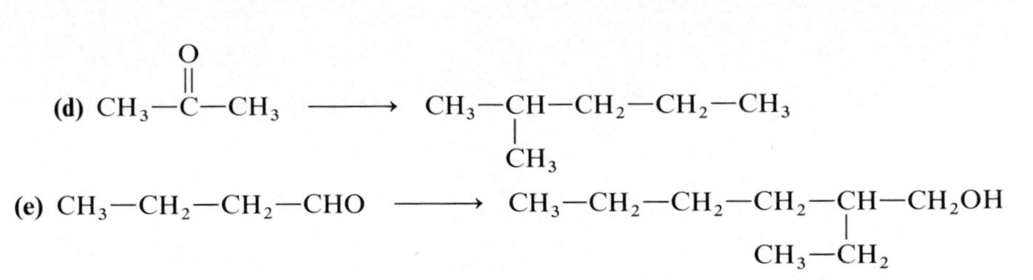

7.33 Show reagents and conditions by which you could prepare each of the following substances from cyclopentanone. In addition to cyclopentanone, use any additional organic or inorganic reagents necessary.

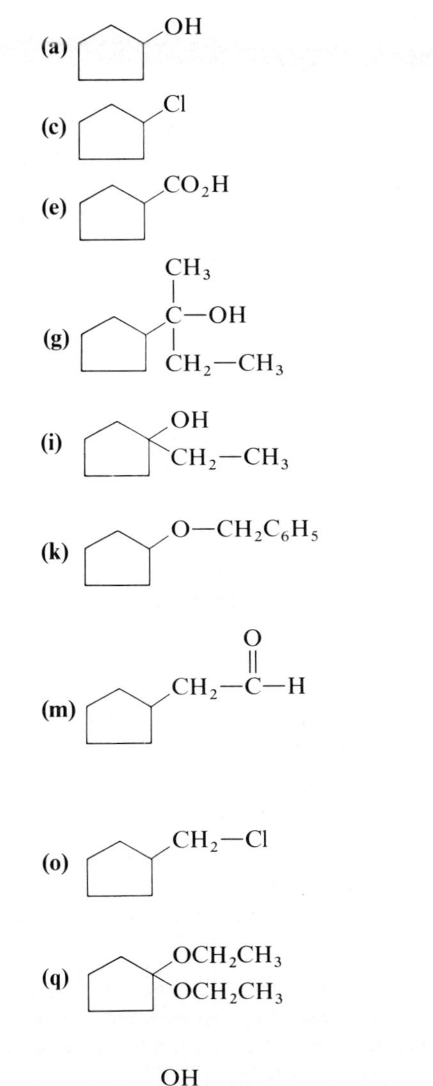

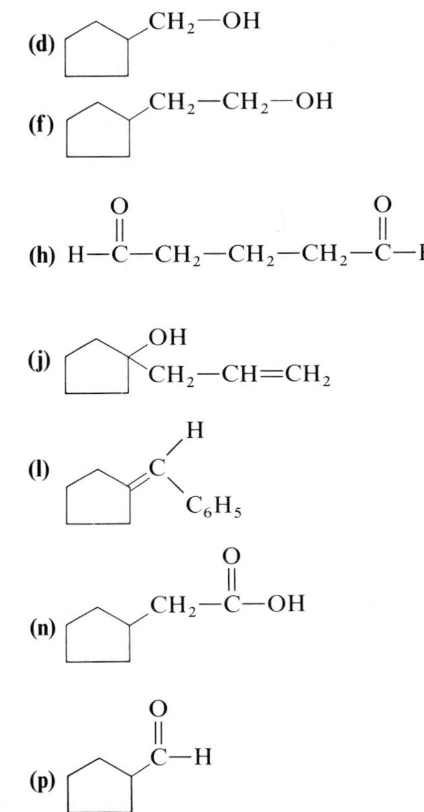

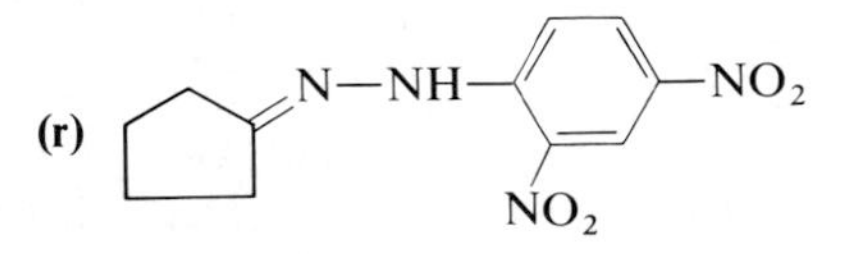

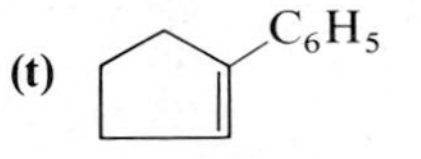

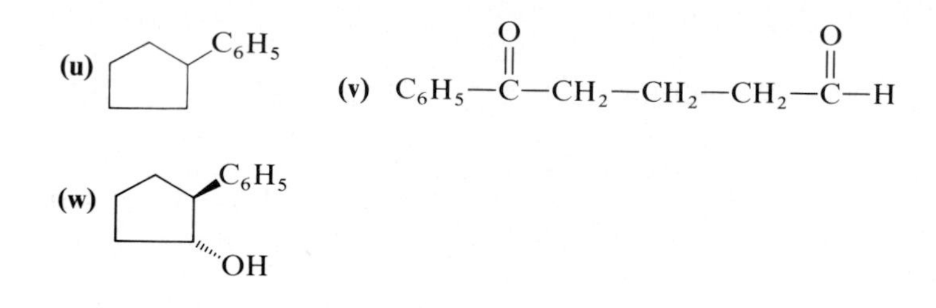

7.34 Cyclohexene can be converted into cyclopentene carboxaldehyde by the following series of steps. Ozonolysis of cyclohexene followed by work up in the presence of zinc and acetic acid forms a compound $C_6H_{10}O_2$. Treatment of this compound with dilute base transforms it into an isomer also of formula $C_6H_{10}O_2$. Warming this isomer in dilute acid yields cyclopentene carboxaldehyde.

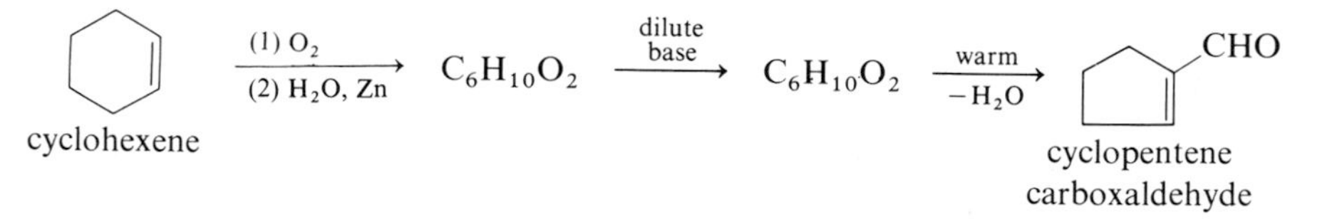

Propose structural formulas for the isomeric compounds of formula $C_6H_{10}O_2$ and account for the conversion of one isomer into the other in the presence of dilute base.

7.35 Compound A ($C_5H_{12}O$) does not give a precipitate with 2,4-dinitrophenylhydrazine. Oxidation of A with potassium dichromate gives B ($C_5H_{10}O$). Compound B reacts with 2,4-dinitrophenylhydrazine but does not give a precipitate with silver nitrate in ammonia. Acid-catalyzed dehydration of compound A gives hydrocarbon C (C_5H_{10}). Ozonolysis of hydrocarbon C gives acetone and acetaldehyde. Propose structural formulas for compounds A, B, and C.

7.36 Make a list of the reagents or types of reagents we have seen so far that add to carbonyl groups of aldehydes and ketones. Do the same for reagents that add to carbon-carbon double bonds.

(a) Which of these reagents add to both carbon-carbon and carbon-oxygen double bonds?

(b) Which of these reagents add to carbon-carbon double bonds but not to carbon-oxygen double bonds? How might you account for these differences in reactivity?

(c) Which of these reagents add to carbon-oxygen double bonds but not to carbon-carbon double bonds? How might you account for these differences in reactivity?

7.37 Make a list of the reactions we have examined so far that lead to the formation of new carbon-carbon bonds. Do not include reactions that merely transform a single bond into a double bond (as for example an alcohol into an aldehyde or ketone), or vice versa.

7.38 Show how you might distinguish between the following pairs of compounds by a simple chemical test. In each case, tell what test you would perform, what you would expect to observe, and write an equation for each positive test.

(a) cyclohexanone and cyclohexanol

(b) benzaldehyde and acetophenone

(c) benzaldehyde and benzyl alcohol

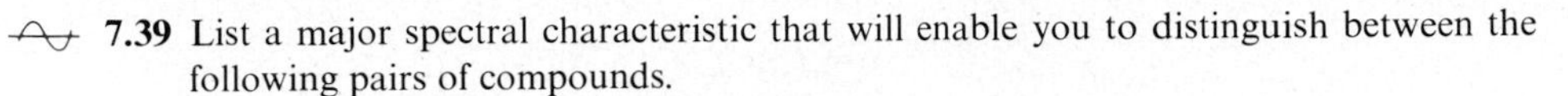

7.39 List a major spectral characteristic that will enable you to distinguish between the following pairs of compounds.

(a) cyclohexanone and cyclohexanol (IR)
(b) benzaldehyde and acetophenone (IR, NMR)
(c) benzaldehyde and benzyl alcohol (IR, NMR)
(d) cyclohexanone and 1,1-dimethoxycyclohexane (IR, NMR)
(e) progesterone (Figure 7.1) and testosterone (IR)
(f) cyclohexane carboxaldehyde (C_6H_{11}—CHO) and benzaldehyde (IR, NMR)

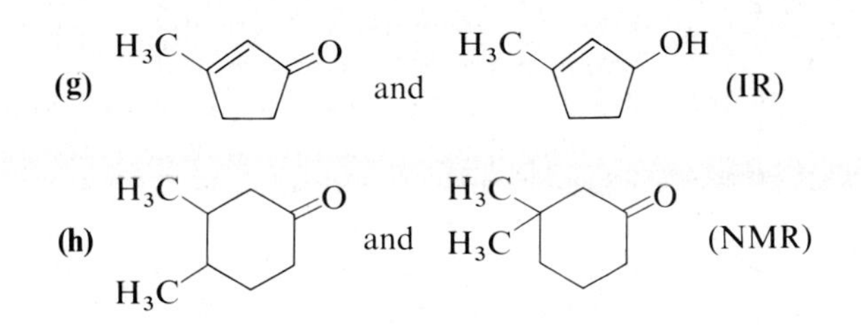

Insect Juvenile Hormones

MINI-ESSAY 3

OUR PRESENT KNOWLEDGE of insect juvenile hormones is the product of basic research on insect physiology and on the chemical events that control insect development. In the four decades since recognizing the existence of insect juvenile hormones, scientists have isolated and determined the structural formulas of several such hormones. They have also discovered extremely potent and selective hormone-mimicking substances in the plant world, synthesized hundreds of hormone analogs in the laboratory, and discovered substances in the plant world with anti-juvenile hormone activity.

To put these results in perspective, let us begin with insect development. Insect growth and metamorphosis generally proceed through four stages—egg to larva to pupa to adult. Internal glands secrete hormones that control each stage. One such hormone, known as juvenile hormone, is secreted by the corpora allata, two tiny glands in the head of insects. At certain stages in development, this hormone must be present; at other stages, it must be absent. For example, juvenile hormone must be present for the immature larva to progress through the usual stages of larval growth. Then, for the mature larva to undergo metamorphosis into an adult, hormone secretion must stop. If juvenile hormone is supplied at this critical time, either by implantation of active corpora allata or by application of the hormone itself, the pupa does not form a viable, mature adult. Juvenile hormone must also be absent for insect eggs to undergo normal embryonic development. If the hormone is applied, eggs either fail to hatch or the immature insects will die without reproducing.

Although the existence of juvenile hormone was recognized as early as 1939 and its site of production in the corpora allata was established, all efforts to extract and isolate it from living insects failed. Then in 1956, Carroll Williams of Harvard University discovered that the abdomen of the adult male Cecropia moth contains a rich storage depot of hormone. In retrospect, discovery of this depot seems most remarkable, for even today the only other known insect from which juvenile hormone can be extracted is the closely related male Cynthia moth.

The first crude preparations of Cecropia juvenile hormone were obtained by ether extraction of excised abdomens. Evaporation of the ether extracts leaves a golden-colored oil, about 0.2 mL per abdomen. Injection of this crude oil produced all the effects achieved by implanting active corpora allata. In fact, injection of the hormone is not even necessary to produce these effects. Merely placing the

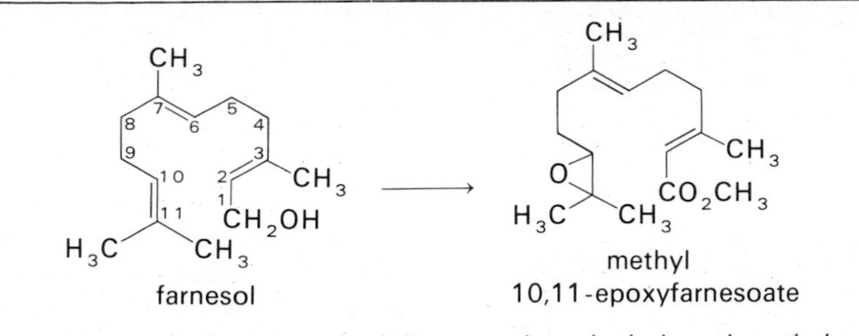

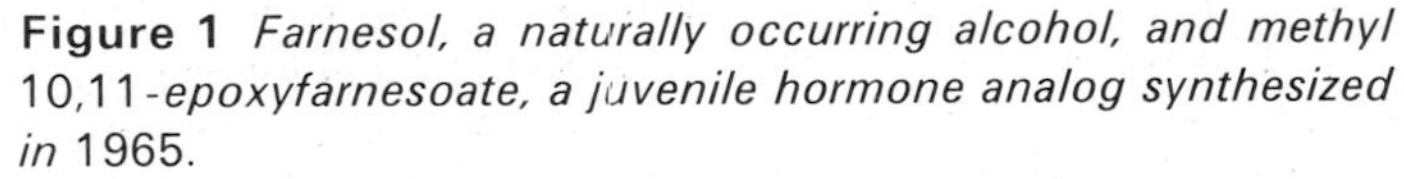
Figure 1 *Farnesol, a naturally occurring alcohol, and methyl* 10,11*-epoxyfarnesoate, a juvenile hormone analog synthesized in* 1965.

oily extract on the insects' bodies produces the same results—derangement of growth and the formation of nonviable adults. It was just this type of disrupted development, coupled with the simplicity of application, that first suggested juvenile hormone as a potential insecticide. Although its activity varies from family to family, natural Cecropia hormone affects such diverse insects as representatives of Coleoptera, Lepidoptera, Hemiptera, and Orthoptera.

Before the structure of Cecropia juvenile hormone was established, scientists had observed some slight degree of juvenile hormone activity in farnesol, a naturally occurring plant sesquiterpene alcohol (Figure 1). Following this lead, William Bowers of the United States Department of Agriculture (USDA) experimental station at Beltsville, Md., began to make systematic structural modifications of farnesol, hoping to synthesize new and even more active compounds. In 1965, he prepared methyl 10,11-epoxyfarnesoate from farnesol by forming an epoxide between carbons 10 and 11 and oxidizing the alcohol on carbon 1 to a carboxylic acid, followed by esterification.

The farnesol derivative synthesized by Bowers was 1600 times more active than farnesol. Although it had only about 0.02% of the Cecropia hormone activity, this suggested that when juvenile hormone was identified, it would bear at least some structural resemblance to farnesol. Bowers speculated that ". . . it is quite possible that juvenile hormone will be synthesized before it is structurally identified from natural sources."

In 1964, another juvenile hormone analog was discovered by what must be regarded as serendipity smiling. Karel Sláma, a young Czech entomologist, arrived at Harvard University to study with Carroll Williams. Sláma brought with him jars of *Pyrrhocoris apterus*, a species of linden bug which he had reared and studied for many years in Prague. To his considerable surprise and mystification, these bugs, when reared in the Harvard laboratory environment failed to metamorphose into sexually mature adults. Instead, they continued to grow as larvae or molted into adultlike forms while retaining many larvalike characteristics. All died without attaining maturity. It appeared that the bugs had been exposed to some unknown source of the hormone because such behavior had previously been observed only upon application of juvenile hormone. Sláma and Williams discovered that the source was none other than the paper towels placed in each rearing jar to give the bugs a surface to walk on. Almost all American paper had the same effect, but surprisingly, paper of European or Japanese manufacture had no effect. For want of a better name, the substance was termed "paper factor." The origin of paper factor was traced to the balsam fir (*Abies balsamea*), a principal source of American paper pulp. Balsam fir synthesizes the active material, which stays with the paper pulp through the entire manufacturing process. Paper factor eventually became known as juvabione (Figure 2), suggesting its relation to juvenile hormone.

Unlike Cecropia juvenile hormone, juvabione is active only on the Pyrrhocorids, an insect family containing some of the most destructive pests of cotton. Closely related families appear to be totally unaffected by it. This exciting discovery was the first evidence for the existence of juvenile hormone-like material with highly selective action on a particular family of insects.

Sláma and his associates in Czechoslovakia prepared a number of compounds structurally related to juvabione but incorporating a benzene rather than a cyclohexene ring (Figure 2). Some of these derivatives are about 100 times more active than juvabione itself, and all retain specific action only on the Pyrrhocorids.

Research came full cycle in 1965 when Herbert Röller of

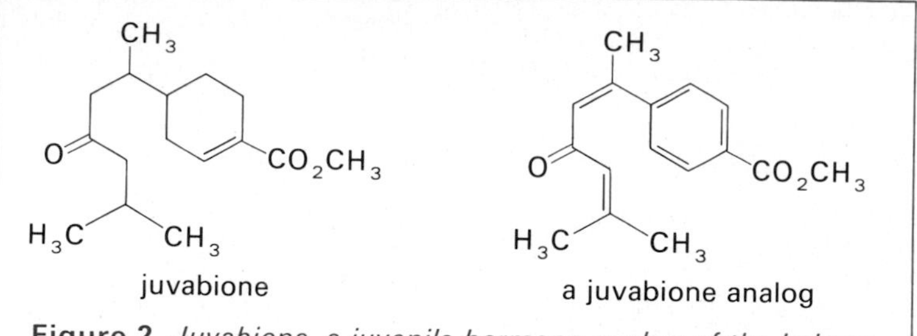

Figure 2 *Juvabione, a juvenile hormone analog of the balsam fir, is active only on insects of the family Pyrrhocoridae. The juvenile hormone analog shown is derived from p-(1,5-dimethylhexyl) benzoic acid.*

the University of Wisconsin isolated the male Cecropia hormone itself and, in 1967, using less than 0.3 mg of pure material, determined its structure. Cecropia hormone is remarkably similar to the farnesol derivative synthesized by Bowers in 1965. The difference is only two carbon atoms: the alkyl groups at carbons 7 and 11 in Cecropia hormone are $—CH_2CH_3$ rather than $—CH_3$. Clearly Bower's prediction in 1963 that Cecropia hormone would bear some structural similarity to farnesol was amply confirmed.

Cecropia hormone contains two *trans* double bonds and a *cis* epoxide. The *trans* configuration of both double bonds appears crucial for biological activity. In contrast, the stereochemistry of the epoxide ring is of secondary importance (Figure 3).

Juvenile hormones and hormone analogs offer promise of a new approach in pest control. In contrast to DDT and other persistent chemical pesticides, these new, third-generation pesticides are extremely potent and so highly selective in their action that they scarcely affect the surrounding biosphere. The trick is to use them at critical times in the insects' life cycle to so derange their normal growth and development that they fail to mate, reproduce, and multiply. Both laboratory and field studies are currently underway to test these substances in insect control programs.

But the story of insect juvenile hormones has not yet ended. Rather, it has taken a new and intriguing turn. Bowers and others reasoned that if plants contain substances with insect

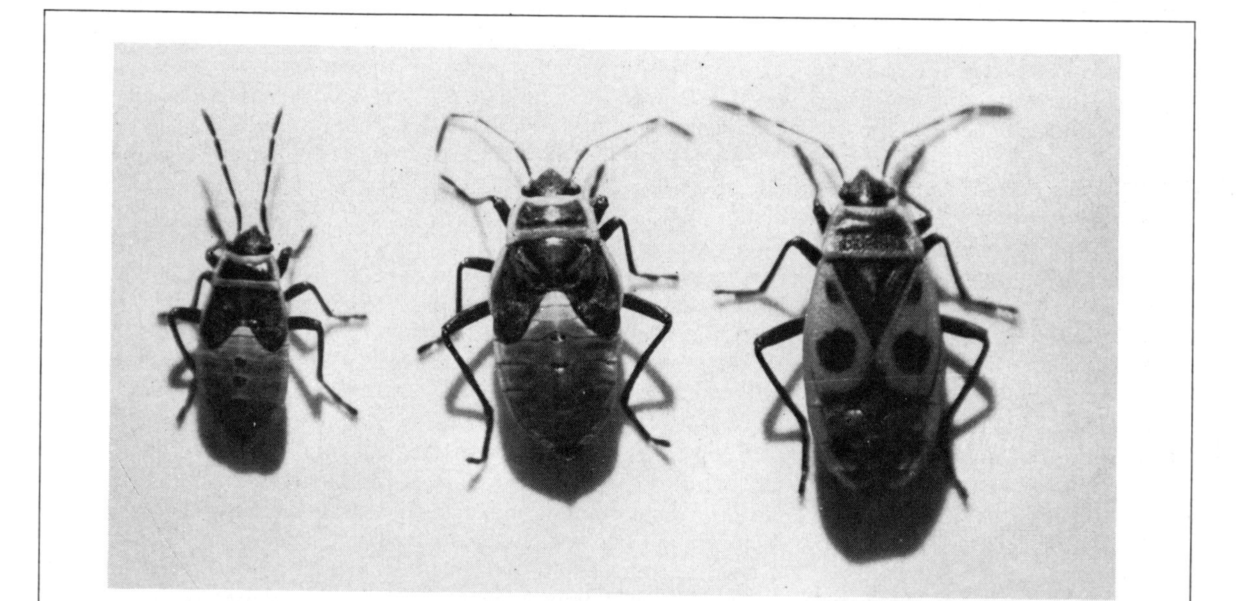

Three linden bugs: left, the insect in the last nymph phase; center, an overgrown nymph that has been treated with juvabione; right, a normal adult. The treated nymph will not develop into an adult and thus it will not reproduce. (USDA)

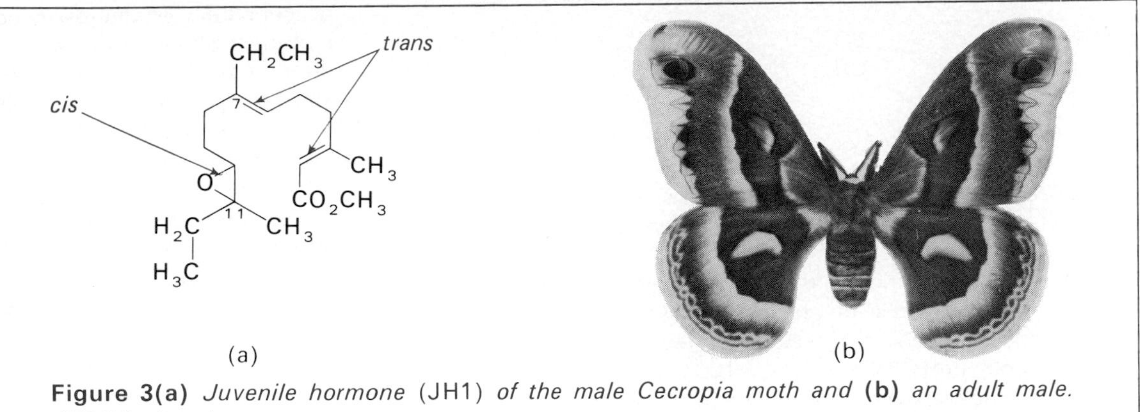

Figure 3(a) *Juvenile hormone* (JH1) *of the male Cecropia moth and* **(b)** *an adult male. (USDA photo)*

juvenile hormone activity, might they not also contain substances with anti-juvenile hormone activity? Bowers began a screening program by extracting plants with nonpolar solvents and testing the effects of the extracts on immature insects. He discovered that an extract from the common ornamental bedding plant *Ageratum houstonianum* possesses anti-juvenile hormone activity. Isolation and structural determination showed that there were two active components in the extract, precocene I and precocene II, substances that differ only in the presence of a methoxyl group (Figure 4).

When milkweed bug nymphs are treated with either precocene I or II, they molt into miniature adults and shortly thereafter they die. These substances show similar anti-juvenile hormone in other insects as well. Precocenes and precocene analogs offer even greater potential for pest control. Juvenile hormones and hormone analogs are effective only during a brief period in the insects' life cycle, but precocene I and precocene II analogs may be effective for much of the insects' lives.

Three yellow mealworms: left, a normal pupa; center, an abnormal adult that had been sprayed with a synthetic juvenile hormone which kept it from developing an adult abdomen; right, a normal adult. (USDA)

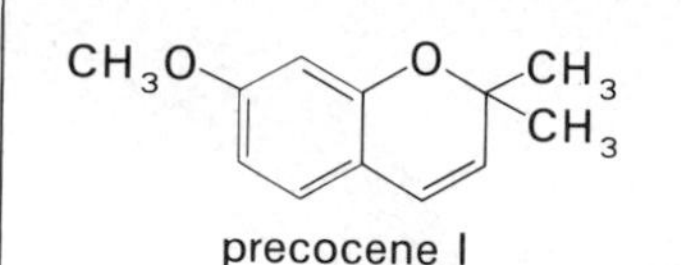

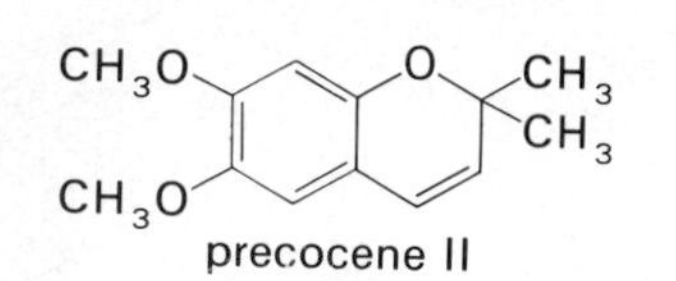

Figure 4 *Precocene I and II, substances with anti-juvenile hormone activity isolated from the bedding plant, ageratum.*

Bowers has continued the search for substances in the plant world that affect insect development. In 1980 he announced the discovery of two remarkably potent juvenile hormone mimics. From 150 g of the oil of sweet basil, *Ocimum basilicum*, he isolated milligram quantities of substances he named juvocimene 1 and juvocimene 2 (Figure 5), names derived in part from their biological activity and in part from their source in the plant world. Juvocimene 1, the more potent of the two, is approximately 1000 times more active than natural juvenile hormone 1. The isolation and identification of the precocenes and juvocimenes by Bowers and his colleagues represent the fruits of a highly effective research strategy, namely, having an idea of what to look for, where to look for it, how to test for it.

Studies such as we have described in this mini-essay add new dimensions to the study of insect growth and development and they give us a glimpse into the incredibly complex interactions between insects and plants. Furthermore, they offer the potential for a new generation of pesticides, compounds so selective in their action that they will attack only certain insects without presenting a hazard to other organisms.

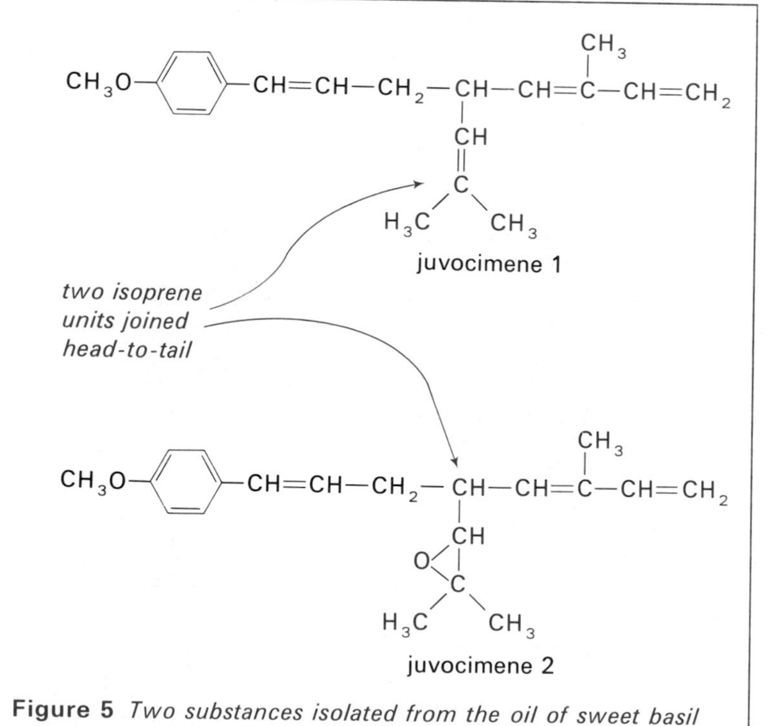

Figure 5 *Two substances isolated from the oil of sweet basil with juvenile hormone activity even more potent than natural juvenile hormone* 1.

References

Bowers, W. S., *Science* 164 (1969): 323.

———, and Nishida, R., "Juvocimenes: Potent Juvenile Hormone Mimics from Sweet Basil," *Science* 209 (1980): 1030–1032.

Bowers, W. S. et al., *Science* 93 (August 13, 1976): 542–547.

Judy, K. J. et al., *Proceedings of the National Academy of Science, U.S.A.* 70 (1973): 1509.

Röller, H. et al., *Angewant Chemie*, International Edition (English), 6 (1967): 179.

Sláma, K. and Williams, C. N., *Proceedings of the National Academy of Science, U.S.A.* 54 (1965): 411.

Williams, C., "Third-Generation Pesticides," *Scientific American* 217 (July 1967): 13.

———, *Chemical Ecology*, Sondheimer, E. and Simeone, J., eds., (N.Y.: Academic Press, 1970).

Pheromones

MINI-ESSAY 4

CHEMICAL communication abounds in nature: the clinging, penetrating odor of the skunk's defensive spray; the hound, nose to the ground, in pursuit of prey; the female dog making known her sexual availability; the female moth attracting males from great distances for mating. As biologists and chemists cooperate to extend our knowledge of other animals, it is becoming increasingly clear that chemical communication is the primary mode of communication in the vast majority of species in the animal world.

Prior to 1950, isolation of enough biologically active material to permit us to decipher any chemical communications seemed an insurmountable task. However, rapid progress in instrumental techniques, particularly in chromatography and spectroscopy, has now made it possible to isolate and carry out structural determinations on as little as a few micrograms of material. Yet, even with these advances, the isolation and identification of the components of pheromones remains a major challenge to technical and experimental expertise. For example, obtaining a mere 12 milligrams of gypsy moth sex attractant required the processing of 500,000 virgin female moths, each yielding only 0.02 micrograms of attractant. In other insect species, it is not uncommon to process at least 20,000 insects to obtain enough material for chemical identification.

The term *pheromone* (from the Greek *pherein*, "to carry" and *horman*, "to excite") is the accepted name for chemicals secreted by an organism of one species to evoke a response in another member of the same species. In this essay, we will look at insect pheromones, for these have been the most widely studied. Pheromones are generally divided into two classes: releaser and primer pheromones, depending on their mode of action. Primer pheromones cause important physiological changes that affect an organism's development and later behavior. The most clearly understood primer pheromones regulate caste systems in social insects (bees, ants, and termites). A typical colony of honey bees (*Apis mellifera*) consists of one queen, several hundred drones (males), and thousands of workers (underdeveloped females). The queen bee is the only fully developed female in the colony. She secretes a "queen substance" that prevents the development of workers' ovaries and promotes the construction of royal colony cells for the rearing of new queens. One of the components of the primer pheromone in the queen substance has been identified as

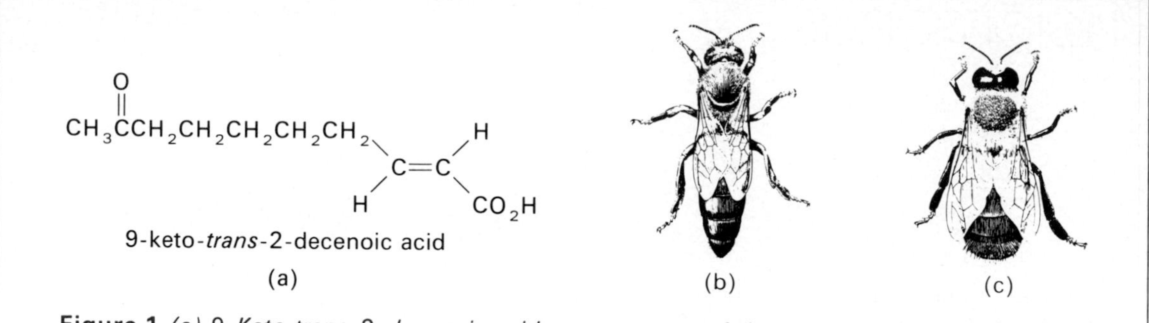

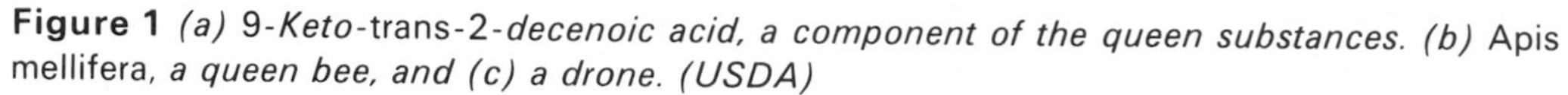
Figure 1 *(a) 9-Keto-*trans*-2-decenoic acid, a component of the queen substances. (b)* Apis mellifera, *a queen bee, and (c) a drone. (USDA)*

9 - keto - *trans* - 2 - decenoic acid (Figure 1). In addition, this same substance serves as a sex pheromone, attracting drones to the queen during her mating flight.

Releaser pheromones produce a rapid, reversible change in behavior, such as sexual attraction and stimulation, aggregation, trail marking, territorial and home range marking, and other social behavior. Some of the earliest observations of releaser pheromones were recorded on the alarm pheromones of the honey bee. Bee-keepers, and perhaps some of the rest of us too, are well aware that the sting of one bee often causes swarms of angry workers to attack the same spot. When a worker stings an intruder, it discharges, along with venom, an alarm pheromone that evokes the aggressive attack by other bees. One component of this alarm pheromone is isoamyl acetate, a sweet-smelling substance with an odor similar to that of banana oil (Figure 2).

One of the aggregating pheromones recently identified and intensively studied is that of the *Ips* family of bark beetles. Bark beetles live in the soil during the winter. In early spring when the temperature begins to rise, a few males emerge from the ground and seek trees in which to construct breeding chambers. The few males bore into trees and during this process, a pheromone produced in the hind gut is emitted triggering a massive secondary invasion of both males and females. After fertilization and hatching, *Ips* larva grow and develop behind the bark. In autumn, they leave the tree and return to the soil to begin another life cycle. In recent years there has been an explosive growth of *Ips typographus*, a species of bark beetle largely confined to the coniferous forests of Europe and Asia. *Ips typographus* is particularly attracted to the commercially valuable Norwegian spruce tree. In 1979, they killed or severely damaged an estimated 5 million trees in Norway and Sweden. Three active substances have been identified in the aggregating and sex pheromone of *I. typographus*. One of these is 2-methyl-3-buten-1-ol, an isomer of a key, five-carbon intermediate in the biosynthesis of terpenes. (See the mini-essay "Terpenes" for the structural formula of 3-methyl-3-buten-1-ol pyrophosphate and the role of this intermediate in terpene biosynthesis.) The other two active substances are 2 - methyl - 6 - methylene - 2,7 - octadien-4-ol (ipsdienol) and verbenol. Both of these substances are monoterpenes (Figure 3).

The governments of Norway and Sweden have initiated

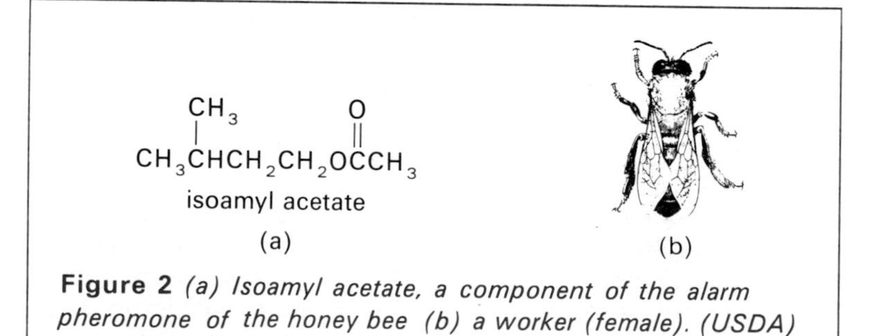

Figure 2 *(a) Isoamyl acetate, a component of the alarm pheromone of the honey bee (b) a worker (female). (USDA)*

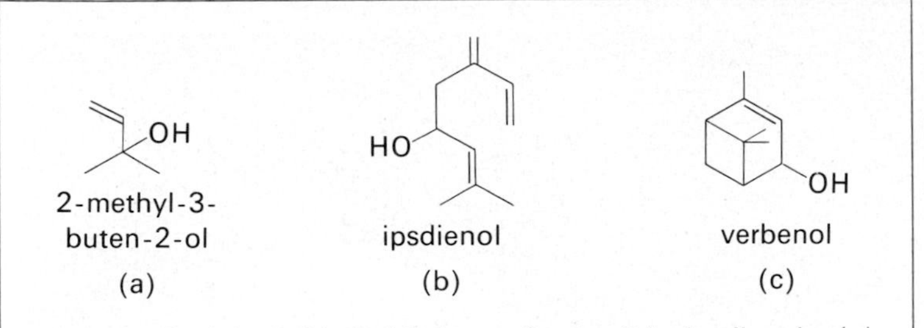

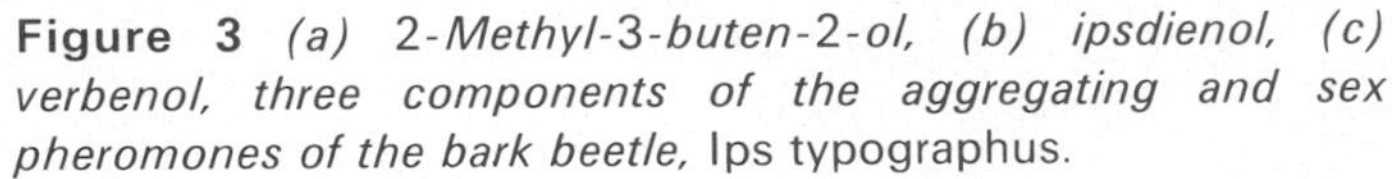

Figure 3 *(a) 2-Methyl-3-buten-2-ol, (b) ipsdienol, (c) verbenol, three components of the aggregating and sex pheromones of the bark beetle,* Ips typographus.

a large scale program to control the population of *I. typographus* and in the summer of 1979 placed more than 900,000 pheromone baited traps in forest areas damaged by bark beetles the previous year.

Of all classes of pheromones, sex pheromones have received the greatest attention in both the scientific community and the popular press. Larvae of certain insects that release them, particularly moths and beetles, are among the most serious agricultural pests. Sex pheromones are commonly referred to as "sex attractants," but this term is misleading because it implies only attraction. Actually, the behavior elicited by the pheromones is considerably more complex. Low levels of sex pheromone stimulation cause orientation and flight of the male toward the female (or in some species, flight of the female to the male). If the level of stimulation is high enough, copulation follows. At least in the case of the cabbage looper (*Trichoplusia mi*), none of these behavioral responses requires the presence of the female. A droplet of female cabbage looper sex pheromone, *cis*-7-dodecenyl acetate, on a piece of filter paper elicits orientation, flight, and even copulatory behavior of the male, all directed toward the spot of the evaporating pheromone (Figure 4).

Grandisol (Figure 5) is one of four components of the sex pheromone of the male cotton boll weevil (*Anthonomus grandis*). Work at the U.S. Department of Agriculture's Boll Weevil Research Laboratory showed that at any given time, a male contains only about 200 nanograms (200×10^{-9} grams) of the attractant, and that about 1.5 micrograms of attractant are released daily. The female of the species does not produce any of the attractant.

Several groups of scientists have studied the components of the sex pheromone of both the Iowa and New York strains of the European corn borer. Females of these closely related species secrete the same sex attractant, 11-tetradecenyl acetate. (See Figure 6.) Males of the Iowa strain show maximum response to a mixture containing about 96% of the *cis* isomer and 4% of the *trans* isomer. When the pure *cis* isomer is used alone, males are only weakly

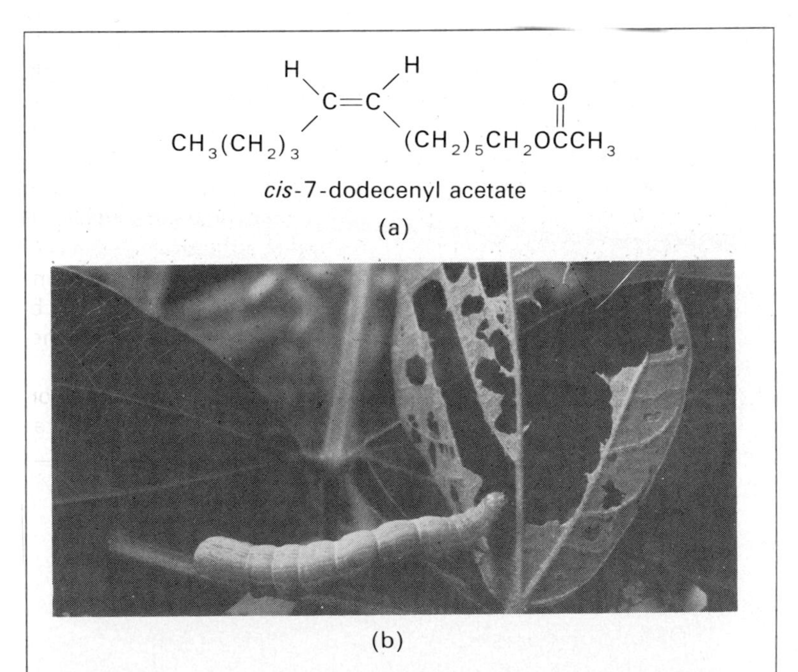

Figure 4 *(a)* Cis-7-*dodecenyl acetate, a component of the sex pheromone of the cabbage looper,* Trichopulsia mi. *(b) The cabbage looper feeding on soybean leaves. (USDA)*

Figure 5 *(a) Grandisol, a component of the sex pheromone of the male cotton boll weevil,* Anthonomus grandis. *(b) A boll weevil on a cotton plant. (USDA)*

attracted. Males of the New York strain show an entirely different response pattern; they respond maximally to a mixture containing 3% of the *cis* isomer and 97% of the *trans* isomer. There is evidence that optimum response to a narrow range of stereoisomers is widespread in nature and that at least some species of insects maintain species isolation (at least for the purposes of mating and reproduction) by the nature of the stereochemistry of their pheromones.

In this essay, we have examined several primer and releaser pheromones. Let us now consider the structural characteristics of the molecules that make them ideally suited for the tasks they must perform. An ordinary community of insects contains hundreds to thousands of different species, each with its own set of pheromones to communicate its needs for food, protection, reproduction, and so on. Such an odor environment must be enormously complex! An insect selects signals from its own species by means of chemoreceptors that are highly sensitive to relevant stimuli and far less sensitive to irrelevant ones. What kinds of structural and physical properties might we expect of organic molecules that meet these demanding requirements of species specificity?

First, insects must be able to synthesize pheromones and store them easily. It is probably for this reason that the majority of insect pheromones so far discovered and identified are related in structure to either fatty

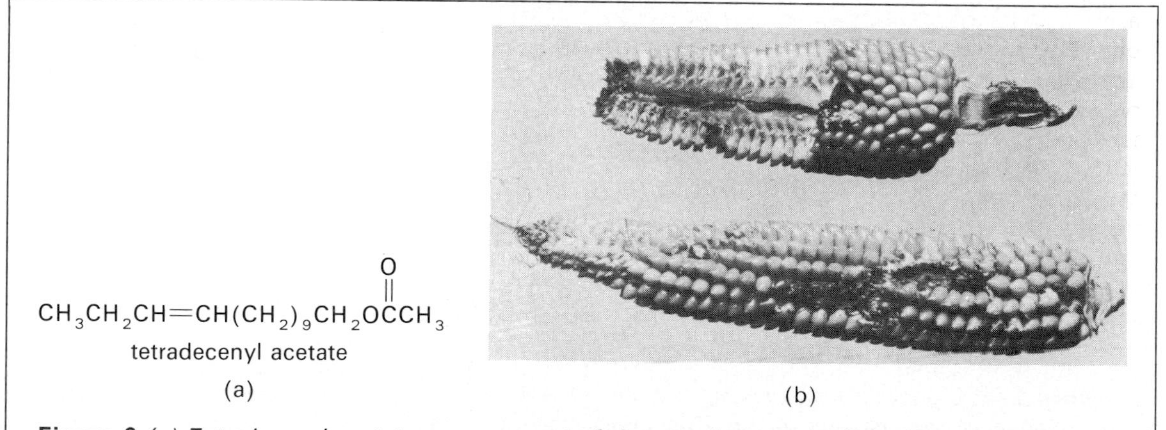

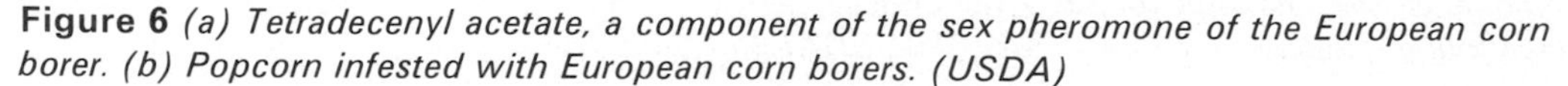
Figure 6 *(a) Tetradecenyl acetate, a component of the sex pheromone of the European corn borer. (b) Popcorn infested with European corn borers. (USDA)*

A field of corn devastated by European corn borers. The larvae damage corn by boring into the stalks and then feeding on the ears. (*USDA*)

acids or terpenes, molecules for whose production insects have elaborate biochemical pathways. Second, a molecule that is to serve as a pheromone must be volatile enough so that the concentration in a given volume of air is high enough to activate the chemoreceptors of males or females of the species, as the case may be. Organic molecules of very low molecular weight are generally too volatile and disperse too rapidly; organic molecules of very high molecular weight are generally not volatile enough. Perhaps for these reasons most airborne pheromones have molecular weights in the range of 100–300. Third, a molecule that is to serve as a pheromone must have a distinctive enough molecular architecture to insure that there is very little or no possibility for cross-attraction between even closely related species. Clearly, there is a tremendous range of diversity and, therefore, of biological individuality available with as few as ten to twenty carbon atoms through variations of structural formula, functional groups, and configuration about the chiral carbons and carbon-carbon double bonds. The potential for biological individuality is further compounded by the use of pheromones containing mixtures of *cis-trans* isomers, enantiomers, functional group isomers, and other species-specific substances.

In this essay we have discussed insect pheromones only. It is now certain that other animals including rodents, some primates, and a few other mammals also use pheromones for communications. There is also some information on fright and alarm pheromones in fish and amphibians. And what of humans?

References

Birch, M. C., ed. "Pheromones," *Frontiers of Biology*. (American Elsevier, 1974) 32.

Hummel, et al., "Clarification of the Chemical Status of the Pink Bollworm Sex Pheromone," *Science* 181 (1973): 873.

Klun, J. A. et al., "Insect Sex Pheromones: Minor Amounts of Opposite Geometrical Isomer Critical to Attraction," *Science* 181 (1973): 661.

O'Sullivan, Dermot A., "Pheromone Lures Help Control Bark Beetles," *Chemical and Engineering News* (July 30, 1979).

Regnier, F. D., *Science* 203 (1979): 559.

Sanders, Howard J., "New Weapons Against Insects: A Special Report," *Chemical and Engineering News* (July 28, 1975).

Shorey, H. H., *Animal Communication by Pheromones*, (N.Y.: Academic Press, 1976).

Sondheimer, E. and Simeone, J. B., eds., *Chemical Ecology*, (N.Y.: Academic Press, 1970).

Wilson, E. O., *Bio-Organic Chemistry*, Clavin, M. and Jorgenson, M., eds., (San Francisco: W. H. Freeman & Co., 1968).

Wood, D. L., Silverstein, R. M., and Nakajama, M., eds., *Control of Insect Behavior by Natural Products*, (N.Y.: Academic Press, 1970).

MINI-ESSAY 5

Footsteps on the Borane Trail

"HERBERT C. BROWN has systematically studied various boron compounds and their chemical reactions. He has shown how various specific reductions can be carried out using borohydrides. One of the simplest of these, sodium borohydride, has become one of the most used chemical reagents. The organoboranes, which he discovered, have become the most versatile reagents in organic synthesis. The exploitation of their chemistry has led to new methods of rearrangements, for addition to double bonds, and for joining carbon atoms to one another."[1]

In this essay, let us trace a part of the path that took Herbert C. Brown from his first encounter with the chemistry of boron compounds in 1936 to Stockholm some 40 years later. As we do so, we will pay particular attention to the origins of some of his discoveries and the sometimes tortuous but always fascinating chain of events through which he so enlarged our knowledge. And we will quote often from his responses to these events in order to catch a glimpse of the person behind the discoveries.

As Brown tells the story, his first encounter with the chemistry of boron compounds came in the form of a gift from Sarah Beylan, a fellow chemistry major later to become his wife. She presented him with a copy of Alfred Stock's book *Hydrides of Boron and Silicon* on the occasion of his graduation from the University of Chicago in 1936. Brown entered graduate school at Chicago that same year and began research under the direction of Professor Hermann Schlesinger on the reactions of diborane, B_2H_6. At that time, diborane was a very rare substance, available in milligram quantities in only two laboratories in the world, that of Stock at Karlsruhe, Germany and that of Schlesinger at Chicago. Brown quickly mastered the high-vacuum techniques developed by Stock for the preparation and handling of diborane and discovered that simple aldehydes and ketones react with this substance and that hydrolysis gives the corresponding alcohols.

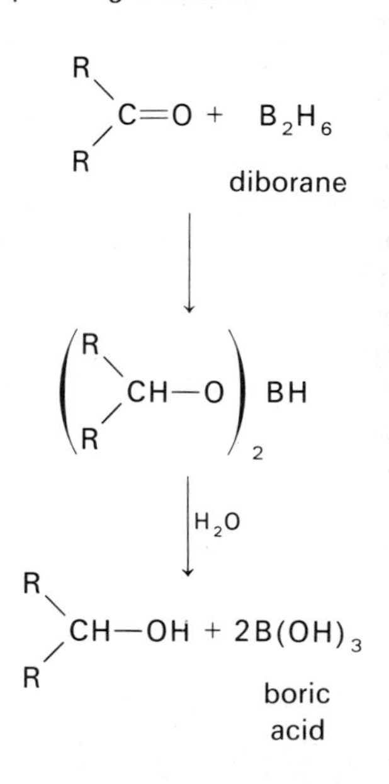

Thus was discovered the fact that diborane is a chemical reducing agent. This study was reported in 1938 in Brown's

Ph.D. thesis from the University of Chicago. Looking back on this discovery from the perspective of 1975, he comments, "Interest in this development among organic chemists was minimal. Diborane was a chemical rarity at that time, available only in milligram quantities through complex procedures. I wish I could say that we had the intelligence to recognize that this was an important synthetic procedure which would require only the development of a practical route to diborane to make it useful. However, we did not. We were later to do so as a result of research forced upon us by World War II."[2]

In 1940, Brown became part of a war-related research program. Uranium-235 is one of the few isotopes to undergo nuclear fission and there was an intensive research effort, funded by the National Defense Research Council, to find a way to separate U-235 (natural abundance 0.71%) from nonfissionable U-238 (natural abundance 99.28%). The most promising route appeared to be enrichment by gaseous diffusion. What was needed was a compound of uranium with a high volatility and, for maximum efficiency of the separation process, one with as low a molecular weight as possible. One such compound under study was uranium hexafluoride, UF_6, but it was highly corrosive and handling it presented what seemed to be insoluble technical problems. In 1940 Schlesinger was asked to undertake a research program in search of new, volatile, and noncorrosive compounds of uranium. Brown was a part of the group Schlesinger assembled for the task. They proceeded as follows. Only a short time before, Schlesinger had prepared aluminum borohydride, $Al(BH_4)_3$, and beryllium borohydride, $Be(BH_4)_2$, and found them to be the most volatile compounds of aluminum and beryllium yet discovered. Accordingly, Schesinger's group attempted to prepare a borohydride of uranium in the hopes that it would have the required high volatility. Within a short time, they were able to prepare uranium(IV) borohydride by the reaction of aluminum borohydride and uranium(IV) fluoride.

$$4\,Al(BH_4)_3 + 3\,UF_4 \longrightarrow 3\,U(BH_4)_4 + 4\,AlF_3$$

The compound had a low molecular weight, adequate volatility, lacked the corrosive properties of UF_6, and the government requested several kilograms for large-scale testing. The problem now was that preparation of uranium(IV) borohydride depended on having available considerably larger quantities of diborane than could be prepared by any method known at that time. There followed an intensive search for methods of preparing this key starting material in considerably larger quantities. They soon discovered such a method; reduction of boron trifluoride, BF_3, by lithium hydride, LiH, in diethyl ether solution.

$$6\,LiH + 8\,BF_3 \xrightarrow{(CH_3CH_2)_2O} B_2H_6 + 6\,LiBF_4$$

Among other substances discovered during the search was sodium borohydride, $NaBH_4$. (We will return to this hydride of boron in a moment.)

While the study of diborane and uranium(IV) borohydride was underway, Schlesinger was informed by the government that the problems encountered in safe handling of UF_6 had been solved and there was no further need for other volatile compounds of uranium! The reason, of course, was the accidental discovery of Teflon (Section 2.11) in 1938 by chemists at du Pont. Du Pont began limited production of Teflon in 1941, and the small quantity of polymer was immediately preempted by the Manhattan Project where it was used in equipment to contain the highly corrosive uranium hexafluoride.

In 1943, Schlesinger's research group was on the verge of being disbanded when he was approached by the Signal Corps who had heard of the preparation of $NaBH_4$ and wondered whether this substance might be used as a portable source for the generation of hydrogen gas in the field. Although $NaBH_4$ had not been used for this purpose, Brown felt certain that it would react with water, just as did B_2H_6, to liberate hydrogen. Accordingly, he set up a demonstration, all conducted behind safety glass and explosion-proof shields, for no one was certain just how violent the reaction between $NaBH_4$ and water might be. As all watched, Brown very cautiously added water to a sample of $NaBH_4$. The substance dissolved in water and to the great amazement of all, nothing else happened. As Brown later comments, "This was one of the greatest shocks of my life and the way we discovered that sodium borohydride possesses a remarkable stability, for a simple boron-hydrogen compound, in water."

Brown continued research on the properties of $NaBH_4$ and subsequently developed a new and commercially feasible route for its preparation by the reaction of sodium hydride with trimethyl borate at 250°C.

$$4\,NaH + \underset{\text{trimethyl borate}}{B(OCH_3)_3} \xrightarrow{250} NaBH_4 + \underset{\text{sodium methoxide}}{3\,NaOCH_3}$$

This route generated a new problem: how to separate sodium borohydride from the by-product sodium methoxide. The most logical strategy seemed to be to find a solvent in which sodium borohydride would dissolve but sodium methoxide would not. Among the solvents tested was acetone. To Brown's surprise, a vigorous reaction took place between sodium borohydride and acetone, and analysis revealed that there were now four moles of 2-propanol per mole of sodium borohydride originally present.

Sodium borohydride is a carbonyl reducing agent! The discovery of $NaBH_4$ in 1942 and that of $LiAlH_4$ in 1945 brought about a revolution in the methods available to chemists for reduction of carbonyl and other functional groups in organic molecules.

Sodium borohydride is now manufactured on a commercial scale by the process discovered by Brown. The crude product containing sodium borohydride and sodium methoxide is quenched with water to give a 12% solution of $NaBH_4$ in aqueous methanol. This solution is used directly as a bleach in the paper and pulp industry. Alternatively, $NaBH_4$ is extracted from the 12% aqueous solution using water-insoluble amines and isolated as a white powder. Purified hydride reducing agents are now a \$10 million per year market in the U.S., a \$50 million per year market worldwide, and growing rapidly. This market is dominated by $NaBH_4$ which accounts for over 70% of the dollar value of all hydride reducing agents used by the chemical industry.

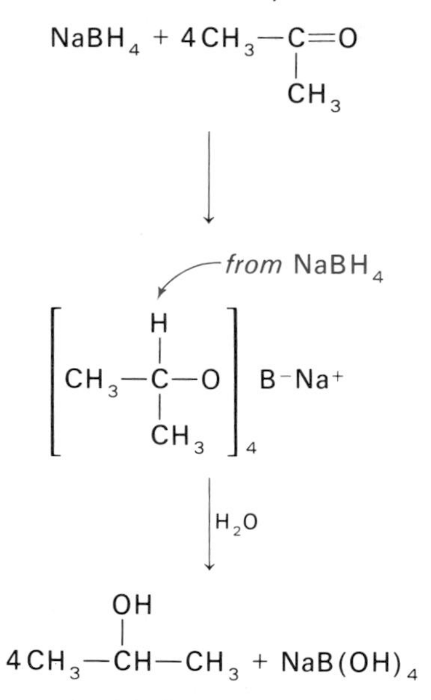

In 1943, Brown moved to Wayne University in Detroit and then in 1947 to Purdue University. At Purdue, he ". . . decided to explore means of increasing the reducing properties of sodium borohydride and of decreasing the reducing properties of lithium aluminum hydride. In this way the organic chemist would have at his disposal a full spectrum of reducing agents—he could select that reagent which would be most favorable for the particular reduction required in a given situation."

During the course of this research, Brown made another remarkable discovery. It was the result of a minor discrepancy between an expected experimental result and the actual result. One of Brown's colleagues, Dr. Subba Rao, was investigating the reaction of $NaBH_4$ in the presence of $AlCl_3$ with a variety of carbonyl-containing compounds. He determined that esters such as ethyl acetate and ethyl stearate are reduced to alcohols by this combination of reagents and that two moles of hydride ion (one-half mole of $NaBH_4$) are consumed per mole of ester.

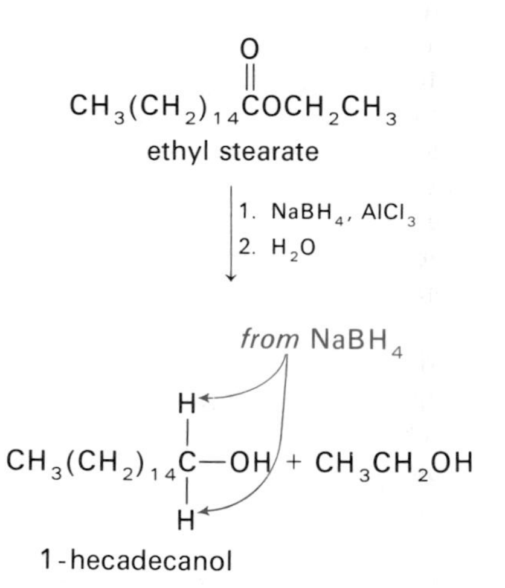

Among the several hundred experiments Subba Rao performed was one with $NaBH_4/AlCl_3$ and ethyl oleate, an unsaturated ester.

$$\underset{\text{ethyl oleate}}{CH_3(CH_2)_7CH{=}CH(CH_2)_7\overset{O}{\overset{\|}{C}}OCH_2CH_3}$$

The ester was reduced as expected. What was unexpected was that 2.4 moles of hydride ion were consumed per mole of ethyl oleate rather than the expected 2.0 moles. Dr. Subba Rao repeated the experiment several times, always with the same result. He soon discovered the reason for the apparent discrepancy. Both the ester and the carbon-carbon double bond were reacting with $NaBH_4/AlCl_3$ and what was being accomplished was simultaneous reduction of the ester and hydroboration of the double bond.

Brown soon discovered that hydroboration of alkenes proceeds even more rapidly and easily with diborane in ether solutions and that the reaction is quantitative and practically instantaneous. Thus was discovered the hydroboration of alkenes. As Brown has pointed out, one is tempted to be impatient with little discrepancies (in this case, in the reduction of esters) but as one gains experience, one tends to appreciate the importance of these discrepancies (in this case, hydroboration of alkenes).

The discovery of a simple and direct way to prepare organoboranes opened a vast field of chemistry and there followed more than two decades of enormously productive and highly creative work during which Brown charted this entirely new field of chemistry.

Brown concluded his Nobel Laureate lecture by putting his more than four decades of research on the chemistry of boron and its derivatives in the following perspective: "In 1938 when I received my Ph.D. degree, I felt that organic chemistry was a relatively mature science, with essentially all of the important reactions and structures known. There appeared to be little new to be done except the working out of reaction mechanisms and the improvement of reaction yields. I now recognize that I was wrong. I have seen major new reactions discovered. Numerous new reagents are available to us. Many new structures are known to us. We have at hand many valuable new techniques. I know that many of the students of today feel the same way that I did in 1938. But I see no reason for believing that the next 40 years will not be as fruitful as in the past."

On December 10, 1979, Herbert C. Brown received the Nobel Prize in Chemistry. He is shown here in Concert Hall, Stockholm, accepting the award from King Carl Gustav of Sweden. (*Courtesy of H. C. Brown*)

Notes

[1] From the Presentation Introduction on the occasion of the award of the Nobel Prize in Chemistry to H. C. Brown on December 8, 1979 in Stockholm, Sweden.

[2] Brown, H. C., "Footsteps on the Borane Trail," Journal of Organometallic Chemistry 100(1975): 3–15.

References

Brown, H. C., "Footsteps on the Borane Trail," *Journal of Organometallic Chemistry* 100 (1975): 3–15.

———, "Organoboranes—The Modern Miracle," *Pure and Applied Chemistry* 47(1976): 49–60.

———, "Hydride Reductions: A 40-Year Revolution in Organic Chemistry," *Chemical and Engineering News* (March 5, 1979).

———, "From Little Acorns to Tall Oaks—from Boranes through Organoboranes," *Science* (October 31, 1980) 485–492.

CHAPTER 8

Carboxylic Acids

In this chapter we shall discuss the structure and acidity of carboxylic acids. These acids and their functional derivatives are widespread both in the biological world and, thanks to the blend of research and technology, in the world of man-made materials. In addition, we will describe a special group of carboxylic acids known as fatty acids. This subject, in turn, will lead us into the chemistry of soaps and synthetic detergents.

8.1 THE STRUCTURE OF CARBOXYLIC ACIDS

The characteristic structural feature of a carboxylic acid is the presence of a carbonyl group (C=O) bonded to a hydroxyl group (—OH). This combination of a carbonyl group and a hydroxyl group is called a carboxyl group (carbonyl + hydroxyl). Shown in Figure 8.1 is a Lewis structure and ball-and-stick model of formic acid, HCO_2H, the simplest compound containing a carboxyl group.

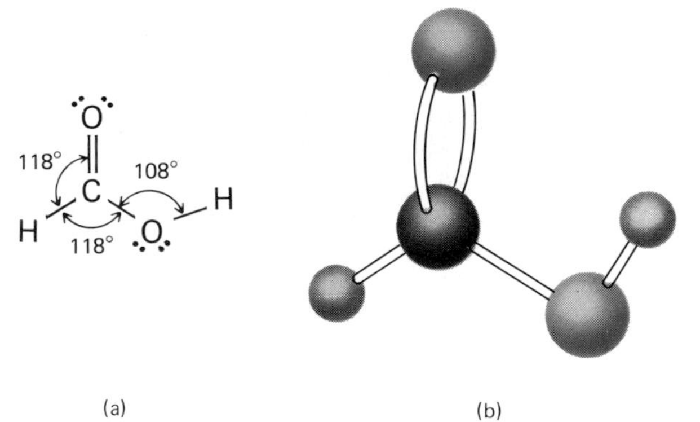

Figure 8.1 *The structure of formic acid. (a) Lewis structure showing bond angles; (b) ball-and-stick model.*

Predicted bond angles in the carboxyl group are 120° about the carbonyl carbon and 109.5° about the hydroxyl oxygen. Notice from Figure 8.1(a) that the observed angles are quite close to those predicted.

8.2 NOMENCLATURE OF CARBOXYLIC ACIDS

The IUPAC system of nomenclature selects as the parent compound the longest chain of carbon atoms that contains the $—CO_2H$ group and indicates the presence of the carboxyl group by changing the -e of the parent compound to -oic acid. The carbon of the carboxyl group is always number 1 of the parent compound and therefore there is no need to give it a number. Following are structural formulas and IUPAC names for several carboxylic acids.

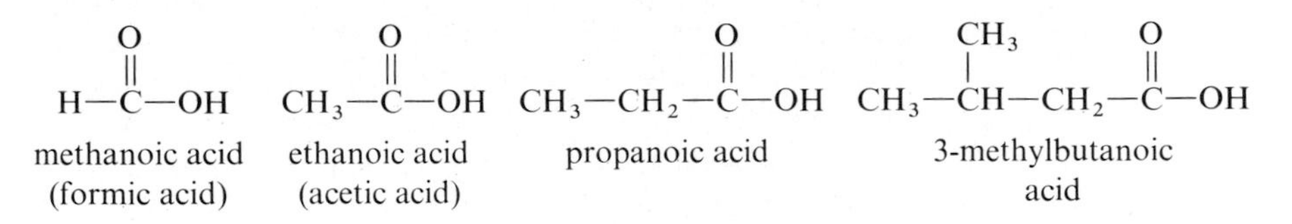

methanoic acid (formic acid) ethanoic acid (acetic acid) propanoic acid 3-methylbutanoic acid

The IUPAC system retains the common names formic acid and acetic acid, and so there are two acceptable IUPAC names for these carboxylic acids. Dicarboxylic acids have the ending -dioic acid to indicate the presence of two —CO_2H groups.

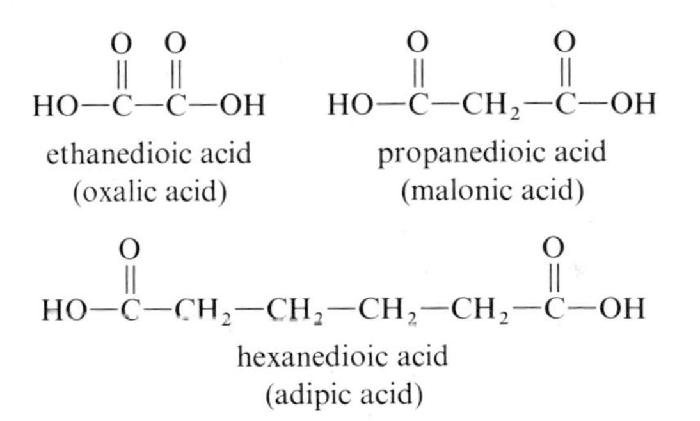

ethanedioic acid (oxalic acid) propanedioic acid (malonic acid)

hexanedioic acid (adipic acid)

Common names for these dicarboxylic acids are given in parentheses under the IUPAC names. Most dicarboxylic acids are known almost exclusively by their common names.

Aromatic carboxylic acids are named as derivatives of the parent aromatic hydrocarbon by changing the ending of the hydrocarbon name to -oic acid. For example, the carboxylic acid derived from benzene is named benzoic acid.

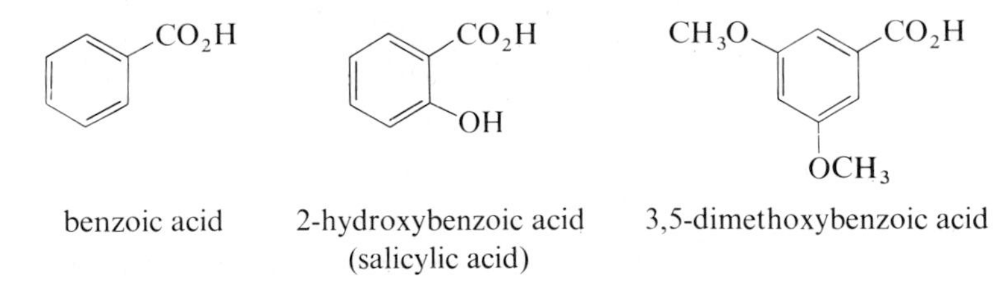

benzoic acid 2-hydroxybenzoic acid (salicylic acid) 3,5-dimethoxybenzoic acid

In more complex structural formulas, the carboxyl group may be named by adding the words -carboxylic acid to the name of the parent hydrocarbon.

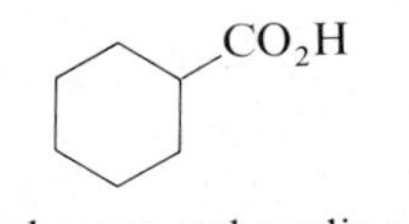

cyclohexanecarboxylic acid

Many lower-molecular-weight mono- and dicarboxylic acids are still known by their common names which generally refer to the natural sources of these acids. They have no relation to any systematic nomenclature. For example, formic acid adds to the sting of the bite of an ant (Latin, *formica*, "ant"); butyric acid gives rancid butter its characteristic smell (Latin, *butyrum*, "butter"); caproic acid is found in goat fat (Latin, *caper*, "goat").

When using common names, Greek letters (α, β, γ, δ) are often used to locate substituents. The α-carbon is the one next to the carboxyl group. Therefore an α-substituent in a common name is equivalent to a 2-substituent in the IUPAC name.

$$\overset{\delta}{C}-\overset{\gamma}{C}-\overset{\beta}{C}-\overset{\alpha}{C}-\overset{\overset{O}{\|}}{C}-OH \qquad HO-CH_2-CH_2-CH_2-\overset{\overset{O}{\|}}{C}-OH \qquad CH_3-\overset{\overset{Cl}{|}}{CH}-\overset{\overset{O}{\|}}{C}OH$$

γ-hydroxybutyric acid $\qquad$ α-chloropropionic acid

Table 8.1 lists IUPAC and common names for the most common straight chain monocarboxylic acids of up to 10 carbons and for dicarboxylic acids of up to six carbons.

Table 8.1 IUPAC and common names of some mono- and dicarboxylic acids.

Structural formula	*IUPAC name*	*Common name*
HCO_2H	methanoic, formic	formic
CH_3CO_2H	ethanoic, acetic	acetic
$CH_3CH_2CO_2H$	propanoic	propionic
$CH_3(CH_2)_2CO_2H$	butanoic	butyric
$CH_3(CH_2)_3CO_2H$	pentanoic	valeric
$CH_3(CH_2)_4CO_2H$	hexanoic	caproic
$CH_3(CH_2)_5CO_2H$	heptanoic	enanthic
$CH_3(CH_2)_6CO_2H$	octanoic	caprylic
$CH_3(CH_2)_8CO_2H$	decanoic	capric
$CH_3(CH_2)_{10}CO_2H$	dodecanoic	lauric
$CH_3(CH_2)_{12}CO_2H$	tetradecanoic	myristic
$CH_3(CH_2)_{14}CO_2H$	hexadecanoic	palmitic
$CH_3(CH_2)_{16}CO_2H$	octadecanoic	stearic
$CH_3(CH_2)_{18}CO_2H$	eicosanoic	arachidic
HO_2CCO_2H	ethanedioic	oxalic
$HO_2CCH_2CO_2H$	propanedioic	malonic
$HO_2C(CH_2)_2CO_2H$	butanedioic	succinic
$HO_2C(CH_2)_3CO_2H$	pentanedioic	glutaric
$HO_2C(CH_2)_4CO_2H$	hexanedioic	adipic

Example 8.1

Give IUPAC names for the following compounds.

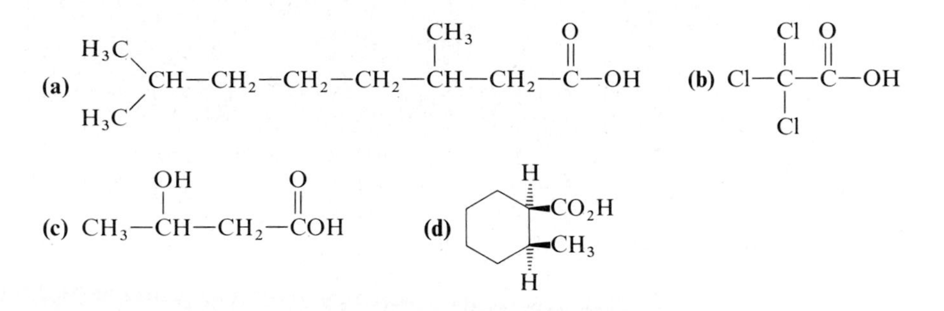

Solution

(a) 3,7-Dimethyloctanoic acid. The longest chain containing the carboxyl group is eight carbons, so this is a disubstituted octanoic acid.
(b) Trichloroacetic acid. Acetic acid is accepted as the IUPAC name for CH_3CO_2H; ethanoic acid is also accepted. It is not necessary to use a numbering system to show the location of the three chlorine substituents for they can only be on the second carbon of the acid.
(c) 3-Hydroxybutanoic acid. The common name of this acid is β-hydroxybutyric acid.
(d) *Cis*-2-methylcyclohexanecarboxylic acid. Numbering of cyclic carboxylic acids begins with the carbon bearing the $—CO_2H$ group. Therefore it is not necessary to use a number to show its location. Both the location of the $—CH_3$ group and the fact that it is *cis* must be indicated in the name.

PROBLEM 8.1 Give IUPAC names for the following compounds:

(a) $CH_3—C(CH_3)_2—CO_2H$

(b) $HO—C(=O)—CH_2—CH(CH_3)—C(=O)—OH$

(c) $CH_3—CH(OH)—CH(OH)—C(=O)—OH$

Although they are only weakly acidic, carboxylic acids do react with strong bases to form salts. Salts of carboxylic acids are named in much the same manner as salts of inorganic acids: the cation is named first and then the anion. The anion derived from a carboxylic acid is named by dropping the terminal -ic from the name of the acid and adding -ate.

Example 8.2

Name the following salts.

(a) CH_3CO_2Na **(b)** $(CH_3CH_2CO_2)_2Ca$ **(c)** $Cl—C_6H_4—CO_2NH_4$

Solution In the following answers, the name and structural formula of the carboxylic acid is listed followed by the name and structural formula of the anion derived from the acid. Finally, the salt is named.

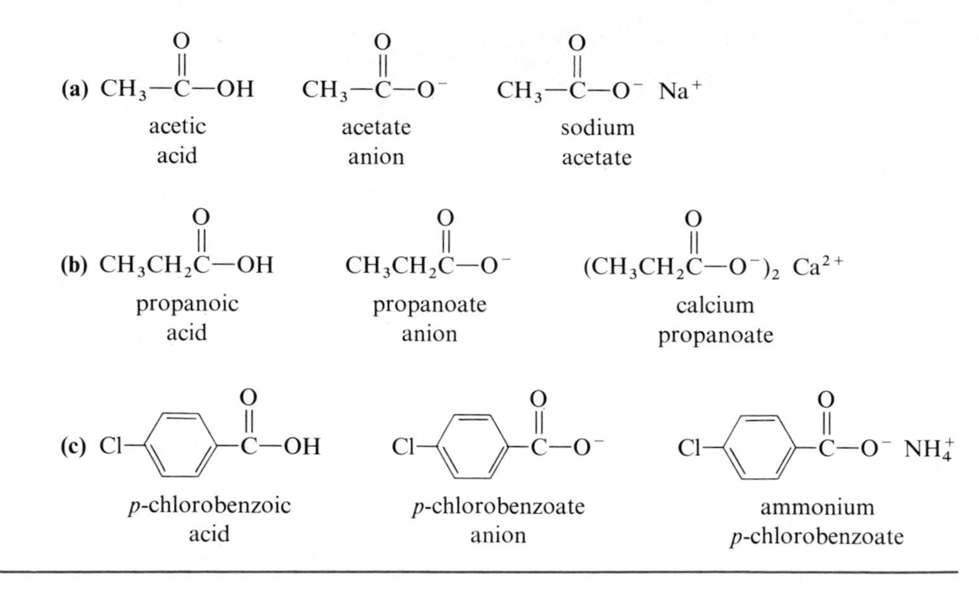

PROBLEM 8.2 Name the following salts.

(a) $CH_3\underset{\displaystyle CH_3}{\underset{|}{C}}HCO_2K$ **(b)** $CH_3(CH_2)_8CO_2Na$ **(c)** $CH_3CO_2NH_4$

8.3 PHYSICAL PROPERTIES OF CARBOXYLIC ACIDS

The carboxyl group contains three polar covalent bonds. Therefore carboxylic acids are polar substances. The carbonyl oxygen and the hydroxyl oxygen each bear partial negative charges, and the carbonyl carbon and the hydroxyl hydrogen each bear partial positive charges as shown in Figure 8.2 for acetic acid.

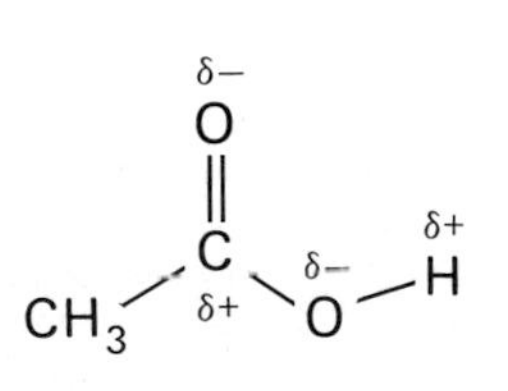

Figure 8.2 *Polarity of the carboxyl group.*

Carboxylic acids can participate in hydrogen bonding through both the C=O and O—H groups, as shown in Figure 8.3 for acetic acid.

Carboxylic acids are even mor extensively hydrogen bonded than alcohols, and therefore they have higher boiling points than alcohols of comparable molecular weight. For example, propanoic acid and 1-butanol have almost identical molecular weights, but the boiling point of propanoic acid is over 20° higher than the boiling point of 1-butanol.

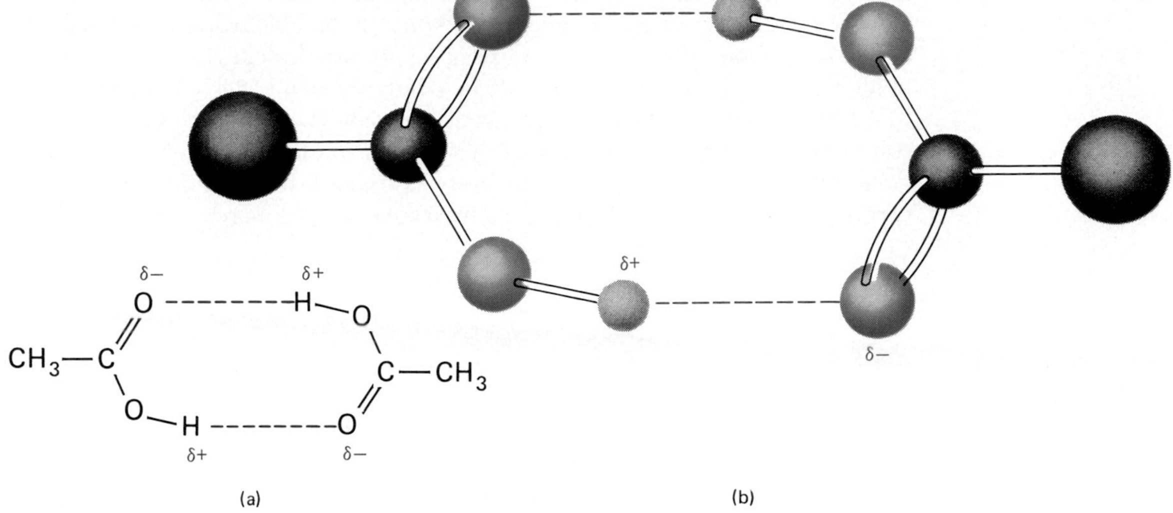

Figure 8.3 *Hydrogen bonding between acetic acid molecules in the pure liquid: (a) Lewis structure, (b) ball-and-stick model.*

$$\underset{\substack{\text{propanoic acid}\\ \text{(MW 74, bp 141°C)}}}{CH_3—CH_2—\overset{\overset{\displaystyle O}{\|}}{C}—OH} \qquad \underset{\substack{\text{1-butanol}\\ \text{(MW 74, bp 117°C)}}}{CH_3—CH_2—CH_2\quad CH_2—OH}$$

Carboxylic acids interact with water molecules by hydrogen bonding through both the carbonyl oxygen and the hydroxyl group.

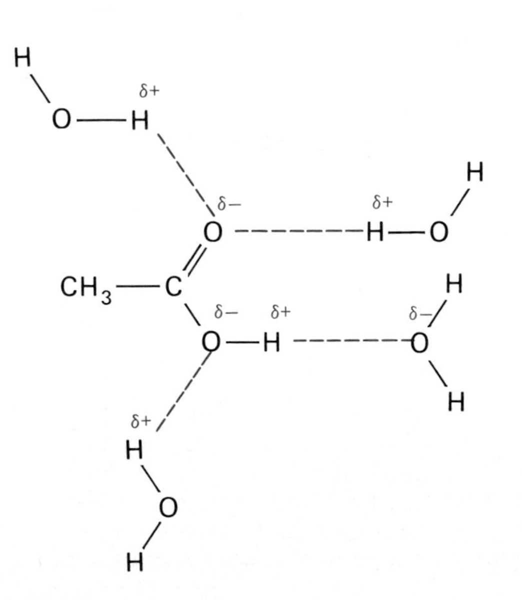

Because of this interaction with water molecules, carboxylic acids are more soluble in water than are alkanes, ethers, alcohols, aldehydes, or ketones of comparable molecular weight. For example, propanoic acid is soluble in water in all proportions, while the solubility of 1-butanol is only 8 g/100 g water. The first four carboxylic acids (formic, acetic, propanoic, and butanoic acids) are soluble in water in all proportions. The higher-molecular-weight carboxylic acids have longer nonpolar hydrocarbon chains, and therefore the solubility of carboxylic acids in water decreases as molecular weight increases.

$$CH_3CH_2-\overset{O}{\overset{\|}{C}}-OH$$

polar carboxyl group

nonpolar hydrocarbon chain

propanoic acid
(soluble in water in all proportions)

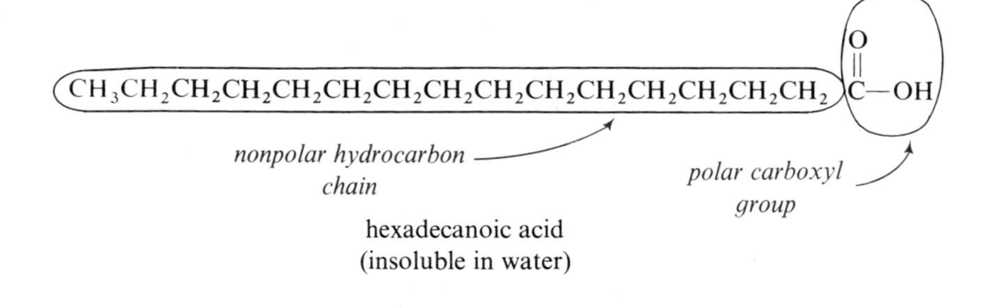

hexadecanoic acid
(insoluble in water)

In propanoic acid, the polar (hydrophilic) carboxyl group is large compared to the nonpolar (hydrophobic) hydrocarbon chain, therefore the acid is soluble in water. In hexadecanoic acid, the polar (hydrophilic) carboxyl group is very small in comparison to the large nonpolar (hydrophobic) chain, therefore this acid is insoluble in water.

Table 8.2 shows the physical properties of some monocarboxylic acids.

Table 8.2 Physical properties of some monocarboxylic acids.

Name	*Structural formula*	*mp(°C)*	*bp(°C)*	*Solubility in water (g/100g H_2O)*	
formic acid	HCO_2H	8	100	∞	soluble in all proportions
acetic acid	CH_3CO_2H	16	118	∞	
propanoic acid	$CH_3CH_2CO_2H$	−22	141	∞	
butanoic acid	$CH_3(CH_2)_2CO_2H$	−6	164	∞	
hexanoic acid	$CH_3(CH_2)_4CO_2H$	−3	205	1.0	slightly soluble
decanoic acid	$CH_3(CH_2)_8CO_2H$	32	—	—	insoluble

Example 8.3

Arrange the following compounds in order of increasing boiling points.

$$\underset{\text{butanoic acid}}{CH_3CH_2CH_2\overset{\overset{O}{\|}}{C}OH} \qquad \underset{\text{pentanal}}{CH_3CH_2CH_2CH_2\overset{\overset{O}{\|}}{C}H} \qquad \underset{\text{1-pentanol}}{CH_3CH_2CH_2CH_2CH_2OH}$$

Solution

All three substances are polar molecules of comparable molecular weight. Both 1-pentanol and butanoic acid have polar —OH bonds, and therefore are extensively hydrogen bonded in the pure liquid. Pentanal has no —OH group and so cannot participate in hydrogen bonding in the pure liquid. Pentanal has the lowest boiling point. 1-Pentanol can participate in hydrogen bonding through the —OH group and is next in boiling point. The carboxyl group of butanoic acid can participate in hydrogen bonding through the —OH group and also the carbonyl oxygen, and therefore has the highest boiling point.

pentanal	1-pentanol	butanoic acid
(MW 82, bp 103°)	(MW 84, bp 137°)	(MW 86, bp 164°)

PROBLEM 8.3 Arrange the following compounds in order of increasing solubility in water.

$$CH_3CH_2OCH_2CH_3 \qquad CH_3CH_2CH_2\overset{\overset{O}{\|}}{C}OH \qquad CH_3(CH_2)_8\overset{\overset{O}{\|}}{C}OH$$

8.4 PREPARATION OF CARBOXYLIC ACIDS

There are several general methods for the synthesis of simple carboxylic acids:

(1) oxidation of primary alcohols and aldehydes;
(2) oxidation of alkenes of the type $RCH{=}CHR$ and $R_2C{=}CHR$;
(3) oxidation of alkyl substituents on benzene rings;
(4) carbonation of Grignard reagents.

Following are examples of each method.

First, oxidation of primary alcohols (Section 5.5C) yields carboxylic acids. In the laboratory, the most commonly used oxidizing agents are potassium dichromate and potassium permanganate.

$$\underset{\text{1-heptanol}}{CH_3(CH_2)_5CH_2OH} + Cr_2O_7^{2-} \xrightarrow{H_3O^+} \underset{\text{heptanoic acid}}{CH_3(CH_2)_5\overset{\overset{O}{\|}}{C}OH} + Cr^{3+}$$

$$\underset{\text{1,4-butanediol}}{HOCH_2CH_2CH_2CH_2OH} + MnO_4^- \xrightarrow{H_3O^+} \underset{\substack{\text{butanedioic acid}\\ \text{(succinic acid)}}}{HO\overset{\overset{O}{\|}}{C}CH_2CH_2\overset{\overset{O}{\|}}{C}OH} + Mn^{2+}$$

Aldehydes are also easily oxidized to carboxylic acids, even by such weak oxidizing agents as $Ag(NH_3)_2^+$ and $Fe(CN)_6^{3-}$.

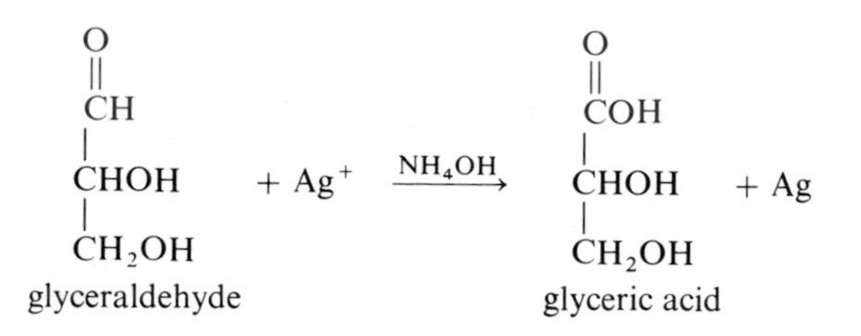

Second, oxidation of disubstituted alkenes results in cleavage of the carbon skeleton and formation of two carboxylic acids.

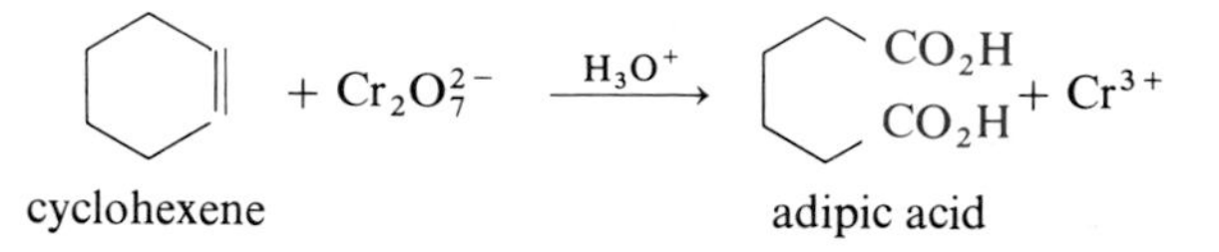

Note that oxidation of a cycloalkene such as cyclohexene results in cleavage of a carbon-carbon double bond but yields a single molecule of a dicarboxylic acid. Potassium dichromate oxidation converts a trisubstituted alkene, $R_2C{=}CHR$, to a ketone and a carboxylic acid. Under these conditions, 2-methyl-2-pentene is oxidized to acetone and propanoic acid.

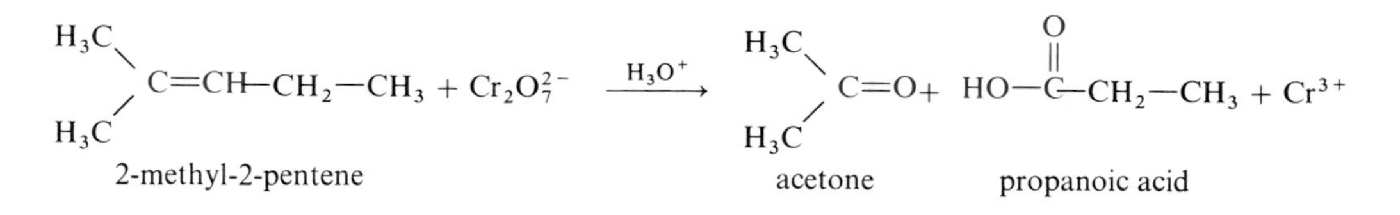

Third, carboxylic acids can be prepared by the oxidation of alkyl substituents on an aromatic ring. Recall from Chapter 2 that most alkanes and cycloalkanes are quite resistant to oxidation by $K_2Cr_2O_7$ or $KMnO_4$. However, when the alkyl group has attached to it a benzene ring and at least one hydrogen atom, the alkyl group is more susceptible to vigorous oxidation. No matter how long the alkyl group is, it is degraded to a substituted benzoic acid.

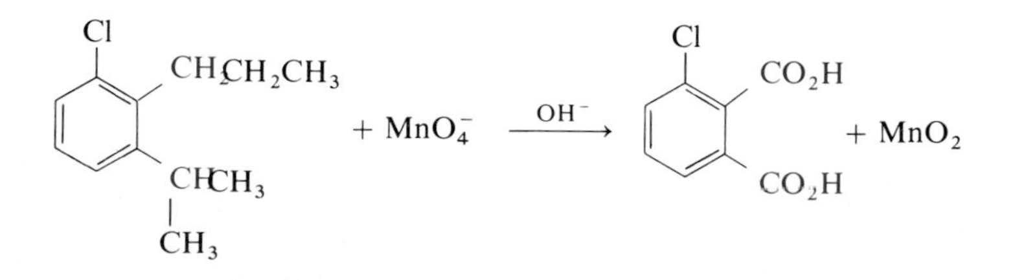

Fourth, carboxylic acids can be prepared by the reaction of Grignard reagents with carbon dioxide (Section 7.5C). This reaction gives good yields and is widely applicable to the preparation of both aliphatic and aromatic carboxylic acids.

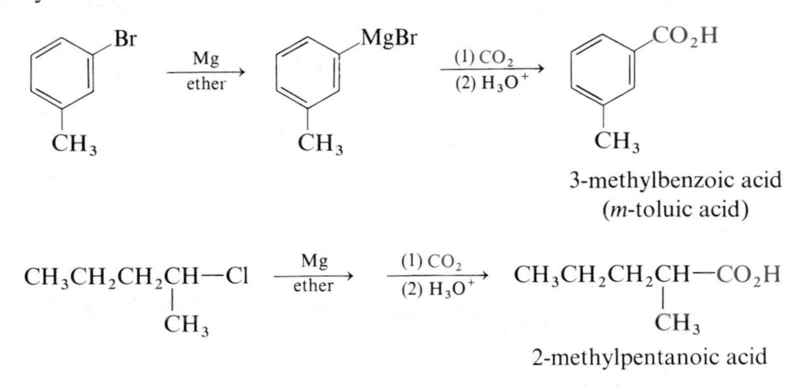

8.5 REACTIONS OF CARBOXYLIC ACIDS

A. ACIDITY OF CARBOXYLIC ACIDS

Carboxylic acids ionize in water to give acidic solutions. However, carboxylic acids are quite different from inorganic acids such as HCl, HBr, HNO_3, and H_2SO_4. These inorganic acids are 100% ionized in aqueous solution and are classified as strong acids.

$$HCl \longrightarrow H^+ + Cl^-$$

Carboxylic acids are only slightly ionized in aqueous solution and are classified as weak acids. When a carboxylic acid is dissolved in water, an equilibrium is established between the carboxylic acid, the carboxylate anion, and H^+.

$$CH_3-\overset{\overset{\displaystyle O}{\|}}{C}-OH \rightleftharpoons CH_3-\overset{\overset{\displaystyle O}{\|}}{C}-O^- + H^+$$

The equilibrium constant for this ionization is called K_a, the acid dissociation or ionization constant, and has the form

$$K_a = \frac{[CH_3CO_2^-][H^+]}{[CH_3CO_2H]} = 1.8 \times 10^{-5}$$

Values of K_a for some representative carboxylic acids are given in Table 8.3. Other than for formic acid, the values of K_a for unsubstituted aliphatic carboxylic acids are essentially the same as that for acetic acid. Therefore, the value of K_a for acetic acid is a useful number to remember.

Substituents, particularly on the α-carbon, exert a marked effect on the acidity of a carboxylic acid. Compare for example the K_a value given in Table 8.3 for acetic acid with those of its mono-, di-, and trichloro- derivatives. You will see that chloroacetic acid is almost 80 times stronger an acid than acetic

Table 8.3 K_a and pK_a for some carboxylic acids.

Name	*Structural formula*	K_a	pK_a
formic	HCOOH	1.8×10^{-4}	3.74
acetic	CH_3COOH	1.8×10^{-5}	4.74
propanoic	CH_3CH_2COOH	1.4×10^{-5}	4.85
butanoic	$CH_3CH_2CH_2COOH$	1.6×10^{-5}	4.80
benzoic	C_6H_5COOH	6.5×10^{-5}	4.19
chloroacetic	$ClCH_2COOH$	1.4×10^{-3}	2.85
dichloroacetic	$Cl_2CHCOOH$	3.3×10^{-2}	1.48
trichloroacetic	Cl_3CCOOH	2.3×10^{-1}	0.64

acid. Trichloroacetic acid is a very much stronger acid, and in dilute aqueous solutions is about as strong as sulfuric acid. The most usual explanation is that due to its greater electronegativity than carbon, chlorine polarizes the electrons of the Cl—C bond and induces a partial positive charge on the carbon atom. This in turn induces a polarization of electrons in the C—C bond and in the C—O bond and consequently the proton is more easily removed. Such polarization of electrons through sigma bonds is called the inductive effect and is indicated by an arrow on a sigma bond with the head of the arrow pointing toward the polarizing atom.

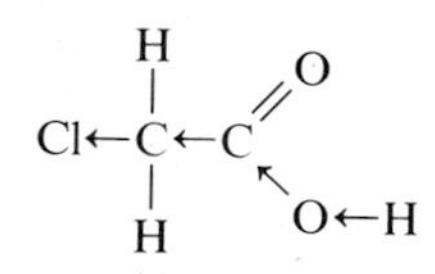

Inductive effects are largest when the electronegative substituent is close to the carboxylic acid group, and becomes progressively weaker as the electronegative substituent is moved farther away. Whereas chloroacetic acid is about 80 times stronger than acetic, 3-chloropropanoic acid is only about six times stronger, and 4-chlorobutanoic acid is only about twice as strong as acetic acid.

chloroacetic acid	$Cl—CH_2—CO_2H$	$K_a = 140 \times 10^{-5}$
3-chloropropanoic acid	$Cl—CH_2—CH_2—CO_2H$	$K_a = 10 \times 10^{-5}$
4-chlorobutanoic acid	$Cl—CH_2—CH_2—CH_2—CO_2H$	$K_a = 3 \times 10^{-5}$

Why are carboxylic acids so much more acidic than water or alcohols—compounds that also contain the —OH functional group? The acidity of ethanol is estimated to be about 10^{-16}. The value of K_a for acetic acid is over ten trillion (10^{10}) times larger than that of ethanol.

$$CH_3—CH_2—O—H \rightleftharpoons CH_3—CH_2—O^- + H^+ \qquad K_a = 10^{-16}$$

$$CH_3—\overset{\overset{\displaystyle O}{\|}}{C}—O—H \rightleftharpoons CH_3—\overset{\overset{\displaystyle O}{\|}}{C}—O^- + H^+ \qquad K_a = 1.8 \times 10^{-5}$$

The question may be stated another way: Why is acetic acid so much more extensively ionized in water than ethanol? We can account for this enhanced acidity (enhanced ionization) by using the resonance model and looking at the relative stabilities of the acetate ion and the ethoxide ion. Recall that this is the approach we used in Section 6.4 to account for the greater acidity of phenol compared to ethanol.

Resonance has no effect on the dissociation of ethanol for there is no possibility for resonance or resonance stabilization in the ethoxide ion. For the acetate ion, however, we can write two equivalent contributing structures.

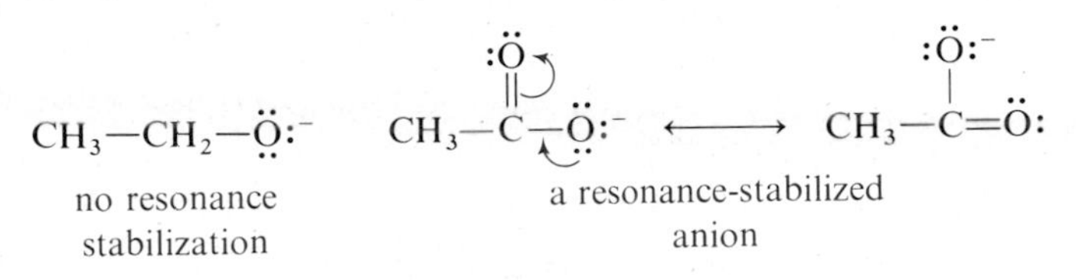

Since acetate is stabilized by resonance compared to ethoxide, the equilibrium for the ionization of acetic acid lies farther to the right than that for the ionization of ethanol, that is, acetic acid is a stronger acid than ethanol.

Before we leave this discussion of the acidity of carboxylic acids, we should compare this class of organic acids with phenols, another class of organic acids. The difference in acidity between a strong acid such as HCl and weak acids such as acetic acid and phenol can be seen by comparing the K_a values and the hydrogen ion concentration of a 0.1M solution of each of these acids in water (Table 8.4). Because they are only partially dissociated in water solution, both carboxylic acids and phenols are classed as weak acids. Note, however, that carboxylic acids are much stronger organic acids than phenols.

All carboxylic acids, whether soluble or insoluble in water, react quantitatively with aqueous solutions of sodium hydroxide to form salts.

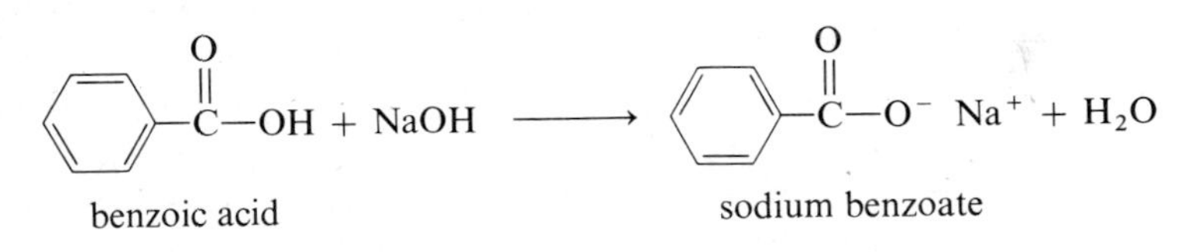

Table 8.4 Relative acidities of 0.1 M solutions of hydrochloric acid, acetic acid, and phenol.

Acid	K_a	*Ionization in water*	$[H^+]$ *of* 0.1 M *solution*
HCl	very large	100%	0.1 M
$CH_3-\overset{O}{\overset{\|\|}{C}}-OH$	1.8×10^{-5}	1.3%	0.0013 M
C_6H_5-OH	3.3×10^{-10}	0.0033%	3.3×10^{-6} M

Sodium, potassium, and ammonium salts of carboxylic acids are ionic substances, and are much more soluble in water than the carboxylic acids from which they are derived. For example, benzoic acid is insoluble in water, while sodium benzoate is soluble.

Carboxylic acids also react with sodium bicarbonate and sodium carbonate. In these reactions, bicarbonate and carbonate ions give carbonic acid, H_2CO_3, which breaks down spontaneously to form CO_2 and H_2O.

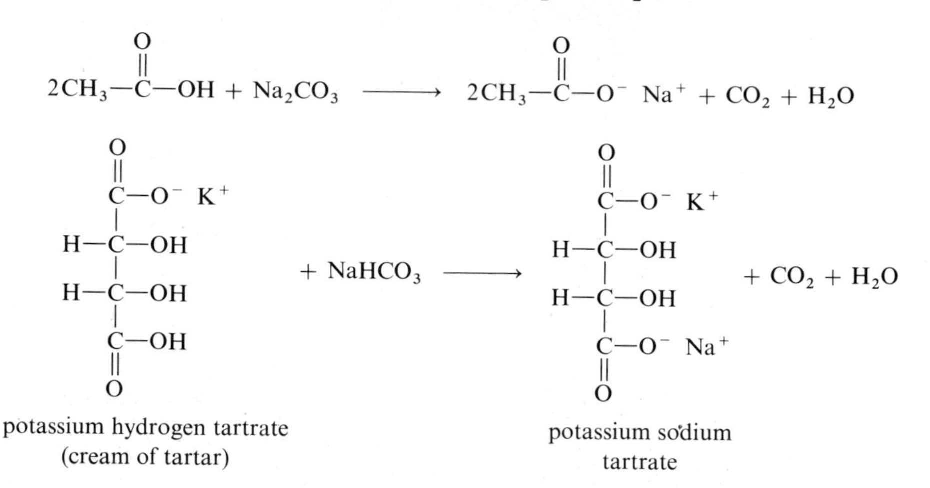

This second reaction explains why sodium carbonate and sodium bicarbonate are used in baking. Baking powder is a combination of sodium carbonate or sodium bicarbonate and potassium hydrogen tartrate. When water is added to baking powder, the acid and base react to liberate CO_2. The CO_2 bubbles in the batter or dough cause it to rise. Baking soda or sodium bicarbonate can also be mixed with vinegar (a solution of acetic acid in water) or lemon juice (a solution containing citric acid) to produce CO_2. Some recipes call for mixing baking soda and sour cream (containing lactic acid), which also react to produce carbon dioxide.

Most phenols are considerably weaker acids than carboxylic acids and do not dissolve in aqueous sodium bicarbonate because the bicarbonate ion is too weakly basic to remove a proton from phenol. This fact permits the ready separation of phenols from the more strongly acidic carboxylic acids. An aqueous solution of sodium bicarbonate dissolves carboxylic acids as their sodium salts but does not dissolve phenols. The water-soluble carboxylate salt can then be physically separated from the water-insoluble phenol.

Example 8.4 Write equations for the following acid-base reactions.

(a) $CH_3-CH(OH)-C(=O)-OH + NaOH \xrightarrow{H_2O}$

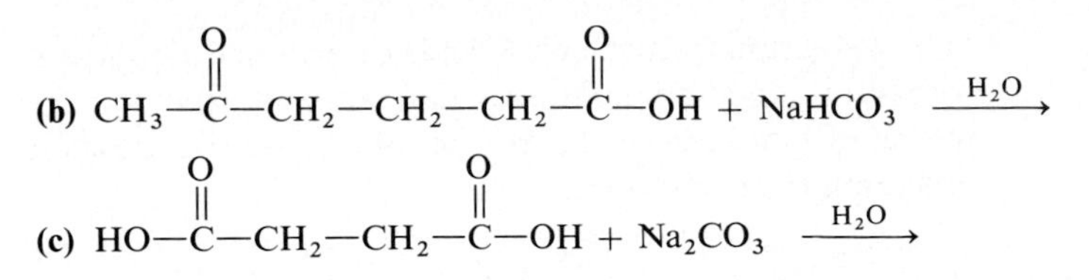

Solution

(a) 2-Hydroxypropanoic acid, or lactic acid as it is more commonly known, is a monocarboxylic acid and reacts in a 1:1 molar ratio with sodium hydroxide to form sodium lactate.

$$\underset{\text{lactic acid}}{CH_3-\underset{\displaystyle OH}{\underset{|}{CH}}-\overset{\displaystyle O}{\overset{\|}{C}}-OH} + NaOH \longrightarrow \underset{\text{sodium lactate}}{CH_3-\underset{\displaystyle OH}{\underset{|}{CH}}-\overset{\displaystyle O}{\overset{\|}{C}}-O^- \; Na^+} + H_2O$$

(b) 5-Ketohexanoic acid is a monocarboxylic acid and reacts in a 1:1 molar ratio with sodium bicarbonate.

$$CH_3\overset{\displaystyle O}{\overset{\|}{C}}CH_2CH_2CH_2\overset{\displaystyle O}{\overset{\|}{C}}OH + NaHCO_3 \longrightarrow CH_3\overset{\displaystyle O}{\overset{\|}{C}}CH_2CH_2CH_2\overset{\displaystyle O}{\overset{\|}{C}}O^- \; Na^+ + CO_2 + H_2O$$

(c) Butanedioic acid, or succinic acid as it is more commonly known, is a dicarboxylic acid. It reacts with sodium carbonate in a 1:1 molar ratio.

$$\underset{\text{succinic acid}}{HO\overset{\displaystyle O}{\overset{\|}{C}}CH_2CH_2\overset{\displaystyle O}{\overset{\|}{C}}OH} + Na_2CO_3 \longrightarrow \underset{\text{sodium succinate}}{Na^+ \; {}^-O\overset{\displaystyle O}{\overset{\|}{C}}CH_2CH_2\overset{\displaystyle O}{\overset{\|}{C}}O^- \; Na^+} + CO_2 + H_2O$$

PROBLEM 8.4 Write equations for the following acid-base reactions.

(a) $CH_3CH_2CH{=}CHCO_2H + NaOH \xrightarrow{H_2O}$

(b) $2C_6H_5CO_2H + Na_2CO_3 \xrightarrow{H_2O}$

(c) $$\underset{\text{citric acid}}{HO-\underset{\displaystyle CH_2-CO_2H}{\underset{|}{\overset{\displaystyle CH_2-CO_2H}{\overset{|}{C}}}}-CO_2H} + 3NaHCO_3 \xrightarrow{H_2O}$$

B. REDUCTION OF CARBOXYLIC ACIDS

Carboxylic acids are reduced to primary alcohols by hydrogen in the presence of a metal catalyst at high pressures (up to 100 atmospheres) and high temperatures (up to 250°C).

$$\underset{\text{hexanoic acid}}{CH_3CH_2CH_2CH_2CH_2\overset{\displaystyle O}{\overset{\|}{C}}-OH} + 2H_2 \xrightarrow[\text{heat, pressure}]{\text{catalyst}} \underset{\text{1-hexanol}}{CH_3CH_2CH_2CH_2CH_2CH_2OH} + H_2O$$

The reduction of carboxylic acids to primary alcohols requires so much higher pressures and temperatures than are required for the reduction of alkenes, aldehydes, or ketones that it is possible to reduce these functional groups without affecting the carboxyl group.

$$\underset{\text{3-ketobutanoic acid}}{CH_3{-}\overset{O}{\overset{\|}{C}}{-}CH_2{-}\overset{O}{\overset{\|}{C}}{-}OH} + H_2 \xrightarrow[25^\circ]{Pt} \underset{\text{3-hydroxybutanoic acid}}{CH_3{-}\overset{OH}{\overset{|}{C}H}{-}CH_2{-}\overset{O}{\overset{\|}{C}}{-}OH}$$

The high hydrogen pressures and temperatures necessary for reduction of carboxylic acids to primary alcohols make this a difficult reaction to do in the laboratory. Fortunately, the discovery of lithium aluminum hydride, $LiAlH_4$, in 1946 made it possible to reduce carboxylic acids to primary alcohols easily. Aldehydes and ketones are also reduced by this reagent (Section 7.4D) but alkenes and alkynes are unaffected. Reaction of lithium aluminum hydride with a carboxylic acid gives the aluminum salt of a primary alcohol (an aluminum alkoxide) which is then treated with water to form the primary alcohol.

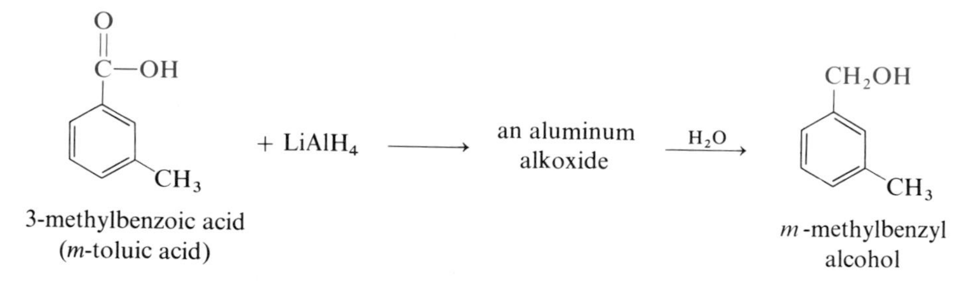

Example 8.5 Draw structural formulas for the products formed by reduction of the following with lithium aluminum hydride.

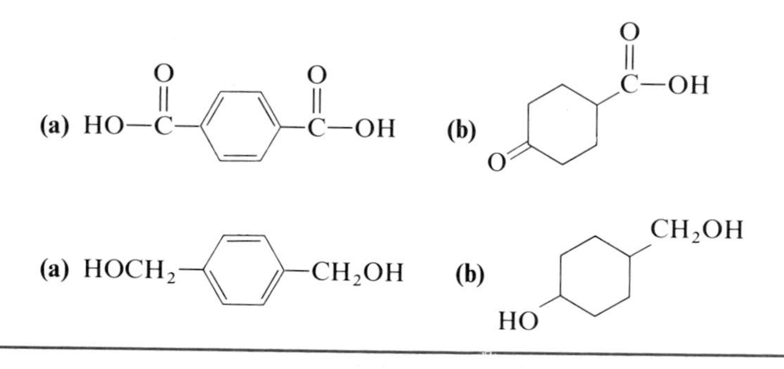

Solution

PROBLEM 8.5 Draw structural formulas for the products formed by reduction of the following with lithium aluminum hydride.

(a) $(H_3C)_2C{=}CH{-}CH_2{-}\overset{CH_3}{\overset{|}{C}H}{-}\overset{O}{\overset{\|}{C}}{-}OH$ (b) $H{-}\overset{O}{\overset{\|}{C}}{-}CH_2{-}\overset{CH_3}{\overset{|}{C}H}{-}\overset{O}{\overset{\|}{C}}{-}OH$

C. DECARBOXYLATION OF β-KETO ACIDS AND β-DICARBOXYLIC ACIDS

Carboxylic acids that have a carbonyl group on the carbon atom beta to the carboxyl group lose CO_2 on heating. This reaction results in loss of the carboxyl group as CO_2 and is called decarboxylation. For example, when heated, 3-ketobutanoic acid decarboxylates to yield acetone and carbon dioxide.

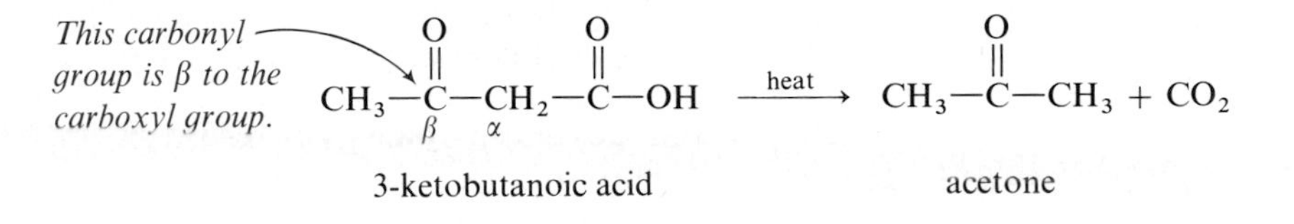

Decarboxylation on heating is a unique property of β-ketocarboxylic acids, and is not observed with other classes of ketocarboxylic acids.

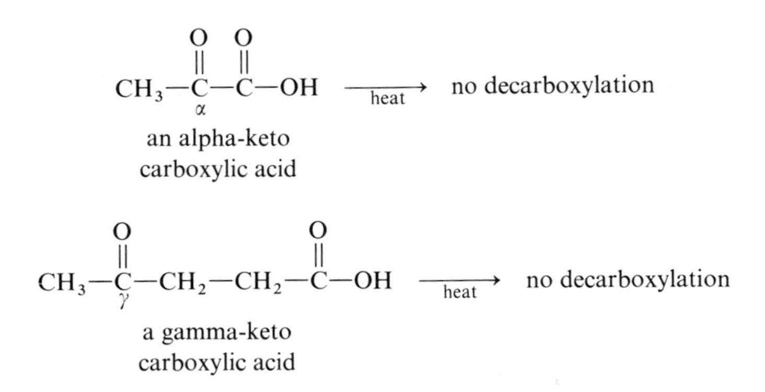

The mechanism for the decarboxylation of β-ketoacids is illustrated by the decarboxylation of 3-ketobutanoic acid. The reaction involves a cyclic six-membered transition state, and by rearrangement of electrons, leads to the enol form of acetone and carbon dioxide. The enol form of acetone is in equilibrium with the keto form.

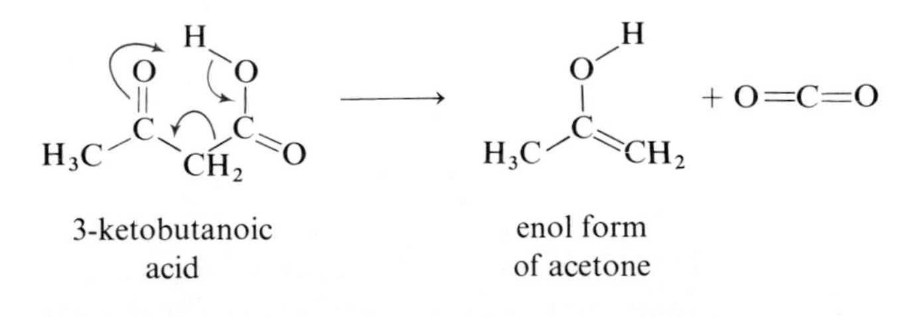

An important example of decarboxylation of a β-ketoacid occurs during the oxidation of foodstuffs in the tricarboxylic acid cycle. Oxalosuccinic acid, a tricarboxylic acid, undergoes decarboxylation to produce α-ketoglutaric acid.

Only one of the three carboxyl groups of oxalosuccinic acid has a carbonyl group in the β position to it. It is this one that is lost as CO_2.

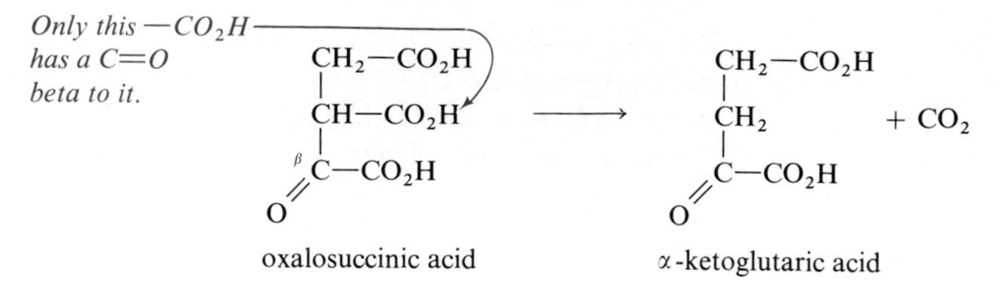

1,3-Dicarboxylic acids also undergo decarboxylation on heating. For example, decarboxylation of isopropylmalonic acid gives 3-methylbutanoic acid.

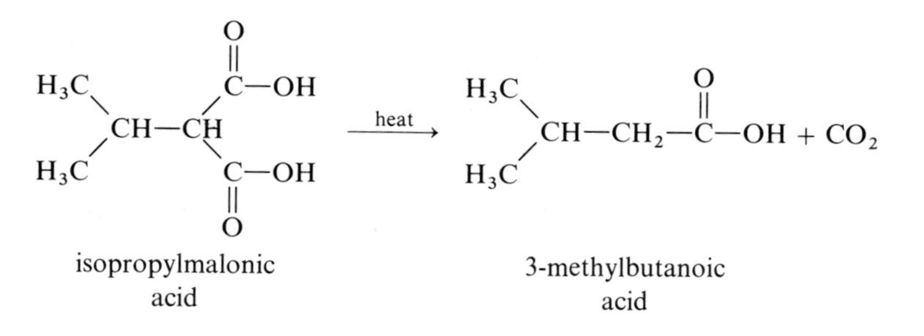

8.6 FATTY ACIDS

Fatty acids are monocarboxylic acids obtained from the hydrolysis of neutral fats and oils.

Neutral fats and oils are triesters of glycerol and their hydrolysis yields one molecule of glycerol and three molecules of fatty acid.

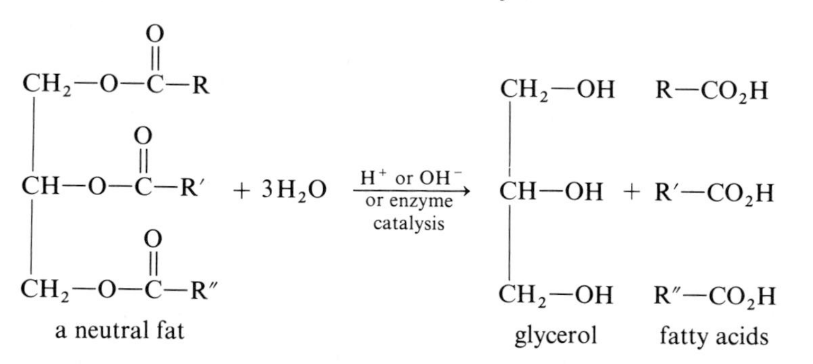

(We shall discuss the structure and properties of fats and oils in Chapter 13). Over 70 fatty acids have been isolated from various cells and tissues. Table 8.5 gives structural formulas of some important fatty acids.

We can make certain generalizations about the more abundant fatty acid components of higher plants and animals.

Table 8.5 Some important naturally occurring fatty acids.

Carbon atoms	*Structural formula*	*Common name*	*mp (°C)*
Saturated fatty acids			
12	$CH_3(CH_2)_{10}COOH$	lauric	44
14	$CH_3(CH_2)_{12}COOH$	myristic	58
16	$CH_3(CH_2)_{14}COOH$	palmitic	63
18	$CH_3(CH_2)_{16}COOH$	stearic	70
20	$CH_3(CH_2)_{18}COOH$	arachidic	77
Unsaturated fatty acids			
16	$CH_3(CH_2)_5CH{=}CH(CH_2)_7COOH$	palmitoleic	−1
18	$CH_3(CH_2)_7CH{=}CH(CH_2)_7COOH$	oleic	16
18	$CH_3(CH_2)_4CH{=}CHCH_2CH{=}CH(CH_2)_7COOH$	linoleic	−5
18	$CH_3CH_2(CH{=}CHCH_2)_3(CH_2)_6COOH$	linolenic	−11
20	$CH_3(CH_2)_3(CH_2CH{=}CH)_4(CH_2)_3COOH$	arachidonic	−49

(1) Nearly all fatty acids have an even number of carbon atoms, usually between 14 and 22 carbons in an unbranched chain. Those having 16 or 18 carbon atoms are the most abundant in nature.

(2) Unsaturated fatty acids have lower melting points than their saturated counterparts. The physical properties of the particular fatty acid components also affect the fats into which they are incorporated and, as we shall see in Section 13.1, fats rich in unsaturated fatty acids have lower melting points than those rich in saturated fatty acids.

(3) In most of the unsaturated fatty acids of higher organisms, there is a double bond between carbon atoms 9 and 10. In these unsaturated fatty acids, *cis* isomers predominate; the *trans* configuration is very rare.

Because of their long hydrophobic hydrocarbon chains, fatty acids are essentially insoluble in water. However, they do interact with water. If a drop of fatty acid is placed on the surface of water, it will spread out to form a thin film one molecule thick (a monomolecular layer), with polar carboxyl groups dissolved in water and nonpolar hydrocarbon chains forming a hydrocarbon layer on the surface of the water (Figure 8.4).

If fatty acids (in the form of fats) are entirely withheld from their diets, rats soon begin to suffer from retarded growth, scaly skin, kidney damage, and eventually, early death. Addition of unsaturated fatty acids (linoleic, linolenic, and arachidonic acids) will cure this condition. Humans and higher animals produce the enzymes necessary to catalyze the introduction of double bonds between carbons 9 and 10 in fatty acid chains. However, they lack the enzymes necessary to introduce unsaturation beyond carbon-10, and therefore fatty acids with unsaturation beyond carbon-10 must be supplied in the diet. Strictly speaking, linoleic acid is the critical fatty acid because it can be converted within

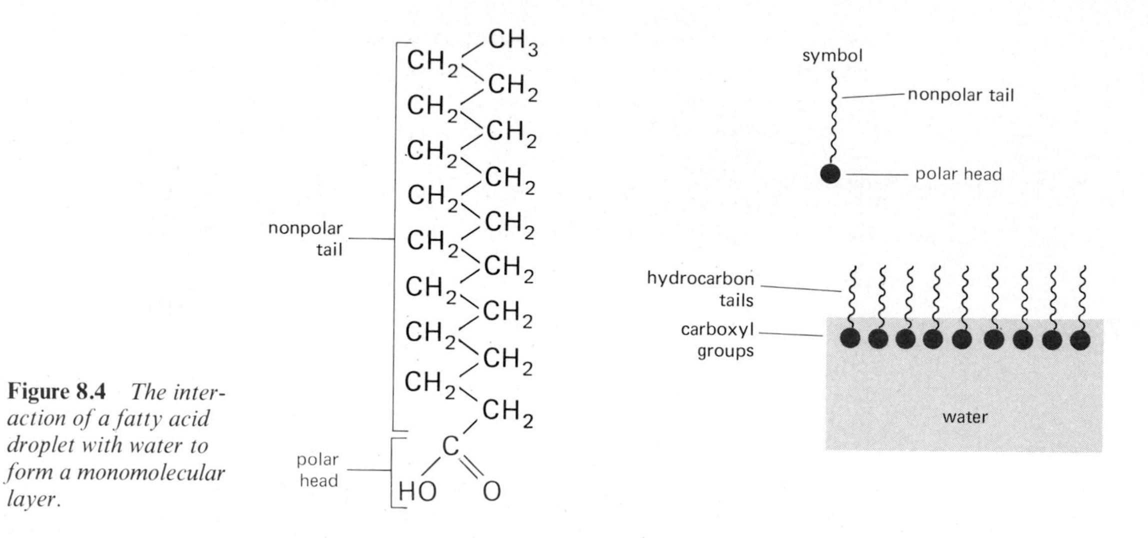

Figure 8.4 *The interaction of a fatty acid droplet with water to form a monomolecular layer.*

the body to linolenic and arachidonic acids. Because linoleic acid must be obtained in the diet for normal growth and well-being of humans and higher animals, it is classified as an essential fatty acid.

Table 8.6 shows the distribution of saturated and unsaturated fatty acids in some foods.

As you can see from Table 8.6, most animal fats are relatively rich in saturated fatty acids, and though the percentage of unsaturated fatty acids is also high, this unsaturation is due mostly to oleic acid. Vegetable fats, on the other hand, generally have a lower content of saturated fatty acids and a relatively

Table 8.6 Distribution of saturated and unsaturated fatty acids in some foods.

	% Fat in edible portion of food	*% Total fat*		
		saturated	*oleic*	*linoleic*
Animal fats				
beef	5–37	43–48	43	0.5–3.0
butter	81	57	33	3
eggs	11.5	35	44	8.7
fish (tuna)	4.1	24.4	24.6	0.5
milk (whole, pasturized)	3.7	57	33	3
pork	52	36.5	42	9.6
Vegetable fats				
coconut oil	100	85	6	0.5
corn oil	100	10	28	53
margarine	81	22.2	58	17.3
peanut oil	100	18	47	29
soybean oil	100	15	20	52
cottonseed oil	100	25	21	50

higher content of unsaturated fatty acids, including linoleic acid. Corn, cottonseed, soybean, and wheat-germ oils are especially rich in linoleic acid. The Committee on Dietary Allowances and the Food and Nutrition Board, both under the auspices of the National Research Council, recommend that linoleic acid consumption should be 2–3% of dietary calories, or about 8-10 grams of linoleic acid per day. The vegetable oils consumed in most diets are particularly rich in linoleic acid so that this requirement is more than met.

8.7 SOAPS

Natural soaps are sodium or potassium salts of fatty acids. One of the most ancient organic reactions known to man is the preparation of soaps by boiling lard or other animal fat with a slight excess of sodium or potassium hydroxide in an open kettle, and the eventual isolation of soap. The reaction that takes place is the hydrolysis of naturally occurring fats and oils and is known as saponification.

$$\underset{\text{a fat}}{\begin{array}{l} CH_2-O-\overset{O}{\overset{\|}{C}}-(CH_2)_{14}CH_3 \\ | \\ CH-O-\overset{O}{\overset{\|}{C}}-(CH_2)_{14}CH_3 \\ | \\ CH_2-O-\overset{O}{\overset{\|}{C}}-(CH_2)_{14}CH_3 \end{array}} + 3NaOH \longrightarrow \underset{\text{glycerol}}{\begin{array}{l} CH_2-OH \\ | \\ CH-OH \\ | \\ CH_2-OH \end{array}} + \underset{\text{a soap}}{3CH_3(CH_2)_{14}-\overset{O}{\overset{\|}{C}}-O^-Na^+}$$

In the present-day industrial manufacture of soap, molten tallow (the fat of cattle, sheep, etc.) is heated with a slight excess of sodium hydroxide. After saponification is complete, sodium chloride is added to precipitate the soap as thick curds. The water layer is drawn off and glycerol is recovered from it by vacuum distillation.

The crude soap curds contain salt, alkali, and glycerol as impurities. These are removed by boiling the curds in water and reprecipitating with salt. After several such purifications, the soap may be used without further processing as an inexpensive industrial soap. Fillers such as sand or pumice may be added to make a scouring soap. Other treatments transform the crude soap into laundry soaps, medicated soaps, cosmetic soaps, liquid soaps, and so on.

Soap owes its remarkable cleansing properties to its ability to act as an emulsifying agent. Regarded from one end, a natural soap is a polar, negatively charged hydrophilic (water-seeking) carboxylate group, $-CO_2^-$, which interacts with surrounding water molecules by hydrogen bonding and ion-dipole interactions. Regarded from the other end, it is a long, nonpolar hydrophobic (water-repelling) hydrocarbon chain, which does not interact at all with the surrounding water molecules. Because the long hydrocarbon chains of natural

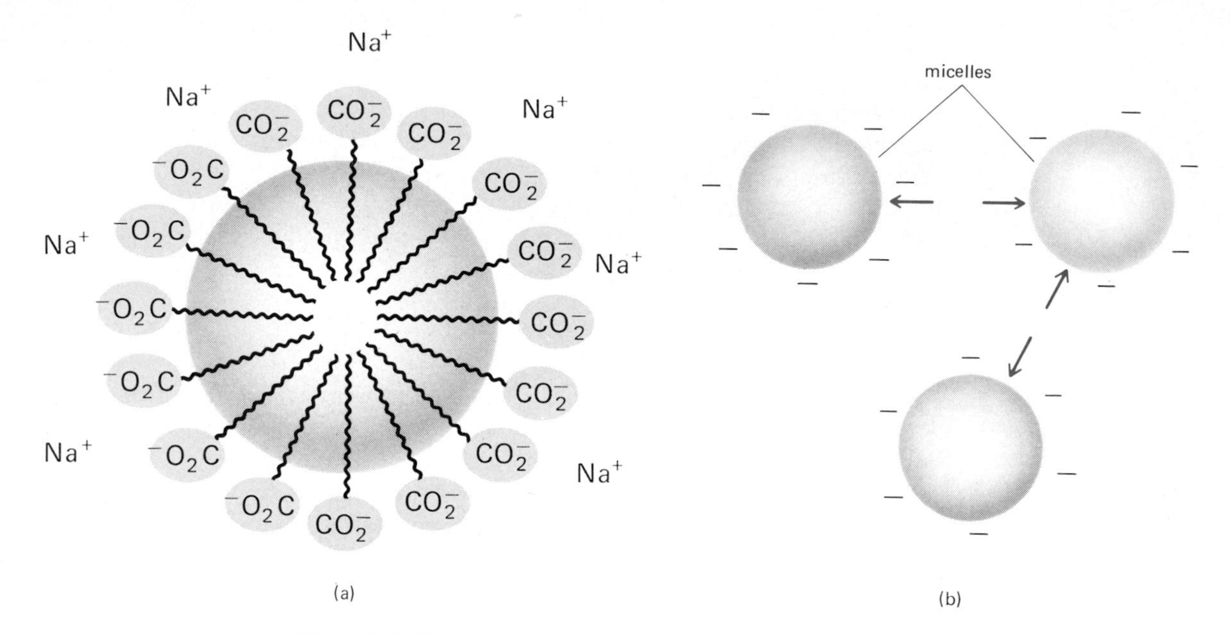

Figure 8.5 *Soap micelles. (a) Diagram of a soap micelle. (b) Because of the negative charges in their surface, soap micelles repel each other.*

soaps are insoluble in water, they cluster together, attracted to each other by dispersion forces. These clusters are called micelles. In soap micelles the carboxylate groups form a negatively charged surface and the water-insoluble hydrocarbon chains lie buried within the center (Figure 8.5). Soap micelles have a net negative charge and remain suspended or dispersed because of mutual repulsion of one for another.

Most of the things we commonly think of as dirt, such as grease, oil, and fat stains, are nonpolar substances, and are insoluble in water. If soap and dirt are mixed together, the nonpolar hydrocarbon ends of the soap micelles "dissolve" the nonpolar dirt molecules. In effect, new soap micelles form, this time with the nonpolar dirt molecules in the center of the micelles. In this way, nonpolar organic oil, grease, or fat is dissolved and washed away in the polar wash water (Figure 8.6).

Soaps are not without their disadvantages. First, soaps are sodium or potassium salts of weak acids, and are converted by mineral acids into free fatty acids.

$$\underset{\text{(soluble in water)}}{CH_3(CH_2)_{16}CO_2^-Na^+} + HCl \longrightarrow \underset{\text{(insoluble in water)}}{CH_3(CH_2)_{16}CO_2H} + Na^+ + Cl^-$$

Fatty acids are far less soluble than their potassium or sodium salts; they precipitate forming a scum. Therefore, soaps cannot be used in acidic solution. Second,

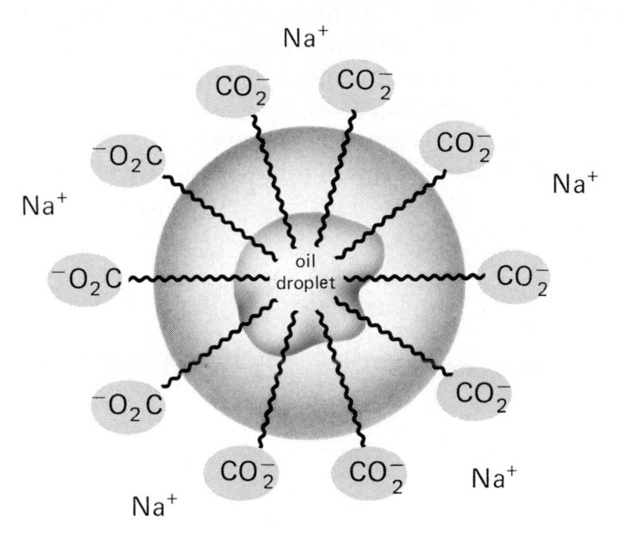

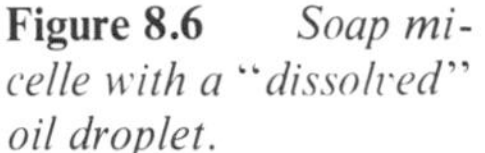
Figure 8.6 *Soap micelle with a "dissolved" oil droplet.*

soaps form insoluble salts when used in water containing calcium (II), magnesium (II), or iron (III) ions ("hard water").

$$\underset{\text{(water soluble)}}{2CH_3(CH_2)_{16}CO_2^- Na^+} + Ca^{2+} \longrightarrow \underset{\text{(water insoluble)}}{[CH_3(CH_2)_{16}CO_2^-]_2Ca^{2+}} + 2Na^+$$

This precipitate or scum formation creates problems including rings around the bathtub, films that spoil the luster of hair, and grayness and harshness of feel that build up on textiles after repeated washing.

Given these limitations on the use of natural soaps, the problem for the chemist was to create a new type of cleansing agent that would be soluble in both acidic and alkaline solutions and would not form insoluble precipitates when used in hard water. Despite considerable effort, there was no significant progress in this problem until early in the 1940s, with the introduction of synthetic detergents.

8.8 SYNTHETIC DETERGENTS

One of the most useful innovations in cleansing has been the development of synthetic detergents (often called syndets). These synthetic products have cleansing power as good or better than ordinary soaps, and at the same time they avoid the two major difficulties already listed for soaps. Given an understanding of the action of soaps, the design criteria for a synthetic detergent were as follows: a molecule with a long hydrocarbon chain (preferably 12 to 18 carbon atoms) and a highly polar group or groups at one end of the molecule that will not form insoluble salts with Ca^{2+}, Mg^{2+}, or other ions present in hard water. It was recognized that these essential characteristics could be produced in a molecule containing a sulfate group rather than a carboxylate group. Such compounds are strong acids, comparable in strength to sulfuric acid. Furthermore, the calcium, magnesium, and iron(III) salts of alkyl sulfate esters are soluble in water.

The first synthetic detergent was produced from 1-dodecanol (lauryl alcohol) by reaction with sulfuric acid, followed by neutralization with sodium hydroxide. The product of this sequence of reactions is sodium dodecyl sulfate (SDS).

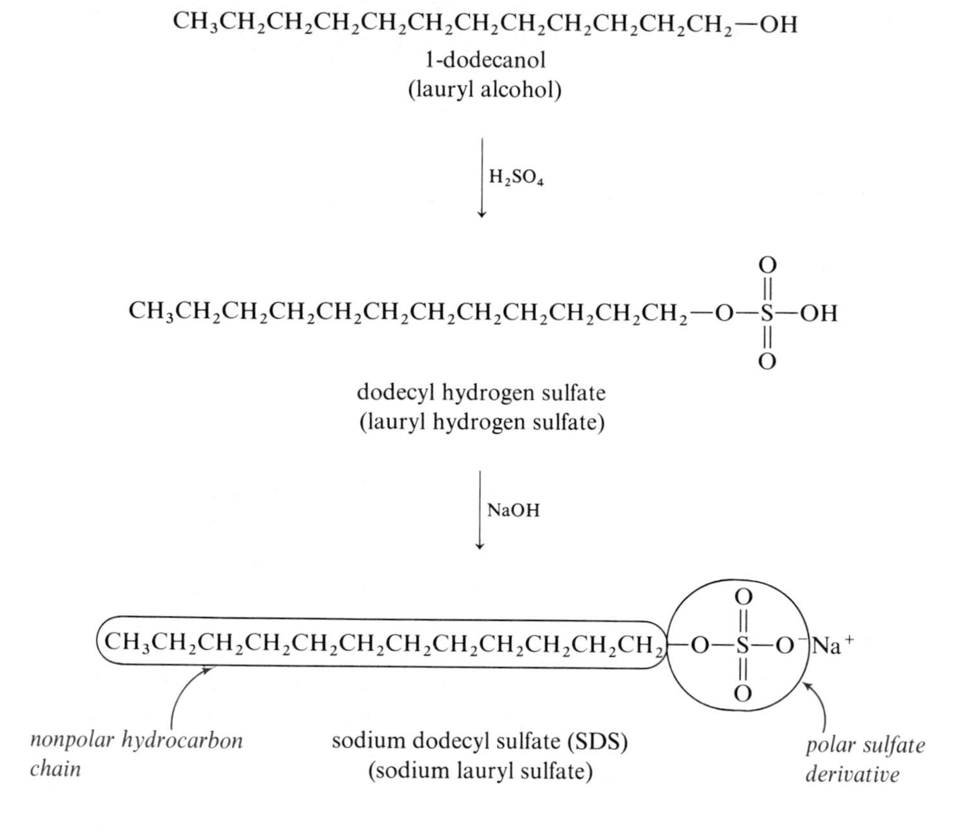

The physical resemblances between this synthetic detergent and ordinary soaps are obvious: a long nonpolar hydrocarbon chain and a highly polar end. Sodium dodecyl sulfate is an excellent detergent. However, the supply of lauryl alcohol was not nearly sufficient to meet the demand for synthetic detergents.

Another major advance in detergents came in the late 1940s, when it became technologically feasible to make the so-called alkylbenzene sulfonate detergents. The raw materials for this synthesis, propene and benzene, had become available from the petroleum refining industry. Propene was polymerized to a tetramer and reacted with benzene. This product was then sulfonated and reacted with sodium hydroxide to yield an alkylbenzene sulfonate salt of the following type.

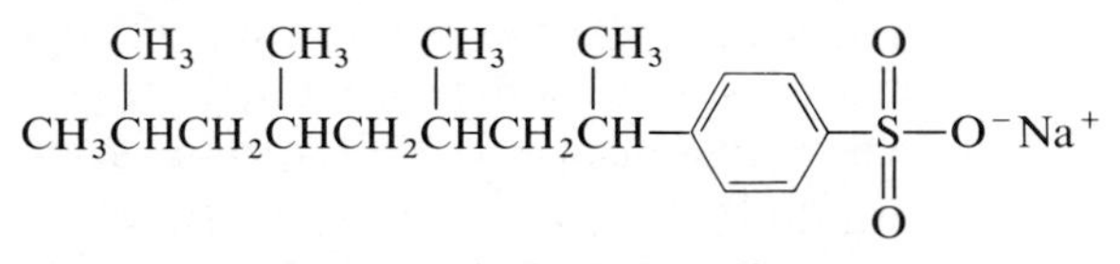

a sodium alkylbenzene sulfonate

Alkylbenzene sulfonate detergents were introduced in the 1950s and were accepted very rapidly. Within a decade, U.S. production of synthetic detergents increased twentyfold. Today they command close to 90% of the market once held by soaps.

The cleansing power of synthetic detergents of this type is enhanced enormously by certain additives. Sodium tripolyphosphate is added to coordinate with calcium, magnesium, copper, iron, and many other ions.

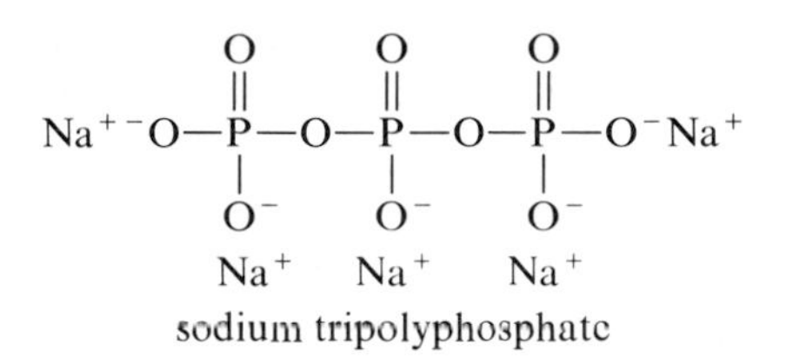

sodium tripolyphosphate

This substance is able to break up and suspend certain clays and pigments by forming water-soluble complexes with the metal ions, thereby facilitating their removal. Phosphates were introduced into cleansing agents in 1948 when Proctor & Gamble Company introduced Tide. By 1953, phosphates were used in more than half of the detergents sold in the United States, and by 1970 almost all detergents contained phosphates, sometimes as much as 60% by weight. Other common additives are whitening agents (optical brighteners), sudsing enhancers or repressors, and granular salts to create a satisfactory consistency for the commercial product.

As useful as the synthetic detergents proved to be, they created two major problems. One of the problems, that of disposability and biodegradability, has been solved. The other problem, the phosphate additives, is now being brought under control.

The first of these problems began to appear as excessive foaming in natural waters and sewage treatment plants. Most of it was found to be caused by the alkylbenzene sulfonate detergents. Soaps are removed from sewage waters by precipitation or by degradation in the treatment plants by microorganisms that are able to metabolize the linear alkyl hydrocarbon chains of the natural soaps derived from fats and oils. Such soaps are said to be biodegradable. It was discovered that the first alkylbenzene sulfonate detergents marketed could not be removed in either of these ways; they could not be precipitated nor could they be degraded by the microorganisms in sewage treatment plants. Instead they remained in suspension, causing sudsing and foaming, polluting streams, and in some cases finding their way into municipal drinking supplies. The solution to the problem was to replace the nonbiodegradable branched-chain hydrocarbon of the alkylbenzene sulfonate by a biodegradable linear-chain hydrocarbon. Fortunately, such linear hydrocarbons had become available through advances in petroleum refining and in 1965 the detergent industry converted entirely to the new linear alkylbenzene sulfonates of the following type.

$$CH_3CH_2CH_2CH_2CH_2CH_2CH_2CH_2CH_2CH_2CH_2CH_2-C_6H_4-S(=O)_2-O^-Na^+$$

sodium dodecylbenzenesulfonate

The second problem is that of the phosphate additives themselves and their contribution to water pollution. Strictly speaking, phosphate is not a pollutant, but rather a fertilizer, and it is as such that phosphates created a problem. The tremendous quantities of phosphates added to lakes, rivers, and streams through the use of detergents (and agricultural phosphate-based fertilizers as well) greatly increased the nutrient quality of the water. According to the House of Representatives' Subcommittee on Conservation and Natural Resources, when a rich stream of fertilizer flows steadily into a lake

> ...overstimulated, the water plants grow to excess. Seasonally they die off and rot.... In the process of decay they exhaust the dissolved oxygen of the water and produce the rotten-egg stench of hydrogen sulfide.... The game fish die of oxygen deficiency.... Intake filters for potable water become clogged, and boat propellers fouled with algae. The lake loses its value as a water supply, as an esthetic and recreational asset, and as an avenue of commerce. Finally the water itself is displaced by the accumulating masses of living and dead vegetation and their decay products and the lake becomes a bog, and eventually, dry land.

Of course, the evolution of a lake is a natural process, but one which has been greatly accelerated. Lake Erie is the most notorious example of an aging American lake. It is said to have "aged" 15,000 years in the last 50. After phosphate-containing detergents were introduced in 1948, the phosphate content of the lake's western basin more than tripled. Fortunately, this situation has not been allowed to continue. Public pressure and intelligent legislation have combined to require much lower phosphate levels in all detergents and, in many instances, to ban the use of phosphate-containing detergents. There are a number of nonphosphate-containing detergents on the market at the present time.

8.9 SPECTROSCOPIC PROPERTIES

The carboxyl group of carboxylic acids gives rise to two characteristic absorptions in the infrared spectrum. One of these occurs in the region 1700 to 1725 cm^{-1} and is associated with the stretching vibration of the carbonyl group. Note that this is essentially the same range of absorption as that for the carbonyl group of simple aldehydes and ketones (Section 7.6). The other infrared absorption characteristic of the carboxyl group is a peak between 2500 to 3000 cm^{-1} due to the stretching vibration of the O—H group. This absorption is generally very broad due to hydrogen bonding between molecules of carboxylic acid. Both C=O and O—H stretching frequencies can be seen in the infrared spectrum of butanoic acid (Figure 8.7).

Like aldehydes and ketones (Section 7.6) simple unconjugated carboxylic acids show only weak absorption in the ultraviolet spectrum.

The hydrogen of the carboxyl group gives rise to a signal in the NMR spectrum in the range $\delta = 10$ to 13.5. Note that the chemical shift of the carboxyl hydrogen is even larger than that of the aldehyde hydrogen. In fact, this chemical shift is so large that it serves to distinguish carboxyl hydrogens from most other types of hydrogens. The NMR signal for carboxyl hydrogens generally falls off the scale of chart papers, most of which are calibrated from $\delta = 0$ to $\delta = 10$. Therefore, in order to record this signal, the pen is set back on the paper at some arbitrary point and a note is made of how much it has been displaced or offset. The signal for the carboxyl proton of 2-methylpropanoic acid appears in the top right corner of the NMR spectrum in Figure 8.8 and has been offset

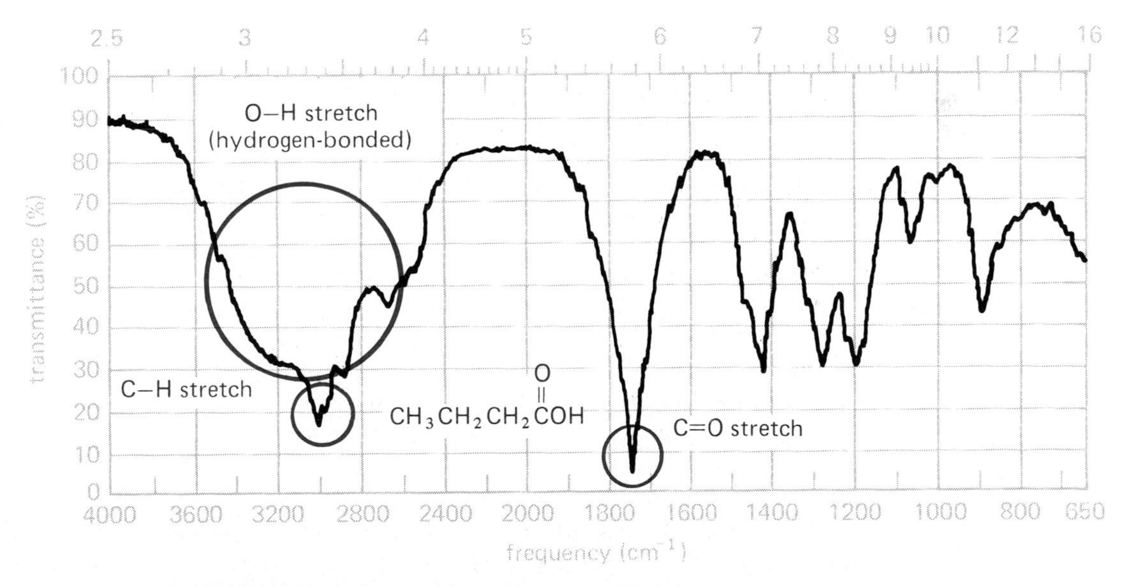

Figure 8.7 *An IR spectrum of butanoic acid.*

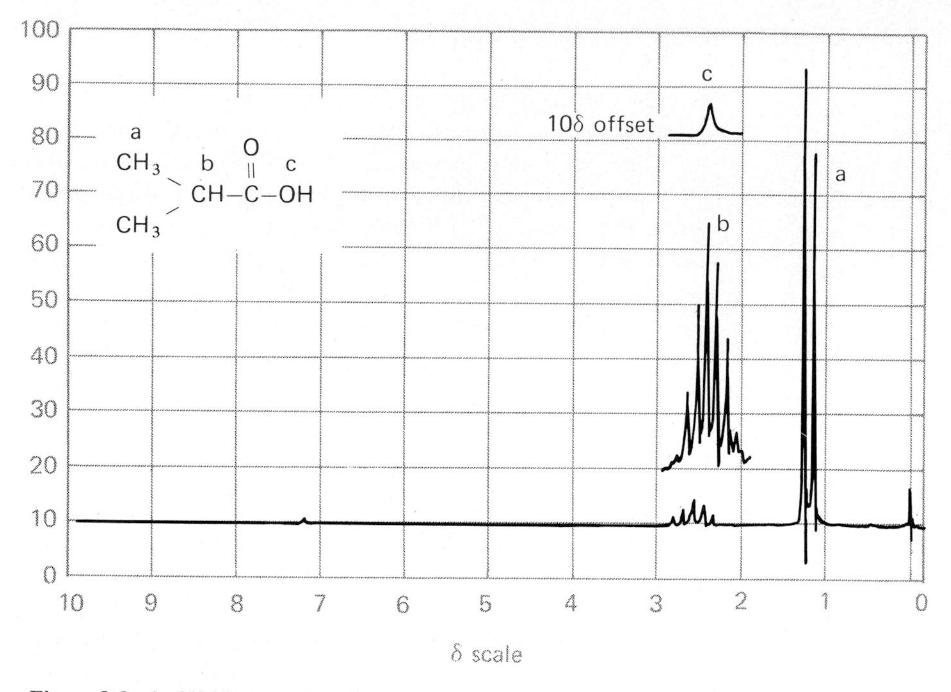

Figure 8.8 *An NMR spectrum of 2-methylpropanoic acid (isobutyric acid).*

by 10δ. Thus, the chemical shift of this proton is $\delta = 12.5$. The multiplet of peaks immediately below the carboxyl signal is an enlargement showing greater detail of the low intensity signal at $\delta = 2.6$.

The signal at $\delta = 1.2$ is due to the six equivalent hydrogens of the two $—CH_3$ groups and is split into a doublet by the single adjacent proton. The signal at $\delta = 2.5$ is due to the single CH hydrogen and is split into a septet (actually only five peaks are clearly visible even in the enlargement) by the six adjacent protons of the methyl groups. The areas of these three signals are in the ratio 6:1:1.

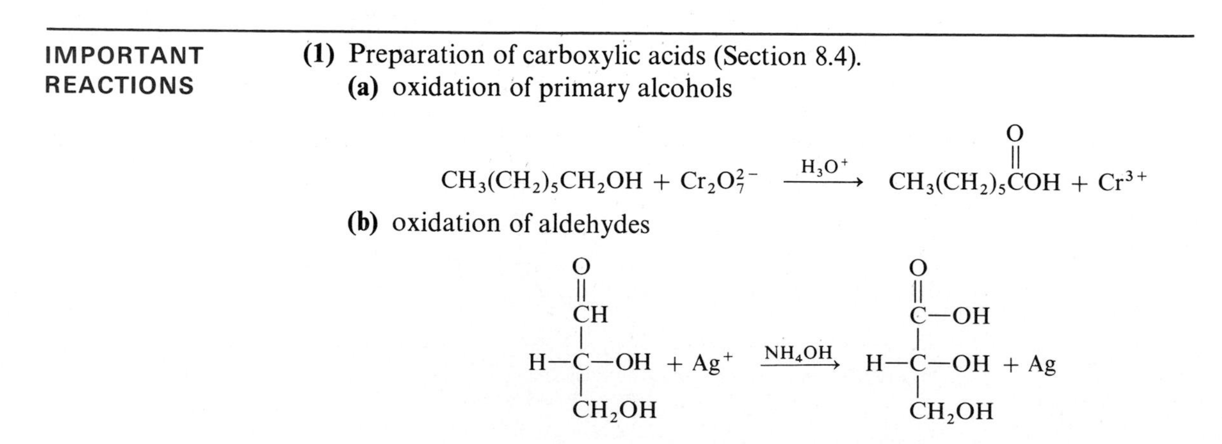

IMPORTANT REACTIONS

(1) Preparation of carboxylic acids (Section 8.4).

(a) oxidation of primary alcohols

$$CH_3(CH_2)_5CH_2OH + Cr_2O_7^{2-} \xrightarrow{H_3O^+} CH_3(CH_2)_5\overset{O}{\overset{\|}{C}}OH + Cr^{3+}$$

(b) oxidation of aldehydes

$$H-\underset{CH_2OH}{\underset{|}{\overset{\overset{O}{\|}{CH}}{\overset{|}{C}}}}-OH + Ag^+ \xrightarrow{NH_4OH} H-\underset{CH_2OH}{\underset{|}{\overset{\overset{O}{\|}{C}-OH}{\overset{|}{C}}}}-OH + Ag$$

(c) oxidation of alkenes

$$\text{cyclohexene} + Cr_2O_7^{2-} \xrightarrow{H^+} HO\overset{O}{\overset{\|}{C}}CH_2CH_2CH_2CH_2\overset{O}{\overset{\|}{C}}OH + Cr^{3+}$$

(d) oxidation of alkyl substituted benzenes

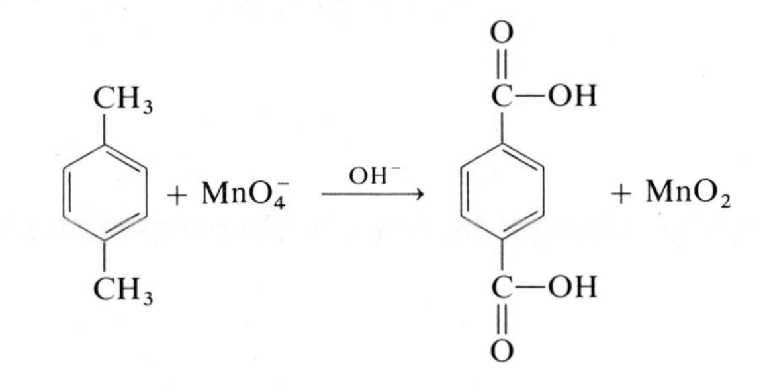

(2) Ionization of carboxylic acids (Section 8.5A).

$$CH_3\overset{O}{\overset{\|}{C}}—OH = CH_3—\overset{O}{\overset{\|}{C}}—O^- + H^+$$

$$K_a = \frac{[CH_3CO_2^-][H^+]}{[CH_3CO_2H]} = 1.8 \times 10^{-5}$$

(3) Reaction of carboxylic acids with bases (Section 8.5A).

$$C_6H_5—\overset{O}{\overset{\|}{C}}—OH + NaHCO_3 \longrightarrow C_6H_5—\overset{O}{\overset{\|}{C}}—O^-Na^+ + CO_2 + H_2O$$

(4) Reduction of carboxylic acids to primary alcohols (Section 8.5B).

$$CH_3(CH_2)_{10}\overset{O}{\overset{\|}{C}}OH + 2H_2 \xrightarrow[\text{heat}]{\text{catalyst}} CH_3(CH_2)_{10}CH_2OH + H_2O$$
high pressure

$$CH_3\overset{O}{\overset{\|}{C}}CH_2CH_2\overset{O}{\overset{\|}{C}}—OH + LiAlH_4 \longrightarrow CH_3\overset{OH}{\overset{|}{C}}HCH_2CH_2CH_2OH$$

(5) Decarboxylation of β-ketocarboxylic acids and β-dicarboxylic acids (Section 8.5C).

$$CH_3—\underset{\beta}{\overset{O}{\overset{\|}{C}}}—\underset{\alpha}{CH_2}—\overset{O}{\overset{\|}{C}}—OH \xrightarrow{\text{heat}} CH_3—\overset{O}{\overset{\|}{C}}—CH_3 + CO_2$$

(6) Reaction of natural soaps with Ca^{2+}, Mg^{2+}, and Fe^{3+} (Section 8.7).

$$2CH_3(CH_2)_{16}CO_2^- + Mg^{2+} \longrightarrow [CH_3(CH_2)_{16}CO_2]_2Mg$$

PROBLEMS

8.6 Name each of the following molecules.

(a) $CH_3CHCH_2CH_2CO_2H$
$\quad\quad|$
$\quad\;\; OH$

(b) 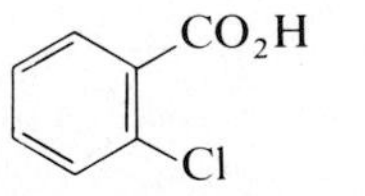

(c) $ClCH_2CO_2H$

(d) CH_3CHCO_2H
$\quad\quad|$
$\quad\;\; CH_3$

(e) $C_6H_5CH_2CO_2H$

(f) 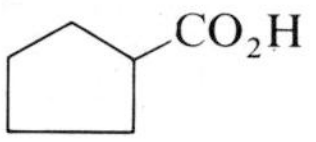

(g) $C_6H_5CO_2Na$

(h) $CH_3CH_2CH_2CH_2CO_2NH_4$

(i) $HO_2CCH_2CH_2CH_2CH_2CO_2H$

(j) CF_3CO_2H

(k) $(CH_3CH_2CO_2)_2Mg$

(l) $CH_3CH_2CH_2CH_2CH_2CH_2CH_2CH{=}CHCO_2H$

(m) 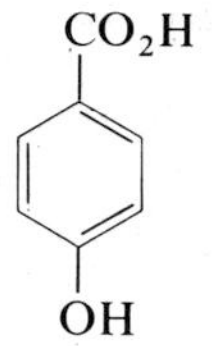

(n) 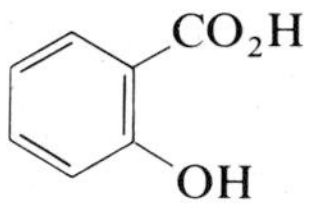

(o) $C_6H_5CH_2CO_2H$

(p) 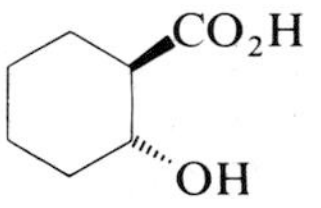

(q) $CH_2{=}CH{-}\overset{\overset{\displaystyle O}{\|}}{C}OH$

(r) 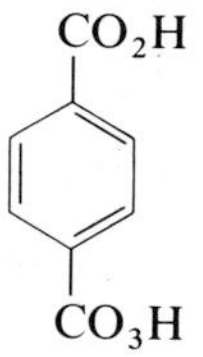

8.7 Draw structural formulas for each of the following molecules.

(a) 3-hydroxybutanoic acid
(b) sodium oxalate
(c) trichloroacetic acid
(d) 4-hydroxybutanoic acid
(e) sodium hexadecanoate
(f) calcium octanoate
(g) potassium phenylacetate
(h) octanoic acid
(i) 2-hydroxypropanoic acid (lactic acid)
(j) 2-aminopropanoic acid (alanine)
(k) *p*-methoxybenzoic acid
(l) potassium 2,4-hexadienoate (the food preservative, potassium sorbate)

8.8 Arrange the following compounds in order of increasing boiling points.

(a) $CH_3CH_2\overset{\overset{\displaystyle O}{\|}}{C}OH \quad CH_3CH_2CH_2CH_2OH \quad CH_3CH_2OCH_2CH_3$

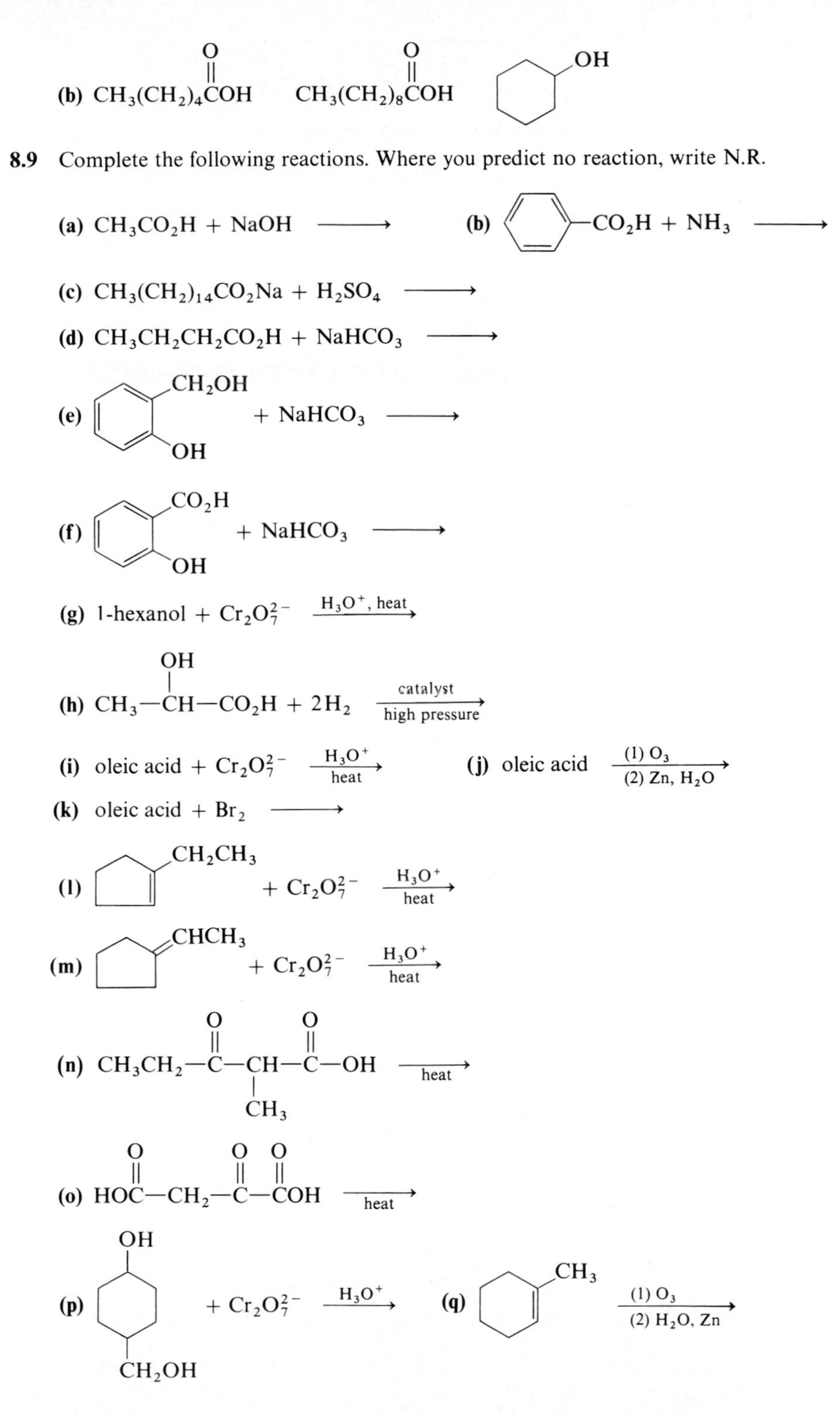
(b) CH3(CH2)4COH
CH3(CH2)8COH
OH
8.9 Complete the following reactions. Where you predict no reaction, write N.R.
(a) CH3CO2H + NaOH
(b) CO2H + NH3
(c) CH3(CH2)14CO2Na + H2SO4
(d) CH3CH2CH2CO2H + NaHCO3
(e) CH2OH OH + NaHCO3
(f) CO2H OH + NaHCO3
(g) 1-hexanol + Cr2O7 2− H3O+, heat
(h) CH3—CH—CO2H + 2H2 catalyst high pressure
(i) oleic acid + Cr2O7 2− H3O+ heat
(j) oleic acid (1) O3 (2) Zn, H2O
(k) oleic acid + Br2
(l) CH2CH3 + Cr2O7 2− H3O+ heat
(m) CHCH3 + Cr2O7 2− H3O+ heat
(n) CH3CH2—C—CH—C—OH heat CH3
(o) HOC—CH2—C—COH heat
(p) OH CH2OH + Cr2O7 2− H3O+
(q) CH3 (1) O3 (2) H2O, Zn

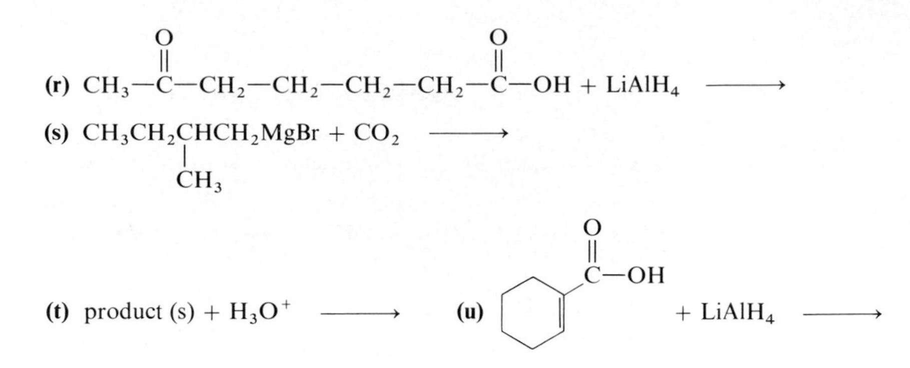

8.10 Arrange the following compounds in order of increasing acidity.

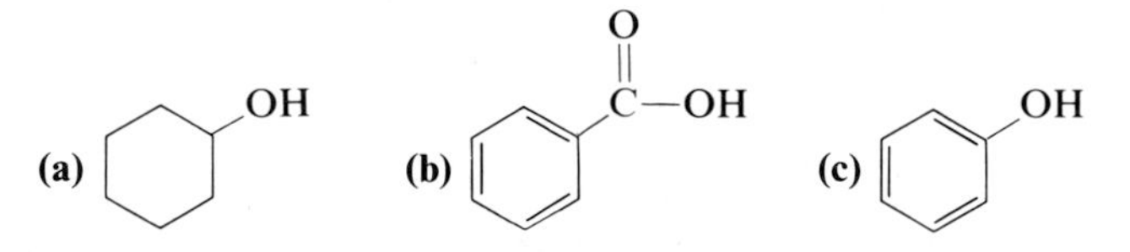

8.11 The following conversions can be carried out in one, two, or three steps. Show the reagents you would use and draw structural formulas for the intermediate formed in any conversion that requires more than one step.

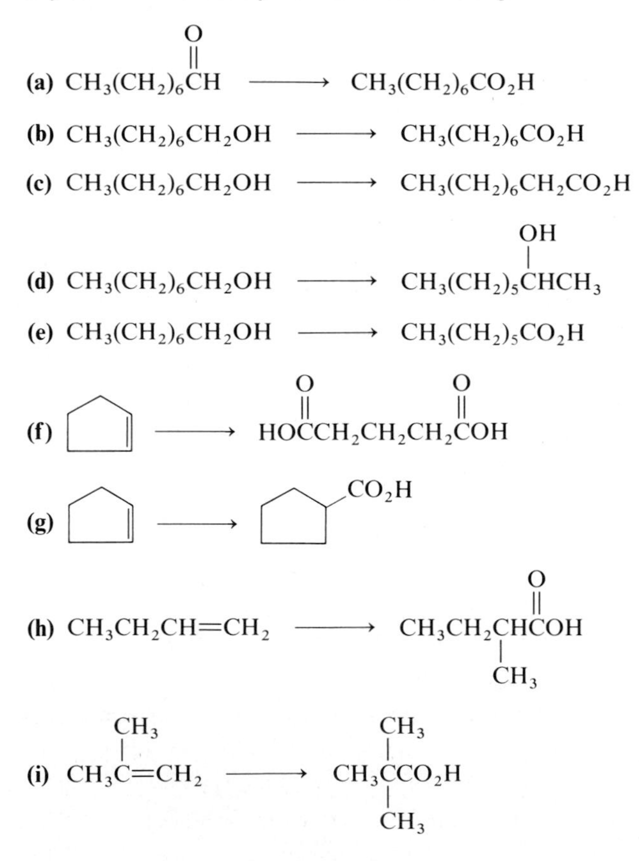

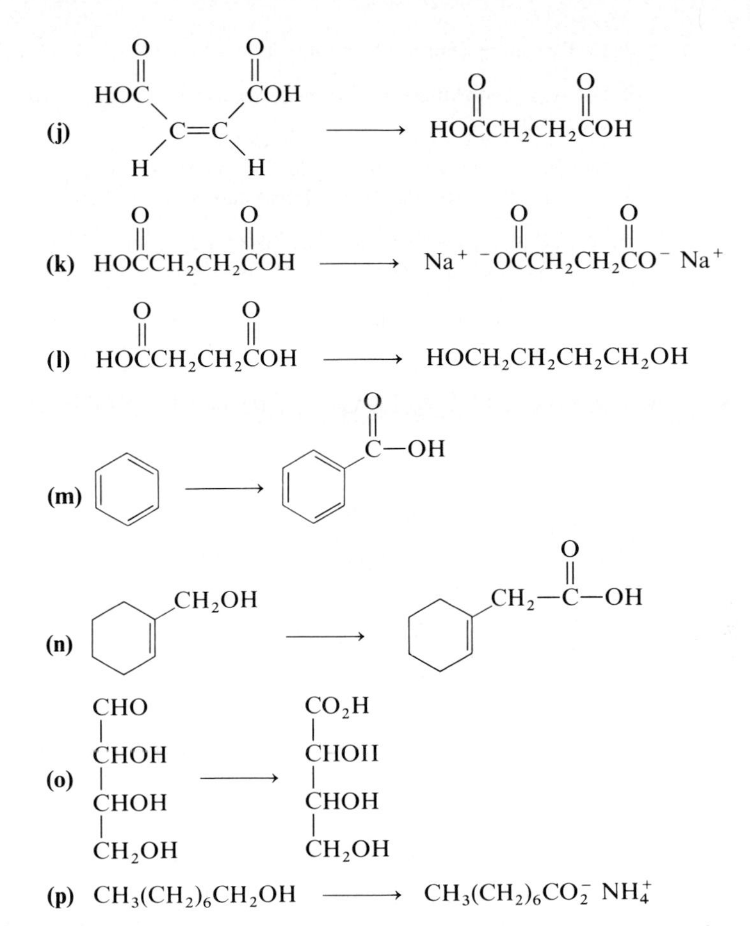

(p) $CH_3(CH_2)_6CH_2OH \longrightarrow CH_3(CH_2)_6CO_2^- \ NH_4^+$

8.12 Show how you might distinguish between the following pairs of compounds by a simple chemical test. In each case, tell what test you would perform, what you would expect to observe, and write an equation for each positive test.

(a) acetic acid and acetaldehyde
(b) hexanoic acid and 1-hexanol
(c) benzoic acid and phenol
(d) sodium salicylate and salicylic acid
(e) oleic acid and stearic acid (see Table 8.5 for structural formulas of these fatty acids)
(f) phenylacetic acid and acetophenone
(g) sodium lauryl sulfate (a synthetic detergent, Section 8.8) and sodium stearate (a natural soap)

8.13 Examine the structural formulas for lauric, palmitic, stearic, oleic, linoleic, and arachidonic acids. For each that shows *cis-trans* isomerism, state the total number of such isomers possible.

8.14 What does it mean to say that linoleic acid is an "essential" fatty acid? Name several dietary sources of linoleic acid.

8.15 By using structural formulas, illustrate how fatty acid molecules interact with water to form a monomolecular layer on the surface of water.

8.16 By using structural formulas, show how a soap "dissolves" fats, oils, and grease.

8.17 Show by balanced equations the reaction of a soap with **(a)** hard water, and **(b)** acidic solution.

8.18 Characterize the structural features necessary to make a good detergent. Illustrate by structural formulas two different classes of synthetic detergents. Name each example.

8.19 Following are structural formulas for a cationic detergent and a nonionic detergent. How would you account for the detergent properties of each?

$$C_6H_5-CH_2-\overset{\overset{\displaystyle CH_3}{|}}{\underset{\underset{\displaystyle C_8H_{17}}{|}}{N^+}}-CH_3\ Cl^-$$

benzyldimethyloctylammonium chloride (a cationic detergent)

$$CH_3(CH_2)_{14}\overset{\overset{\displaystyle O}{\|}}{C}-O-CH_2-\overset{\overset{\displaystyle CH_2OH}{|}}{\underset{\underset{\displaystyle CH_2OH}{|}}{C}}-CH_2OH$$

pentaerythrityl palmitate (a nonionic detergent)

8.20 Compound *A* ($C_5H_{10}O_3$) readily dissolves in water to give an acidic solution and can be titrated with aqueous sodium hydroxide. Compound *A* is also optically active and contains an alcohol group. Oxidation of compound *A* by potassium dichromate gives compound *B* ($C_5H_8O_4$). Compound *B* is a dicarboxylic acid and is optically inactive. Deduce structures for compounds *A* and *B* consistent with these observations.

8.21 There are four isomeric alcohols of molecular formula $C_4H_{10}O$. Draw and name each. Which of these alcohols is indicated by the following experimental observations? Compound *D* ($C_4H_{10}O$), on oxidation by potassium dichromate in acid solution, gives compound *E* ($C_4H_8O_2$), a carboxylic acid. Treatment of compound *D* with warm phosphoric acid brings about dehydration and yields compound *F* (C_4H_8). Treatment of compound *F* with warm aqueous sulfuric acid gives *G* ($C_4H_{10}O$), a new alcohol isomeric with compound *D*. Compound *G* is resistant to oxidation. Propose structures for compounds *D*, *E*, *F*, and *G* consistent with these observations.

8.22 List one major spectral characteristic that will enable you to distinguish between the following compounds.

(a) acetic acid and acetaldehyde (IR and NMR)
(b) hexanoic acid and 1-hexanol (IR and NMR)
(c) benzoic acid and phenol (IR)
(d) phenylacetic acid and acetophenone (IR and NMR)
(e) 4-ketohexanoic acid and hexanoic acid (NMR)
(f) benzoic acid and cyclohexanecarboxylic acid (UV and NMR)
(g) 2-methylpropanoic acid (isobutyric acid) and butanoic acid (*n*-butyric acid)

Prostaglandins

MINI-ESSAY 6

THE PROSTAGLANDINS are a group of naturally occurring substances all of which have the 20-carbon skeleton of prostanoic acid.

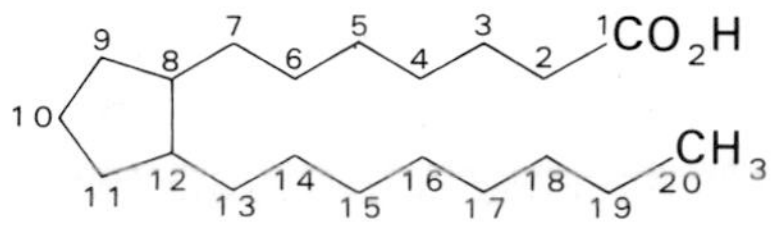

prostanoic acid

Prostaglandins and protaglandin-derived materials are intimately involved in a host of bodily processes. They have been found in virtually all human tissues examined thus far. They are involved in almost every phase of reproductive physiology. Certain prostaglandin-derived substances stimulate blood clotting by promoting the aggregation of blood platelets, and others inhibit the clotting process. Prostaglandins are involved in both the induction of the inflammatory response and in its relief. The medical significance of their involvement in the process of inflammation is obvious when we realize that more than five million Americans suffer from rheumatoid arthritis, an inflammatory disease. Certain prostaglandins also appear to stimulate the enzyme adenyl cyclase, which in turn catalyzes the conversion of adenosine triphosphate (ATP) into cyclic adenosine monophosphate (cAMP). Cyclic-AMP is a compound of major importance in the regulation of cellular metabolism, and it may well be that the ability of prostaglandins to influence the metabolism of this substance is the key to their wide range of physiological activities.

The discovery and structure determination of prostaglandins began in 1930, when Raphael Kurzrok and Charles Lieb, both gynecologists practicing in New York, observed that human seminal fluid stimulates contraction of isolated human uterine muscle. A few years later, in Sweden, Ulf von Euler confirmed this report and noted that human seminal fluid also produces contraction of intestinal smooth muscle and lowers blood pressure when injected into the bloodstream. Von Euler proposed the name prostaglandin for the mysterious substance or substances responsible for such diverse effects, for at that time it was believed that they originated in the prostate gland. We now know that prostaglandin production is by no means limited to the prostate gland. However, the name has stuck. By 1960, several prostaglandins had been isolated in pure crystalline form and their structural formulas had been determined. Structural formulas for three common prostaglandins are given in Figure 1.

Prostaglandins are abbreviated PG, with an additional letter and numerical subscript to indicate the type and series. The various types differ in the functional groups present in the five-membered ring. Those of the A-type are α,β-unsaturated ketones; those of the B-type are

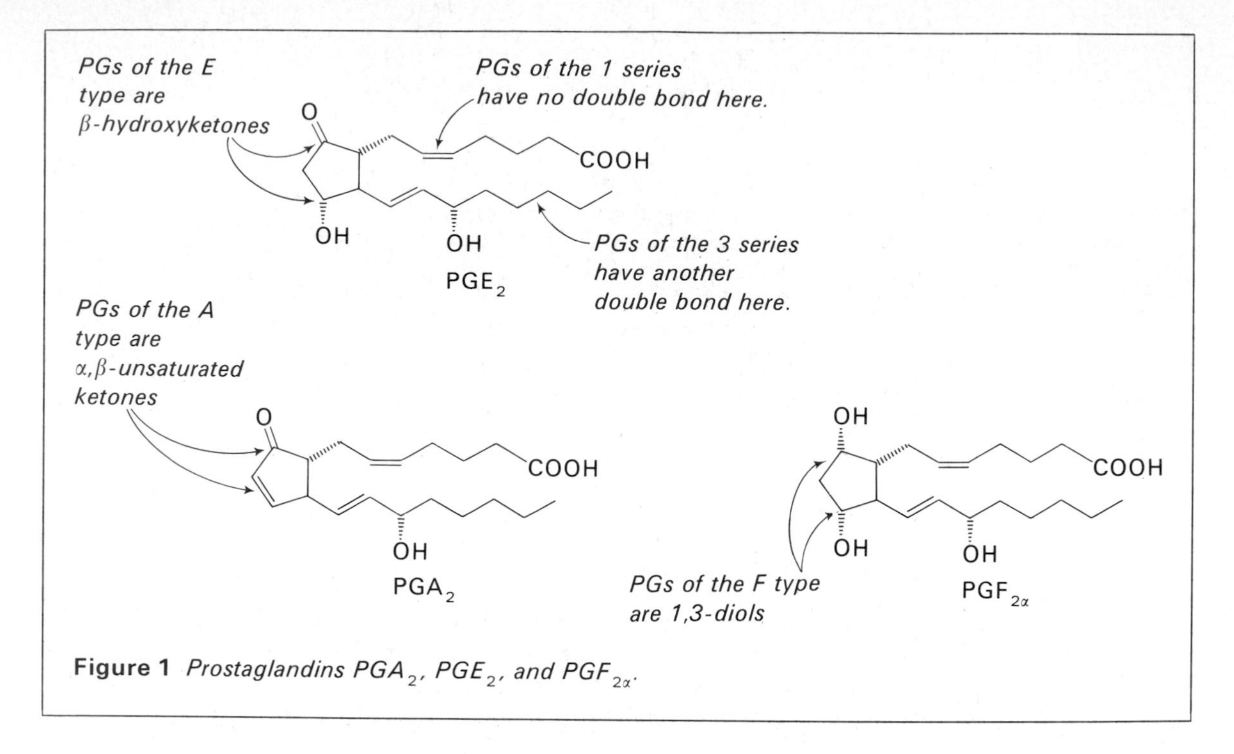

Figure 1 *Prostaglandins PGA_2, PGE_2, and $PGF_{2\alpha}$.*

β-hydroxyketones; and those of the F-type are 1,3-diols. The subscript α in the F-type indicates that the hydroxyl group at carbon 9 is below the plane of the five-membered ring and on the same side as the hydroxyl at carbon 11. The various series of prostaglandins differ in the number of double bonds on the two side chains. Those of the 1-series have only one double bond; those of the 2-series have two double bonds; and those of the 3-series have three double bonds.

Coincident with the investigations of the chemical structure of prostaglandins, clinical scientists began to study the biochemistry of these remarkable substances and their potential as drugs. Initially, research was hampered by the high cost and great difficulty of isolating and purifying the substances. If they could not be isolated easily, could they be synthesized instead? The first totally synthetic prostaglandins became available in 1968 when Dr. John Pike of the Upjohn Company and Professor E. J. Corey of Harvard University each announced laboratory synthesis of several prostaglandins and prostaglandin analogs. However, costs were still high. Then, in 1969, the price of prostaglandins dropped dramatically with the discovery that the gorgonian sea whip or sea fan, *Plexaura homomalla*, which grows on the coral reefs off the coast of Florida and in the Caribbean, is a rich source of prostaglandin-like materials (Figure 2). The concentration of PG-like substances in this marine organism is about 100 times the normal concentration found in most mammalian sources. In the laboratory the PG-like compounds were extracted and then transformed into prostaglandins and prostaglandin analogs. At the present time, however, there is no need to depend on this natural source, for chemists have developed highly effective and stereospecific laboratory schemes for the synthesis of almost any prostaglandin or prostaglandin-like substance.

Prostaglandins are not stored as such in tissues, but they are synthesized in response to specific environmental or physiological triggers. Starting materials for prostaglandin synthesis are unsaturated fatty acids of twenty carbon atoms. Those of the 2-series are derived from arachidonic acid (5, 8, 11, 14-eicosatetraenoic acid), an unsaturated fatty acid containing four carbon-carbon double bonds. Steps in the biochemical pathways by which arachidonic acid is converted to several key prostaglandins are summarized in

A molecular model of prostaglandin E_1 is examined by Dr. John Pike, Head of Experimental Chemistry Research, The Upjohn Company. (Upjohn International Inc.)

Found in the Caribbean Sea, Plexaura homomalla, *known as the sea whip or gorgonian, contains the highest concentration of prostaglandin-like substances so far found in nature. Before synthetic production of prostaglandins, Upjohn extracted the rare substances from this soft coral. (Upjohn International Inc.)*

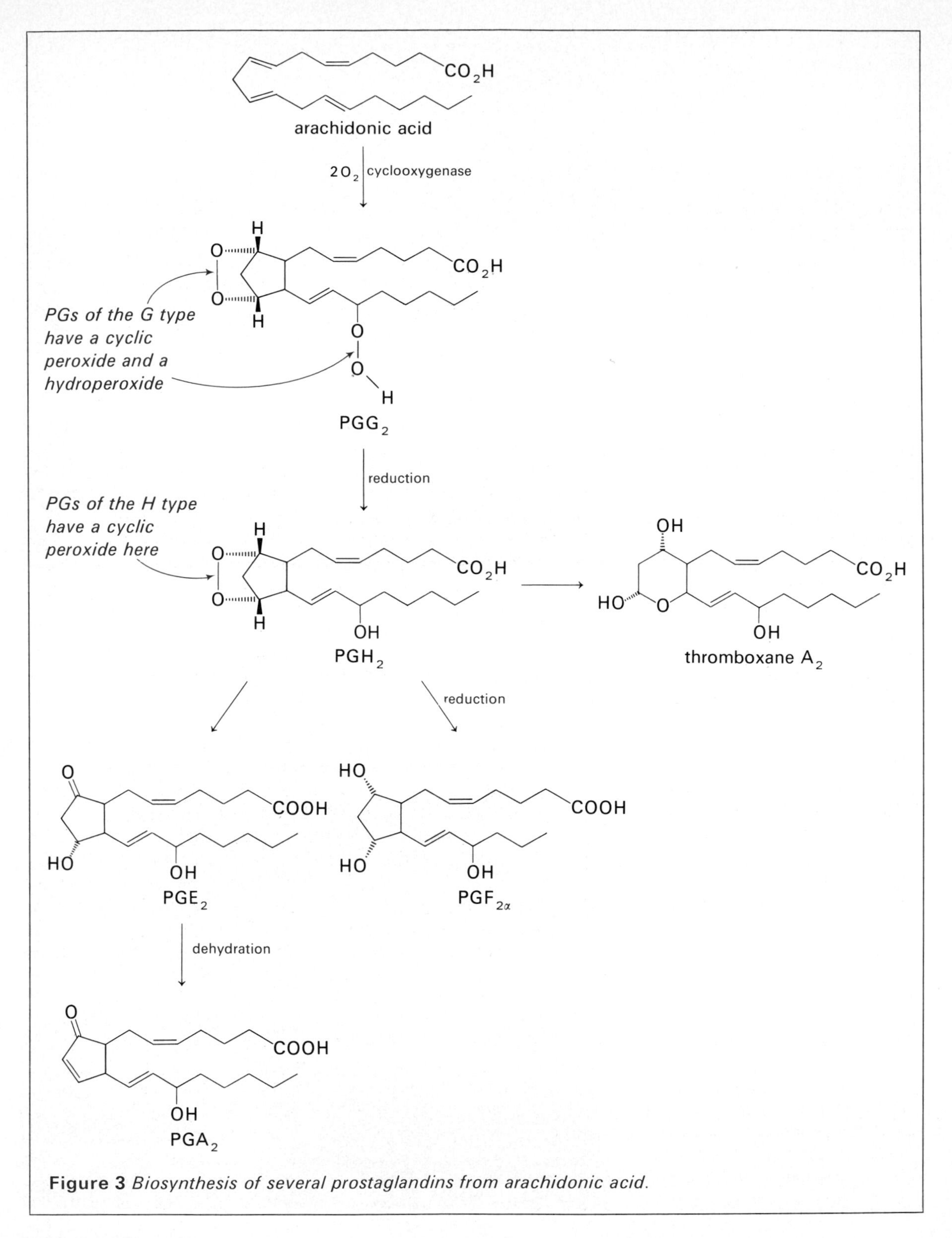

Figure 3 *Biosynthesis of several prostaglandins from arachidonic acid.*

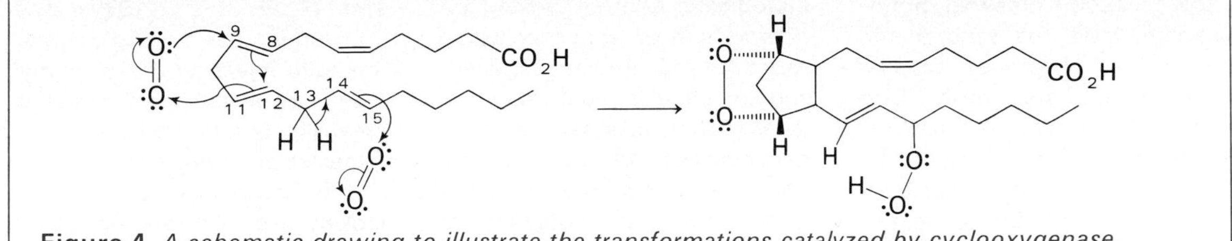

Figure 4 *A schematic drawing to illustrate the transformations catalyzed by cyclooxygenase.*

Figure 3. Arachidonic acid is drawn in this figure to show the relationship between its structural formula and that of the prostaglandins derived from it.

The first step in the biosynthesis of prostaglandins of the 2-series is reaction of arachidonic acid with two molecules of oxygen, O_2. This complex reaction, catalyzed by the enzyme cyclooxygenase, involves transformations in two different parts of the molecule. In one part, a hydrogen atom is removed from carbon 13, a carbon-carbon double bond is formed between carbons 13 and 14, and formation of a carbon-oxygen bond between carbon 15 and molecular oxygen generates a hydroperoxide (—O—O—H). In another part of the molecule, formation of a bond between carbons 8 and 11 generates a five-membered ring characteristic of the prostaglandins, and reaction with a second molecule of O_2 generates a peroxide (—O—O—) linkage between carbons 9 and 12. These transformations are shown in detail in Figure 4, with arrows to show you where bonds are broken and new ones formed. The actual mechanism of the reaction catalyzed by cyclooxygenase is considerably more complicated than is indicated by these arrows.

Enzyme-catalyzed reduction of the hydroperoxide at carbon 15 of PGG_2 gives PGH_2. This substance, a key intermediate from which all other prostaglandins of the 2-series are synthesized, has a very short half-life. Within minutes it is converted into other prostaglandins and prostaglandin-derived substances. Enzyme-catalyzed reduction of the peroxide between carbons 9 and 11 of PGH_2 forms the diol characteristic of prostaglandins of the F series. Alternatively, rearrangement of the —O—O— linkage of PGH_2 leads to the formation of the α,β-unsaturated ketone system characteristic of prostaglandins of the E series. Dehydration of PGE_2 forms PGA_2. In another, more complex enzyme-catalyzed transformation, PGH_2 is transformed into thromboxane A_2. This substance is not a true prostaglandin for it does not have the carbon skeleton characteristic of prostanoic acid. Nonetheless, it is very important in human physiology and is grouped with the prostaglandins. Notice that carbon 11 of thromboxane A_2 is a hemiacetal. Figure 3 shows the biosynthesis of types A, E, F, G, and H. There are other types, also derived from the key intermediate PGH_2. Precisely which prostaglandins or prostaglandin-like substances are produced depends on the enzymes present in a particular tissue. Platelets for example make primarily thromboxane A_2.

Now that we have seen the types of transformations by which the body synthesizes prostaglandins, let us look at several functions of these substances within the body.

First is the participation of prostaglandins in blood clotting. There are three distinct phases to the physiological mechanisms which come into play within the body to stop bleeding from a ruptured blood vessel. The first phase is called platelet aggregation which is initiated by agents such as thrombin. During platelet aggregation, blood platelets become sticky and form a platelet plug at the site of the injury. If damage is minor and the blood vessel is small, this platelet plug may be sufficient to stop loss of blood from the vessel. If it is not sufficient, the platelets are stimulated to release a group of substances (the platelet release reaction) which in turn promote a second wave of platelet aggregation and constriction of the injured vessel. The third phase is the triggering of the actual blood coagulation process.

We have learned within the past few years that among the substances released in platelet release reactions is thromboxane A_2. This prostaglandin-derived molecule is a very potent vasoconstrictor and the key substance that triggers platelet aggregation. Just as we have known for some time that thrombin stimulates the second and irreversible phase of platelet aggregation, we have also known that aspirin and aspirin-like drugs such as indomethacin, inhibit this second phase. How these drugs are able to do this remained a mystery until it was discovered that aspirin inhibits cyclooxygenase, the enzyme that initiates the synthesis of thromboxane A_2.

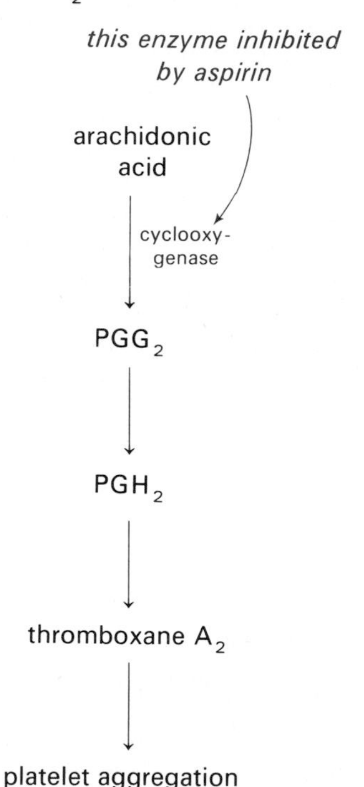

There is now good evidence that the ability of aspirin to reduce inflammation is also related to its ability to inhibit prostaglandin synthesis. Further research on the prostaglandins may help us understand even more about inflammatory diseases such as rheumatoid arthritis, asthma, and other allergic responses.

As indicated in the introduction to this essay, the first recorded observations on the biological activity of prostaglandins were those of gynecologists Kurzrok and Lieb. The first widespread clinical application of these substances was also by gynecologists and obstetricians. The observation that prostaglandins stimulate contraction of uterine smooth muscle led to the suggestion that these substances might be used for termination of second-trimester pregnancy. One problem with the use of naturally occurring prostaglandins for this purpose is that they are very rapidly degraded within the body. Therefore, their use required repeated administration over a period of hours. In the search for less rapidly degraded prostaglandins, a number of semi-synthetic prostaglandin analogs were synthesized. One of the most effective of these was 15-methyl prostaglandin $F_{2\alpha}$ which is longer acting and has 10 to 20 times the potency of $PGF_{2\alpha}$.

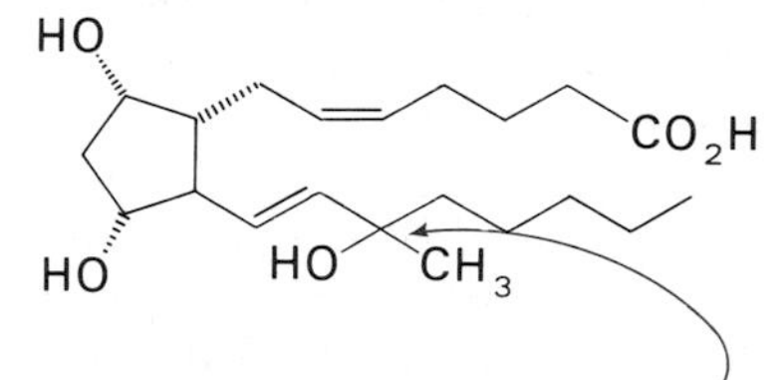

15-methyl prostaglandin $F_{2\alpha}$

The potential clinical use of prostaglandins and prostaglandin analogs for termination of second trimester pregnancy was explored in a study designed and conducted by the World Health Organization Task Force on the Use of Prostaglandins for the Regulation of Fertility. This multicenter, multinational study, entitled "Prostaglandins and Abortion," is described in the *American Journal of Obstetrics and Gynecology* (1977) and concludes that a single intra-amniotic injection of 15-methyl $PGF_{2\alpha}$ is a safe and effective means for termination of second-trimester pregnancy.

In this mini-essay we have looked at only a few aspects of the biosynthesis and importance in human physiology of prostaglandins. From even this brief encounter, it should be clear that we are only beginning to understand the chemistry and biochemistry of this group of substances. And it should also be clear that the enormous prostaglandin research effort now under way offers great promise for even deeper insight into human physiology and for the development of new and highly effective drugs for use in clinical medicine.

References

Bergstrom, S., "Prostaglandins: Members of a New Hormonal System," *Science* 157 (1967): 382.

———, Carlson, L. A., and Weeks, J. R., "The Prostaglandins: A Family of Biologically Active Lipids," *Pharmacological Review* 20 (1968): 1.

Goodman and Gilman, *The Pharmacological Basis of Theraputics,* 5th ed., (N.Y.: Macmillan, 1975).

Kuehl, F. A. and Egan, R. W., "Prostaglandins, Arachidonic Acid, and Inflammation," *Science* 210 (1980): 978–984.

Needleman, P. et al., *Nature* 261 (London, 1976): 558–560.

"Prostaglandins and Abortion," *The American Journal of Obstetrics and Gynecology* 129 (1977): 593–606.

The Prostaglandins, (N.Y.: Plenum Press).

Samuellson, B. et al., "Prostaglandins," *Annual Review of Biochemistry* 44 (1975): 669–695.

Acetic Acid: From What and How

IF YOU WERE ASKED as a part of a laboratory assignment in organic chemistry to prepare acetic acid, how would you do it? What reagents would you begin with and what reaction or series of reactions would you choose? Perhaps you might start with the fact that a common laboratory method for the preparation of carboxylic acids is oxidation of primary alcohols and choose ethanol as the source of carbons and hydrogens and either potassium dichromate or potassium permanganate as the oxidizing agent.

$$\underset{\text{ethanol}}{CH_3—CH_2—OH} + Cr_2O_7^{2-} \xrightarrow{H^+ \mid H_2O} \underset{\text{acetic acid}}{CH_3—\overset{\overset{\displaystyle O}{\|}}{C}—OH} + Cr^{3+}$$

While oxidation of ethanol by potassium dichromate in aqueous sulfuric acid is a practical method for the preparation of acetic acid in the laboratory, it is not a practical method for production on an industrial scale. First, there are other starting materials less expensive than ethanol and potassium dichromate. In this mini-essay we will discuss two alternative sources of carbon atoms (ethylene and methanol) and one alternative oxidizing agent (molecular oxygen, O_2). Second, oxidation of ethanol by potassium dichromate is most commonly carried out in aqueous solution and the problems associated with handling and removing the massive amounts of water involved make this an impractical way to manufacture the several billion pounds of acetic acid produced in the United States each year. Third, what happens to Cr^{3+} salts formed in the reaction and to any unreacted $K_2Cr_2O_7$? Are they to be recovered and recycled, and if so how? Or are they to be disposed of in a manner consistent with protection of the environment? And how is the spent sulfuric acid to be treated?

In this mini-essay, we will look at several methods that have been used over the years to synthesize acetic acid on an industrial scale. In 1979, acetic acid production in the United States totalled 3.3 billion pounds, a volume that ranks it thirty-third on the list of all chemicals manufactured by the U.S. chemical industry and nineteenth on the list of organics. The synthesis of acetic acid and products derived directly from it is a multi-billion dollar business and as we shall see, the cost of energy and of raw material feedstocks (sources of carbon and hydrogen) are of first importance in process design.

MINI-ESSAY 7

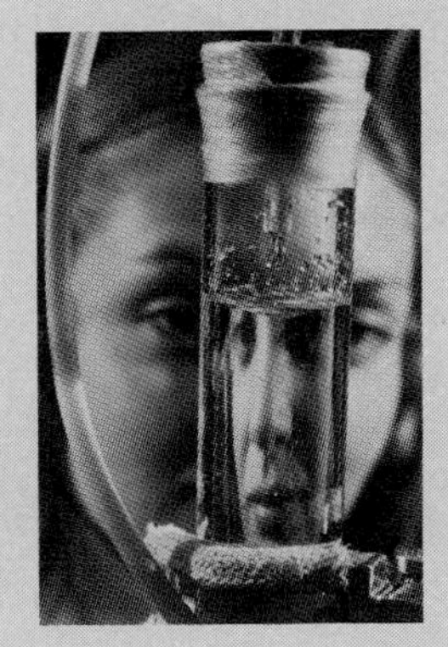

For about 75 years, wood was the principal raw material for the production of acetic acid. Actually, wood was the raw material for the production of charcoal, at one time the chief fuel for iron refinery furnaces, and acetic acid was a by-product that could be recovered and sold for profit. In the process known as carbonization, wood is heated for prolonged periods of time at 400–500°C and the vapors given off are collected and cooled. The portion that condenses consists of a mixture of at least 100 different substances. Chief among these, and the ones separated and recovered for their commercial value, are acetic acid and methanol. Typical yields per cord of dry, hard wood are 140–185 pounds of acetic acid, 55–65 pounds of methanol, and 1000–1050 pounds of charcoal. As the iron and steel industry converted to other fuels, the demand for charcoal decreased to the point where carbonization of wood ceased to be an important source of acetic acid.

The first industrial synthesis of acetic acid was commercialized in 1916 in Canada and Germany, using acetylene as the feedstock. The process involved two stages: (1) hydration of acetylene to acetaldehyde followed by (2) oxidation of acetaldehyde to acetic acid.

In Stage 1, high purity acetylene was blown through a hot solution of 20% sulfuric acid containing small amounts of mercury (II) sulfate, $HgSO_4$, as a catalyst. The reaction is exothermic and the optimum temperature range of 70–80°C was maintained by circulating cold water through coils within the reaction chamber. Acetylene was added at a rate sufficient to sweep acetaldehyde out of the chamber before it could undergo further reaction. Acetaldehyde and unreacted acetylene were then separated by bubbling the mixture into water at 0°C. Acetaldehyde dissolved to form an aqueous solution while acetylene was collected in gaseous form and recycled. Distillation of the aqueous solution yielded high purity (99.9%) acetaldehyde.

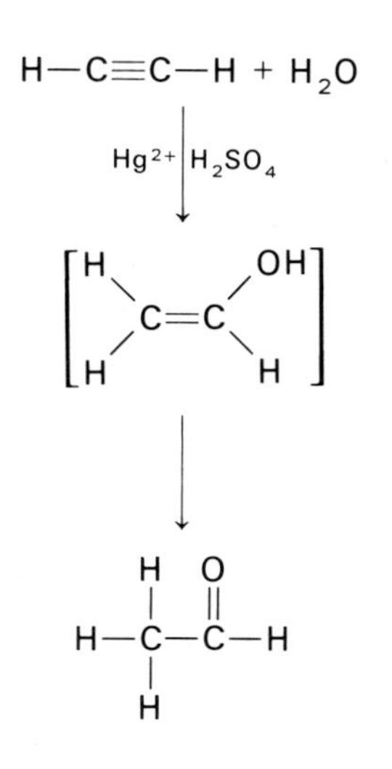

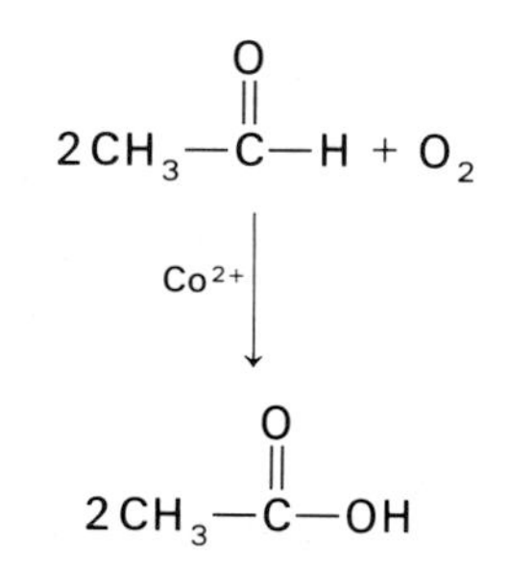

Stage 2 of the acetylene-to-acetic-acid process involved O_2 oxidation of acetaldehyde, catalyzed by small amounts of cobalt (II) acetate, $Co(CH_3CO_2)_2$. The chemistry of the aldehyde oxidation involves two steps. The first is reaction of acetaldehyde and oxygen by a free radical mechanism to form peracetic acid. In the second step, peracetic acid oxidizes a molecule of acetaldehyde to acetic acid and is itself reduced to a second molecule of acetic acid.

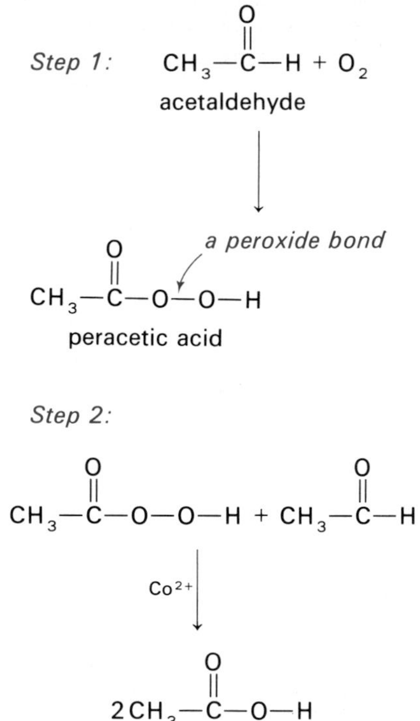

The technology of the acetylene based process is simple and the yields are high, factors that made this the major route to acetic acid for over 50 years. A problem was that the process required a huge input of energy, not for the hydration or oxidation stages themselves, but for the synthesis of acetylene, the feedstock on which the entire process was based. At that time, acetylene was prepared by reaction of calcium carbide, CaC_2, with water.

$$CaC_2 + 2H_2O \text{ (calcium carbide)} \longrightarrow H{-}C{\equiv}C{-}H \text{ (acetylene)} + Ca(OH)_2 \text{ (calcium hydroxide)}$$

Calcium carbide, in turn, was prepared by heating calcium oxide (from limestone, $CaCO_3$) with coke (from coal) to 2000–2500°C in an electric furnace.

$$\text{coal} \xrightarrow{\text{heat}} \underset{\text{coke}}{C}$$

$$\underset{\text{calcium carbonate}}{CaCo_3} \xrightarrow{\text{heat}} \underset{\text{calcium oxide}}{CaO} + CO_2$$

$$\underset{\text{calcium oxide}}{CaO} + \underset{\text{coke}}{3C} \xrightarrow{2500°C} \underset{\text{calcium carbide}}{CaC_2} + CO$$

Production of calcium carbide by fusion of calcium oxide and coke in an electric furnace required enormous amounts of energy, and as the cost of energy rose, acetylene, based as it was on calcium carbide, ceased to be an economical feedstock from which to manufacture acetic acid.

As an alternative feedstock, chemists and chemical engineers turned to ethylene, already available in huge quantities by refining natural gas and petroleum (see the mini-essay "Ethylene") and at a considerably lower price than that of acetylene. The ethylene-to-acetic-acid process depends on the fact, known since 1894, that in the presence of catalytic amounts of Pd^{2+} and Cu^{2+} salts, ethylene is oxidized by molecular oxygen to acetaldehyde.

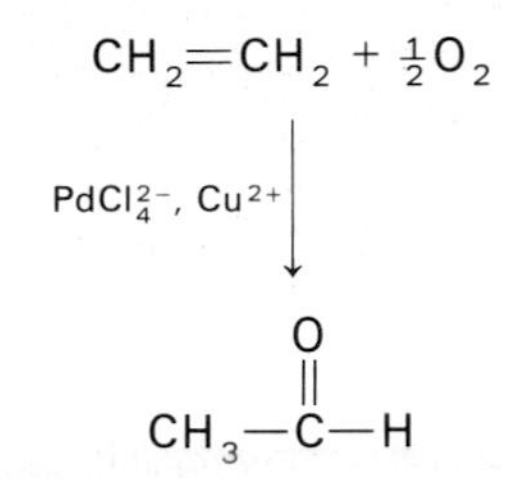

The first plant to use ethylene oxidation for the manufacture of acetaldehyde was built by Wacker-Chemie in 1959 and the process itself became known as the Wacker process. The net equation of the Wacker process is simple to write but its chemistry is complex. Because of its economic importance, there has been a great deal of research on the chemistry of the Wacker process and its mechanism is now quite well understood. Following is a brief outline of key steps in the mechanism. Under the conditions of the reaction, palladium is present initially in the form $PdCl_4^{2-}$, a complex ion in which the oxidation state of the metal is +2. As one way to simplify the mechanism, we will write palladium as Pd^{2+} rather than as the complex ion. In Step 1 of the Wacker process, Pd^{2+} interacts with the pi electrons of the carbon-carbon double bond in much the same way as does H^+ during acid-catalyzed hydration of an alkene. Reaction of Pd^{2+} (an electrophile) at one carbon of the double bond is followed by reaction of H—OH (a nucleophile) at the other carbon. This reaction forms a substance with a carbon-palladium single bond ($C{-}Pd^+$) and is a specific example of electrophilic addition to a carbon-carbon double bond. Note that the formal charge on palladium initially is +2 and in the product of Step 1, it is +1. Rearrangement in Step 2 followed by elimination of H^+ and Pd^0 in Step 3 gives acetaldehyde. As a result of Steps 1–3, ethylene is oxidized to acetaldehyde (a two-electron oxidation) and Pd^{2+} is reduced to Pd^0 (a two-electron reduction). In Step 4, Pd^0 is reoxidized by Cu^{2+} and in the process Cu^{2+} is reduced to Cu^+. Finally, in Step 5, Cu^+ is reoxidized to Cu^{2+} by molecular oxygen (Figure 1).

The mechanism of the Wacker process as sketched here involves five steps and if you add them and then simplify the resulting equation by cancelling substances that appear on both the left and right sides, you will discover that the net reaction involves one molecule of ethylene and one-half molecule of oxygen and gives one molecule of acetaldehyde. Neither palladium nor copper appears in the balanced equation. The function of these substances is to undergo a continuous cycle of oxidation-reduction reactions, the effect of which is to catalyze the oxidation of ethylene and the reduction of molecular oxygen. The Wacker process was introduced in 1959 and within a few years, ethylene completely replaced acetylene as a feedstock for the synthesis of both acetaldehyde and acetic acid.

At the same time the Wacker process was introduced, Badische Anilin und Soda Fabrik (BASF) in Germany began production of acetic acid by a process that at first sight must

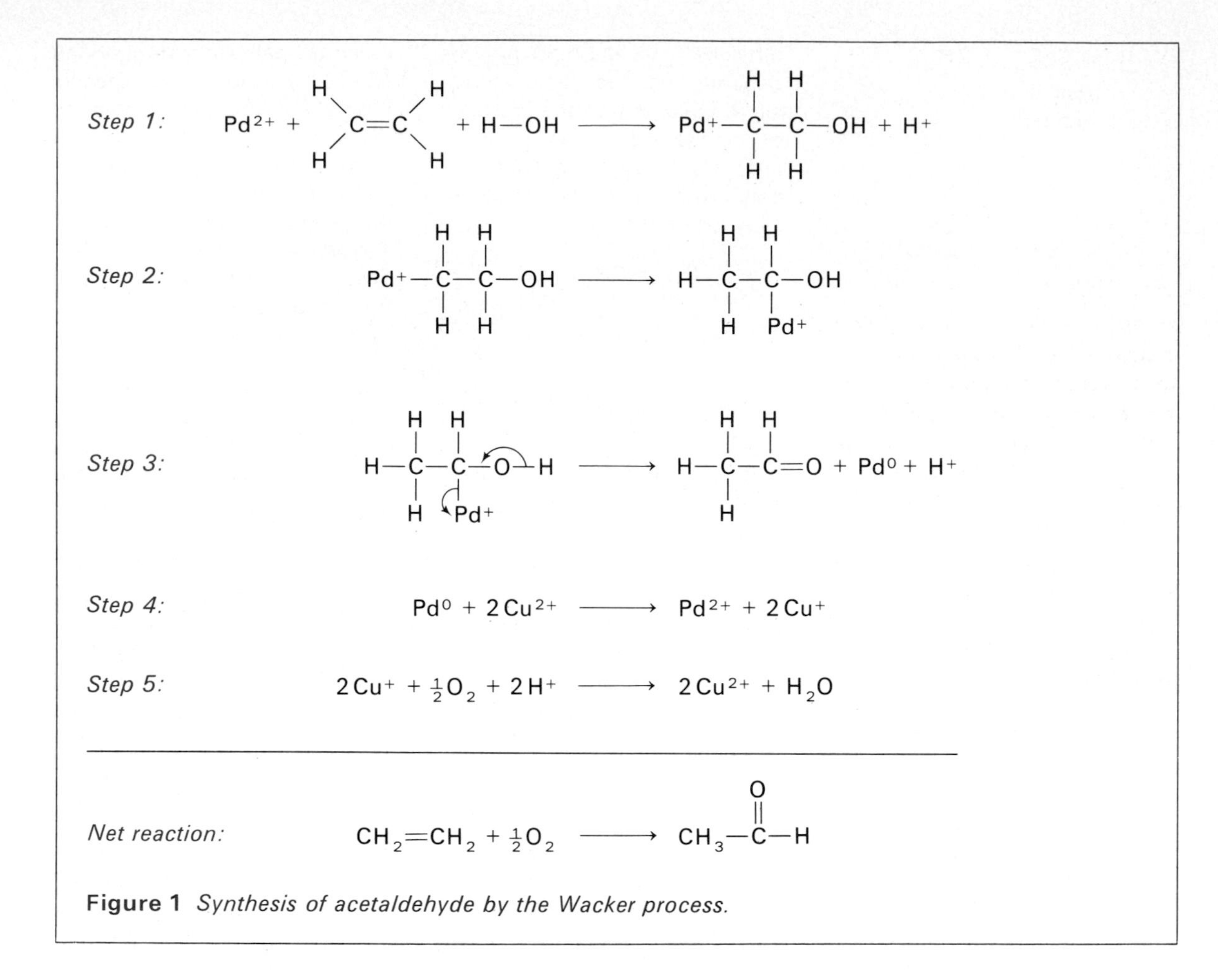

Figure 1 *Synthesis of acetaldehyde by the Wacker process.*

seem both remarkable and mysterious. The BASF process is based on the fact, first observed in 1880, that reaction of sodium methoxide with carbon monoxide gives small amounts of sodium acetate.

$$CH_3{-}O^-Na^+ + CO \longrightarrow CH_3{-}\overset{\overset{\displaystyle O}{\|}}{C}{-}O^-Na^+$$

Yields of sodium acetate, however, were so low that the reaction was of no immediate commercial value. In 1925 Henry Dreyfus of British Celanese began systematic studies of this reaction and in particular sought catalysts that would improve yields of sodium acetate. He discovered that silver and copper salts were effective. However, conditions of temperature, pressure, and alkalinity were so corrosive that few materials withstood them. For a time in the mid-1920s a pilot plant using a gold-lined autoclave as a reaction chamber produced 250 pounds of acetic acid per day, but the sodium methoxide-carbon monoxide process was in no way competitive with other routes to acetic acid. Chemists and chemical engineers next turned to the reaction of carbon monoxide with methanol and in 1965 BASF built a plant to produce acetic acid from CH_3OH and CO in the presence of HI and soluble salts of Co^{2+} and Cu^{2+}. The BASF process operates at 250°C and pressures of up to 700 atmospheres and is referred to as high-pressure carbonylation of methanol. In 1973 the Monsanto Company developed an entirely different process for the conversion of methanol and carbon monoxide to acetic acid based on the observation that these reagents in the presence

In 1973 Monsanto developed a process for the conversion of methanol and carbon monoxide to acetic acid in yields approaching 100%. Shown here is Monsanto's acetic acid plant at Texas City, Texas. (Monsanto Chemical Intermediates Co.)

of small amounts of soluble rhodium (III) salts, HI, and H_2O give acetic acid in yields approaching 100%. Again, catalysts provided the key to development of an economical and technologically feasible process.

$$CH_3OH + CO \xrightarrow{Rh^{3+},\ HI,\ H_2O} CH_3-\overset{O}{\overset{\|}{C}}-OH$$

The Monsanto process operates at 175°C and 15 atmospheres and is referred to as low-pressure carbonylation of methanol. The mechanism of the Monsanto process is complex and in common with the palladium-catalyzed oxidation of ethylene in the Wacker process, involves formation of several different substances containing carbon-metal bonds. Among these are Rh—CH_3, Rh—CO, and Rh—$COCH_3$ bonds. Key features of the Monsanto process are reaction of CH_3OH and HI to form methyl iodide, reaction of methyl iodide and carbon monoxide in the presence of Rh^{3+} to form acetyl iodide, and hydrolysis of acetyl iodide to form acetic acid and regenerate HI.

$$CH_3-OH + HI \longrightarrow CH_3-I + H-OH$$

$$CH_3-I + CO \xrightarrow{Rh^{3+}} CH_3-\overset{O}{\overset{\|}{C}}-I$$

$$CH_3-\overset{O}{\overset{\|}{C}}-I + H-OH \longrightarrow CH_3-\overset{O}{\overset{\|}{C}}-OH + HI$$

Both the high-pressure and low-pressure carbonylations are based on methanol, a substance readily available by catalytic reduction of carbon monoxide.

$$CO + 2H_2 \longrightarrow CH_3-OH$$

Carbon monoxide and hydrogen in turn are available from the reaction of water with methane, coal, various petroleum products, or biological wastes.

$$CH_4 + H_2O \rightleftharpoons CO + 3H_2$$

$$\underset{\text{coal}}{C} + H_2O \rightleftharpoons CO + H_2$$

The mixture of hydrogen and carbon monoxide produced by these reactions is termed "synthesis gas" and each year the quantity of it manufactured is enormous. For example, 70 billion pounds of synthesis gas was produced in the United States in 1979. Thus, with synthesis gas as the feedstock, the raw material base for the synthesis of acetic acid can be shifted from natural gas and petroleum to coal, a substance available in the United States in vast quantities.

In this mini-essay, we have seen that the critical elements on which to base the synthesis of acetic acid are sources of carbon and hydrogen, and of course, energy. Over the last several decades, sources of these elements have changed and through a combination of basic and applied research, the chemical industry has developed new technologies based on different feedstocks. As the cost of energy rose, ethylene replaced acetylene as a source of carbon and hydrogen. Ethylene as a feedstock depends on the availability at attractive prices of natural gas and petroleum. The cost of these raw materials has risen dramatically in recent years and ethylene is now being replaced by synthesis gas, a mixture of carbon monoxide and hydrogen that can be manufactured from a variety of carbon sources, including coal. Synthesis gas is now a major source of methanol and acetic acid and it is likely that it will be a major feedstock for the production of other organics in the decades ahead.

References

Parshall, G. S., "Organometallic Chemistry in Homogeneous Catalysis," *Science* 208 (1980): 1221–1224.

Pruett, R. L., "Synthesis Gas: A Raw Material for Industrial Chemicals," *Science* 211 (1980): 1116.

Wagner, F. S., Jr., "Acetic Acid," *Encyclopedia of Chemical Technology* 1 (N.Y.: Wiley, 1978), 124–147.

Witcoff, H. A. and Reuben, B. G., *Industrial Organic Metals in Perspective*, (N.Y.: Wiley, 1980).

CHAPTER 9

Functional Derivatives of Carboxylic Acids

In this chapter we shall describe the structure and chemical properties of esters, amides, anhydrides, and acid chlorides, all functional derivatives of carboxylic acids in which the —OH of the carboxyl group has been replaced by —OR, $—NH_2$, OCOR, or —Cl.

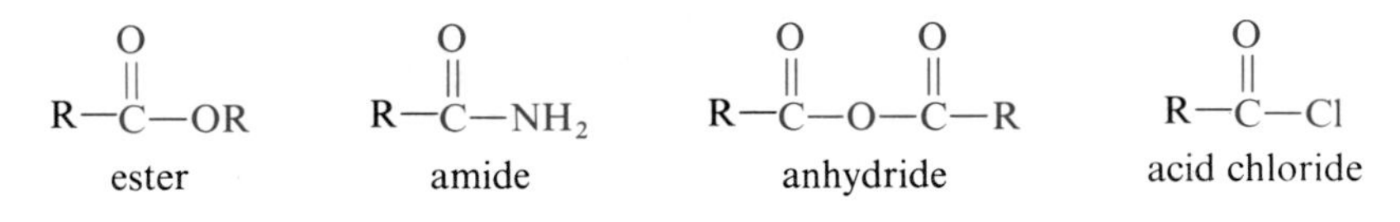

9.1 NOMENCLATURE

In the IUPAC system of nomenclature, esters are named as derivatives of carboxylic acids. The alkyl or aryl group attached to oxygen is named first, followed by the name of the acid from which the ester is derived. The acid is named by dropping the suffix -ic from the IUPAC name or common name of the acid and adding -ate.

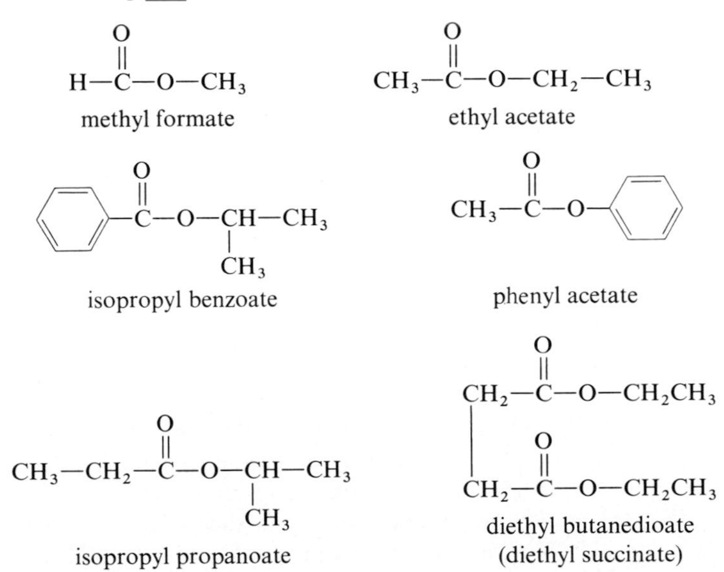

Example 9.1

Name the following esters.

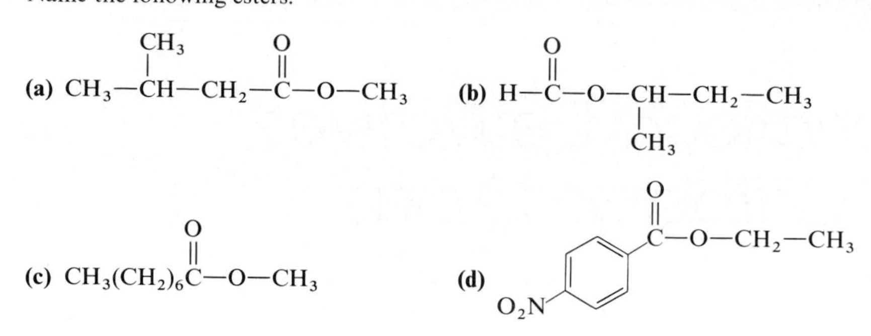

Solution

The names of these esters are derived in a stepwise manner. First the name of the alkyl group attached to oxygen is given; then the IUPAC name of the carboxylic acid from which the ester is derived; and finally, the name of the ester.

alkyl group attached to oxygen	*IUPAC name of carboxylic acid*	*name of ester*
(a) methyl	3-methylbutanoic acid	methyl 3-methylbutanoate
(b) *sec*-butyl	formic acid	*sec*-butyl formate
(c) methyl	octanoic acid	methyl octanoate
(d) ethyl	*p*-nitrobenzoic acid	ethyl *p*-nitrobenzoate

PROBLEM 9.1 Name the following esters.

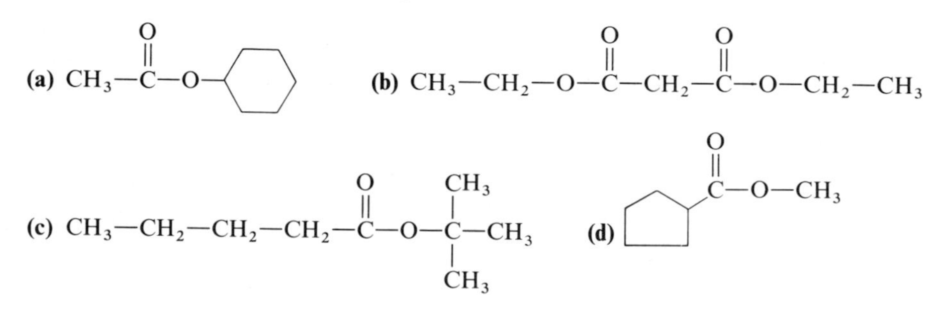

Esters of thiols are named by adding the prefix thio- to the name of the acid to indicate the presence of the sulfur atom.

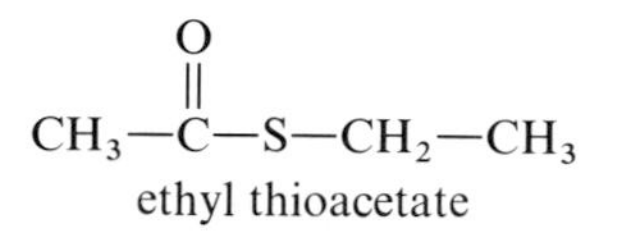

One of the most important thioesters in biological systems is derived from acetic acid and a thiol called coenzyme A. Coenzyme A is a complex biomolecule with the molecular formula $C_{21}H_{36}O_{16}N_7P_3S$. One structural feature of coenzyme A is the presence of a thiol group. For this reason, coenzyme A is sometimes written CoA—SH to emphasize this functional group.

The thioester derived from acetic acid and coenzyme A is called acetyl coenzyme A.

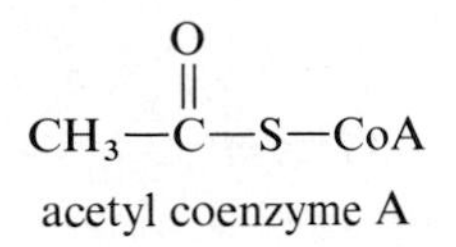

acetyl coenzyme A

Acetyl coenzyme A is an essential intermediate in the metabolism of fatty acids, carbohydrates, and amino acids.

Amides are named as derivatives of carboxylic acids by dropping the suffix -oic from the IUPAC name of the acid, or the suffix -ic from the common name of the acid, and adding -amide.

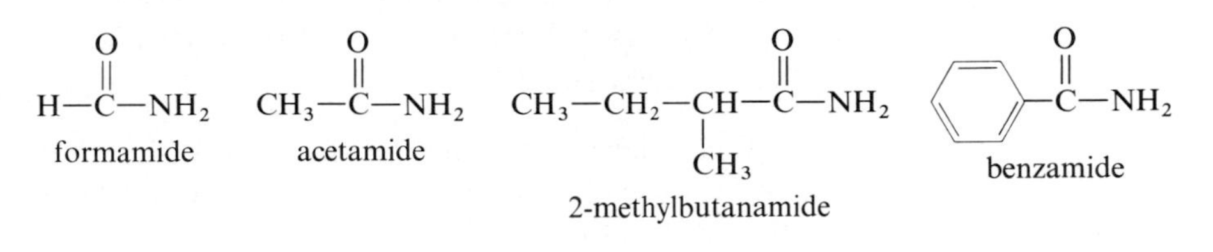

formamide acetamide 2-methylbutanamide benzamide

If the nitrogen atom is substituted with an alkyl or aryl group, the substituent is named and its location on nitrogen is indicated by a capital N.

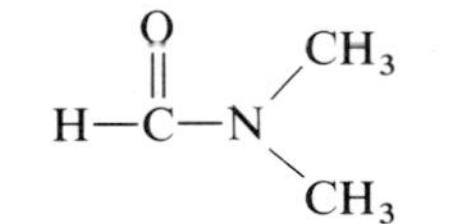

N,N-dimethylformamide

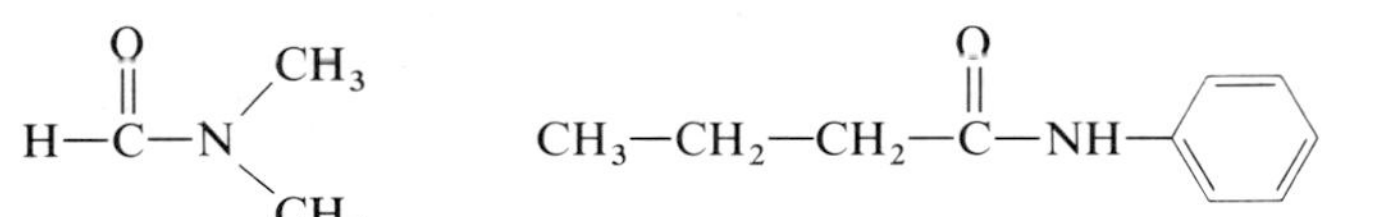

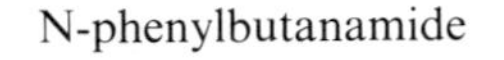

N-phenylbutanamide

Example 9.2 Name the following amides.

(a) $CH_3-C(=O)-NH-CH_3$ (b) $(H_3C)_2CH-CH_2-C(=O)-NH_2$ (c) $CH_3-C(CH_3)_2-C(=O)-NH_2$

Solution The names of these amides are derived in a stepwise manner just as we derived those of the esters in Example 9.1. First is listed the name of the alkyl or aryl group (if any) attached to nitrogen, then the name of the carboxylic acid from which the amide is derived; and finally, the name of the amide.

alkyl or aryl group attached to nitrogen	*IUPAC name of carboxylic acid*	*name of amide*
(a) N-methyl	acetic acid	N-methylacetamide
(b) ——	3-methylbutanoic acid	3-methylbutanamide
(c) ——	2,2-dimethylpropanoic acid	2,2-dimethylpropanamide

PROBLEM 9.2 Name the following amides.

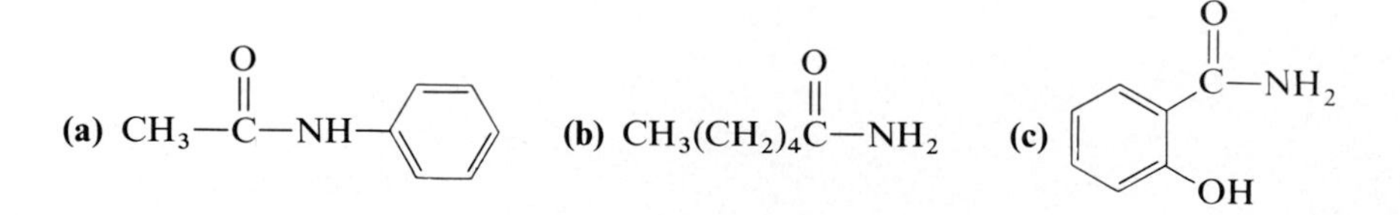

Anhydrides are named by adding the word anhydride to the name of the acid from which it is derived. For our purposes, the most important organic anhydride is acetic anhydride.

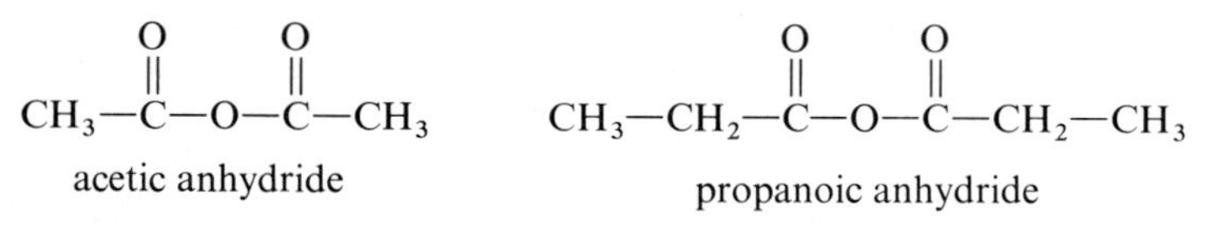

acetic anhydride propanoic anhydride

Acid halides are named as derivatives of carboxylic acids by dropping the suffix -ic acid from the IUPAC or common name and adding -yl followed by the name of the halogen atom.

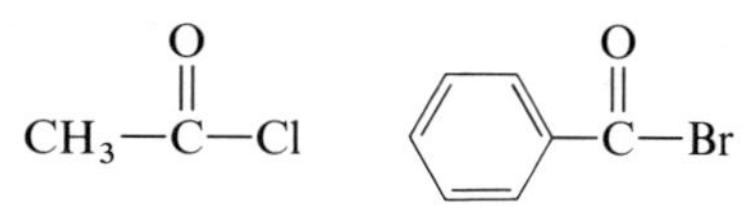

acetyl chloride benzoyl bromide

9.2 ESTERS, ANHYDRIDES, AND AMIDES OF INORGANIC ACIDS

In esters of carboxylic acids, the —OH of the carboxylate group is replaced by —OR. Inorganic acids form esters in which the —OH of the acid is replaced by —OR.

Phosphoric acid, H_3PO_4, has three —OH groups and can form mono-, di-, and triesters. Following are examples of each.

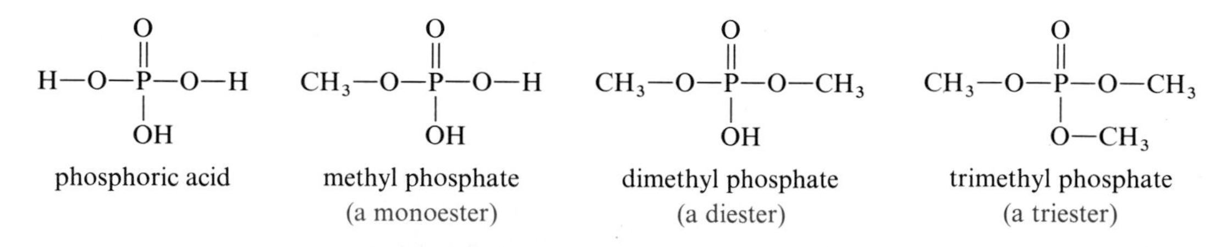

phosphoric acid; methyl phosphate (a monoester); dimethyl phosphate (a diester); trimethyl phosphate (a triester)

Esters of phosphoric acid are especially important in biological chemistry because many organic molecules, for example glyceraldehyde and dihydroxyacetone, can be metabolized only in the phosphorylated state. As another example, vitamin B_6 or pyridoxal, is metabolically active only after it is phosphorylated to pyridoxal phosphate.

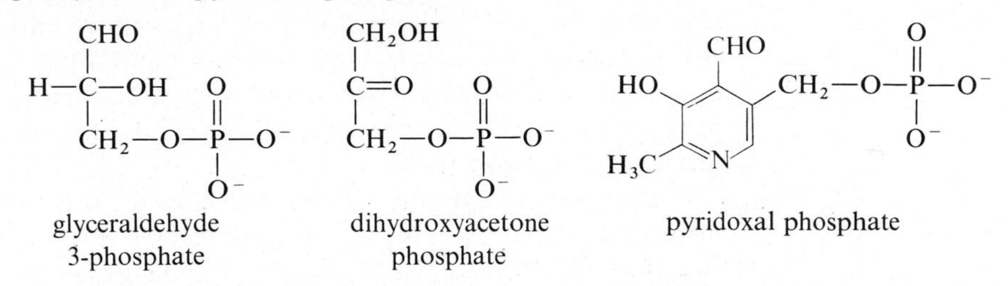

When drawing structural formulas for phosphate esters and anhydrides found in living systems, we will show the state of ionization at pH 7.4. Phosphoric acid is a triprotic acid, and at pH 7.4 two protons of H_3PO_4 are completely ionized. Glyceraldehyde 3-phosphate is a monoester of phosphoric acid, and at pH 7.4 this substance has a net charge of −2.

Phosphoric acid also forms anhydrides in which the characteristic structural feature is the presence of a —P(=O)(OH)—O—P(=O)(OH)— group. Following is the structural formula of phosphoric acid anhydride. The common name for this substance is pyrophosphoric acid.

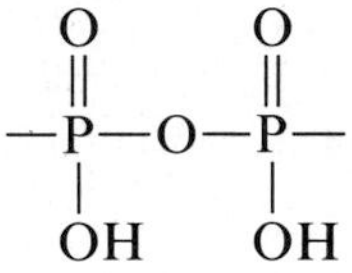

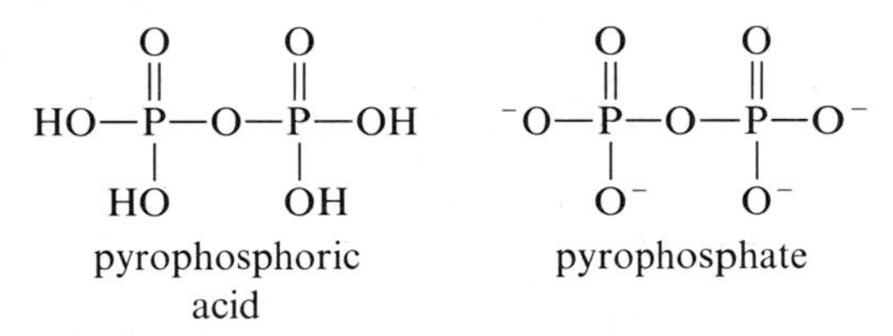

At pH 7.4, pyrophosphoric acid has a net charge of −4. The anion is named pyrophosphate.

Several organic esters of nitric acid and nitrous acid have been used as drugs for more than 100 years. Two of these are glycerol trinitrate (more commonly known as nitroglycerine) and 3-methybutyl nitrite (more commonly known as isopentyl nitrite or isoamyl nitrite).

H—O—NO_2
nitric acid

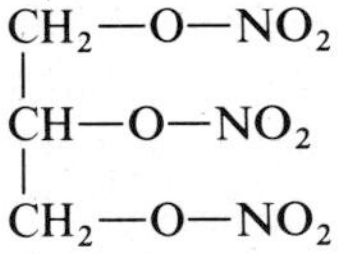

CH_2—O—NO_2
|
CH—O—NO_2
|
CH_2—O—NO_2
glyceryl trinitrate
(nitroglycerine)

H—O—NO
nitrous acid

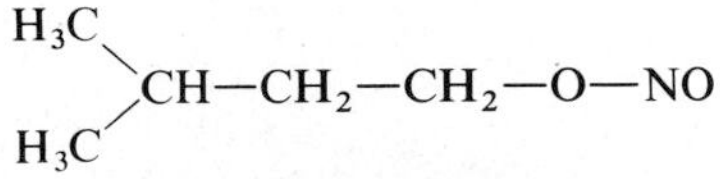

$(H_3C)_2$CH—CH_2—CH_2—O—NO
3-methylbutyl nitrite
(isopentyl nitrite)
(isoamyl nitrite)

Both nitroglycerine and isoamyl nitrite produce rapid relaxation of most smooth muscles of the body. Medically, their more important action is relaxation of the smooth muscle of blood vessels and dilation of all large and small arteries of the heart, for which reason they are called vasodilators. Both esters are used extensively for the treatment of angina pectoris, a heart disease characterized by agonizing pain.

Phosphoric acid and sulfuric acid also form amides in which an —OH group is replaced by $—NH_2$ or a derivative of $—NH_2$. Following is the structural formula of creatine phosphate, an amide of phosphoric acid.

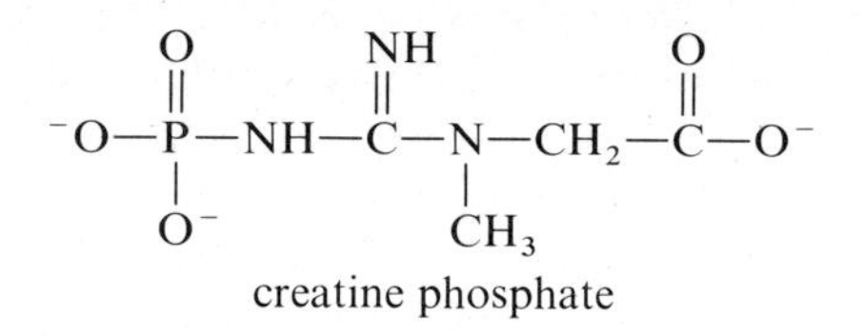

creatine phosphate

Creatine phosphate is found in muscle tissue and is a source of energy to drive muscle contraction.

Sulfonic acids also form amides in which the —OH is replaced by $—NH_2$. Following are structural formulas of benzene sulfonic acid, benzenesulfonamide, and *p*-aminobenzenesulfonamide.

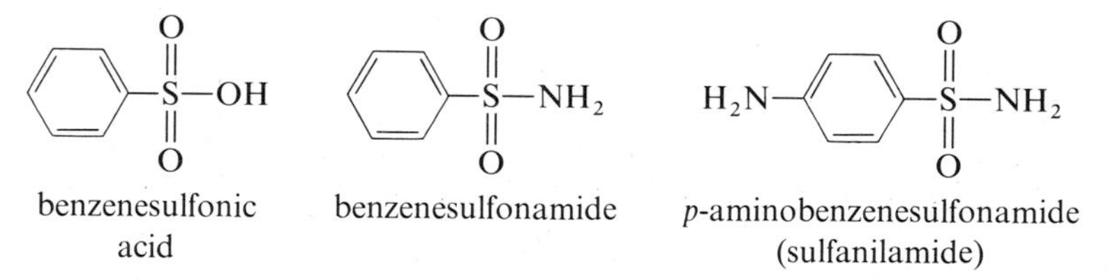

benzenesulfonic acid   benzenesulfonamide   *p*-aminobenzenesulfonamide (sulfanilamide)

The discovery of the medicinal use of sulfanilamide and its derivatives was a milestone in the history of chemotherapy, because it represents one of the first rational investigations of synthetic organic molecules as potential drugs to fight infection. Sulfanilamide was first prepared in 1908 in Germany, but it was not until 1932 that its possible therapeutic value was realized. In that year, the dye Protonsil was prepared and in research over the next two years, the German scientist G. Domagk observed that mice with streptococcal septicemia (an infection of the blood) could be treated with Protonsil. He also observed the remarkable effectiveness of Protonsil in curing experimental streptococcal and staphylococcal infections in other experimental animals. Domagk further discovered that Protonsil is rapidly reduced in the cell to sulfanilamide, and that it is sulfanilamide and not Protonsil that is the actual antibiotic. His discoveries were honored in 1939 by the Nobel Prize in Medicine.

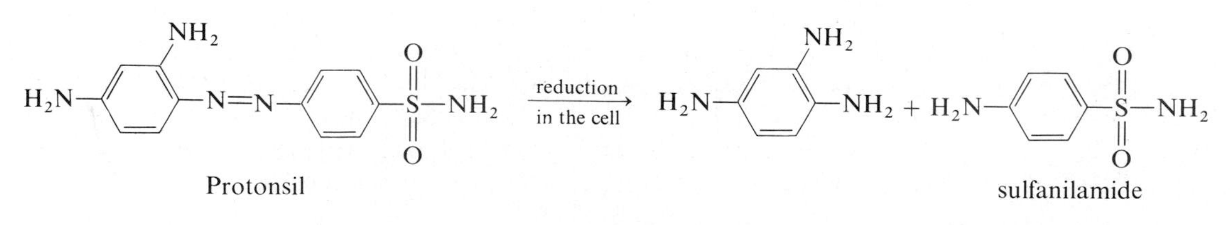

Protonsil   sulfanilamide

The key to understanding the action of sulfanilamide came in 1940 with the observation that the inhibition of bacterial growth caused by sulfanilamide can be reversed by adding large amounts of *p*-aminobenzoic acid (PABA). From this experiment, it was recognized that *p*-aminobenzoic acid is a growth factor for certain bacteria, and that in some way not then understood, sulfanilamide interferes with the bacteria's ability to use PABA.

There are obvious structural similarities between *p*-aminobenzoic acid and sulfanilamide.

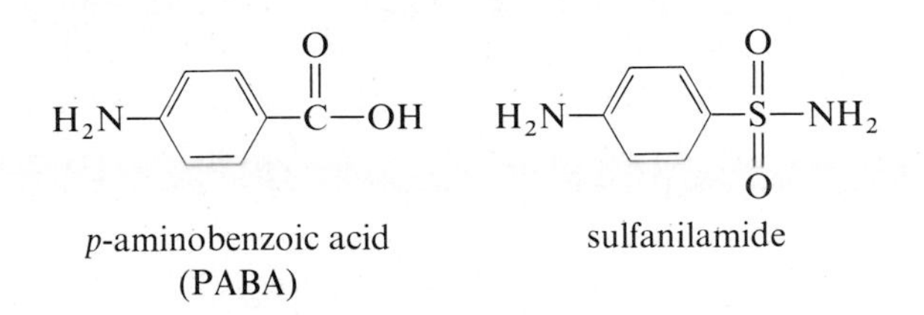

p-aminobenzoic acid (PABA)   sulfanilamide

It now appears that sulfanilamide drugs inhibit one or more enzyme-catalyzed steps in the synthesis of folic acid from *p*-aminobenzoic acid. The ability of sulfanilamide to combat infections in humans depends on the fact that humans also require folic acid, but do not make it from *p*-aminobenzoic acid. For humans, folic acid is a vitamin and must be supplied in the diet.

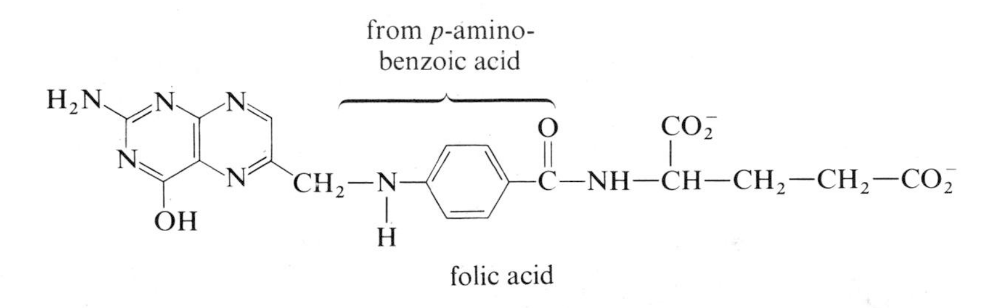

folic acid

In the search for even better sulfa drugs, literally thousands of derivatives of sulfanilamide have been synthesized. Two of the more effective sulfa drugs are

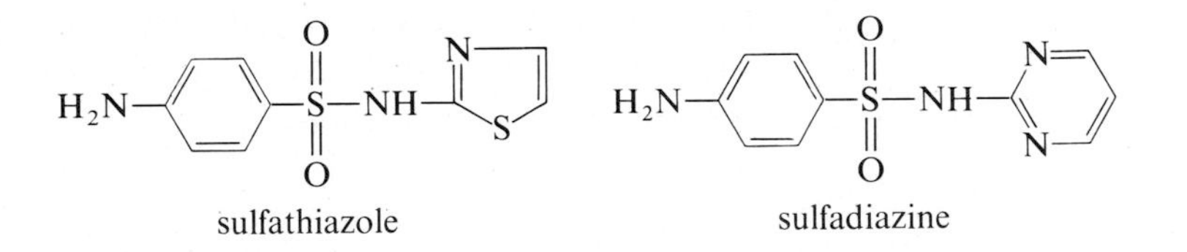

sulfathiazole   sulfadiazine

Sulfa drugs were found to be effective in the treatment of tuberculosis, pneumonia, and diphtheria, and they helped usher in a new era in public health in the United States in the 1930s. During World War II, they were routinely sprinkled on wounds to prevent infection. These drugs were among the first of the new "wonder drugs." As a footnote in history, the use of sulfa drugs to fight bacterial infection has been largely supplanted by an even newer wonder drug, the penicillins.

9.3 PHYSICAL PROPERTIES OF ESTERS AND AMIDES

Esters are polar substances, and are attracted to each other in the pure state by a combination of dipole-dipole interactions between polar $—CO_2—$ groups and by dispersion forces between nonpolar hydrocarbon groups. Most esters are insoluble in water due to the hydrophobic character of the hydrocarbon portion of the molecule. Esters are soluble in polar organic solvents such as ether and acetone.

The low-molecular-weight esters have rather pleasant odors. The characteristic fragrances of many flowers and fruits are in many instances due to the presence of esters, either singly or in mixtures. Some of the more familiar esters are ethyl formate (artificial rum flavor), methyl butanoate (apples), octyl acetate (oranges), and ethyl butanoate (pineapples). Artificial fruit flavors are made largely from mixtures of low-molecular-weight esters.

Amides are polar substances; and because of the polar character of the C=O and N—H bonds, there is the possibility for hydrogen bonding between a partially positive hydrogen atom of one amide group and a partially negative oxygen atom of another amide.

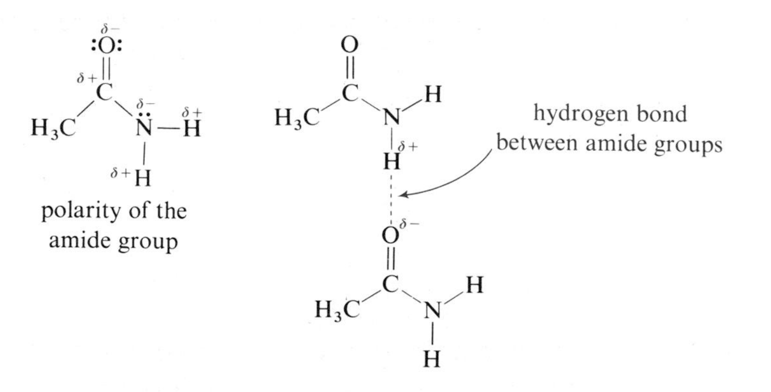

Because of this polarity and association by hydrogen bonding, amides have higher boiling points and are more soluble in water than esters of comparable molecular weight. Virtually all amides are solids at room temperature. However, formamide, the amide of lowest molecular weight, has a melting point of 3° and is a liquid at room temperature.

9.4 NUCLEOPHILIC SUBSTITUTION AT AN UNSATURATED CARBON

The basic reaction theme common to the carbonyl group of aldehydes, ketones, carboxylic acids, esters, amides, anhydrides, and acid halides is nucleophilic addition to the carbonyl group. In the case of aldehydes and ketones, the carbonyl addition product is sometimes isolated as such. For example, in aldol condensations (Section 7.5F), the carbonyl group undergoing reaction is transformed into an alcohol, and this addition product is the final product of the reaction.

$$CH_3\overset{O}{\overset{\|}{C}}H + CH_3\overset{O}{\overset{\|}{C}}H \xrightarrow{\text{base}} CH_3\overset{OH}{\overset{|}{C}}HCH_2\overset{O}{\overset{\|}{C}}H$$

In other reactions of aldehydes and ketones, the carbonyl addition product is formed but then undergoes loss of H_2O or other small molecule to yield a new functional group. For example, the reaction of an aldehyde or ketone with a primary amine forms a carbonyl addition product that then loses a molecule of H_2O to form a Schiff base; in this reaction, a $>C{=}O$ group is transformed into a $C{=}N{-}$ group.

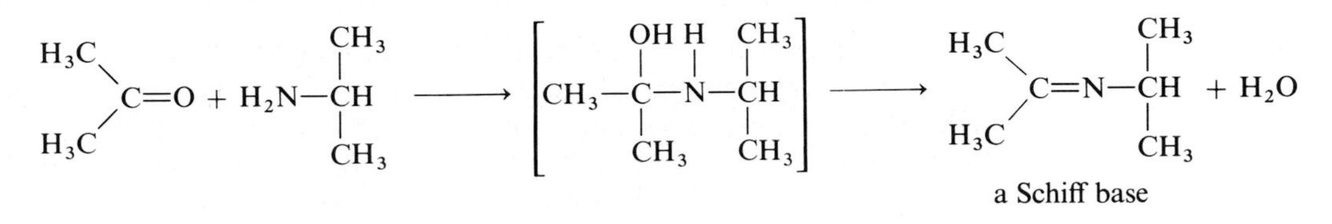

a Schiff base

With the new functional groups to be studied in this chapter, the carbonyl addition product collapses to regenerate the carbonyl group.

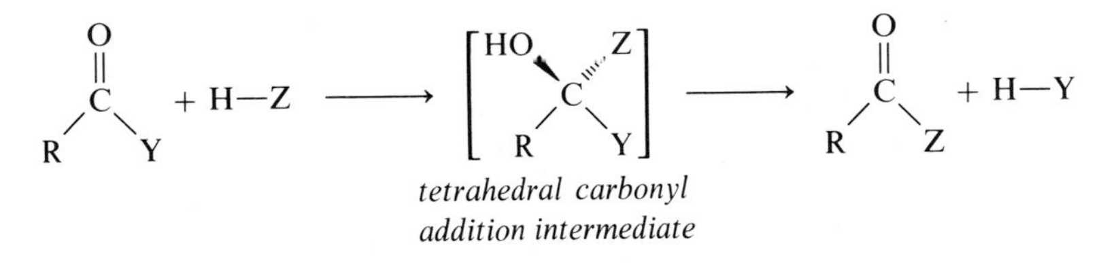

tetrahedral carbonyl addition intermediate

The effect of this reaction is to substitute a new atom or group of atoms for one already attached to the carbonyl group. For this reason, we characterize these reactions of the carbonyl group as nucleophilic substitution at an unsaturated carbon.

9.5 PREPARATION OF ESTERS

A carboxylic acid can be converted into an ester by heating with an alcohol in the presence of an acid catalyst, usually dry hydrogen chloride, concentrated sulfuric acid, or an ion-exchange resin in the hydrogen ion form.

Direct esterification of alcohols and acids in this manner is called Fischer esterification. As an example, reaction of acetic acid and ethanol in the presence of concentrated sulfuric acid produces ethyl acetate and water.

$$\underset{\text{acetic acid}}{CH_3{-}\overset{O}{\overset{\|}{C}}{-}OH} + \underset{\text{ethanol}}{HO{-}CH_2{-}CH_3} \overset{H^+}{\rightleftharpoons} \underset{\text{ethyl acetate}}{CH_3{-}\overset{O}{\overset{\|}{C}}{-}O{-}CH_2{-}CH_3} + H{-}O{-}H$$

Acid-catalyzed esterification is reversible, and generally at equilibrium there are appreciable quantities of both ester and alcohol present. If 60.0 g (one mole) of acetic acid and 60.0 g (one mole) of 1-propanol are refluxed in the presence of a few drops of concentrated sulfuric acid, the reaction mixture at equilibrium contains about 68.0 g (0.67 mole) of propyl acetate, 12.0 g (0.67 mole) of water, and 20.0 g (0.33 mole) each of acetic acid and 1-propanol. At equilibrium there is about 67% conversion of acid and alcohol into ester.

$$CH_3\overset{O}{\overset{\|}{C}}-OH + HOCH_2CH_2CH_3 \overset{H^+}{\rightleftharpoons} CH_3\overset{O}{\overset{\|}{C}}-OCH_2CH_2CH_3 + H_2O$$

initial:	1.00 mole	1.00 mole	0.00 mole	0.00 mole
equilibrium:	0.33 mole	0.33 mole	0.67 mole	0.67 mole

Direct esterification can be used to prepare esters in high yields by careful control of reaction conditions. For example, if the alcohol is inexpensive, a large excess of it can be used to drive the reaction to the right and achieve a high conversion of acid into ester. Or it may be possible to take advantage of a situation in which the boiling points of the reactants and ester are higher than that of water. In this case heating the reaction mixture somewhat above 100°C removes water as it is formed and shifts the equilibrium toward the production of higher yields of ester.

Fischer esterification is but one of the general methods of preparing esters. From the standpoint of the organic chemist interested in the laboratory synthesis of esters, other important methods are the reaction of hydroxyl compounds (alcohols, phenols) with acid anhydrides and acid halides. We shall examine these methods of ester preparation in Sections 9.8 and 9.10.

Example 9.3 Name and draw structural formulas for the esters produced in the following reactions.

(a) $HCO_2H + CH_3CH_2OH \xrightarrow{H^+}$

(b) $C_6H_4(CO_2H)(OH)$ (ortho) $+ CH_3OH \xrightarrow{H^+}$

(c) $HO\overset{O}{\overset{\|}{C}}CH_2CH_2CH_2\overset{O}{\overset{\|}{C}}OH + 2CH_3CH_2OH \xrightarrow{H^+}$

Solution Ester formation follows the general reaction shown in Section 9.4.

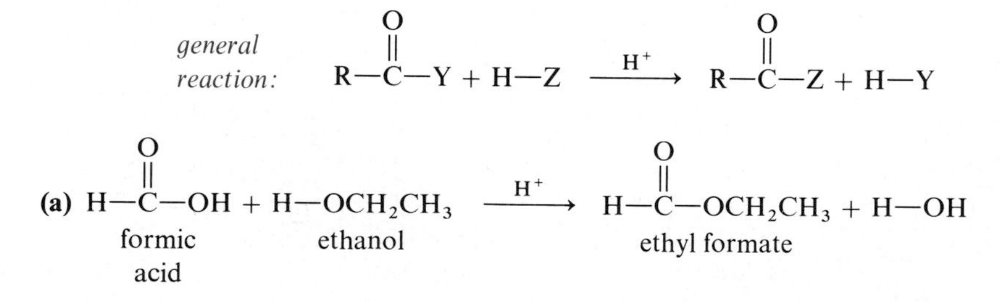

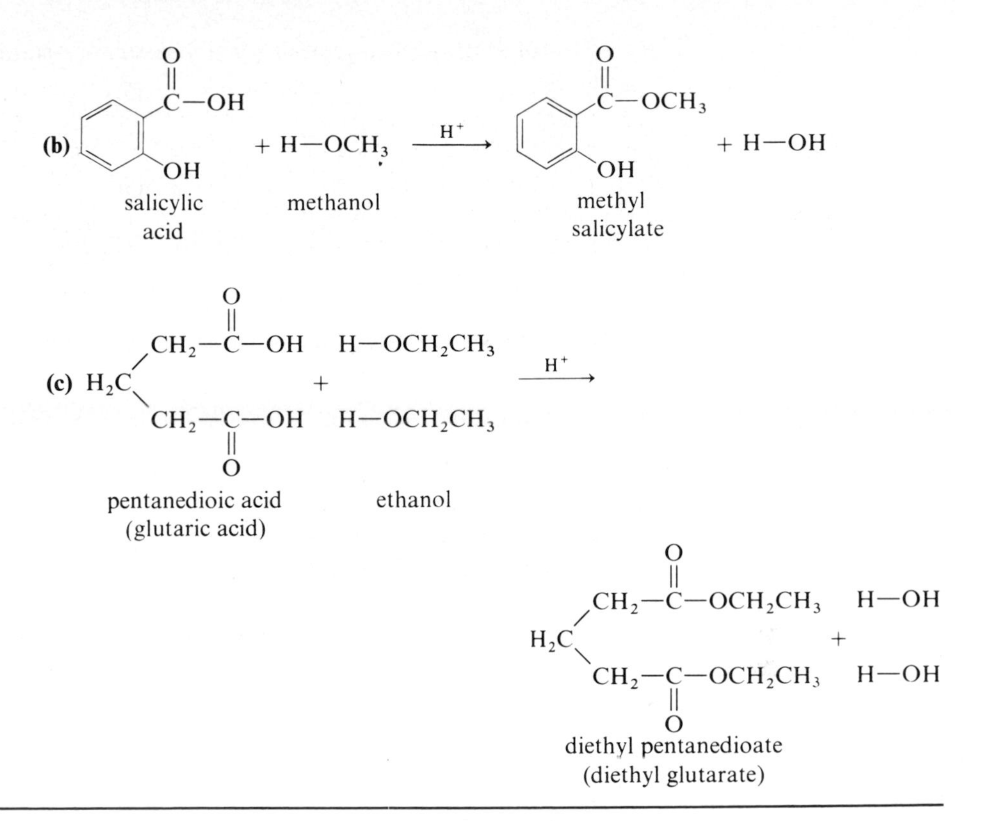

PROBLEM 9.3 Name and draw structural formulas for the esters produced in the following reactions.

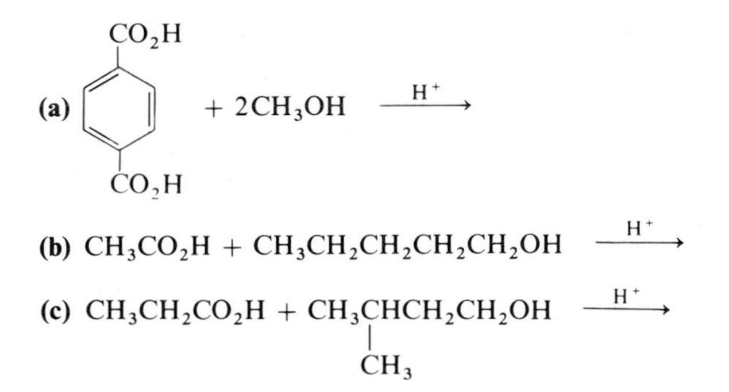

(b) $CH_3CO_2H + CH_3CH_2CH_2CH_2CH_2OH \xrightarrow{H^+}$

(c) $CH_3CH_2CO_2H + CH_3\underset{\displaystyle CH_3}{\underset{|}{C}}HCH_2CH_2OH \xrightarrow{H^+}$

Following is a mechanism for acid-catalyzed esterification. Protonation of the carbonyl oxygen in Step 1 gives a resonance-stabilized oxonium ion in which the carbonyl carbon bears an increased positive charge and is more susceptible to attack by nucleophiles. Addition of alcohol in Step 2 followed by loss of a proton in Step 3 gives a tetrahedral carbonyl addition intermediate. Protonation of this intermediate in Step 4 followed by loss of H_2O in Step 5 gives a second resonance-stabilized cation. Loss of a proton in Step 6 gives the ester.

Step 1: Protonation of oxygen to form a resonance-stabilized cation.

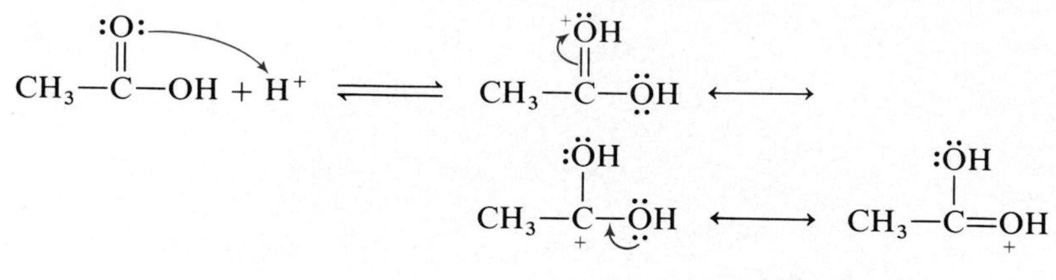
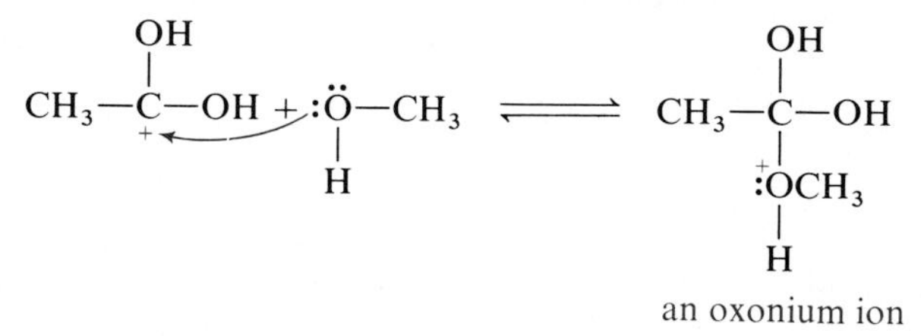
resonance-stabilized cation

Step 2: Addition of nucleophile to the carbonyl carbon to form an oxonium ion.

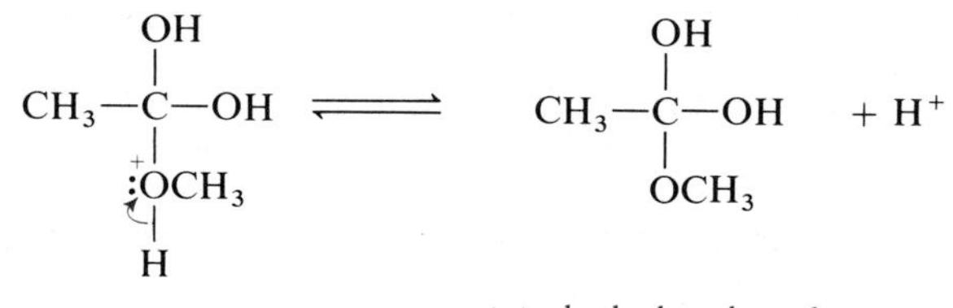
an oxonium ion

Step 3: Loss of H^+ to form a tetrahedral carbonyl addition intermediate.

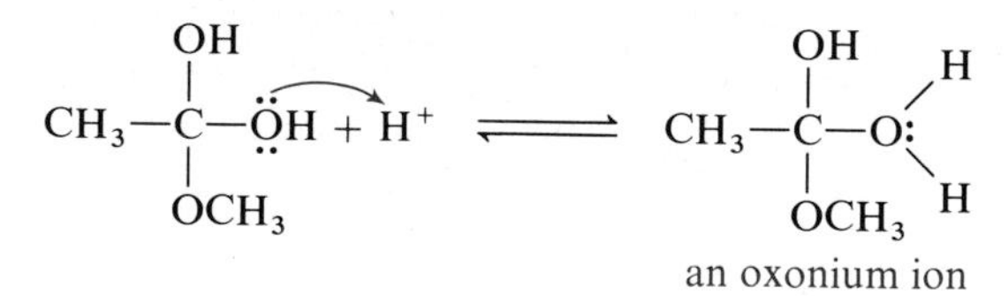
tetrahedral carbonyl addition intermediate

Step 4: Protonation of oxygen to form an oxonium ion.

an oxonium ion

Step 5: Loss of H_2O to form a resonance-stabilized cation.

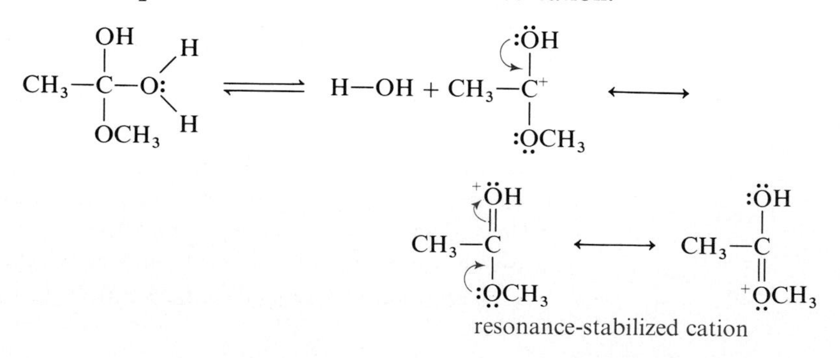
resonance-stabilized cation

Step 6: Loss of H^+ to form the ester.

$$CH_3-\overset{\overset{+}{:OH}}{\overset{\|}{C}}-OCH_3 \rightleftharpoons CH_3-\overset{\overset{O}{\|}}{C}-OCH_3 + H^+$$

The mechanism for Fischer esterification is a specific example of the general mechanism shown in the previous section for nucleophilic substitution at a carbonyl carbon. A key step is formation of a tetrahedral carbonyl addition intermediate.

$$CH_3-\overset{\overset{O}{\|}}{C}-OH + HO-CH_3 \rightleftharpoons \left[CH_3-\underset{\underset{OH}{|}}{\overset{\overset{OH}{|}}{C}}-O-CH_3 \right]$$

tetrahedral carbonyl addition intermediate

$$\rightleftharpoons CH_3-\overset{\overset{O}{\|}}{C}-O-CH_3 + H-OH$$

The mechanism we have proposed for acid-catalyzed esterification predicts that the molecule of water formed during the reaction is derived from the —OH of the carboxylic acid and the —H of the alcohol. This prediction has been tested in the following way. Oxygen in nature is a mixture of three isotopes: 99.7% ^{16}O, 0.04% ^{17}O, and 0.20% ^{18}O. Through the use of modern techniques of separating isotopes, it is possible to prepare oxygen-containing compounds significantly enriched in oxygen-18. One of these is $CH_3{}^{18}OH$. When methanol enriched with oxygen-18 reacts with acetic acid containing only ordinary oxygen, all of the oxygen-18 is found in the ester. The water formed in the reaction contains no oxygen-18. This fact is consistent with the mechanism we have proposed.

$$CH_3\overset{\overset{O}{\|}}{C}-OH + H-{}^{18}OCH_3 \longrightarrow CH_3-\overset{\overset{O}{\|}}{C}-{}^{18}OCH_3 + H_2O$$

Example 9.4

Draw structural formulas for the tetrahedral carbonyl addition intermediates formed in the following acid-catalyzed esterifications.

(a) $$CH_3CH_2\overset{\overset{O}{\|}}{C}OH + HOCH_2CH_3 \xrightarrow{H^+} CH_3CH_2\overset{\overset{O}{\|}}{C}OCH_2CH_3 + H_2O$$

(b) $$C_6H_5-\overset{\overset{O}{\|}}{C}OH + HOCH_3 \xrightarrow{H^+} C_6H_5-\overset{\overset{O}{\|}}{C}OCH_3 + H_2O$$

Solution In the answers, the starting materials are drawn to emphasize that the oxygen atom of the —OH group of the alcohol adds to the carbon atom of the C=O group of the carboxylic acid.

(a) $$CH_3{-}CH_2{-}\overset{\overset{\displaystyle O}{\|}}{\underset{\underset{\displaystyle OH}{|}}{C}} + \overset{\overset{\displaystyle H}{|}}{O}{-}CH_2CH_3 \longrightarrow \left[CH_3{-}CH_2{-}\overset{\overset{\displaystyle O{-}H}{|}}{\underset{\underset{\displaystyle OH}{|}}{C}}{-}O{-}CH_2CH_3 \right]$$

(b) $$C_6H_5{-}\overset{\overset{\displaystyle O}{\|}}{\underset{\underset{\displaystyle OH}{|}}{C}} + \overset{\overset{\displaystyle H}{|}}{O}{-}CH_3 \longrightarrow \left[C_6H_5{-}\overset{\overset{\displaystyle O{-}H}{|}}{\underset{\underset{\displaystyle OH}{|}}{C}}{-}O{-}CH_3 \right]$$

PROBLEM 9.4 Draw structural formulas for the tetrahedral carbonyl addition intermediates formed in the following acid-catalyzed esterification reactions.

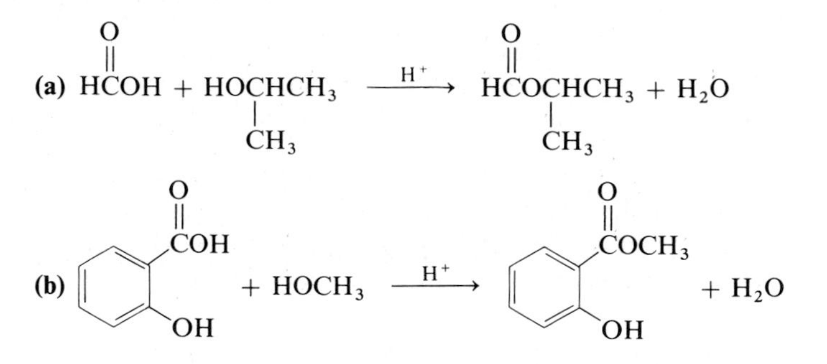

9.6 REACTIONS OF ESTERS

A. REDUCTION

Esters can be reduced by hydrogen in the presence of a catalyst, but as in the case of carboxylic acids, catalytic reduction of esters requires high temperatures and pressures of hydrogen. However, esters are reduced smoothly at room temperature by lithium aluminum hydride. Reduction of an ester gives two alcohols, one derived from the carboxylic acid portion of the ester, the other derived from the alkyl group on oxygen.

$$\underset{\text{diethyl adipate}}{CH_3CH_2O\overset{\overset{\displaystyle O}{\|}}{C}(CH_2)_4\overset{\overset{\displaystyle O}{\|}}{C}OCH_2CH_3} + 4H_2 \xrightarrow[\text{heat, pressure}]{\text{catalyst}}$$

$$\underset{\text{1,6-hexanediol}}{HOCH_2(CH_2)_4CH_2OH} + \underset{\text{ethanol}}{2CH_3CH_2OH}$$

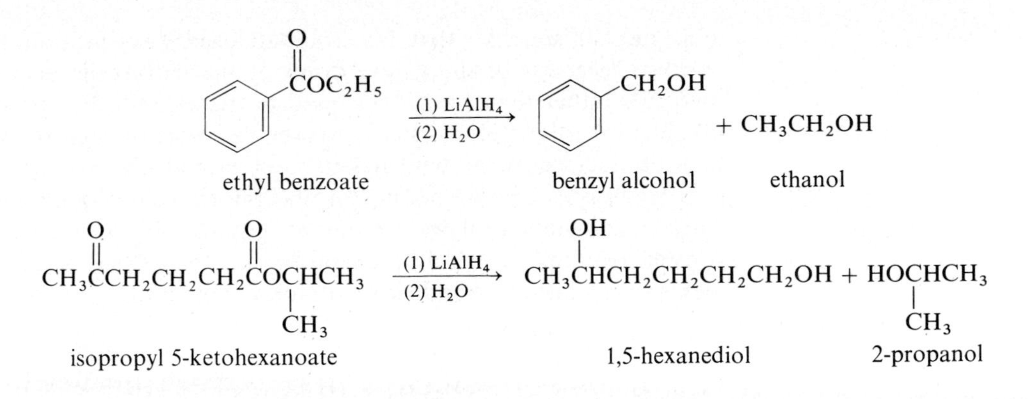

B. HYDROLYSIS

Esters of carboxylic acids, as well as those of inorganic acids such as phosphoric acid, are converted to the corresponding acids and alcohols by hydrolysis in either aqueous acid or base.

$$CH_3-\overset{\displaystyle O}{\overset{\|}{C}}-O-CH_2CH_3 + H-O-H \underset{}{\overset{H^+}{\rightleftharpoons}} CH_3-\overset{\displaystyle O}{\overset{\|}{C}}-OH + HO-CH_2CH_3$$

Since the mechanism proposed in Section 9.5 for acid-catalyzed esterification is reversible, the formation of the same tetrahedral carbonyl addition intermediate accounts equally well for acid-catalyzed hydrolysis.

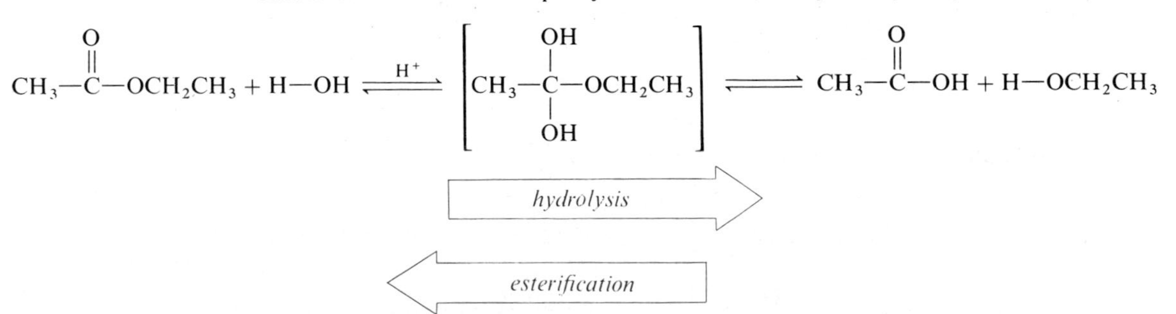

By carrying out hydrolysis in a large excess of water, the position of the equilibrium is shifted to favor the formation of carboxylic acid and alcohol.

In alkaline hydrolysis of esters, hydroxide ion adds to the carbonyl carbon to form a tetrahedral carbonyl addition intermediate. This intermediate then eliminates a molecule of alcohol to form the carboxylate anion, $RCOO^-$.

$$CH_3-\overset{\displaystyle O}{\overset{\|}{C}}-OCH_2CH_3 + OH^- \rightleftharpoons \left[CH_3-\underset{\displaystyle OCH_2CH_3}{\underset{|}{\overset{\displaystyle O^-}{\overset{|}{C}}}}-OH \right] \longrightarrow CH_3-\overset{\displaystyle O}{\overset{\|}{C}}-O^- + HOCH_2CH_3$$

tetrahedral carbonyl addition intermediate

This mechanism, like that for acid-catalyzed esterification and hydrolysis, involves cleavage of the C—O bond of the carboxylic acid portion of the molecule rather than the C—O bond of the alcohol. For practical purposes, alkaline hydrolysis of an ester is irreversible because a carboxylate anion, the final product, shows no tendency to react with alcohol.

Hydrolysis of a monoester of phosphoric acid requires one molecule of water, while the hydrolysis of a diester requires $2H_2O$, and the hydrolysis of a triester requires $3H_2O$. For example, the hydrolysis of diethyl phosphate yields phosphoric acid and two molecules of ethanol.

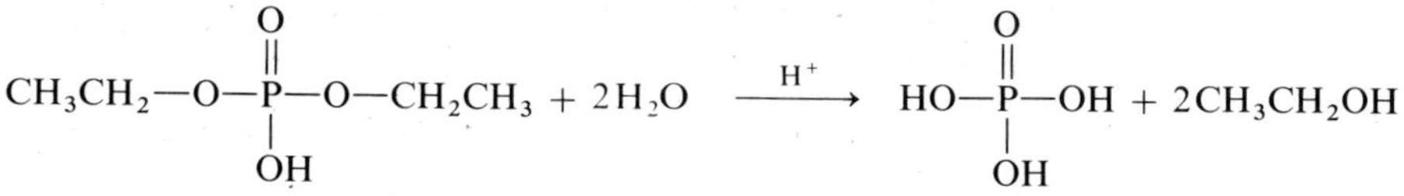

$$\underset{\text{diethyl phosphate}}{CH_3CH_2{-}O{-}\overset{O}{\overset{\|}{\underset{OH}{\underset{|}{P}}}}{-}O{-}CH_2CH_3} + 2H_2O \xrightarrow{H^+} HO{-}\overset{O}{\overset{\|}{\underset{OH}{\underset{|}{P}}}}{-}OH + 2CH_3CH_2OH$$

Example 9.5

Write equations for hydrolysis of the following esters.

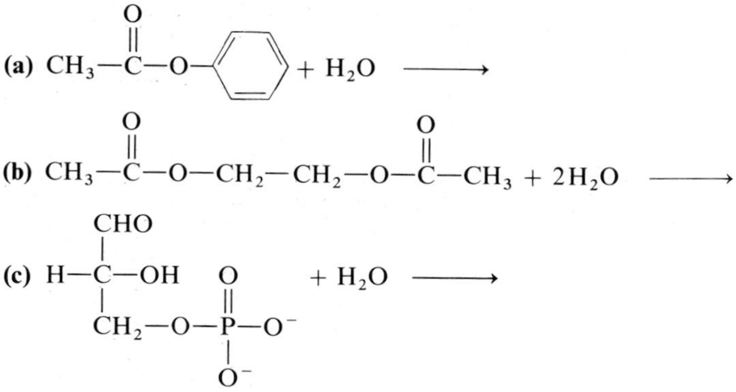

(a) $CH_3{-}\overset{O}{\overset{\|}{C}}{-}O{-}C_6H_5 + H_2O \longrightarrow$

(b) $CH_3{-}\overset{O}{\overset{\|}{C}}{-}O{-}CH_2{-}CH_2{-}O{-}\overset{O}{\overset{\|}{C}}{-}CH_3 + 2H_2O \longrightarrow$

(c) $H{-}\overset{CHO}{\overset{|}{\underset{CH_2{-}O{-}\overset{O}{\overset{\|}{\underset{O^-}{\underset{|}{P}}}}{-}O^-}{\underset{|}{C}}}}{-}OH + H_2O \longrightarrow$

Solution

(a) $CH_3{-}\overset{O}{\overset{\|}{C}}{-}O{-}C_6H_5 + H_2O \longrightarrow CH_3{-}\overset{O}{\overset{\|}{C}}{-}OH + HO{-}C_6H_5$

(b) $CH_3{-}\overset{O}{\overset{\|}{C}}{-}O{-}CH_2{-}CH_2{-}O{-}\overset{O}{\overset{\|}{C}}{-}CH_3 + 2H_2O \longrightarrow$

$$2CH_3{-}\overset{O}{\overset{\|}{C}}{-}OH + HO{-}CH_2{-}CH_2{-}OH$$

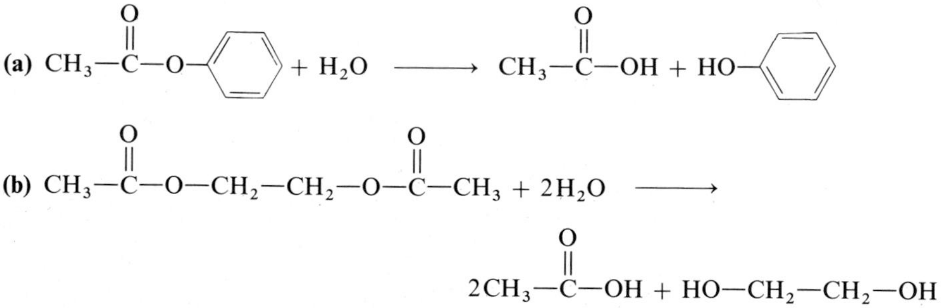

This substance is a diester, and on hydrolysis yields two molecules of acetic acid and one of ethylene glycol.

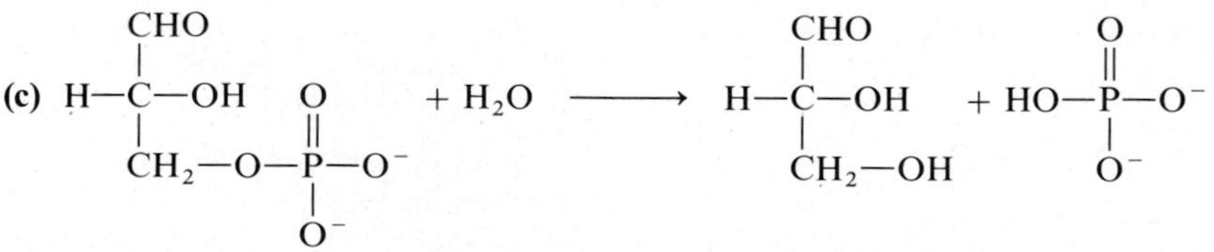

(c) $H{-}\overset{CHO}{\overset{|}{\underset{CH_2{-}O{-}\overset{O}{\overset{\|}{\underset{O^-}{\underset{|}{P}}}}{-}O^-}{\underset{|}{C}}}}{-}OH + H_2O \longrightarrow H{-}\overset{CHO}{\overset{|}{\underset{CH_2{-}OH}{\underset{|}{C}}}}{-}OH + HO{-}\overset{O}{\overset{\|}{\underset{O^-}{\underset{|}{P}}}}{-}O^-$

PROBLEM 9.5 Write equations for hydrolysis of the following esters.

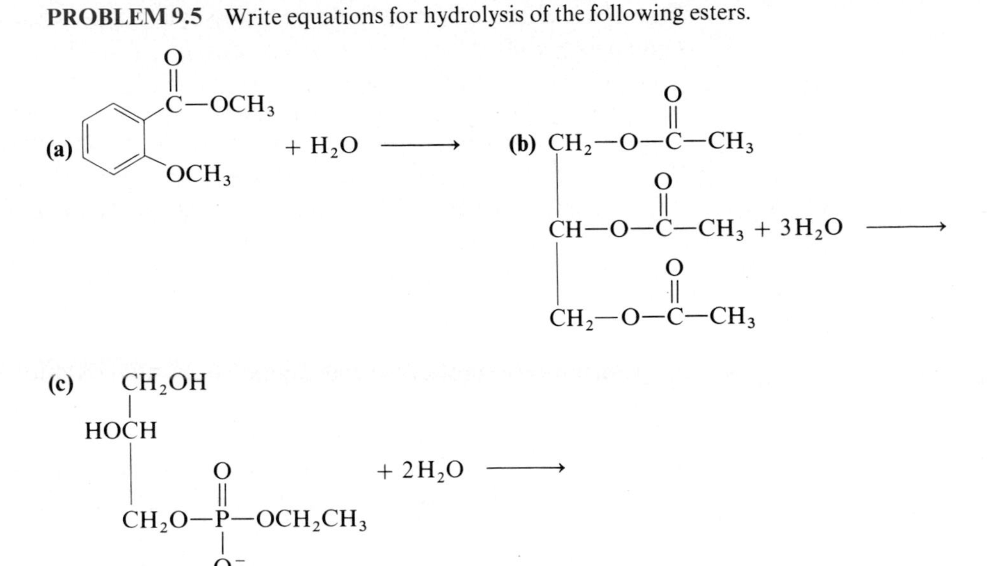

C. AMMONOLYSIS

Reaction with ammonia converts an ester into an amide. This reaction is similar to hydrolysis and is called ammonolysis. Ammonia is a strong nucleophile and adds directly to the carbonyl carbon. No catalyst is necessary.

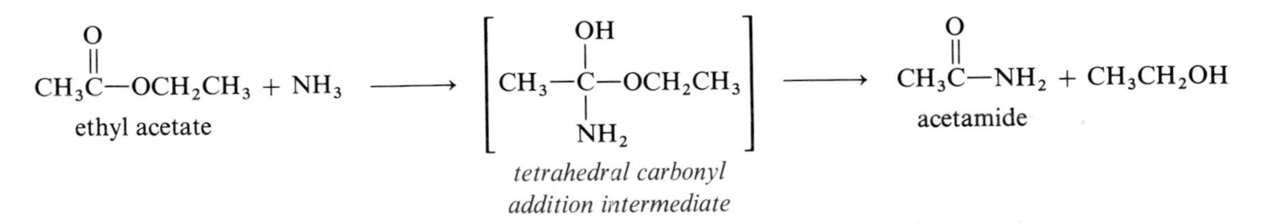

Although ammonolysis is an equilibrium reaction, the concentration of ester present at equilibrium is so small that it may be regarded as zero. Therefore, it is not possible to prepare an ester by treating an amide with alcohol.

Another example of ammonolysis of an ester is the laboratory synthesis of barbituric acid and barbiturates. Heating urea and diethyl malonate at 110°C in the presence of sodium ethoxide yields barbituric acid.

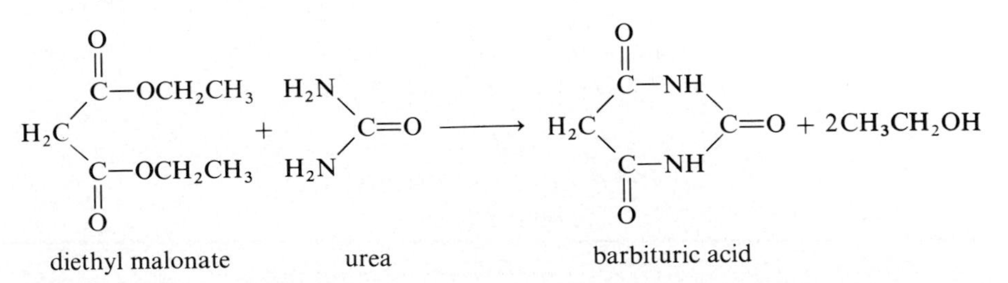

Mono- and disubstituted malonic esters yield substituted barbituric acids known as barbiturates.

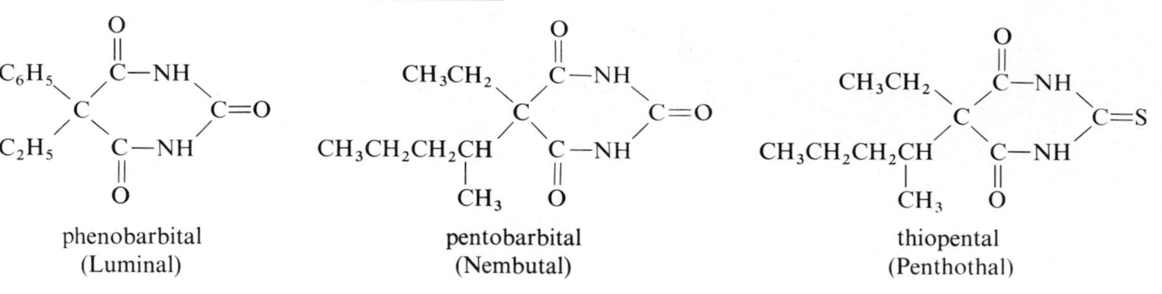

phenobarbital (Luminal) | pentobarbital (Nembutal) | thiopental (Penthothal)

Barbiturates produce effects ranging from mild sedation to deep anesthesia, and even death, depending on the dose and the particular barbiturate. Sedation, long or short acting, depends on the structure of the barbiturate. Phenobarbital is long acting while pentobarbital acts for a shorter time, about three hours. Thiopental is very fast acting and is used as an anesthetic for producing deep sedation quickly. With barbiturates in general, sleep can be produced with as little as 0.1 g (1 capsule) and toxic symptoms and even death can result from 1.5 g (15 capsules).

Example 9.6 Write equations for the following ammonolysis reactions.

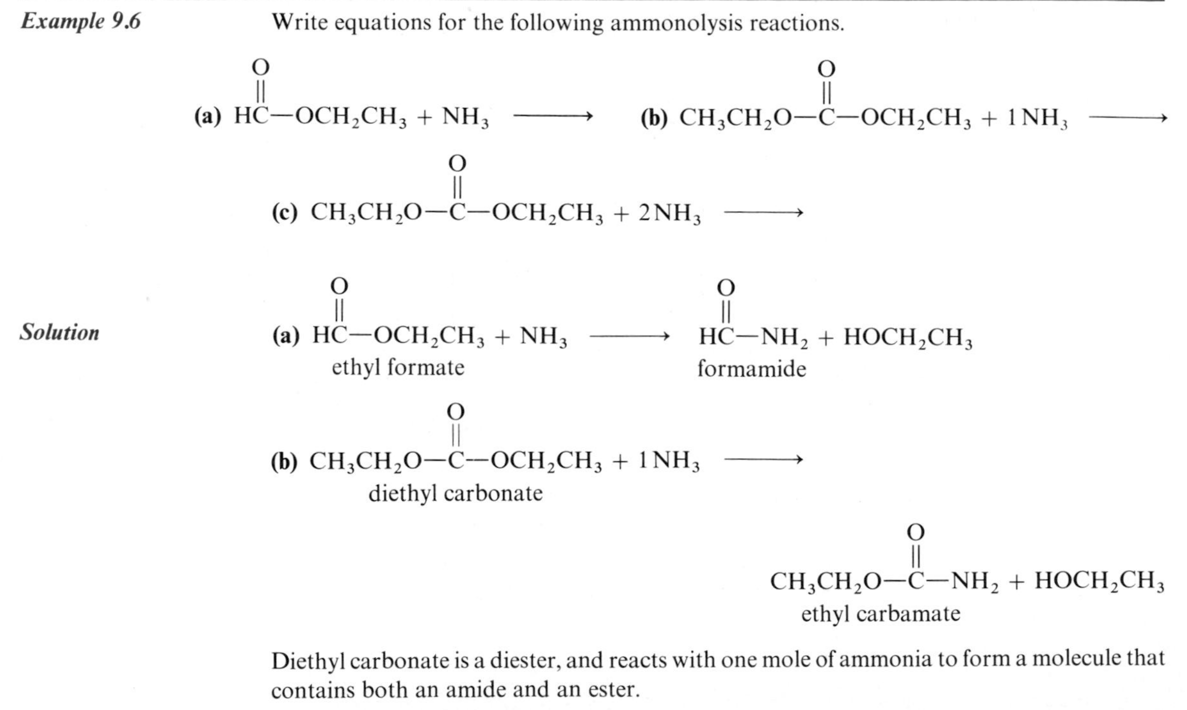

(a) $HC(=O)-OCH_2CH_3 + NH_3 \longrightarrow$ (b) $CH_3CH_2O-C(=O)-OCH_2CH_3 + 1NH_3 \longrightarrow$

(c) $CH_3CH_2O-C(=O)-OCH_2CH_3 + 2NH_3 \longrightarrow$

Solution

(a) $HC(=O)-OCH_2CH_3 + NH_3 \longrightarrow HC(=O)-NH_2 + HOCH_2CH_3$
ethyl formate → formamide

(b) $CH_3CH_2O-C(=O)-OCH_2CH_3 + 1NH_3 \longrightarrow CH_3CH_2O-C(=O)-NH_2 + HOCH_2CH_3$
diethyl carbonate → ethyl carbamate

Diethyl carbonate is a diester, and reacts with one mole of ammonia to form a molecule that contains both an amide and an ester.

(c) $CH_3CH_2O-C(=O)-OCH_2CH_3 + 2NH_3 \longrightarrow H_2N-C(=O)-NH_2 + 2HOCH_2CH_3$
diethyl carbonate → urea

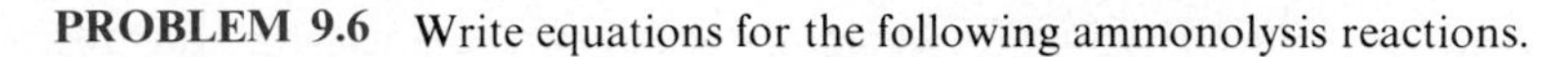

PROBLEM 9.6 Write equations for the following ammonolysis reactions.

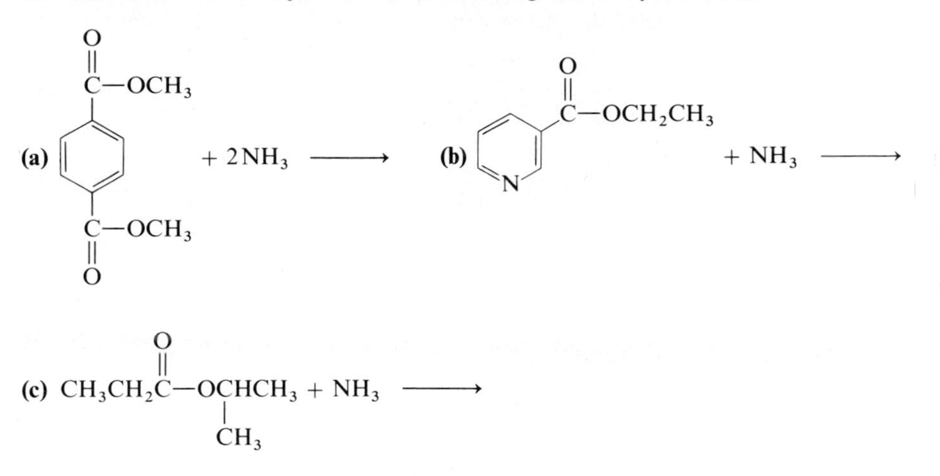

D. TRANSESTERIFICATION

Reaction of an ester with an alcohol in the presence of an acid catalyst results in the interchange of alkyl groups on the carboxyl oxygen.

$$\underset{\text{methyl acetate}}{CH_3\overset{O}{\overset{\|}{C}}—OCH_3} + \underset{\text{ethanol}}{HOCH_2CH_3} \overset{H^+}{\rightleftharpoons} \underset{\text{ethyl acetate}}{CH_3\overset{O}{\overset{\|}{C}}—OCH_2CH_3} + \underset{\text{methanol}}{HOCH_3}$$

Because the reaction of an ester with an alcohol results in the formation of a different ester, the process is called transesterification. Transesterification is an equilibrium reaction. An excess of alcohol can be used to drive the equilibrium to the right.

Example 9.7

Complete the following transesterification reactions. (The stoichiometry of each is given in the problem.)

$$\textbf{(a)}\ CH_3\overset{O}{\overset{\|}{C}}—OCH_2(CH_2)_8CH_3 + CH_3OH \xrightarrow{H^+}$$

$$\textbf{(b)}\ CH_3\overset{O}{\overset{\|}{C}}—OCH_2CH_2O—\overset{O}{\overset{\|}{C}}CH_3 + 2CH_3OH \xrightarrow{H^+}$$

Solution

$$\textbf{(a)}\ CH_3\overset{O}{\overset{\|}{C}}—OCH_3 + HOCH_2(CH_2)_8CH_3$$

$$\textbf{(b)}\ 2CH_3\overset{O}{\overset{\|}{C}}—OCH_3 + HOCH_2CH_2OH$$

PROBLEM 9.7 Complete the following transesterification reactions. (The stoichiometry of each is given in the problem.)

(a)

$$\begin{array}{l} CH_2O-\overset{O}{\overset{\|}{C}}(CH_2)_{14}CH_3 \\ | \\ CHO-\overset{O}{\overset{\|}{C}}(CH_2)_{14}CH_3 \quad + 3CH_3OH \xrightarrow{H^+} \\ | \\ CH_2O-\overset{O}{\overset{\|}{C}}(CH_2)_{14}CH_3 \end{array}$$

(b)

$$2\,C_6H_5-\overset{O}{\overset{\|}{C}}-OCH_3 + HOCH_2CH_2OH \xrightarrow{H^+}$$

E. REACTION WITH GRIGNARD REAGENTS

Carboxylate esters react readily with two moles of Grignard reagents to form alcohols. The reaction of Grignard reagents with formate esters yields secondary alcohols as illustrated by the preparation of 4-heptanol from ethyl formate and propyl magnesium bromide.

$$H\overset{O}{\overset{\|}{C}}-OCH_2CH_3 + 2CH_3CH_2CH_2MgBr \longrightarrow CH_3CH_2CH_2-\overset{OH}{\overset{|}{C}H}-CH_2CH_2CH_3$$

4-heptanol

This reaction proceeds in two steps. First, one mole of propyl magnesium bromide adds to the carbonyl group to give a tetrahedral carbonyl addition intermediate. This intermediate collapses to regenerate the carbonyl group and lose the magnesium salt of ethanol.

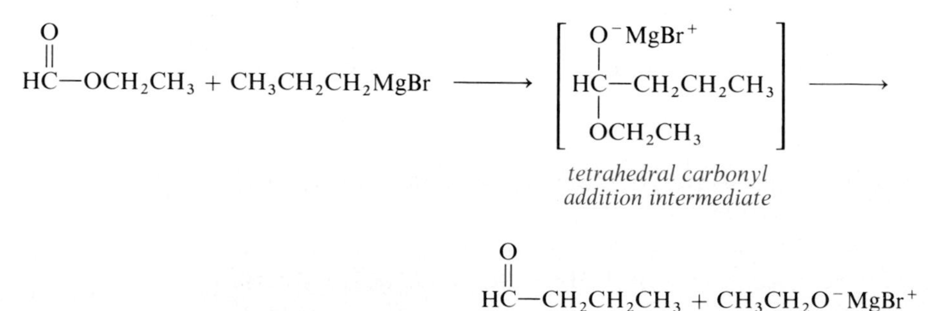

$$H\overset{O}{\overset{\|}{C}}-CH_2CH_2CH_3 + CH_3CH_2O^-MgBr^+$$

The aldehyde formed by collapse of the tetrahedral carbonyl addition intermediate then adds a second mole of Grignard reagent to form a secondary alcohol.

$$\mathrm{H\overset{\overset{\displaystyle O}{\|}}{C}{-}CH_2CH_2CH_3 + CH_3CH_2CH_2MgBr \longrightarrow CH_3CH_2CH_2{-}\overset{\overset{\displaystyle OH}{|}}{C}H{-}CH_2CH_2CH_3}$$

Esters other than those of formic acid give tertiary alcohols.

$$\mathrm{CH_3CH_2CH_2\overset{\overset{\displaystyle O}{\|}}{C}{-}OCH_3 + 2C_6H_5MgBr \longrightarrow CH_3CH_2CH_2\underset{\underset{\displaystyle C_6H_5}{|}}{\overset{\overset{\displaystyle OH}{|}}{C}}{-}C_6H_5}$$

1,1-diphenyl-1-butanol

9.7 PREPARATION OF ACID CHLORIDES

Acid chlorides are most often prepared by the reaction of a carboxylic acid with either thionyl chloride or phosphorus pentachloride in much the same way that alkyl chlorides are prepared from alcohols (Section 5.5E).

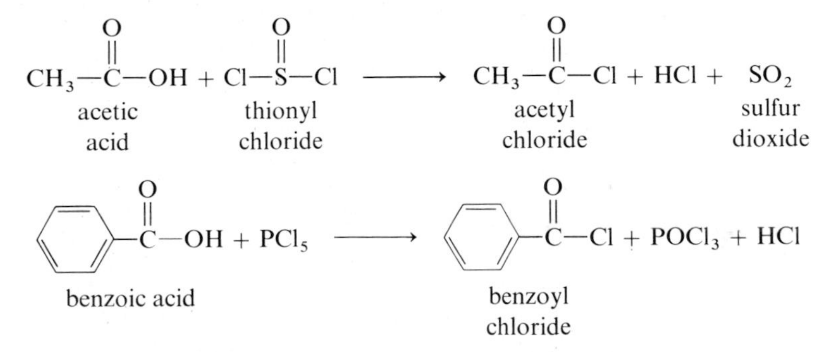

9.8 REACTIONS OF ACID CHLORIDES

Acid chlorides react readily with a wide variety of nucleophiles including water, ammonia, amines, alcohols, and phenols.

A. HYDROLYSIS

Reaction of an acid chloride with water regenerates the parent carboxylic acid.

$$\mathrm{CH_3{-}\overset{\overset{\displaystyle O}{\|}}{C}{-}Cl + H_2O \longrightarrow CH_3{-}\overset{\overset{\displaystyle O}{\|}}{C}{-}OH + HCl}$$

Acetyl chloride and many other low molecular weight aliphatic acid halides react so readily with water that they must be protected from atmospheric moisture during storage.

B. REACTION WITH ALCOHOLS

Acid chlorides react with alcohols and phenols to give esters. Such reactions are most often carried out in the presence of an organic base such as pyridine, which both catalyzes the reaction and neutralizes the HCl formed as a by-product.

$$CH_3-\overset{\overset{\displaystyle O}{\|}}{C}-Cl + CH_3-\underset{\underset{\displaystyle CH_3}{|}}{\overset{\overset{\displaystyle CH_3}{|}}{C}}-OH \xrightarrow{\text{base}} CH_3-\overset{\overset{\displaystyle O}{\|}}{C}-O-\underset{\underset{\displaystyle CH_3}{|}}{\overset{\overset{\displaystyle CH_3}{|}}{C}}-CH_3 + HCl$$

tert-butyl acetate

Recall that esters can also be prepared by acid-catalyzed Fischer esterification (Section 9.5). Of these two methods for the preparation of esters, the use of acid chlorides is often more convenient primarily because both the preparation of the acid chlorides and their reaction with alcohols are rapid and irreversible reactions. In contrast, Fischer esterification is a slow, reversible reaction, and yields of ester depend on the position of the equilibrium.

C. REACTION WITH AMMONIA AND AMINES

Acid chlorides react with ammonia and amines to give amides. In this reaction it is necessary to use two moles of ammonia or amine for each mole of acid chloride, the first to form an amide and the second to neutralize the HCl produced in the reaction.

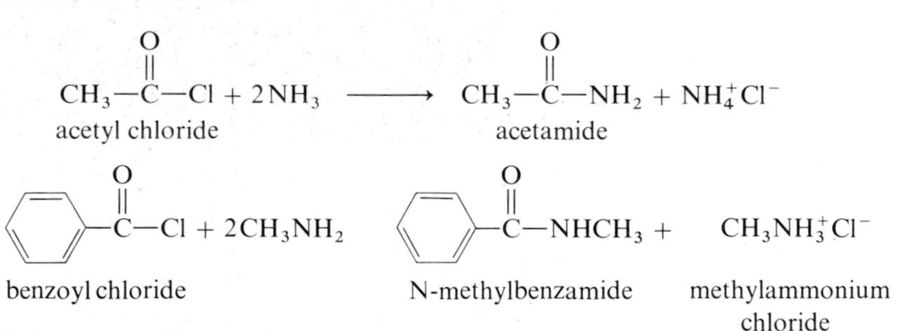

9.9 PREPARATION OF ACID ANHYDRIDES

Acid anhydrides are not normally prepared in the laboratory. The most commonly used acid anhydride is acetic anhydride and this substance is commercially available.

9.10 REACTIONS OF ACID ANHYDRIDES

The most important uses of anhydrides are reaction with alcohols to form esters and reaction with amines to form amides. Anhydrides also react with water to give two molecules of carboxylic acid.

A. HYDROLYSIS

Anhydrides of carboxylic acids, as well as those of inorganic acids such as phosphoric acid, are converted to the corresponding acids by hydrolysis in either aqueous acid or base. For example, hydrolysis of acetic anhydride gives two molecules of acetic acid.

$$\underset{\text{acetic anhydride}}{CH_3{-}\overset{O}{\overset{\|}{C}}{-}O{-}\overset{O}{\overset{\|}{C}}{-}CH_3} + H{-}OH \longrightarrow CH_3{-}\overset{O}{\overset{\|}{C}}{-}OH + HO{-}\overset{O}{\overset{\|}{C}}{-}CH_3$$

Acetic anhydride reacts so readily with water that it must be protected from moisture during storage.

B. REACTION WITH ALCOHOLS

Anhydrides react with alcohols to form an ester and a carboxylic acid.

$$\underset{\text{acetic anhydride}}{CH_3{-}\overset{O}{\overset{\|}{C}}{-}O{-}\overset{O}{\overset{\|}{C}}{-}CH_3} + \underset{\text{ethanol}}{HOCH_2CH_3} \longrightarrow$$

$$\underset{\text{ethyl acetate}}{CH_3{-}\overset{O}{\overset{\|}{C}}{-}OCH_2CH_3} + \underset{\text{acetic acid}}{HO{-}\overset{O}{\overset{\|}{C}}{-}CH_3}$$

Aspirin is prepared by the reaction of acetic anhydride and the —OH group of salicylic acid.

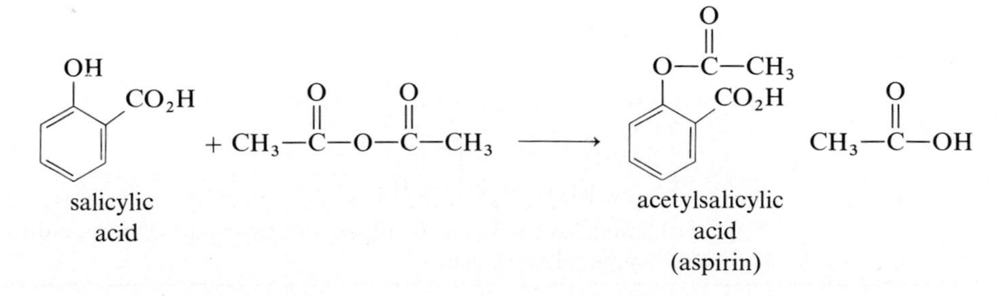

C. REACTION WITH AMMONIA AND AMINES

Anhydrides react very rapidly with ammonia to form one molecule of amide and one of carboxylate salt. In this reaction, the anhydride and ammonia first react to form a tetrahedral carbonyl addition intermediate, which then breaks apart to give the amide and a carboxylic acid. The carboxylic acid then reacts with a second molecule of ammonia to form an ammonium salt. For example, reaction of one mole of acetic anhydride with two moles of ammonia gives one mole of acetamide and one of ammonium acetate.

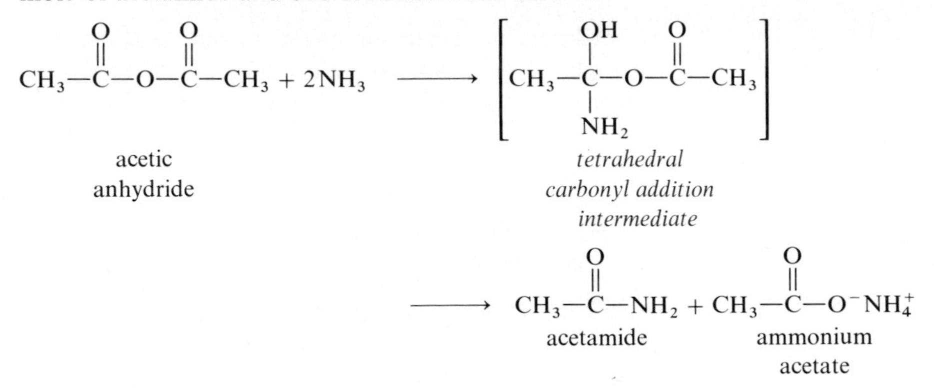

Example 9.8 Complete the following reactions. (The stoichiometry of each is given in the problem.)

(a) $2CH_3{-}\overset{O}{\overset{\|}{C}}{-}O{-}\overset{O}{\overset{\|}{C}}{-}CH_3 + HO{-}CH_2{-}CH_2{-}OH \longrightarrow$

(b) $\overset{O}{\overset{\|}{C}}{-}H$ / $H{-}C{-}OH$ / $CH_2{-}OH$ (glyceraldehyde) $+ 2CH_3{-}\overset{O}{\overset{\|}{C}}{-}O{-}\overset{O}{\overset{\|}{C}}{-}CH_3 \longrightarrow$

Solution

(a) $CH_3{-}\overset{O}{\overset{\|}{C}}{-}O{-}CH_2{-}CH_2{-}O{-}\overset{O}{\overset{\|}{C}}{-}CH_3 + 2CH_3{-}\overset{O}{\overset{\|}{C}}{-}OH$

The starting material, ethylene glycol is a diol, and reacts with two moles of acetic anhydride to produce a diester and two moles of acetic acid.

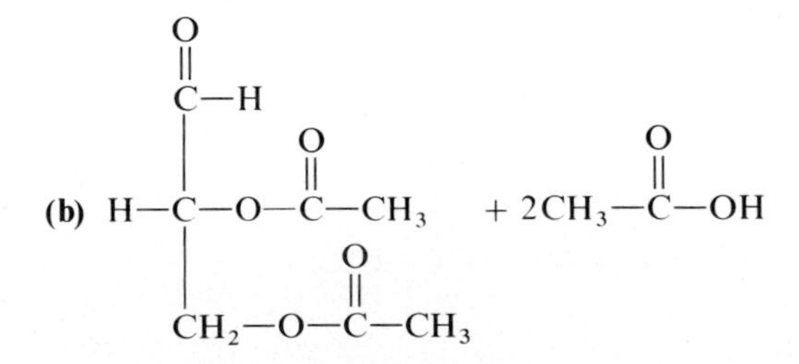

Glyceraldehyde is a diol and reacts with two moles of acetic anhydride to give a diester and two moles of acetic acid.

PROBLEM 9.8 Complete the following reactions.

$$\text{(a)}\; CH_2{=}CH-CH_2-OH + CH_3-\overset{O}{\overset{\|}{C}}-O-\overset{O}{\overset{\|}{C}}-CH_3 \longrightarrow$$

$$\text{(b)}\; 2CH_3-\overset{O}{\overset{\|}{C}}-O-\overset{O}{\overset{\|}{C}}-CH_3 + HO-CH_2-CH_2-NH_2 \longrightarrow$$

9.11 PREPARATION OF AMIDES

Amides are most commonly synthesized in the laboratory by reaction of an anhydride with ammonia (Section 9.10C), reaction of an acid halide with ammonia (Section 9.8C), or reaction of an ester with ammonia (Section 9.6C). We have already seen several examples of each of these reactions.

$$CH_3-\overset{O}{\overset{\|}{C}}-O-CH_2CH_3 + NH_3 \longrightarrow CH_3-\overset{O}{\overset{\|}{C}}-NH_3 + CH_3CH_2OH$$

$$CH_3-\overset{O}{\overset{\|}{C}}-O-\overset{O}{\overset{\|}{C}}-CH_3 + 2NH_3 \longrightarrow CH_3-\overset{O}{\overset{\|}{C}}-NH_2 + CH_3-\overset{O}{\overset{\|}{C}}-O^-NH_4^+$$

$$CH_3-\overset{O}{\overset{\|}{C}}-Cl + 2NH_3 \longrightarrow CH_3-\overset{O}{\overset{\|}{C}}-NH_2 + NH_4^+Cl^-$$

Amides can also be prepared by heating the ammonium salt of a carboxylic acid above its melting point.

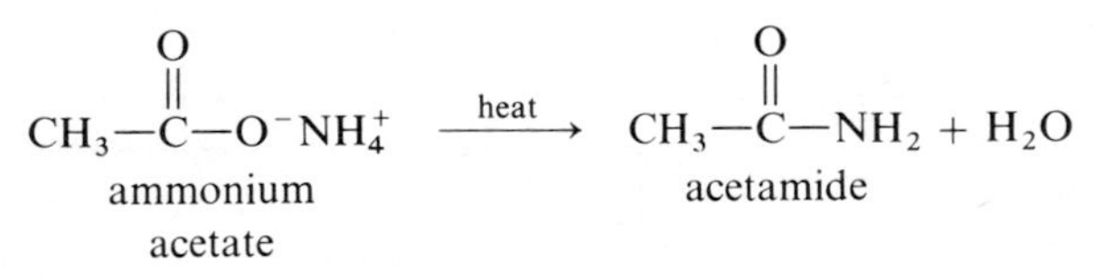

9.12 HYDROLYSIS OF AMIDES

Amides are very resistant to hydrolysis. However, in the presence of concentrated aqueous acid or base, hydrolysis does occur, though not as rapidly as in the case of esters. Often, it is necessary to reflux an amide for several hours with concentrated hydrochloric acid to bring about hydrolysis.

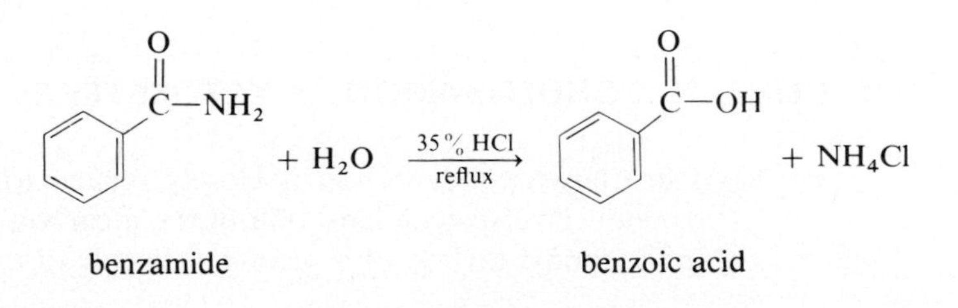

9.13 RELATIVE REACTIVITIES OF ACID HALIDES, ANHYDRIDES, ESTERS, AND AMIDES

The four common functional derivatives of carboxylic acid we have described in this chapter show marked differences in reactivities. For example, consider the ease of hydrolysis of an acid chloride, an anhydride, an ester, and an amide. Both acetyl chloride and acetic anhydride react so readily with water that they must be protected from atmospheric moisture during storage. Ethyl acetate reacts slowly with water at room temperature but hydrolyzes readily on heating in the presence of an acid or base catalyst. Acetamide is very resistant to hydrolysis except in the presence of moderately strong acid or base and heat. The reactivity of these functional derivatives of carboxylic acids decreases in the following order.

$$\underset{\text{acid chloride}}{CH_3\overset{O}{\overset{\|}{C}}{-}Cl} \approx \underset{\text{acid anhydride}}{CH_3\overset{O}{\overset{\|}{C}}{-}O{-}\overset{O}{\overset{\|}{C}}CH_3} > \underset{\text{ester}}{CH_3\overset{O}{\overset{\|}{C}}{-}OCH_2CH_3} > \underset{\text{amide}}{CH_3\overset{O}{\overset{\|}{C}}{-}NH_2}$$

Any less reactive derivative of a carboxylic acid may be prepared directly from a more reactive derivative, but not vice versa. For example, an ester can be synthesized from an acid anhydride plus an alcohol or phenol. However, an ester cannot be synthesized from an amide plus an alcohol. These interconversions are summarized in Table 9.1.

Table 9.1 Interconversion of functional derivatives of carboxylic acids.

	Acyl derivative		*Can be converted to*	*By reaction with*	*Name of process*
increasing reactivity (↑)	acid anhydride or acid chloride	⟶	carboxylic acid	water	hydrolysis
			ester	alcohol or phenol	alcoholysis
			amide	NH_3, primary or secondary amine	ammonolysis
	ester	⟶	carboxylic acid	water	hydrolysis
			amide	NH_3, primary or secondary amine	ammonolysis
	amide	⟶	carboxylic acid	water	hydrolysis

9.14 THE CLAISEN CONDENSATION: β-KETOESTERS

The Claisen condensation is closely related to the aldol condensation. The reaction involves condensation of the α-carbon of one molecule of ester with the carbonyl carbon of a second molecule of ester. For example, when ethyl acetate is heated with sodium ethoxide in ethanol, the α-carbon of one molecule of ethyl acetate condenses with the carbonyl carbon of a second molecule to form a new carbon-carbon bond.

$$CH_3\overset{O}{\overset{\|}{C}}{-}OCH_2CH_3 + CH_3\overset{O}{\overset{\|}{C}}OCH_2CH_3 \xrightarrow[CH_3CH_2OH]{CH_3CH_2O^-\ Na^+} CH_3\overset{O}{\overset{\|}{C}}{-}CH_2\overset{O}{\overset{\|}{C}}OCH_2CH_3 + CH_3CH_2OH$$

carbonyl carbon (arrow to the C of the first ester); α-carbon (arrow to the CH_3 of the second ester); new C–C bond formed (arrow to the bond in the product)

ethyl 3-ketobutanoate
(ethyl acetoacetate)

The characteristic structural feature of the product of a Claisen condensation is a ketone on carbon-3 of an ester chain. In the common name system of nomenclature, carbon-3 of a carboxylic acid is called a beta-carbon (β-carbon), thus the products of Claisen condensation reactions are often called β-ketoesters. In naming β-ketoesters, the prefix keto- is used to indicate the presence of the ketone group on the ester chain.

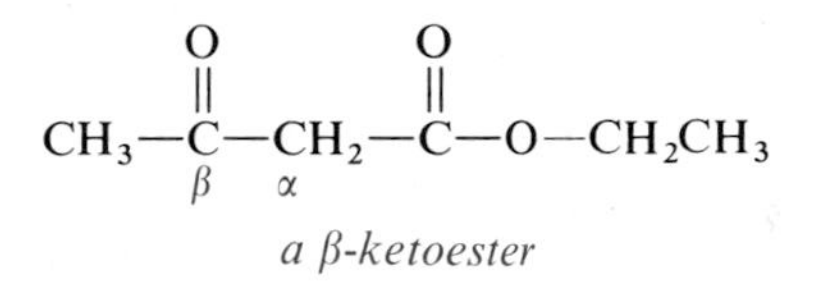

a β-ketoester

Claisen condensation of ethyl propanoate yields the following β-ketoester.

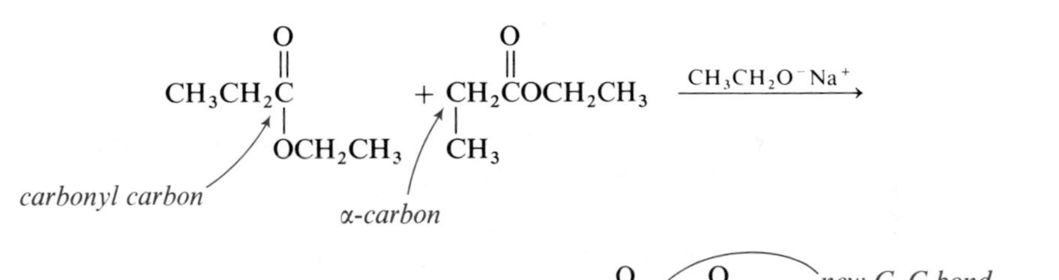

$$CH_3CH_2\overset{O}{\overset{\|}{C}}{-}\underset{CH_3}{\underset{|}{C}}H\overset{O}{\overset{\|}{C}}OCH_2CH_3 + CH_3CH_2OH$$

new C–C bond

ethyl 2-methyl-3-ketopentanoate
(a β-ketoester)

In this reaction, the structural formulas of the two ester molecules are written to emphasize that the step forming the new carbon-carbon bond involves the α-carbon of one ester and the carbonyl carbon of the other.

The mechanism of the Claisen condensation is similar to the three-step mechanism for the aldol condensation (Section 7.5F) in that it begins with the formation of an anion on the carbon alpha to the carbonyl group. Recall from the discussion in Section 7.5F that hydrogen atoms on α-carbons show acidity in the presence of a strong base such as sodium ethoxide. The first step in the

Claisen condensation of ethyl acetate is the formation of an anion. This anion is a nucleophile, and in Step 2 it attacks the carbonyl carbon of a second molecule of ethyl acetate, forming a tetrahedral carbonyl addition intermediate. Elimination of ethoxide ion in Step 3 gives ethyl acetoacetate.

Step 1:

$$CH_3CH_2O^- + H{-}CH_2{-}\overset{O}{\overset{\|}{C}}{-}OCH_2CH_3 \rightleftharpoons CH_3CH_2OH + :\bar{C}H_2{-}\overset{O}{\overset{\|}{C}}{-}OCH_2CH_3$$

Step 2:

$$CH_3{-}\overset{\ddot{O}:}{\overset{\|}{C}}{-}OCH_2CH_3 + :\bar{C}H_2{-}\overset{O}{\overset{\|}{C}}{-}OCH_2CH_3 \rightleftharpoons CH_3{-}\overset{:\ddot{O}:^-}{\overset{|}{C}}{-}CH_2{-}\overset{O}{\overset{\|}{C}}{-}OCH_2CH_3$$
$$\qquad\qquad\qquad\qquad\qquad\qquad\qquad\qquad\quad |$$
$$\qquad\qquad\qquad\qquad\qquad\qquad\qquad\qquad OCH_2CH_3$$

tetrahedral carbonyl addition intermediate

Step 3:

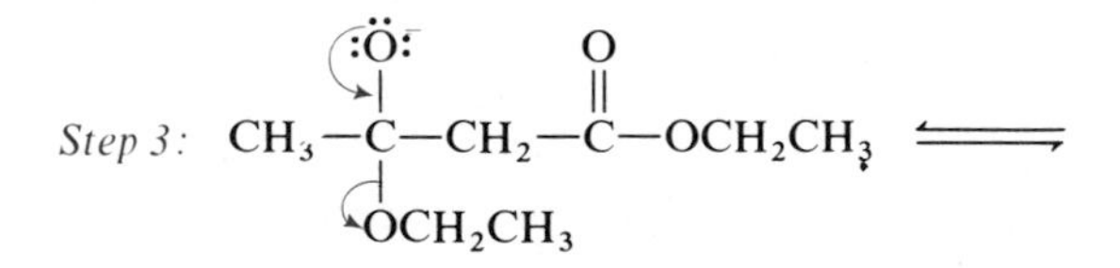

$$\rightleftharpoons CH_3{-}\overset{O}{\overset{\|}{C}}{-}CH_2{-}\overset{O}{\overset{\|}{C}}{-}OCH_2CH_3 + CH_3CH_2O^-$$

This reaction is reversible and the position of equilibrium is quite unfavorable for the formation of ethyl acetoacetate. However, the yield of β-ketoester can be increased by adding a mole of sodium ethoxide per mole of product formed. The K_a of ethyl acetoacetate is approximately 10^{-10} which means that it is about as strong an acid as phenol. Ethyl acetoacetate reacts with sodium ethoxide to form a sodium salt that precipitates from solution and thus is removed from the equilibrium.

$$CH_3CH_2O^-Na^+ + \underset{\text{ethyl acetoacetate}}{CH_3{-}\overset{O}{\overset{\|}{C}}{-}CH_2{-}\overset{O}{\overset{\|}{C}}{-}OCH_2CH_3} \longrightarrow$$

$$\underset{\text{a sodium salt of ethyl acetoacetate}}{Na^+\ CH_3{-}\overset{O}{\overset{\|}{C}}{-}\underset{H}{\underset{|}{\ddot{C}^-}}{-}\overset{O}{\overset{\|}{C}}{-}OCH_2CH_3} + CH_3CH_2OH$$

Diesters of six- and seven-carbon diacids undergo intramolecular Claisen condensation to form five- and six-membered rings.

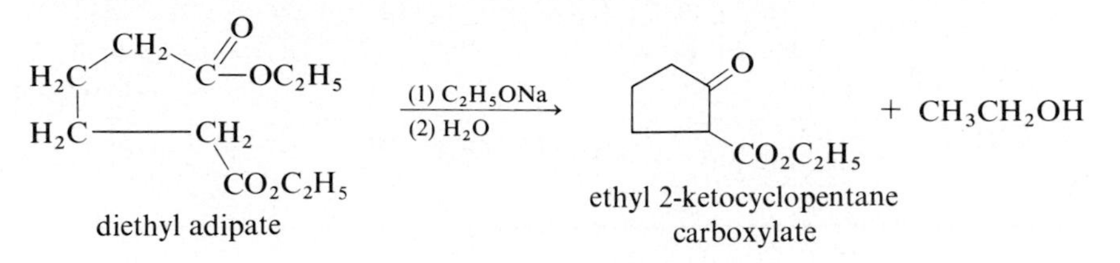

In the case of a mixed Claisen condensation, that is, a condensation between two different esters, a mixture of four possible products results unless there is an appreciable difference in reactivity between one ester and the other. One such difference is if one of the esters has no α-hydrogens and therefore, cannot serve as a carbanion. Examples of esters without α-hydrogens are ethyl formate, ethyl benzoate, and diethyl carbonate.

$$H-\overset{O}{\overset{\|}{C}}-OCH_2CH_3 \qquad C_6H_5-\overset{O}{\overset{\|}{C}}-OCH_2CH_3 \qquad CH_3CH_2O-\overset{O}{\overset{\|}{C}}-OCH_2CH_3$$

ethyl formate — ethyl benzoate — diethyl carbonate

Example 9.9

Draw structural formulas for the products of the following Claisen condensations.

(a) $$C_6H_5-\overset{O}{\overset{\|}{C}}-OCH_2CH_3 + CH_3CH_2\overset{O}{\overset{\|}{C}}-OCH_2CH_3 \xrightarrow{\text{base}}$$

(b) $$\text{cyclohexanone} + H-\overset{O}{\overset{\|}{C}}-OCH_2CH_3 \xrightarrow{\text{base}}$$

Solution

(a) Ethyl benzoate has no α-hydrogen atoms, and therefore can function only as a carbanion acceptor in a Claisen condensation. Ethyl propanoate has two α-hydrogens and reacts with a mole of base to form a carbanion that then reacts with the carbonyl group of ethyl benzoate.

$$C_6H_5-\overset{O}{\overset{\|}{C}}-OCH_2CH_3 + \underset{CH_3}{\underset{|}{CH_2}}\overset{O}{\overset{\|}{C}}OCH_2CH_3 \xrightarrow{\text{base}}$$

new C–C bond

$$C_6H_5-\overset{O}{\overset{\|}{C}}-\underset{CH_3}{\underset{|}{CH}}\overset{O}{\overset{\|}{C}}OCH_2CH_3 + CH_3CH_2OH$$

ethyl 2-methyl-3-keto-3-phenylpropanoate

(b) Only cyclohexanone has an α-hydrogen that can be removed by base to form an anion. The carbonyl group of ethyl formate functions as the anion acceptor.

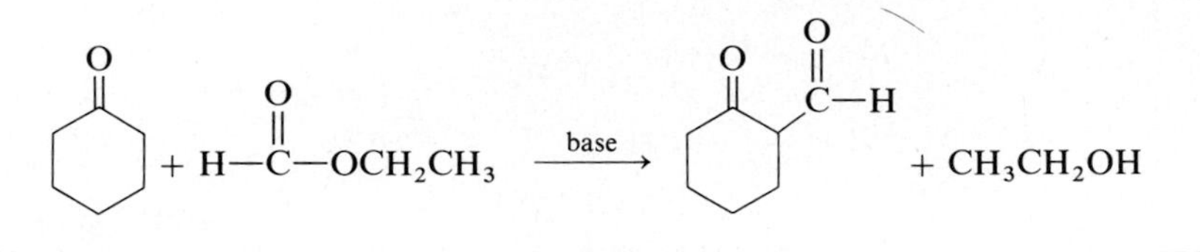

PROBLEM 9.9 Draw structural formulas for the products of the following Claisen condensations

(a) $CH_3CH_2O{-}\overset{O}{\overset{\|}{C}}(CH_2)_5\overset{O}{\overset{\|}{C}}{-}OCH_2CH_3 \xrightarrow{\text{base}}$

(b) cyclohexanone $+ CH_3CH_2O{-}\overset{O}{\overset{\|}{C}}{-}OCH_2CH_3 \xrightarrow{\text{base}}$

Hydrolysis of a β-ketoester in aqueous acid gives the corresponding β-ketoacid.

$$CH_3\overset{O}{\overset{\|}{C}}CH_2\overset{O}{\overset{\|}{C}}OCH_2CH_3 + H_2O \xrightarrow[\text{heat}]{H^+} CH_3\overset{O}{\overset{\|}{C}}CH_2\overset{O}{\overset{\|}{C}}OH + CH_3CH_2OH$$

3-ketobutanoic acid
(β-ketobutyric acid)

β-Ketoacids lose carbon dioxide on heating to give ketones (Section 8.5C).

$$CH_3\overset{O}{\overset{\|}{C}}CH_2\overset{O}{\overset{\|}{C}}OH \xrightarrow[\text{heat}]{} CH_3\overset{O}{\overset{\|}{C}}CH_3 + O{=}C{=}O$$

In the more usual reaction, ester hydrolysis and decarboxylation occur together, and only the ketone is isolated.

$$CH_3CH_2\overset{O}{\overset{\|}{C}}\underset{CH_3}{\underset{|}{C}H}\overset{O}{\overset{\|}{C}}OCH_2CH_3 + H_2O \xrightarrow[\text{heat}]{H^+}$$

ethyl 2-methyl-3-keto-
pentanoate

$$CH_3CH_2\overset{O}{\overset{\|}{C}}CH_2CH_3 + CO_2 + CH_3CH_2OH$$

3-pentanone

9.15 SPECTROSCOPIC PROPERTIES

All carbonyl-containing organic molecules show characteristic absorption in the infrared spectrum due to the stretching vibration of the carbonyl group. These absorption ranges for unconjugated aldehydes, ketones, carboxylic acids, esters, and amides are summarized below in Table 9.2.

Notice that while these carbonyl stretching frequencies fall within a relatively narrow range, there is some variation and it is sometimes possible to distinguish between these functional groups by the position of the carbonyl stretching absorption. Of greater use, however, in distinguishing between carbonyl-containing functional groups is the presence or absence of peaks in the spectrum corresponding to aldehyde C—H stretching (Section 7.6), carboxyl O—H stretching (Section 8.9), or amide N—H stretching.

The most characteristic feature of the NMR spectrum of esters is the chemical shift of hydrogens on the carbon atom attached to the carbonyl group and to the ester oxygen. The presence of the carbonyl group and the oxygen shifts the signals of adjacent protons to larger delta values. For example, the NMR spectrum of methyl propanoate (Figure 9.1) shows three signals; a triplet at $\delta = 1.1$, a quartet at $\delta = 2.3$, and a singlet at $\delta = 3.6$. These signals are in the ratio 3:2:3.

Compare the chemical shifts of the protons on the two methyl groups in this molecule. The methyl protons of CH_3CH_2— appear as a triplet at $\delta = 1.1$, while the methyl protons of CH_3O— appear as a singlet at $\delta = 3.6$. Note that

Table 9.2 Characteristic infrared absorptions of carbonyl-containing groups.

Functional group	*Carbonyl stretching frequency* (cm^{-1})	*Other characteristic stretching frequency* (cm^{-1})	
$R{-}C({=}O){-}H$	1710–1740	2720–2830	C—H
$R{-}C({=}O){-}R$	1710–1740		
$R{-}C({=}O){-}OH$	1700–1725	2500–3000	O—H
$R{-}C({=}O){-}OR$	1725–1750	1000–1300	O—R
$R{-}C({=}O){-}NH_2$	1680–1690	3400–3500	N—H

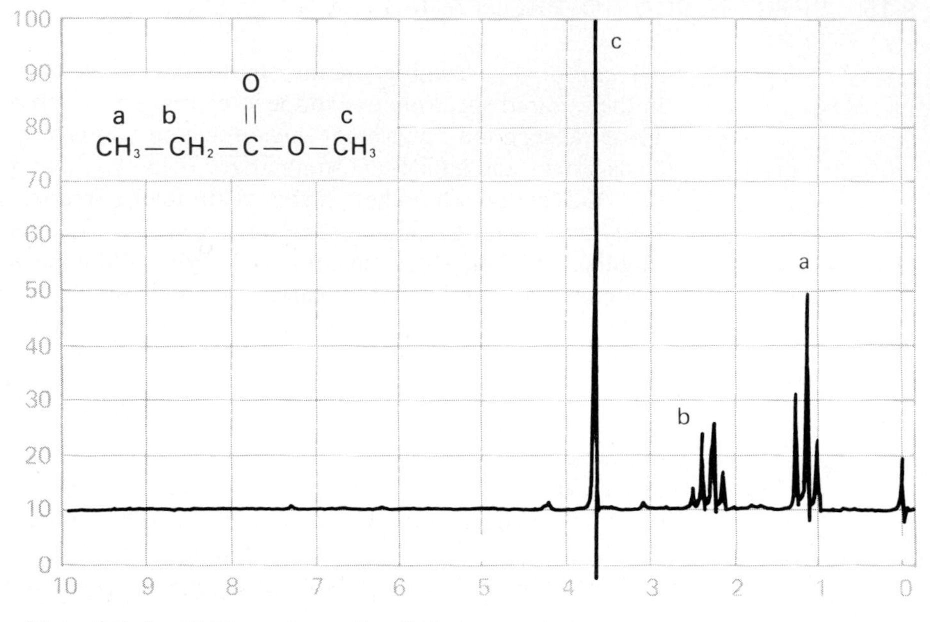

Figure 9.1 An NMR spectrum of methyl propanoate.

this same shift to larger delta values is seen in the protons attached to oxygen atoms in alcohols and ethers. (Compare Table 15.3.)

IMPORTANT REACTIONS

(1) Fischer esterification (Section 9.5).

$$CH_3\overset{O}{\overset{\|}{C}}-OH + H-OCH_2CH_2CH_3 \overset{H^+}{\rightleftharpoons} CH_3\overset{O}{\overset{\|}{C}}-OCH_2CH_2CH_3 + H_2O$$

(2) Hydrolysis

(a) of esters (Section 9.6B).

$$CH_3\overset{O}{\overset{\|}{C}}-OCH_2CH_2CH_3 + H_2O \overset{H^+}{\rightleftharpoons} CH_3\overset{O}{\overset{\|}{C}}-OH + HOCH_2CH_2CH_3$$

$$CH_3\overset{O}{\overset{\|}{C}}-OCH_2CH_2CH_3 + NaOH \longrightarrow CH_3\overset{O}{\overset{\|}{C}}-O^-Na^+ + HOCH_2CH_2CH_3$$

$$CH_3CH_2O-\overset{O}{\overset{\|}{\underset{OCH_2CH_3}{\underset{|}{P}}}}-OCH_2CH_3 + 3H_2O \xrightarrow{H_3O^+} 3CH_3CH_2OH + HO-\overset{O}{\overset{\|}{\underset{OH}{\underset{|}{P}}}}-OH$$

(b) of amides (Section 9.12).

$$CH_3\overset{\overset{\displaystyle O}{\|}}{C}{-}NH_2 + H_2O + HCl \longrightarrow CH_3\overset{\overset{\displaystyle O}{\|}}{C}{-}OH + NH_4Cl$$

(c) of anhydrides (Section 9.10A).

$$CH_3\overset{\overset{\displaystyle O}{\|}}{C}{-}O{-}\overset{\overset{\displaystyle O}{\|}}{C}CH_3 + H_2O \longrightarrow 2CH_3\overset{\overset{\displaystyle O}{\|}}{C}{-}OH$$

(d) of acid halides (Section 9.8A).

$$CH_3\overset{\overset{\displaystyle O}{\|}}{C}{-}Cl + H_2O \longrightarrow CH_3\overset{\overset{\displaystyle O}{\|}}{C}{-}OH + HCl$$

(3) Ammonolysis

(a) of esters (Section 9.6C).

$$CH_3\overset{\overset{\displaystyle O}{\|}}{C}{-}OCH_2CH_3 + NH_3 \longrightarrow CH_3\overset{\overset{\displaystyle O}{\|}}{C}{-}NH_2 + CH_3CH_2OH$$

(b) of anhydrides (Section 9.10C).

$$CH_3\overset{\overset{\displaystyle O}{\|}}{C}{-}O{-}\overset{\overset{\displaystyle O}{\|}}{C}CH_3 + 2NH_3 \longrightarrow CH_3\overset{\overset{\displaystyle O}{\|}}{C}{-}NH_2 + CH_3\overset{\overset{\displaystyle O}{\|}}{C}{-}O^-\ NH_4^+$$

(c) of acid halides (Section 9.8C).

$$CH_3\overset{\overset{\displaystyle O}{\|}}{C}{-}Cl + 2NH_3 \longrightarrow CH_3\overset{\overset{\displaystyle O}{\|}}{C}{-}NH_2 + NH_4Cl$$

(4) Alcoholysis

(a) of esters (Section 9.6D).

$$CH_3\overset{\overset{\displaystyle O}{\|}}{C}{-}OCH_3 + HOCH_2CH_2CH_3 \xrightarrow{H^+} CH_3\overset{\overset{\displaystyle O}{\|}}{C}{-}OCH_2CH_2CH_3 + CH_3OH$$

The reaction of an ester with an alcohol is also called transesterification.

(b) of anhydrides (Section 9.10B).

$$CH_3\overset{\overset{\displaystyle O}{\|}}{C}{-}O{-}\overset{\overset{\displaystyle O}{\|}}{C}CH_3 + CH_3CH_2OH \longrightarrow CH_3\overset{\overset{\displaystyle O}{\|}}{C}{-}OCH_2CH_3 + HO{-}\overset{\overset{\displaystyle O}{\|}}{C}CH_3$$

(c) of acid halides (Section 9.8B).

$$CH_3\overset{\overset{\displaystyle O}{\|}}{C}{-}Cl + HOCH_2CH_2CH_3 \longrightarrow CH_3\overset{\overset{\displaystyle O}{\|}}{C}{-}OCH_2CH_2CH_3 + HCl$$

(5) Reduction of esters (Section 9.6A).

$$C_6H_5\overset{O}{\overset{\|}{C}}-OCH_2CH_3 \xrightarrow[(2)\ H_2O]{(1)\ LiAlH_4} C_6H_5CH_2OH + HOCH_2CH_3$$

(6) Reaction of esters with Grignard reagents (Section 9.6E).

(a) Reaction with formate esters gives secondary alcohols.

$$2C_6H_5MgBr + H-\overset{O}{\overset{\|}{C}}-OCH_2CH_3 \longrightarrow C_6H_5-\underset{C_6H_5}{\underset{|}{CH}}-OH$$

(b) Reaction with other esters gives tertiary alcohols.

$$2C_6H_5MgBr + C_6H_5\overset{O}{\overset{\|}{C}}-OCH_2CH_3 \longrightarrow C_6H_5-\overset{OH}{\overset{|}{\underset{C_6H_5}{\underset{|}{C}}}}-C_6H_5$$

(7) The Claisen condensation (Section 9.14).

$$2CH_3\overset{O}{\overset{\|}{C}}-OCH_2CH_3 \xrightarrow{CH_3CH_2ONa} CH_3\underset{\beta}{\overset{O}{\overset{\|}{C}}}-\underset{\alpha}{CH_2}-\overset{O}{\overset{\|}{C}}-OCH_2CH_3 + CH_3CH_2OH$$

The product of a Claisen condensation is a β-ketoester.

PROBLEMS

9.10 Name the following compounds.

(a) $CH_3CH_2\overset{O}{\overset{\|}{C}}O\underset{CH_3}{\underset{|}{C}}HCH_3$

(b) $CH_3\overset{O}{\overset{\|}{C}}-NH_2$

(c) $C_6H_5-\overset{O}{\overset{\|}{C}}-O-\overset{O}{\overset{\|}{C}}-C_6H_5$

(d) $CH_3CH_2O-\overset{O}{\overset{\|}{\underset{OCH_2CH_3}{\underset{|}{P}}}}-OCH_2CH_3$

(e) $CH_3CH_2CH_2CH_2\overset{O}{\overset{\|}{C}}-NH-CH_3$

(f) $CH_3CH_2O-\overset{O}{\overset{\|}{C}}CH_2CH_2\overset{O}{\overset{\|}{C}}-OCH_2CH_3$

(g) $CH_3CH_2\overset{O}{\overset{\|}{C}}-S-CH_3$

(h) $CH_3\overset{O}{\overset{\|}{C}}-S-CoA$

(i) $C_6H_{11}-O-\overset{O}{\overset{\|}{C}}-CH_3$ (cyclohexyl)

(j) $C_6H_5-O-\overset{O}{\overset{\|}{C}}-CH_3$ (phenyl)

(k) 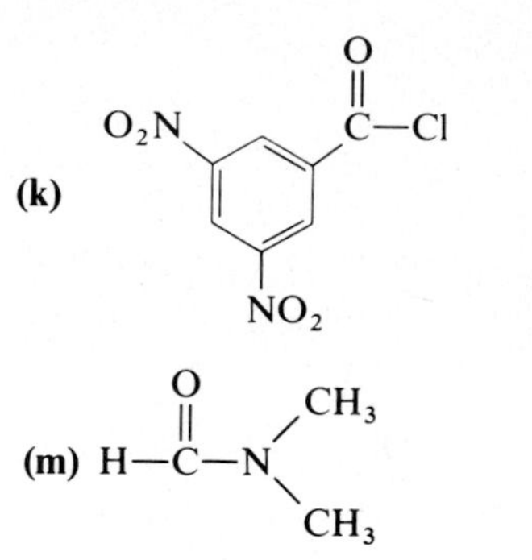

(l) $CH_2{=}CH{-}\overset{\overset{\large O}{\|}}{C}{-}OCH_3$

(m) $H{-}\overset{\overset{\large O}{\|}}{C}{-}N(CH_3)_2$

(n) $CH_2{-}ONO_2$ | $CH{-}ONO_2$ | $CH_2{-}ONO_2$

9.11 Draw structural formulas for the following:

(a) phenyl benzoate
(b) diethyl carbonate
(c) benzamide
(d) cyclobutyl butanoate
(e) methyl 3-methylbutanoate
(f) isopropyl 3-methylhexanoate
(g) diethyl oxalate
(h) ethyl *cis*-2-pentenoate
(i) acetamide
(j) *N,N*-dimethylacetamide
(k) acetic anhydride
(l) *N*-phenylbutanamide
(m) diethyl malonate
(n) formamide
(o) ethyl 3-hydroxybutanoate
(p) acetyl chloride
(q) methyl diethyl phosphate
(r) methyl formate
(s) *p*-nitrophenyl acetate
(t) acetyl salicylate
(u) methyl *p*-hydroxybenzoate
(v) urea

9.12 Draw structural formulas for the nine isomeric esters of molecular formula $C_5H_{10}O_2$. Give each an IUPAC name.

9.13 Both acetic acid and methyl formate have the same molecular formula $C_2H_4O_2$. Both are liquids, one with a boiling point of 32°C, the other with a boiling point of 118°C. Which of the two compounds would you predict to have the boiling point of 32°C? Explain your reasoning.

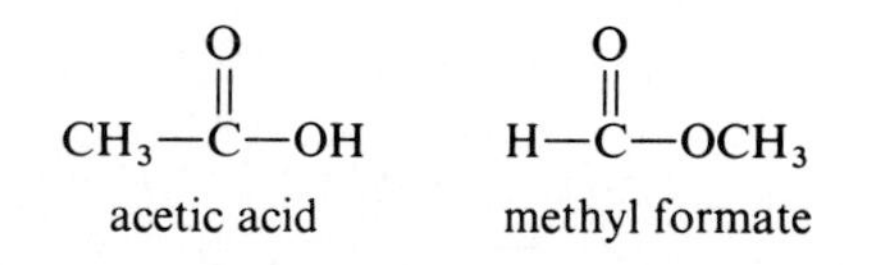

9.14 Write structural formulas for the products of hydrolysis of the following esters, amides, and anhydrides.

(a) $CH_3\overset{\overset{\large O}{\|}}{C}OCH_2CH_3 + H_2O \longrightarrow$

(b) 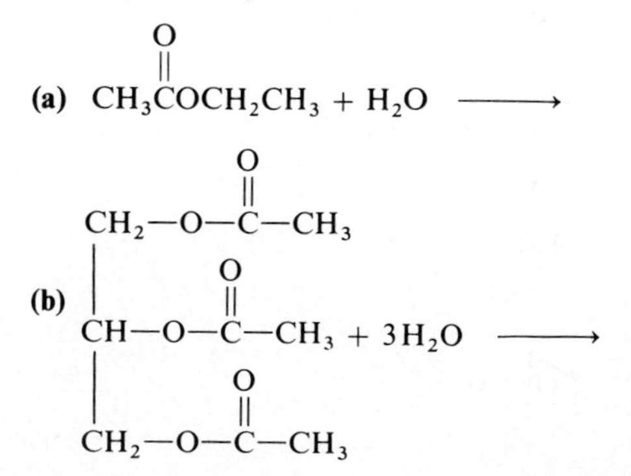

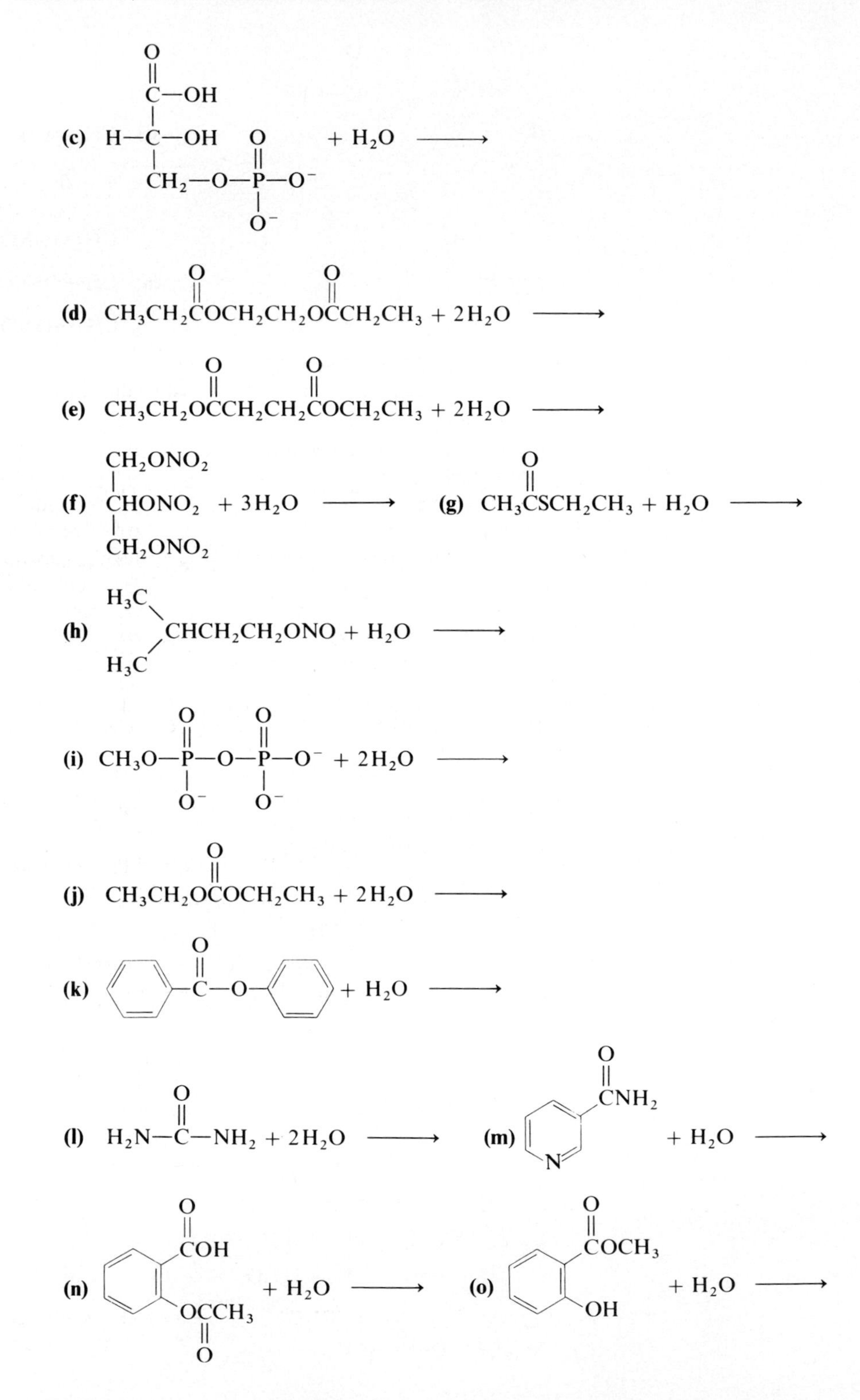
(c)
+ H2O
(d) CH3CH2COCH2CH2OCCH2CH3 + 2H2O
(e) CH3CH2OCCH2CH2COCH2CH3 + 2H2O
(f) + 3H2O
(g) CH3CSCH2CH3 + H2O
(h) CHCH2CH2ONO + H2O
(i) CH3O—P—O—P—O⁻ + 2H2O
(j) CH3CH2OCOCH2CH3 + 2H2O
(k) + H2O
(l) H2N—C—NH2 + 2H2O
(m) + H2O
(n) + H2O
(o) + H2O

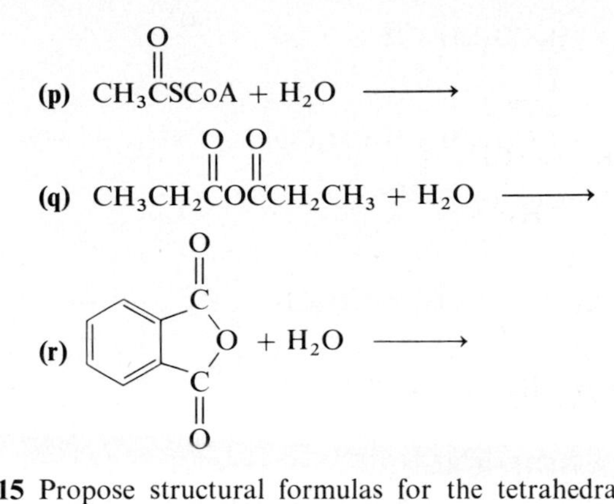

(p) $CH_3\overset{O}{\overset{\|}{C}}SCoA + H_2O \longrightarrow$

(q) $CH_3CH_2\overset{O}{\overset{\|}{C}}O\overset{O}{\overset{\|}{C}}CH_2CH_3 + H_2O \longrightarrow$

(r) (phthalic anhydride) $+ H_2O \longrightarrow$

9.15 Propose structural formulas for the tetrahedral carbonyl addition intermediates formed during the following reactions.

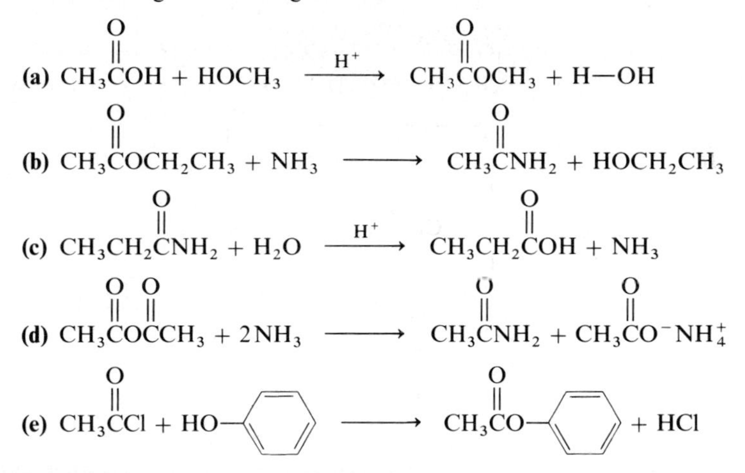

(a) $CH_3\overset{O}{\overset{\|}{C}}OH + HOCH_3 \xrightarrow{H^+} CH_3\overset{O}{\overset{\|}{C}}OCH_3 + H{-}OH$

(b) $CH_3\overset{O}{\overset{\|}{C}}OCH_2CH_3 + NH_3 \longrightarrow CH_3\overset{O}{\overset{\|}{C}}NH_2 + HOCH_2CH_3$

(c) $CH_3CH_2\overset{O}{\overset{\|}{C}}NH_2 + H_2O \xrightarrow{H^+} CH_3CH_2\overset{O}{\overset{\|}{C}}OH + NH_3$

(d) $CH_3\overset{O}{\overset{\|}{C}}O\overset{O}{\overset{\|}{C}}CH_3 + 2NH_3 \longrightarrow CH_3\overset{O}{\overset{\|}{C}}NH_2 + CH_3\overset{O}{\overset{\|}{C}}O^-NH_4^+$

(e) $CH_3\overset{O}{\overset{\|}{C}}Cl + HO{-}C_6H_5 \longrightarrow CH_3\overset{O}{\overset{\|}{C}}O{-}C_6H_5 + HCl$

9.16 Draw structural formulas for the products of the following reactions.

(a) $CH_3CH_2CO_2CH_3 + CH_3CH_2NH_2 \longrightarrow$

(b) $o\text{-}HOC_6H_4CO_2H + CH_3\overset{O}{\overset{\|}{C}}O\overset{O}{\overset{\|}{C}}CH_3 \longrightarrow$

(c) $CH_3CH_2O{-}C_6H_4{-}NH_2 + CH_3\overset{O}{\overset{\|}{C}}O\overset{O}{\overset{\|}{C}}CH_3 \longrightarrow$ (phenacetin, a pain reliever)

(d) $C_6H_5CO_2CH_3 + 2C_6H_5MgBr \longrightarrow$

(e) $C_6H_5CO_2CH_3 + LiAlH_4 \longrightarrow$

(f) $C_6H_5CO_2CH_3 + NH_3 \longrightarrow$

(g) $o\text{-}HOC_6H_4CO_2H + CH_3OH \xrightarrow{H^+}$ (oil of wintergreen)

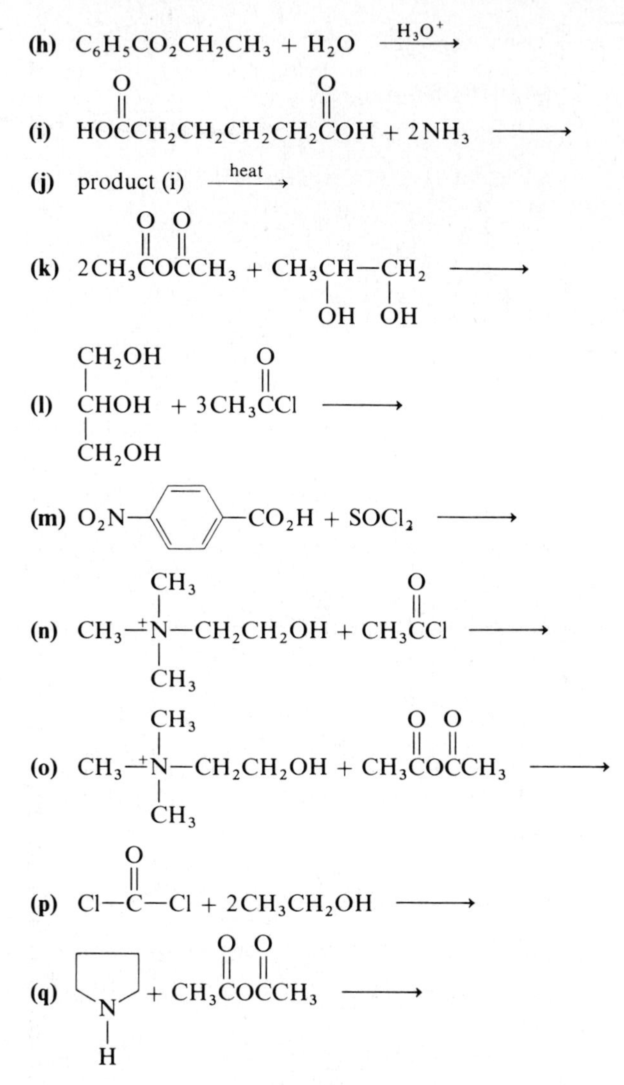

9.17 Following are structural formulas of two drugs widely used in clinical medicine. The first is a tranquilizer. Miltown is one of the several trade names for this substance. The second drug, phenobarbital, is a long acting sedative, hypnotic, and central nervous system depressant. Phenobarbital is used to treat mild hypertension and temporary emotional strain.

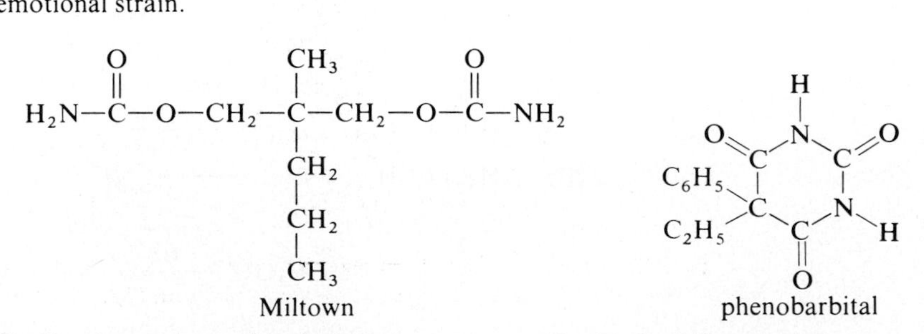

Miltown phenobarbital

Predict the products of hydrolysis in aqueous acid of each of these substances.

9.18 The following conversions can be done in either one, two, or three steps. Show the reagents you would use and draw structural formulas for any intermediates involved in two- or three-step conversions. In addition to the indicated starting material, use any necessary inorganic and organic compounds.

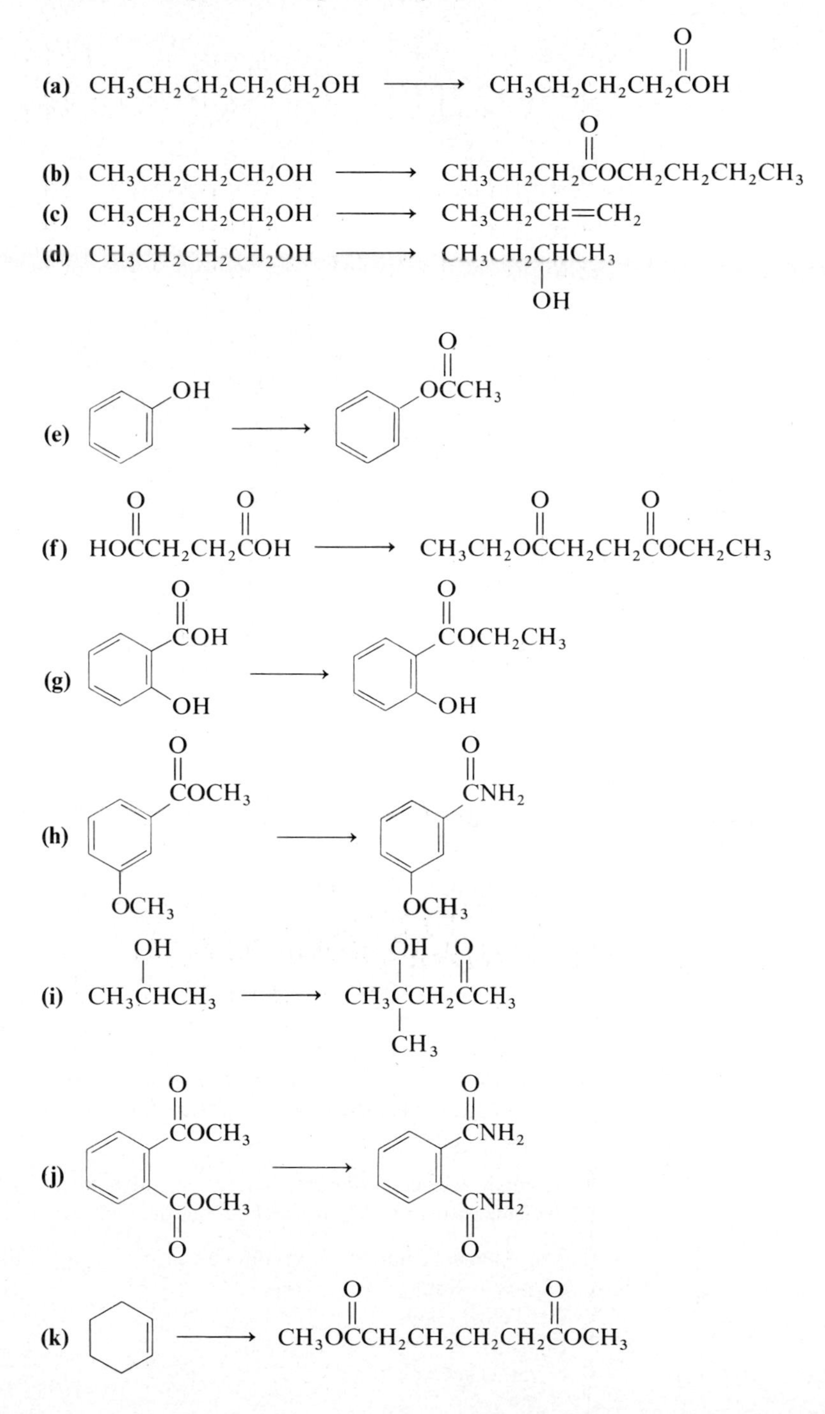

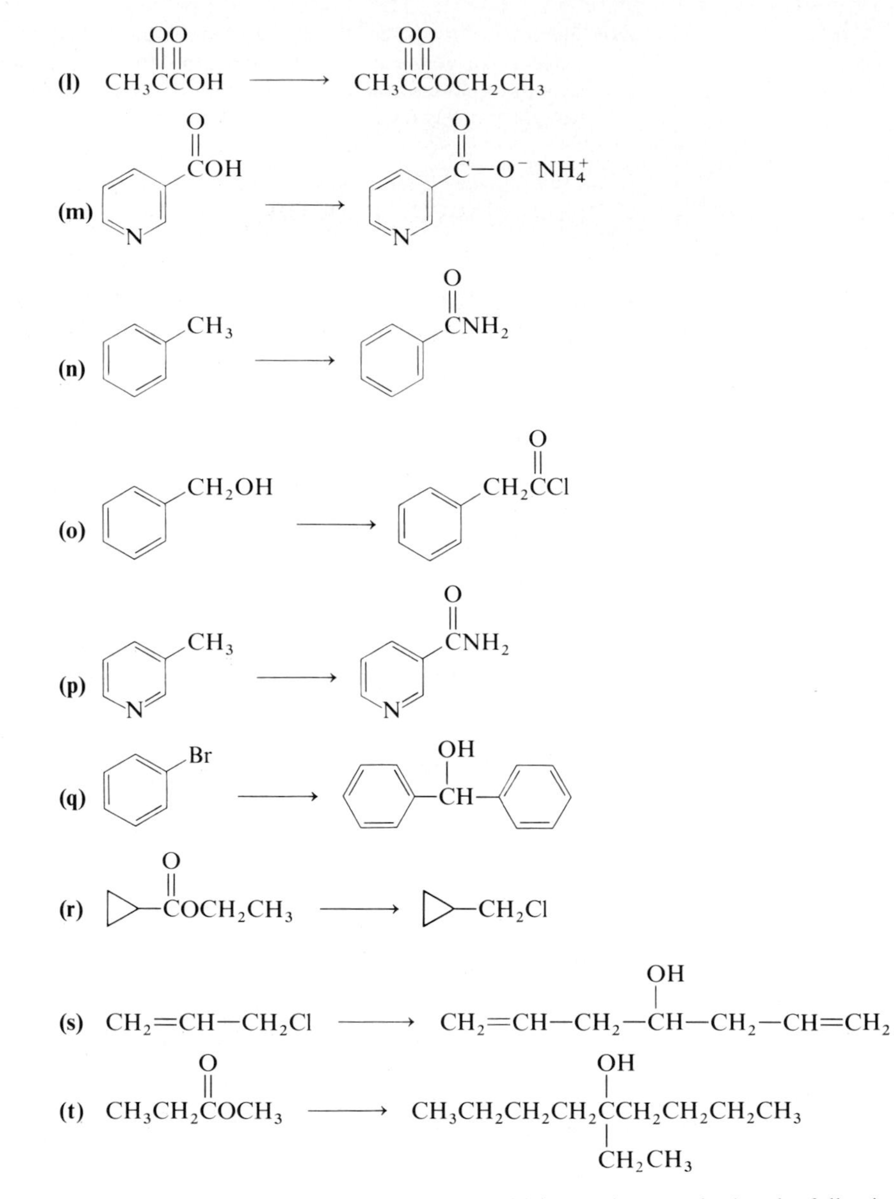

9.19 Suggest a combination of reagents that could be used to synthesize the following barbiturates: phenobarbital, thiopental. (Structural formulas for each are given in Section 9.6C).

9.20 Compare acid-catalyzed ester formation with acid-catalyzed acetal formation, and list the similarities and differences between the two reactions.

9.21 The following compounds all contain a carbon-oxygen linkage. Compare the reactivity of each to water.

(a) diethyl ether
(b) acetic anhydride
(c) ethyl acetate

9.22 The following compounds are derivatives of carboxylic acids. Compare the reactivities of each to water.

(a) ethyl acetate
(b) acetic anhydride
(c) acetamide
(d) acetyl chloride

9.23 Explain why it is preferable to hydrolyze esters in aqueous base rather than in aqueous acid.

9.24 **(a)** Write an equation for the equilibrium established when acetic acid and 1-propanol are refluxed in the presence of a few drops of concentrated sulfuric acid.
(b) Using the data in Section 9.5, calculate the equilibrium constant for this reaction.

9.25 If 15g of salicylic acid is reacted with an excess of acetic anhydride, how many grams of aspirin could be formed?

9.26 Draw structural formulas for the products of the following carbonyl condensations.

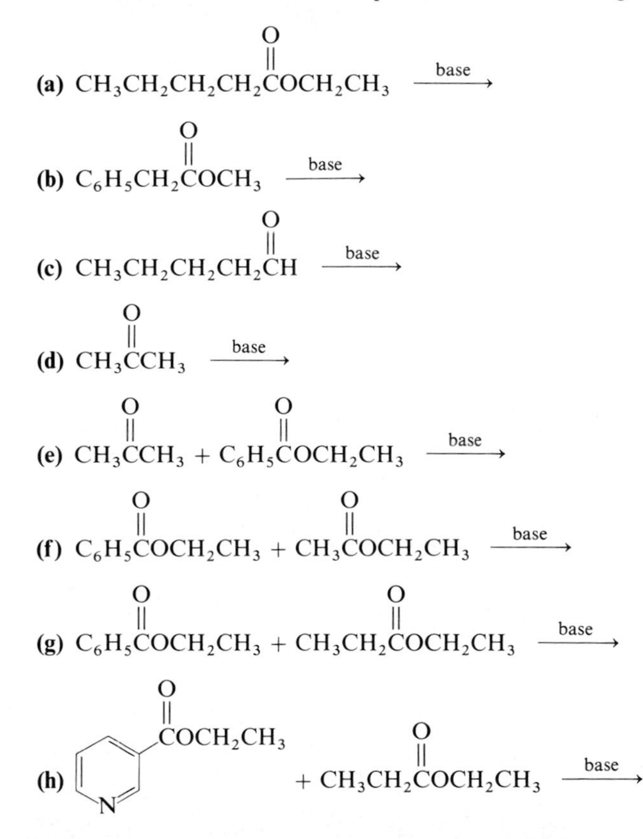

9.27 Write equations to show how you could convert ethyl propanoate into the following substances.

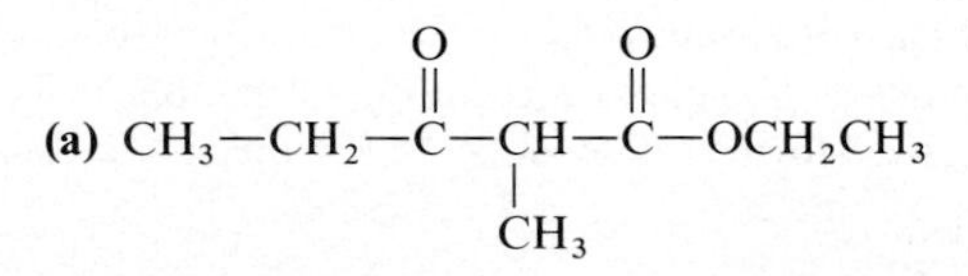

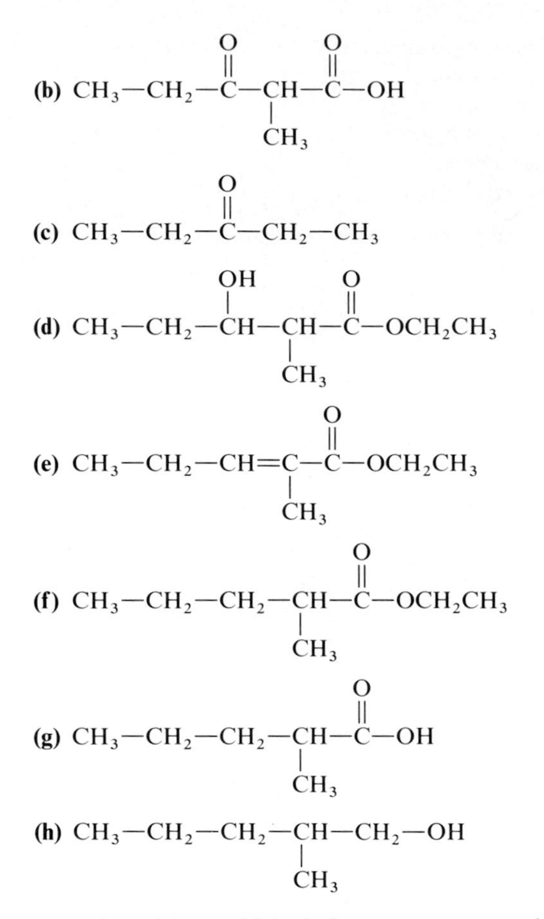

9.28 Starting with any aldehyde, ketone, or ester, synthesize each of the following by using the aldol or Claisen condensation and any necessary subsequent steps.

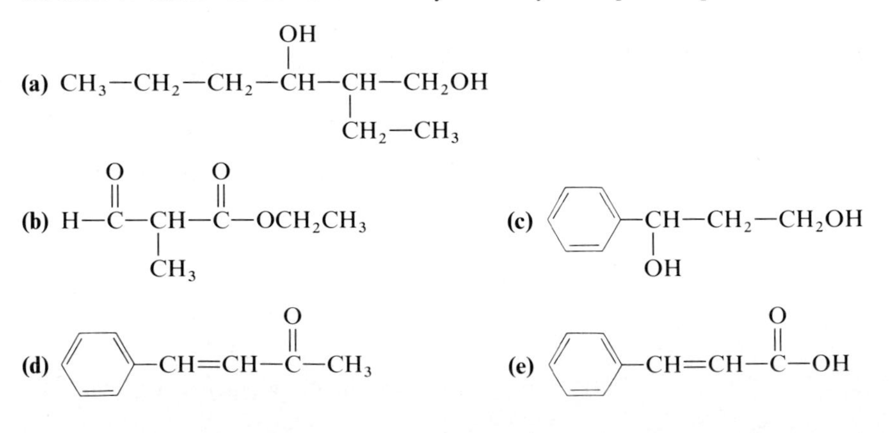

9.29 One biosynthetic route to the formation of aromatic compounds, including derivatives of benzene, is thought to involve the formation and cyclization of a polycarbonyl chain derived form acetate units. An example is the biosynthesis of orsellinic acid by the mold *Penicillium urticae*. If acetate supplied to the mold is labeled at the carbonyl carbon with carbon-14, orsellinic acid has the indicated label pattern.

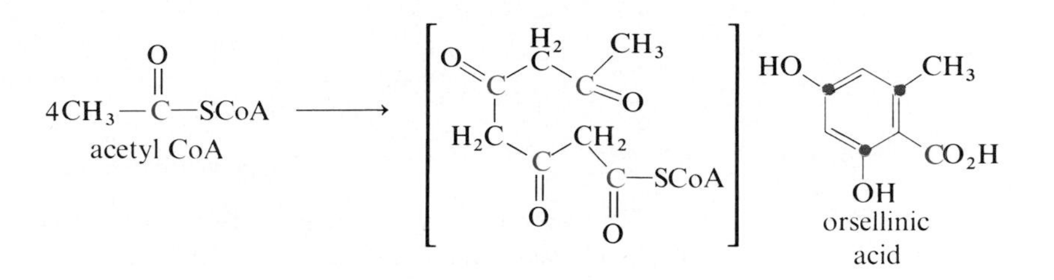

(a) Propose a series of reactions involving successive Claisen-like condensations for biosynthesis of the polycarbonyl compound shown in brackets.

(b) Propose a series of reactions for conversion of the bracketed polycarbonyl compound into orsellinic acid. (*Hint*: use a combination of aldol condensation, dehydration, ketoenol tautomerism, and ester hydrolysis.)

9.30 The same polycarbonyl compound shown in brackets in problem 9.29 can also cyclize in an alternative manner to give phloroacetophenone.

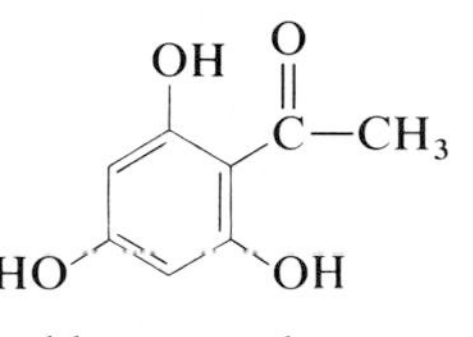

phloroacetophenone

(a) Propose a series of reactions for conversion of the bracketed polycarbonyl compound to phloroacetophenone.

(b) If acetate supplied as a starting material for the synthesis of phloroacetophenone is labeled in the carbonyl carbon with carbon-14, predict the position of carbon-14 labels in phloroacetophenone.

9.31 Carboxylic acids and alcohols may be converted to esters by a variety of chemical methods. When the acid contains more than one —COOH group and the alcohol contains more than one —OH group, then under the appropriate experimental conditions, hundreds of molecules may be linked together to give a polyester. Dacron is a polyester of terephthalic acid and ethylene glycol.

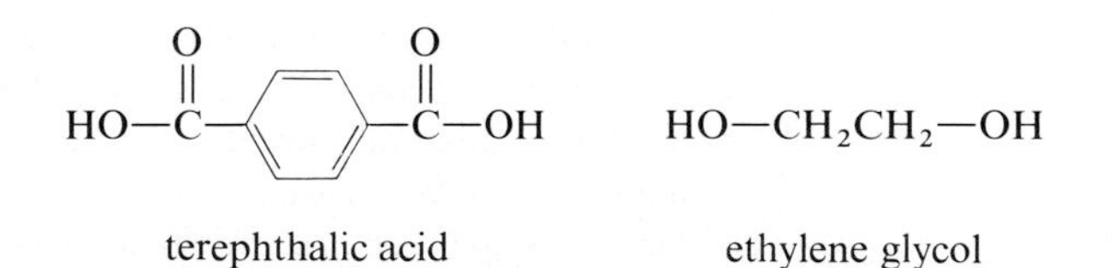

(a) Formulate a structure for Dacron polyester. Be certain to show in principle how several hundred molecules can be hooked together to form the polyester.

(b) Write an equation for the chemistry involved when a drop of concentrated hydrochloric acid makes a hole in a Dacron polyester shirt.

(c) From what starting materials do you think the condensation fiber Kodel Polyester is made?

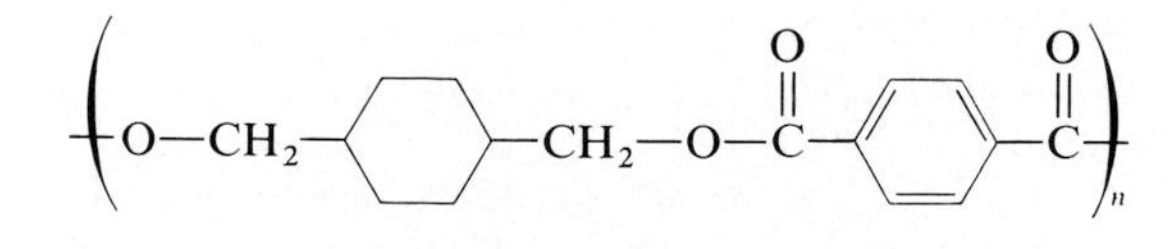

Kodel Polyester

9.32 Show how you might distinguish between the following pairs of compounds by a simple chemical test. In each case, tell what test you would perform, what you would expect to observe, and write an equation for each positive test.

(a) isopropyl formate and 2-methylpropanoic acid (isobutyric acid)
(b) butanoic acid and butanamide
(c) methyl benzoate and acetophenone (methyl phenyl ketone)
(d) methyl hexanoate and hexanal

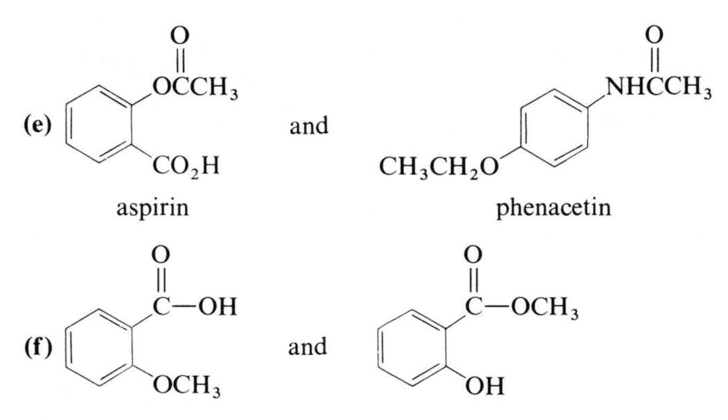

9.33 List one major spectral characteristic that will enable you to distinguish between the following compounds.

(a) isopropyl formate and 2-methylpropanoic acid (IR, NMR)
(b) butanoic acid and butanamide (IR)
(c) methyl benzoate and acetophenone (NMR)
(d) methyl hexanoate and hexanal (IR, NMR)
(e) aspirin and phenacetin (IR, NMR)
(f) phenyl acetate and cyclohexyl acetate (IR, NMR, UV)
(g) methyl benzoate and benzamide (IR, NMR)

9.34 There are four isomeric esters of molecular formula $C_4H_8O_2$.

(a) The NMR spectrum of one of these esters shows signals at $\delta = 1.3$ (doublet), $\delta = 5.3$ (septet), and $\delta = 8.0$ (singlet). Draw the structural formula of this ester.
(b) Draw the structural formula of the other ester whose NMR spectrum will also show a singlet at $\delta = 8.0$.
(c) The NMR spectra of the remaining two esters each show three signals; a singlet, a triplet, and a quartet. Show how these esters can be distinguished by the chemical shift of the singlet; by the chemical shift of the quartet.

Nylon and Dacron

MINI-ESSAY 8

A NUMBER OF CHEMISTS recognized the need to develop a basic knowledge of polymer chemistry, shortly after World War I. One of the most creative of these pioneers was Wallace M. Carothers of E. I. du Pont de Nemours & Co., Inc. In the early 1930s, Carothers and his associates began fundamental research into the reactions of aliphatic dicarboxylic acids and dialcohols. From adipic acid and ethylene glycol they obtained a polyester of high molecular weight that could be drawn into fibers. However, melting points of the first polyester fibers obtained by Carothers were too low for them to be used as textile fibers and they were not investigated further. (That is, not for another decade.) Carothers then turned his attention to the reactions of dicarboxylic acids and diamines. In 1934 he synthesized Nylon 66, the first purely synthetic fiber. Nylon 66 is so named because it is synthesized from two different organic starting materials, each of six carbon atoms.

In the chemical synthesis of Nylon 66, adipic acid and hexamethylene diamine (HMDA) dissolved in aqueous alcohol react to form a one-to-one salt, commonly called nylon salt.

Nylon salt is then heated in an autoclave to 250°C. As the temperature increases in the closed system, the internal pressure rises to about 250 pounds per square inch (about 15 atmospheres). Under these conditions $—CO_2^-$ and $—NH_3^+$ groups react to form amides and water is formed as a by-product.

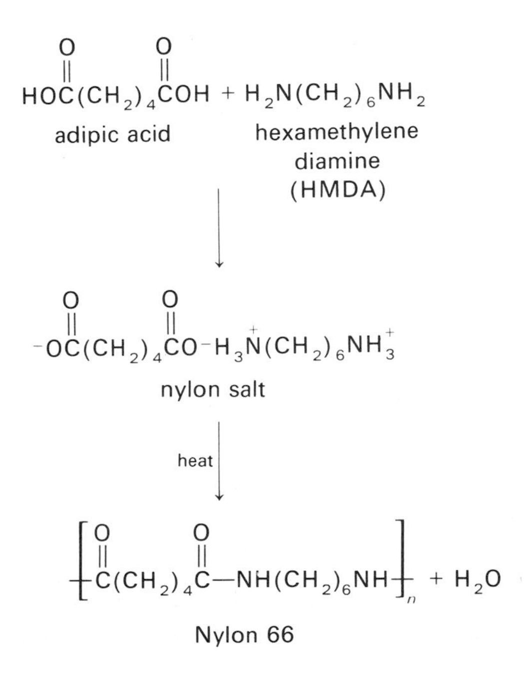

Synthetic textile fiber is being rolled from a storage hopper into a baling press for shipment to mills which will then process it into spun yarns on cotton, woolen, worsted, or other established spinning systems. (E. I. du Pont de Nemours & Co.)

is benzene. During the synthesis of Nylon 6, caprolactam is partially hydrolyzed and heated to 250°C to drive off water and bring about polymerization.

Why is Nylon 66 the primary nylon fiber produced in the United States and Canada while Nylon 6 is the primary nylon fiber produced in Germany, Italy, and Japan? The answer lies chiefly in the availability of different raw material bases. In the United States and Canada, butadiene is readily available from thermal cracking of light hydrocarbons extracted from natural gas, itself a vast natural resource. Natural gas is not as plentiful in Europe and Japan. Therefore, these countries are forced to depend on petrochemicals as a raw material base for their synthesis of nylon. It is more economical for them to synthesize Nylon 6 from caprolactam than it is to synthesize Nylon 66 from adipic acid and hexamethylene diamine.

By the 1940s, scientists were beginning to understand some of the relationships between molecular structure and bulk physical properties, and the polyester condensations were reexamined. Recall that Carothers and his associates had already concluded that the polyester fibers from aliphatic dicarboxylic acids and ethylene glycol were not suitable for textile use because they were too low melting. Winfield and Dickson at the Calico Printers Association in England reasoned, quite correctly as it turned out, that a greater resistance to rotation in the polymer backbone would stiffen the polymer, raise its melting point, and thereby lead to a more acceptable polyester fiber. To create stiffness in the polymer chain, they used terephthalic acid, an aromatic dicarboxylic acid (see Figure 3).

The crude polyester is first spun into fibers and then cold-

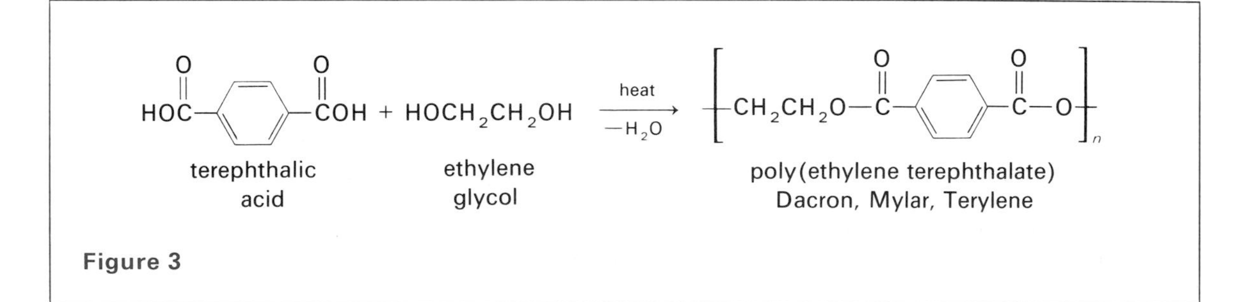

Figure 3

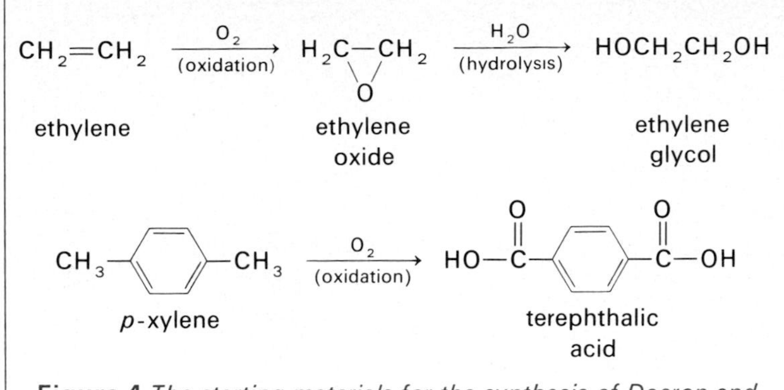

Figure 4 *The starting materials for the synthesis of Dacron and Mylar are derived from petroleum and natural gas.*

drawn to form a textile fiber tradenamed Dacron polyester. The outstanding feature of Dacron polyester is its stiffness (about four times that of Nylon 66), very high strength, and a remarkable resistance to creasing and wrinkling. Because Dacron polyesters are harsh to the touch (because of their stiffness), they are usually blended with cotton or wool to make acceptable textile fibers. Crude polyester is also fabricated into films and marketed under the trade name Mylar.

The raw material bases for ethylene glycol and terephthalic acid, starting materials for the synthesis of Dacron and Mylar, are petroleum and natural gas. Ethylene glycol is prepared by air oxidation of ethylene followed by hydrolysis. Terephthalic acid is obtained by oxidation of *p*-xylene, an aromatic hydrocarbon obtained with benzene and toluene from catalytic cracking and reforming of naphtha and other petroleum fractions (Figure 4).

In 1979 production of man-made fibers in the United States exceeded 10 billion pounds. Heading the list were polyester fibers (4.18 billion pounds) and polyamide fibers (2.72 billion pounds).

The fruits of the research into the relationships between molecular structure and physical properties of polymers, and of advances in fabrication techniques are no where better illustrated than by the polyaromatic amides or aramids introduced by du Pont in the early 1960s. The goal was a polyamide with increased heat resistance, flammability resistance, and that could be drawn into fibers even tougher and stronger than Nylon 66. The results were the aramids marketed under the tradename Kevlar. Figure 5 gives the structural formula for the repeating unit of the aramid synthesized from polymerization of terephthalic acid and 1,4-diaminobenzene (*p*-phenylene diamine). Aramids can be drawn into fibers and then woven into cables as strong as steel but with only one-fifth the weight of steel. One use of aramid cables is to anchor off-shore drilling platforms.

An innovative use of two of the polymers described in this essay is the manufacture of a new material for the construction of sails. The material, manufactured by du Pont, consists of a Mylar polyester film laminated to a woven Kevlar aramid fabric. The high-strength and low-stretch properties of this material provide a sailcloth equal in performance to other more conventional materials but at a much lower weight.

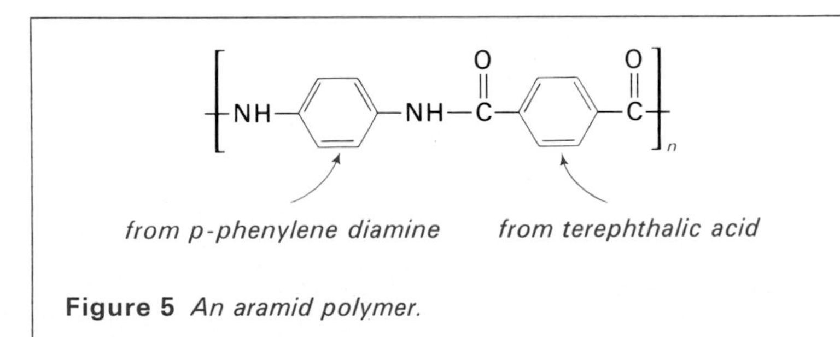

Figure 5 *An aramid polymer.*

References

Anderson, B. C., Barton, L. R. and Collette, J. W., "Trends in Polymer Development," *Science* 208 (1980): 807–812.

Deanin, R. D., *New Industrial Polymers*, (Washington, D. C.: Am. Chemical Society, 1979).

Encyclopedia of Polymer Science and Technology, (N.Y.: Wiley, 1976).

Stille, J. K., *Industrial Organic Chemistry*, (Englewood Cliffs, N.J.: Prentice-Hall, 1968).

Wittcoff, H. A. and Reuben, B. G., *Industrial Organic Chemicals in Perspective*, (N.Y.: Wiley, 1980).

CHAPTER 10

Amines

In Chapter 5 we discussed the structure of alcohols and ethers—substances in which first one and then both hydrogens of H_2O are replaced by alkyl or aromatic groups. In this chapter, we will discuss amines, derivatives of ammonia in which one, two, or three hydrogens of NH_3 are replaced by alkyl or aromatic groups. The most important chemical property of amines is their basicity.

10.1 THE STRUCTURE OF AMINES

The structure of simple alkyl amines is much like that of ammonia, NH_3. Shown in Figure 10.1 are structural formulas and ball-and-stick models of methylamine, dimethylamine, and trimethylamine.

Amines are classified as primary, secondary, or tertiary depending on the number of hydrogen atoms that have been replaced by carbon atoms. In a primary amine, one hydrogen is replaced by a carbon atom. In a secondary amine, two hydrogens are replaced by carbon atoms. In a tertiary amine, three hydrogens are replaced by carbon atoms.

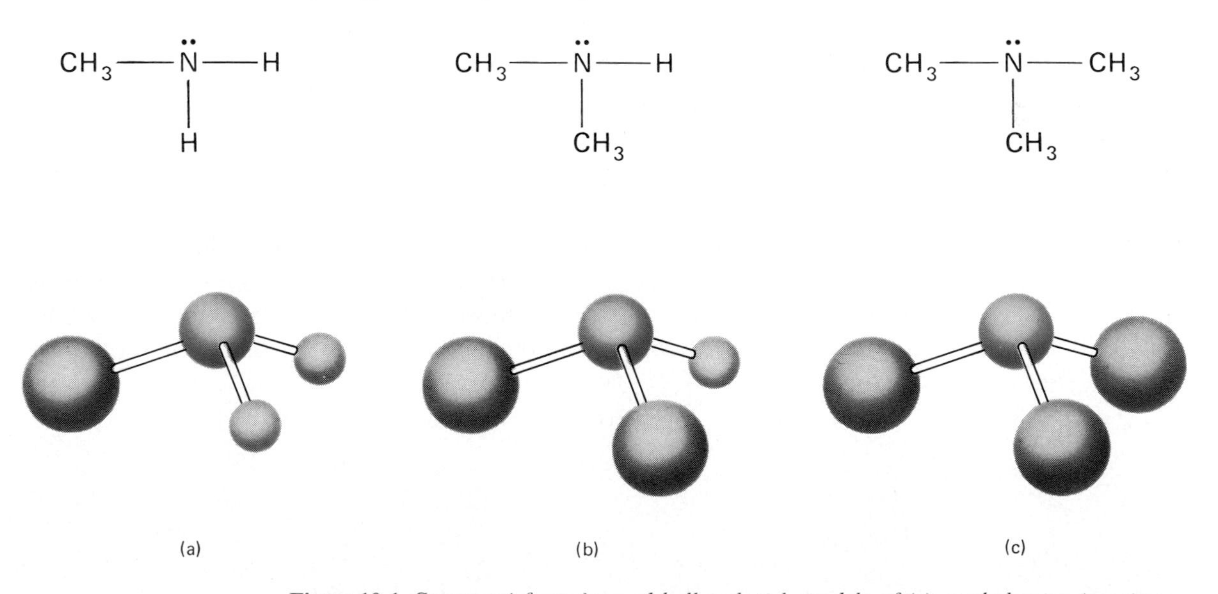

Figure 10.1 *Structural formulas and ball-and-stick models of (a) methylamine (a primary amine); (b) dimethylamine (a secondary amine), and (c) trimethylamine (a tertiary amine).*

In amines, nitrogen uses three sp^3 hybrid orbitals to form sigma bonds with three other atoms. The unshared pair of electrons on the nitrogen atom is in the remaining sp^3 hybrid orbital. Therefore, we predict bond angles of 109.5° about the nitrogen atom.

10.2 NOMENCLATURE OF AMINES

The simple aliphatic amines are named by specifying the alkyl groups attached to the nitrogen atom and adding the ending -amine.

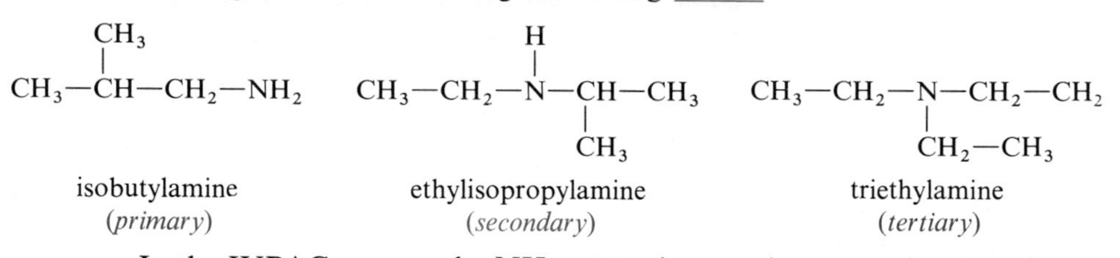

isobutylamine (*primary*) ethylisopropylamine (*secondary*) triethylamine (*tertiary*)

In the IUPAC system, the NH_2 group is named as an amino substituent (like chloro-, nitro-, etc.).

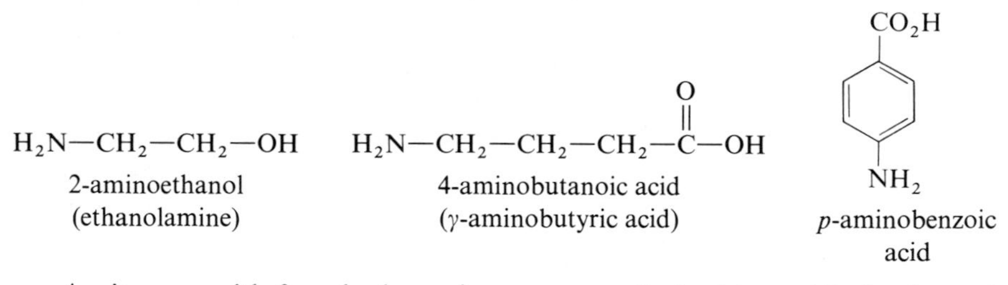

2-aminoethanol (ethanolamine) 4-aminobutanoic acid (γ-aminobutyric acid) *p*-aminobenzoic acid

A nitrogen with four hydrocarbon groups attached is positively charged and is known as a quaternary ammonium ion.

tetramethylammonium chloride tetramethylammonium hydroxide

Compounds containing an $—NH_2$ group on a benzene ring are named as derivatives of aniline. Many simple substitution derivatives are also known by common names, for example, anisidine (methxoyaniline) and toluidine (methylaniline).

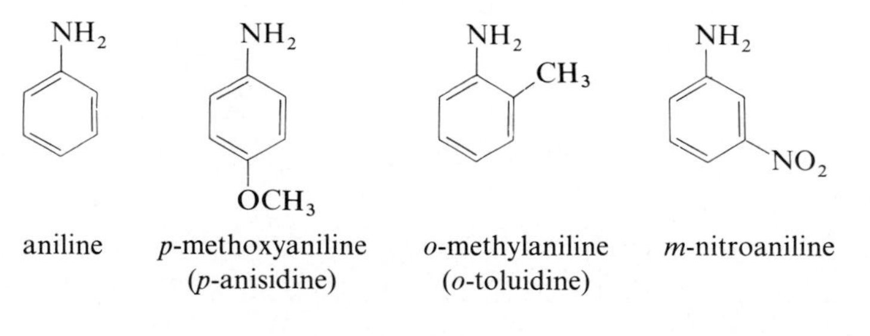

aniline *p*-methoxyaniline (*p*-anisidine) *o*-methylaniline (*o*-toluidine) *m*-nitroaniline

Heterocyclic amines, that is, cyclic compounds in which the amine nitrogen is one of the atoms of a ring, are given special names. Following are structural formulas for piperidine and pyrrolidine, both secondary amines. N-Methylpyrrolidine is an example of a heterocyclic amine in which the amine is tertiary.

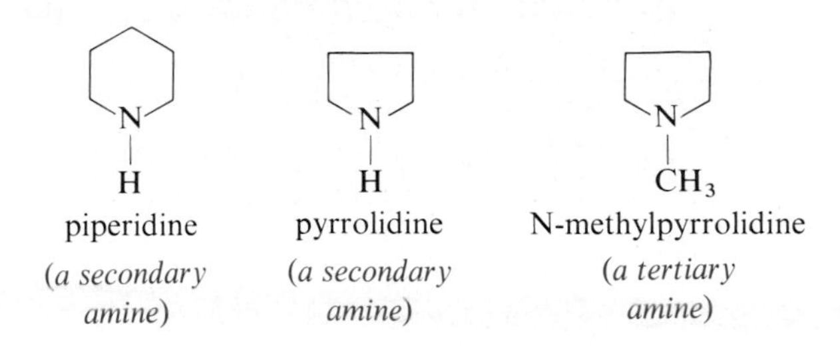

Of special interest are amines in which a nitrogen atom replaces a carbon atom in an aromatic ring. These amines are known as heterocyclic aromatic amines. In Section 6.8 we examined the structure and nomenclature of several heterocyclic aromatic amines including pyridine, pyrimidine, pyrrole, and imidazole.

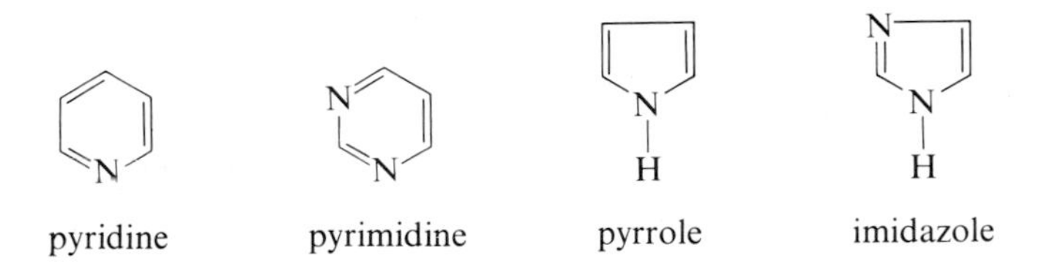

These amines show chemical properties more like those of aromatic hydrocarbons and we will not discuss them further in this chapter.

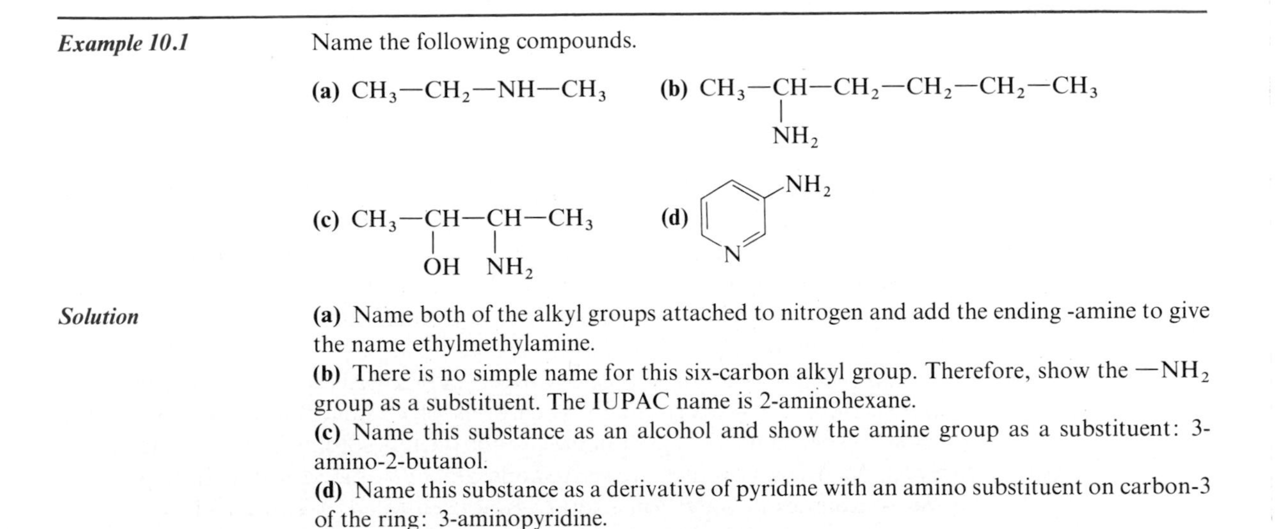

Example 10.1 Name the following compounds.

(a) $CH_3—CH_2—NH—CH_3$

(b) $CH_3—CH(NH_2)—CH_2—CH_2—CH_2—CH_3$

(c) $CH_3—CH(OH)—CH(NH_2)—CH_3$

(d) [pyridine ring with NH_2 substituent]

Solution

(a) Name both of the alkyl groups attached to nitrogen and add the ending -amine to give the name ethylmethylamine.

(b) There is no simple name for this six-carbon alkyl group. Therefore, show the $—NH_2$ group as a substituent. The IUPAC name is 2-aminohexane.

(c) Name this substance as an alcohol and show the amine group as a substituent: 3-amino-2-butanol.

(d) Name this substance as a derivative of pyridine with an amino substituent on carbon-3 of the ring: 3-aminopyridine.

PROBLEM 10.1 Name the following compounds.

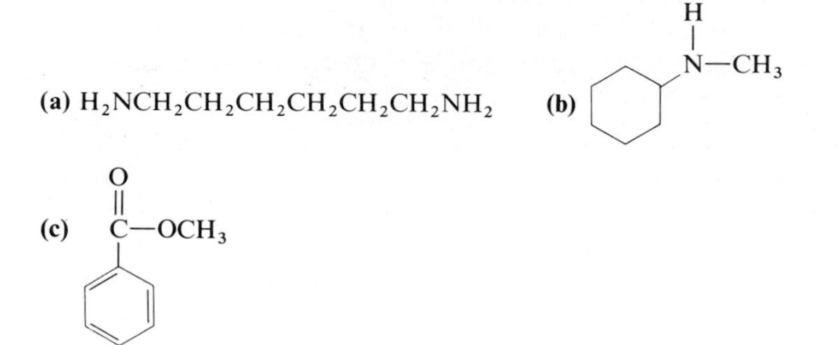

10.3 PHYSICAL PROPERTIES OF AMINES

Amines are polar substances and both primary and secondary amines can form intermolecular hydrogen bonds.

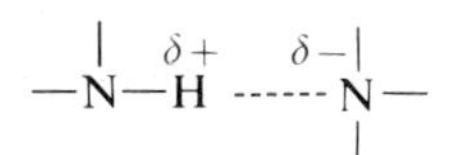

However, because the difference in electronegativity between nitrogen and hydrogen ($3.0 - 2.1 = 0.9$) is not as great as that between oxygen and hydrogen ($3.5 - 2.1 = 1.4$), the N—H···N hydrogen bond is not nearly as strong as is the O—H···O hydrogen bond. The boiling points of ethane, methylamine, and methanol, all compounds of comparable molecular weight, are

	CH_3—CH_3	CH_3—NH_2	CH_3—OH
mol. wt.	30	31	32
bp °C	−88	−7	65

Ethane is a nonpolar hydrocarbon, and the only interaction between its molecules in the pure liquid are very weak dispersion forces. Therefore, it has the lowest boiling point of the three. Both methylamine and methanol are polar molecules, and they interact in the pure liquid by hydrogen bonding. Hydrogen bonding is weaker in methylamine than methanol, and therefore methylamine has a lower boiling point than does methanol.

All classes of amines form hydrogen bonds with water and therefore are more soluble in water than hydrocarbons of comparable molecular weight. Most low-molecular-weight amines are completely soluble in water. The higher-molecular-weight amines are only moderately soluble in water. Boiling points and solubilities in water for several amines are listed in Table 10.1.

Table 10.1 Physical properties of amines.

Name	*bp*(°C)	*Solubility* (g/110 g H_2O)	K_b	pK_b	pK_a
ammonia	−33	90	1.8×10^{-5}	4.74	9.26
methylamine	−7	∞	4.4×10^{-4}	3.34	10.66
ethylamine	17	∞	6.3×10^{-4}	3.20	10.80
diethylamine	55	∞	3.1×10^{-4}	3.51	10.49
triethylamine	89	1.5	1.0×10^{-3}	3.00	11.00
cyclohexylamine	134	slightly	5.5×10^{-4}	3.34	10.66
benzylamine	185	∞	2.1×10^{-5}	4.67	9.33
aniline	184	3.7	4.2×10^{-10}	9.37	4.63
pyridine	116	∞	1.8×10^{-9}	8.75	5.25
imidazole	257	∞	8.9×10^{-8}	7.05	6.95

10.4 PREPARATION OF AMINES

A. ALKYLATION OF AMMONIA AND AMINES

One of the characteristic reactions of ammonia and amines is participation in nucleophilic substitution (S_N2) reactions, as illustrated by the synthesis of methylamine from methyl chloride and ammonia.

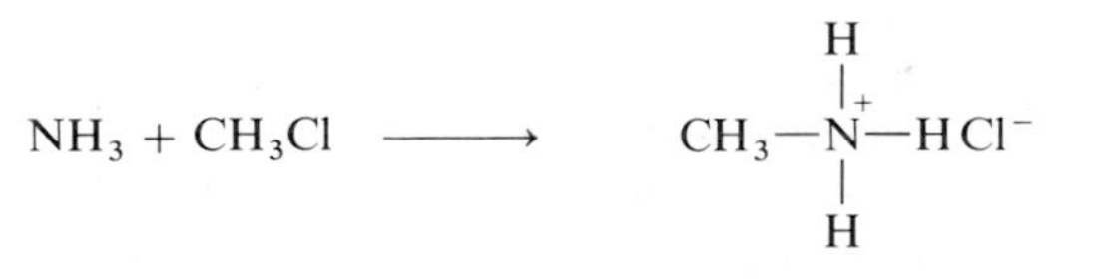

methylammonium chloride

The reaction forms a new bond between nitrogen and an alkyl group and therefore is called alkylation. Note that the initial product is a substituted ammonium salt. This salt is in equilibrium with other protonated species through transfer of a proton.

$$CH_3NH_3^+ + NH_3 \longrightarrow CH_3NH_2 + NH_4^+$$

Methylamine may undergo further alkylation to yield dimethylamine, trimethylamine, or a tetramethylammonium salt. The alkylation proceeds in stages and it is possible to maximize the yield of the primary, secondary, or tertiary amine, or quaternary ammonium ion through control of the concentrations of the reactants, reaction times, and other experimental conditions.

Intramolecular alkylation to produce a pyrrolidine ring is illustrated by the following reaction in one laboratory synthesis of nicotine.

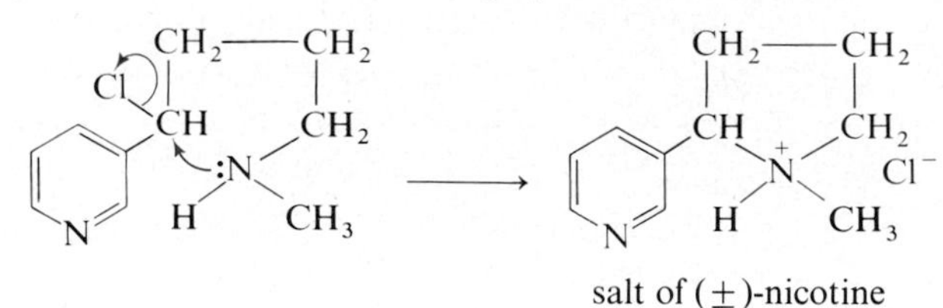

salt of (±)-nicotine

B. REDUCTION OF NITRO GROUPS

Aromatic amines are almost always made from reduction of aromatic nitro compounds. Commonly used reducing agents are iron and steam, zinc and hydrochloric acid, or hydrogen gas in the presence of a heavy metal catalyst such as platinum, palladium, or nickel.

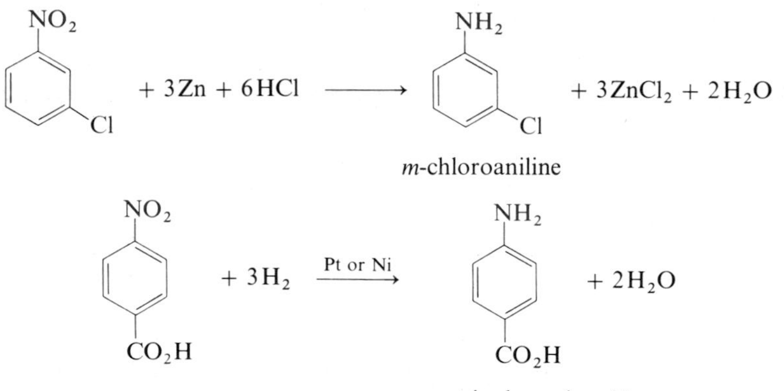

m-chloroaniline

p-aminobenzoic acid

Since nitro groups can be introduced easily into an aromatic ring, nitration followed by reduction is one of the best methods for preparing aromatic amines.

C. REDUCTION OF AMIDES

An especially valuable method for preparing pure primary, secondary, or tertiary aliphatic amines is the reduction of amides in the presence of a heavy metal catalyst or by lithium aluminum hydride.

$$H_2N\overset{\overset{O}{\|}}{C}(CH_2)_4\overset{\overset{O}{\|}}{C}NH_2 + 4H_2 \xrightarrow[\text{heat, pressure}]{\text{catalyst}} H_2NCH_2(CH_2)_4CH_2NH_2 + 2H_2O$$

hexanediamide (adipamide) → 1,6-hexanediamine (hexamethylenediamine)

$$CH_3CH_2\overset{\overset{O}{\|}}{C}NH_2 \xrightarrow[\text{(2) } H_2O]{\text{(1) } LiAlH_4} CH_3CH_2CH_2NH_2$$

propanamide → propylamine

Catalytic reduction of amides requires high temperatures and pressures of hydrogen. However, reduction by lithium aluminum hydride proceeds smoothly at room temperature. The starting amides are available from ammonolysis of esters (Section 9.6C), anhydrides (Section 9.10C), and acid chlorides (Section 9.8C). This synthesis of amines is particularly useful for the preparation of secondary and tertiary amines having different alkyl or aryl groups on nitrogen.

10.5 REACTIONS OF AMINES

A. BASICITY

Like ammonia, all primary, secondary, and tertiary amines are weak bases and aqueous solutions of amines are basic.

$$NH_3 + H_2O \rightleftharpoons NH_4^+ + OH^-$$

$$CH_3NH_2 + H_2O \rightleftharpoons CH_3NH_3^+ + OH^-$$

The equilibrium constant expression for the reaction of methylamine and water is

$$K_b = \frac{[CH_3NH_3^+][OH^-]}{[CH_3NH_2]} = 4.4 \times 10^{-4}$$

Values for K_b for some primary, secondary, tertiary, and aromatic amines are given in Table 10.1. All aliphatic amines, whether primary, secondary, tertiary, or saturated heterocyclic amines, have about the same base strength as ammonia.

Aromatic amines such as aniline are significantly less basic than aliphatic amines. The K_b of aniline is less than that of cyclohexylamine by a factor of 10^6.

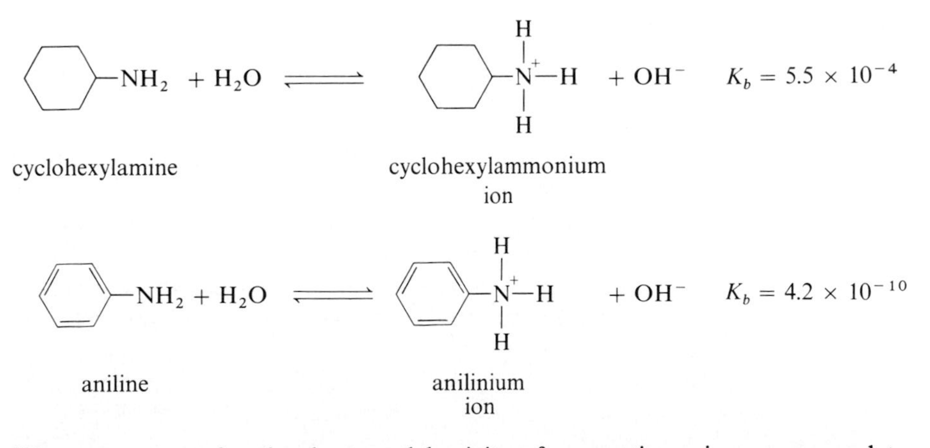

We can account for the decreased basicity of aromatic amines compared to aliphatic amines by using the resonance theory. When comparing two similar equilibria, the position of equilibrium is shifted toward the side favored by resonance. Both aniline and the anilinium ion are stabilized by resonance. For each, two Kekulé structures can be drawn. There are three additional contributing structures that can be drawn for aniline, and although they involve the separation of unlike charge, they nevertheless make a contribution to the hybrid. No such structures involving the separation of unlike charge can be drawn for the anilinium ion because all electron pairs on nitrogen are used in forming sigma bonds.

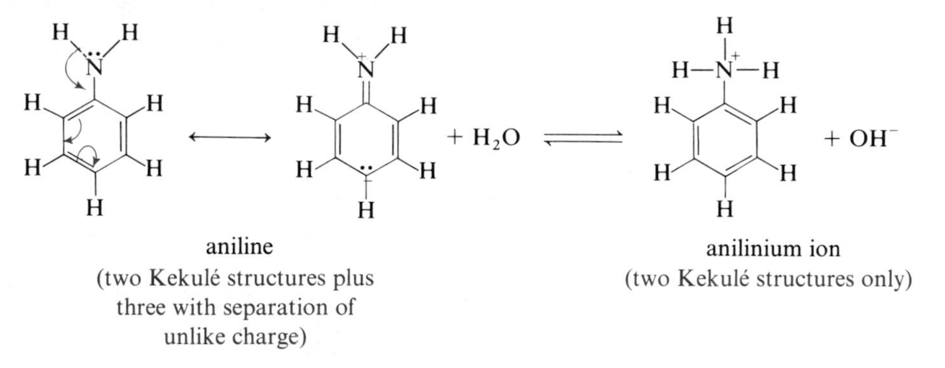

Because of the greater resonance stabilization of aniline compared to that of the anilinium ion, the position of equilibrium is shifted to the left compared to that of cyclohexylamine; consequently, aniline is a weaker base than cyclohexylamine.

To extend our discussion of the basicity of the $-NH_2$ group one step further, let us compare the basicities of cyclohexylamine, aniline, and acetamide—all compounds containing the $-NH_2$ group.

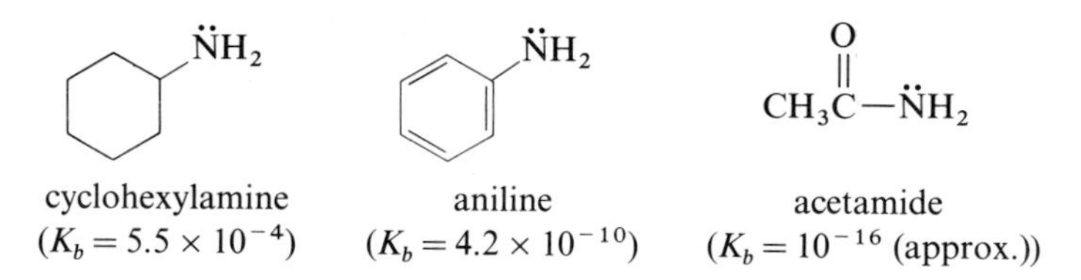

Although amines are bases (and aliphatic amines are stronger bases than aromatic amines), amides are neutral molecules. The basicity of $-NH_2$ containing molecules depends on the availability of the electron pair on nitrogen for bonding with a proton. In the case of aniline compared to cyclohexylamine, the electron pair is less available for bonding with a proton because of resonance interaction with the benzene ring. Much the same argument applies to amides. The amide group is best represented as a hybrid of two contributing structures.

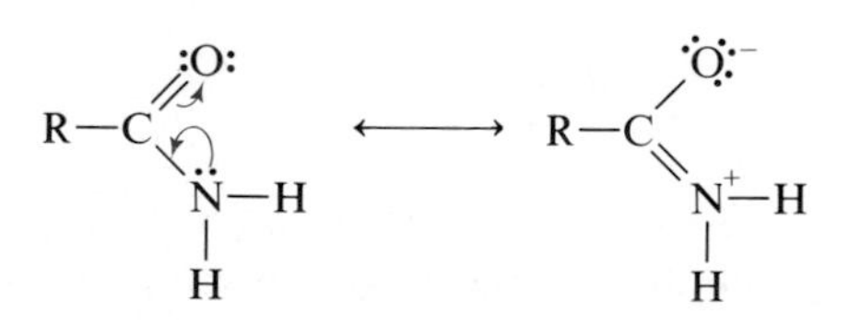

While we cannot predict in advance just how important this resonance stabilization might be, the fact that amides are neutral molecules indicates that resonance stabilization is very significant.

One final note about the basicity of amines. Until recently it was common practice to list only K_b of pK_b values for amines. Now, however, it is more and more common to list only K_a and pK_a values for amines. For example, only K_a and pK_a values are used in discussing the acid-base properties of the amino group of amino acids (Chapter 12). For this reason, Table 10.1 also lists pK_a values for a variety of common amines. Note that $pK_a + pK_b = 14$.

To illustrate the difference between pK_b and pK_a for an amine, consider methylamine.

$$CH_3NH_2 + H_2O \rightleftharpoons CH_3NH_3^+ + OH^- \qquad K_b = 4.4 \times 10^{-4} \qquad pK_b = 3.34$$

$$CH_3NH_3^+ \rightleftharpoons CH_3NH_2 + H^+ \qquad K_a = 2.7 \times 10^{-11} \qquad pK_a = 10.66$$

pK_b measures directly the strength of CH_3NH_2 as a base while pK_a measures directly the strength of $CH_3NH_3^+$ as an acid. For perspective, you might compare the pK_a values for acetic acid and for methylamine.

$$CH_3CO_2H \rightleftharpoons CH_3CO_2^- + H^+ \qquad K_a = 1.8 \times 10^{-5} \qquad pK_a = 4.74$$

$$CH_3NH_3^+ \rightleftharpoons CH_3NH_2 + H^+ \qquad K_a = 2.7 \times 10^{-11} \qquad pK_a = 10.66$$

By using K_a or pK_a values for carboxylic acids and amines, we can compare them directly, for in each case we are looking at the dissociation of an acid to form a base and a proton. It is obvious that acetic acid is a much stronger acid than methylammonium ion.

All amines, whether soluble or insoluble in water, react quantitatively with acids to form substituted ammonium ions.

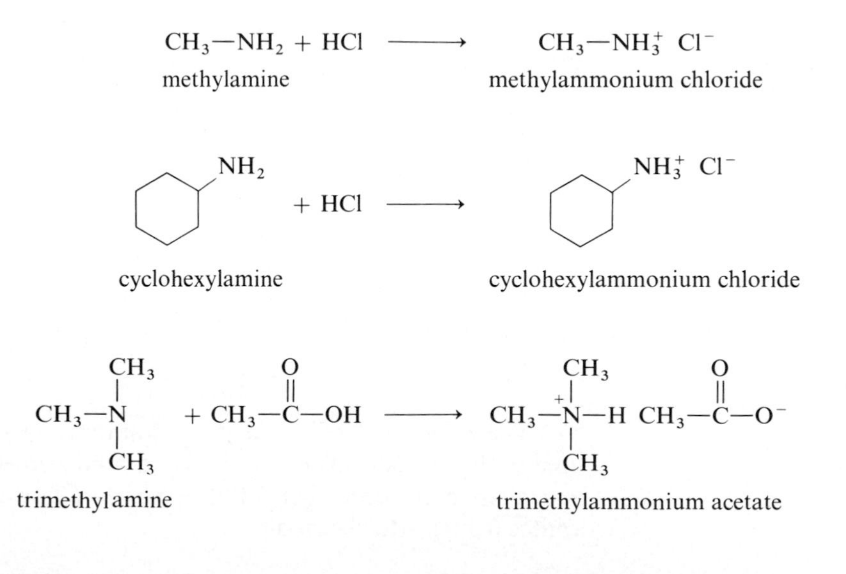

Example 10.2 Complete the following acid-base reactions.

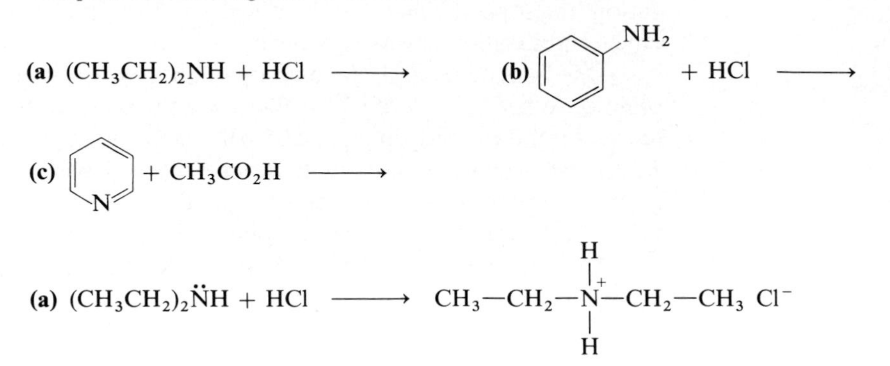

Solution

$$(CH_3CH_2)_2\ddot{N}H + HCl \longrightarrow CH_3{-}CH_2{-}\overset{+}{N}H_2{-}CH_2{-}CH_3\ Cl^-$$

Diethylamine, a secondary amine, reacts with HCl to form the salt diethylammonium chloride.

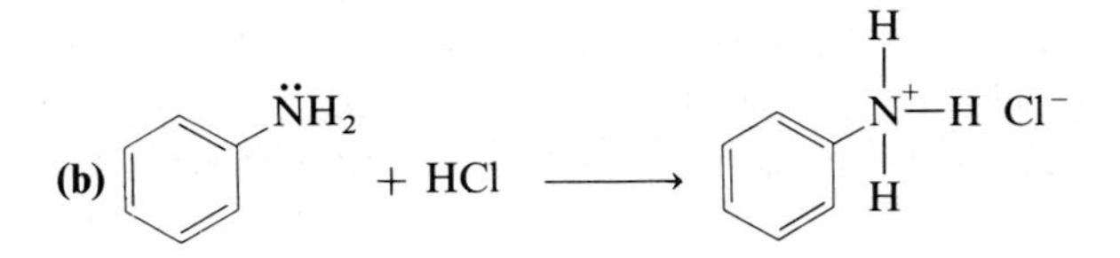

Aniline, a primary aromatic amine, reacts with hydrochloric acid to form the salt anilinium chloride. Another name for this salt is aniline hydrochloride.

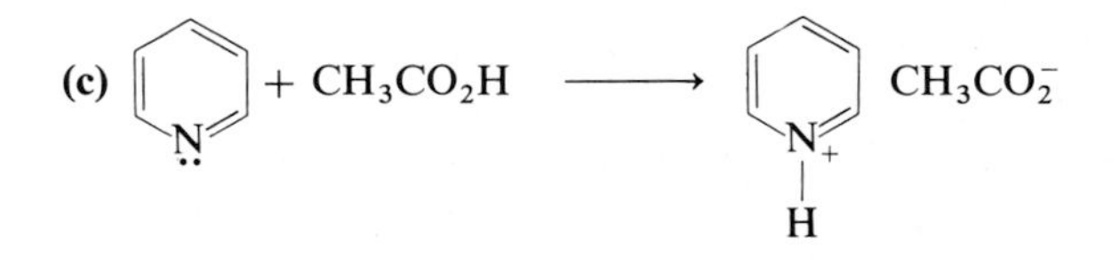

Pyridine, a heterocyclic aromatic amine, reacts with acetic acid to form the salt pyridinium acetate.

PROBLEM 10.2 Complete the following acid-base reactions.

(a) $HO{-}CH_2{-}CH_2{-}N(CH_3){-}CH_3 + HCl \longrightarrow$

(b) $(CH_3CH_2)_3N + C_6H_5CO_2H \longrightarrow$

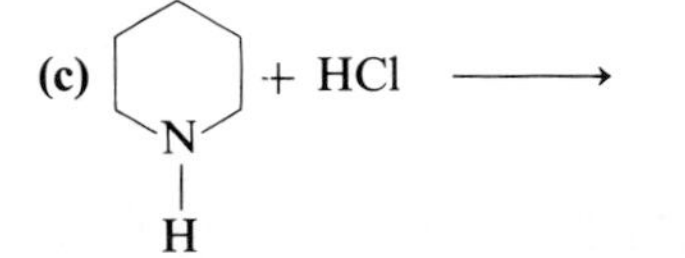

The basicity of amines and the solubility of amine salts in water can be used to distinguish between amines and nonbasic, water-insoluble compounds, and also to separate them. Following is a flowchart for the separation of aniline from methyl benzoate.

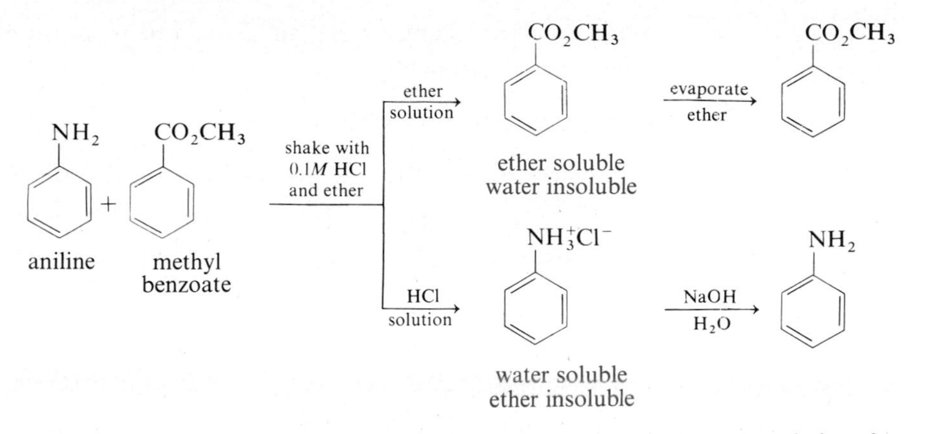

Aniline and methyl benzoate are only slightly soluble in water, and therefore cannot be separated on the basis of their water solubility. However, both dissolve in ether. When an ether solution of the two substances is shaken with dilute aqueous hydrochloric acid, aniline reacts to form a water-soluble salt. Methyl benzoate remains in the ether layer. Separation of the ether layer and evaporation of the ether yields methyl benzoate. Treatment of the aqueous solution with sodium hydroxide converts the water-soluble salt to free aniline which then separates as a water-insoluble layer.

B. REACTION WITH CARBOXYLIC ACID DERIVATIVES TO GIVE AMIDES

As we have already seen in Chapter 9, primary and secondary amines react with esters, acid chlorides, and acid anhydrides to give amides. Following are examples of each of these reactions.

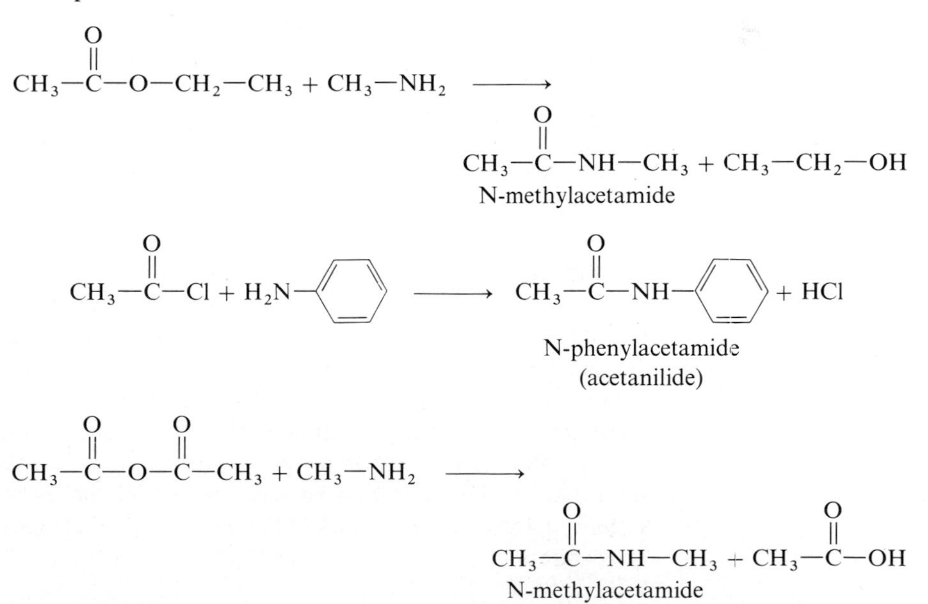

We also saw in Chapter 9 that amides can be prepared by heating ammonium salts of carboxylic acids.

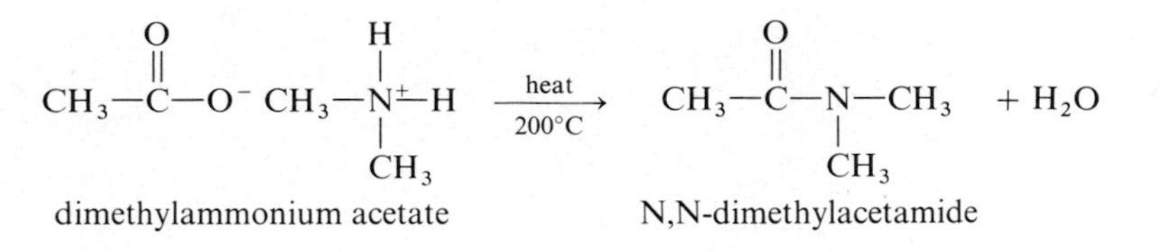

C. REACTION WITH NITROUS ACID

Tertiary amines react with nitrous acid in a typical acid-base reaction to form ammonium salts.

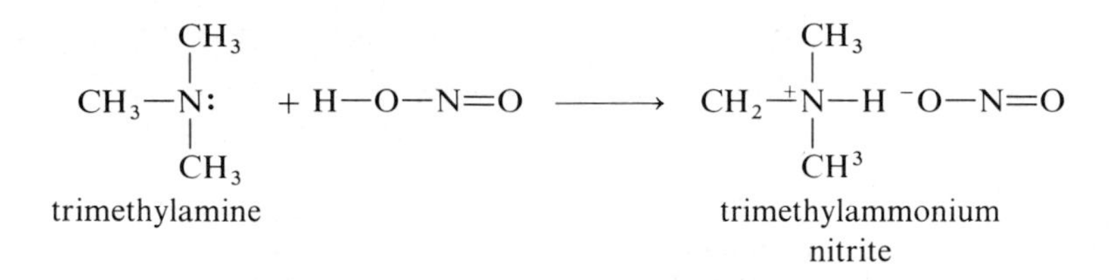

With secondary amines, the reaction is somewhat more complex and leads to the formation of N-nitrosoamines, commonly called nitrosamines. For example, the reaction of dimethylamine and nitrous acid gives N-nitrosodimethylamine (dimethylnitrosamine).

$$CH_3{-}\underset{\displaystyle CH_3}{\underset{|}{N}}{-}H + H{-}O{-}N{=}O \longrightarrow CH_3{-}\underset{\displaystyle CH_3}{\underset{|}{N}}{-}N{=}O + H{-}OH$$

dimethylamine → N-nitrosodimethylamine (dimethylnitrosamine)

Sodium nitrite, a substance that reacts with acids to give nitrous acid, is used in many industrial processes and commercial products. For example, it was common in years past, especially before the development of adequate refrigeration for the transport and storage of meats, to add $NaNO_2$ to retard spoilage, prevent botulism food poisoning, and to preserve the red color of processed meats. Thus, there are many potential sources of nitrite in the environment. However, for humans, the environment is not the only source of nitrite. Many foods contain nitrate as natural constituents. Nitrate is reduced to nitrite in the saliva and by bacterial action in the intestine. Thus there is the potential for internally generated nitrites to react with amines and produce nitrosoamines.

Nitrosoamines are also formed by the reaction of amines with certain oxides of nitrogen as, for example, N_2O_3.

$$CH_3{-}\underset{\substack{|\\ CH_3}}{N}{-}H + O{=}N{-}\overset{+}{N}\!\begin{smallmatrix} {\diagup\!\!\diagup} O \\ \diagdown O^- \end{smallmatrix} \longrightarrow CH_3{-}\underset{\substack{|\\ CH_3}}{N}{-}N{=}O + H{-}O{-}N{=}O$$

Because of the prevalence of low-molecular-weight secondary amines in the environment and of nitrites and oxides of nitrogen, N-nitrosoamines have been found in such places as in the air in tannery plants, rubber tire plants, iron foundries, new car interiors, and in such consumer products as beer, whiskey, lotions, shampoos, and cooked nitrite-cured bacon.

What makes N-nitrosoamines of special concern is that more than 130 of these substances have been tested as chemical carcinogens in animals and over 100 have produced tumors in one or more species and in one or more organs. Most studies have been done on rodents. The possible link between nitrosoamines and human cancer is being studied in several ways. First, improvements are being made in analytical techniques to detect these substances in the environment. Second, expanded studies of the occurrence and distribution of various forms of cancer are being carried out by both government and private research groups, and third, the chemistry and biochemistry of N-nitrosoamines are being studied intensively.

Primary amines react with nitrous acid to form diazonium ions. In the case of a primary aliphatic amine, the diazonium ion is unstable and decomposes to give nitrogen gas and a carbocation. The carbocation then reacts with water to form an alcohol or loses a proton to form an alkene.

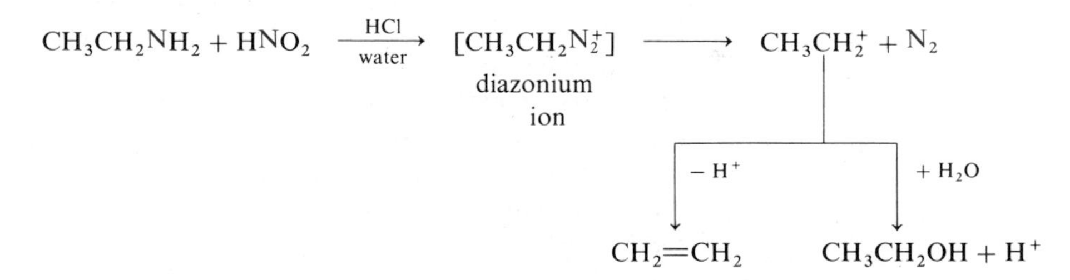

For each mole of primary amine, one mole of N_2 is formed, and therefore the reaction of a primary amine with nitrous acid can be used as a quantitative measure of the number of moles of primary amine present in a sample of organic material.

Primary aromatic amines also react with cold nitrous acid to give diazonium salts. Aromatic diazonium salts are stable at 0°C, but they too decompose with the evolution of nitrogen gas when they are warmed to room temperature. In water, phenol is the organic product of the decomposition.

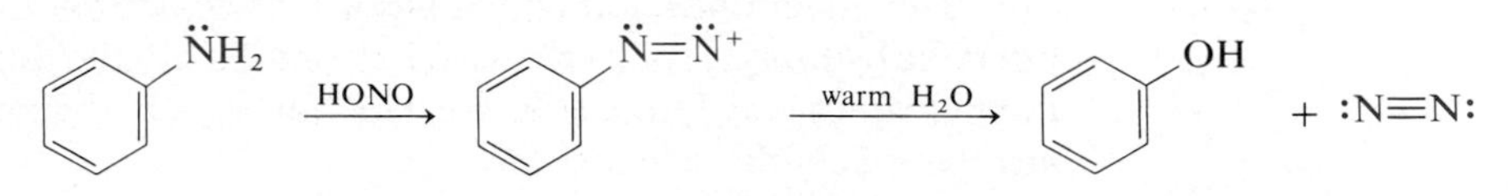

If the diazonium salt is warmed in the presence of cuprous chloride, cuprous bromide, or potassium iodide, an aryl halide is formed.

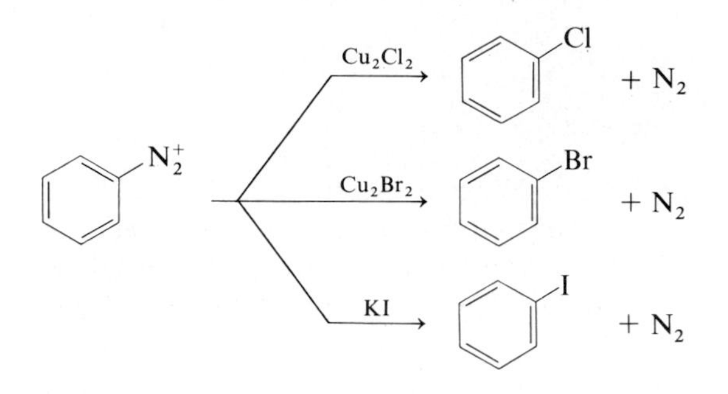

Thus, aromatic amines and the aromatic nitro compounds from which they can be synthesized are a very convenient synthetic route to a large number of aromatic substances.

Diazonium ions also react with aromatic rings by electrophilic aromatic substitution.

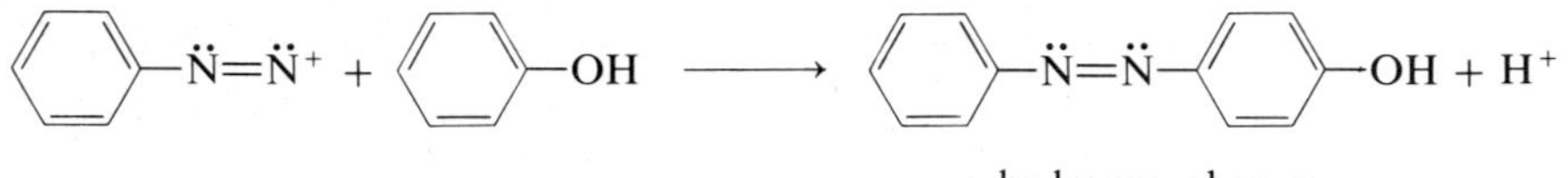

p-hydroxyazobenzene

Such reactions are often referred to as diazo coupling reactions. Notice that in this reaction both nitrogen atoms are retained. Because the diazonium ion is not a powerful electrophile, diazo coupling reactions are limited to phenols, aromatic amines, and a few other aromatic compounds.

The products of diazo coupling reactions are called azo compounds. The simple azo compounds are yellow or orange. With increasing complexity and substitution on either of the aromatic rings, an almost unlimited range of colors and shades can be produced. Azo compounds have found extensive use as dyes and indicators. Following is shown the structural formula of the commercial dye, congo red.

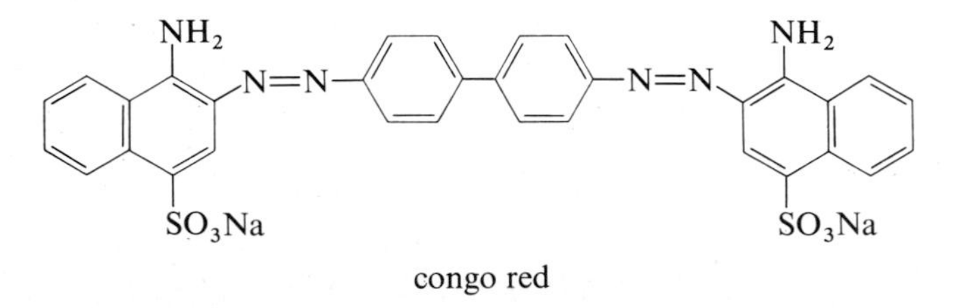

congo red

Azo compounds containing weakly basic amino groups or weakly acidic phenolic hydroxyl groups are useful as acid-base indicators. One such indicator, methyl orange, is formed by the reaction of the diazonium ion of sulfanilic acid with N,N-dimethylaniline.

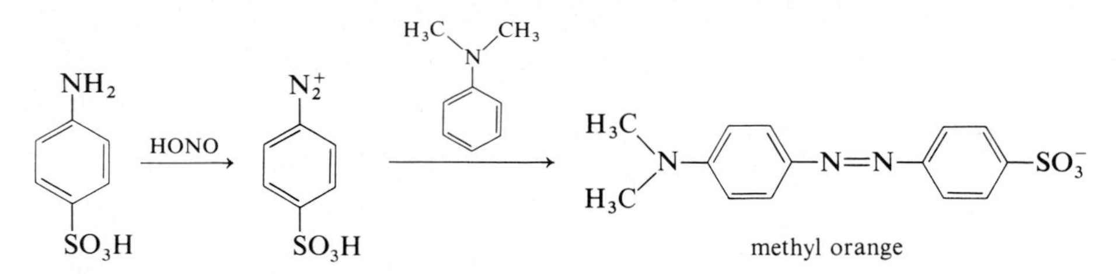

At pH 3.1 methyl orange is red; at pH 4.4 it is yellow-orange. Therefore this azo dye is widely used as an indicator in acid-base titrations where the end point of the titration is in the pH range 3.1 to 4.4.

10.6 SOME NATURALLY OCCURRING AMINES

Structural formulas of a variety of naturally occurring amines of both plant and animal origin are shown in Figure 10.2. These are chosen to illustrate something of the structural diversity and range of physiological activity of amines, their value as drugs, and their importance in nutrition. Coniine from the water hemlock is highly toxic. It can cause weakness, labored respiration, paralysis, and eventually death. The levorotatory isomer of coniine is the toxic substance in the "poison hemlock" used by Socrates to commit suicide. Nicotine is one of the chief heterocyclic amines of the tobacco plant. In small doses, nicotine is a stimulant. However, in larger doses it causes depression, nausea, and vomiting. In still larger doses, nicotine acts as a poison. Solutions of nicotine are often used as insecticides. Note that nicotinic acid, an oxidation product of nicotine, is one of the water-soluble vitamins required by humans for proper nutrition. You should also note that ingested or inhaled nicotine does not give rise to nicotinic acid in the body for humans have no enzyme systems capable of catalyzing this conversion. Smoking will not supply any vitamins! Quinine, isolated from the bark of the cinchona tree in South America, has been long used as a medicine for the treatment of malaria.

Histamine, formed by the decarboxylation of the amino acid histidine (Table 12.1), is a toxic substance present in all tissues of the body, combined in some manner with proteins. Extensive production of histamine occurs during hypersensitive allergic reactions, and the symptoms of this release are unfortunately familiar to most of us, in particular those who suffer from hay fever. The search for antihistamines—drugs that inhibit the effects of histamine—has led to the synthesis of several drugs whose trade names are well known. Structural formulas for three of the more widely used antihistaminic drugs are shown in Figure 10.3.

Observe the structural similarity in these three drugs: each has two aromatic rings and a dimethylaminoethyl, $-CH_2CH_2N(CH_3)_2$, group. Dexbrompheniramine is one of the most potent of the available antihistaminics. Notice that it has one chiral carbon atom. Pharmacological studies have shown that the (+)-isomer is nearly twice as potent as the racemic mixture and about thirty times as potent as the (−)-isomer.

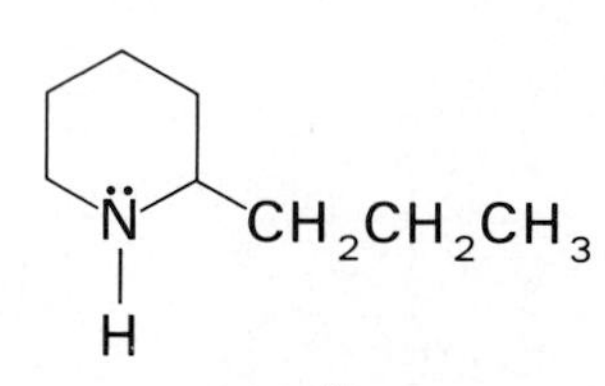

coniine
(from poison hemlock)

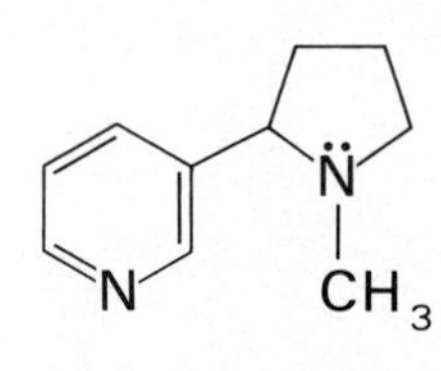

nicotine
(from tobacco)

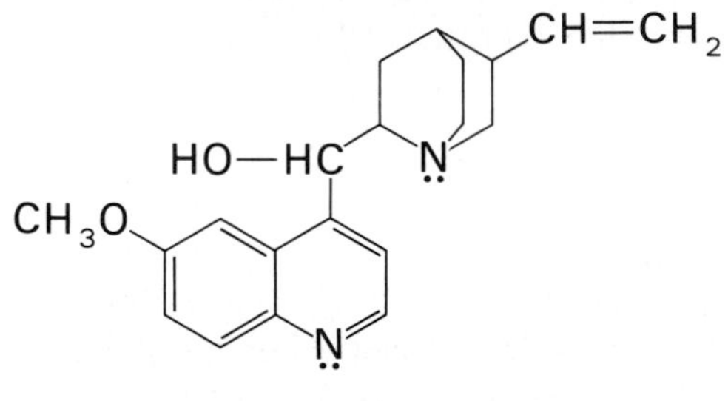

quinine

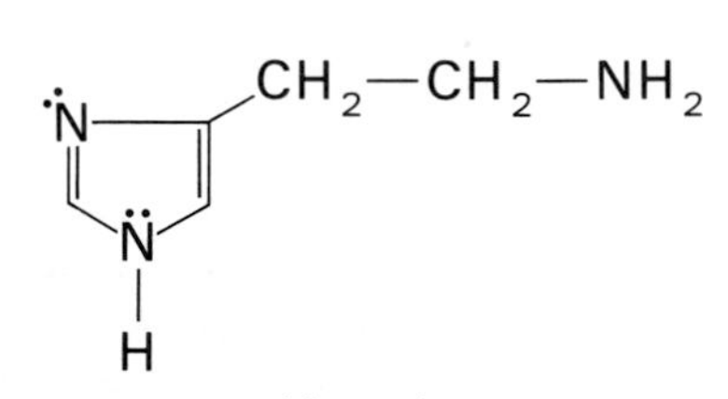

histamine

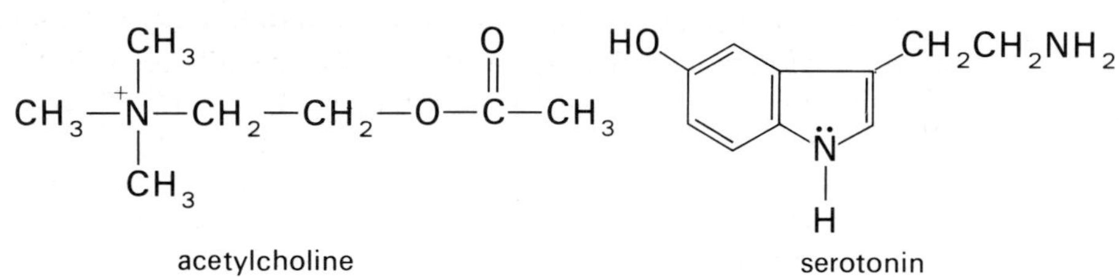

acetylcholine

serotonin
(5-hydroxytryptamine)

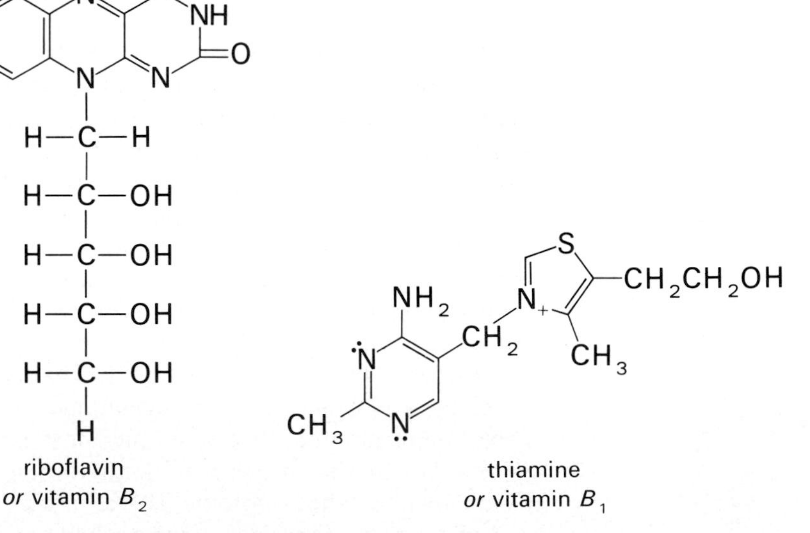

riboflavin
or vitamin B_2

thiamine
or vitamin B_1

Figure 10.2 *Several naturally occurring amines of plant and animal origin.*

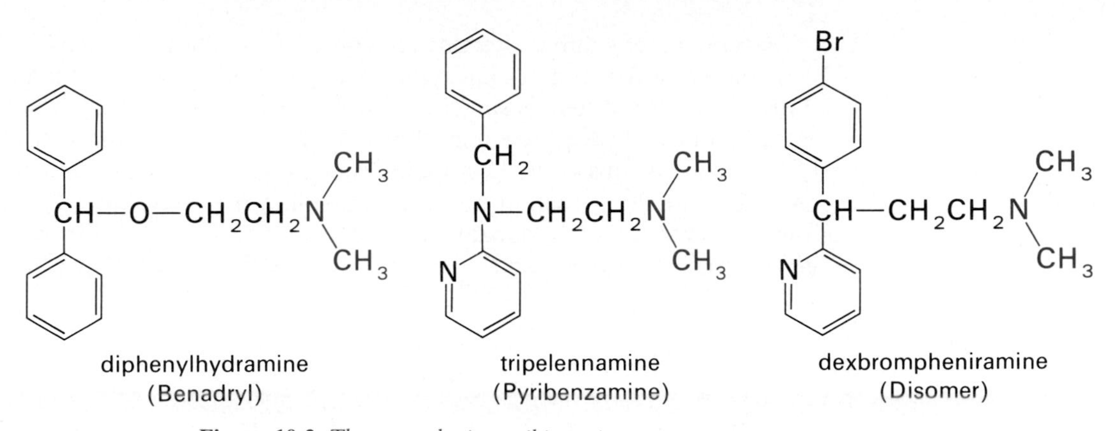

Figure 10.3 *Three synthetic antihistamines.*

Serotonin and acetylcholine are both neurotransmitter substances important in human physiology; serotonin in parts of the central nervous system mediating affective behavior, acetyl choline in certain motor neurons responsible for causing contraction of voluntary muscles. Acetylcholine is stored in synaptic vesicles and released in response to electrical activity in the neuron. It diffuses across the synapse and interacts with receptor sites on a neighboring neuron to cause transmission of a nerve impulse. After interacting with receptor sites, acetylcholinc is dcactivatcd through hydrolysis catalyzed by the enzyme acetylcholinesterase.

$$\underset{\text{acetylcholine}}{CH_3-\overset{\overset{CH_3}{|}}{\underset{\underset{CH_3}{|}}{N^+}}-CH_2-CH_2-O-\overset{\overset{O}{\|}}{C}-CH_3} + H_2O \xrightarrow[\text{esterase}]{\text{acetylcholin-}} \underset{\text{choline}}{CH_3-\overset{\overset{CH_3}{|}}{\underset{\underset{CH_3}{|}}{N^+}}-CH_2-CH_2-OH} + \underset{\text{acetate}}{{}^-O-\overset{\overset{O}{\|}}{C}-CH_3}$$

There are several other classes of synthetic compounds that affect acetylcholine-mediated nerve transmission. Among the most widely known of these are the nerve gases and related insecticides. The nerve gases diisopropyl fluorophosphate (DFP) and Tabun are both potent inhibitors of acetylcholinesterase and a few milligrams of either can kill a person in a few minutes through paralysis and respiratory failure.

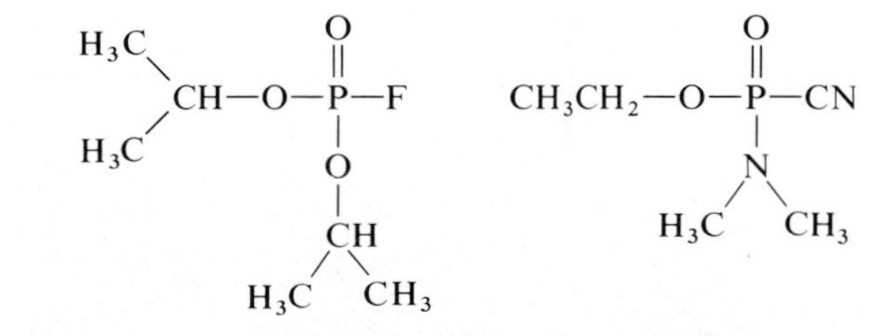

diisopropyl fluorophosphate (DFP) Tabun

Several water-soluble vitamins contain cyclic amines. Riboflavin (Figure 10.2) contains a fused three-ring system known as flavin. One of the nitrogen atoms of this ring system is substituted with a five-carbon chain derived from the sugar ribose, hence the name riboflavin. Thiamine or vitamin B_1 (Figure 10.2) contains a substituted pyrimidine ring as well as a five-membered ring containing one atom each of nitrogen and sulfur. You have already seen the structural formula of pyridoxal phosphate in Section 9.2. This substance, derived from pyridoxine, or vitamin B_6, contains a substituted pyridine ring.

10.7 SPECTROSCOPIC PROPERTIES

All primary and secondary amines show characteristic absorption between 3200 and 3500 cm^{-1} in the infrared region of the spectrum associated with stretching of the N—H bond. For primary amines, the N—H stretching frequency appears as a split peak at about 3500 cm^{-1}. Compare for example, the infrared spectrum of isopropylamine (Figure 10.4). For secondary amines, the N—H stretching frequency appears as a single peak at about 3500 cm^{-1}. Tertiary amines lack an N—H bond and therefore show no absorption in this region.

Saturated amines, whether they be primary, secondary, or tertiary, show no absorption in the ultraviolet-visible region.

Hydrogens bound to nitrogen give NMR signals in the range $\delta = 1.0$ to 3.0. These signals are often broad and very difficult to detect. As is also

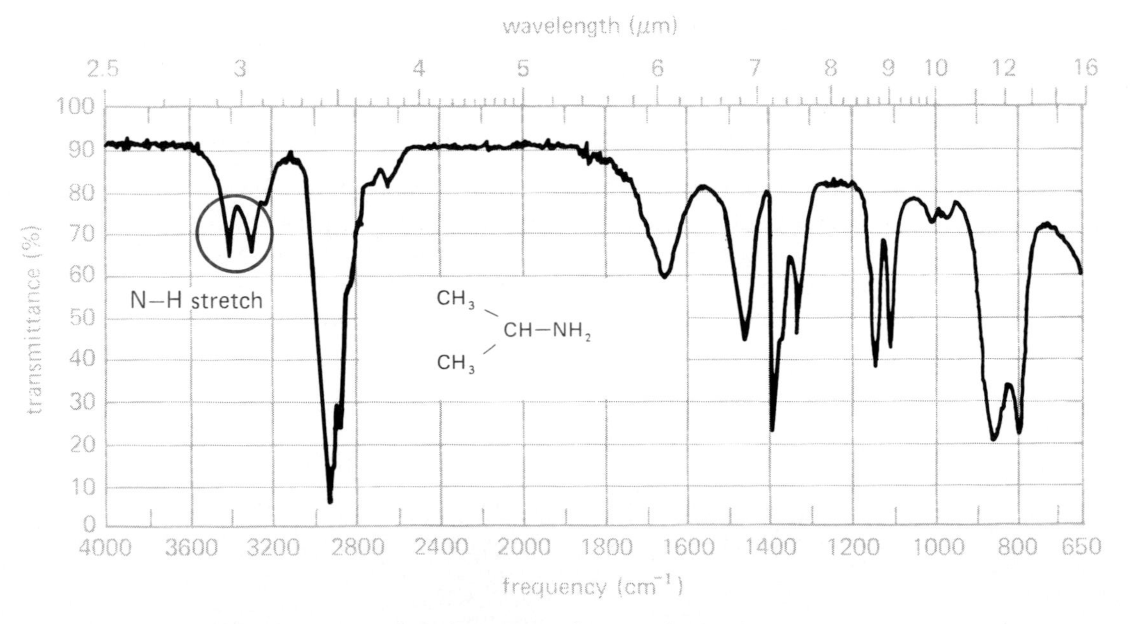

Figure 10.4 *An infrared spectrum of isopropylamine.*

the case with protons on O—H groups (Section 5.12), the signals of N—H protons of primary and secondary amines are not split by protons on adjacent carbon atoms and, therefore, generally appear as singlets.

IMPORTANT REACTIONS

(1) Alkylation of ammonia and amines (Section 10.4A).

$$(CH_3)_3N + CH_3Cl \longrightarrow (CH_3)_4N^+ \; Cl^-$$

(2) Reduction of nitro groups (Section 10.4B).

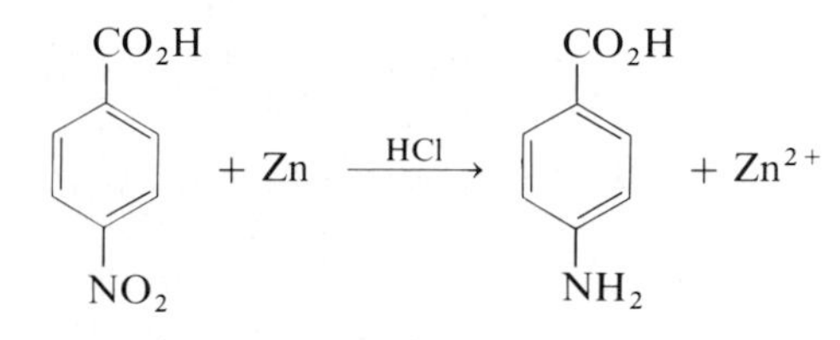

Nitro groups are also reduced by H_2 in the presence of a catalyst.

(3) Basicity of amines (Section 10.5A).

$$CH_3NH_2 + H_2O = CH_3NH_3^+ + OH^-$$

$$K_b = \frac{[CH_3NH_3^+][OH^-]}{[CH_3NH_2]} = 4.4 \times 10^{-4}$$

$$pK_b = 3.34$$

(4) Acidity of amine salts (Section 10.5A).

$$CH_3NH_3^+ = CH_3NH_2 + H^+$$

$$K_a = \frac{[CH_3NH_2][H^+]}{[CH_3NH_3^+]} = 2.7 \times 10^{-11}$$

$$pK_a = 10.66$$

(5) Formation of salts with acids (Section 10.5A).

$$CH_3NH_2 + HCl \longrightarrow CH_3NH_3^+Cl^-$$

$$CH_3NH_2 + CH_3\overset{O}{\overset{\|}{C}}OH \longrightarrow CH_3NH_3^+CH_3\overset{O}{\overset{\|}{C}}O^-$$

(6) Ammonolysis of esters (Section 10.5B; see also Section 9.6C).

$$CH_3\overset{O}{\overset{\|}{C}}OCH_2CH_3 + CH_3NH_2 \longrightarrow CH_3\overset{O}{\overset{\|}{C}}NHCH_3 + CH_3CH_2OH$$

(7) Ammonolysis of anhydrides (Section 10.5B; see also Section 9.10C).

$$CH_3\overset{O}{\overset{\|}{C}}O\overset{O}{\overset{\|}{C}}CH_3 + 2CH_3NH_2 \longrightarrow CH_3\overset{O}{\overset{\|}{C}}NHCH_3 + CH_3\overset{O}{\overset{\|}{C}}O^- \; CH_3NH_3^+$$

(8) Ammonolysis of acid chlorides (Section 10.5B; see also Section 9.8).

$$CH_3\overset{\overset{\large O}{\|}}{C}Cl + 2CH_3NH_2 \longrightarrow CH_3\overset{\overset{\large O}{\|}}{C}NHCH_3^+ + CH_3NH_3^+\ Cl^-$$

(9) Reaction with nitrous acid (Section 10.5C).
(a) Tertiary amines.

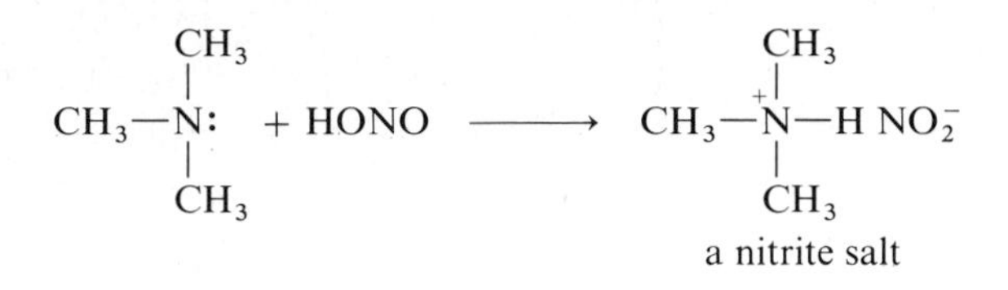

a nitrite salt

(b) Secondary amines.

$$CH_3CH_2\underset{\underset{\large CH_3}{|}}{N}H + HONO \longrightarrow CH_3CH_2\underset{\underset{\large CH_3}{|}}{N}{-}N{=}O + H_2O$$

an N-nitrosoamine

(c) Primary amines.

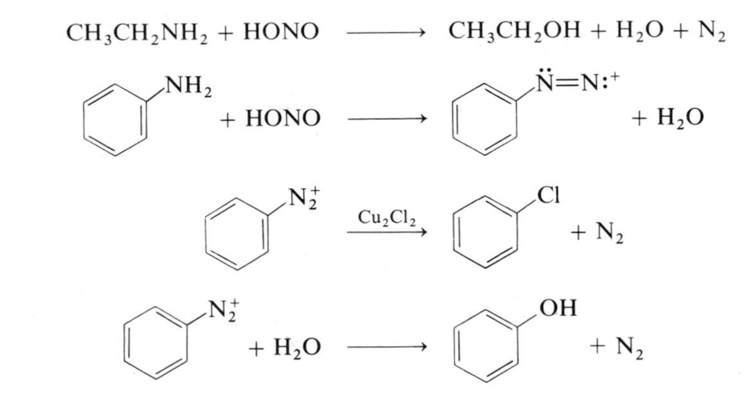

(10) Formation of azo dyes and indicators (Section 10.5C).

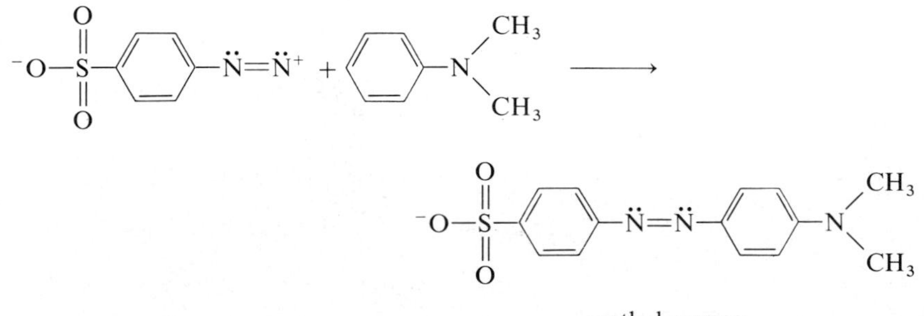

methyl orange

PROBLEMS

10.3 Write structural formulas for the following compounds. Classify each as a primary amine, a secondary amine, a tertiary amine, an aromatic amine, a heterocyclic aromatic amine, or an ammonium salt.

(a) diethylamine
(b) aniline
(c) cyclohexylamine
(d) pyrrole
(e) pyridine
(f) tetraethylammonium iodide
(g) 2-aminoethanol (ethanolamine)
(h) 2-aminopropanoic acid (alanine)
(i) pyrimidine
(j) trimethylammonium benzoate
(k) *p*-methoxyaniline
(l) N,N-dimethylacetamide
(m) N-methylaniline
(n) acetylcholine
(o) choline acetate
(p) ethyl *p*-aminobenzoate
(q) pyridine 3-carboxylic acid (nicotinic acid)
(r) 3,5-dichloroaniline
(s) 3,5-diaminochlorobenzene
(t) *p*-aminobenzenesulfonic acid

10.4 Give an acceptable name for each of the following compounds.

(a) 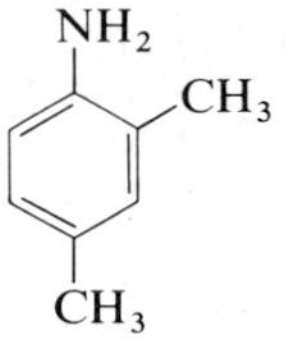 NH_2, CH_3, CH_3

(b) 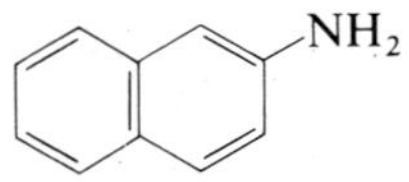 NH_2

(c) 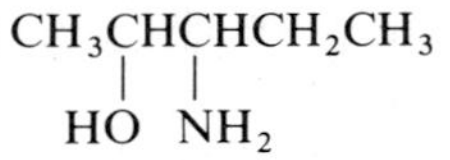 $CH_3CHCHCH_2CH_3$ (with HO and NH_2 on carbons 2 and 3)

(d) 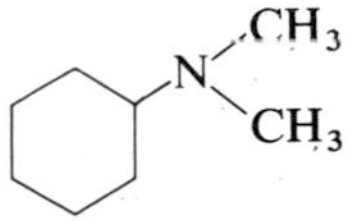 N, CH_3, CH_3

(e) $CH_3CH_2CH_2CH_2NH_2$

(f) $(CH_3CH_2)_2NCH_3$

(g) 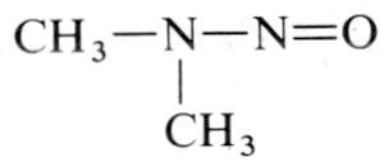 $CH_3—N—N=O$ (with CH_3 on N)

(h) 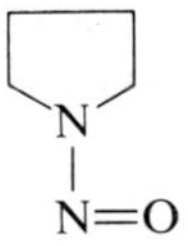 N, N=O

10.5 Draw structural formulas for the eight isomeric amines of molecular formula $C_4H_{11}N$. Name each. Label each as primary, secondary, or tertiary.

10.6 Arrange the following compounds in order of increasing basicity.

(a) 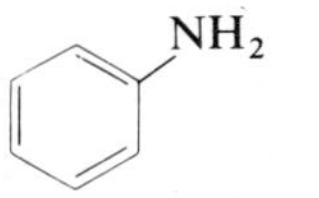 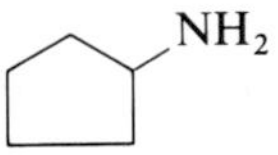 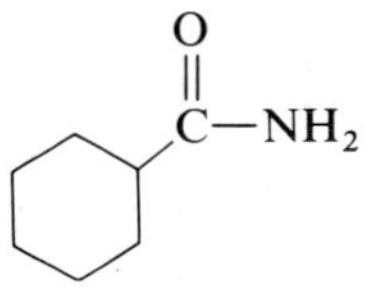 (i) NH_2 (ii) NH_2 (iii) O, C—NH_2

(b) 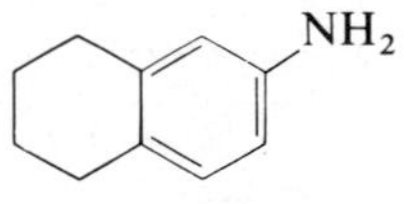 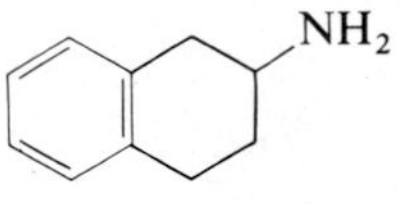 (i) NH_2 (ii) NH_2

10.7 Arrange the following compounds in order of increasing boiling points.

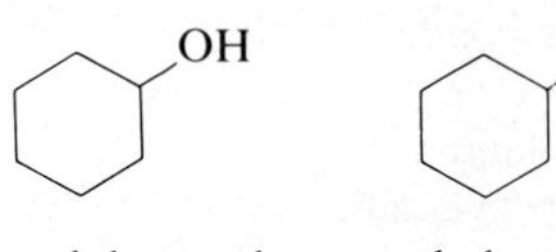

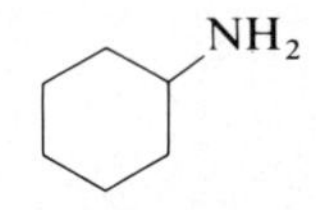

cyclohexanol methylcyclohexane cyclohexylamine

10.8 Select the stronger acid.

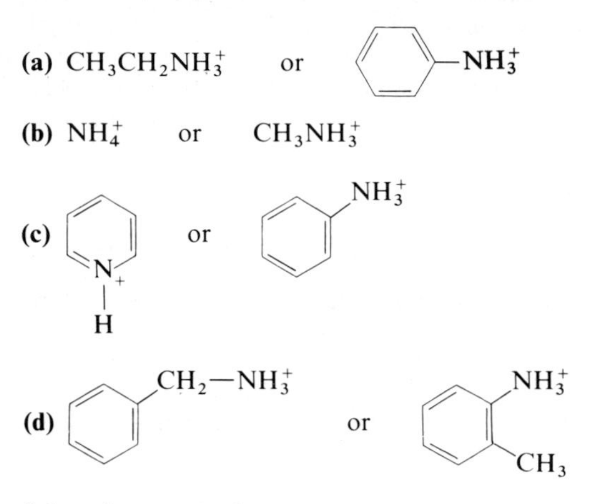

10.9 Select the stronger base.

(a) $CH_3CH_2NH_2$ or $C_6H_5NH_2$ **(b)** NH_3 or CH_3NH_2

(c) CH_3NH_2 or $CH_3CO_2^-$

(d) $C_6H_5CH_2NH_2$ or $C_6H_5NH_2$

(e) (pyridine) or (aniline, NH_2)

10.10 Using the theory of resonance, account for the fact that
(a) phenol is a stronger acid than cyclohexanol.
(b) aniline is a weaker base than cyclohexylamine.

10.11 How would you account for the fact that pyrrole is considerably less basic than pyridine?

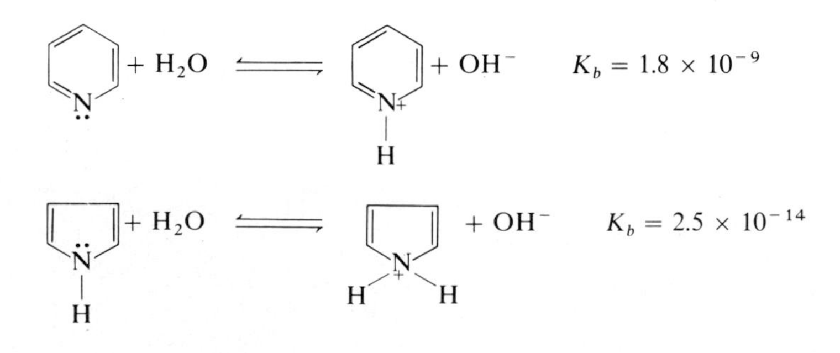

10.12 Suppose you are given a mixture of the following three compounds. Describe a procedure you could use to separate and isolate each in pure state.

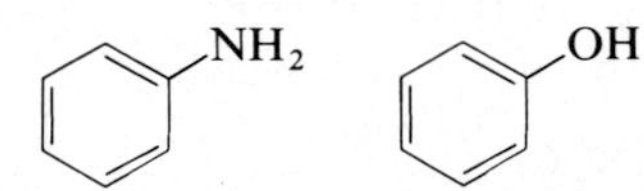

$CH_3CH_2CH_2CH_2CH_2CH_2OH$

10.13 Alanine (2-aminopropanoic acid) is one of the important amino acids found in proteins. Would you expect the structural formula of alanine to be represented better by (I) or (II)? Explain.

$$\underset{\text{(I)}}{CH_3{-}\underset{\displaystyle NH_2}{\underset{|}{CH}}{-}CO_2H} \qquad \underset{\text{(II)}}{CH_3{-}\underset{\displaystyle NH_3^+}{\underset{|}{CH}}{-}CO_2^-}$$

10.14 Both 1-aminobutane and 1-butanol are liquids. One of these compounds has a boiling point of 117°C, the other a boiling point of 78°C. Which compound has the boiling point of 78°C? which the boiling point of 117°C? Explain your reasoning.

10.15 Would you expect aniline to be more soluble in water or in 0.1 M HCl? Explain.

10.16 Draw structural formulas for the products of the following reactions.

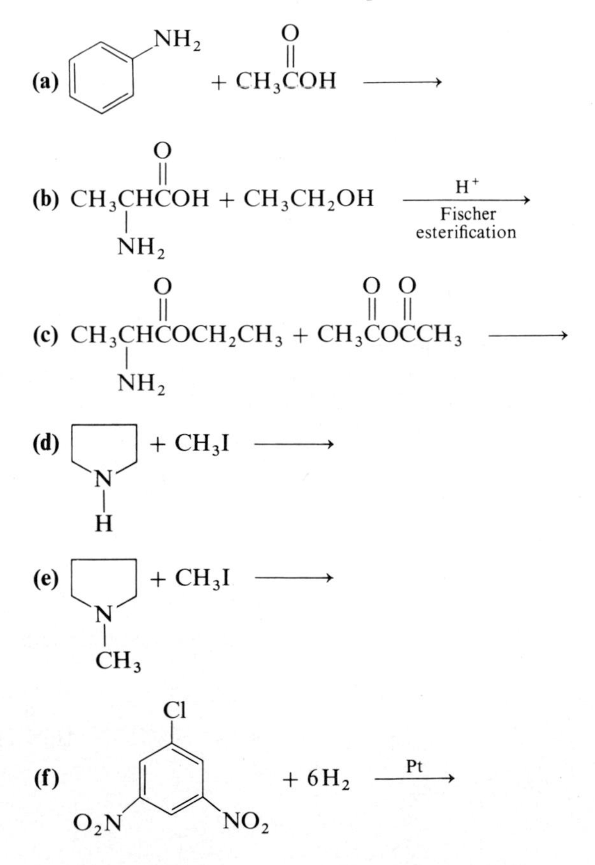

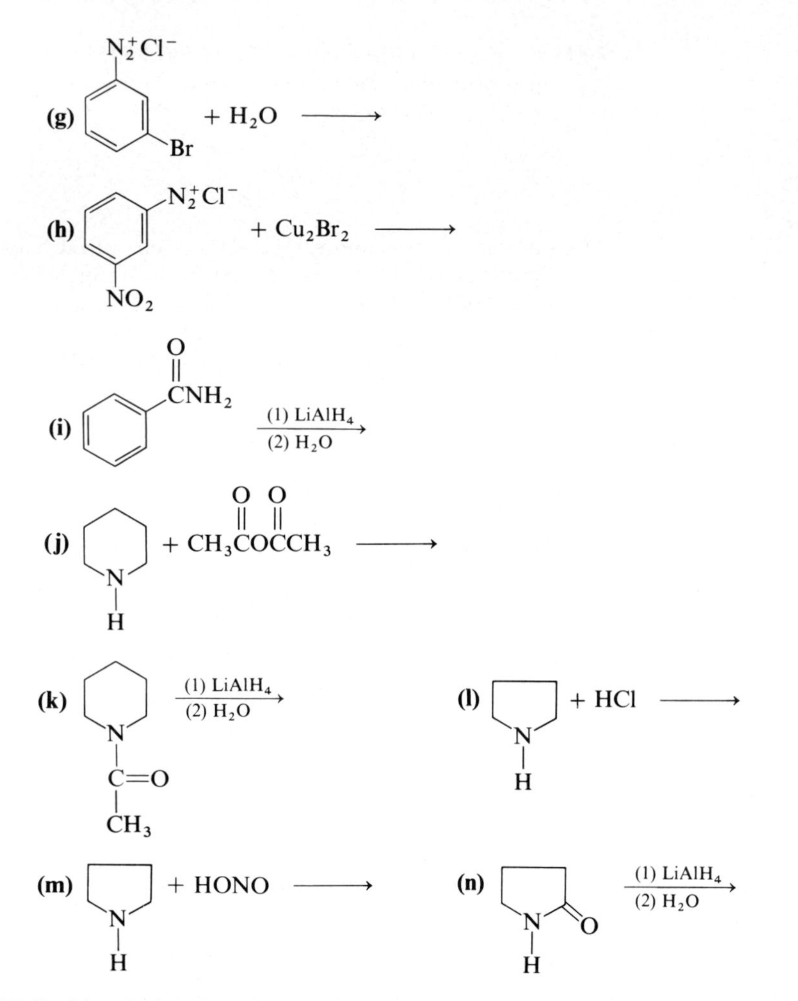

10.17 Pyridoxal phosphate is one of the metabolically active forms of vitamin B_6. Draw structural formulas for the Schiff bases formed by reaction of pyridoxal phosphate with **(a)** tyrosine and **(b)** glutamic acid.

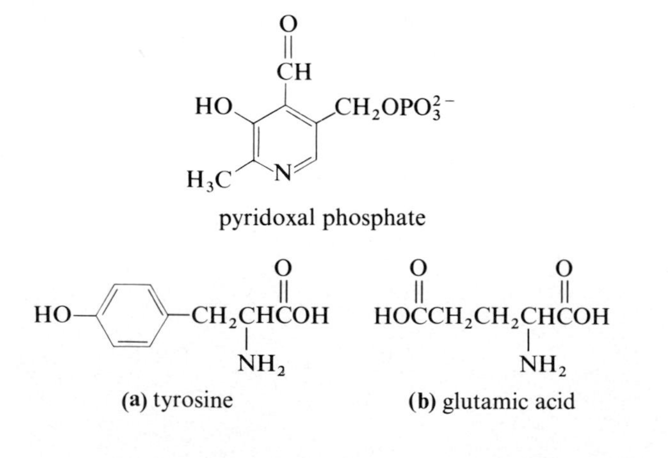

(a) tyrosine **(b)** glutamic acid

10.18 Another of the metabolically active forms of vitamin B_6 is pyridoxamine phosphate. Draw structural formulas for the Schiff bases formed by the reaction of pyridoxamine phosphate with the ketone of **(a)** pyruvic acid and **(b)** oxaloacetic acid.

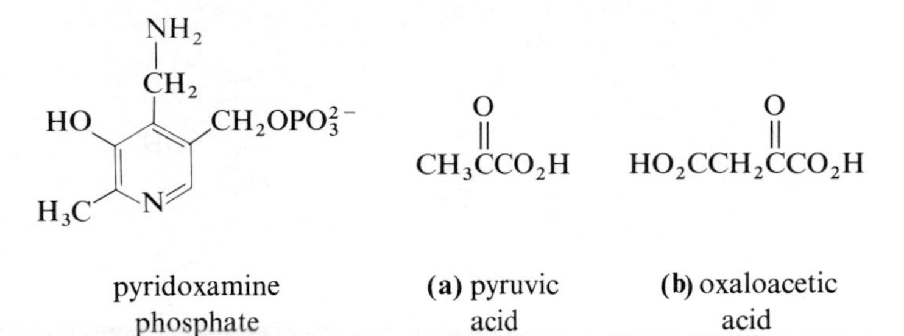

pyridoxamine phosphate — **(a)** pyruvic acid — **(b)** oxaloacetic acid

10.19 Write equations to show how the following amides can be prepared by the reaction of (1) an ester and an amine, (2) an acid anhydride and an amine, and (3) an acid chloride and an amine.

(a) $CH_3CH(CH_3)CH_2CH_2C(=O)NH_2$

(b) $C_6H_5CH_2C(=O)NHCH_3$

(c) $CH_3C(=O)N$ (pyrrolidine ring)

10.20 Write structural formulas for the products formed when each amide in the previous problem is reduced with $LiAlH_4$.

10.21 Write structural formulas for an amide, which on reduction with $LiAlH_4$, will give the following amines.

(a) $CH_3CH_2CH_2CH_2NH_2$

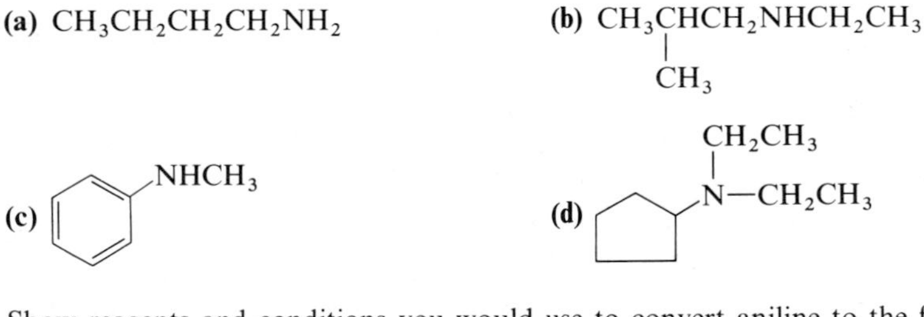

10.22 Show reagents and conditions you would use to convert aniline to the following compounds.

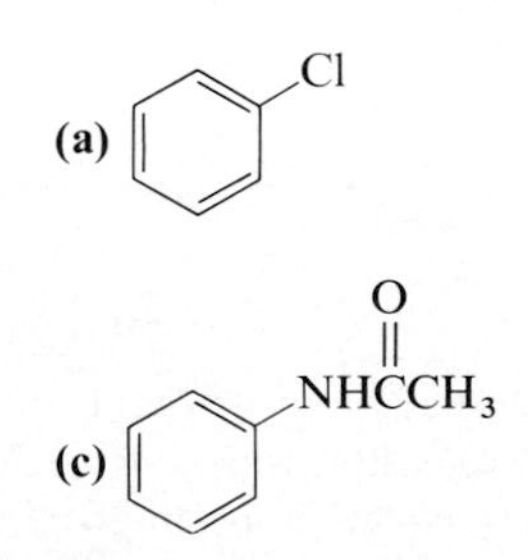

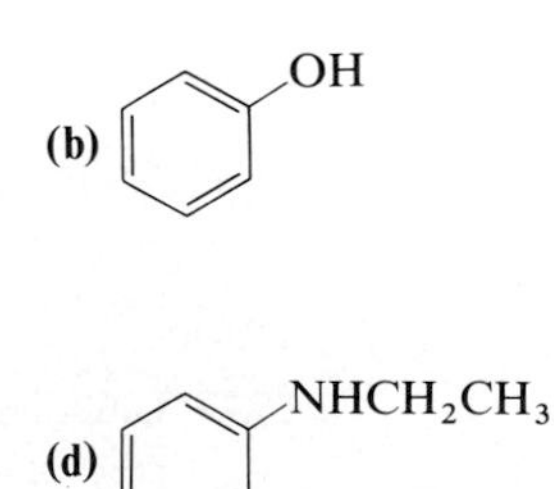

10.23 Show reagents and conditions you would use to carry out the following conversions.

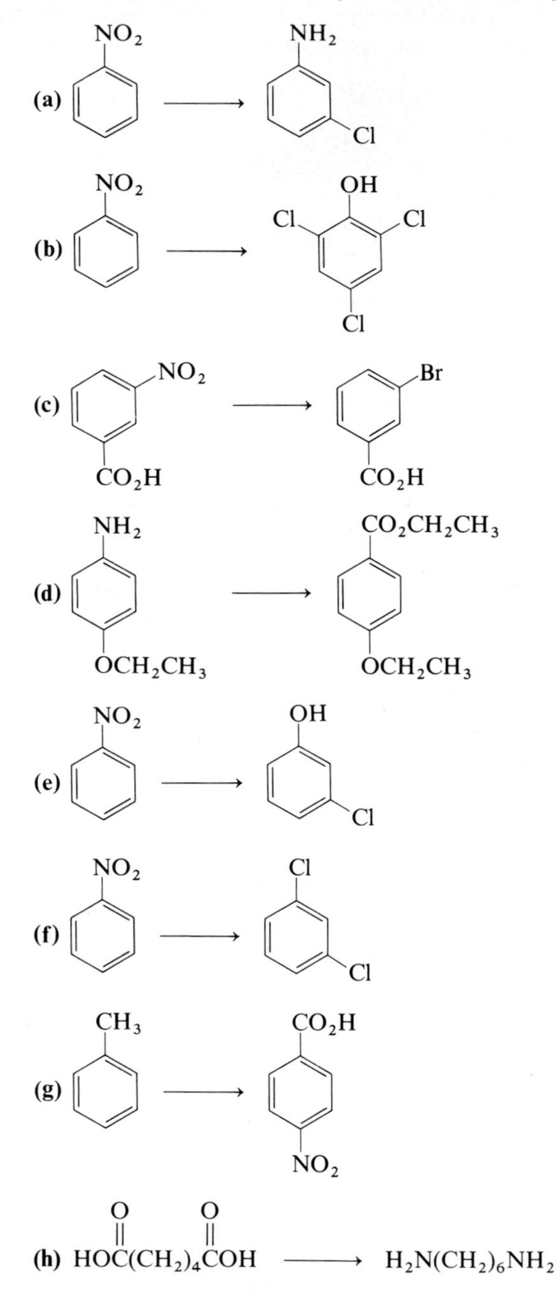

(h) $HO\overset{O}{\overset{\|}{C}}(CH_2)_4\overset{O}{\overset{\|}{C}}OH \longrightarrow H_2N(CH_2)_6NH_2$

10.24 Describe a simple chemical test by which you could distinguish between the following pairs of compounds. In each case, state what you would do, what you would expect to observe, and write a balanced equation for all positive results.

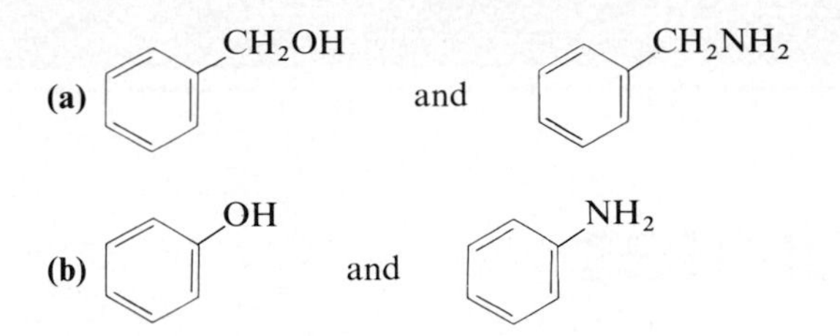

(c) cyclohexanol and cyclohexylamine
(d) ammonium butanoate and butanamide
(e) trimethylacetic acid and 2,2-dimethylpropylamine
(f) propylamine and methylethylamine
(g) N-methylpiperidine and cyclohexylamine
(h) methylpropylamine and dimethylethylamine
(i) N-methylaniline and *p*-toluidine

10.25 List one major spectral characteristic by which you could distinguish between the following compounds.
(a) 1-aminohexane (*n*-hexylamine) and triethylamine (IR)
(b) aniline and benzamide (IR)
(c) aniline and cyclohexylamine (UV)
(d) methylpropylamine and dimethylethylamine (IR)
(e) *tert*-butylamine and dimethylethylamine (IR, NMR)

10.26 In Problem 10.5 you drew structural formulas for the eight isomeric amines of molecular formula $C_4H_{11}N$. Which of these isomeric amines has the following NMR spectra?
(a) a singlet (9H) at $\delta = 1.1$ and a singlet (2H) at $\delta = 1.3$
(b) a singlet (1H) at $\delta = 0.8$, a triplet (6H) at $\delta = 1.1$, and a quartet (4H) at $\delta = 2.6$

CHAPTER 11

Carbohydrates

Carbohydrates are among the most abundant constituents of the plant and animal worlds. They serve many vital functions such as storehouses of chemical energy (glucose, starch, glycogen); supportive structural components in plants (cellulose); and essential components in the mechanisms of genetic control of development and growth of living cells (D-ribose and 2-deoxyribose).

The name carbohydrate was derived from early observations that many members of this class have the empirical formula $C_n(H_2O)_m$ and were termed "hydrates of carbon." For example, grape sugar (glucose) has the molecular formula $C_6H_{12}O_6$. Cane sugar (sucrose) has the formula $C_{12}H_{22}O_{11}$. Starch and cellulose are very large molecules of variable molecular weight having the empirical formula $(C_6H_{10}O_5)_x$, where x may be as large as several hundred thousand.

It soon became clear that not all compounds having the properties of carbohydrates have this general formula. For example, we now know of carbohydrates that contain nitrogen in addition to the elements of carbon, hydrogen, and oxygen. Although the term carbohydrate is not fully descriptive, it is firmly rooted in the chemical nomenclature and has persisted as the name of this class of compounds.

These substances are often referred to as saccharides because of the sweet taste of the simpler members of the family, the sugars (Latin, *saccharum*, "sugar").

Carbohydrates are polyhydroxyaldehydes or polyhydroxyketones or substances that after hydrolysis yield polyhydroxyaldehydes and/or polyhydroxyketones. Therefore the chemistry of carbohydrates is essentially the chemistry of two functional groups, the hydroxyl group and the carbonyl group.

11.1 MONOSACCHARIDES

Monosaccharides are carbohydrates that cannot be hydrolyzed to simpler compounds. They have the general formula $C_nH_{2n}O_n$, where n varies from three to eight. The terms triose, tetrose, pentose, etc., refer to the number of carbon atoms in the monosaccharide. A triose contains three carbon atoms, a tetrose four carbon atoms, and so on.

$C_3H_6O_3$	triose	$C_6H_{12}O_6$	hexose
$C_4H_8O_4$	tetrose	$C_7H_{14}O_7$	heptose
$C_5H_{10}O_5$	pentose	$C_8H_{16}O_8$	octose

There are only two trioses: glyceraldehyde and dihydroxyacetone. Glyceraldehyde contains two —OH groups and an aldehyde; dihydroxyacetone contains two —OH groups and a ketone.

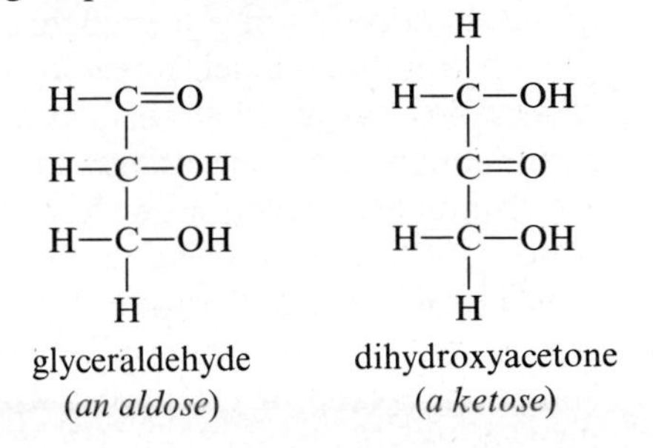

All monosaccharides contain a carbonyl group. One type of monosaccharide has a carbonyl in the form of an aldehyde; these are known as aldoses (aldehyde + -ose). Glyceraldehyde is the simplest aldose. A second type of monosaccharide has a carbonyl group in the form of a ketone; these are known as ketoses (ketone + -ose). Dihydroxyacetone is the simplest ketose.

Dihydroxyacetone has no chiral carbon, and therefore cannot show stereoisomerism. Glyceraldehyde, however, has one chiral carbon and shows stereoisomerism. Drawn in Figure 11.1 are structural formulas for the enantiomers of glyceraldehyde. The compound of structural formula Ia is named D-glyceraldehyde and has a specific rotation of +13.5°. Its enantiomer, Ib,

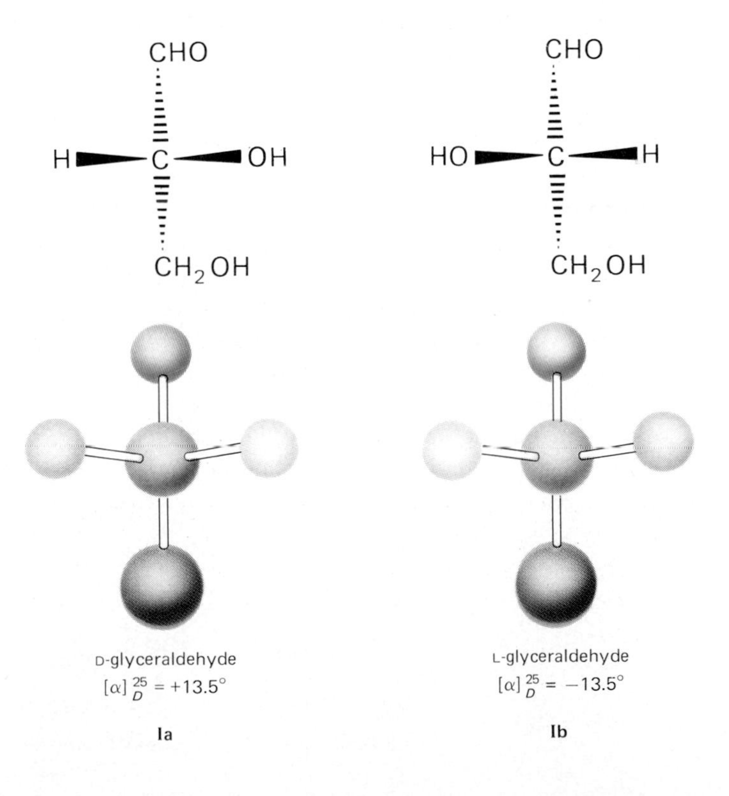

Figure 11.1 *The enantiomers of glyceraldehyde (Ia and Ib) and ball-and-stick models of each.*

is named L-glyceraldehyde and has a specific rotation of $-13.5°$. Note that the magnitude of the specific rotation for the enantiomers of glyceraldehyde is equal but opposite in sign. Also shown in Figure 11.1 are ball-and-stick models of the enantiomers of glyceraldehyde.

In the three-dimensional formulas in Figure 11.1, the configuration of each atom or group of atoms bonded to the chiral carbon atom is indicated by a combination of dashed wedges and solid wedges. Fischer projection formulas are a simplified way to show the configuration about such chiral carbon atoms. According to this convention, the carbon chain is written vertically, with the most highly oxidized carbon atom at the top. Horizontal lines show groups projecting above the plane of the paper; vertical lines show groups projecting behind the plane of the paper. Applying the Fischer projection rules to the three-dimensional drawings of the enantiomers of glyceraldehyde gives formulas Ia′ and Ib′.

```
     CHO               CHO
      |                 |
  H—C—OH           HO—C—H
      |                 |
     CH2OH             CH2OH
     Ia′               Ib′
D-glyceraldehyde   L-glyceraldehyde
```

The configurations of D-glyceraldehyde and L-glyceraldehyde serve as reference points for the assignment of configuration to all other aldoses and ketoses. All those that have the same configuration as D-glyceraldehyde about the chiral carbon farthest from the aldehyde or ketone group are called D-monosaccharides; all those that have the same configuration as L-glyceraldehyde about the chiral carbon farthest from the aldehyde or ketone group are called L-monosaccharides.

According to the Cahn-Ingold-Prelog convention for specifying absolute configuration about a chiral carbon (Section 4.7), D-glyceraldehyde becomes R-glyceraldehyde and L-glyceraldehyde becomes S-glyceraldehyde. While the newer R, S convention is more general, the older D, L convention continues to be used in carbohydrate chemistry.

Tables 11.1 and 11.2 show the names and structural formulas for all trioses, tetroses, pentoses, and hexoses of the D-series.

D-Ribose and 2-deoxy-D-ribose, the most important pentoses, are building blocks of nucleic acids; D-ribose in ribonucleic acid (RNA) and 2-deoxy-D-ribose in deoxyribonucleic acid (DNA)

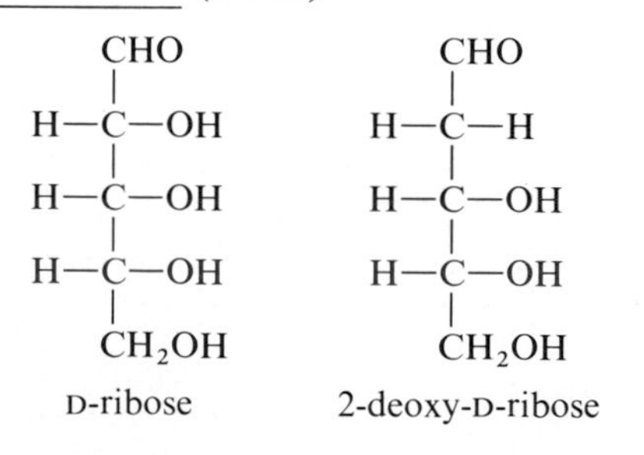

D-ribose 2-deoxy-D-ribose

Table 11.1 The isomeric D-aldotetroses, D-aldopentoses, and D-aldohexoses derived from D-glyceraldehyde.

D-aldotriose	CHO H—C—OH CH_2OH D-glyceraldehyde			
D-aldotetroses	CHO H—C—OH H—C—OH CH_2OH D-erythrose	CHO HO—C—H H—C—OH CH_2OH D-threose		
D-aldopentoses	CHO H—C—OH H—C—OH H—C—OH CH_2OH D-ribose	CHO HO—C—H H—C—OH H—C—OH CH_2OH D-arabinose	CHO H—C—OH HO—C—H H—C—OH CH_2OH D-xylose	CHO HO—C—H HO—C—H H—C—OH CH_2OH D-lyxose
D-aldohexoses	CHO H—C—OH H—C—OH H—C—OH H—C—OH CH_2OH D-allose	CHO HO—C—H H—C—OH H—C—OH H—C—OH CH_2OH D-altrose	CHO H—C—OH HO—C—H H—C—OH H—C—OH CH_2OH D-glucose	CHO HO—C—H HO—C—H H—C—OH H—C—OH CH_2OH D-mannose
	CHO H—C—OH H—C—OH HO—C—H H—C—OH CH_2OH D-gulose	CHO HO—C—H H—C—OH HO—C—H H—C—OH CH_2OH D-idose	CHO H—C—OH HO—C—H HO—C—H H—C—OH CH_2OH D-galactose	CHO HO—C—H HO—C—H HO—C—H H—C—OH CH_2OH D-talose

Table 11.2 The isomeric D-ketopentoses and D-ketohexoses derived from dihydroxyacetone and D-erythrulose.

ketotriose	CH_2OH CO CH_2OH dihydroxyacetone			
D-ketotetrose	CH_2OH CO H—C—OH CH_2OH D-erythrulose			
D-ketopentoses	CH_2OH CO H—C—OH H—C—OH CH_2OH D-ribulose	CH_2OH CO HO—C—H H—C—OH CH_2OH D-xylulose		
D-ketohexoses	CH_2OH CO H—C—OH H—C—OH H—C—OH CH_2OH D-psicose	CH_2OH CO HO—C—H H—C—OH H—C—OH CH_2OH D-fructose	CH_2OH CO H—C—OH HO—C—H H—C—OH CH_2OH D-sorbose	CH_2OH CO HO—C—H HO—C—H H—C—OH CH_2OH D-tagatose

The most abundant hexoses in the biological world are D-glucose, D-galactose, and D-fructose. The first two of these are D-aldohexoses; the third, fructose, contains a keto group on carbon-2 and is a D-ketohexose.

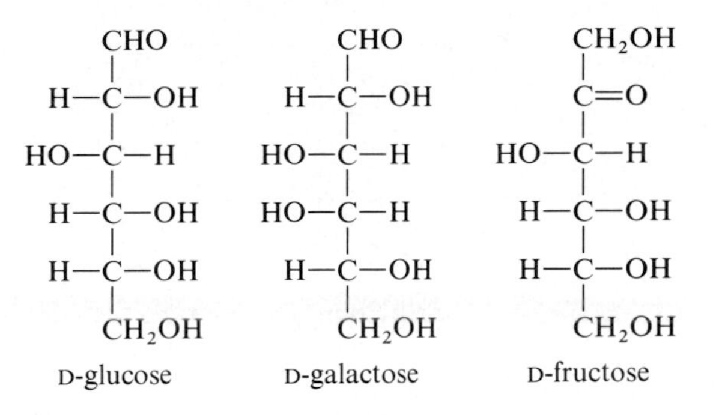

Glucose is by far the most common hexose monosaccharide. It is also known as dextrose because it is dextrorotatory. Other names for this monosaccharide are grape sugar, blood sugar, and corn sugar, names that clearly indicate its sources in an uncombined state. (Human blood normally contains 65–110 mg of glucose per 100 mL).

Fructose is found combined with glucose in the disaccharide sucrose, or as it is more commonly known, table sugar (Section 11.8). D-Galactose is found combined with glucose in the disaccharide lactose (Section 11.8). Lactose appears nowhere else except in milk, where it is present to the extent of about 5%. The enzyme systems of glycolysis and the tricarboxylic acid cycle cannot metabolize galactose and it must first be converted by isomerization at carbon-4 into glucose. There is an inherited disease galactosemia among human infants that manifests itself by the inability to metabolize galactose, which then accumulates in various tissues, including the central nervous system, and causes damage to cells. Without treatment, the infant either suffers permanent damage to its organs, including the brain, or at worst, succumbs. This disease occurs because one of the enzymes involved in the isomerization of galactose to glucose is not synthesized in the liver of these infants due to a lack of the necessary gene. Since we are unable to alter the genetic mechanism to restore a non-functioning gene, it is not possible to correct this enzyme deficiency. Yet it is possible to treat the disease in a surprisingly simple manner providing diagnosis is made early enough. Because milk is the only source of the galactose, it is excluded from the infants' diet. Our ability to remedy this disease marked an important milestone, for it is the first hereditary disease to be controlled through a knowledge of genetics and biochemistry.

Example 11.1

(a) Draw Fischer projection formulas for all aldoses of four carbon atoms.
(b) Label those which are D-monosaccharides and those which are L-monosaccharides.
(c) Indicate which are pairs of enantiomers.
(d) Refer to Table 11.1 and give names to the aldoses you have drawn.

Solution

(a) Aldoses have an aldehyde on carbon-1 and an —OH group on each of the other carbons. Start by writing the aldehyde group and the four carbon chains. By convention, the carbon chains are written vertically. In aldotetroses, carbon-2 and carbon-3 are chiral carbons.

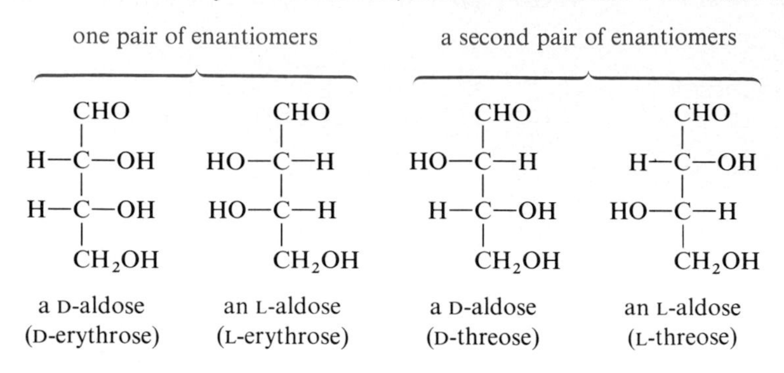

(b) D and L configurations refer to the arrangement of groups attached to the next-to-last carbon of the monosaccharide. In the aldotetroses, the next-to-last carbon is number 3 of the chain. This —OH is on the right in a D-aldose and on the left in a L-aldose.

PROBLEM 11.1 **(a)** Draw Fischer projection formulas for all 2-ketoses with five carbon atoms.

(b) Label those which are D-ketopentoses and those which are L-ketopentoses.

(c) Indicate which are pairs of enantiomers.

(d) Refer to Table 11.2 and give names to the ketoses you have drawn.

11.2 ASCORBIC ACID

The structural formula of ascorbic acid (vitamin C) resembles that of a monosaccharide. In fact, this vitamin is synthesized biochemically from D-glucose (Figure 11.2). Enzyme-catalyzed oxidation of D-glucose by NAD^+ gives D-glucuronic acid. This four-electron oxidation requires two moles of NAD^+ per mole of D-glucose. A selective enzyme-catalyzed reduction of the —CHO group of D-glucuronic acid gives L-gulonic acid. Carbon atom 1 of what was D-glucose is now $—CH_2OH$, and carbon atom 6 is now $—CO_2H$. According to the Fischer convention, the carbon chain should be renumbered and turned 180° in the plane of the paper so that the $—CO_2H$ group is uppermost and now appears as carbon-1. This substance, L-gulonic acid, forms a cyclic ester, and is then oxidized by molecular oxygen in an enzyme-catalyzed reaction to L-ascorbic acid. Humans, other primates, and guinea pigs lack the enzyme system necessary to convert L-gulonic acid to L-ascorbic acid and, therefore, are unable to synthesize ascorbic acid. Vitamin C is readily oxidized to L-dehydroascorbic acid. Both forms are physiologically active and are found in body fluids.

The biochemical function of ascorbic acid is not completely understood. Its most clearly established biological role is in the maintenance of collagen

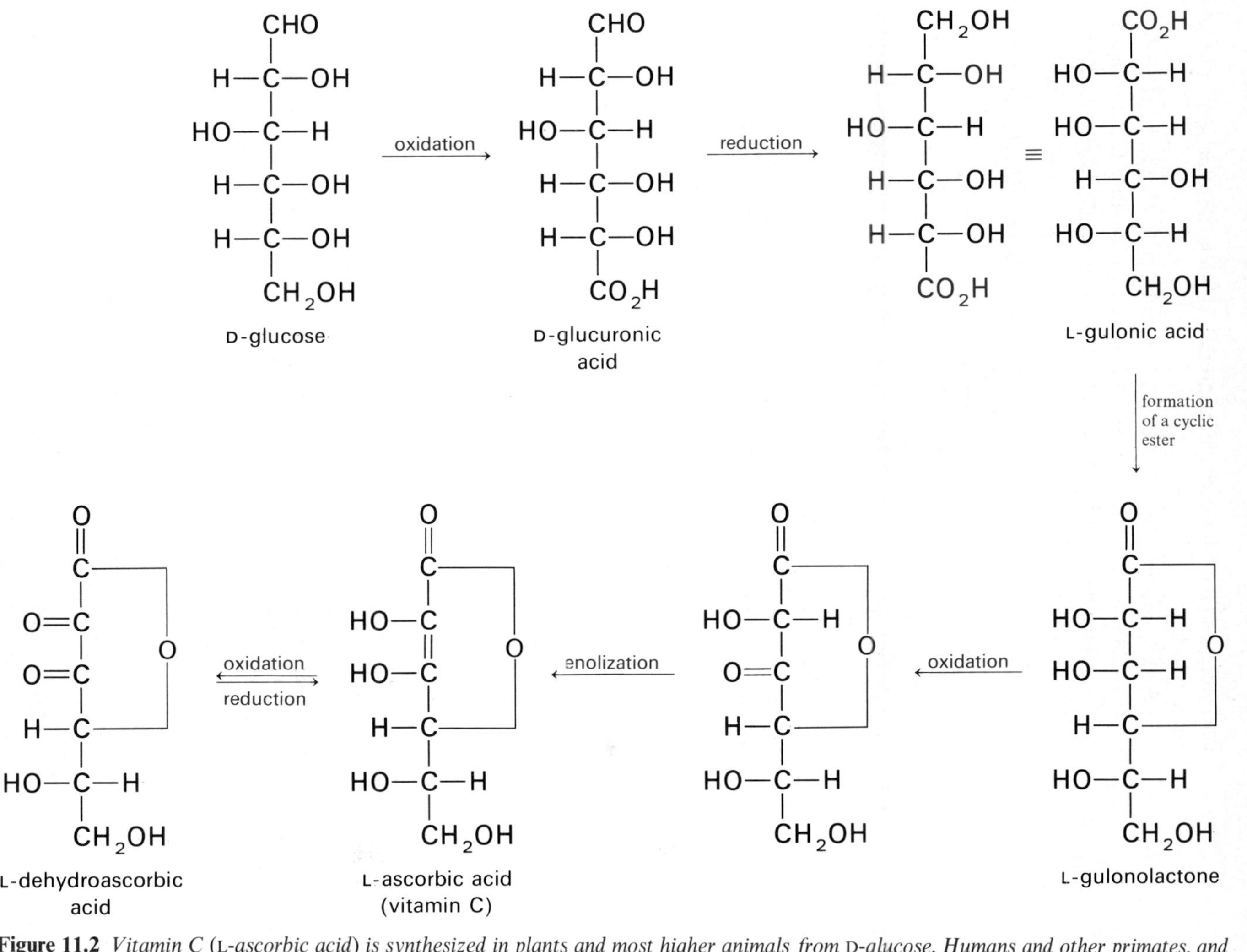

Figure 11.2 *Vitamin C (L-ascorbic acid) is synthesized in plants and most higher animals from D-glucose. Humans and other primates, and guinea pigs have a genetic defect that prevents them from converting L-gulonic acid to L-gulonolactone, and therefore from synthesizing vitamin C. These organisms require vitamin C in the diet.*

(Section 12.11), an intercellular material of bone, dentine, cartilage, and connective tissue.

Severe deficiency of ascorbic acid produces scurvy. A most vivid description of this vitamin deficiency disease was given by Jacques Cartier in 1536 when it afflicted his men during the exploration of the Saint Lawrence River.

> Some did lose all their strength, and could not stand on their feet. . . . Others also had all their skin spotted with spots of blood of a purple color; then did it ascend up to their ankles, knees, thighs, shoulders, arms, and necks. Their mouths became stinking, their gums so rotten that all of the flesh did fall off, even to the roots of the teeth, which did also almost fall out.

This is scurvy! Green or fresh vegetables and ripe fruits are the best remedies and the best means of preventing the disease. (These foods are of course sources of vitamin C.) This was recognized in 1753 by the English physician James Lind, who urged the inclusion of lemon or lime juice in the diet of British sailors, hence the nickname "limeys."

There is good evidence that severe ascorbic acid deficiency as seen in scurvy is the result of impaired synthesis of collagen. Collagen synthesized in the absence of ascorbic acid cannot properly form fibers, thereby resulting in such typical symptoms as fragility of blood vessels and hemorrhaging, loosening of the teeth, poor wound healing, and so on. We will discuss the structure and function of collagen more fully in Chapter 12.

11.3 AMINO SUGARS

Amino sugars contain an $—NH_2$ group in place of an $—OH$ group. Only three amino sugars are common and widely distributed in nature: D-glucosamine, D-mannosamine, and D-galactosamine.

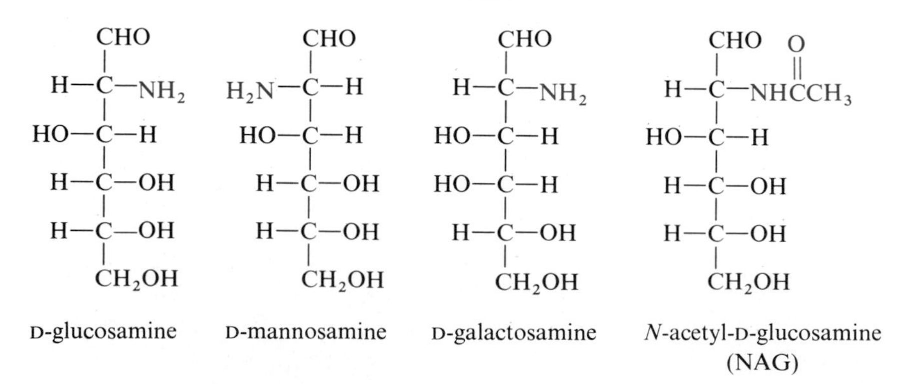

In most cases where these monosaccharides occur in nature, the $—NH_2$ group is acetylated. *N*-Acetylglucosamine is a component of many polysaccharides including chitin, the hard, shell-like exoskeleton of lobsters, crabs, shrimps, and other crustaceans.

11.4 THE CYCLIC STRUCTURE OF MONOSACCHARIDES

As we saw in Section 7.5A, aldehydes and ketones react with alcohols to form hemiacetals and hemiketals.

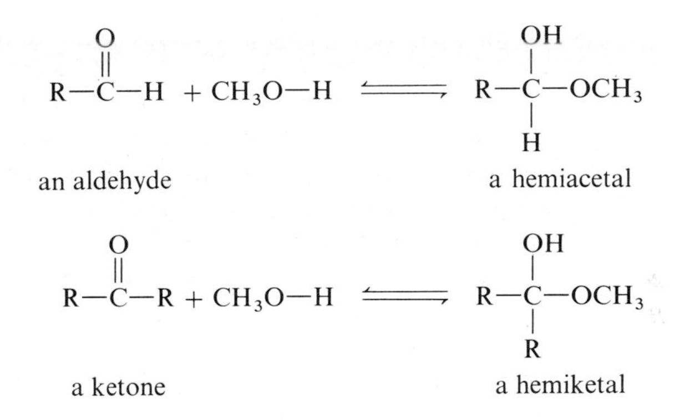

We saw also that cyclic hemiacetals and hemiketals form when hydroxyl and carbonyl groups are part of the same molecule. For example, 4-hydroxypentanal forms a five-membered cyclic hemiacetal.

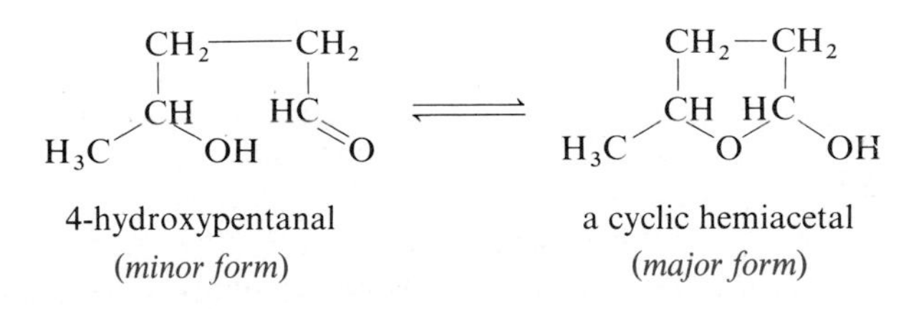

4-hydroxypentanal (*minor form*) a cyclic hemiacetal (*major form*)

Monosaccharides have hydroxyl and carbonyl groups in the same molecule, and they too can form cyclic hemiacetals and hemiketals. One such example is glucose.

In the cyclic hemiacetal of glucose, the —OH on carbon-5 bonds to carbon-1 to form a six-membered cyclic hemiacetal. When this cyclic hemiacetal forms, carbon-1 becomes a chiral center and is called an anomeric carbon. Because of this new chiral center, there are two isomeric forms of D-glucose, each with different physical properties. One form (α), isolated by crystallization from alcohol-water below 30°C, has a specific rotation of +112°. The other form (β), obtained when an aqueous solution of D-glucose is evaporated above 98°C, has a specific rotation of +19°.

Following are Fischer projection formulas for α- and β-D-glucose, along with that of the open-chain form.

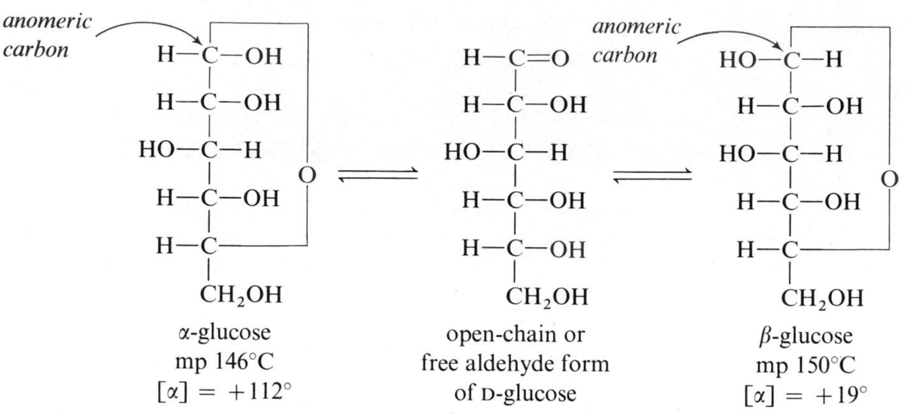

In Fischer projection formulas, the —OH on the anomeric carbon is on the right in α-D-glucose and on the left in β-D-glucose. We can also draw structural formulas showing these cyclic hemiacetals as planar hexagons. Such representations are called <u>Haworth structures</u>. Haworth structures for α-D-glucose and β-D-glucose are

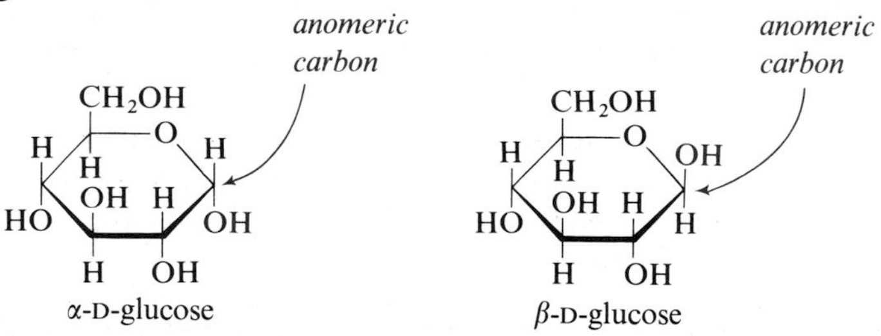

In these Haworth structures, the —OH on the anomeric carbon is below the plane of the ring in the α-anomer and above the plane of the ring in the β-anomer.

Other monosaccharides also form cyclic hemiacetals and hemiketals. Following are Haworth structures for the cyclic hemiketals formed by D-fructose. Note that these cyclic hemiketals contain five-membered rings.

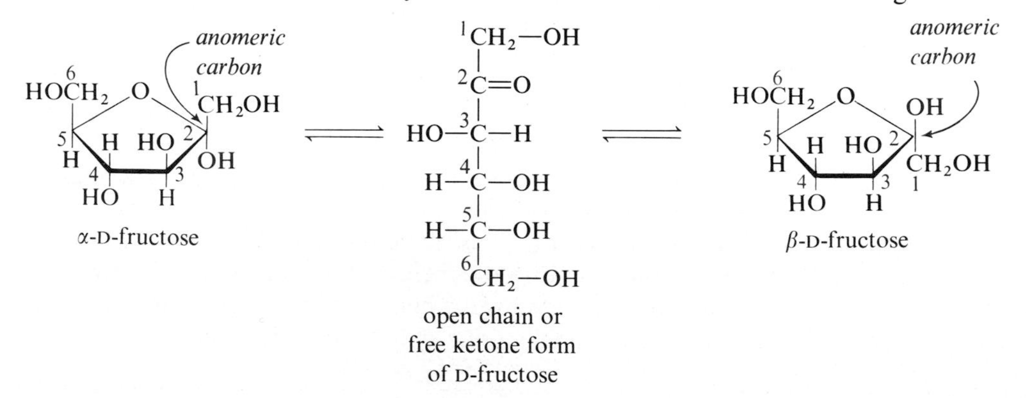

In the cyclic hemiketals of fructose, the —OH on the anomeric carbon (carbon-2 of fructose) is below the plane of the ring in the α-anomer and above the plane of the ring in the β-anomer.

Example 11.2

D-Galactose forms a cyclic hemiacetal containing a six-membered ring. Draw Haworth structures for α-D-galactose and β-D-galactose. Label the anomeric carbon in each cyclic hemiacetal.

Solution

One way to draw Haworth structures for six-membered cyclic hemiacetals is to use the α- and β-forms of D-glucose as reference points. D-galactose differs from D-glucose in the configuration of carbon-4. Therefore the α- and β-forms of D-galactose differ from the α- and β-forms of D-glucose only in the orientation of the —OH group on carbon-4.

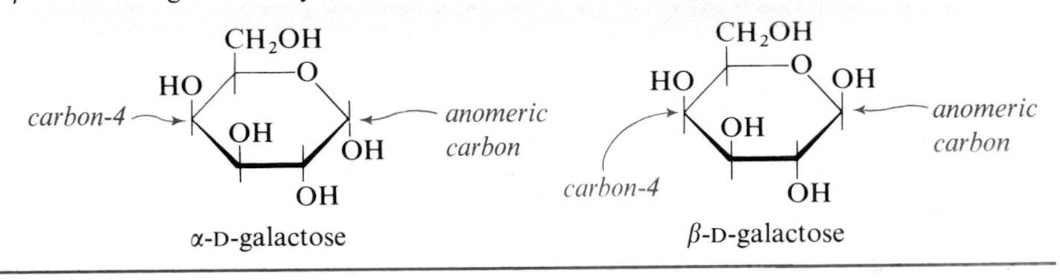

PROBLEM 11.2 D-Mannose forms a cyclic hemiacetal containing a six-membered ring. Draw Haworth structures for α-D-mannose and β-D-mannose. Label the anomeric carbon atom in each of these cyclic hemiacetals.

The α- and β-forms of monosaccharides are interconvertible in aqueous solution, and the change in specific rotation that accompanies this interconversion is known as mutarotation. For example, a freshly prepared solution of α-D-glucose shows an initial rotation of +112°, which gradually decreases to +52° as α-D-glucose reaches an equilibrium with β-D-glucose. A solution of β-glucose also undergoes mutarotation during which the specific rotation changes from an initial value of +19° to the same equilibrium value of +52°.

Mutarotation is common to all monosaccharides that exist in α- and β-forms. Shown in Table 11.3 are specific rotations for α- and β-forms of D-galactose and D-fructose with equilibrium values for the specific rotation of each after mutarotation.

Table 11.3 Specific rotations of D-galactose and D-fructose before and after mutarotation.

Monosaccharide	*Specific rotation*	*Specific rotation after mutarotation*
α-D-galactose	+151°	+84°
β-D-galactose	−53°	
α-D-fructose	21°	−92°
β-D-fructose	−133°	

11.5 STEREOREPRESENTATION OF MONOSACCHARIDES

There are three common ways for representing the stereochemistry of glucose and other carbohydrates. The first is the Fischer projection. While this type of representation shows clearly the configuration at each carbon, it is inaccurate in its representation of bond angles and the geometry of the molecule. In Haworth structures, cyclic forms of carbohydrates are represented as regular hexagons or pentagons, as shown for the α-anomers of glucose and ribose.

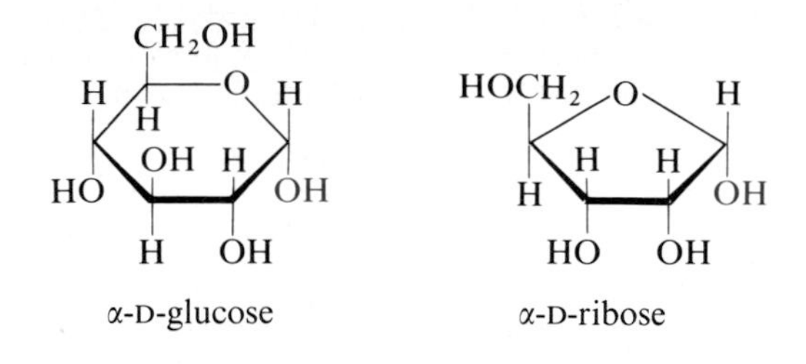

A third stereorepresentation for monosaccharides is to show six-membered cyclic hemiacetals as strain-free chair conformations. Following are chair conformations of α-D-glucose and β-D-glucose along with the open-chain form of D-glucose.

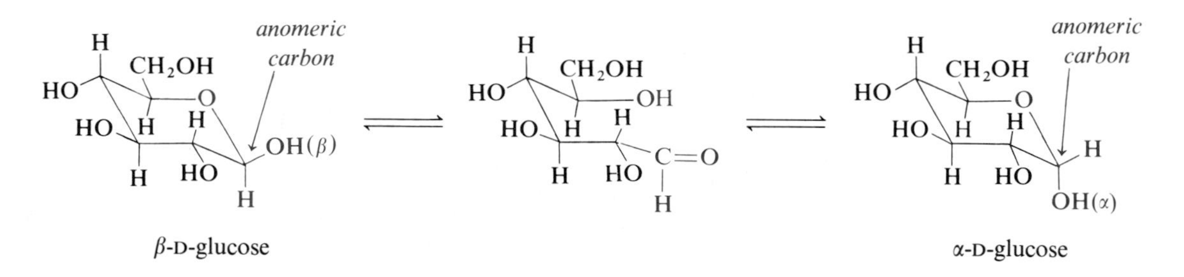

Notice that in the chair conformations of α- and β-D-glucose, substituents on carbons 2, 3, 4, and 5 of each ring are equatorial. The —OH on the anomeric carbon of α-D-glucose is axial; the —OH on the anomeric carbon of β-D-glucose is equatorial.

11.6 PHYSICAL PROPERTIES OF MONOSACCHARIDES

Monosaccharides are colorless crystalline solids. Because of the possibility of hydrogen bonding between polar —OH groups and water, monosaccharides are very soluble in water. They are only slightly soluble in alcohol, and are insoluble in nonhydroxylic solvents such as ether, chloroform, and benzene.

Although all monosaccharides are sweet to the taste, some are sweeter than others (Table 11.4). Of the monosaccharides, D-fructose is the sweetest

Table 11.4 Relative sweetness of some monosaccharides, disaccharides, and other sweetening agents.

Monosaccharides		**Disaccharides**	
D-fructose	174	sucrose (table sugar)	100
D-glucose	74	lactose (milk sugar)	0.16
D-xylose	0.40		
D-galactose	0.22		
Other Sweetening Agents			
saccharine	3,000		
honey	97		
molasses	74		
corn syrup	74		

to the taste, even sweeter than sucrose (table sugar). Saccharine is a noncarbohydrate, artificial sweetener. The sweet taste of honey is due largely to the presence of D-fructose, that of corn syrup is due largely to the presence of glucose. Molasses is a by-product of table sugar production. In the production of table sugar, sugar cane is boiled with water and then cooled. As the mixture cools, both sucrose crystals and light molasses separate and are collected. Subsequent boilings and coolings yield a very dark, thick syrup known as blackstrap molasses.

11.7 REACTIONS OF MONOSACCHARIDES

A. OXIDATION

Monosaccharides (and carbohydrates in general) are classified as reducing or nonreducing sugars according to their behavior toward Cu^{2+} (Benedict's solution, Fehling's solution) or toward Ag^+ in ammonium hydroxide (Tollens' solution). To understand the chemical basis for this classification, remember that all monosaccharides contain either an aldehyde or an α-hydroxyketone and that in dilute base (i.e., the conditions of Benedict's, Tollens' and Fehling's tests) ketoses are in equilibrium with aldoses via an enediol intermediate (Problems 7.26 and 7.27).

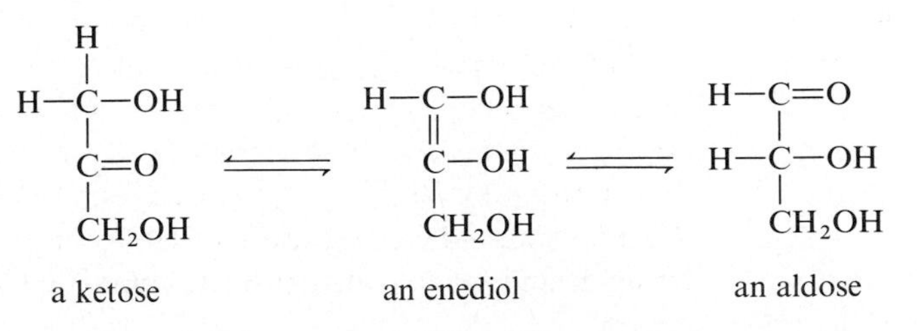

Both Ag(I) and Cu(II) oxidize aldehydes to carboxylic acids.

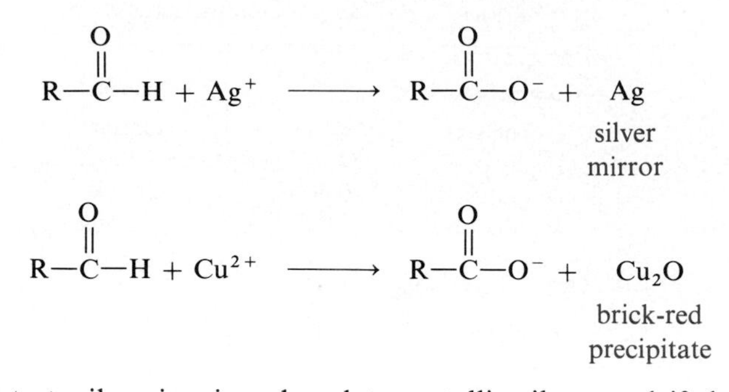

In Tollens' test, silver ion is reduced to metallic silver, and if the reaction is done properly, silver precipitates as a mirror-like coating on the surface of the container. For this reason, this test is called the silver mirror test. In Benedict's or Fehling's test, copper(II) is reduced to copper(I) which precipitates as Cu_2O, a brick-red solid. Any carbohydrate that reduces silver ion to metallic silver or copper(II) ion to Cu_2O is classified as a reducing sugar. Sugars that do not reduce these reagents are classified as nonreducing sugars.

Even though monosaccharides are predominantly in the cyclic hemiacetal form, the cyclic forms are in equilibrium with the free aldehyde and are therefore susceptible to oxidation by the solutions mentioned. All monosaccharides are reducing sugars.

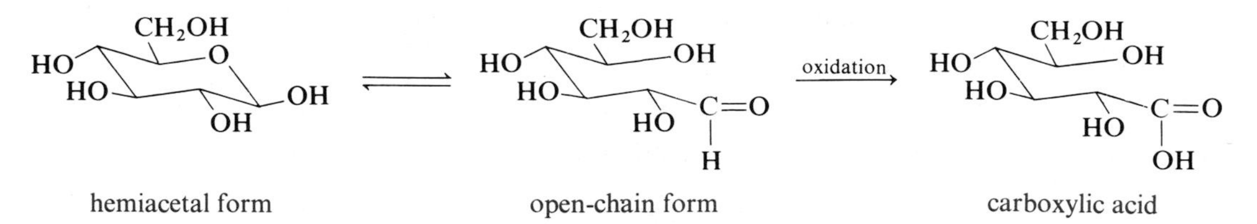

Oxidation of aldoses by Fehling's, Benedict's, or Tollens' solutions forms monocarboxylic acids known as aldonic acids. For example, D-glucose forms D-gluconic acid, D-threose yields D-threonic acid.

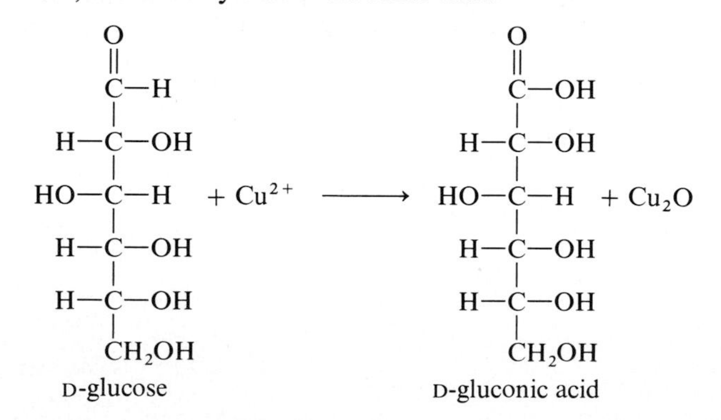

Warm nitric acid oxidizes the —CHO group of a monosaccharide to a carboxylic acid, and also oxidizes the terminal $—CH_2OH$ (a primary alcohol) to a carboxylic acid. Dicarboxylic acids derived from monosaccharides are known as aldaric acids. For example, oxidation of D-glucose by nitric acid yields D-glucaric acid.

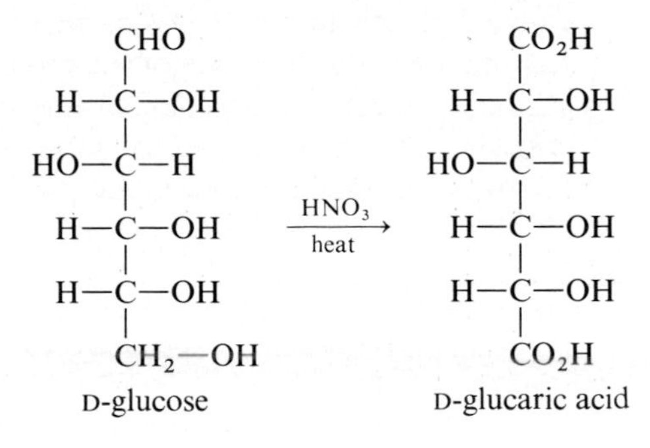

B. REDUCTION

The carbonyl group of a monosaccharide is reduced to an alcohol by a variety of reducing agents, including hydrogen in the presence of a heavy-metal catalyst and $NaBH_4$. These reduction products are known as alditols. Reduction of D-glucose gives D-glucitol, known more commonly as sorbitol.

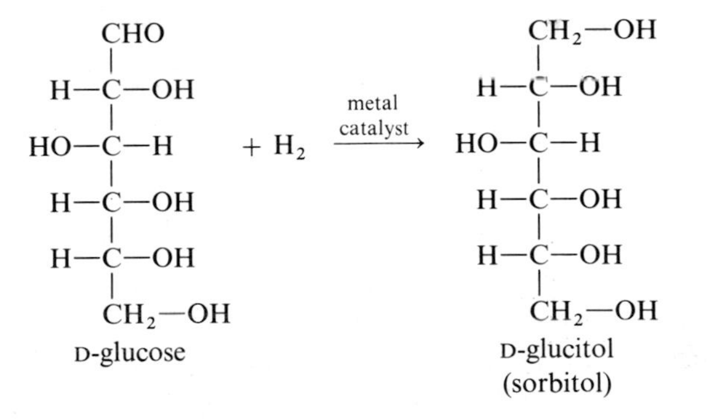

C. FORMATION OF GLYCOSIDES (ACETALS)

We have already seen in Section 11.4 that monosaccharides of five or more carbon atoms form cyclic hemiacetals or hemiketals. Reaction of a monosaccharide hemiacetal or hemiketal with a second molecule of alcohol forms an acetal or ketal. Following is a Haworth structure for the acetal formed by the reaction of β-D-glucose with methanol.

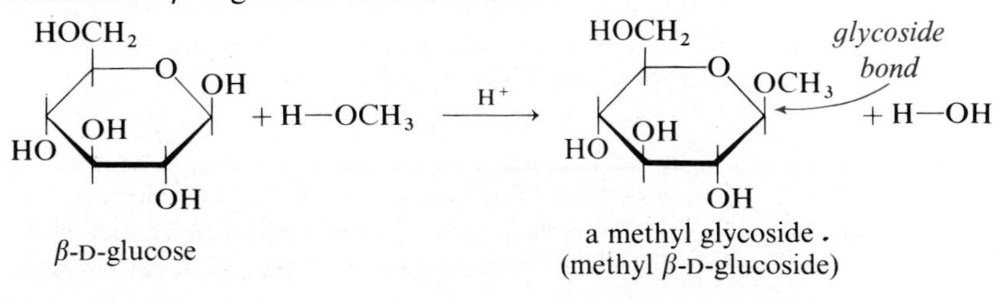

In this reaction, the —OH on the anomeric carbon is replaced by an —OR group. The cyclic acetal or ketal derived from a monosaccharide is called a glycoside, and the bond from the anomeric carbon to the —OR group is called a glycoside bond. The name of the glycoside is derived by dropping the terminal -e from the name of the monosaccharide from which it is derived and adding -ide. For example, glycosides derived from D-glucose are named D-glucosides; those derived from D-ribose are named D-ribosides; etc. Glycosides of monosaccharides are nonreducing sugars.

Example 11.3

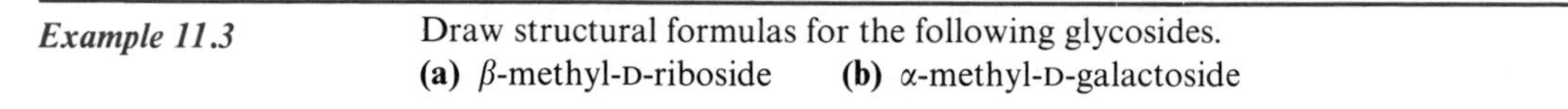

Draw structural formulas for the following glycosides.
(a) β-methyl-D-riboside **(b)** α-methyl-D-galactoside

Solution

(a) D-Ribose forms a five-membered cyclic hemiacetal. This hemiacetal reacts with methanol to form a cyclic acetal. The $—OCH_3$ group on the anomeric carbon is above the plane of the ring in the β-riboside.

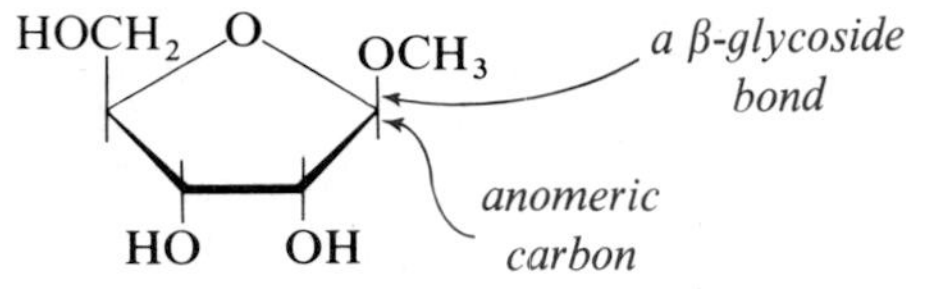

(b) D-Galactose forms a six-membered cyclic hemiacetal which reacts with methanol to form a six-membered cyclic acetal. The $—OCH_3$ group is below the plane of the ring in the α-galactoside. Following are shown Haworth and chair structures for α-methyl-D-galactoside.

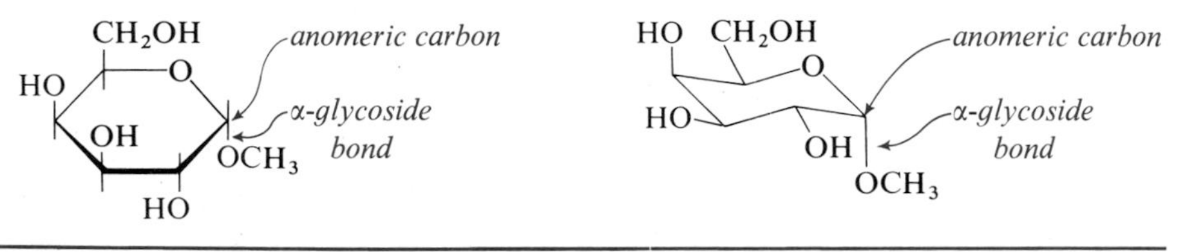

PROBLEM 11.3 Draw structural formulas for the following glycosides. In each, label the anomeric carbon and the glycoside bond.
(a) β-methyl-D-fructoside **(b)** α-methyl-D-mannoside

The anomeric carbon of a cyclic hemiacetal reacts with an N—H group to form an N-glycoside. Especially important in the biological world are the N-glycosides formed between D-ribose and 2-deoxy-D-ribose and the heterocyclic aromatic amines uracil, cytosine, thymine, adenine, and guanine (Figure 11.3). N-Glycosides of these purine and pyrimidine bases are structural units of DNA and RNA (Chapter 14).

Example 11.4

Draw structural formulas for the N-glycosides formed between the following compounds. In each, label the anomeric carbon and the glycoside bond.
(a) β-D-ribose and cytosine **(b)** β-2-deoxy-D-ribose and adenine

Solution

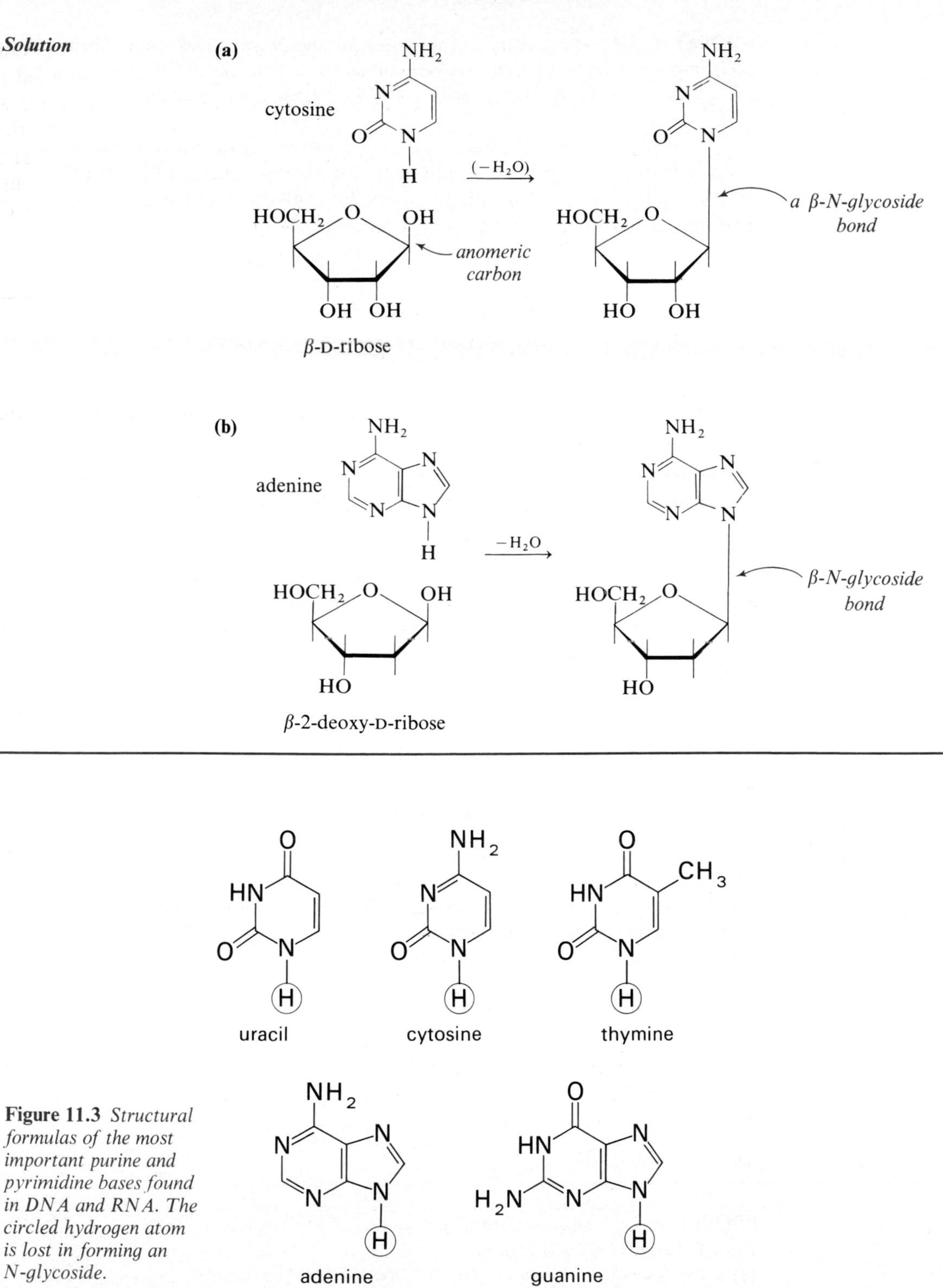

Figure 11.3 *Structural formulas of the most important purine and pyrimidine bases found in DNA and RNA. The circled hydrogen atom is lost in forming an N-glycoside.*

PROBLEM 11.4 Draw structural formulas for the N-glycosides formed between the following compounds. In each, label the anomeric carbon and the N-glycoside bond.
(a) β-2-deoxy-D-ribose and uracil **(b)** β-D-ribose and guanine

Phosphate esters of monosaccharides are important intermediates in the metabolism of carbohydrates and they are also structural units in DNA and RNA. Following are phosphate esters of D-glyceraldehyde and dihydroxyacetone, each shown as it would be ionized at pH 7.4.

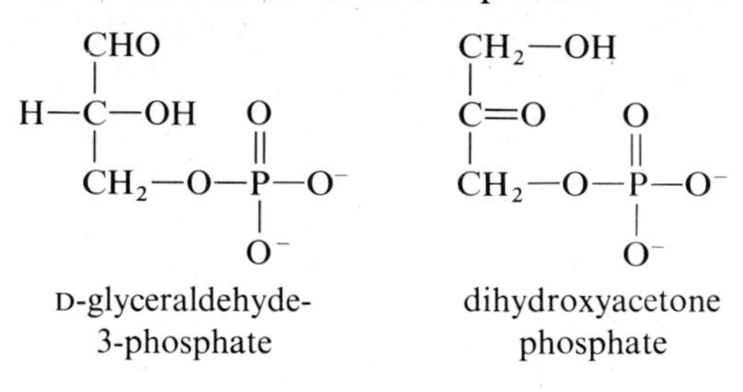

Example 11.5

Draw structural formulas for the following phosphate esters. Calculate the net charge on each at pH 7.4.
(a) α-D-glucose-6-phosphate **(b)** β-D-fructose-1,6-diphosphate

Solution

(a) Draw D-glucose as a six-membered cyclic hemiacetal with the —OH on the anomeric carbon atom below the plane of the ring. Show the phosphate ester bond between the oxygen atom of carbon 6 and phosphoric acid.

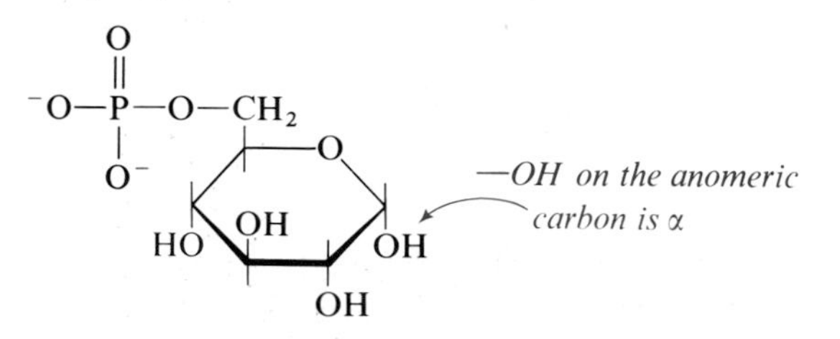

The net charge at pH 7.4 is −2.

(b) Draw D-fructose in the five-membered cyclic hemiketal form. Then draw one phosphate ester between the —OH on carbon-6 and phosphate and another between the —OH on carbon-1 and phosphate.

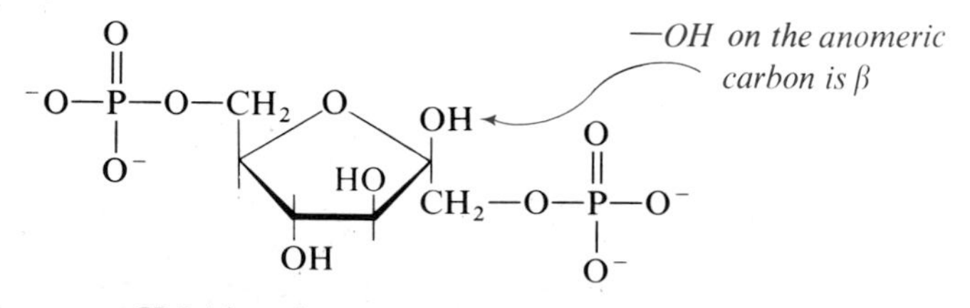

The net charge at pH 7.4 is −4.

PROBLEM 11.5 Draw structural formulas for the following phosphate esters. Calculate the net charge on each at pH 7.4.
(a) α-D-glucose-1-phosphate **(b)** β-D-ribose-5-phosphate

11.8 DISACCHARIDES

Most carbohydrates in nature contain more than one monosaccharide. Those that contain two monosaccharide units are called disaccharides, those that contain three monosaccharide units are called trisaccharides, and those that contain many monosaccharide units are called polysaccharides. In a disaccharide, two monosaccharide units are joined together by a glycoside bond between the anomeric carbon of one unit and an —OH of the other. In this section, we will look at the structure and natural occurrence of three disaccharides: maltose, lactose (milk sugar), and sucrose (cane sugar).

Maltose derives its name from the fact that it occurs mainly in malt liquors, the juice from sprouted barley and other cereal grains. Maltose consists of two molecules of glucose joined by a glycoside bond between carbon-1 (the anomeric carbon) of one glucose unit and carbon-4 of the second glucose unit. The oxygen atom on the anomeric carbon of the first glucose unit is α (below the plane of the ring), and therefore the bond joining the two glucose units is called an α-1,4-glycoside bond. Following are Haworth and chair formulas for maltose.

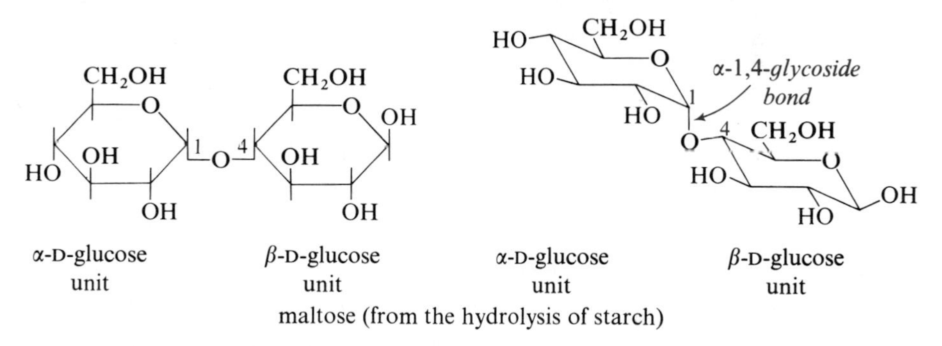

maltose (from the hydrolysis of starch)

Hydrolysis of maltose yields two molecules of glucose. Maltose is a reducing sugar because the anomeric carbon on the right unit of D-glucose is in equilibrium with the free aldehyde that can be oxidized to a carboxylic acid.

Another disaccharide of special interest is lactose, the major sugar of milk. It is present to the extent of 5–8% in human milk and 4–6% in cow's milk. Hydrolysis of lactose affords D-glucose and D-galactose. In lactose, galactose is joined to glucose by a β-1,4-glycoside bond.

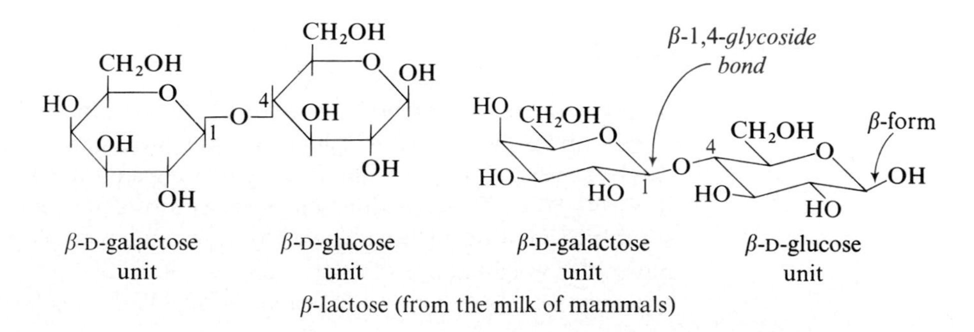

β-lactose (from the milk of mammals)

Sucrose (table sugar) is the most abundant disaccharide. It is obtained from the juice of sugar cane and the sugar beet.

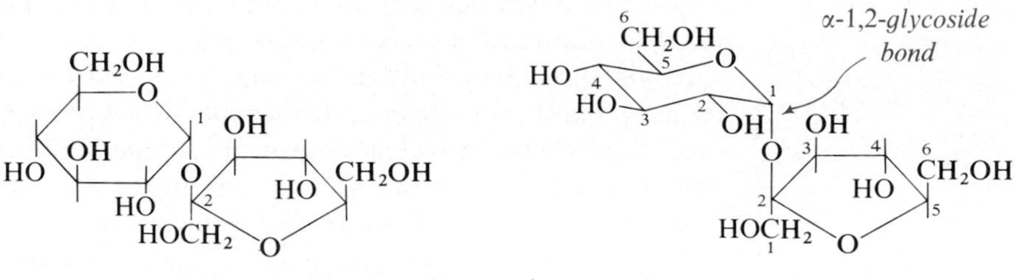

sucrose (cane or beet sugar)

In sucrose, carbon-1 of glucose is joined by an α-1,2-glycoside bond to carbon-2 of fructose. Glucose is in a six-membered ring form while fructose is in a five-membered ring form. Sucrose is a nonreducing sugar because the anomeric hemiacetal carbons of both glucose and fructose are involved in the formation of the glycoside bond in sucrose.

Example 11.6 Draw Haworth and chair formulas of a disaccharide in which two units of D-glucose are joined by an α-1,6-glycoside bond.

Solution First draw the structural formulas of α-D-glucose. Then connect the anomeric carbon of this monosaccharide by an α-glycoside bond to carbon-6 of the second glucose unit.

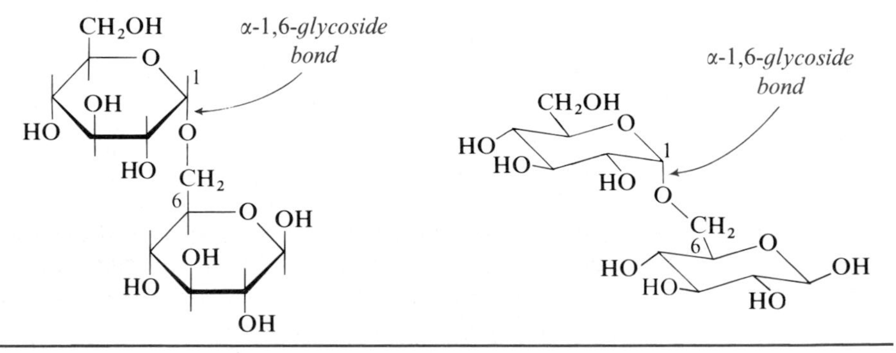

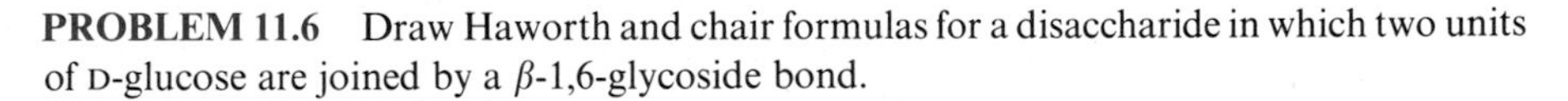

PROBLEM 11.6 Draw Haworth and chair formulas for a disaccharide in which two units of D-glucose are joined by a *β*-1,6-glycoside bond.

11.9 POLYSACCHARIDES

Starch is the reserve carbohydrate for plants. It is found in all plant seeds and tubers, and is the form in which glucose is stored for later use by plants. Starch can be separated into two fractions by making a paste with water and warming it to 60–80°C. One fraction, amylose, comprising about 20% of starch, is soluble in hot water. The water-insoluble fraction is amylopectin. Amylose has a molecular weight range of 10,000–50,000 (60 to 300 glucose units), and amylopectin has a molecular weight range of 50,000–1,000,000 (300 to 6000 glucose units). X-ray diffraction studies of amylose show a continuous, unbranched chain of glucose units joined by α-1,4-glycoside bonds.

Amylopectin has a highly branched structure. It contains the same type of repetitive sequence of α-1,4-glycoside bonds as does amylose, but chain lengths vary only from about 24 to 30 units. In addition, there is considerable branching from this linear network. At branch points a new chain is started by an α-1,6-glycoside linkage between carbon-1 of one glucose unit and carbon-6 of another glucose unit.

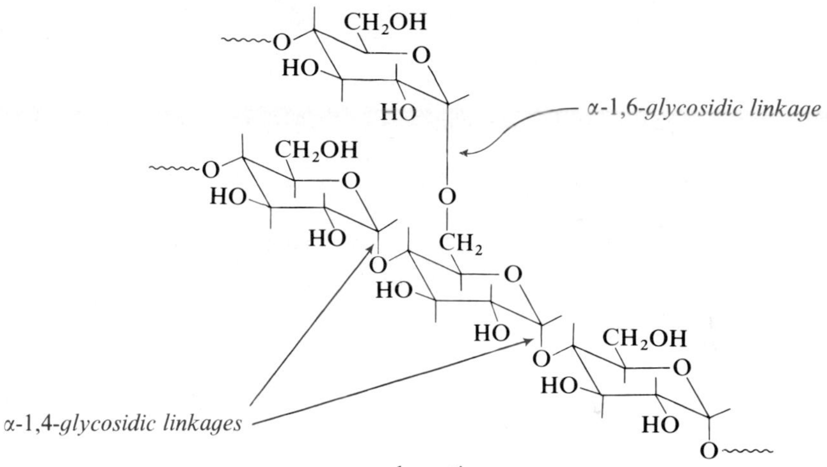

amylopectin

Why are carbohydrates stored in plants as polysaccharides rather than as monosaccharides? The answer has to do with osmotic pressure. Osmotic pressure is proportional to the molar concentration, not the molecular weight, of a solute. If we assume that 1000 molecules of glucose are assembled into one starch macromolecule, then we can predict that a solution containing 1 gram of starch per 10 mL will have only 1/1000 the osmotic pressure of a solution of 1 gram of glucose in the same volume of solution. This feat of packaging is of tremendous advantage because it reduces the strain on various membranes enclosing such macromolecules.

Glycogen is the reserve carbohydrate for animals. Like amylopectin, glycogen is a nonlinear polymer of glucose units joined by α-1,4- and α-1,6-glycoside bonds, but it has a lower molecular weight and a more highly branched structure. The total amount of glycogen in the body of a well-nourished adult is about 350 grams divided almost equally between liver and muscle. Figure 11.4 illustrates the highly branched structure of glycogen.

Cellulose is the most widely distributed skeletal polysaccharide. It constitutes almost half of the cell wall material of wood. Cotton is almost pure cellulose. Cellulose is a linear polymer of glucose units joined together by β-1,4-glycoside linkages and has a molecular weight of approximately 400,000, corresponding to 2800 glucose units. X-ray analysis indicates that cellulose fiber consists of bundles of parallel polysaccharide chains held together by hydrogen bonding between the hydroxyls of adjacent chains. This type of

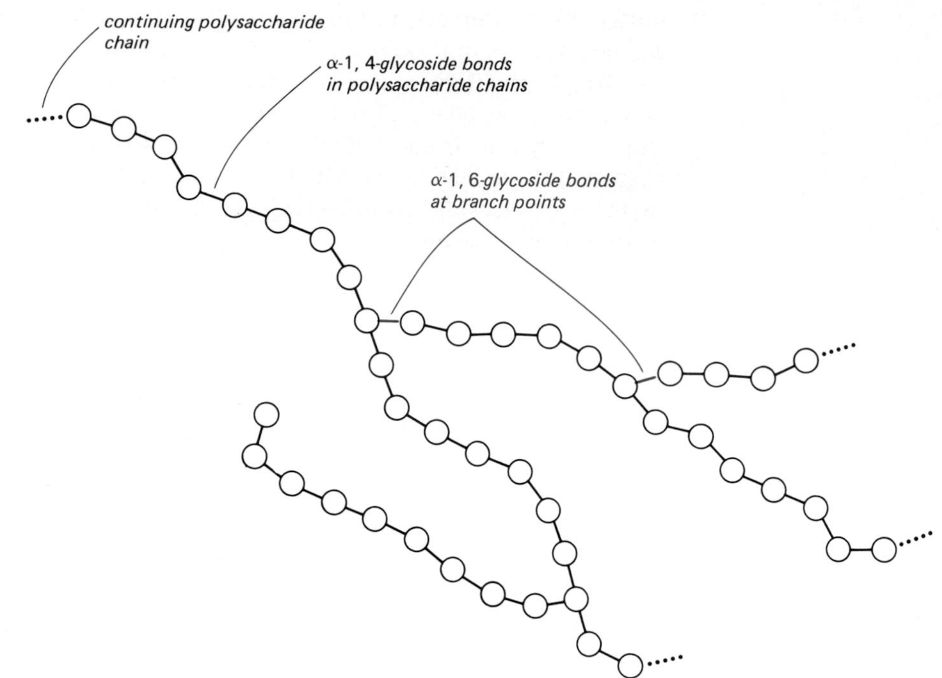

Figure 11.4 *Glycogen.*

arrangement of parallel chains into bundles and the resultant hydrogen bonding gives cellulose fibers a high mechanical strength and a chemical stability.

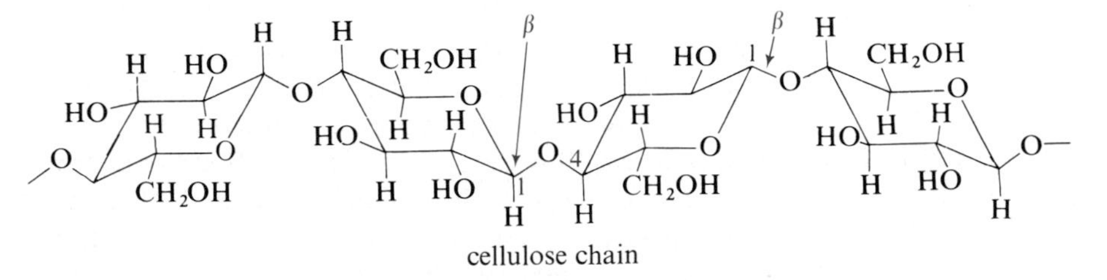

Humans and other animals cannot use cellulose as a food because our digestive systems do not contain β-glycosidases (enzymes that catalyze hydrolysis of the β-glycoside bonds). Our digestive systems contain only α-glycosidases and hence the polysaccharides we use as sources of glucose are starch and glycogen. On the other hand, many bacteria and microorganisms do contain β-glycosidases and are able to digest cellulose. Termites are fortunate to have such bacteria in their intestine, and can use wood as their principal food. Ruminants (cud-chewing animals) can also digest grasses and wood because of the presence of microorganisms within their specially constructed alimentary system.

Cellulose is the raw material for the production of certain widely used semi-synthetic materials, the most important of which are cellulose nitrate and cellulose acetate.

Nitration of cellulose by a mixture of nitric and sulfuric acids yields a polynitrate ester called cellulose nitrate or nitrocellulose. The extent of nitration varies with the conditions of the process used. Following is a partial structural formula of fully nitrated cellulose.

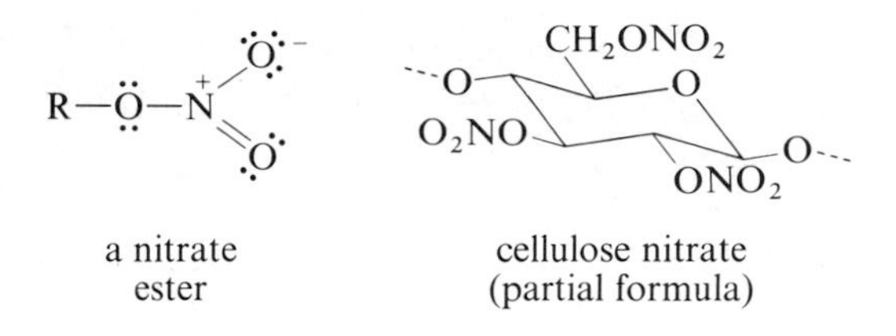

a nitrate ester

cellulose nitrate (partial formula)

Note that there are three nitrate esters per glucose unit. Guncotton approaches this degree of nitration. Less completely nitrated celluloses are used in the manufacture of the common moldable plastic celluloid. In this process, partially nitrated cellulose is mixed with alcohol and camphor and then molded or rolled into sheets. The molded or rolled article is then heated to evaporate the alcohol. The finished material is a tough, hard plastic. One of the first uses for this plastic was as a substitute for ivory billiard balls. Celluloid deserves special mention in the chemistry and technology of plastics for it was the first (and until about 1920 the only) plastic to be manufactured on an industrial scale. Its major disadvantage is that it is highly flammable.

Fibers made from regenerated and chemically modified cellulose were the first of the man-made fibers to become commercially important. In one industrial process, cellulose is reacted with carbon disulfide to form an alkali-soluble xanthate ester. A solution of cellulose xanthate is then extruded into dilute sulfuric acid to hydrolyze the xanthate ester and precipitate free or regenerated cellulose. Extruding regenerated cellulose as a filament produces "viscose rayon" threads; extruding it as sheets produces cellophane.

In another industrial process, cellulose is acetylated with acetic anhydride. Acetylated cellulose is dissolved in a suitable solvent, precipitated, and then drawn into fibers. This material is known as acetate rayon. Cellulose acetate, acetylated to the extent of about 80%, became commercial in Europe about 1920 and in the United States a few years later. Cellulose triacetate, which has about 97% of the hydroxyls acetylated, became commercial in the United States in 1954. Acetate fibers rank fourth in production in the United States, surpassed only by polyester, nylon, and rayon fibers.

11.10 BLOOD GROUP SUBSTANCES

Plasma membranes of animal cells have large numbers of relatively small carbohydrates bound to them. In fact, it now appears that the outsides of most plasma membranes are literally sugar-coated. Typically, these membrane-bound sugars contain from 4 to 15 monosaccharides, and are built from just a few monosaccharides, including D-galactose, D-mannose, L-fucose, N-acetyl-D-glucosamine, and N-acetyl-D-galactosamine.

Following is the structural formula of L-fucose.

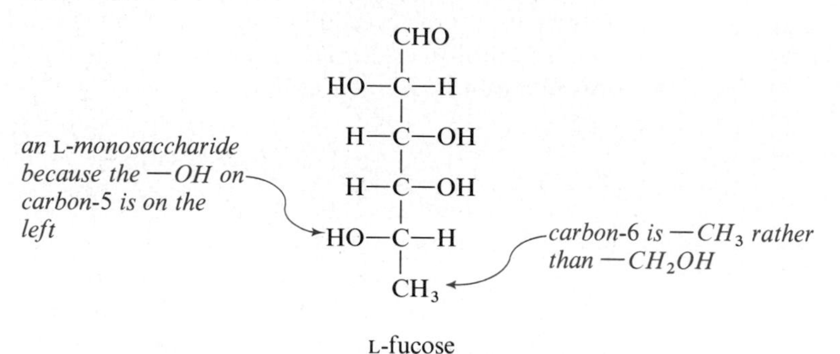

L-fucose

One of the first discovered and best understood of these membrane-bound carbohydrates are the so-called blood group substances. Although blood group substances or "markers" are found chiefly on the surface of erythrocytes (red blood cells), they are also found on proteins and lipids in other parts of the body as well. In the ABO system, first described in 1900, individuals are classified according to four blood types: A, B, AB, and O. Blood from individuals of the same type can be mixed without clumping of erythrocytes. However, if serum from a type A individual is mixed with type B blood, or vice versa, the erythrocytes will clump. Serum from a type O individual causes clumping of both type A and type B blood. At the cellular level, the chemical basis for this classification is a relatively small, membrane-bound carbohydrate. Following is the terminal tetrasaccharide portion of the sugar found on erythrocytes of individuals with type A blood. The stereochemistry of the glycoside bonds between the monosaccharides are shown in parentheses.

$$N\text{-acetyl-D-galactosamine} \xrightarrow{(\alpha\text{-}1,3)} \text{D-galactose} \xrightarrow{(\beta\text{-}1,3)} N\text{-acetyl-D-glucosamine} \xrightarrow{(\beta\text{-}1,\ldots)}$$

$$\uparrow (\alpha\text{-}1,2) \text{ from L-fucose (to D-galactose)}$$

This tetrasaccharide is shown using planar hexagons for the monosaccharide units and indicating the stereochemistry of the glycoside bonds by the symbols written over the glycoside bonds themselves.

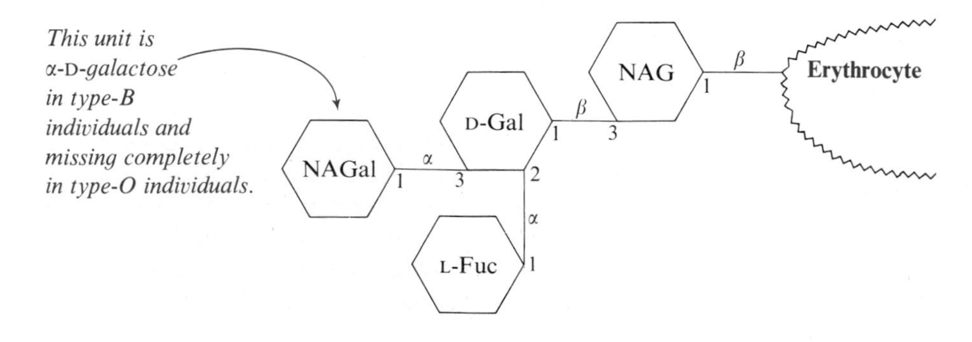

There are several distinctive features of this tetrasaccharide. First, it contains a monosaccharide of the "unnatural" or L-series, namely L-fucose. Second, it contains D-galactose to which are bonded two other monosaccharides, one by an α-1,2-glycoside bond, the other by an α-1,3-glycoside bond.

It is this last monosaccharide that determines the ABO classification. In type-A individuals, the chain terminates in N-acetyl-D-galactosamine (NAGal); in type-B individuals it terminates instead in D-galactose (D-Gal); and in type-O individuals, the last monosaccharide is missing completely. The saccharides of type-AB individuals contain both kinds of carbohydrate chains.

IMPORTANT REACTIONS

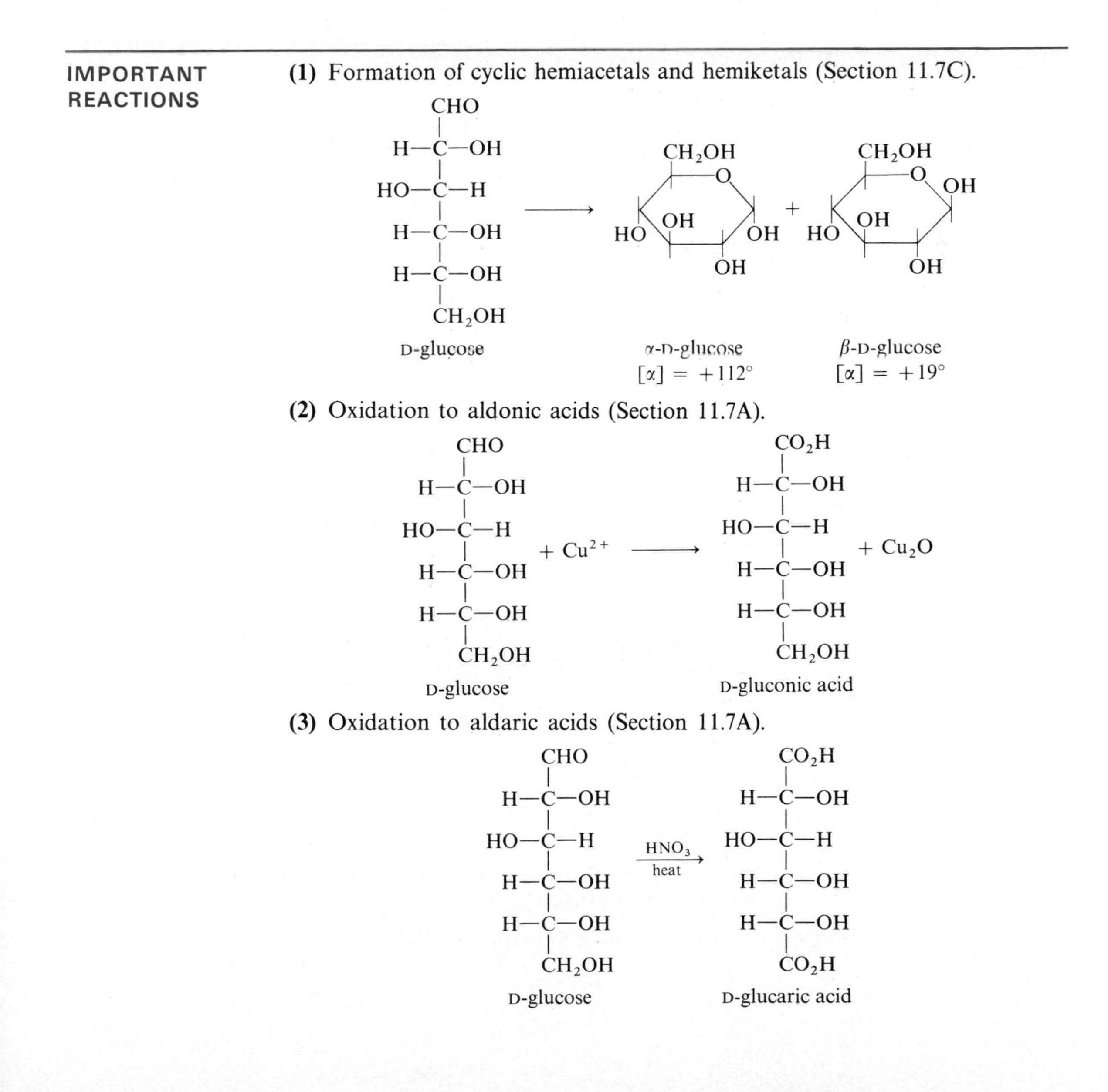

(4) Reduction to alditols (Section 11.7B).

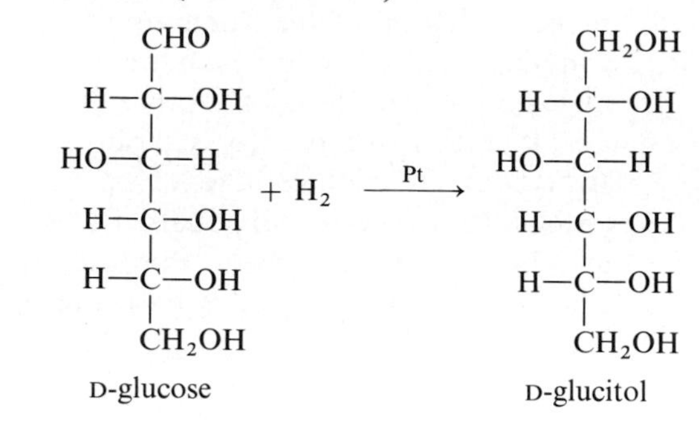

(5) Formation of glycosides (Section 11.7C).

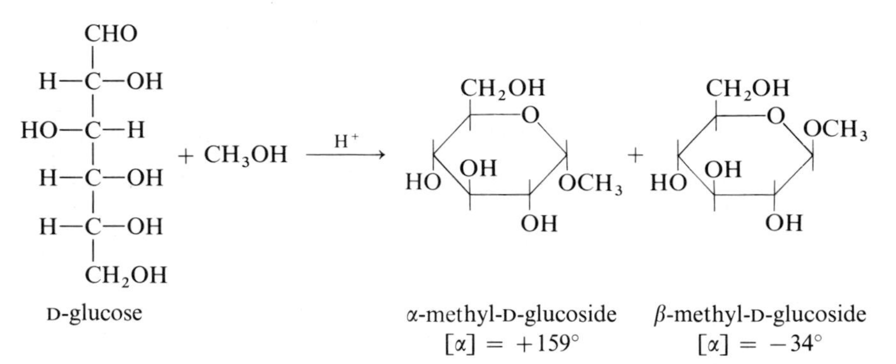

(6) Formation of N-glycosides (Section 11.7C).

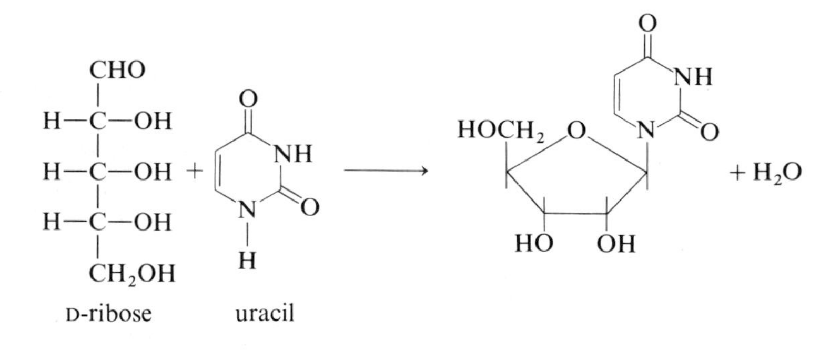

PROBLEMS

11.7 Define "carbohydrate" in terms of the functional groups present. Literally, the term carbohydrate is derived from "hydrates of carbon." Show, by reference to molecular formulas, the origin of this term.

11.8 Explain the meaning of the designations D- and L- as used to specify the stereochemistry of monosaccharides.

11.9 List the rules for drawing Fischer projection formulas.

11.10 Draw the four stereoisomers of 2,3,4-trihydroxybutanal. Label them A, B, C, and D. Which are pairs of enantiomers?

11.11 What is the difference in structure between D-ribose and 2-deoxy-D-ribose?

11.12 Both D-ribose and 2-deoxy-D-ribose form five-membered cyclic hemiacetals. Draw structural formulas for the α- and β-forms of each.

11.13 Draw a Fischer projection formula of D-glucose. State the total number of chiral carbons and the total number of stereoisomers possible for this structural formula. Of these, only D-glucose, D-galactose, and D-mannose are common in nature. Draw Fischer projection formulas for D-galactose and D-mannose.

11.14 Table 11.1 shows a Fischer projection formula of D-arabinose. Draw a Fischer projection formula of L-arabinose, a naturally occurring aldopentose of the "unnatural" L-configuration.

11.15 There are three common conventions for representing the stereochemistry of carbohydrates. They are (1) Fischer projections, (2) Haworth structures, and (3) chair structures. Draw α-D-glucose according to the rules of each of these conventions. Do the same for β-D-glucose.

11.16 **(a)** Build a molecular model of D-glucose in the open-chain form.
(b) Using this molecular model, show the reaction of the —OH on carbon-5 with the aldehyde of carbon-1 to form a cyclic hemiacetal. Show that either α-D-glucose or β-glucose can be formed, depending on the direction from which the —OH group interacts with the aldehyde group.

11.17 Explain the convention α- and β- as used to designate the stereochemistry of cyclic forms of monosaccharides.

11.18 Draw structural formulas for the open-chain and cyclic forms of D-fructose.

11.19 D-Glucosamine, D-mannosamine, and D-galactosamine form six-membered cyclic hemiacetals. Draw Haworth and chair structures for the α- and β-forms of each.

11.20 Explain the phenomenon of mutarotation with reference to carbohydrates. By what means is it detected?

11.21 A solution of α-D-glucose has a specific rotation of $+112°$; one of β-D-glucose has a specific rotation of $+19°$. On mutarotation, the specific rotation of each solution changes to an equilibrium value of $+52°$. Calculate the percentage of α-D-glucose in the equilibrium mixture.

11.22 Treatment of D-glucose in dilute aqueous base at room temperature yields an equilibrium mixture of D-glucose, D-mannose, and D-fructose. Account for this transformation. *Hint*: review Section 7.5E and your answer to Problem 7.27.

11.23 Ketones are not oxidized by mild oxidizing agents. However, both dihydroxyacetone and fructose give a positive Benedict's test and therefore are classed as reducing sugars. Using structural formulas, show how these monosaccharides react to give a positive test. *Hint*: these tests are done in the presence of dilute aqueous base.

11.24 Fischer attempted to convert D-glucose into its dimethyl acetal according to the following reaction:

$$\text{(i)}\quad \underset{\text{D-glucose}}{C_6H_{12}O_6} + 2CH_3OH \xrightarrow{H^+} \underset{\text{D-glucose dimethyl acetal}}{C_8H_{18}O_7} + H_2O$$

However, the reaction that takes place is instead:

$$\text{(ii)}\quad \underset{\text{D-glucose}}{C_6H_{12}O_6} + CH_3OH \xrightarrow{H^+} \underset{\text{methyl D-glucoside}}{C_7H_{14}O_6} + H_2O$$

Draw a structural formula for the expected product, $C_8H_{18}O_7$. Reaction (ii) gives isomeric methyl glucosides, designated as α-methyl-D-glucoside and β-methyl-D-glucoside. Draw Fischer, Haworth, and chair structures for each of these and label them accordingly.

11.25 Classify the following as reducing or nonreducing sugars.

(a) sucrose
(b) lactose
(c) maltose
(d) 2-deoxy-D-ribose
(e) α-D-glucose
(f) α-D-glucose-6-phosphate
(g) D-ribose
(h) α-methyl-D-glucose

11.26 Draw structural formulas for the following phosphate esters.

(a) β-D-galactose-6-phosphate
(b) β-D-ribose-3-phosphate
(c) β-2-deoxy-D-ribose-5-phosphate
(d) β-D-ribose-1,3-diphosphate

11.27 What is the major difference in structure between cellulose and starch? Why are humans unable to digest cellulose?

11.28 Trehalose is found in young mushrooms and is the chief carbohydrate in the blood of certain insects. Trehalose is a disaccharide consisting of two glucose units joined by an α-1,1-glycoside bond.

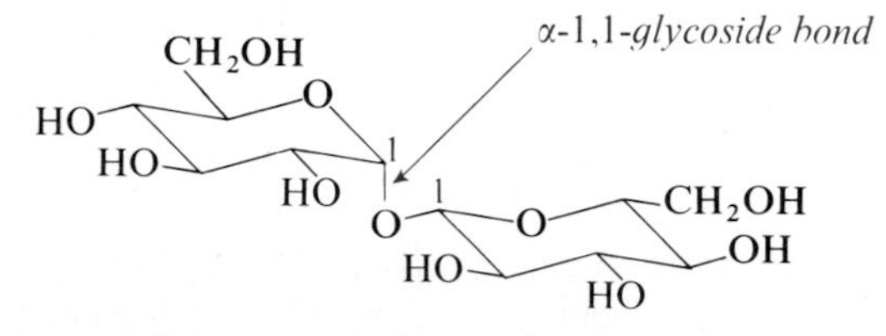

On the basis of its structural formula, would you expect trehalose **(a)** to be a reducing sugar? **(b)** to undergo mutarotation?

11.29 Raffinose is the most abundant trisaccharide in nature.

(a) Name the three monosaccharide units in raffinose.

(b) There are two glycoside bonds in raffinose. Describe each as you have already done for other disaccharides (i.e., an α-1,2-glycoside bond in sucrose).

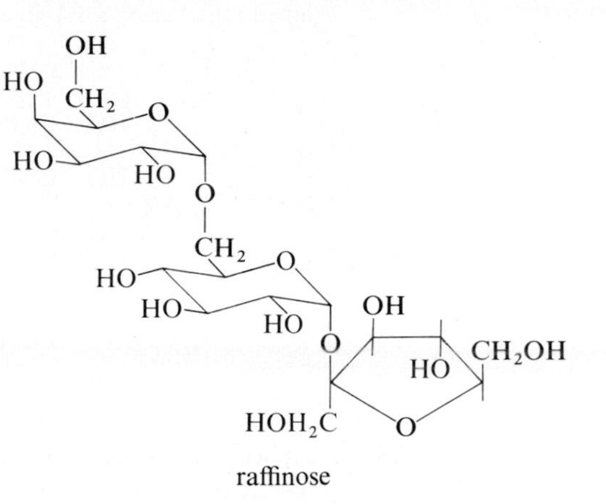

raffinose

11.30 Following is the Fischer projection formula for N-acetyl-D-glucosamine. This substance forms a six-membered cyclic hemiacetal.

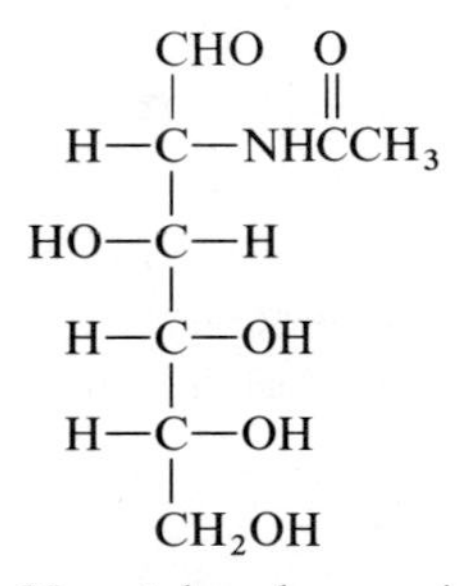

N-acetyl-D-glucosamine

(a) Draw Haworth and chair structures for the α- and β-forms of this monosaccharide.

(b) Draw Haworth and chair structures for the disaccharide formed by joining two units of N-acetyl-D-glucosamine by a β-1,4-glycoside bond. (If you have done this correctly, you have drawn the structural formula of the repeating dimer of chitin, the polysaccharide component of the shells of lobster and other crustacea.)

11.31 **(a)** Draw a structural formula for the trisaccharide found in the membrane-bound carbohydrate of individuals with type-O blood. Use Haworth structures for the monosaccharide units.

(b) Show how N-acetyl-D-galactosamine is attached to this trisaccharide to form the type-A blood group substance.

(c) Show how D-galactose is attached to this trisaccharide to form the type-B blood group substance.

11.32 Propose a likely structure for the following polysaccharides.

(a) Alginic acid, isolated from seaweed, is used as a thickening agent in ice cream and other foods. Alginic acid is a polymer of D-mannuronic acid units joined together by β-1,4-glycoside bonds.

(b) Pectic acid is the main constituent of pectin responsible for the formation of jellies from fruits and berries. Pectic acid is a polymer of D-galacturonic acid units joined together by α-1,4-glycoside bonds.

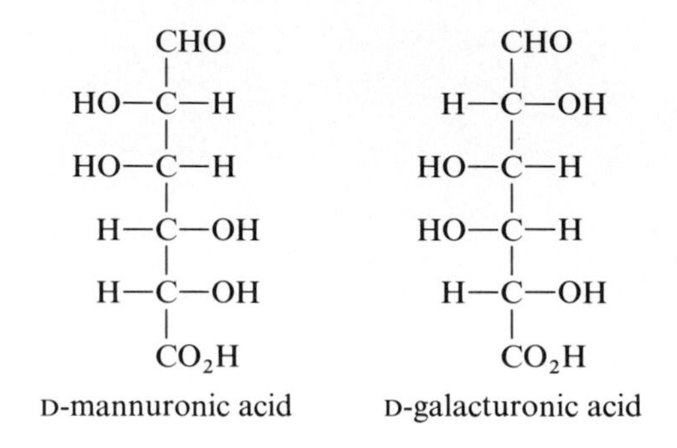

11.33 An important technique for establishing relative configurations among isomeric aldoses is to convert both terminal carbon atoms into the same functional group. This can be done either by selective oxidation or reduction. As a specific example, nitric acid oxidation of D-erythrose gives *meso*-tartaric acid. Oxidation of D-threose under similar conditions gives D-tartaric acid.

$$\text{D-threose} \xrightarrow[\text{oxidation}]{HNO_3} \text{D-tartaric acid}$$

$$\text{D-erythrose} \xrightarrow[\text{oxidation}]{HNO_3} \textit{meso}\text{-tartaric acid}$$

Using this information, show which of the structural formulas **(a)** or **(b)** is D-erythrose and which is D-threose. Check you answer by referring to Table 11.1.

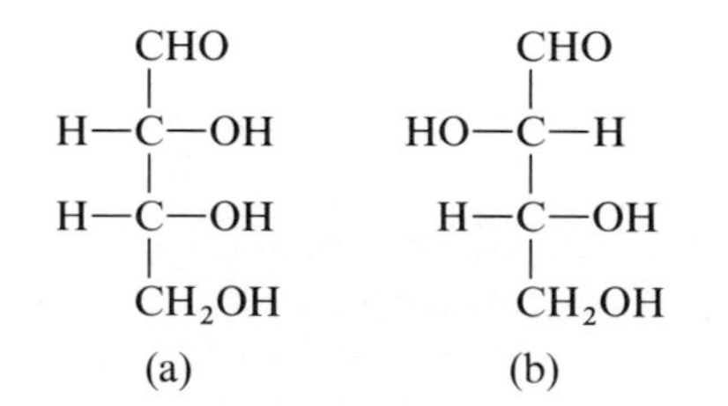

11.34 There are four D-aldopentoses (Table 11.1). Suppose that each is reduced with $NaBH_4$. Which of the four yield optically inactive D-alditols? which yield optically active D-alditols?

11.35 L-Fucose is one of several monosaccharides commonly found in the surface polysaccharides of animal cells. This 6-deoxyaldohexose is synthesized from D-mannose in the series of eight reactions shown opposite.

(a) Describe the type of reaction (i.e. oxidation, reduction, hydration, etc.) involved in each of the eight steps. We have already studied all of these reaction types, and you should be able to find several additional examples of each in this or previous chapters.

(b) Explain why this monosaccharide now belongs to the L-series even though it is derived biochemically from a D-sugar.

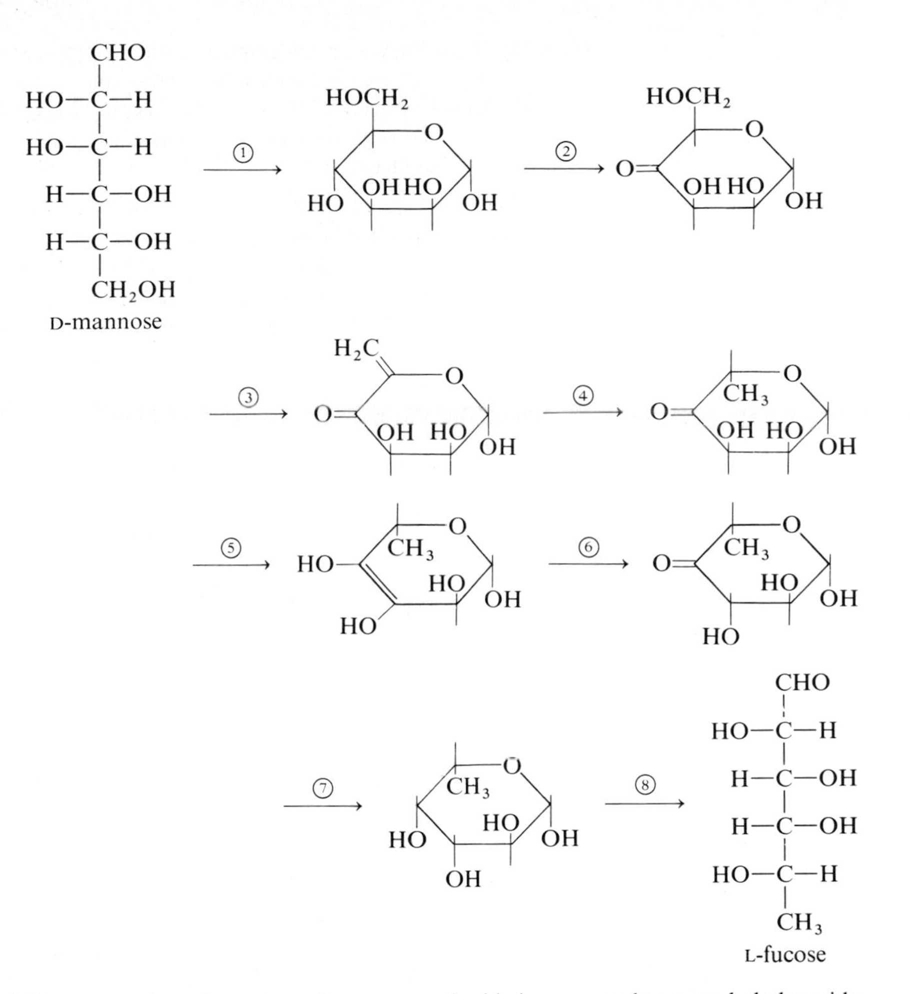

11.36 The anomeric carbon atom of a monosaccharide is converted to a methyl glycoside by refluxing the monosaccharide with methanol in the presence of an acid catalyst. The remaining —OH groups can then be converted to ethers by an application of the Williamson ether synthesis (Section 5.7) by reacting the monosaccharide methyl glycoside with methyl iodide in the presence of a strong base. Under these conditions, —OH groups are converted to $—O^-$, strong nucleophiles which then displace iodide ion from methyl iodide to give methyl ethers. The new functional groups formed are commonly called O-methyl ethers. Treatment of α-methyl-D-glucoside with methyl iodide under these conditions gives α-methyl-2,3,4,6-tetra-O-methyl-D-glucoside. Draw the structural formula of this substance.

11.37 Hydrolysis of α-methyl-2,3,4,6-tetra-O-methyl-D-glucoside (Problem 11.36) in aqueous acid gives 2,3,4,6-tetra-O-methyl-D-glucose.

(a) Draw the structural formula for the open-chain, free aldehyde form of this substance.

(b) How do you account for the fact that the methyl glucoside bond is hydrolyzed under these conditions but the methyl ether bonds on carbons 2, 3, 4, and 6 are not hydrolyzed?

11.38 **(a)** Draw the structural formula of the methyl glycoside of β-lactose. The name of this substance is β-methyl lactoside.

(b) Draw the structural formula of the hepta-O-methyl ether of this methyl glycoside.

(c) Hydrolysis of the hepta-O-methyl ether of β-methyl lactoside gives the 2,3,4,6-tetra-O-methyl ether of D-galactose and the 2,3,6-tri-O-methyl ether of D-glucose. Draw the structural formulas of each of these substances.

11.39 Certain types of streptococci found in the mouth, especially *streptococcus mutans*, have an enzyme system that uses sucrose as a starting material for the synthesis of high molecular weight polysaccharides known as dextrans. About 10% of the dry weight of dental plaque is composed of dextran. In one study of the dextran components of dental plaque, dextran was methylated with methyl iodide in the presence of a strong base and then the fully methylated polysaccharide was hydrolyzed in aqueous acid. The only monosaccharides found were four O-methyl derivatives of D-glucose.

D-glucose derivative	*Mole %*
2,3,4,6-tetra-O-methyl-D-glucose	14.6
2,4,6-tri-O-methyl-D-glucose	50.5
2,3,4-tri-O-methyl-D-glucose	20.9
2,4-di-O-methyl-D-glucose	14.0
	100%

(a) Draw the structural formula of the open-chain, free aldehyde form of each of these glucose derivatives.

(b) The isolation of one of these glucose derivatives is evidence for the presence of 1,3-glycoside bonds in this dextran. Explain. What is the percent of 1,3-glycoside bonds?

(c) The isolation of a second of these glucose derivatives is evidence for the presence of 1,6-glycoside bonds in this dextran. Explain. What is the percent of 1,6-glycoside bonds?

(d) The isolation of a third of these glucose derivatives is evidence that some monosaccharide units participate in both 1,3- and 1,6-glycoside bonds and serve therefore as branch points in the polysaccharide chain. Explain. What is the percent of chain branching?

(e) The fourth derivative isolated represents the monosaccharide at the end of the branched chains. Compare the percent yield of this terminal monosaccharide unit with the percent of chain branching you determined in part (d).

(f) From all of this evidence, sketch the polysaccharide of dextran in the same manner as glycogen is sketched in Figure 11.4.

Clinical Chemistry: The Search for Specificity

MINI-ESSAY 9

THE ANALYTICAL procedure that is most often performed in the clinical chemistry laboratory is the determination of glucose in blood, urine, or other biological fluid. The need for a rapid and reliable test for blood glucose stems from the high incidence of the disease diabetes mellitus. There are approximately 2 million known diabetics in the United States and it is estimated that another 2 million more are undiagnosed.

Diabetes mellitus is characterized by insufficient blood levels of the polypeptide hormone insulin (Section 12.10). Deficiency of insulin results in the inability of glucose to enter muscle and liver cells, leading to increased levels of blood glucose (hyperglucosemia); impaired metabolism of fats and proteins; ketosis; and possibly diabetic coma. Thus, it is critical for the early diagnosis and effective management of this disease to have a rapid and reliable procedure for the determination of blood glucose.

Over the past 60 years or more, a great many such tests have been developed. We will discuss only three of these, each chosen to illustrate something of the problems involved in developing suitable clinical laboratory tests and how these problems can be solved. Furthermore, these tests will illustrate the use of both chemical and enzymatic techniques in the modern clinical chemistry laboratory.

The first widely used glucose test was based on the fact that glucose is a reducing sugar. Specifically, the aldehyde group of glucose is oxidized by ferricyanide ion to a carboxyl group. In the process Fe^{3+} in ferricyanide ion (yellow in hot alkaline solution) is reduced to Fe^{2+} in ferrocyanide ion (colorless in hot alkaline solution). See Figure 1.

The decrease in absorbance of the test solution, measured at 420 nm, is proportional to the concentration of glucose in the sample. A modification of this method introduced in 1928 and for many years the standard procedure for the determination of blood glucose, involved carrying out the reaction in the presence of excess ferric ion. Under these conditions, the ferrocyanide ion formed on oxidation of glucose reacts further with Fe^{3+} to form ferric ferrocyanide, or as it is more commonly known, Prussian blue. In this procedure, the absorbance is directly proportional to the concentration of glucose in the test sample:

$$3\,Fe(CN)_6^{4-} + 4\,Fe^{3+} \longrightarrow \underset{\text{ferric ferrocyanide (Prussian blue)}}{Fe_4[Fe(CN)_6]_3}$$

Although this method can be used to measure glucose

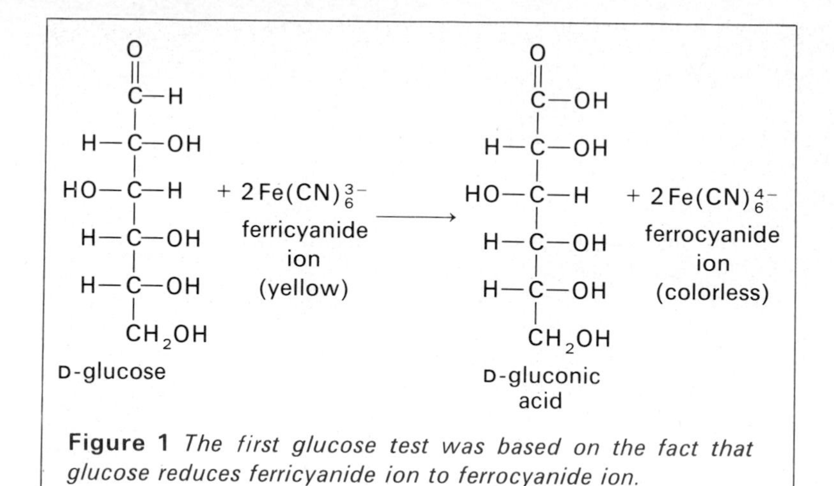

Figure 1 *The first glucose test was based on the fact that glucose reduces ferricyanide ion to ferrocyanide ion.*

concentration, it has the disadvantage that ferricyanide also oxidizes several other reducing substances found in blood, including ascorbic acid, uric acid, certain amino acids, and phenols. In addition, any other aldoses present in blood will also reduce ferricyanide. All of these substances are said to give false positive results. The ferricyanide and other oxidative tests first developed often gave values as much as 30% or more higher than the so-called "true glucose" value.

How could the determination of blood glucose be made more specific? Quite naturally, a considerable effort was devoted, with varying degrees of success, to eliminating or at least minimizing the interference of other reducing substances. However, a more satisfactory approach in the search for specificity lay in attacking the problem in a completely different way, namely, by taking advantage of a chemical reactivity of glucose other than its property as a reducing sugar. One of the most successful and widely used of these nonoxidative methods involves reaction of glucose with *o*-toluidine to form a blue-green Schiff base. The absorbance of this Schiff base can be measured at 625 nm and is directly proportional to glucose concentration (Figure 2). The procedure calls for mixing a test sample with *o*-toluidine, heating in a boiling-water bath for 8 minutes, and then measuring the absorbance at 625 nm. The *o*-toluidine method can be applied directly to serum, plasma, cerebro-spinal fluid, and urine, and to samples as small as 20 microliters (20×10^{-6} liter). In addition, it does not give false positive results with other reducing substances because the procedure itself does not involve oxidation. However, galactose and mannose, and to a lesser extent lactose and xylose, are potential sources of false positive results, for they also react with *o*-toluidine to give colored Schiff bases. This is generally not a problem because these mono- and disaccharides are normally present in serum and plasma only in very low concentrations.

In recent years the search for even greater specificity in glucose determinations has led to the introduction of enzyme-based glucose assay procedures. What was needed was an enzyme that catalyzes a specific reaction of glucose, but does not catalyze comparable reactions of any other substance normally present in biological fluids. The enzyme glucose oxidase meets these requirements. It catalyzes the oxidation of β-D-glucose to D-gluconic acid (Figure 3).

Glucose oxidase is highly specific for β-D-glucose. Therefore, complete oxidation of any

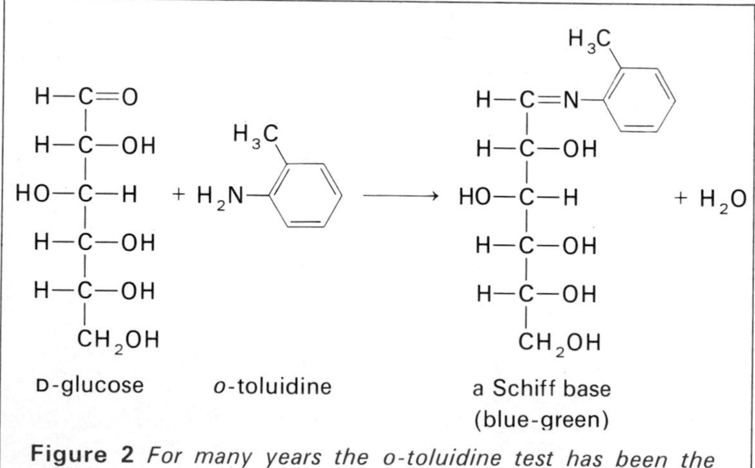

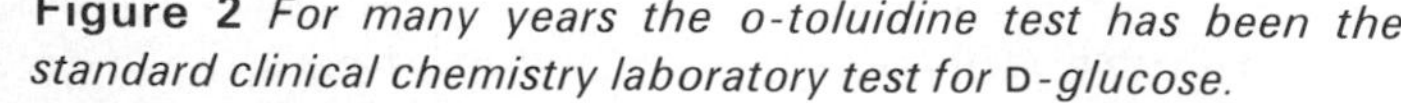
Figure 2 *For many years the o-toluidine test has been the standard clinical chemistry laboratory test for D-glucose.*

The insulin pump worn by this diabetes patient mimics the normal pancreas by continuously injecting insulin under the skin in tiny, predetermined amounts. However, the patient must measure his own glucose levels and then "tell" the machine how much insulin to release. (*Joslin Diabetes Foundation, Inc.*)

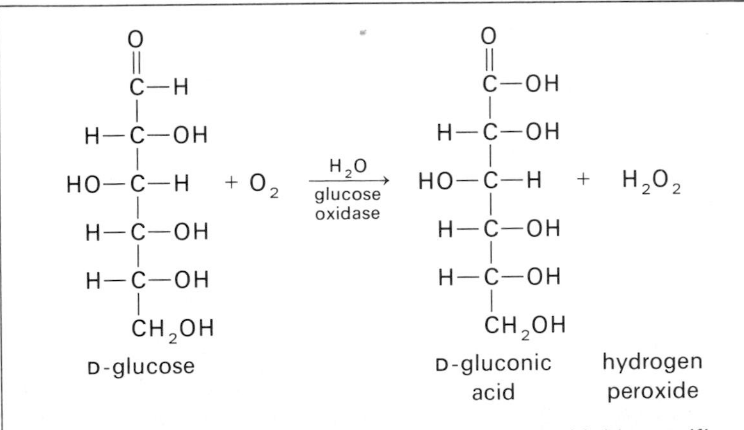

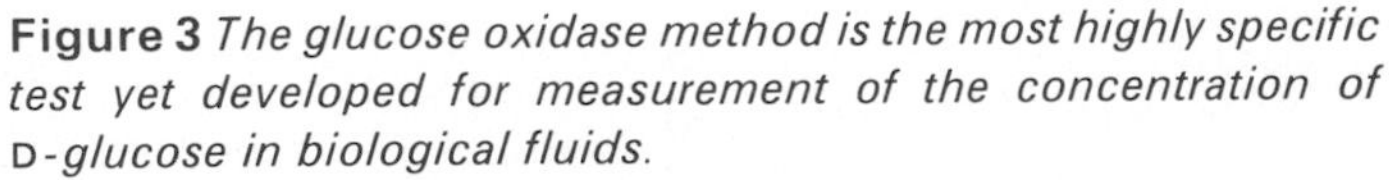

Figure 3 *The glucose oxidase method is the most highly specific test yet developed for measurement of the concentration of* D-*glucose in biological fluids.*

sample containing both β-D-glucose and α-D-glucose requires conversion of the α-form to the β-form. This interconversion is rapid and complete even in the short time required for the test itself. Notice that the oxidizing agent, O_2, is reduced to hydrogen peroxide, H_2O_2. The hydrogen peroxide generated in the glucose oxidase reaction is in turn used to oxidize another substance whose concentration can then be determined spectrophotometrically. In one procedure, H_2O_2 is reacted with iodide ion to form molecular iodine, I_2.

$$2I^- + H_2O_2 + 2H^+ \longrightarrow I_2 + 2H_2O$$

The absorbance at 420 nm is used to calculate iodine concentration and then glucose concentration. In aother procedure, H_2O_2 is used to oxidize the aromatic amine *o*-toluidine to a colored product. The enzyme peroxidase catalyzes this second oxidation. The concentration of the colored oxidation product is determined spectrophotometrically.

$$o\text{-toluidine} + H_2O_2 \xrightarrow{\text{peroxidase}} \text{oxidized } o\text{-toluidine (colored)} + H_2O$$

A number of commercially available test kits employ the glucose oxidase reaction for qualitative determination of glucose in urine. One of these, Clinistix, (produced by the Ames Co., Elkhart, Indiana), consists of a filter paper strip impregnated with glucose oxidase, peroxidase, and *o*-toluidine. The test end of the paper is dipped in urine, removed, and examined after 10 seconds. A blue color develops if the concentration of glucose in the urine exceeds about 100 mg/mL.

Recall that the first of the assay methods we looked at in this mini-essay was also an oxidative procedure. However, it gave positive errors in the presence of other reducing substances. In a sense, the search for specificity has now turned full circle, for at the present time the most highly specific and accurate assay, and the one which is said to give "true" glucose values, is also an oxidative method. Unlike the earlier ferricyanide method, this newer oxidative method is highly specific because it is catalyzed by a highly specific enzyme, β-D-glucose oxidase.

References

Henry, R. J., Cannon, D. O., and Winkleman, J. W. eds., *Clinical Chemistry, Principles and Techniques*, 2nd ed., (N.Y.: Harper & Row, 1974).

Tietz, N., ed., *Fundamentals of Clinical Chemistry*, (N.Y.: W. B. Saunders, 1976).

CHAPTER 12

Amino Acids and Proteins

Proteins, as much as any other class of compounds, are inseparable from life itself. These remarkable molecules are classified into two broad categories depending on their physical properties. Fibrous proteins are stringy, physically tough, and generally insoluble in water and most solvents. There are three major classes of fibrous proteins: the keratins of skin, wool, claws, horn, scales, and feathers; the silks; and the collagens of tendons and hides. Globular proteins, the second broad category, are generally spherical in shape, and soluble in water and aqueous solutions. Nearly all enzymes, antibodies, hormones, and transport proteins are globular.

In this chapter we will examine the structure and properties of proteins and of the amino acids from which they are constructed.

12.1 AMINO ACIDS

Amino acids are substances that contain both a carboxyl group and an amino group. While many types of amino acids are known, it is the α-amino acids that are most significant in the biological world because they are the units from which proteins are constructed. The general formula of an α-amino acid is shown in Figure 12.1. Although Figure 12.1a is a common way of writing structural formulas for amino acids, it is not accurate because it shows an acid ($—CO_2H$) and a base ($—NH_2$) within the same molecule. These acidic and basic groups react with each other to form an internal salt called a zwitterion (Figure 12.1b). Note that the zwitterion has no net charge; it contains one positive and one negative charge.

Although pK_a values for carboxyl and amino groups of particular amino acids vary, the average value of pK_a for the α-CO_2H group is 2.2; that for the α-NH_3^+ group is 9.5. Using these values, we can calculate the ratio of [α-CO_2H] to [α-CO_2^-] and of [α-NH_3^+] to [α-NH_2] at any given pH. As an example,

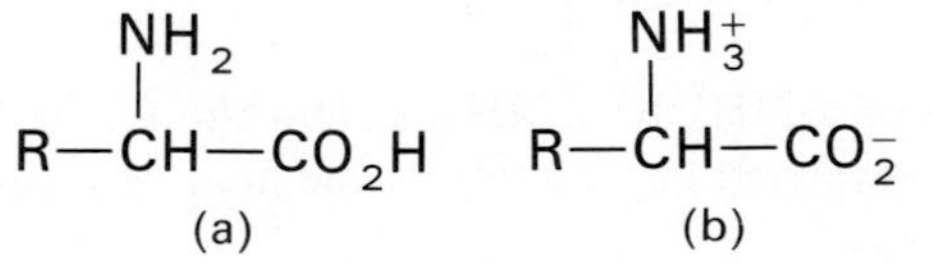

$$\underset{\text{(a)}}{R—\overset{\overset{\displaystyle NH_2}{|}}{CH}—CO_2H} \qquad \underset{\text{(b)}}{R—\overset{\overset{\displaystyle NH_3^+}{|}}{CH}—CO_2^-}$$

Figure 12.1 *General formula for an α-amino acid: (a) un-ionized form; (b) zwitterion form.*

let us calculate these ratios at pH 7.4, the physiological pH. Consider first the ionization of the weak acid α-CO_2H to form H^+ and its conjugate base α-CO_2^-.

$$\underset{\text{weak acid}}{\alpha\text{-}\overset{\overset{\displaystyle O}{\|}}{C}\text{—OH}} \rightleftharpoons \underset{\text{conjugate base}}{\alpha\text{-}\overset{\overset{\displaystyle O}{\|}}{C}\text{—O}^-} + H^+ \qquad K_a = 6.3 \times 10^{-3} \qquad pK_a = 2.2$$

The equilibrium constant for this ionization is called the acid dissociation constant and is given by the expression:

$$K_a = \frac{[H^+][\alpha\text{-}CO_2^-]}{[\alpha\text{-}CO_2H]}$$

Rearranging this expression gives

$$\frac{[\alpha\text{-}CO_2^-]}{[\alpha\text{-}CO_2H]} = \frac{K_a}{[H^+]}$$

Substituting the value of K_a for the α-CO_2H group (6.3×10^{-3}) and the hydrogen ion concentration at pH 7.4 (4.0×10^{-8}) in this equation gives

$$\frac{[\alpha\text{-}CO_2^-]}{[\alpha\text{-}CO^2H]} = \frac{6.3 \times 10^{-3}}{4.0 \times 10^{-8}} = 1.6 \times 10^5$$

Thus we see that at pH 7.4, the physiological pH, the ratio of [α-CO_2^-] to [α-CO_2H] is over 160,000 to 1. It is clear that at pH 7.4 an α-carboxyl group is virtually 100% in the conjugate base or ionized form and has a charge of -1.

We can also calculate the ratio of acid to conjugate base for the α-amino group. The average value of pK_α for α-amino groups is 9.5.

$$\underset{\text{weak acid}}{\alpha\text{-}NH_3^+} \rightleftharpoons \underset{\text{conjugate base}}{\alpha\text{-}NH_2} + H^+ \qquad K_a = 3.2 \times 10^{-10} \qquad pK_a = 9.5$$

The acid dissociation constant for the ionization of α-NH_3^+ can be rearranged to give the following expression.

$$\frac{[\alpha\text{-}NH_2]}{[\alpha\text{-}NH_3^+]} = \frac{K_a}{[H^+]}$$

Substituting the value of K_a for the α-NH_3^+ group (3.2×10^{-10}) and the hydrogen ion concentration at pH 7.4 (4.0×10^{-8}) gives

$$\frac{[\alpha\text{-}NH_2]}{[\alpha\text{-}NH_3^+]} = \frac{3.2 \times 10^{-10}}{4.0 \times 10^{-8}} = 8 \times 10^{-3}$$

Thus, the ratio of α-NH_2 to α-NH_3^+ is less than 1 to 100, and at pH 7.4, an α-amino group is almost completely in the acid or protonated form and has a charge of $+1$.

We can see from these calculations that the zwitterion form of an amino acid predominates at pH 7.4. In the remainder of the text, we shall use the zwitterion form for amino acids.

If the R-group of an α-amino acid is something other than hydrogen, the amino acid contains a chiral carbon atom adjacent to the carboxyl group and shows stereoisomerism. Figure 12.2 shows stereorepresentations and Fischer projection formulas for the enantiomers of serine.

With D-glyceraldehyde as a standard, it has been established that all amino acids occurring naturally in proteins have the opposite, or L-configuration, about the chiral carbon. D-Amino acids are not found in proteins and are not part of the metabolism of higher organisms. However, several D-amino acids are important in the structure and metabolism of lower forms of life. As an example, both D-alanine and D-glutamic acid are structural components of the cell walls of certain bacteria.

Table 12.1 shows names, structural formulas, and standard three-letter abbreviations for the 20 common amino acids found in proteins. Amino acids in this table are grouped into three categories according to the nature of their side chain. The nonpolar side chain category includes 8 amino acids: of these, glycine, alanine, and proline have small nonpolar side chains and are weakly hydrophobic. The other five amino acids in this category (phenylalanine, valine, leucine, isoleucine, and methionine) have larger side chains and are more strongly hydrophobic.

The polar but uncharged side chain category includes 8 amino acids: serine and threonine with hydroxyl groups; asparagine and glutamine with amide groups; tyrosine with a phenolic side chain; tryptophan and histidine with heterocyclic aromatic amine side chains; and cysteine with a sulfhydryl group. Three amino acids included in the polar uncharged category have side chains that show some degree of ionization, depending on the pH. These are the sulfhydryl group of cysteine, the imidazole group of histidine, and the phenolic hydroxyl of tyrosine.

There are four amino acids in the charged side chain category: aspartic acid, glutamic acid, lysine, and arginine. Aspartic and glutamic acids have carboxyl groups on the side chain. The pK_a values for these side-chain $—CO_2H$ groups are approximately 4.0, and therefore each is completely ionized at

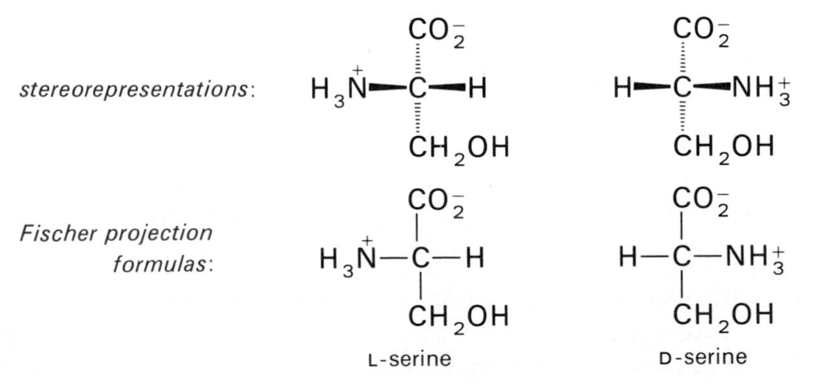

Figure 12.2 *The enantiomers of serine.*

Table 12.1 The 20 common amino acids of protein origin, grouped by categories. Each is shown as it would be ionized at pH 7.4

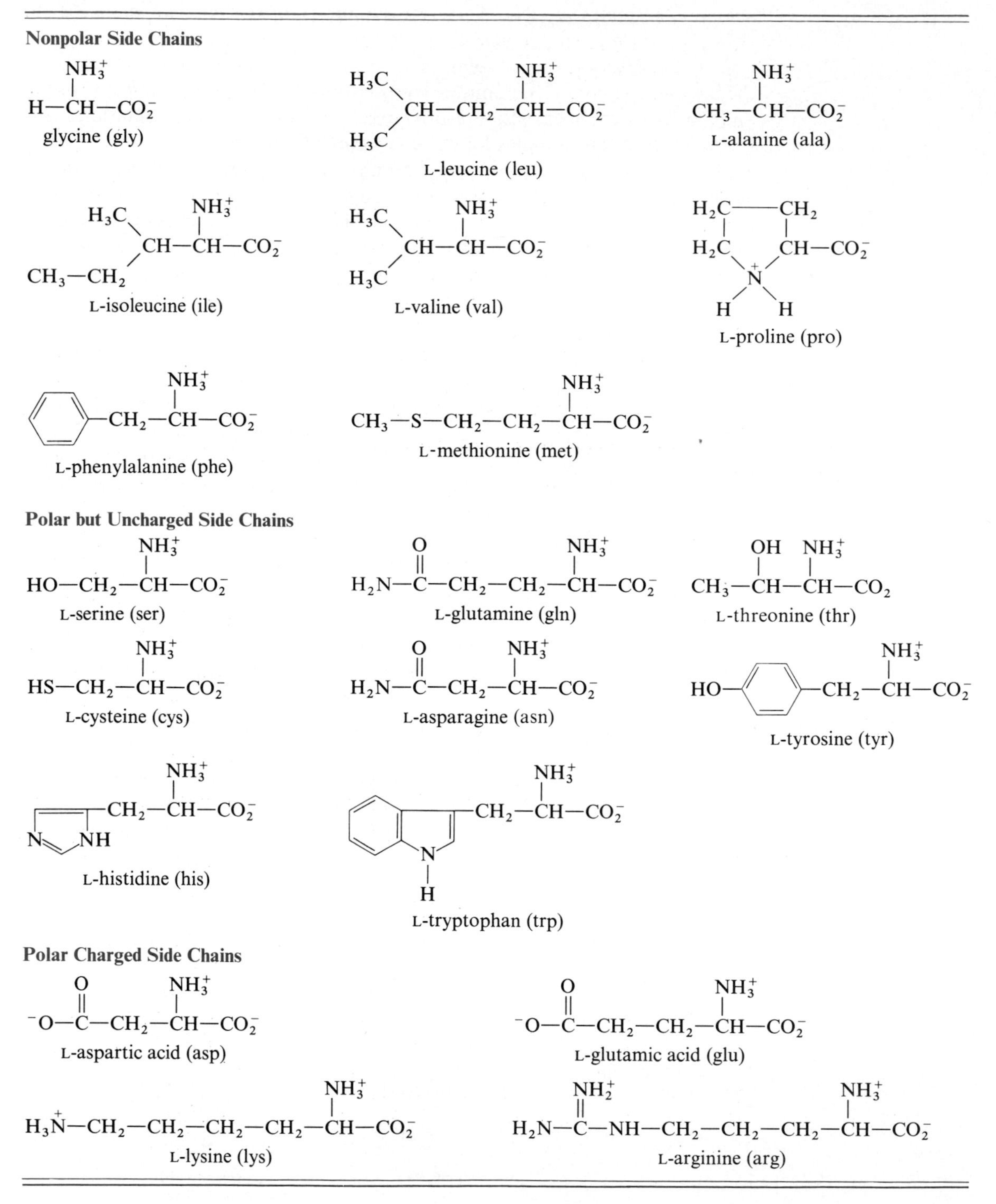

pH 7.4. The side chains of lysine and arginine have amino groups. The pK_a of the lysine side chain is 10.5 and that of arginine is 12.5. Therefore these side chains are fully protonated at pH 7.4 and lysine and arginine have a net charge of +1 at pH 7.4.

The pK_a values for the seven amino acids with ionizable side chains are given in Table 12.2.

It is possible to estimate the degree of ionization of an alpha carboxyl group, alpha amino group, or ionizable side chain at any pH using the following guidelines.

(1) If the pH of the solution is 2.0 or more units greater (more basic) than the pK_a of the ionizable group, then the group is almost entirely in the conjugate base form.

(2) If the pH of the solution is 1.0 unit greater (more basic) than the pK_a of the ionizable group, then the group is 90% in the conjugate base form and 10% in the acid form.

(3) If the pH of the solution is equal to the pK_a of the ionizable group, then the group is 50% in the conjugate base form and 50% in the acid form.

(4) If the pH of the ionizable group is 1.0 unit less (more acidic than the pK_a of the ionizable group, then the group is 10% in the conjugate base form and 90% in the acid form.

(5) If the pH of the solution is 2.0 or more units less (more acidic) than the pK_a of the ionizable group, then the group is almost entirely in the acid form.

Table 12.2 pK_a values of ionizable amino acid side chains.

Amino acid	*Side chain (acid form)*	*pK_a of side chain*	*Predominant form at pH 7.4*
aspartic acid	$—CO_2H$	3.86	$—CO_2^-$
glutamic acid	$—CO_2H$	4.07	$—CO_2^-$
histidine	imidazolium ring (N—H, N^+—H)	6.10	imidazole ring (N—H, N)
cysteine	—SH	8.00	—SH
tyrosine	$—C_6H_4—OH$	10.07	$—C_6H_4—OH$
lysine	$—NH_3^+$	10.53	$—NH_3^+$
arginine	$—NH—C(=NH_2^+)—NH_2$	12.48	$—NH—C(=NH_2^+)—NH_2$

Example 12.1

Draw structural formulas for the following amino acids and estimate the net charge on each at pH 1.0, 6.0, and 12.0.
(a) serine **(b)** glutamic acid

Solution

(a) The pK_a for the α-carboxyl group of serine is approximately 2.2 and that of the α-amino group is 9.5. A pH of 1.0 is 1.2 units below the pK_a of the α-CO_2H group, thus it is more than 90% in the acid form. The α-amino group is also completely in the acid form at this pH, and therefore serine has a net charge of +1. At pH 6.0, the α-carboxyl group is fully ionized to —CO_2^- and the α-amino group is fully protonated to α-NH_3^+. Therefore, at pH 6.0, serine has a net charge of zero. At pH 12.0, both the α-carboxyl and α-amino groups are fully ionized to α-CO_2^- and α-NH_2, and therefore serine has a net charge of −1.

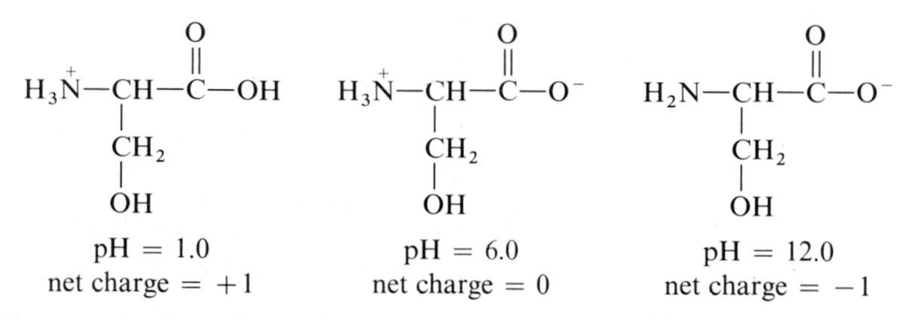

(b) The degree of ionization and the charged character of the α-carboxyl and α-amino groups in glutamic acid are the same as those in serine. The pK_a of the side chain carboxyl group is 4.07 (Table 12.2). A pH of 1.0 is more than 2 units less (more acidic) than this pK_a and pHs of 6.0 and 12.0 are each more than 2 units greater (more basic) than this pK_a. Therefore, the side-chain carboxyl group of glutamic acid is fully protonated at pH 1.0 and fully ionized at pHs of 6.0 and 12.0.

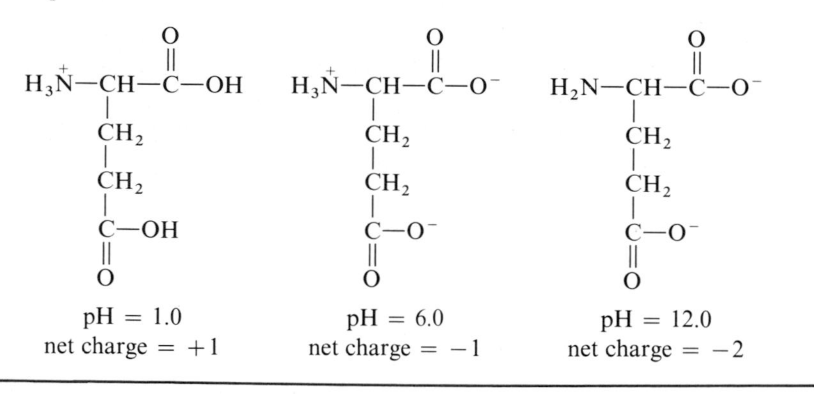

PROBLEM 12.1 Draw structural formulas for the following amino acids and estimate the net charge on each at pH 1.0, 6.0, and 12.0.
(a) histidine **(b)** lysine

As you can see from this description of amino acids and from your solutions to Example 12.1 and Problem 12.1, the net charge on an amino acid in solution becomes more positive as the pH decreases (the solution becomes more acidic), and conversely, the net charge becomes more negative as the pH increases (the solution becomes more basic). The pH at which an amino acid has a net charge of zero is called the isoelectric point. This pH is given the symbol pI.

Amino acids with non-ionizing side chains have isoelectric points in the pH range 5.5–6.5 (for example, the pIs of glycine and serine are 6.0). The isoelectric points for amino acids with ionizable side chains are given in Table 12.3.

Given values for the isoelectric point of these amino acids, you can estimate the charge on each at any pH. For example, the charge on tyrosine is zero at pH 5.63. A small fraction of the molecules are positively charged at pH 5.0. Virtually all molecules of tyrosine are positively charged at pH 3.63 (2 units less than pI). As another example, the net charge on lysine is zero at pH 9.47. At pH values smaller than 9.47, some fraction of lysine molecules are positively charged, and at pH values greater than 9.47, some fraction are negatively charged.

An understanding of the acid-base behavior of amino acids, and also of proteins, is important for two reasons. First, it helps us to understand the solubility of these molecules as a function of pH. While amino acids generally are quite soluble in water, solubility is a minimum at the isoelectric point. To crystallize an amino acid or protein, the pH of an aqueous solution is adjusted to the pI and the substance is precipitated, filtered, and collected. This process is known as isoelectric precipitation. Second, a knowledge of isoelectric points enables us to predict the way components of mixtures of amino acids or proteins migrate in an electrical field. This process of separating substances on the basis of their electrical charges is called electrophoresis.

In paper electrophoresis, a paper strip saturated with an aqueous buffer of predetermined pH serves as a bridge between two electrode vessels. A sample of amino acid or protein is applied as a spot. When an electrical potential is applied to the electrode vessels, amino acid or protein molecules migrate toward the electrode carrying the opposite charge. Molecules having a high charge density move more rapidly than those with a low charge density. Any molecule already at its isoelectric point remains at the origin. After separation

Table 12.3 Isoelectric points of amino acids with ionizable side-chain groups.

Amino acid	*Ionizable side chain group*	*pI (isoelectric point)*
aspartic acid	carboxyl	2.98
glutamic acid	carboxyl	3.08
histidine	imidazole	7.64
cysteine	thiol	5.02
tyrosine	phenol	5.63
lysine	amino	9.47
arginine	amino	10.76

is complete, the strip is dried, and the paper is sprayed with a dye to make the separated components visible. The dye most commonly used for amino acids is ninhydrin.

Ninhydrin consists of a benzene ring fused to a five-membered ring. In aqueous solution, the middle ketone of the five-membered ring reacts with a molecule of water (it undergoes hydration) to give a substance called ninhydrin hydrate.

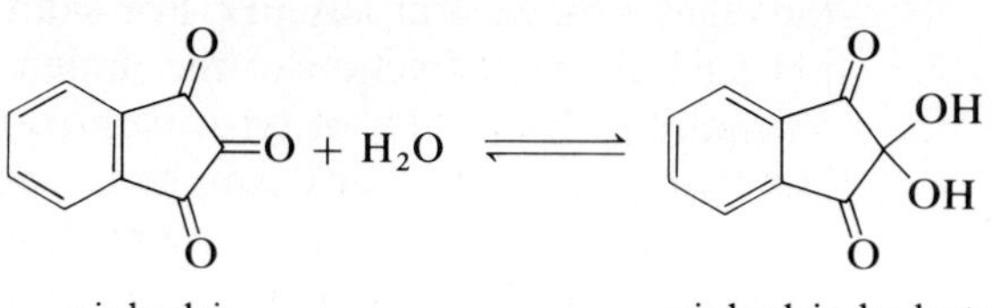

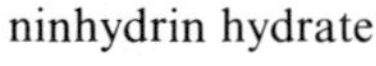

Ninhydrin hydrate reacts with α-amino acids to produce a purple-colored anion, an aldehyde, and carbon dioxide.

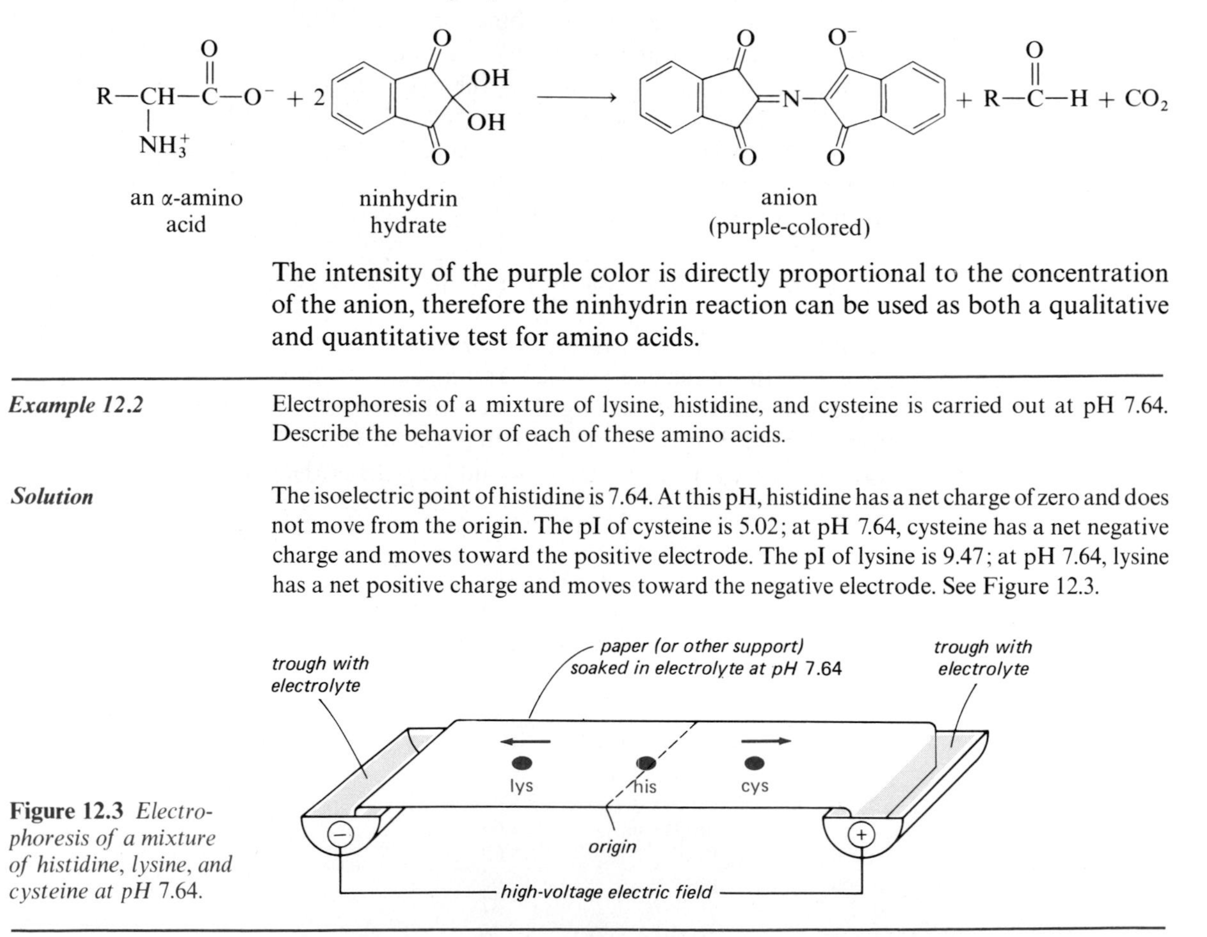

The intensity of the purple color is directly proportional to the concentration of the anion, therefore the ninhydrin reaction can be used as both a qualitative and quantitative test for amino acids.

Example 12.2 Electrophoresis of a mixture of lysine, histidine, and cysteine is carried out at pH 7.64. Describe the behavior of each of these amino acids.

Solution The isoelectric point of histidine is 7.64. At this pH, histidine has a net charge of zero and does not move from the origin. The pI of cysteine is 5.02; at pH 7.64, cysteine has a net negative charge and moves toward the positive electrode. The pI of lysine is 9.47; at pH 7.64, lysine has a net positive charge and moves toward the negative electrode. See Figure 12.3.

Figure 12.3 *Electrophoresis of a mixture of histidine, lysine, and cysteine at pH 7.64.*

PROBLEM 12.2 Describe the behavior of a mixture of glutamic acid, arginine, and valine on paper electrophoresis at pH 6.0.

Electrophoretic separations also can be carried out using starch, agar, certain plastics, and cellulose acetate as solid supports. This technique is extremely important in biochemical research, and is also an invaluable tool in the clinical chemistry laboratory. For a discussion of the electrophoretic screening of blood samples for sickle-cell anemia, see the mini-essay "Abnormal Human Hemoglobins."

Thus far in this section we have concentrated on the 20 amino acids derived from proteins. A few proteins contain special amino acids. For example, L-hydroxyproline and L-5-hydroxylysine are important components of collagen, but are found in very few other proteins. These and all other special amino acids are formed after proteins are constructed by modification of one of the 20 common amino acids already incorporated into the protein.

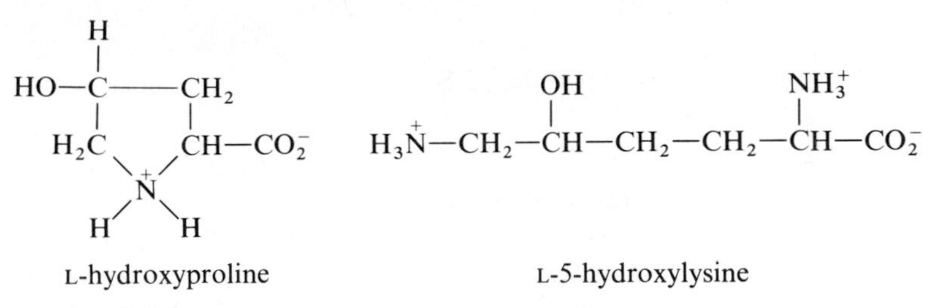
L-hydroxyproline L-5-hydroxylysine

In addition to those amino acids listed in Table 12.1, there are a number of important nonprotein-derived amino acids, many of which are either metabolic

Table 12.4 Several nonprotein-derived amino acids.

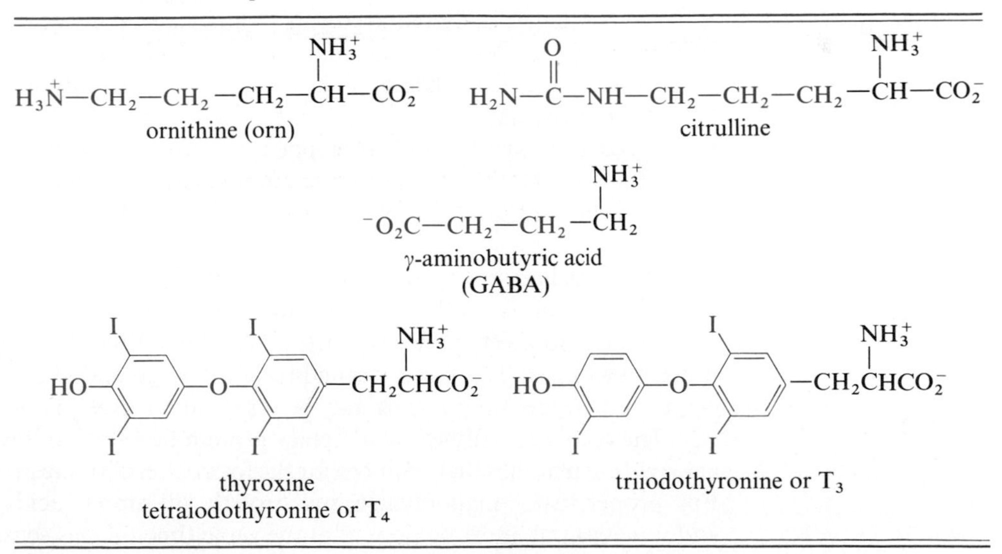

intermediates or parts of nonprotein biomolecules. Several of these are shown in Table 12.4. Ornithine and citrulline are part of the metabolic pathway that converts excess ammonia to urea. Gamma-aminobutyric acid (GABA) is present in the free state in brain tissue. Its function in the brain is as yet largely unknown.

Thyroxine, one of several hormones derived from the amino acid tyrosine, was first isolated from thyroid tissue in 1914. In 1952, triiodothyronine, a compound identical to thyroxine except that it contains only three atoms of iodine, was also discovered in the thyroid. Triiodothyronine is even more potent than thyroxine. Although the exact mechanisms of action of these thyroid hormones are not known, they are essential for the proper regulation of cellular metabolism. The levorotatory isomer of each is significantly more active than the dextrorotatory isomer.

12.2 ESSENTIAL AMINO ACIDS

Of the 20 amino acids required by the body for the production of proteins, adequate amounts of 12 can be synthesized by enzyme-catalyzed reactions starting from carbohydrates or lipid fragments and a source of nitrogen. For the remaining amino acids, either there are no biochemical pathways available for their synthesis, or the available pathways do not provide adequate amounts for proper nutrition. Accordingly, these amino acids must be supplied in the diet, and are called "essential" amino acids. In reality, all amino acids are essential for normal tissue growth and development. However, the term "essential" is reserved for those that must be supplied in the diet.

The estimated daily requirements of eight essential amino acids are given in Table 12.5. Anyone who consumes about 40–55 grams of protein daily in the form of meat, fish, cheese, milk, or eggs satisfies daily needs for these essential amino acids.

Tyrosine is synthesized from phenylalanine in the body. Therefore the requirements for these two aromatic amino acids are combined in Table 12.5. Similarly the sulfur containing amino acids methionine and cysteine are combined.

Histidine is essential for growth in infants, and it may be needed by adults as well. Arginine is synthesized by adults, but the rate of internal synthesis is not adequate to meet the needs of the body during periods of rapid growth and protein synthesis. Therefore, depending on the age and state of health, either eight, nine, or ten amino acids may be essential for humans.

The relative usefulness of a dietary protein depends on how well its amino acid pattern matches that required for the formation of tissue protein in humans. For proper tissue maintenance and growth, all amino acids, both essential and nonessential, must be present at the same time. In this sense, tissue growth

Table 12.5 Estimated amino acid requirements of humans.

	Requirement, mg/kg *body weight/day*		
Amino acid	*Infant* (4–6 *months*)	*Child* (10–12 *years*)	*Adult*
Histidine	33	—	—
Isoleucine	83	28	12
Leucine	135	42	16
Lysine	99	44	12
Total *S*-containing amino acids (methionine and cysteine)	49	22	10
Total aromatic amino acids (phenylalanine and tyrosine)	141	22	16
Threonine	68	28	8
Tryptophan	21	4	3
Valine	92	25	14

is an all-or-nothing process; if even one amino acid is missing, no protein is made.

The biological value of a dietary protein is a measure of the percentage that is absorbed and used to build body tissue. Some of the first information on the biological value of dietary proteins came from studies on rats. In one series of experiments, young rats were fed diets containing 18% protein in the form of either casein (a milk protein), gliadin (a wheat protein), or zein (a corn protein). With casein as the sole source of protein, the rats remained healthy and grew normally. Those fed gliadin maintained their weight but did not grow much. Those fed zein not only failed to grow but lost weight and, if kept on this diet, eventually died. Since casein evidently supplies all required amino acids in the correct proportions needed for growth, it is called a complete protein. Analysis revealed that gliadin contains too little lysine, and that zein is low in both lysine and tryptophan. When the gliadin diet was supplemented with lysine, or the zein diet with lysine and tryptophan, the test animals grew normally.

Table 12.6 shows the biological value for rats of some common dietary sources of protein. The proteins in egg are the best-quality natural protein. The proteins of milk rank 84, those of meats and soybeans about 74. The legumes, vegetables, and cereal grains are in the range 50–70.

Plant proteins generally vary more from the amino acid pattern required by humans than do animal proteins. Fortunately, however, not all plant proteins are deficient in the same amino acids. For example, beans are low in the sulfur-containing amino acids cysteine and methionine, yet are high in lysine. Wheat has just the opposite pattern. By eating wheat and beans together, it is possible to increase by 33% the usable protein you would get by eating either of these foods separately.

Table 12.6 The biological value for rats of some common sources of dietary protein. (Biological value is the portion of absorbed protein retained as body tissue.)

Food	*Protein as % of dry solid*	*Biological value of protein, %*
hen's egg, whole	48	94
cow's milk, whole	27	84
fish	72	83
beef	45	74
soybeans	41	73
rice, brown	9	73
potato, white	9	67
wheat, whole grain	14	65
corn, whole grain	11	59
dry beans, common	25	58

12.3 THE PEPTIDE BOND

In 1902, Emil Fischer proposed that proteins are long chains of amino acids joined together by amide bonds between the α-carboxyl group of one amino acid and the α-amino group of another. For these amide bonds Fischer proposed the special name "peptide bond." Figure 12.4 shows the peptide bond formed between glycine and alanine in the peptide glycylalanine.

A molecule containing two amino acids joined by an amide bond is called a dipeptide. Those containing larger numbers of amino acids are called tri-

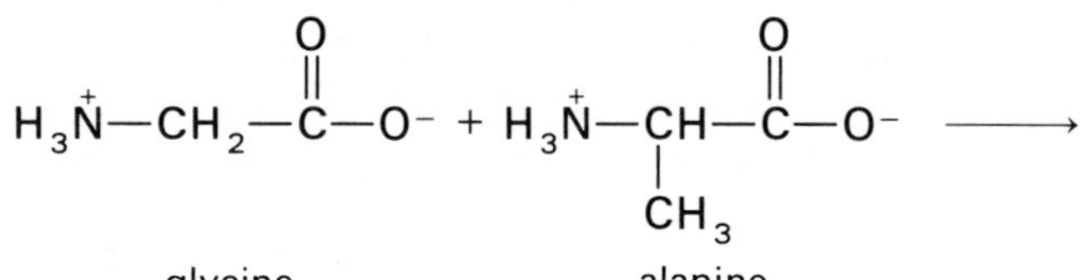

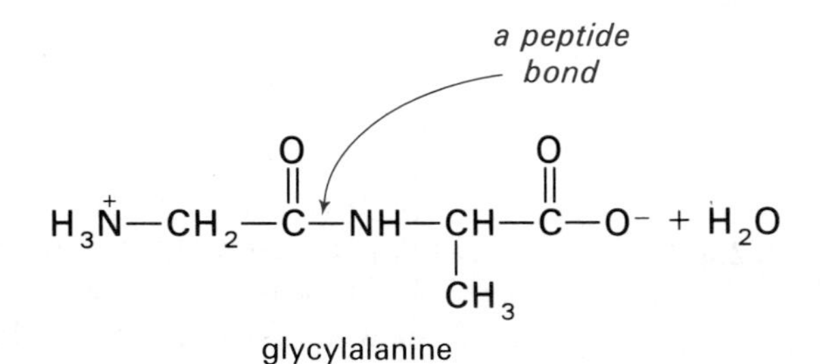

Figure 12.4 *The peptide bond in glycylalanine.*

peptides, tetrapeptides, pentapeptides, etc. Molecules containing 10 or more amino acids are generally called polypeptides. Proteins are biological macromolecules of molecular weight 5,000 or greater, consisting of one or more polypeptide chains.

By convention, polypeptides are written from the left, beginning with the amino acid having the free H_3N^+— group, and proceeding to the right toward the amino acid with the free —CO_2^- group. The amino acid with the free H_3N^+— group is called the N-terminal amino acid and that with the free —CO_2^- group is called the C-terminal amino acid. The structural formula for a polypeptide sequence may be written out in full, or the sequence of amino acids may be indicated using the standard abbreviation for each.

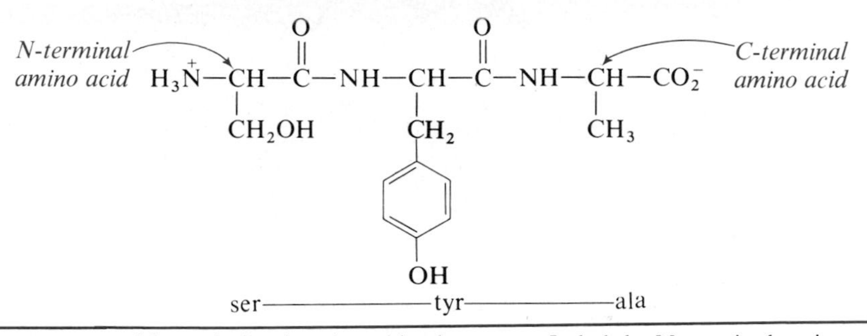

Example 12.3 Draw a structural formula for the tripeptide gly-ser-asp. Label the N-terminal amino acid and the C-terminal acid. What is the net charge on this tripeptide at pH 6.0?

Solution In writing the formula for this tripeptide, begin with glycine on the left. Then connect glycine to serine by a peptide bond. Finally connect serine to aspartic acid by a peptide bond. The net charge on this tripetpide at pH 6.0 is −1.

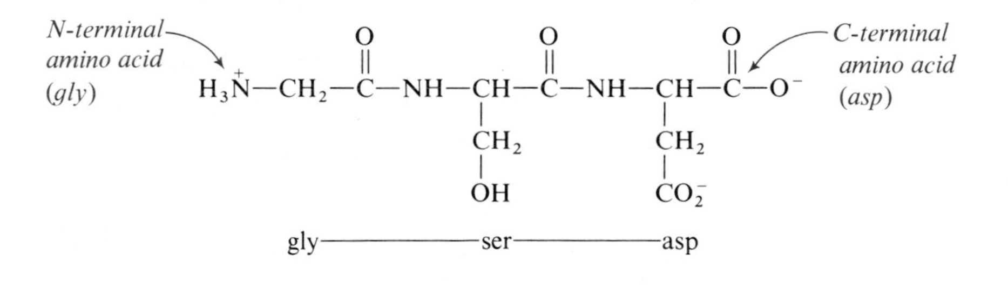

PROBLEM 12.3 Draw a structural formula for lys-phe-ala. Label the N-terminal amino acid and the C-terminal amino acid. What is the net charge on this tripeptide at pH 6.0?

12.4 THE GEOMETRY OF THE PEPTIDE BOND

In the late 1930s, Linus Pauling began a series of studies to determine the geometry of the peptide bond. One of his first discoveries was that the peptide bond itself is planar. As shown in Figure 12.5, the four atoms of the peptide

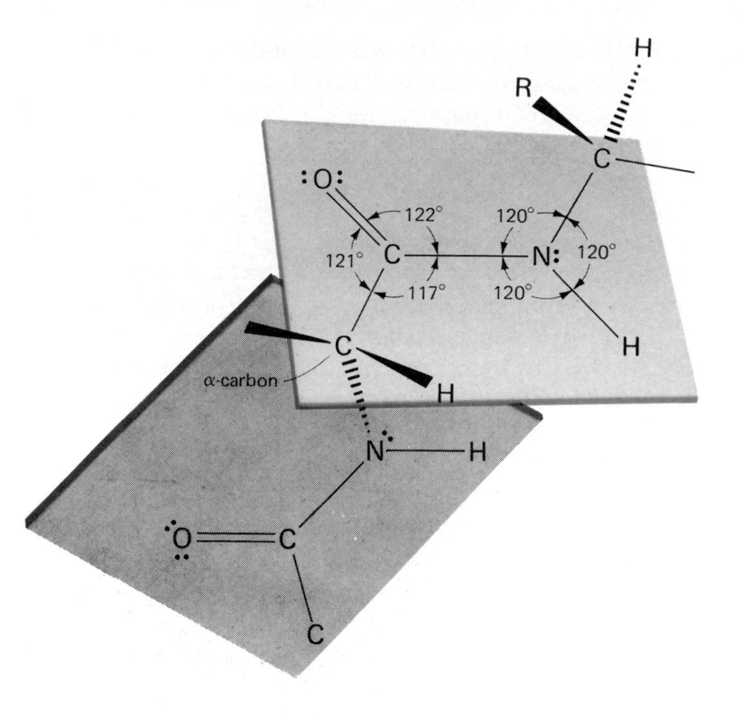

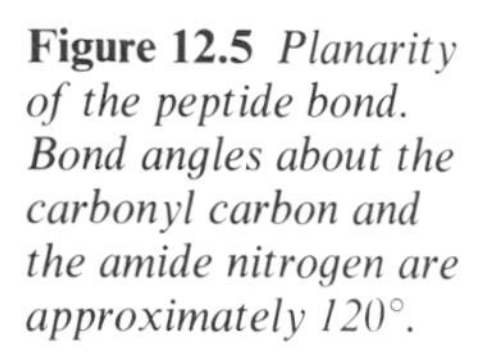
Figure 12.5 *Planarity of the peptide bond. Bond angles about the carbonyl carbon and the amide nitrogen are approximately 120°.*

bond and the two alpha carbon atoms joined to the peptide bond all lie in the same plane.

Had you been asked in Chapter 1 to describe the geometry of the peptide bond, you would have predicted bond angles of approximately 120° about the carbonyl carbon and 109.5° about the amide nitrogen. This prediction agrees with the observed bond angle of 120° about the carbonyl carbon. However, the geometry of the amide nitrogen is unexpected. To account for this observed geometry, Pauling proposed that the peptide bond is more accurately represented as a resonance hybrid of two important contributing structures:

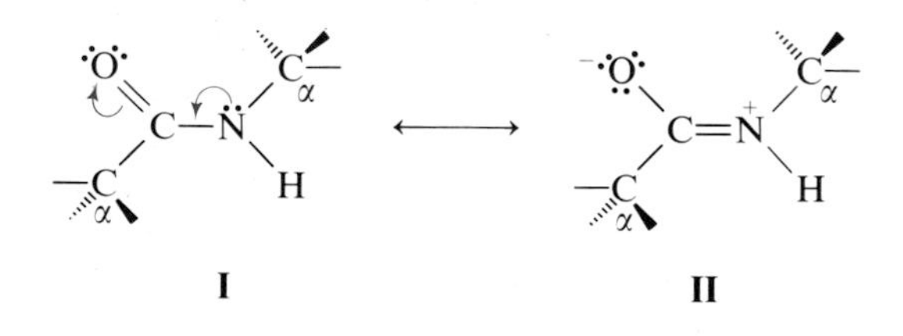

Contributing structure I shows a carbon-oxygen double bond, while structure II shows a double bond between the carbonyl carbon and the nitrogen atom of the peptide bond. The hybrid, of course, is neither of these, but in the real structure, the carbon-nitrogen bond has at least partial double-bond character. Accordingly, in the hybrid, the six-atom group is planar.

There are two possible planar arrangements of the atoms of the peptide bond. In one configuration, the two α-carbons are *cis* to each other, and in the second configuration they are *trans*:

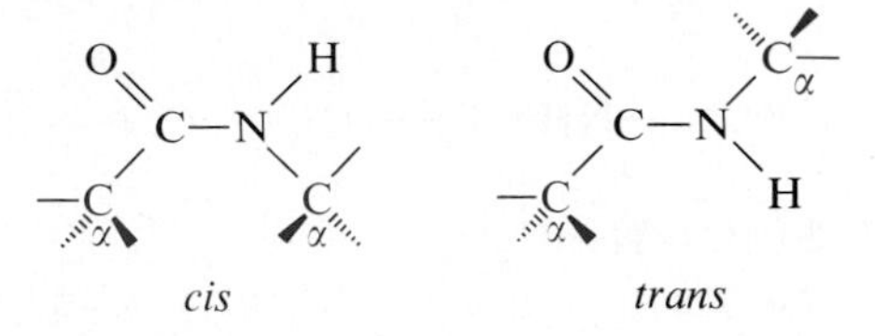

The *trans* configuration is more favorable because the bulky α-carbons are farther from each other than they are in the *cis* configuration. Virtually all peptide bonds in naturally occurring proteins have the *trans* configuration.

12.5 AMINO ACID SEQUENCE: PRIMARY STRUCTURE

Primary structure refers to the sequence of amino acids in a polypeptide chain and also to the location of any disulfide bonds. In this sense, then, primary structure is a complete description of all covalent bonding in a polypeptide chain or protein.

To appreciate the problem of deciphering the primary structure of a polypeptide chain, just imagine the incredibly large number of different chemical words (polypeptides) that can be constructed with a 20-letter alphabet, where words can range from under 10 letters to hundreds of letters. With only three amino acids, there are six entirely different tripeptides possible. For glycine, alanine, and serine, the six tripeptides are

gly–ala–ser	ala–gly–ser	ser–gly–ala
gly–ser–ala	ala–ser–gly	ser–ala–gly

For a polypeptide containing one each of the 20 different amino acids, the number of possible polypeptides is $20 \times 19 \times 18 \times \cdots \times 2 \times 1$ or about 2×10^{18}. With larger polypeptides and proteins, the number of possible arrangements becomes truly countless!

Of the various chemical methods developed for determining the amino acid sequence of a polypeptide chain, the one most widely used today was introduced in 1950 by Pehr Edman of the University of Lund, Sweden. In this procedure, a polypeptide chain is reacted with phenylisothiocyanate, $C_6H_5—N{=}C{=}S$, a substance that reacts selectively with the $—NH_3^+$ group of the N-terminal amino acid. The effect of the Edman degradation is to cleave the N-terminal amino acid as a substituted phenylthiohydantoin (Fig. 12.6) that is then separated and identified.

The special value of the Edman degradation is that it cleaves the N-terminal amino acid from a polypeptide chain without affecting any other bonds in the chain. Furthermore, Edman degradation can be repeated on the now-shortened polypeptide chain and the next amino acid in the sequence cleaved and identified. In practice, it is now possible to sequence as many as the first 40 or so amino acids in a polypeptide chain by this method.

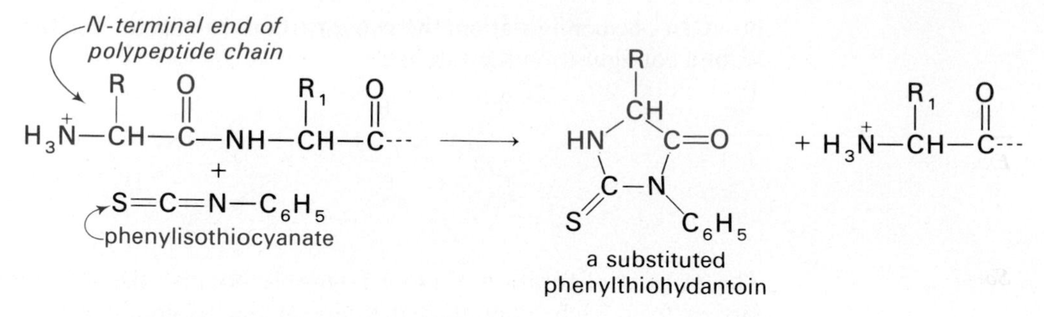

Figure 12.6 *The Edman degradation. Reaction of a polypeptide chain with phenylisothiocyanate selectively removes the N-terminal amino acid as a substituted phenylthiohydantoin.*

If it is not possible to sequence an entire polypeptide chain by repeated Edman degradation, the chain is partially hydrolyzed to yield a series of smaller fragments and each fragment is sequenced separately. Such partial hydrolysis is most often carried out using enzymes that catalyze the hydrolysis of specific peptide bonds.

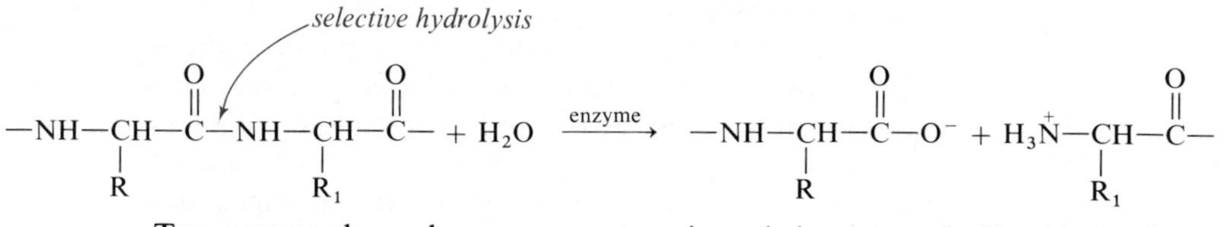

Two commonly used enzymes are trypsin and chymotrypsin. Trypsin catalyzes the hydrolysis of peptide bonds in which the carboxyl group is contributed

Table 12.7 Specific cleavage of peptide bonds catalyzed by trypsin and chymotrypsin.

Enzyme	*Catalyzes the hydrolysis of peptide bonds formed by the carboxyl group of*	*Side chain (R-group) of the amino acid undergoing selective cleavage*
trypsin	arginine	$-CH_2-CH_2-CH_2-NH-C(=NH_2^+)-NH_2$
	lysine	$-CH_2-CH_2-CH_2-CH_2-NH_3^+$
chymotrypsin	phenylalanine	$-CH_2-C_6H_5$
	tyrosine	$-CH_2-C_6H_4-OH$
	tryptophan	$-CH_2-$(indol-3-yl, N–H)

by either arginine or lysine; chymotrypsin catalyzes the hydrolysis of peptide bonds in which the carboxyl group is contributed by either phenylalanine, tyrosine, or tryptophan (Table 12.7).

Example 12.4

Following are amino acid sequences for several tripeptides. Which are hydrolyzed by trypsin? which by chymotrypsin?

(a) arg-glu-ser **(b)** phe-gly-lys **(c)** phe-lys-met

Solution

(a) Trypsin catalyzes hydrolysis of peptide bonds between the carboxyl groups of lysine and arginine and the α-amino group of another amino acid. Therefore, the peptide bond between arginine and glutamic acid is hydrolyzed in the presence of trypsin.

$$\text{arg–glu–ser} + H_2O \xrightarrow{\text{trypsin}} \text{arg} + \text{glu–ser}$$

Chymotrypsin catalyzes the hydrolysis of peptide bonds formed by the carboxyl groups of phenylalanine, tyrosine, and tryptophan. Since none of these three is present, tripeptide (a) is not affected by chymotrypsin.

(b) Tripeptide (b) is not affected by trypsin. While there is a lysine present, its carboxyl group is at the C-terminal end and is not involved in peptide bond formation. Tripeptide (b) is hydrolyzed in the presence of chymotrypsin.

$$\text{phe–gly–lys} + H_2O \xrightarrow{\text{chymotrypsin}} \text{phe} + \text{gly–lys}$$

(c) Tripeptide (c) is hydrolyzed both by trypsin and chymotrypsin.

$$\text{phe–lys–met} + H_2O \xrightarrow{\text{trypsin}} \text{phe–lys} + \text{met}$$

$$\text{phe–lys–met} + H_2O \xrightarrow{\text{chymotrypsin}} \text{phe} + \text{lys–met}$$

PROBLEM 12.4 Following are amino acid sequences for three tripeptides. Which are hydrolyzed by trypsin? which by chymotrypsin?

(a) tyr-gln-val **(b)** thr-phe-ser **(c)** thr-ser-phe

Example 12.5

Deduce the amino acid sequence of a pentapeptide from the following experimental results.

amino acid composition:	arg, glu, his, phe, ser (alphabetical order)	
Edman degradation:	glu	
digestion with chymotrypsin:	Fragment A:	glu, his, phe
	Fragment B:	arg, ser
digestion with trypsin:	Fragment C:	arg, glu, his, phe
	Fragment D:	ser

Solution

Edman degradation cleaves glu from the pentapeptide chain; therefore, glutamic acid must be the N-terminal amino acid. Now write the following partial sequence.

glu–(arg, his, phe, ser)

Next, digestion with chymotrypsin gives fragments A and B and fragment A contains phe. Fragment A also contains glu which you already know is the N-terminal amino acid. From this, conclude that the first three amino acids in the chain must be glu-his-phe, and now write the following fuller partial sequence.

glu–his–phe–(arg, ser)

The fact that trypsin cleaves the pentapeptide means that the single arg must be within the pentapeptide chain; it cannot be the C-terminal amino acid. Therefore the complete sequence must be

glu–his–phe–arg–ser

PROBLEM 12.5 Deduce the amino acid sequence of a hexapeptide from the following experimental results:

amino acid composition:	lys, pro, pro, tyr, val, val	
Edman degradation:	pro	
digestion with trypsin:	Fragment E:	pro, lys, val
	Fragment F:	pro, tyr, val
digestion with chymotrypsin:	Fragment G:	lys, pro, tyr, val, val
	Fragment H:	pro

12.6 CONFORMATIONS OF POLYPEPTIDE CHAINS: SECONDARY STRUCTURE

In addition to studying the geometry of the peptide bond, Linus Pauling also investigated folding patterns of polypeptide chains. He assumed that in folding patterns of greatest stability, (1) all atoms in a peptide bond lie in the same plane, and (2) each amide group is hydrogen bonded between the N—H of one peptide bond and the C=O of another. Pauling was the first to recognize the importance of hydrogen bonding in stabilizing folding patterns. There is appreciable polarity to the C=O and the N—H bonds and when two amide groups lie close to each other, the two peptide bonds interact by hydrogen bonding, as shown in Figure 12.7.

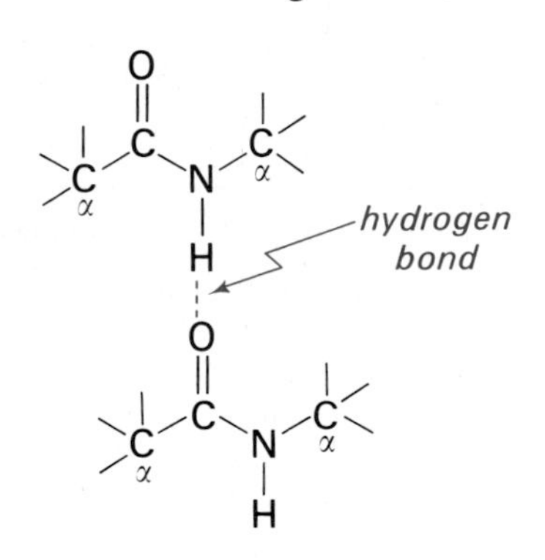

Figure 12.7 *Hydrogen bonding between amide groups.*

On the basis of model building, Pauling proposed that two folding patterns should be particularly stable: the α-helix and the antiparallel β-pleated sheet. In the helix pattern shown in Figure 12.8, the polypeptide chain is coiled in a spiral. As you study this drawing, note the following:

(1) The helix is coiled in a clockwise or right-handed manner. Right-handed means that if you turn the helix clockwise, it twists away from you. In this sense, a right-handed helix is analogous to the right-hand thread of a common wood or machine screw.

(2) There are 3.6 amino acids per turn of the helix.

(3) Each peptide bond is *trans* and planar.

(4) The N—H group of each peptide bond points roughly upward, parallel to the axis of the helix, and the C=O of each peptide bond points roughly downward.

(5) The carbonyl group of each peptide bond is hydrogen bonded to the N—H group of the peptide bond four amino acid units away from it. Hydrogen bonds are shown as dotted lines.

(6) All R— groups point outward from the helix.

Almost immediately after Pauling proposed the α-helix structure, other workers proved the presence of the α-helix in hair keratin. It soon became obvious that the α-helix is one of the fundamental folding patterns of polypeptide chains.

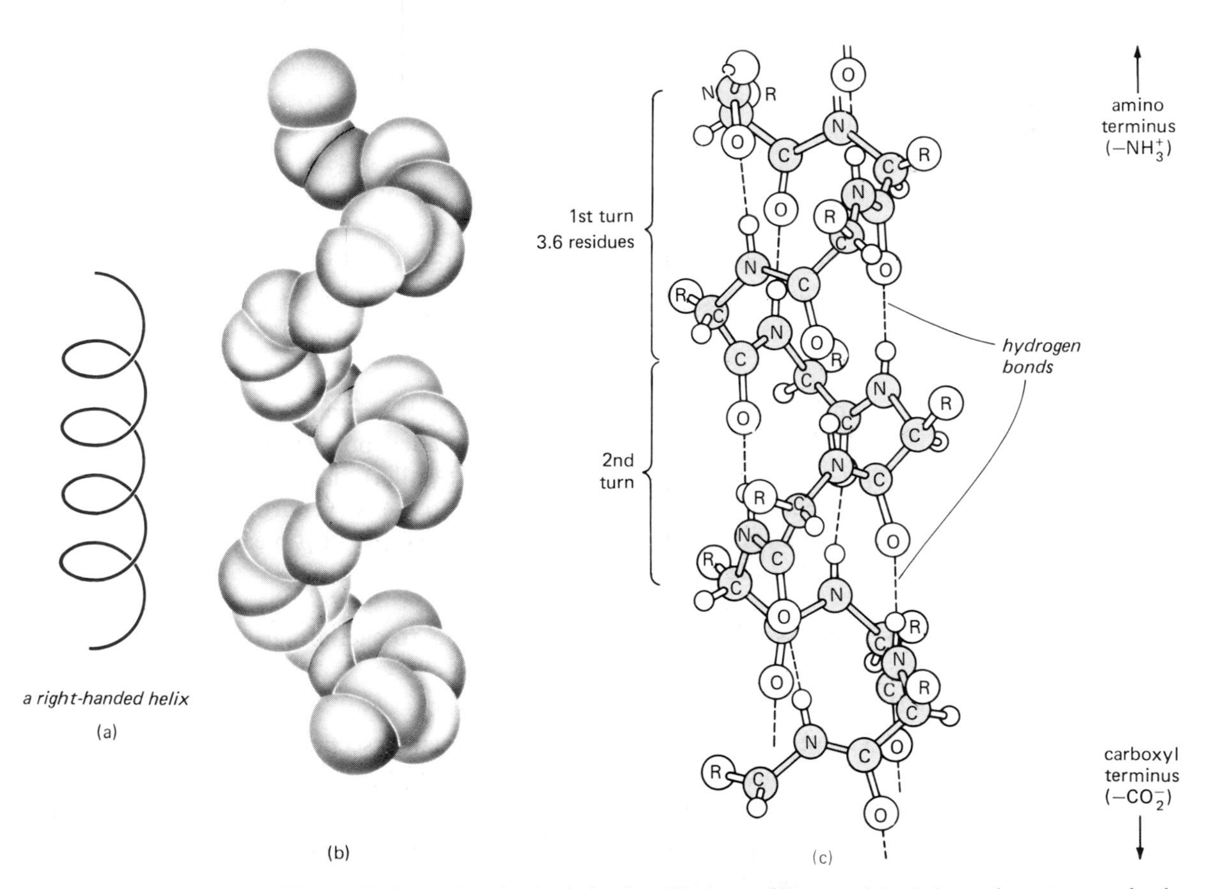

Figure 12.8 *(a) A right-handed helix. (b) Space-filling model of the carbon-nitrogen backbone of an α-helix. (c) Ball-and-stick model of the α-helix showing intrachain hydrogen bonding. There are* 3.6 *amino acid residues per turn (a distance of* 5.4×10^{-8} cm *along the axis of the helix).*

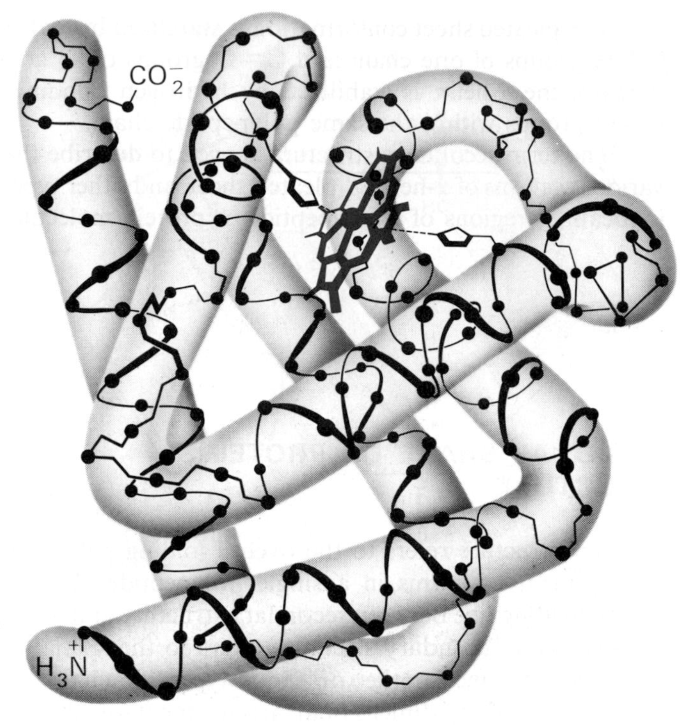

Figure 12.10 *The three-dimensional structure of myoglobin. The heme group is shown in color. The N-terminal amino acid (indicated by* $—NH_3^+$*) is at the lower left and the C-terminal amino acid (indicated by* $—CO_2^-$*) is at the upper left. Reproduced from R. E. Dickerson, in H. Neurath, ed., The Proteins, Vol. II (Academic Press, New York, 1964).*

in diving mammals such as seals, whales, and porpoises. Myoglobin and its structural relative, hemoglobin (Section 12.8), are the oxygen transport and storage molecules of vertebrates. Hemoglobin binds molecular oxygen in the lungs and transports it to myoglobin in muscles. Myoglobin stores the molecular oxygen until it is required for metabolic oxidation.

Myoglobin consists of a single polypeptide chain of 153 amino acids. The complete amino acid sequence (primary structure) of the chain is known. Myoglobin also contains a single heme unit (Fig. 12.17). The determination of the three-dimensional structure of myoglobin represented a milestone in the study of molecular architecture. For his contribution to this research, J. C. Kendrew shared the Nobel Prize in Chemistry in 1963. The secondary and tertiary structure of myoglobin are shown in Figure 12.10. The single polypeptide chain is folded in a complex, almost box-like shape.

A more detailed analysis shows the exact location of the atoms of the peptide backbone and also the location of all side chains. The important structural features of myoglobin are

(1) The backbone consists of eight relatively straight sections of α-helix, each separated by a bend in the polypeptide chain. The longest section of α-helix has 23 amino acids, the shortest has 7. Some 75% of the amino acids are found in these eight regions of α-helix.

PROBLEM 12.6 At pH 7.4, with what amino acid side chains can the side chain of lysine form salt linkages?

12.8 QUATERNARY STRUCTURE

Most proteins of molecular weight greater than 50,000 consist of two or more noncovalently linked polypeptide chains. This arrangement of protein monomers into an aggregation is known as quaternary structure. A good example is hemoglobin, a protein that consists of four separate protein monomers: two α-chains of 141 amino acids each, and two β-chains of 146 amino acids each. The quaternary structure of hemoglobin is shown in Figure 12.11.

The major factor stabilizing the aggregation of protein subunits is hydrophobic interaction. When separate monomers fold into compact three-dimensional shapes to expose polar side chains to the aqueous environment, and shield nonpolar side chains from water, there are still hydrophobic "patches" on the surface and in contact with water. These patches can be shielded from water if two or more monomers assemble so that their hydrophobic patches are in contact. The molecular weights, number of subunits, and biological functions of several proteins with quaternary structure are shown in Table 12.8.

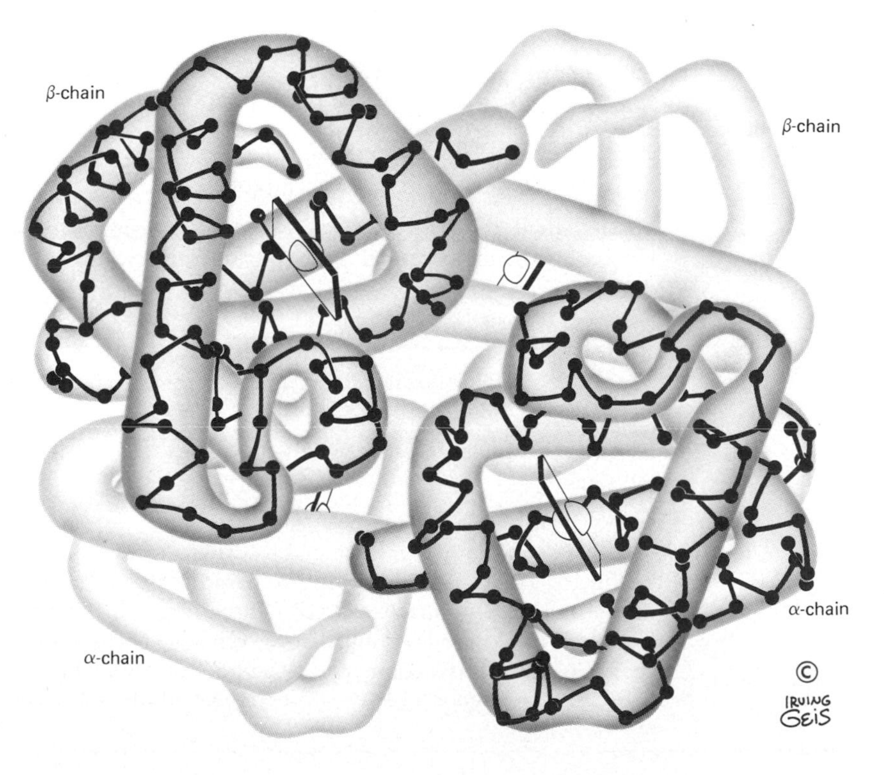

Figure 12.11 *The quaternary structure of hemoglobin, showing the four subunits packed together. The flat disks represent four heme units. From R. E. Dickerson and I. Geis, The Structure and Action of Proteins, W. A. Benjamin, Inc., Menlo Park, CA. © Copyright 1969. All rights reserved. Used by permission.*

(2) Strongly hydrophobic side chains such as those of phenylalanine, v leucine, isoleucine, and methionine cluster together in the interior c molecule where they are shielded from contact with water. Hydropl interactions between nonpolar side chains are a major factor in dire the folding of the polypeptide chain of myoglobin into this tl dimensional shape.

(3) The outer surface of myoglobin is coated with hydrophilic side chains s as those of lysine, arginine, serine, glutamic acid, histidine, and glutan which interact with the aqueous environment by hydrogen bonding. ' only polar side chains that point to the interior of the myoglobin molec are those of two histidines. These side chains can be seen in Figure 12 as five-membered rings pointing inward toward the heme group.

(4) Oppositely charged amino acids close to each other in the three-dimension structure interact by electrostatic attractions called salt linkages. A example of a salt linkage is the attraction of the side chains of lysine ($-NH_3$ and glutamic acid ($-CO_2^-$).

The three-dimensional structures of several other globular proteins hav also been determined and their secondary and tertiary structures analyzed It is clear that globular proteins contain α-helix and β-pleated sheet structure, but that there is wide variation in the relative amounts of each. Lysozyme, with 129 amino acids in a single polypeptide chain, has only about 25% of its amino acids in α-helix regions. Cytochrome, with 104 amino acids in a single polypeptide chain, has no α-helix structure but does contain several regions of β-pleated sheet. Yet whatever the proportions of α-helix and β-pleated sheet structure, virtually all nonpolar side chains of globular proteins are directed toward the interior of the molecule, while polar side chains are on the surface of the molecule and in contact with the aqueous environment. Note that this arrangement of polar and nonpolar groups in globular proteins very much resembles the arrangement of polar and nonpolar groups of soap molecules in micelles (Fig. 8.5).

Example 12.6

With which of the following amino acid side chains can the side chain of threonine form hydrogen bonds?

(a) valine **(b)** phenylalanine **(c)** tyrosine **(d)** asparagine **(e)** histidine **(f)** alanine

Solution

The side chain of threonine contains a hydroxyl group that can participate in hydrogen bonding in two ways: the negatively charged oxygen can function as a hydrogen bond acceptor and the positively charged hydrogen can function as a hydrogen bond donor. Therefore, the side chain of threonine can function as a hydrogen bond acceptor for the side chains of tyrosine, asparagine, and histidine. The side chain of threonine can function as a hydrogen bond donor for hydrogen bonding with the side chains of tyrosine, asparagine, and histidine.

Table 12.8 Quaternary structure of selected proteins.

Protein	*Mol wt*	*Number of subunits*	*Subunit mol wt*	*Biological function*
insulin	11,466	2	5,733	a hormone regulating glucose metabolism
hemoglobin	64,500	4	16,100	oxygen transport in blood plasma
alcohol dehydrogenase	80,000	4	20,000	an enzyme of alcoholic fermentation
lactic dehydrogenase	134,000	4	33,500	an enzyme of anaerobic glycolysis
aldolase	150,000	4	37,500	an enzyme of anaerobic glycolysis
fumarase	194,000	4	48,500	an enzyme of the tricarboxylic acid cycle
tobacco mosaic virus	40,000,000	2200	17,500	plant virus coat

12.9 DENATURATION

Globular proteins found in living organisms are remarkably sensitive to changes in their environment. Relatively mild changes in pH, temperature, or solvent composition, even for only a short period of time, may cause them to denature. Denaturation is a physical change, the most observable result of which is loss of biological activity. With the exception of cleavage of disulfide bonds, denaturation stems from changes in secondary, tertiary, or quaternary structure through disruption of noncovalent interactions, e.g., hydrogen bonds, salt linkages, and hydrophobic interactions. Common denaturing agents include

(1) Heat. Most globular proteins denature when heated above 50–60°C. For example, boiling or frying an egg causes egg-white protein to denature and form an insoluble mass.

(2) Large changes in pH. Adding concentrated acid or alkali to a protein in aqueous solution causes changes in the charged character of ionizable side chains and interferes with salt linkages. For example, in certain clinical chemistry tests where it is necessary to first remove any protein material, trichloroacetic acid (a strong organic acid) is added to denature and precipitate any protein present.

(3) Detergents. Treatment of a protein with sodium dodecylsulfate (SDS), a common detergent, causes the native conformation to unfold and exposes the nonpolar protein side chains. These side chains are then stabilized by hydrophobic interaction with hydrocarbon chains of the detergent.

(4) Organic solvents such as alcohols, acetone, or ether. These solvents disrupt hydrogen bonding in the native protein.

(5) Mechanical treatment. Most globular proteins denature in aqueous solution if they are stirred or shaken vigorously. An example is the whipping of egg whites to make a meringue.

(6) Urea and guanidine hydrochloride.

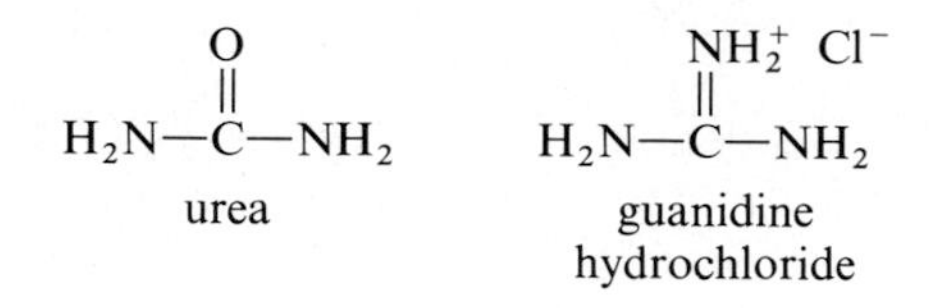

These reagents cause disruption of protein hydrogen bonding and hydrophobic interactions.

Denaturation may be partial or it may be complete. It may also be reversible or irreversible. For example, the hormone insulin can be denatured with 8M urea and the three disulfide bonds reduced to —SH groups. If urea is then removed and the disulfide bonds reformed, the resulting molecule has less than 1% of its former biological activity. In this case, the denaturation has been both complete and irreversible. As another example, consider ribonuclease, an enzyme that consists of a single polypeptide chain of 124 amino acids folded into a compact, three-dimensional structure stabilized in part by four disulfide bonds. Treatment of ribonuclease with urea causes the molecule to unfold and the disulfide bonds can then be reduced to thiol groups. At this point, the protein is completely denatured—it has no biological activity. If urea is removed from solution and the thiol groups reoxidized to disulfide bonds, the protein regains its full biological activity. In this instance, denaturation has been complete but reversible.

12.10 SECONDARY, TERTIARY, AND QUATERNARY STRUCTURE ARE DETERMINED BY THE PRIMARY STRUCTURE

The primary structure of a protein is determined by information coded within the genes. Once the primary structure of a polypeptide is established, it then directs folding of the polypeptide chain into a three-dimensional structure. In other words, information inherent in the primary structure of a protein determines the secondary, tertiary, and quaternary structure.

If the three-dimensional shape of a polypeptide or protein is determined by the primary structure, how can we account for the observation that denaturation of some proteins is reversible while that of others is irreversible?

The reason for this difference in behavior from one protein to another is that some proteins, like ribonuclease, are synthesized as single polypeptide chains that then fold into unique three-dimensional structures that possess full biological activity. Others, like insulin, are synthesized as larger molecules that possess no biological activity, but which are "activated" at some later point by specific enzyme-catalyzed peptide bond cleavage. Insulin is synthesized in the β-cells of the pancreas as a single polypeptide chain of 84 amino acids. This molecule is called proinsulin and has no biological activity. When insulin is needed, a section of 33 amino acids is hydrolyzed from proinsulin in an enzyme-catalyzed reaction to produce the active hormone (Figure 12.12). Bovine insulin contains 51 amino acids in two polypeptide chains. The A chain contains 21 amino acids and has glycine (gly) at the $-NH_3^+$ terminus and asparagine (asn) at the $-CO_2^-$ terminus. Chain B contains 30 amino acids with phenylalanine (phe) at the $-NH_3^+$ terminus and alanine (ala) at the $-CO_2^-$ terminus.

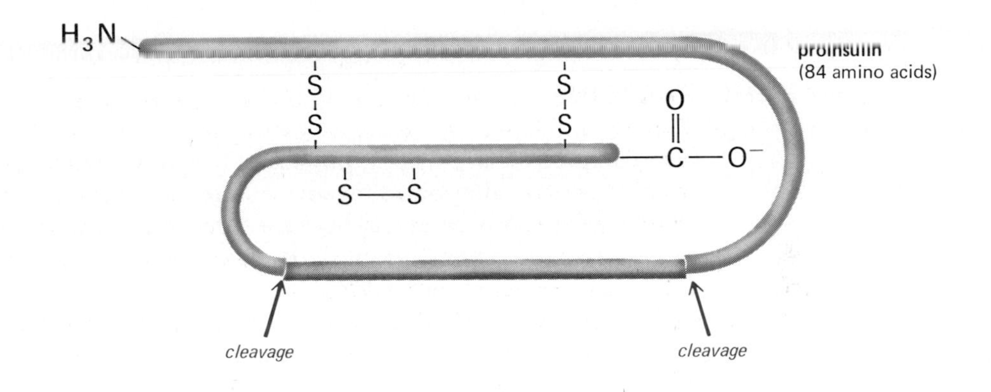

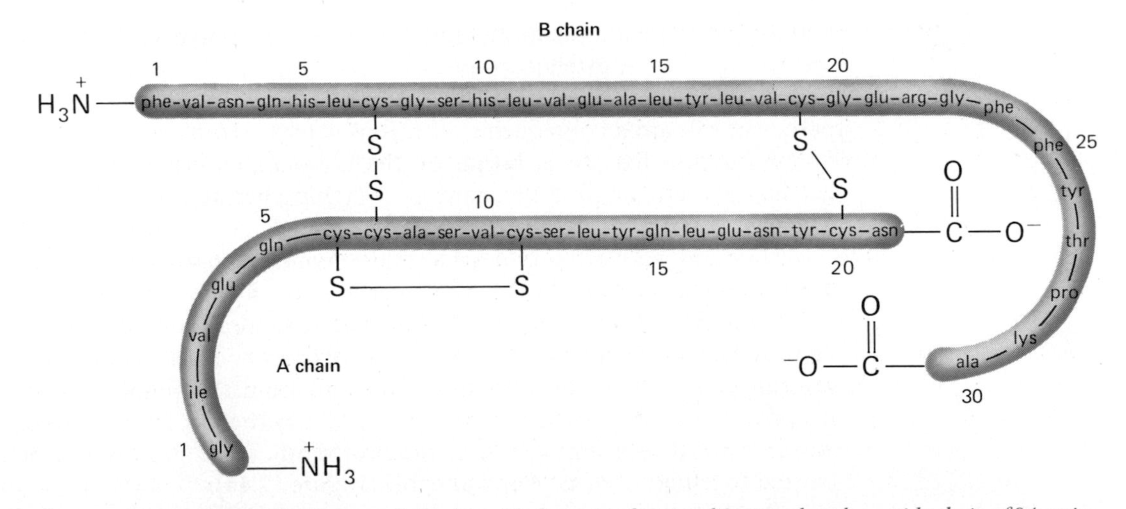

Figure 12.12 *(upper) A schematic diagram of proinsulin, a single polypeptide chain of 84 amino acids. (lower) The amino acid sequence of bovine insulin.*

The information directing the original folding of the single polypeptide chain of proinsulin is no longer present in the A and B chains of the active hormone, and for this reason refolding of the denatured protein is random and the denaturation is irreversible.

The process of producing a protein in an inactive storage form is quite common. For example, the digestive enzymes trypsin and chymotrypsin and the blood clotting enzyme thrombin are produced as inactive proteins. Enzymes produced as inactive proteins that are then activated by cleavage of one or more polypeptide bonds are called zymogens.

12.11 FIBROUS PROTEINS

Fibrous proteins are stringy, physically tough substances composed of elongated, rod-like chains joined together by several types of cross-linkages to form stable, insoluble structures. There are three major classes of fibrous proteins: the silks; the keratins of skin, wool, claws, horn, scales, and feathers; and the collagens of tendons and hides.

In terms of bulk properties, silk is characterized by great strength, flexibility, and resistance to stretching. The distribution of amino acids in silk is unique in that it is largely built from only three amino acids: glycine (45%), alanine (30%), and serine (12%). In addition, there are small amounts of most other amino acids. Chemical studies of silk have revealed that the hexapeptide unit

—(gly-ser-gly-ala-gly-ala)—

repeats for long distances in the polypeptide chain. Notice that every other amino acid in this hexapeptide unit is glycine. The polypeptide chains of silk are arranged in the extended β-pleated sheet (Figure 12.13).

In silk fiber, β-sheets are stacked one on top of another. The bulk properties of silk are a consequence of this molecular structure. Silk fiber is very strong because tension is borne by the covalent bonds of the polypeptide chains themselves. Silk is resistant to stretching because the chain is already extended as far as it can go.

Hair and wool are very flexible, and in contrast to silk, they stretch. They are also elastic, so that when tension is released the fibers revert to their original condition. At the molecular level, the basic structural unit of hair is a polypeptide chain wound in an α-helix conformation. Furthermore, there are several levels of structural organization built from the simple α-helix. First, it appears that three α-helices are arranged together to form a larger interwoven coil called a protofibril. In the protofibril, three strands of α-helix are twisted together to form a rope or cable (Figure 12.14).

Protofibrils are then arranged in bundles to form an 11-stranded cable called a microfibril. These in turn are imbedded in a larger matrix that

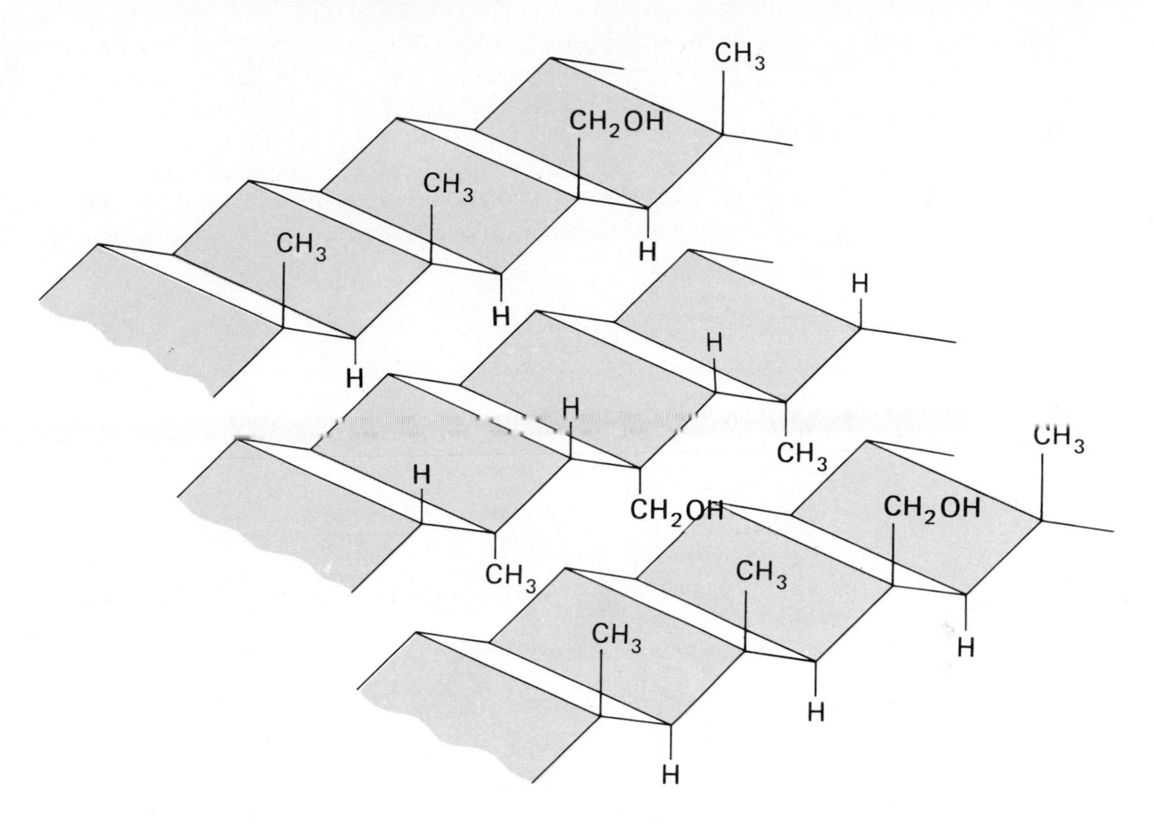

Figure 12.13 *Schematic representation of three polypeptide chains in the β-pleated sheet conformation of silk. The conformation of each β-sheet is as represented in Figure 12.9. In this perspective, each sheet is viewed from the side looking through the plane formed by the atoms of the peptide bonds.*

ultimately forms the hair fiber (Figure 12.15). The α-helices themselves are cross-linked by disulfide bonds between cysteine side chains.

You have surely noted that hair can be stretched and springs back on release of tension. What happens in the stretching process is an elongation of hydrogen bonds along turns of the α-helix. The major force causing the stretched hair fibers to return to their original length is reformation of hydrogen bonds in the α-helices. The α-keratins of horns and claws have essentially the same structure as hair but with a much higher content of cysteine and a greater degree of disulfide bridge cross-linking between the helices. These additional

Figure 12-14 *The supracoiling of three α-helices in hair and wool to form a protofibril.*

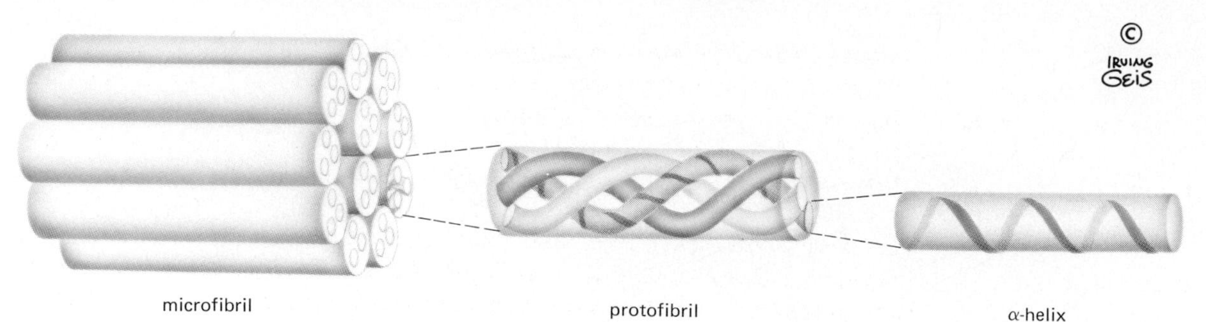

Figure 12.15 *The detailed structure of hair fiber. From R. E. Dickerson and I. Geis, The Structure and Action of Proteins, W. A. Benjamin, Inc., Menlo Park, CA.*

disulfide bonds greatly increase the resistance to stretch and produce the hard keratins of horn and claw.

The third major class of fibrous proteins are the collagens. Collagens are constituents of skin, bone, teeth, blood vessels, tendons, cartilage, and connective tissue. In fact, they are the most abundant of all proteins in higher vertebrates, making up almost 30% of the total body protein mass in humans. The distinctive physical property of collagen is that it forms long, insoluble fibers of very high tensile strength. Table 12.9 lists the collagen content of several tissues. Note that bone, Achilles' tendon, skin, and the cornea of the eye are largely collagen.

Because of its abundance and wide distribution in vertebrates, and because it is associated with a variety of diseases and problems of aging, more is known about collagen than probably any other fibrous protein. The collagen molecule is very large and has a distinctive amino acid composition. One-third of the

Table 12.9 Collagen content of some body tissues.

Tissue	*Collagen (% dry weight)*
bone, mineral-free	88
Achilles' tendon	86
skin	72
cornea	68
cartilage	46–63
ligament	17
aorta	12–24

amino acids in collagen are glycine, and another 20% are proline and hydroxyproline. Tyrosine is present in very small amounts and the essential amino acid tryptophan is absent entirely. Cysteine is also absent, so there are no disulfide cross-links in collagen. When collagen fibers are boiled in water, they are converted into soluble gelatins. Gelatin itself has no biological food value because it lacks the essential amino acid tryptophan.

The polypeptide chains of collagen fold into a conformation that is particularly stable and unique to collagen. In this conformation, three protein strands wrap around each other to form a left-handed superhelix called the collagen triple helix. This unit is called tropocollagen and looks much like a three-stranded rope (Figure 12.16).

Collagen fibers are formed when many tropocollagen molecules line up side by side in a regular pattern and are then cross-linked by the formation of new covalent bonds. One of the effects of severe ascorbic acid deficiency is impaired synthesis of collagen. Without adequate supplies of vitamin C, cross-linking of tropocollagen strands is inhibited, with the result that they do not unite to form stable, physically tough fibers.

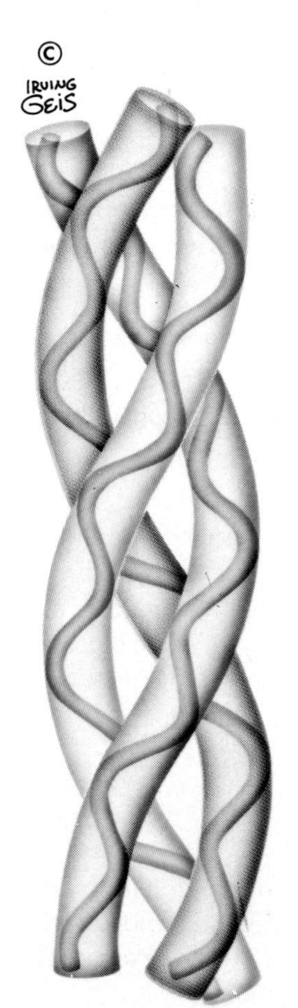

Figure 12.16 *Collagen triple helix. From R. E. Dickerson and I. Geis, The Structure and Action of Proteins, W. A. Benjamin, Inc., Menlo Park, CA. © Copyright 1969. All rights reserved. Used by permission.*

12.12 ENZYMES

One of the unique characteristics of the living cell is its ability to carry out complex reactions rapidly and with remarkable specificity. The agents responsible for these transformations are a group of proteins called enzymes, each designed to catalyze a specific reaction. The first enzyme isolated in pure crystalline form was urease which catalyzes the hydrolysis of urea to ammonia and carbon dioxide.

$$\underset{\text{urea}}{H_2N{-}\overset{\overset{\displaystyle O}{\|}}{C}{-}NH_2} + H_2O \xrightarrow{\text{urease}} 2NH_3 + CO_2$$

It is now clear that all enzymes are proteins and that the one feature which distinguishes them from other proteins is that they are catalysts.

In the case of enzyme-catalyzed reactions, reactants are referred to as substrates. The enzyme and substrate(s) combine to form an activated complex given the special name of enzyme-substrate complex (ES). This complex then undergoes a chemical change to form products and to regenerate the enzyme.

$$E + S \rightleftharpoons ES$$

$$ES \longrightarrow E + P$$

As catalysts, enzymes are far superior to their nonbiological laboratory counterparts in three major ways.

(1) Enzymes have enormous catalytic power.

(2) Enzymes are able to discriminate between very closely related molecules.

(3) The activity of many enzymes is regulated.

Let us look at each of these unique characteristics in more detail.

First, enzymes have enormous power to increase the rate of chemical reactions. In fact, most reactions that occur readily in living cells would occur too slowly to support life in the absence of these biocatalysts. As an example, the enzyme carbonic anhydrase catalyzes the reaction of carbon dioxide and water to produce carbonic acid.

$$CO_2 + H_2O \xrightarrow[\text{anhydrase}]{\text{carbonic}} H_2CO_3$$

Carbonic anhydrase increases the rate of hydration of carbon dioxide by almost 10^7 times compared to the uncatalyzed reaction. Red blood cells are especially rich in this enzyme, and are able to promote the rapid interconversion of carbon dioxide and bicarbonate ion.

Second among their unique properties is that enzymes are highly specific in the reactions they catalyze. A given enzyme will generally catalyze only one reaction or a single type of reaction. Competing reactions and by-products such as we find under laboratory conditions are not observed in enzyme-catalyzed reactions. We have already discussed in Chapter 4 how an enzyme might catalyze a reaction of (+)-glyceraldehyde but not of its enantiomer (−)-glyceraldehyde.

Third among their unique properties is the fact that the activities of many enzymes can be regulated. There are mechanisms in living cells to both increase and decrease the activities of many enzymes. Most often, these regulation mechanisms involve certain small molecules that bind to an enzyme and change its activity.

The fact that enzymes have enormous catalytic power, are highly specific, and can be regulated has important consequences for the living cell. Any living cell contains literally thousands of different molecules, and there is an almost infinite number of chemical reactions that are possible in this mix. Yet the cell, by virtue of its enzymes, selects which chemical reactions take place, and by regulating the activities of key enzymes, it also controls the rates of these reactions and how much of any given product is formed. In this regard, enzymes are truly remarkable catalysts.

12.13 ENZYME PROTEINS AND COFACTORS

Among the enzymes that act as biocatalysts, there is a considerable diversity of structure. Many enzymes are simple proteins, which means that the protein itself is the true catalyst. Still other enzymes catalyze reactions of their substrates only in the presence of specific nonprotein molecules or ions. Nonprotein molecules or metal ions required for enzyme activity are called cofactors. Cofactors are divided into three groups: metal ions, coenzymes, and prosthetic groups.

A. METAL ION COFACTORS

Metal ion cofactors function primarily by forming complexes with the enzyme itself or with other nonprotein groups required by the enzyme for catalytic activity. In some cases, metal ions appear to be loosely associated with the active enzyme and can be removed from them easily. In other instances, metal ions are integral parts of the enzyme structure and are retained throughout normal isolation and purification procedures. As an example, virtually all reactions that involve adenosine triphosphate (ATP) and the hydrolysis of

Table 12.10 Some metal ion cofactors.

Metal ion cofactor	*Function*
Na^+	May be important in controlling the activity of certain enzymes.
Mg^{2+}	Essential for the activity of enzymes catalyzing reactions involving ATP. Required for the stability of quaternary structure of certain enzymes and other proteins as well.
K^+	Required for the activity of certain enzymes. The principal intracellular cation.
Ca^{2+}	Required for the activity of certain enzymes. Binds with the protein of muscle and is required for muscle contraction.
Mn^{2+}	Required for the activity of certain enzymes.
Fe^{2+}, Fe^{3+}	Required for the activity of many enzymes catalyzing oxidation/reduction.
Co^{2+}	Required for activity of all vitamin B_{12}-requiring enzymes.
Cu^+, Cu^{2+}	Essential for the activity of some enzymes catalyzing oxidation/reduction. Essential for oxygen transport in marine organisms.
Zn^{2+}	Essential for the activity of some enzymes.
$Mo^{(?)}$	Essential for enzymes involved in nitrogen fixation and certain oxidation/reduction enzymes.

phosphate anhydride bonds require the divalent cation Mg^{2+} as a cofactor. In these reactions, the positively charged cation coordinates with negatively charged oxygens of the phosphate groups. As another example, carbonic anhydrase requires one Zn^{2+} ion per molecule of enzyme for activity. The Zn^{2+} is an integral part of the active form of this enzyme.

Table 12.10 lists ten metal ions essential for normal growth and development. In recent years we have come to realize that several other metals, present only in trace amounts, are also essential for good health. In many instances we have little or no understanding of what the role in the body is of these elements or why they are essential.

B. COENZYMES AND VITAMINS

A coenzyme is a small organic molecule that binds reversibly to an enzyme and that is required for the activity of the enzyme. Many coenzymes are second substrates for the enzyme. For example, the enzyme lactate dehydrogenase (LDH) requires nicotinamide adenine dinucleotide (NAD^+) for activity.

$$\underset{\text{lactate}}{CH_3-\overset{\overset{\displaystyle OH}{|}}{C}H-CO_2^-} + NAD^+ \xrightarrow{\text{lactate dehydrogenase}} \underset{\text{pyruvate}}{CH_3-\overset{\overset{\displaystyle O}{\|}}{C}-CO_2^-} + NADH + H^+$$

It should be obvious why NAD^+ is required: NAD^+ is the organic molecule that oxidizes lactate to pyruvate and itself is reduced to NADH. Humans and many other organisms cannot synthesize certain coenzymes and so they must obtain in their diet either the coenzyme itself or a substance from which the coenzyme can be synthesized.

These essential coenzymes or coenzyme precursors are called vitamins. Vitamins are divided into two classes based on their physical properties: those that are water-soluble and those that are fat-soluble. As might be expected from these two classifications, water-soluble vitamins are highly polar substances, while fat-soluble vitamins are nonpolar. Most water-soluble vitamins are either coenzymes themselves or are small molecules from which coenzymes are synthesized within the body. The function of the fat-soluble vitamins is less well understood.

Table 12.11 lists twelve essential coenzymes, their vitamin precursors, and the function of each.

Table 12.11 Twelve essential coenzymes, their vitamin precursors, and biological functions.

Coenzyme	*Vitamin precursor*	*Function*
nicotinamide adenine dinucleotide (NAD^+)	nicotinic acid (niacin)	Oxidation–reduction
nicotinamide adenine dinucleotide phosphate ($NADP^+$)	nicotinic acid (niacin)	Oxidation–reduction
flavin adenine dinucleotide (FAD)	riboflavin (vitamin B_2)	Oxidation–reduction
flavin mononucleotide (FMN)	riboflavin (vitamin B_2)	Oxidation–reduction
ascorbic acid (vitamin C)	none	Oxidation–reduction
lipoic acid	none	Oxidation–reduction
thiamine pyrophosphate (TPP)	thiamine (vitamin B_1)	Oxidative decarboxylation and nonoxidative decarboxylation
pyridoxal phosphate	pyridoxine (vitamin B_6)	Transaminations
coenzyme A	pantothenic acid	Transfer of $CH_3{-}\overset{\overset{O}{\|}}{C}$ groups
tetrahydrofolic acid (THF)	folic acid	Transfer of $-CH_3$, $-CH_2OH$, $-CHO$ groups
biotin	none	Carboxylation reactions
cobamide (B_{12})	cobalamine	Transfer of $-\overset{\overset{O}{\|}}{C}-S-CoA$ groups

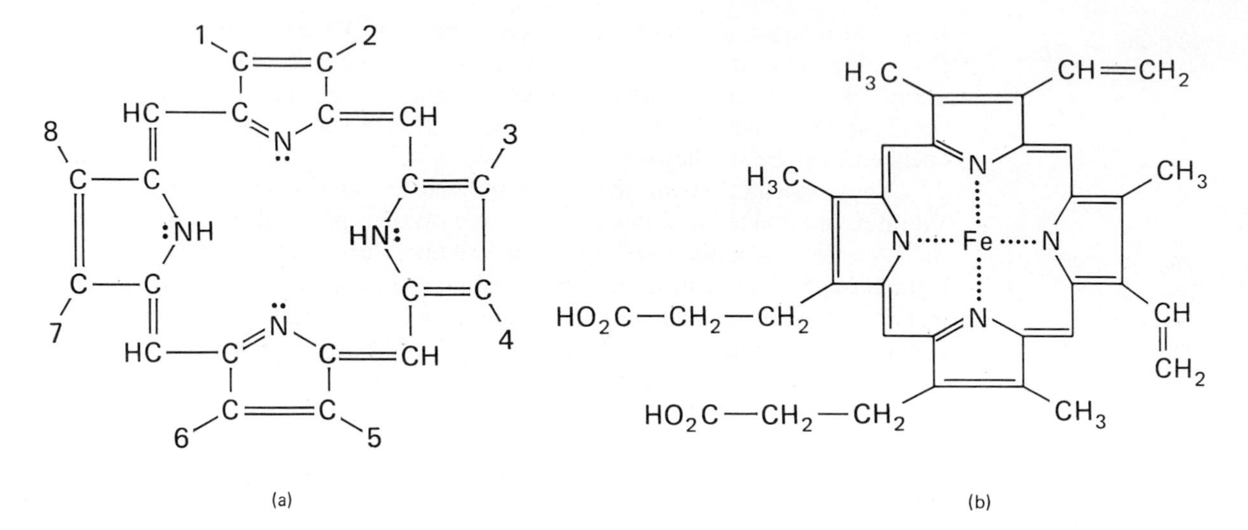

Figure 12.17 *Heme coenzymes: (a) the porphyrin ring system; (b) the heme prosthetic group of hemoglobin, myoglobin, and certain enzymes.*

C. PROSTHETIC GROUPS

Coenzymes and prosthetic groups are similar in that both are organic molecules that bind to enzymes and are necessary for biological activity of the enzyme. The difference is one of degree. Organic molecules that bind reversibly to enzymes are classified as coenzymes. Those which are strongly bound and are an integral part of an enzyme are called prosthetic groups. Metal ions that are an integral part of the protein may also be referred to as prosthetic groups. Perhaps the most important prosthetic groups are the hemes. The structure of heme consists of four substituted pyrrole rings joined by one-carbon bridges into a larger ring called porphyrin (Figure 12.17a). Shown in Figure 12.17b is the heme group found in hemoglobin and myoglobin. Hemoglobin and myoglobin are involved in oxygen transport and storage in mammals. Note that there is an atom of iron in the center of the heme group. In hemoglobin and myoglobin, the iron atom occurs as Fe^{2+}. Other types of heme prosthetic groups have different substituents on the porphyrin ring, for example, Mg^{2+} in chlorophyll.

12.14 MECHANISM OF ENZYME CATALYSIS

The formation of an enzyme-substrate complex (Section 12.12) is the first and crucial step in enzyme catalysis. Virtually all enzymes are globular proteins, and even the simplest have molecular weights ranging from 12,000 to 40,000 (i.e., they consist of 100–400 amino acids). Because enzymes are so large com-

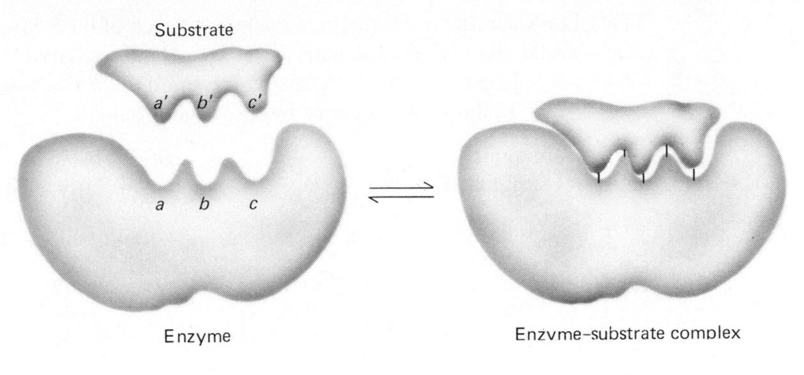

Figure 12.18 *Lock-and-key model of the interaction of enzyme and substrate.*

pared to molecules whose reactions they catalyze, it has been proposed that substrate and enzyme interact at only a small portion or region of the enzyme surface. We call this region of interaction the active site.

To date, a large number of enzymes have been isolated in pure crystalline form and studied by X-ray crystallography. In all cases where enzyme three-dimensional structures have been determined and the interactions between enzyme and substrate studied, the active site has been found to be a relatively small portion of the enzyme surface with a unique arrangement of amino acid side chains. Often these side chains, so close together in the active site, are contributed by amino acids quite far apart in the linear sequence of the polypeptide chain.

In 1890, Emil Fischer likened the binding of an enzyme and its substrate to the interaction of a lock and its key. According to this lock-and-key model, shown schematically in Figure 12.18, the substrate and enzyme have complementary shapes and "fit" together. Recall from our discussion of the significance of chirality in the biological world (Section 4.10) that we accounted for the remarkable ability of enzymes to distinguish between enantiomers by proposing that an enzyme and its substrate must interact through at least three specific binding sites on the surface of the enzyme. In Figure 12.18, these three binding sites are labeled *a*, *b*, and *c*. The complementary regions of the substrate are labeled *a*′, *b*′, and *c*′. Groups on the enzyme surface that participate in binding enzyme and substrate to form an enzyme-substrate complex are called binding groups.

Once a substrate molecule has been recognized and bound to the active site of the enzyme, certain functional groups in the active site participate directly in the making and breaking of chemical bonds. These are called catalytic groups. In a sense, then, the active site on an enzyme is a unique combination of binding groups and catalytic groups.

PROBLEMS

12.7 Explain the meaning of the designation L- as it is used to indicate the stereochemistry of protein-derived amino acids. What is the structural relationship between L-serine and D-glyceraldehyde? and L-glyceraldehyde?

12.8 Draw a structural formula of the form of each of the following amino acids that you would expect to predominate at pH 1.0, pH 6.0, and pH 12.0. Use values of 2.2 for the pK_a of the α-carboxyl group, 9.5 for the pK_a of the α-amino group, and refer to Table 12.2 for pK_a values of ionizable side chains.

(a) alanine
(b) tyrosine
(c) arginine
(d) aspartic acid

12.9 pK_a Values for the three ionizable groups of glutamic acid are

$$\text{p}K_a = 4.1 \rightarrow \text{HO}-\overset{\text{O}}{\overset{\|}{\text{C}}}-\text{CH}_2-\text{CH}_2-\underset{\text{NH}_3^+}{\underset{|}{\text{CH}}}-\overset{\text{O}}{\overset{\|}{\text{C}}}-\text{OH} \leftarrow \text{p}K_a = 2.2$$

$$\text{p}K_a = 9.7 \longrightarrow \text{NH}_3^+$$

(a) Which of these ionizable groups is the strongest acid; the weakest acid?
(b) Estimate the net charge on glutamic acid at pH 1.0; at pH 6.0; at pH 11.0.

12.10 Would you expect an aqueous solution of lysine to be acidic, basic, or neutral? Explain your reasoning. (*Hint*: In thinking about this problem, first consider the effect on pH of the carboxyl and the α-amino groups together in the zwitterion form and then the effect of the terminal amino group.)

12.11 Draw structural formulas for glutamic acid and glutamine. One of these amino acids has an isoelectric point of 5.7, the other has an isoelectric point of 3.08. Which amino acid has which isoelectric point? Explain your reasoning.

12.12 Write a balanced equation for the oxidation of two molecules of cysteine by hydrogen peroxide to form a disulfide bond.

12.13 Name the essential amino acids for humans. Why are they termed "essential?" Compare meat, fish, and the cereal grains in terms of their ability to supply the essential amino acids.

12.14 Write a structural formula for the tripeptide glycylserylaspartic acid. Write a structural formula for an isomeric tripeptide. Calculate the net charge on each at pH 6.0.

12.15 Write the structural formula for the tripeptide lys-asp-val. Calculate the net charge on this tripeptide at pH 6.0.

12.16 How many tetrapeptides can be constructed from the 20 amino acids **(a)** if each of the amino acids is used only once in the tetrapeptide? **(b)** if each amino acid can be used up to four times in the tetrapeptide?

12.17 Examine the amino acid composition of bovine insulin (Figure 12.12) and list all asp, glu, lys, and arg in the molecule. Would you predict the isoelectric point of insulin to be nearer that of the acidic amino acids (pI 2.0–3.0), the neutral amino acids (5.5–6.5), or the basic amino acids (9.5–11.0)?

12.18 Following is the primary structure of glucagon, a polypeptide hormone that helps to regulate glycogen metabolism. Glucagon is secreted by the alpha cells of the pancreas during the fasting state, when blood glucose levels are decreasing. This hormone stimulates the enzymes that catalyze the hydrolysis of glycogen to glucose and thus helps to maintain blood glucose levels within a normal concentration range. Glucagon contains 29 amino acids and has a molecular weight of approximately 3500.

1 5 10 15
his-ser-glu-gly-thr-phe-thr-ser-asp-tyr-ser-lys-tyr-leu-asp-

16 20 25 29
-ser-arg-arg-ala-gln-asp-phe-val-gln-trp-leu-met-asn-thr

glucagon

(a) Estimate the net charge on glucagon at pH 6.0.
(b) Would you predict the isoelectric point of glucagon to be nearer that of the acidic amino acids (pI 2.0–3.0), the neutral amino acids (5.5–6.5), or the basic amino acids (9.5–11.0)?

12.19 Which of the following peptides and proteins migrate to the (+) electrode, and which migrate to the (−) electrode on electrophoresis at pH 6.0? at pH 8.6?

(a) ala-glu-ile
(b) gly-asp-lys
(c) val-ala-leu
(d) tyr-trp-arg
(e) thyroglobulin (pI 4.6)
(f) hemoglobin (pI 6.8)

12.20 The following proteins have approximately the same molecular weight and size. The pI of each is given.

carboxypeptidase	pI 6.0
pepsin	pI 1.0
human growth hormone	pI 6.9
ovalbumin	pI 4.6

(a) State whether each is positively charged, negatively charged, or uncharged at pH 6.0.
(b) Draw a diagram showing the results of electrophoresis of a mixture of the four at pH 6.0.

12.21 Following are amino acid sequences for several tripeptides. Indicate which are hydrolyzed by trypsin, which by chymotrypsin.

(a) ala-asp-lys
(b) glu-arg-ser
(c) phe-met-trp
(d) lys-ala-asp
(e) arg-tyr-gly
(f) asp-lys-phe

12.22 Deduce the amino acid sequence of an octapeptide from the following experimental results.

amino acid composition:	ala, arg, leu, lys, met, phe, ser, tyr	
Edman degradation:	ala	
digestion with trypsin:	Fragment I:	ala, arg
	Fragment J:	leu, met, tyr
	Fragment K:	lys, phe, ser
digestion with chymotrypsin:	Fragment L:	ala, arg, phe, ser
	Fragment M:	leu
	Fragment N:	lys, met, tyr

12.23 Using structural formulas, show how the theory of resonance accounts for the fact that the peptide bond is planar.

12.24 In constructing models of polypeptide chains, Pauling assumed that for maximum stability, (1) all amide bonds are *trans* and coplanar, and (2) there is a maximum of hydrogen bonding between amide groups. Examine the β-pleated sheet (Figure 12.9) and the α-helix (Figure 12.8) and convince yourself that in each of these, the amide bonds are planar and that each carbonyl is hydrogen bonded to an amide hydrogen.

12.25 Examine the structure of the α-helix. Are the amino acid side chains arranged all inside the helix, all outside the helix, or randomly oriented?

12.26 Draw structural formulas to illustrate the noncovalent interactions indicated.
(a) Hydrogen bonding between the side chains of thr and asn.
(b) Salt linkage between the side chains of lys and glu.
(c) Hydrophobic interactions between the side chains of two phenylalanines.

12.27 Consider a typical globular protein in an aqueous medium of pH 6.0. Which of the following amino acid side chains would you expect to find on the outside and in contact with water? which on the inside and shielded from contact with water?

(a) glutamic acid
(b) glutamine
(c) arginine
(d) serine
(e) valine
(f) phenylalanine
(g) lysine
(h) isoleucine
(i) threonine

12.28 Account for the fact that solubility of globular proteins is a function of pH, and that solubility is a minimum when the pH of the solution equals the isoelectric point (pI) of the protein.

12.29 Insulin is a water-soluble globular protein. Calculate the percentage of amino acids with polar side chains (both uncharged and charged) in this protein. Also calculate the percentage of polar amino acid side chains in the repeating hexapeptide of silk (Section 12.11). What generalization might you make about the relative percentages of polar side chains in water-soluble globular proteins compared to water-insoluble fibrous proteins?

12.30 Myoglobin and hemoglobin are globular proteins. Myoglobin consists of a single polypeptide chain of 153 amino acids. Hemoglobin is composed of four polypeptide chains, two of 141 amino acids and two of 146 amino acids. The three-dimensional structure of myoglobin and hemoglobin polypeptide chains is very similar, yet myoglobin exists as a monomer in aqueous solution, while the four polypeptide chains of hemoglobin self-assemble to form a tetramer. Which polypeptide chains, those of myoglobin or hemoglobin, would you predict to have a higher percentage of nonpolar amino acids?

12.31 What is irreversible denaturation and how does it differ from reversible denaturation?

12.32 What is the most characteristic type of secondary structure in the protein components of:

(a) hair
(b) silk
(c) collagen
(d) hooves

12.33 Characterize the amino acid composition of silk. Silk fiber is quite resistant to stretching. Account for this property in terms of molecular structure.

12.34 What is the function of collagen? Describe **(a)** the macroscopic physical properties and **(b)** the molecular structure of collagen.

Abnormal Human Hemoglobins

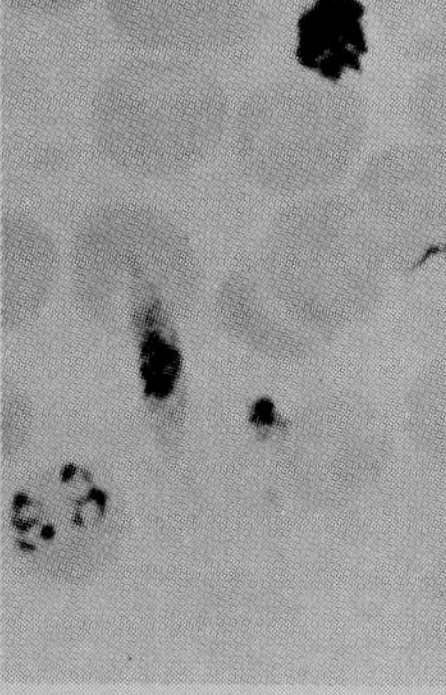

THERE ARE AN ESTIMATED 5 billion red blood cells (erythrocytes) in the bloodstream of an adult, each packed with about 270 million molecules of hemoglobin. In terms of sheer numbers, hemoglobin is one of the most plentiful proteins in the body. Hemoglobin's role is to pick up molecular oxygen in the lungs and deliver it to all parts of the body for metabolic oxidation. Normal adult hemoglobin (hemoglobin A or Hb A) is composed of four polypeptide chains; two α chains, each of 141 amino acids; and two β chains, each of 146 amino acids. Each polypeptide chain surrounds one iron porphyrin or heme group (Section 12.14C) that reversibly binds oxygen. The tetrameric structure of hemoglobin is stabilized principally by hydrophobic interactions. Shown in Figure 1 is the three-dimensional shape of a single β chain. The N-terminal amino acid is indicated by $-NH_3^+$ and the C-terminal amino acid is indicated by $-CO_2^-$.

The 3-dimensional structure of hemoglobin A was determined by Max Perutz. For this pioneering work he shared in the Nobel Prize in 1963.

It is the so-called abnormal human hemoglobins that have

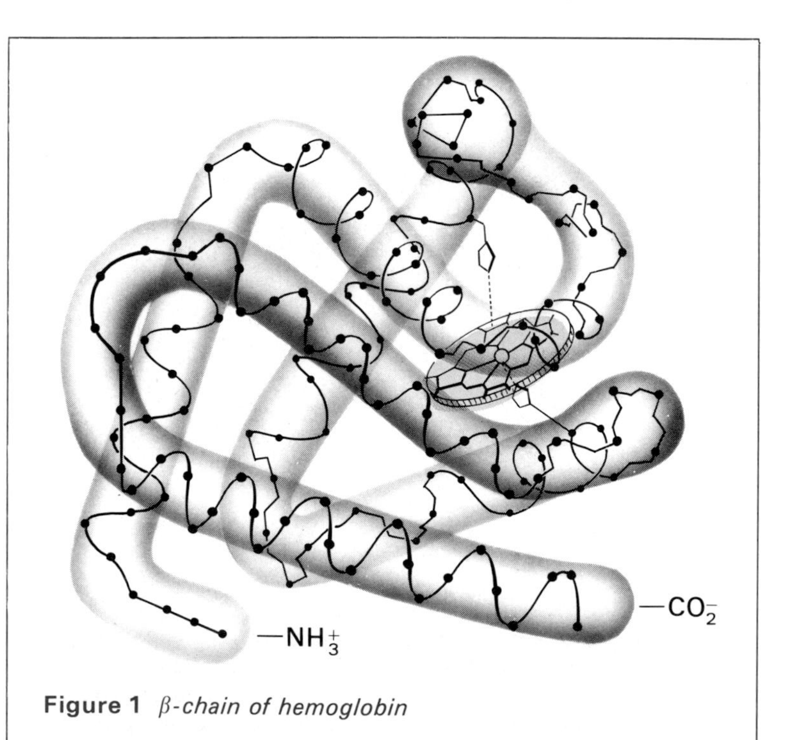

Figure 1 *β-chain of hemoglobin*

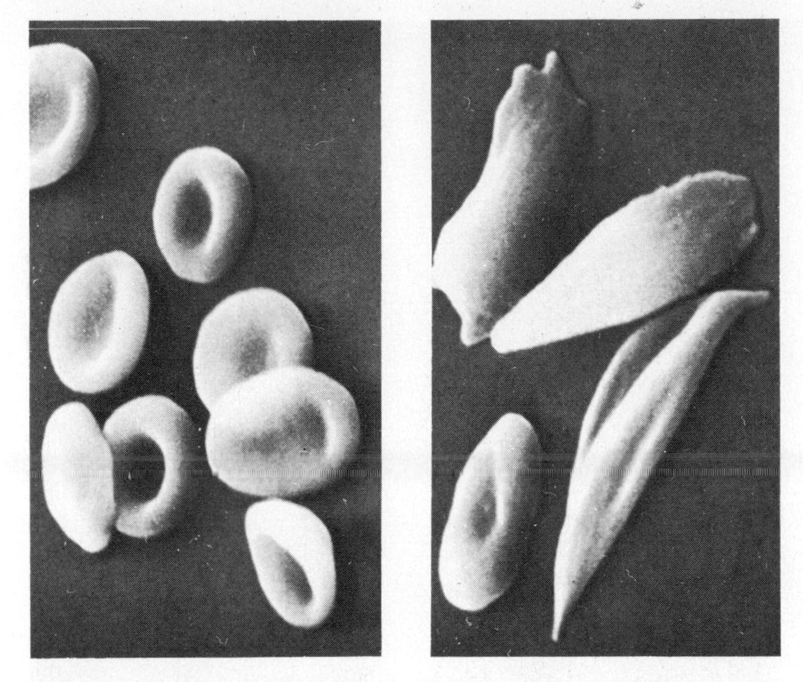

Figure 2 *Normal red blood cells (left) magnified ×6750 and cells that have sickled (right) after discharging oxygen, magnified ×8700. (Courtesy of Dr. Marion I. Barnhart, Wayne State University School of Medicine.)*

attracted particular attention because of the diseases associated with them. The best known of these diseases is sickle-cell anemia, a name derived from the characteristic sickle shape of affected red blood cells when they are deoxygenated. When combined with oxygen, red blood cells of persons with sickle-cell anemia have the flat, disc-like conformation of normal erythrocytes. However, when oxygen pressure is reduced, affected cells become distorted and considerably more rigid and inflexible than normal cells (Figure 2). Because they are larger than some of the blood channels through which they must pass, sickled cells tend to become wedged in capillaries, blocking the flow of blood. Surprisingly, little is known about the reason why some organs and tissues are affected more than others by the disease, the normal age of onset of the disease, and male versus female susceptibility. Some persons afflicted with sickle-cell anemia die at an early age, often due to childhood infections complicated by the disease. Others lead productive lives until advanced ages.

In 1949 Linus Pauling made a discovery that opened the way to an understanding of this disease at the molecular level. He observed that there is a significant difference between normal adult hemoglobin (Hb A) and sickle-cell hemoglobin (Hb S). At pH 6.9, Hb A has a net negative charge and Hb S has a net positive charge and on paper electrophoresis at this pH, Hb A moves toward the positive electrode and Hb S toward the negative electrode. Vernon Ingram pursued this discovery and in 1956 showed that sickle-cell hemoglobin differs from normal hemoglobin only in the amino acid at the sixth position of the β chain. Alpha chains of both are identical, but glutamic acid at position 6 of each β chain of Hb A is replaced by valine in Hb S. A result of the valine-glutamic acid substitution is replacement of two negatively charged side chains by two nonpolar, hydrophobic side chains.

How is the substitution of Hb S for Hb A in red blood cells related to the process of sickling? We know that hemoglobin S functions perfectly normally in transporting molecular oxygen from the lungs to cells. In this regard it is indistinguishable from hemoglobin A. However, when it gives up its oxygen, Hb S tends to form polymers which separate from solution in crystalline form. There is now good evidence that the basic unit of crystalline Hb S polymer is a double-stranded fiber stabilized by hydrophobic interactions including that between the valine at position 6 of one β chain and a hydrophobic patch on other Hb S molecules. Then the double-stranded Hb S polymer molecules interact to form multi-stranded cable-like structures. This is a remarkable phenomenon—polymerization of Hb S is facilitated by the presence of valine at β-6, but comparable polymerization of Hb A is prevented by the presence of glutamic acid at β-6.

Now that we have an understanding of sickle-cell anemia at the molecular level,

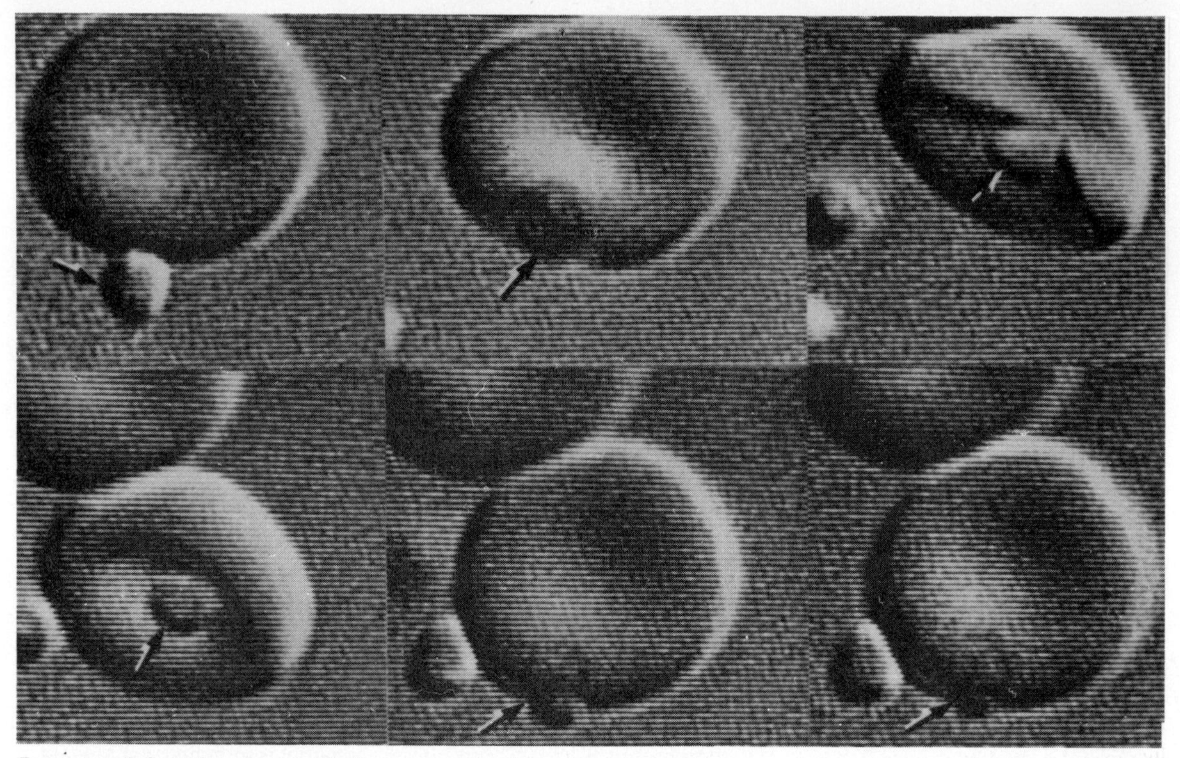

From top left to bottom right, the arrows in this microscope sequence follow the invasion of a red-blood cell by a malaria parasite. (*World Health Organization*)

the challenge is to devise specific medical treatments to prevent or at least inhibit the sickling process. One strategy being actively pursued is the search for substances that will inhibit the polymerization of Hb S by disrupting or preventing hydrophobic interactions of β-6 valines.

Sickle-cell anemia is a genetic disease. Persons with an Hb S gene from only one parent are said to have sickle-cell trait. About 40% of the hemoglobin in these individuals is Hb S. There are generally no ill effects associated with sickle-cell trait except under extreme conditions. Persons with Hb S genes from both parents are said to have sickle-cell disease, and all of their hemoglobin is Hb S. The mutant gene coding for Hb S occurs in about 10% of black Americans and in about 20% of African blacks. The gene is also present in significant numbers of the populations of countries bordering the Mediterranean Sea and parts of India.

The fact that there seems to be so much natural selection pressure against the Hb S gene raises the question of why it has

Table 1 Abnormal human hemoglobins. Many of these names are drived from the location of their discovery.

Hemoglobin variant	*Amino acid substitution*		
	Position	*From*	*To*
	alpha chain		
J-Paris	12	ala	asp
G-Philadelphia	68	asn	lys
M-Boston	58	his	tyr
Dakar	112	his	gln
	beta chain		
S	6	glu	val
J-Trinidad	16	gly	asp
E	26	glu	lys
M-Hamburg	63	his	tyr

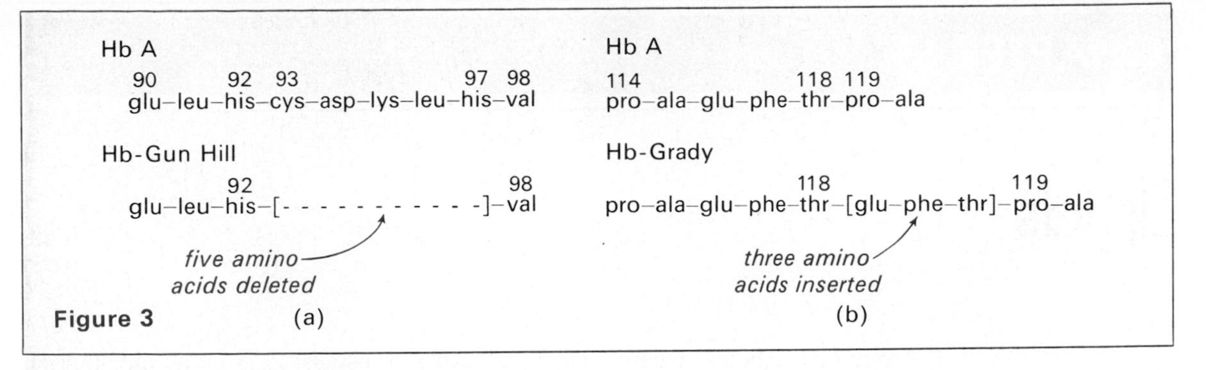

Figure 3 (a) (b)

persisted so long in the gene pool and why sickle-cell trait is so common in populations of specific parts of the world. Several explanations have been offered, but the most likely, first advanced in 1949, is that sickle-cell trait provides some protection against *Plasmodium falciparum*, the parasite responsible for the most severe form of malaria. The falciparum parasite lives part of its life cycle in red blood cells and grows equally well in oxygenated cells containing either Hb S or Hb A. However, when infected cells containing Hb S are deoxygenated and they sickle, the parasites living in them are killed. Not all infected Hb S cells sickle at any one time, but the approximately 40% that do sufficiently reduce the parasite population to reduce the severity of the malaria and prevent death.

The dramatic success in discovering the genetic and molecular basis for sickle-cell anemia spurred interest in searching for other abnormal hemoglobins. To date, several hundred have been isolated and the changes in primary structure have been determined. In the vast majority, there is but a single amino acid residue change in either the alpha or the beta chain, and each substitution is consistent with the change of a single nucleotide in one DNA codon (Section 14.8). Several abnormal hemoglobins are listed in Table 1.

Although most of the abnormal hemoglobins differ from Hb A by only a single amino acid substitution, several have been discovered in which there are either insertions or deletions of amino acids. For example, in hemoglobin-Leiden, discovered in 1968, glutamic acid at position 6 in each beta chain is missing altogether. In hemoglobin-Gun Hill, discovered in 1967 in a 41-year-old man and one of his three daughters, there is a deletion of five amino acids in each beta chain. Thus, each beta chain is shortened to 141 amino acid residues (Figure 3a).

In hemoglobin-Grady, discovered in 1974 in a 25-year-old woman and her father, there is insertion of three amino acids in each alpha chain. Thus, each alpha chain is elongated to 144 amino acid residues (Figure 3b).

References

Antonine, E. and Brunori, M., "Hemoglobin," *Annual Review of Biochemistry* 39 (1970): 977.

Beale, D. and Lehmann, H., "Abnormal Hemoglobins and the Genetic Code," *Nature* 207 (1965): 259.

Dayhoff, M. O., *Atlas of Protein Sequence and Structure* 5 (National Biomedical Research Foundation: 1972).

Harkness, D. R., "Trends," *Biochemical Science* 1 (1976): 73–76.

Huisman, T. H. J. et al., "Hemoglobin Grady: The First Example of a Variant with Elongated Chains Due to the Insertion of Residues," *Proceedings of the National Academy of Science* 71 (1974): 3270.

Ingram, V., "Gene Mutations in Human Hemoglobin: The Chemical Difference Between Normal and Sickle-Cell Hemoglobin," *Nature* 180 (1957): 326.

Maugh, T. H., II, "A New Understanding of Sickle Cell Emerges," *Science* 211 (1981): 265–267.

———, II, "Sickel Cell (II): Many Agents Near Trials," *Science* 211 (1981): 468–470.

Morimoto, H., Lehmann, H., and Perutz, M. F., "Molecular Pathology of Human Hemoglobins," *Nature* 232 (1971): 408.

Pauling, L. et al., "Sickle-Cell Anemia, a Molecular Disease," *Science* 110 (1949): 543.

Perutz, M. F., "The Hemoglobin Molecule," *Scientific American* (Nov. 1964).

CHAPTER 13

Lipids

Lipids are a heterogeneous class of naturally occurring organic substances, grouped together not by the presence of a distinguishing functional group or structural feature, but rather on the basis of common solubility properties. Lipids are all insoluble in water and highly soluble in one or more organic solvents, including ether, chloroform, benzene, and acetone. In fact, these four solvents are often referred to as "lipid-solvents" or "fat-solvents." Proteins, carbohydrates, and nucleic acids are largely insoluble in these solvents.

Lipids are widely distributed in the biological world, and play a variety of roles in both plant and animal tissue. In the human body, lipids function as storage forms of energy, metabolic fuels, structural components of biological membranes, emulsifying agents, vitamins, and regulators of metabolism.

In this chapter we will describe the structure and biological function of representative members of the five major types of lipids: fats and oils, waxes, phospholipids, the fat-soluble vitamins, and steroids. In addition, we will describe the structure of biological membranes.

13.1 FATS AND OILS

You certainly are familiar with fats and oils since you encounter them every day in such things as milk, butter, oleomargarine, corn oil and other liquid vegetable oils, and many other foods. Fats and oils are triesters of glycerol and are called triglycerides. Triglycerides are the most abundant naturally occurring lipids. Complete hydrolysis of a triglyceride yields one molecule of glycerol and three molecules of fatty acid.

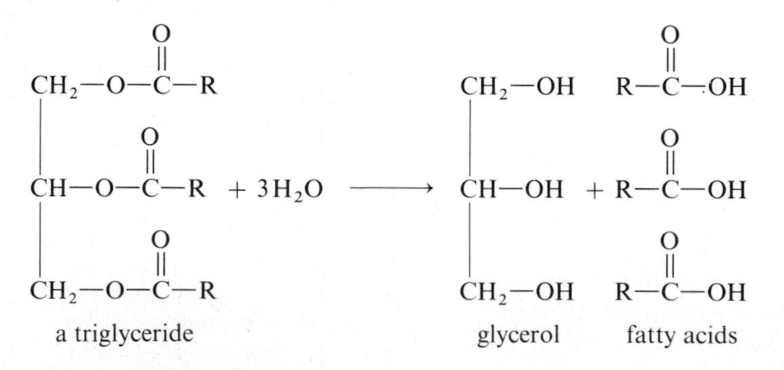

A triglyceride in which all three fatty acids are identical is called a simple triglyceride. An example is tristearin. Simple triglycerides are rare in nature, mixed triglycerides are much more common. Following is a mixed triglyceride formed from glycerol and molecules of stearic acid, oleic acid, and linoleic acid, three of the most abundant fatty acids.

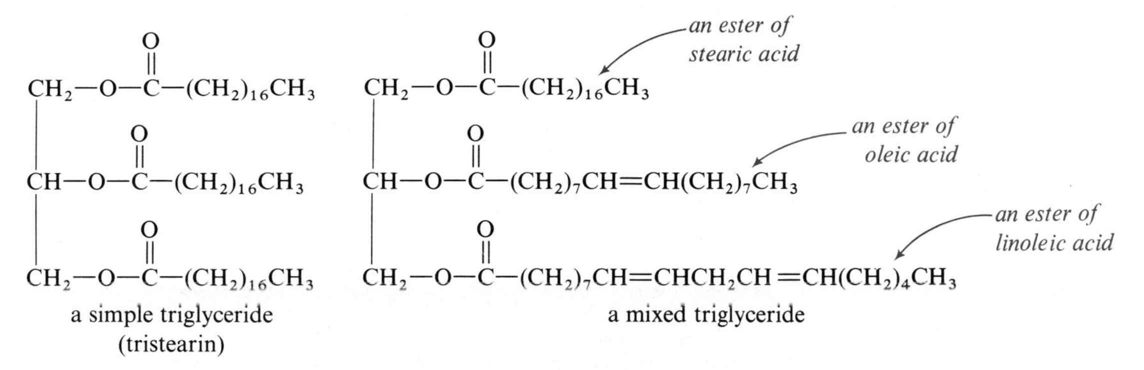

a simple triglyceride (tristearin)

a mixed triglyceride

The physical properties of a triglyceride depend on its fatty acid components. In general, the melting point of a triglyceride increases as the number of carbons in the hydrocarbon chains increases and decreases as the degree of unsaturation increases. Triglycerides rich in oleic acid, linoleic acid, and other unsaturated fatty acids are generally liquid at room temperature and are called oils. Triglycerides rich in palmitic, stearic, and other saturated fatty acids are generally semisolid or solid at room temperature and are called fats. Table 13.1 lists the percent composition in grams of fatty acid per 100 grams of triglyceride for several common fats and oils. Notice that beef tallow is approximately 41.5% by weight saturated fatty acids and 53.1% unsaturated fatty acids. Vegetable oils such as corn oil, soybean oil, and wheat germ oil, however, are all approximately 80% by weight unsaturated fatty acids. Butter fat is distinctive in that it contains significant amounts of lower-molecular-weight fatty acids.

Table 13.1 Fatty acid composition by weight of several triglycerides. Percentages are given for the most abundant fatty acids; other fatty acids are present in lesser amounts.

		Saturated fatty acids		*Unsaturated fatty acids*		
Fat or oil	*mp(°C)*	*palmitic*	*stearic*	*palmitoleic*	*oleic*	*linoleic*
depot fat (human)	15	24.0	8.4	5.0	46.9	10.2
beef tallow	—	27.4	14.1	—	49.6	2.5
corn oil	−20	10.2	3.0	1.5	49.6	34.3
peanut oil	3.0	8.3	3.1	—	56.0	26.0
soybean oil	−16	9.8	2.4	0.4	28.9	50.7
wheat germ oil	—	← 16.0 →		—	28.1	52.3
butter fat	32	29.0	9.2	4.6	26.7	3.6

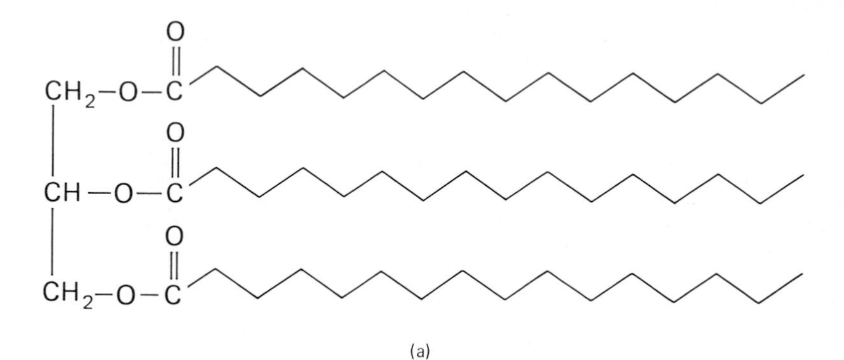

(a)

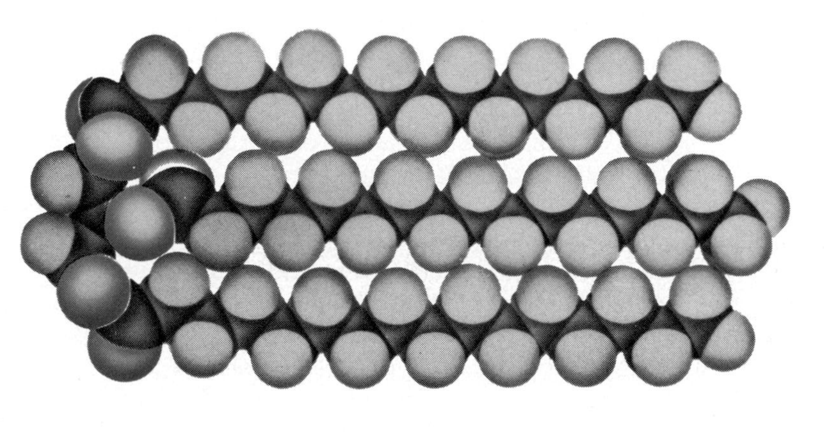

(b)

Figure 13.1 *A saturated triglyceride. (a) Structural formula; (b) space-fiilling model. Space-filling models show relative sizes and shapes of atoms and molecules.*

The lower melting points of triglycerides rich in unsaturated fatty acids are related to differences in three-dimensional shape between the hydrocarbon chains of unsaturated and saturated fatty acid components. Shown in Figure 13.1 is the structural formula of tripalmitin and a space-filling model of this saturated triglyceride. Notice that the three saturated hydrocarbon chains of tripalmitin lie parallel to each other and that the molecule has a very ordered, compact shape. Dispersion forces between these hydocarbon chains are strong. Because of their compactness and interaction by dispersion forces, triglycerides rich in saturated fatty acids have melting points above room temperature.

The three-dimensional shape of an unsaturated fatty acid is quite different from that of a saturated fatty acid. Recall from Section 8.6 that in the unsaturated fatty acids of higher organisms, *cis*-isomers predominate, and *trans*-isomers are very rare. Figure 13.2 shows the structural formula of a triglyceride derived from one molecule each of palmitic acid, oleic acid, and linoleic acid. Notice the *cis* configuration about the double bonds in the hydrocarbon chains of oleic and linoleic acids. Also shown in Figure 13.2 is a space-filling model of this unsaturated triglyceride.

In an unsaturated triglyceride the hydrocarbon chains cannot lie parallel to each other. Compared to saturated triglycerides, molecules of unsaturated

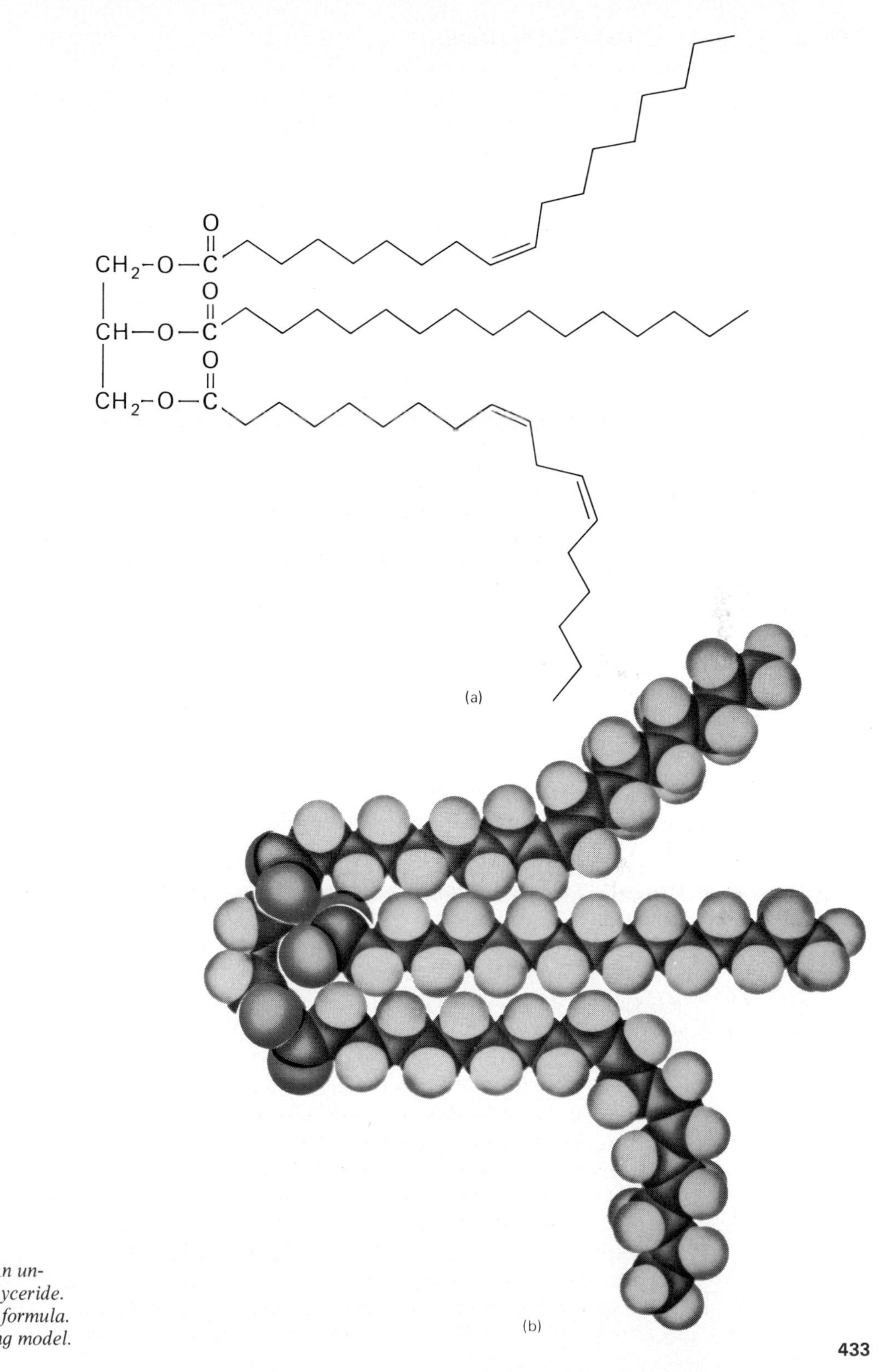

Figure 13.2 *An unsaturated triglyceride. (a) Structural formula. (b) Space-filling model.*

triglycerides are more bulky, pack together less well, and dispersion forces between them are weaker. For these reasons, unsaturated triglycerides have lower melting points than saturated triglycerides.

Example 13.1

Following is given the fatty acid composition by percent of two triglycerides. Predict which triglyceride has the lower melting point.

	palmitic acid	*stearic acid*	*palmitoleic acid*	*oleic acid*	*linoleic acid*
triglyceride A	24.0	8.4	5.0	46.9	10.2
triglyceride B	9.8	2.4	0.4	28.9	50.7

Solution

Triglyceride A is composed of approximately 32% saturated fatty acids and 62% unsaturated fatty acids. Triglyceride B is composed of 12% saturated fatty acids and 80% unsaturated fatty acids. Of the unsaturated fatty acids in B, more than 50% are linoleic acid, a fatty acid with two double bonds. Because of its higher degree of unsaturation, triglyceride B has the lower melting point. Refer to Table 13.1 and you will see that triglyceride A is human depot fat (mp 15°C) and triglyceride B is soybean oil (mp −16°C).

PROBLEM 13.1 How do you account for the fact that both beef tallow and corn oil are composed of approximately 50% oleic acid and yet they have such different melting points?

On exposure to air, most triglycerides develop an unpleasant odor and flavor, and are said to become rancid. In part, rancidity is the result of slight hydrolysis of the fat and oil and production of low-molecular-weight fatty acids. The odor of rancid butter is due largely to the presence of butanoic acid formed by the hydrolysis of butterfat. These same low-molecular-weight fatty acids can be formed by air oxidation of the unsaturated fatty-acid side chains. The rate of rancidification varies with individual triglycerides, largely because of the presence of certain naturally occurring substances called antioxidants that inhibit the process. One of the most common lipid antioxidants is vitamin E (Section 13.4).

For a variety of reasons, in part convenience and in part dietary preference, a major industry has developed for the conversion of oils to fats. The process, called "hardening," involves reaction of an oil with hydrogen in the presence of a catalyst and reduction of some or all of the double bonds of the fatty-acid constituents of a triglyceride. If all of the double bonds are saturated, the resulting triglyceride is hard and brittle. In practice, the degree of hardening is carefully controlled to produce fat of a desired consistency. The resulting fats are sold for kitchen use (Crisco, Spry, etc.). Oleomargarine and other butter substitutes are prepared by hydrogenation of cottonseed, soybean, corn, or peanut oils. The resulting product is often churned with milk and artificially colored to give it a flavor and consistency resembling that of butter.

13.2 WAXES

Plant waxes are esters of fatty acids and alcohols, each having from 16 to 34 carbon atoms. Carnauba wax, which coats the leaves of the carnauba palm native to Brazil, is largely myricyl cerotate, $C_{25}H_{51}CO_2C_{30}H_{61}$. Beeswax, secreted from the wax glands of the bee, is largely myricyl palmitate, $C_{15}H_{31}$-$CO_2C_{30}H_{61}$.

$$CH_3(CH_2)_{14}\overset{\overset{\displaystyle O}{\|}}{C}{-}O(CH_2)_{29}CH_3$$

myricyl palmitate
(a major component of beeswax)

Waxes are harder, more brittle, and less greasy to the touch than fats. Applications are found in polishes, cosmetics, ointments and other pharmaceutical preparations.

13.3 PHOSPHOLIPIDS

Phospholipids are the second most abundant kind of naturally occurring lipid. They are found almost exclusively in plant and animal membranes, which typically consist of about 40–50 % phospholipid and 50–60 % protein.

The most abundant phospholipids contain glycerol and fatty acids, as do the simple fats. In addition, they also contain phosphoric acid and a low-molecular-weight alcohol. The most common of these low-molecular-weight alcohols are choline, ethanolamine, serine, and inositol.

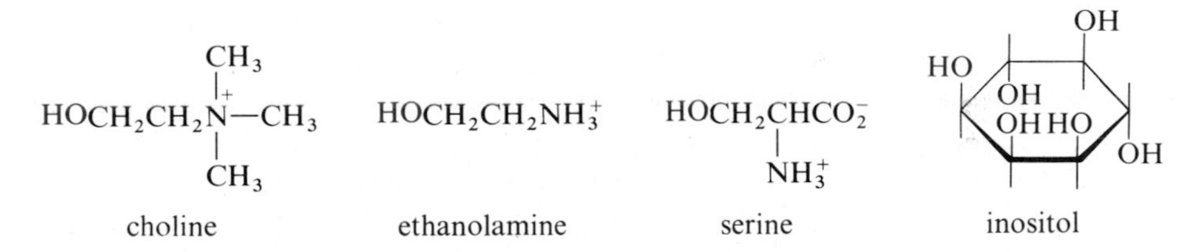

choline ethanolamine serine inositol

The most abundant phospholipids in higher plants and animals are the lecithins and the cephalins.

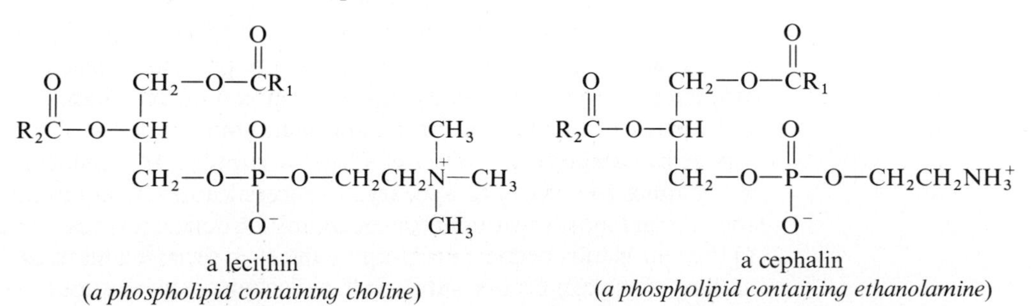

a lecithin
(a phospholipid containing choline)

a cephalin
(a phospholipid containing ethanolamine)

The lecithins are phosphate esters of choline and the cephalins are phosphate esters of ethanolamine. Lecithin and cephalin are shown as they would be ionized at pH 7.4. The fatty acids most common in these membrane phospholipids are palmitic and stearic acids (both fully saturated) and oleic acid (one double bond in the hydrocarbon chain).

13.4 FAT-SOLUBLE VITAMINS

Vitamins are divided into two broad classes on the basis of solubility: those that are water-soluble, and those that are fat-soluble. As we discussed in Section 12.13B, the water-soluble vitamins serve as coenzymes or as building blocks for coenzymes, and the role of the coenzymes is well understood. The fat-soluble vitamins include vitamins A, D, E, and K. At the present time, the molecular basis of their function is only poorly understood.

Vitamin A, or retinol, is a primary alcohol of molecular formula $C_{20}H_{30}O$. Vitamin A alcohol occurs only in the animal world, where the best sources are cod-liver oil and other fish-liver oils, animal liver, and dairy products. Vitamin A in the form of a precursor, or provitamin, is found in the plant world in plant pigments called carotenes. The most common of these, β-carotene ($C_{40}H_{56}$), has an orange-red color and is commonly used as a food coloring. The carotenes have no vitamin A activity; however, after ingestion, β-carotene is cleaved at the central carbon–carbon double bond to give a vitamin-A-related molecule.

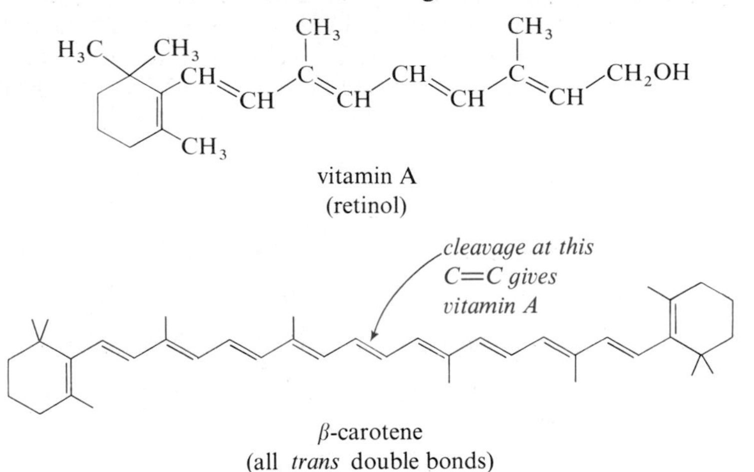

vitamin A
(retinol)

β-carotene
(all *trans* double bonds)

A deficiency of vitamin A or vitamin-A precursors leads to a slowing or stopping of growth. Probably the major action of this vitamin is on epithelial cells, particularly those of the mucous membranes of the eye, respiratory tract, and genitourinary tract. Without adequate supplies of vitamin A, these mucous membranes become hard and dry, a process known as keratinization. One of the first and most obvious effects of vitamin-A deficiency is on the eye. The cells of the tear glands become keratinized and stop secreting tears, and the external surface of the eye becomes dry, dull, and often scaly. Without tears to remove bacteria, the eye is much more susceptible to serious infection. If this condition is not treated in time, blindness results. The mucous membranes of the respiratory, digestive, and urinary tracts also become keratinized in vitamin-A deficiency and become susceptible to infection.

A less serious condition, but one frequently seen in humans whose diets contain insufficient vitamin A, is night blindness. This is the inability to see in dim light or to adapt to a decrease in light intensity.

The term vitamin D is a generic name for a group of structurally related substances produced by the action of ultraviolet light on certain provitamins (vitamin precursors). Vitamin D_3, also called cholecalciferol, is produced in the skin of mammals by the action of sunlight on 7-dehydrocholesterol. Sunlight causes opening of one of the six-membered rings and the formation of a triene. With normal exposure to sunlight, enough 7-dehydrocholesterol is converted to vitamin D_3 so that no dietary vitamin D is necessary. Only when skin manufacture is inadequate is there a need to supplement the diet with artificially fortified foods or multivitamins. Vitamin D_3 has little or no biological activity, but must be metabolically activated before it can function in its target tissues. In the liver, vitamin D_3 undergoes a selective enzyme-catalyzed oxidation at carbon-25 of the side chain to form 25-hydroxyvitamin D_3. This oxidation corresponds to the conversion of a C—H bond to a C—OH group, and the enzyme catalyzing the oxidation is given the trivial name of hydroxylase. Although 25-hydroxyvitamin D_3 is the most abundant form in the circulatory system, it has only modest biological activity, and undergoes further oxidation in the kidneys to form 1,25-dihydroxyvitamin D_3, the hormonally active form of the vitamin. Notice that the first oxidation of the activation process takes place in the liver; the second oxidation takes place in the kidneys.

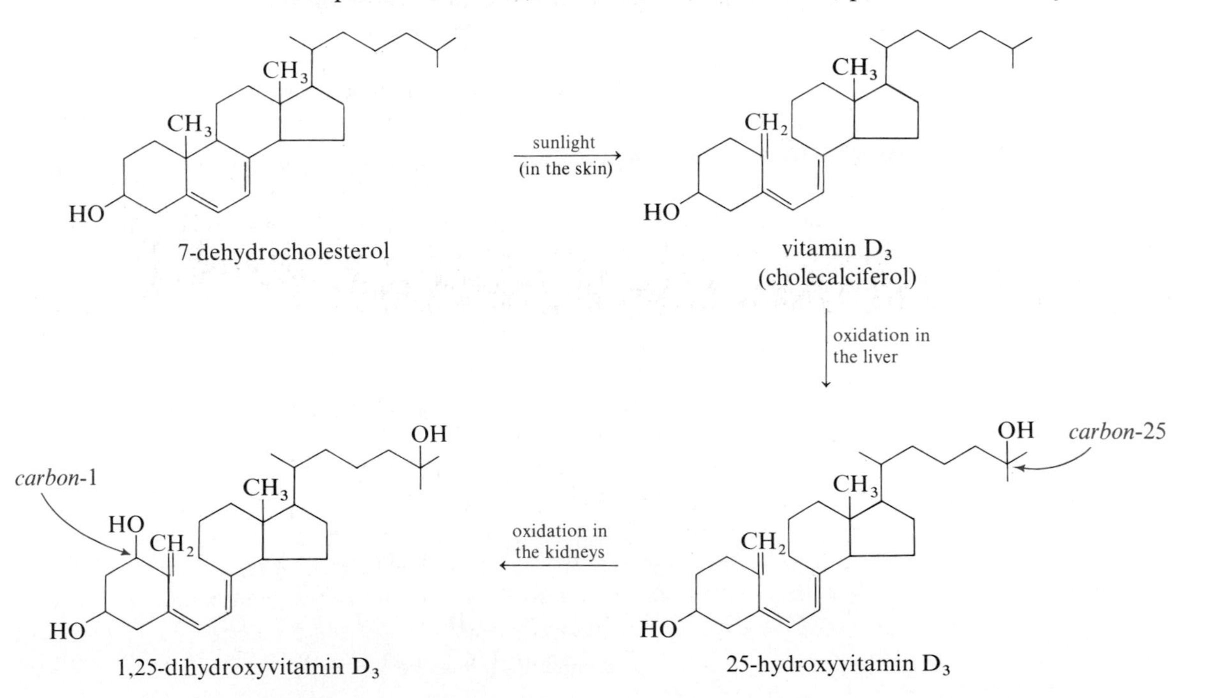

The principal function of vitamin D metabolites is to regulate calcium metabolism. 1,25-Dihydroxyvitamin D_3 acts in the small intestine to facilitate absorption of calcium and phosphate ions; it acts in the kidneys to stimulate reabsorption of filtered calcium ions; and it acts in bone to dissolve bone and release calcium and phosphate ions into the bloodstream. A deficiency of vitamin D in childhood is associated with rickets, a mineral-metabolism disease that leads to bowlegs, knock-knees, and enlarged joints. Adults with damaged kidneys or liver often suffer severe demineralization of bones. Several different methods are used in the clinical chemistry laboratory to measure the concentration of vitamin D metabolites in the blood. The most commonly used method measures the concentration of 25-hydroxyvitamin D_3. Normal values for humans of this substance are approximately 20 nanograms per mL of plasma.

Vitamin E is actually a group of about seven compounds of similar structure. Of these, α-tocopherol has the greatest potency.

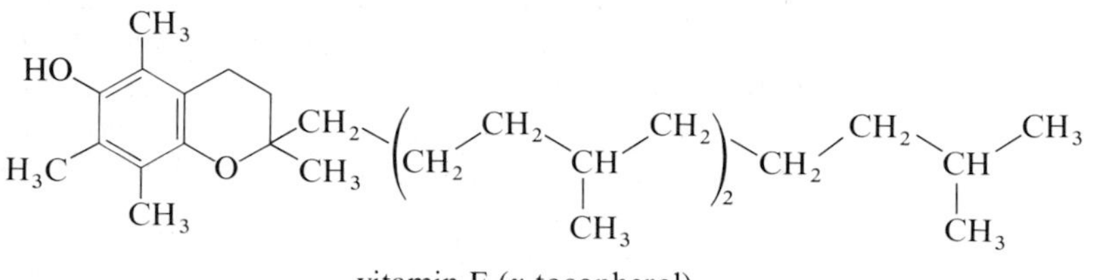

vitamin E (α-tocopherol)

Vitamin E (tocopherol) was first recognized in 1922 as a dietary substance essential for normal reproduction in rats. Its name comes from Greek, *tocopherol*, "promoter of childbirth." Vitamin E occurs in fish oil, in other oils such as cottonseed and peanut oil, and in leafy green vegetables. The richest source of vitamin E is wheat germ oil. In the body, vitamin E functions as an antioxidant in that it inhibits the oxidation of unsaturated lipids by molecular oxygen. In addition, it is necessary for the proper development and functioning of membranes in red blood cells, muscle cells, etc.

Vitamin K was discovered in 1935 as a result of a study of newly hatched chicks that had a fatal disease in which the blood was slow to clot. This condition could be prevented and cured by the administration of a substance found in hog liver and in alfalfa. It was later discovered that the delayed clotting time of the blood was caused by a deficiency of prothrombin, and it is now known that vitamin K is essential for the synthesis of prothrombin in the liver. The natural vitamin has a long, branched alkyl chain of usually 20 to 30 carbon atoms.

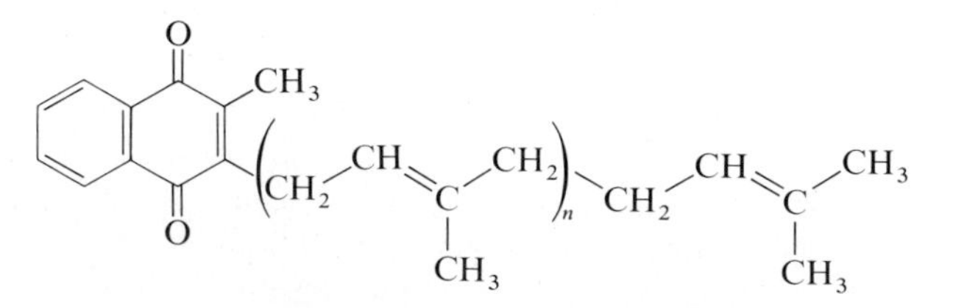

vitamin K_2 (*n* may be 5, 6, or 8)

The natural vitamins of the K family have for the most part been replaced by synthetic preparations. Menadione, one such synthetic material with vitamin-K activity, has only hydrogen in the place of the alkyl chain.

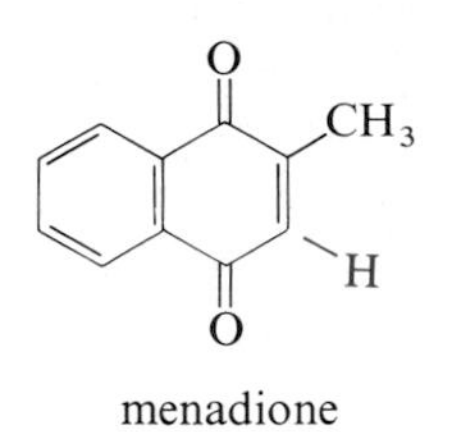

menadione

13.5 STEROIDS

All steroids contain four fused carbon rings: three rings of six carbon atoms and one ring of five carbon atoms. These 17 carbon atoms make up the structural unit known as the steroid nucleus. Figure 13.3 shows both the numbering system and the letter designation for the steroid nucleus.

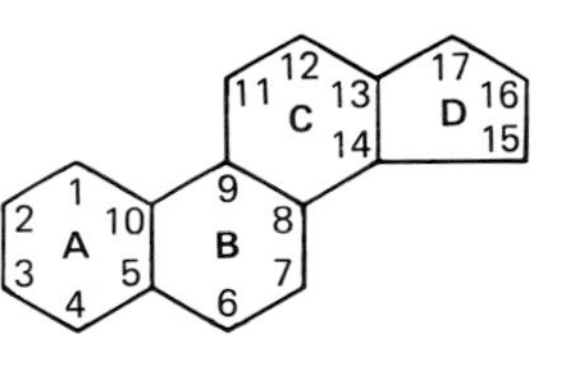

Figure 13.3 *The steroid nucleus.*

The steroid nucleus is found in a number of extremely important biomolecules. For our discussion, we will divide these into four groups: cholesterol, the adrenocorticoid hormones, the sex hormones, and bile acids.

A. CHOLESTEROL

Cholesterol is a white, water-insoluble substance found in varying amounts in practically all living organisms except bacteria. In animal cells it serves as (1) an essential component of membrane structures, and (2) the precursor of bile acids, steroid hormones, and vitamin D. In humans, the central and peripheral nervous systems have a very high cholesterol content (about 10% of dry brain weight). Human plasma contains an average of 50 mg of free cholesterol per 100 mL and about 170 mg of cholesterol esterified with fatty acids. Gallstones are almost pure cholesterol.

Since it is relatively easy to measure the concentration of cholesterol in serum, a great deal of information has been collected in attempts to correlate serum levels with various diseases. One of these diseases, arteriosclerosis, or hardening of the arteries, is among the most common diseases of aging. With increasing age, humans normally develop decreased capacity to metabolize

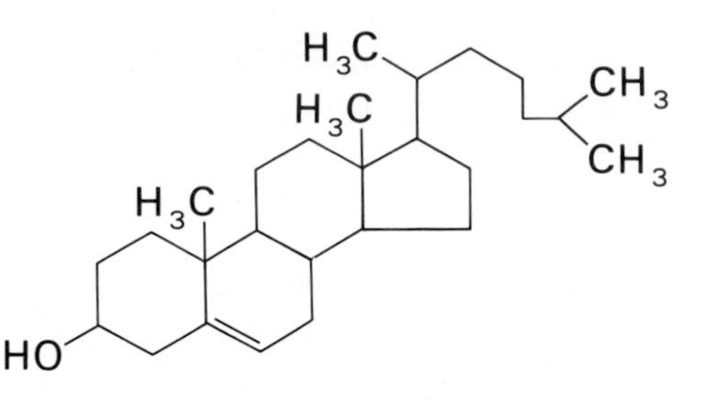

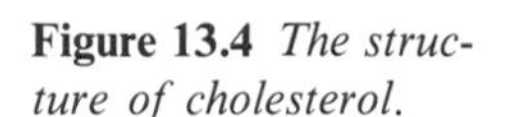

Figure 13.4 *The structure of cholesterol.*

fat, and therefore cholesterol concentration in tissues increases. When arteriosclerosis is accompanied by build-up of cholesterol and other lipids on the inner surfaces of arteries, the condition is known as atherosclerosis and results in a decrease in the diameter of the channels through which blood must flow. This decreased diameter, together with increased turbulence, leads to a greater probability of clot formation within the channel. If the channel is blocked by a clot, cells may be deprived of oxygen and die. The death of tissue in this way is called infarction.

Infarction can occur in many different tissues, and the clinical symptoms depend upon which vessels and tissues are involved. Myocardial infarction, which is the most common, involves the arteries of the heart.

B. ADRENOCORTICOID HORMONES

The cortex of the adrenal gland synthesizes several hormones that affect (1) water and electrolyte balance, and (2) carbohydrate and protein metabolism. They are called adrenocorticoid hormones because they are synthesized in the adrenal cortex. Those that control mineral balance are called mineralocorticoids; those that control glucose and carbohydrate balance are called glucocorticoid hormones. Both groups of hormones are derived from cholesterol and have the four-ring nucleus common to all steroids.

Aldosterone (Figure 13.5) is the most effective mineralocorticoid hormone secreted by the adrenal cortex. This hormone acts on kidney tubules to stimulate

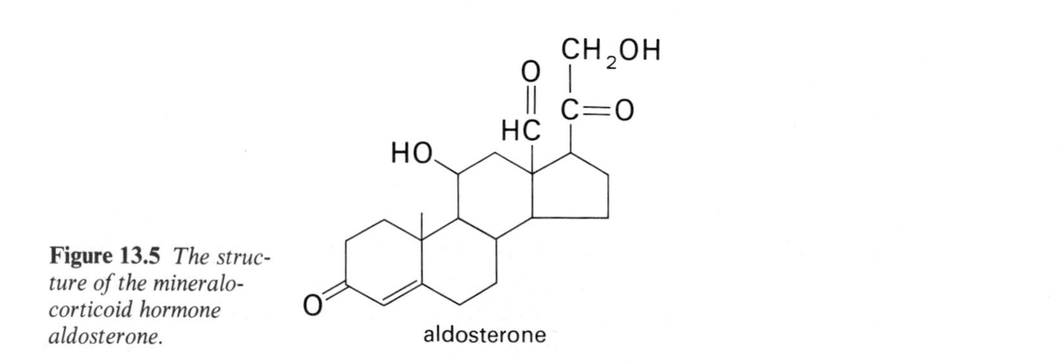

Figure 13.5 *The structure of the mineralocorticoid hormone aldosterone.*

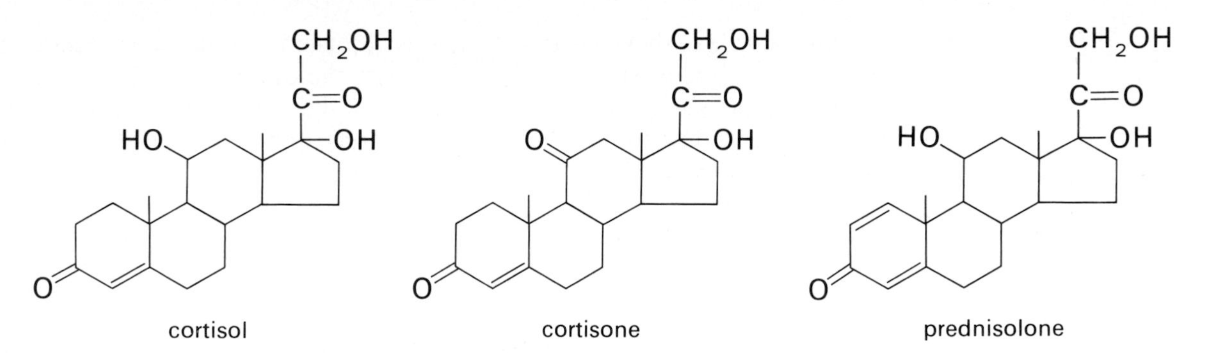

Figure 13.6 *Some glucocorticoid hormones.*

the resorption of sodium ions, and thus regulates water and electrolyte metabolism. An adult on a diet of normal sodium content secretes about 0.1 mg per day of aldosterone.

Cortisol (Figure 13.6) is the principal glucorticoid hormone of the adrenal cortex, which secretes about 25 mg of this substance per day. Cortisol affects the metabolism of carbohydrates, proteins, and fats; it affects water and electrolyte balance; and it affects inflammatory processes within the body. In the presence of cortisol, the synthesis of protein in muscle tissue is depressed, protein degradation is increased, and there is an increase in the supply of free amino acids in both muscle cells and blood plasma. The liver, in turn, is stimulated to use the carbon skeletons of certain amino acids for the synthesis of glucose and glycogen. Thus, cortisol and other glucocorticoid hormones act to increase the supply of glucose and liver glycogen at the expense of body protein. Cortisol also has some mineralocorticoid action; that is, it promotes resorption of sodium ions and water retention by the tubules of the kidney. However, it is far less potent as a mineralocorticoid than is aldosterone.

Cortisol and its oxidation product, cortisone (Figure 13.6), are probably best known for their use in clinical medicine as remarkably effective anti-inflammatory agents. They are used in the treatment of a host of inflammatory diseases, including acute attacks of rheumatoid arthritis and bronchial asthma, and inflammations of the eye, colon, and other organs. Laboratory research has produced a series of semisynthetic steroid hormones (for example, prednisolone, Figure 13.6) that are even more potent than cortisone in treating inflammatory diseases. Many of these semisynthetic hormones have an additional advantage over cortisone in that they do not at the same time stimulate sodium retention and fluid accumulation.

C. SEX HORMONES

The testes in the male and ovaries in the female, besides producing spermatozoa or ova, also produce steroid hormones which control secondary sex characteristics, the reproductive cycle, and the growth and development of accessory reproductive organs.

Of the male sex hormones or androgens, testosterone is the most important (Figure 13.7). It is produced in the testes from cholesterol. The chief function of testosterone is to promote normal growth of the male reproductive organs and development of the characteristic deep voice, pattern of facial and body hair, and male musculature.

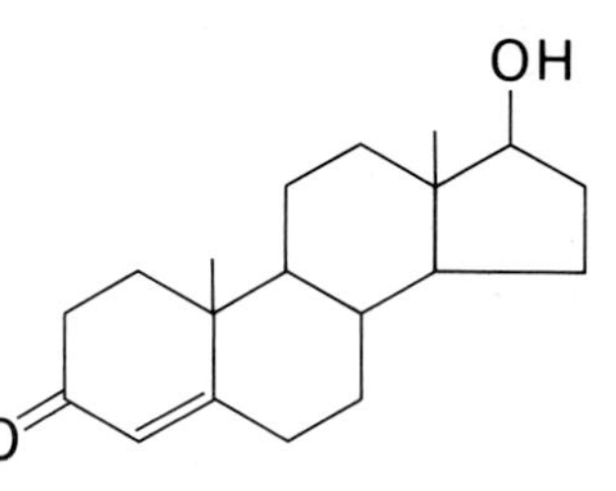

Figure 13.7 *Testosterone.*

In the female there are two types of sex hormones of particular importance, progesterone and a group of hormones known as estrogens (Figure 13.8). Changing rates of secretion of these hormones cause the periodic change in the ovaries and uterus known as the menstrual cycle. Immediately following menstrual flow, increased estrogen secretion causes growth of the lining of the uterus and ripening of the ovum. Estradiol is one of the most important estrogens, which are also responsible for development of the female secondary sex characteristics.

Progesterone is synthesized by the oxidation of cholesterol. Its secretion just prior to ovulation prevents other ova from ripening and also prepares the uterus for implantation and maintenance of a fertilized egg. If conception does not occur, progesterone production decreases and menstruation occurs. If fertilization and implantation do occur, production of progesterone continues and helps to maintain the pregnancy. One of the consequences of continued progesterone production is prevention of ovulation during pregnancy.

Once the role of progesterone in inhibiting ovulation was understood, its potential as a possible contraceptive drug was realized. Unfortunately, progesterone itself is relatively ineffective when taken orally and injection often produces local irritation. As a result of massive research programs, a large

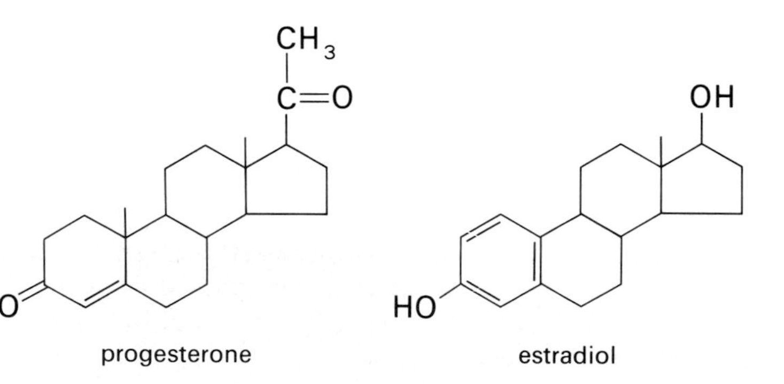

Figure 13.8 *Progesterone and estradiol, two female sex hormones.*

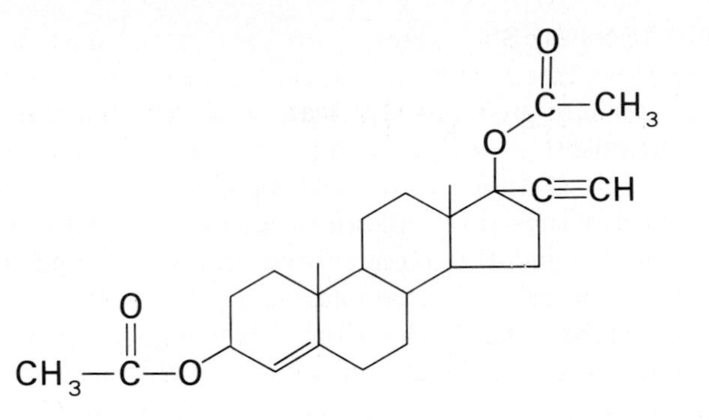

Figure 13.9 *Ethynodiol diacetate, a progesterone analog widely used in oral contraceptive preparations.*

number of synthetic steroids that could be administered orally became available in the early 1960s. When taken regularly, these drugs prevent ovulation, yet allow most women a normal menstrual cycle. Some of the most effective contain a progesterone-like analog such as ethynodiol diacetate (Figure 13.9) combined with a smaller amount of an estrogen-like material. The small amount of estrogen prevents irregular menstrual flow ("break-through bleeding") during prolonged use of contraceptive pills.

D. BILE ACIDS

Bile acids are synthesized in the liver from cholesterol and then stored in the gallbladder. During digestion, the gallbladder contracts and supplies bile to the small intestine by way of the bile duct. The primary bile acid in humans is cholic acid (Figure 13.10).

Bile acids have several important functions. First, they are products of the breakdown of cholesterol and thus are a major pathway for the elimination of cholesterol from the body via the feces. Second, because they are able to emulsify fats in the intestine, bile acids aid in the digestion and absorption of dietary fats. Third, they can dissolve cholesterol by the formation of cholesterol–bile salt micelles or cholesterol–lecithin–bile salt micelles. In this way cholesterol, whether it is from the diet, synthesized in the liver, or removed from circulation by the liver, can be solubilized.

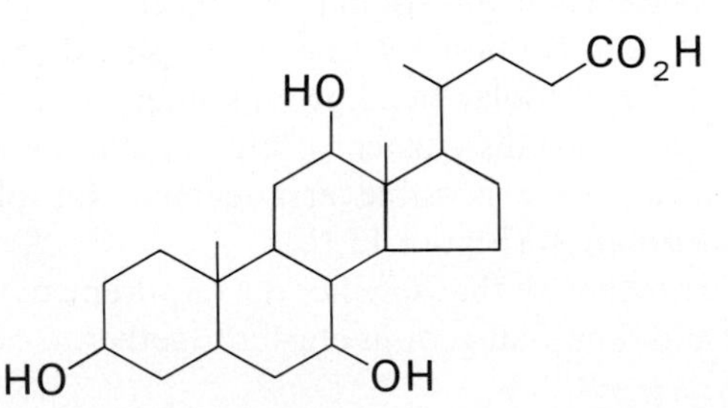

Figure 13.10 *Cholic acid, an important constituent of human bile.*

13.6 BIOLOGICAL MEMBRANES

Membranes are an important feature of cell structure and are vital for all living organisms. Some of the most important functions of membranes can be illustrated by considering the cell membrane. First, the cell membrane is a mechanical barrier that separates the contents of a cell from its external environment. Second, the cell membrane controls the passage of substances into and out of the cell. For example, essential nutrients are transported into the cell and metabolic wastes out of the cell through the membrane. Cell membranes also help to regulate the concentrations of molecules and ions within the cell. Third, the cell membrane provides structural support for certain proteins. Some of these proteins are "receptors" for hormone-carried messages; others are specific enzyme complexes.

Obviously, membranes are more than impervious, mechanical barriers separating the cell and its organelles from the external environment. They are highly specialized structures that perform a multitude of tasks with great precision and accuracy.

We know that the membranes of plant and animal cells are typically composed of 40 to 50% phospholipid and 50 to 60% protein. Yet there are wide variations in phospholipid–protein content even between different types of cells within the same organism. For example, myelin, the membrane that surrounds specific types of nerve cells and serves as an insulator, contains only about 18% protein. Membranes that are active in transporting specific molecules into and out of cells contain about 50% protein. Membranes actively involved in the transformation of energy, such as those of mitochondria and chloroplasts, contain up to 75% protein.

The question of the detailed molecular structure of membranes is one of the most challenging problems in biochemistry today. Despite intensive research, many aspects of membrane structure and activity still are not understood. Before we discuss a model for membrane structure, let us first consider the shapes of phospholipid molecules and the organization of phospholipid molecules in aqueous solution.

Shown in Figure 13.11 is a structural formula and space-filling model of a lecithin, a major type of membrane phospholipid.

Lecithin and other phospholipids are elongated, almost rod-like molecules, with the nonpolar (hydrophobic) hydrocarbon chains lying essentially parallel to one another and with the polar (hydrophilic) phosphate ester group pointing in the opposite direction.

What happens when phospholipid molecules are placed in an aqueous medium? Recall from Section 8.7 that when placed in water, soap molecules form micelles in which polar head groups interact with water molecules and nonpolar hydrocarbon tails cluster within the micelle and are removed from contact with water. One possible arrangement for phospholipids in water also is micelle formation (Figure 13.12).

Another arrangement that satisfies the requirement that polar groups interact with water and nonpolar groups cluster together to exclude water is the lipid

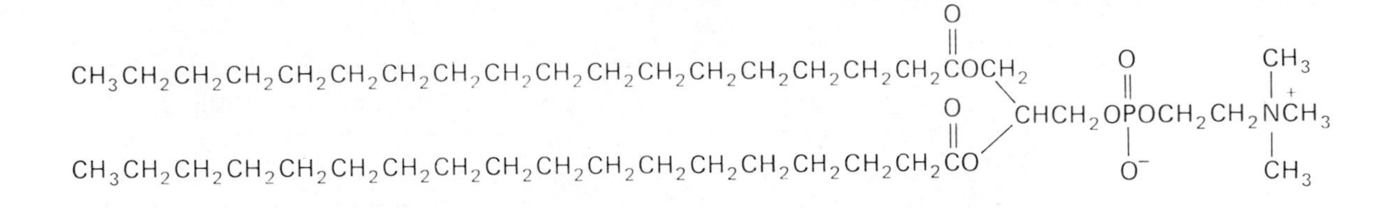

(a)

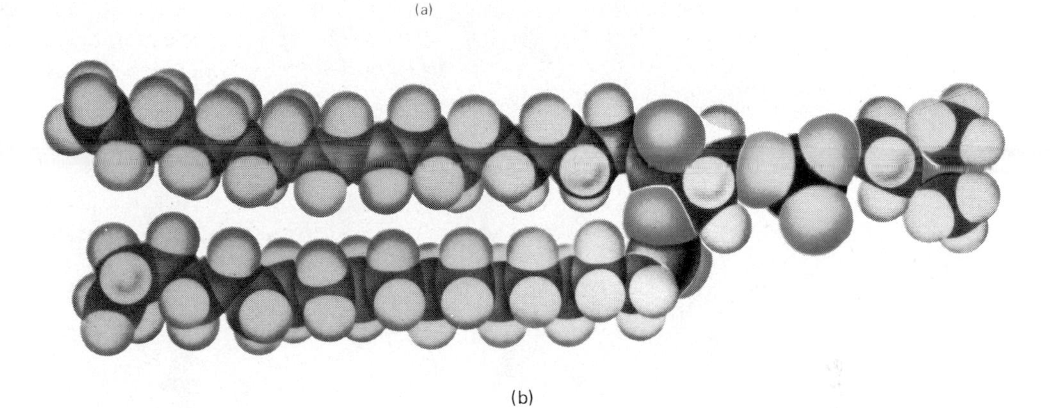

(b)

Figure 13.11 *A lecithin. (a) Structural formula. (b) Space filling-model.*

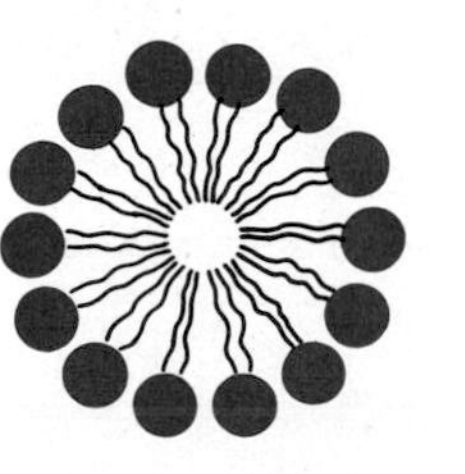

Figure 13.12 *Micelle formation of phospholipids in an aqueous medium.*

bilayer. A schematic diagram of a lipid bilayer is shown in Figure 13.13. The favored structure for phospholipids in aqueous solution is the lipid bilayer rather than the micelle because micelles can only grow to a limited size before holes begin to appear in the outer polar surface. Lipid bilayers can grow to almost infinite extent and provide a boundary surface for a cell or organelle whatever its size.

It is important to realize that the self-assembly of phospholipid molecules into a bilayer is a spontaneous process driven by two types of noncovalent forces: (1) hydrophobic interactions, which result when the nonpolar hydrocarbon chains cluster together and exclude water molecules, and (2) electrostatic interactions and hydrogen bonding, which result when the polar head groups interact with water molecules.

As you might expect from their structural characteristics, lipid bilayers are highly impermeable to ions and most polar molecules, for it would take a great deal of energy to transport an ion or a polar molecule through the nonpolar interior of the bilayer. However, water readily passes in and out of the

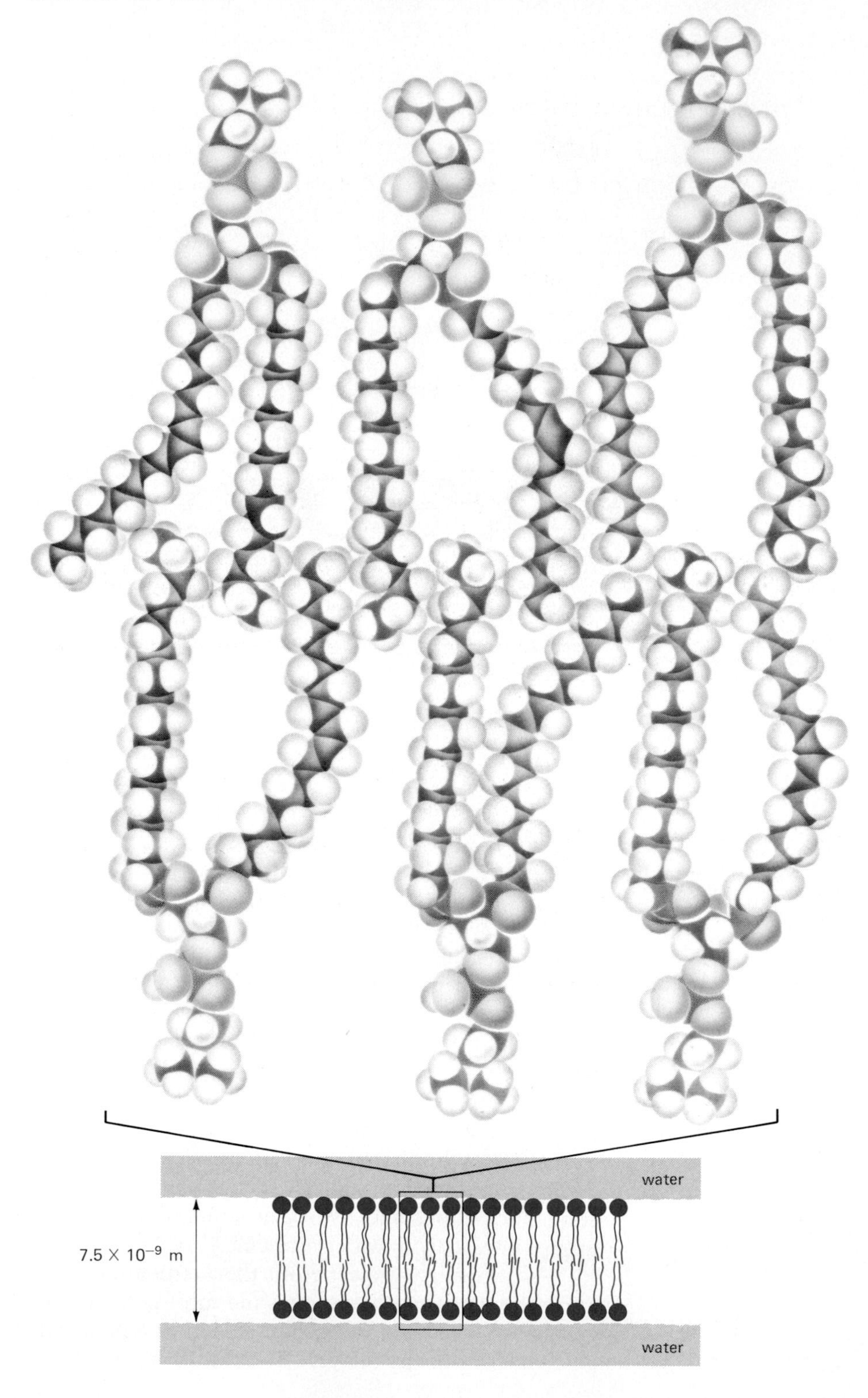

Figure 13.13 *A section of lipid bilayer (lower part). Enlarged (above) is a section of six phospholipid molecules in the bilayer. Note in the enlargement that 50% of the hydrocarbon chains contain unsaturation.*

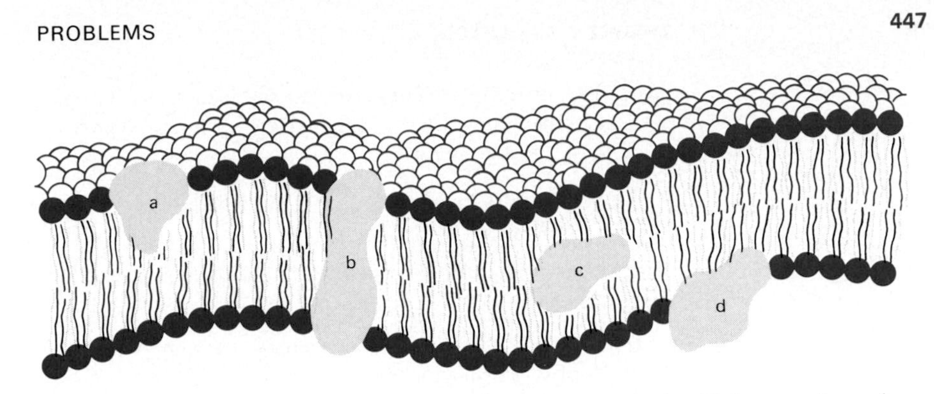

Figure 13.14 *Fluid-mosaic model of a membrane, showing the lipid bilayer and membrane proteins oriented (a) on the outer surface, (b) penetrating the entire thickness of the membrane, (c) embedded within the membrane, and (d) on the inner surface of the membrane.*

lipid bilayer. Glucose passes through lipid bilayers 10,000 times more slowly than water, and sodium ion 1,000,000,000 times more slowly than water.

The most satisfactory current model for the arrangement of proteins and phospholipids in plant and animal membranes is the fluid-mosaic model. According to this model, the membrane phospholipids form a lipid bilayer and membrane proteins are imbedded in this bilayer. Some proteins are exposed to the aqueous environment on the outer surface of the membrane; others provide channels that penetrate from the outer to the inner surface of the membrane; while still others are imbedded within the lipid bilayer. Four possible protein arrangements are shown schematically in Figure 13.14.

The fluid-mosaic model is consistent with the evidence provided by chemical analysis and electron microscope pictures of cell membranes. However, this model does not explain just how membrane proteins act as pumps and gates for the transport of ions and molecules across the membrane, or how they act as receptors for hormone-borne messages and communications between one cell and another. Nor does it explain how enzymes bound on membrane surfaces catalyze the reactions they do. All of these questions are very active areas of research today.

PROBLEMS

13.2 List six major functions of lipids in the human body. Name and draw a structural formula for a lipid representing each function.

13.3 How many isomers (including stereoisomers) are possible for a triglyceride containing one molecule each of palmitic, stearic, and oleic acid.

13.4 What is meant by the term "hardening" as applied to fats and oils?

13.5 Saponification is the alkaline hydrolysis of naturally occurring fats and oils. A saponification number is the number of mg of potassium hydroxide required to saponify 1g of a fat or oil. Calculate the saponification number of tristearin, molecular weight 890.

13.6 The saponification number of butter is approximately 230; that of oleomargarine is approximately 195. Calculate the average molecular weight of butter fat and of oleomargarine.

13.7 The percentage of unsaturated fatty acids in butter fat is approximately 35%. Compare this with the percentage of unsaturated fatty acids in corn oil, soybean oil, and wheat germ oil.

13.8 Draw a structural formula for a phospholipid containing serine; a phospholipid containing inositol.

13.9 Draw structural formulas for the products of complete hydrolysis of a lecithin; a cephalin.

13.10 Examine the structural formula of vitamin A and state the number of *cis-trans* isomers possible for this molecule.

13.11 In fish liver oils, vitamin A is present as esters of fatty acids. The most common of these esters is vitamin A palmitate. Draw a structural formula for this substance.

13.12 Describe the symptoms of severe vitamin A deficiency.

13.13 Examine the structural formulas of vitamins A, D_3, E, and K_2. Based on their structural formulas, would you expect them to be more soluble in water or in olive oil? Would you expect them to be soluble in blood plasma?

13.14 Explain why vitamin E is added to some processed foods.

13.15 Draw the structural formula of cholesterol, label all chiral carbons, and state the total number of stereoisomers possible for this structural formula.

13.16 Esters of cholesterol and fatty acids are normal constituents of blood plasma. The fatty acids esterified with cholesterol are generally unsaturated. Draw the structural formula for cholesteryl oleate.

13.17 Cholesterol is an important component of the lipid fraction of cell membranes. How do you think a cholesterol molecule might be oriented in a biological membrane according to the fluid-mosaic model?

13.18 Name the six functional groups in cortisol; in aldosterone.

13.19 Examine the structural formulas of testosterone, a male sex hormone, and progesterone, a female sex hormone. What are the similarities in structure between the two? the differences?

13.20 Why are progesterone and the estrogens called "sex hormones"?

13.21 Describe how a combination of progesterone and estrogen analogs functions as an oral contraceptive.

13.22 Examine the structural formula of cholic acid and account for the fact that this and other bile acids are able to emulsify fats and oils.

13.23 Two of the major noncovalent forces directing the organization of biomolecules in aqueous solution are the tendencies to (1) arrange polar groups so that they interact with water by hydrogen bonding, and (2) arrange nonpolar groups so that they are shielded from water. Show how these forces direct micelle formation by soap molecules, folding of globular proteins, and lipid bilayer formation by phospholipids.

13.24 Describe the major features of the fluid-mosaic model of the structure of biological membranes.

CHAPTER 14

Nucleic Acids

Nucleic acids are a third great class of biopolymers which, like proteins and polysaccharides, are vital components of living materials. In this chapter we will look at the structure of nucleosides and nucleotides and the manner in which these small building blocks are bonded together to form giant nucleic acid molecules. Then we will consider the three-dimensional structure of nucleic acids. Finally, we will examine the manner in which genetic information coded on deoxyribonucleic acids is expressed in protein biosynthesis.

14.1 THE COMPONENTS OF DEOXYRIBONUCLEIC ACID (DNA)

Controlled hydrolysis breaks DNA molecules into three components: (1) phosphoric acid; (2) 2-deoxy-D-ribose; and (3) four heterocyclic aromatic amine bases. The heterocyclic bases fall into two classes, those derived from pyrimidine (cytosine and thymine), and those derived from purine (adenine and guanine).

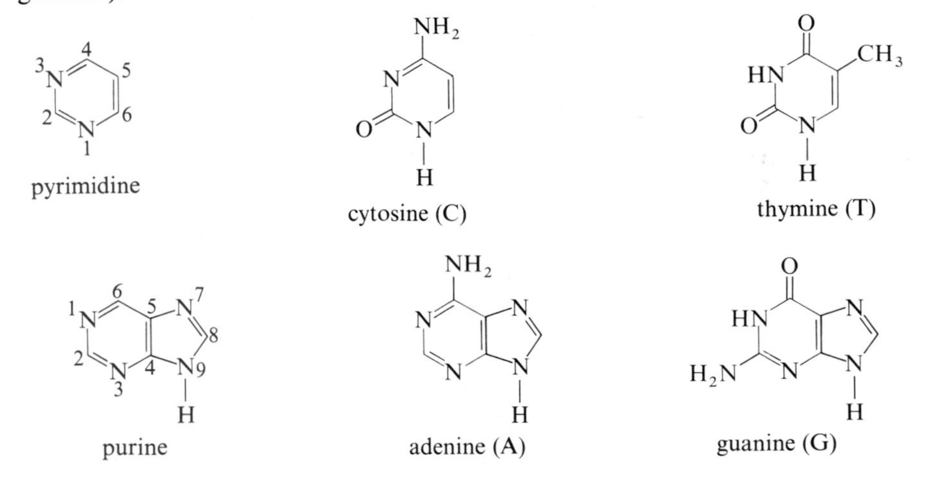

14.2 NUCLEOSIDES

A nucleoside is a glycoside in which nitrogen-9 of a purine base or nitrogen-1 of a pyrimidine base is bonded to 2-deoxyribose by a β-N-glycoside bond (Section 11.7C). Two nucleosides, 2′-deoxyadenosine and 2′-deoxycytidine, are shown in Figure 14.1.

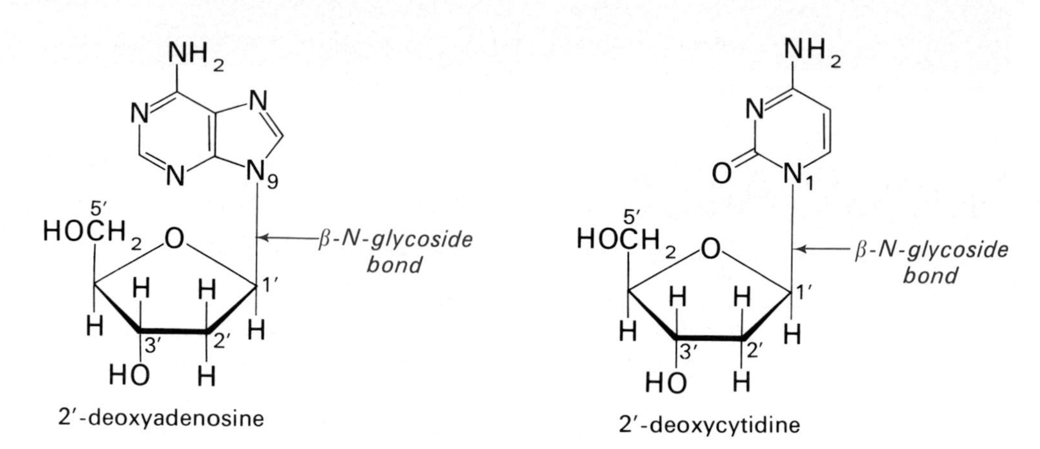

Figure 14.1 *Nucleosides: 2′-deoxyadenosine and 2′-deoxycytidine. Unprimed numbers are used to designate atoms of the purine and pyrimidine bases; primed numbers are used to designate atoms of 2-deoxyribose.*

The other two nucleosides found in DNA are 2′-deoxythymidine and 2′-deoxyguanosine.

14.3 NUCLEOTIDES

A nucleotide is a nucleoside monophosphate ester in which a molecule of phosphoric acid is esterified with one of the free hydroxyl groups of 2-deoxyribose. Nucleoside monophosphates are illustrated in Figure 14.2 by the

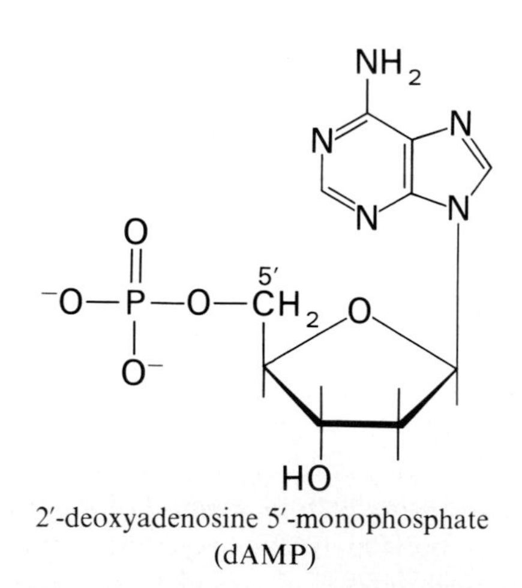

2′-deoxyadenosine 5′-monophosphate (dAMP)

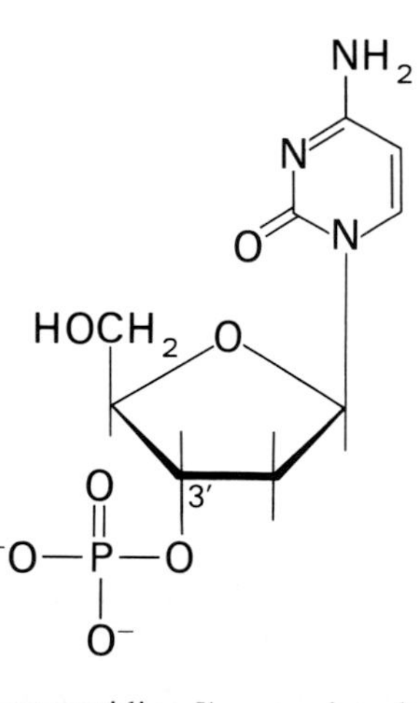

2′-deoxycytidine 5′-monophosphate (dCMP)

Figure 14.2 *Nucleoside monophosphates.*

Table 14.1 The major mononucleotides derived from DNA. Each is named as a monophosphate; as an acid; and by a four-letter abbreviation.

deoxyadenosine monophosphate,	deoxyadenylic acid,	dAMP
deoxyguanosine monophosphate,	deoxyguanylic acid,	dGMP
deoxycytidine monophosphate,	deoxycytidylic acid,	dCMP
deoxythymidine monophosphate,	deoxythymidylic acid,	dTMP

5′-monophosphate ester of 2′-deoxyadenosine and the 3′-monophosphate ester of 2′-deoxycytidine. Note that at pH 7.0, the two protons of the phosphate ester are ionized giving this group a net charge of -2.

Mononucleotides are named either as phosphate esters (e.g., deoxyadenosine 5′-monophosphate); as acids (e.g., deoxyadenylic acid); or by using four-letter abbreviations (e.g., dAMP). In the four-letter abbreviations for mononucleotides, the letter d indicates 2-deoxy-D-ribose, the second letter indicates the nucleoside, and the third and fourth letters indicate that the substance is a monophosphate (MP) ester. Table 14.1 lists names of the major mononucleotides derived from DNA.

All nucleoside monophosphates may be further phosphorylated to form nucleoside diphosphates and nucleoside triphosphates. In the case of the diphosphates and triphosphates, the second and third phosphate groups are joined by anhydride bonds.

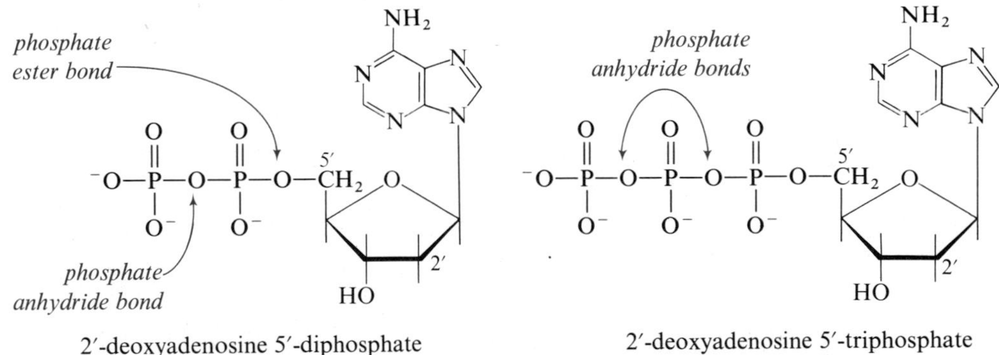

2′-deoxyadenosine 5′-diphosphate (dADP)

2′-deoxyadenosine 5′-triphosphate (dATP)

At pH 7.4, all protons of the diphosphate and triphosphate groups are fully ionized giving these groups net charges of -3 and -4, respectively.

Example 14.1

Draw structural formulas for the following compounds.

(a) 2′-deoxycytidine 5′-monophosphate (dCMP)

(b) 2′-deoxyguanosine 5′-triphosphate (dGTP)

Solution

(a) Cytosine is joined by a β-N-glycoside bond between N-1 of cytosine and carbon-1 of the cyclic hemiacetal form of 2-deoxy-D-ribose. The 5′-hydroxyl of the pentose is bonded to phosphate by an ester bond.

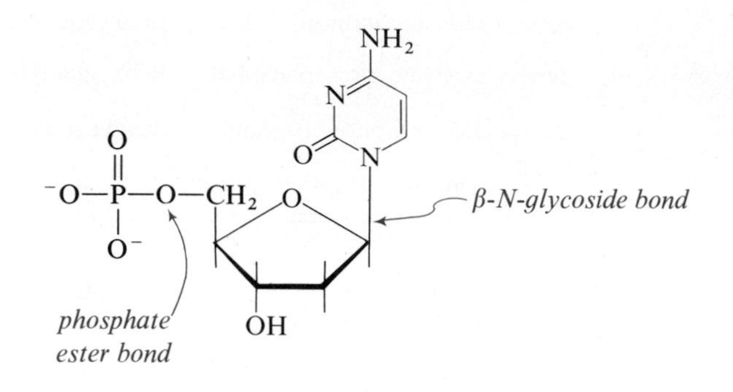

(b) Guanine is joined by a β-N-glycoside bond between N-9 of guanine and carbon-1 of the cyclic hemiacetal form of 2-deoxy-D-ribose. The 5′-hydroxyl group of the pentose is joined to three phosphate groups by a combination of one ester bond and two anhydride bonds.

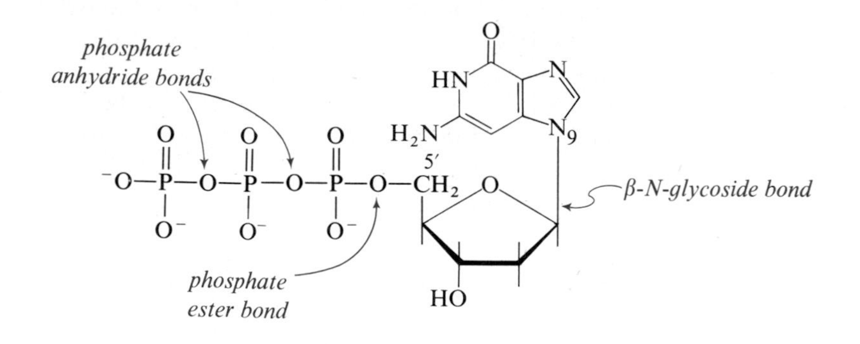

PROBLEM 14.1 Draw structural formulas for
(a) dTPP **(b)** dGMP

14.4 THE STRUCTURE OF DNA

Deoxyribonucleic acid (DNA) consists of a backbone of alternating units of deoxyribose and phosphate in which the 3′-hydroxyl of one deoxyribose is joined to the 5′-hydroxyl of the next deoxyribose by a phosphodiester bond (Figure 14.3). This backbone is constant throughout the entire DNA molecule. A heterocyclic base, either adenine, guanine, thymine, or cytosine, is attached to each deoxyribose by a β-N-glycoside bond.

The sequence of bases in the DNA molecule is indicated by the first letter abbreviation of each base beginning from the free 5′-hydroxyl end of the chain.

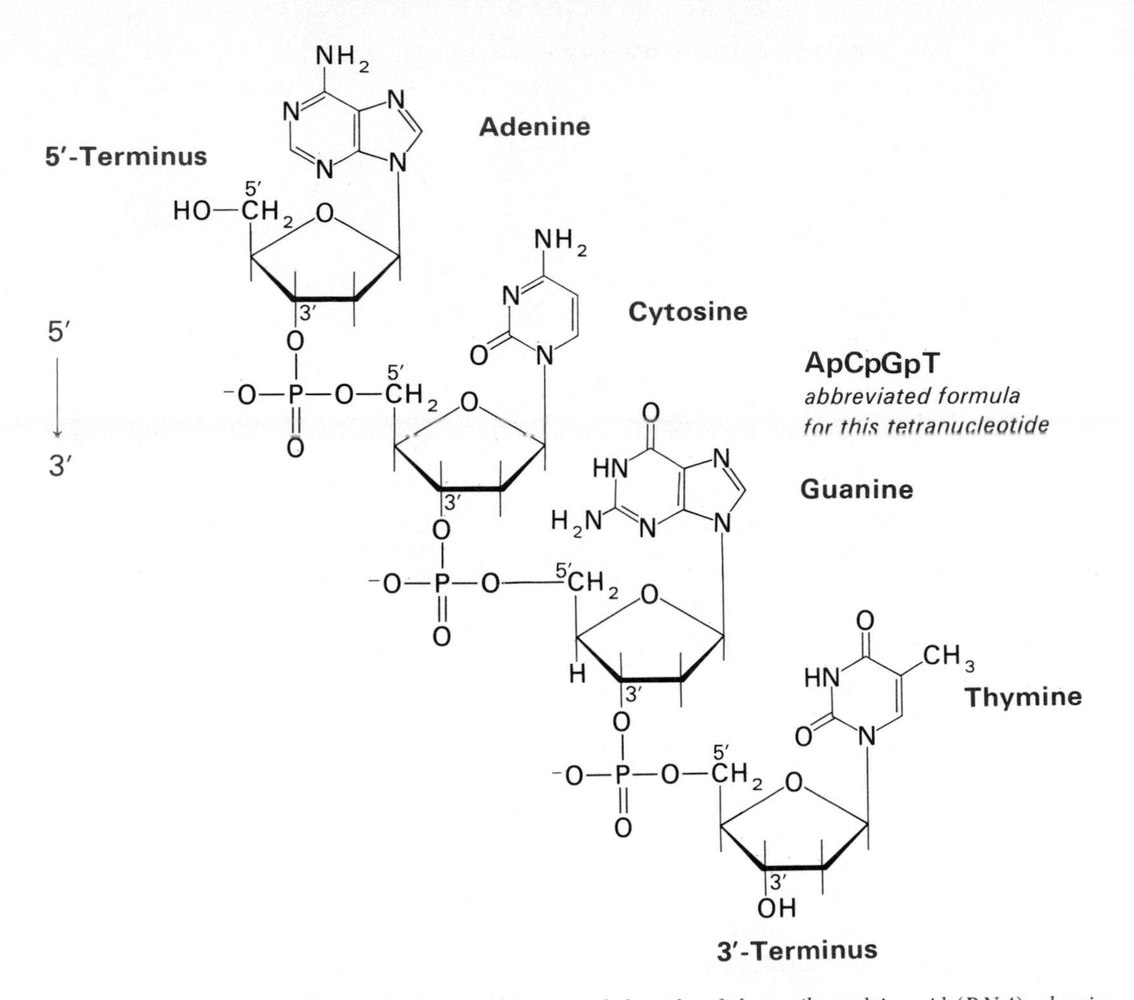

Figure 14.3 *Partial structural formula of deoxyribonucleic acid (DNA), showing a tetranucleotide sequence. In the abbreviated sequence (right), the bases of the tetranucleotide are read from the 5′ end of the chain to the 3′ end, as, indicated by the arrows (far left).*

According to this convention, the base sequence of the section of DNA shown in Figure 14.3 is written ApCpGpT where the letter p indicates the phosphodiester bonds in the backbone of the molecule. Alternatively, the symbol for the phosphodiester bond can be omitted and the base sequence written more simply as ACGT.

Example 14.2 Draw the structural formula of pApC.

Solution The first symbol in the shorthand formula of this dinucleotide is p indicating that the 5′-hydroxyl is bonded to phosphate by an ester bond. The last symbol is C which shows that the 3′-hydroxyl is free, i.e., it is not esterified with phosphate.

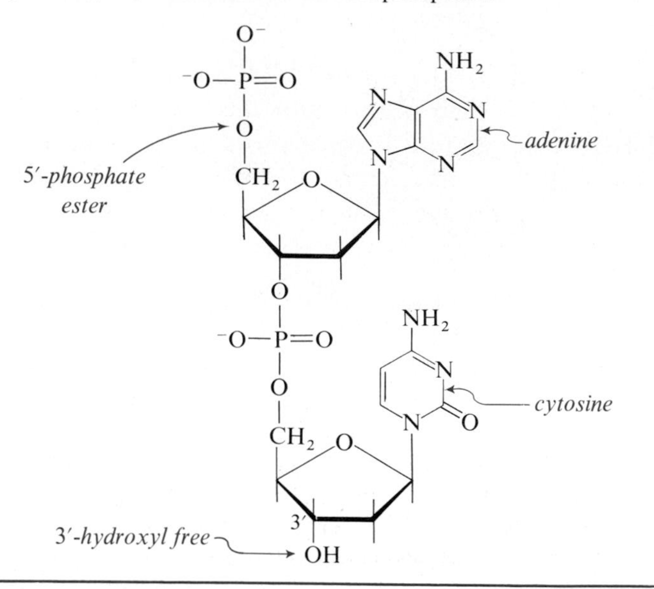

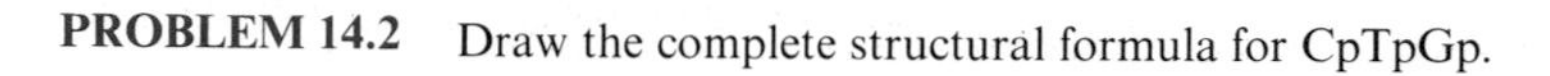

PROBLEM 14.2 Draw the complete structural formula for CpTpGp.

By 1950, it was clear that DNA molecules consist of chains of alternating units of deoxyribose and phosphate linked by phosphodiester bonds and with a base attached to each deoxyribose by a β-N-glycoside bond. However, the precise sequence of bases along the chain of any particular DNA molecule was completely unknown. At one time, it was thought that the four major bases occurred in equal ratios and perhaps repeated in a regular pattern along the pentose-phosphate backbone of the molecule. However, more precise determinations of base composition (Table 14.2) revealed that the bases do not occur in equal ratios.

Table 14.2 Comparison of base composition (in mole percent) of DNAs from several organisms.

Organism	A	G	C	T	$\frac{A}{T}$	$\frac{G}{C}$	$\frac{\textit{purines}}{\textit{pyrimidines}}$
human	30.9	19.9	19.8	29.4	1.05	1.00	1.04
sheep	29.3	21.4	21.0	28.3	1.03	1.02	1.03
sea urchin	32.8	17.7	17.3	32.1	1.02	1.02	1.02
marine crab	47.3	2.7	2.7	47.3	1.00	1.00	1.00
yeast	31.3	18.7	17.1	32.9	0.95	1.09	1.00
E. coli	24.7	26.0	26.0	23.6	1.04	1.01	1.03

From consideration of data such as these, the following conclusions emerged.

(1) The mole percent base composition of DNA in any organism is the same in all cells and is characteristic of the organism.
(2) In nearly all DNAs, the mole percent of adenine equals that of thymine, and the mole percent of guanine equals that of cytosine.
(3) The total number of purine residues (A + G) equals the total number of pyrimidine residues (C + T).

Additional information on the structure of DNA emerged from analysis of X-ray diffraction photographs of DNA fibers taken by Rosalind Franklin and Maurice Wilkins. These photographs showed that DNA molecules are long, fairly straight, and not more than a dozen atoms thick. Furthermore, despite the fact that the base composition of DNAs isolated from different organisms varies over a rather wide range, the DNA molecules themselves are remarkably uniform in thickness. Herein lay one of the major problems to be solved. How could the molecular dimensions of DNA be so regular even though the relative percentages of the various bases differ so widely? There was also another problem to be solved: In what form is genetic information stored in DNA molecules, and how is it transmitted from one generation or cell to the next?

With this accumulated information, the stage was set for the development of a hypothesis about DNA conformation. In 1953, F. H. C. Crick, a British physicist, and James D. Watson, an American biologist, postulated a precise model of the three-dimensional structure of DNA. The model not only accounted for many of the observed physical and chemical properties of DNA, but also suggested a mechanism by which genetic information could be repeatedly and accurately replicated. Watson, Crick, and Wilkins shared the 1962 Nobel Prize in Physiology and Medicine for "their discoveries concerning the molecular structure of nucleic acids, and its significance for information transfer in living material."

The heart of the Watson–Crick model is the postulate that a molecule of DNA consists of two antiparallel polynucleotide strands coiled in a right-handed manner about the same axis to form a double helix. To account for the observed base ratios and the constant thickness of DNA, Watson and Crick postulated that purine and pyrimidine bases project inward toward the axis of the helix and are always paired in a very specific manner.

According to scale models, the dimensions of a thymine-adenine base pair are identical with those of a cytosine-guanine base pair, and the length of each pair is consistent with the thickness of the DNA strand (Figure 14.4). This fact gives rise to the principle of complementarity. In DNA, adenine is always paired by hydrogen bonding with thymine, i.e., adenine and thymine are complementary bases. Similarly, guanine and cytosine are complementary bases. A significant fact arising from Watson and Crick's model building is that no other base pairing is consistent with the observed thickness of a DNA

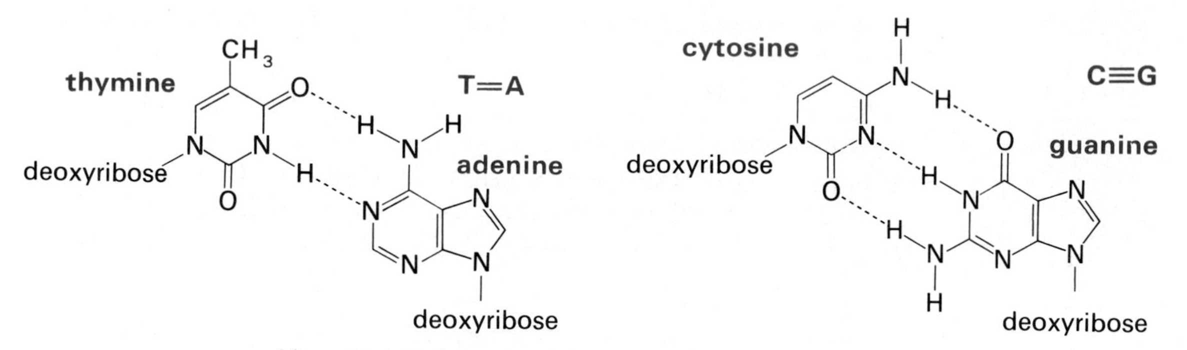

Figure 14.4 *Hydrogen-bonded interaction between thymine and adenine and between cytosine and guanine. The first couple is abbreviated as T=A (showing two hydrogen bonds) and the second couple is abbreviated as C≡G (showing three hydrogen bonds).*

molecule. A pair of pyrimidine bases is too small to account for the observed thickness of a DNA molecule, while a pair cf purine bases is too large. Thus, according to the Watson–Crick model, the repeating units in the double stranded DNA molecule are not single bases of differing dimensions, but specific base pairs of identical dimensions.

To account for the periodicity observed from X-ray data, Watson and Crick postulated that base pairs are stacked one on top of the other with a distance of 3.4×10^{-8} cm between base pairs and exactly ten base pairs are stacked in one complete turn of the helix. There is one complete turn of the helix every 34×10^{-8} cm (Figure 14.5).

Example 14.3

One chain of a double-stranded DNA molecule has the base sequence of —ACTTGCCA—. Write the base sequence for the complementary strand.

Solution

Remember that base sequence is always written from the 5′ end of the strand to the 3′ end. Remember also that A is always paired by hydrogen bonding with its complement T, and that G is always paired by hydrogen bonding with its complement C. In double-stranded DNA, the strands run in opposite (antiparallel) directions so that the 5′ end of one strand is associated with the 3′ end of the other strand. Hydrogen bonds between base pairs are shown by dashed lines.

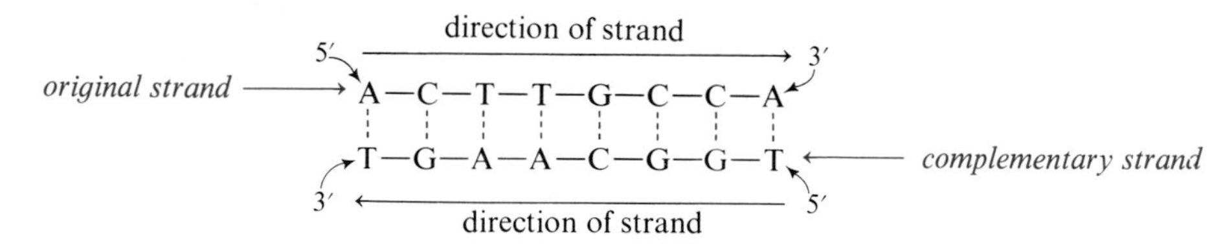

The complement of 5′—ACTTGCCA—3′ is shown under it in the solution. Writing this strand poses a communication problem. DNA strands are always written from the 5′ to 3′ end. Therefore, if the original strand is 5′—ACTTGCCA—3′, its complement is 5′—TGGCAAGT—3′.

PROBLEM 14.3 Write the complementary base sequence for

—C—C—G—T—A—C—G—A—.

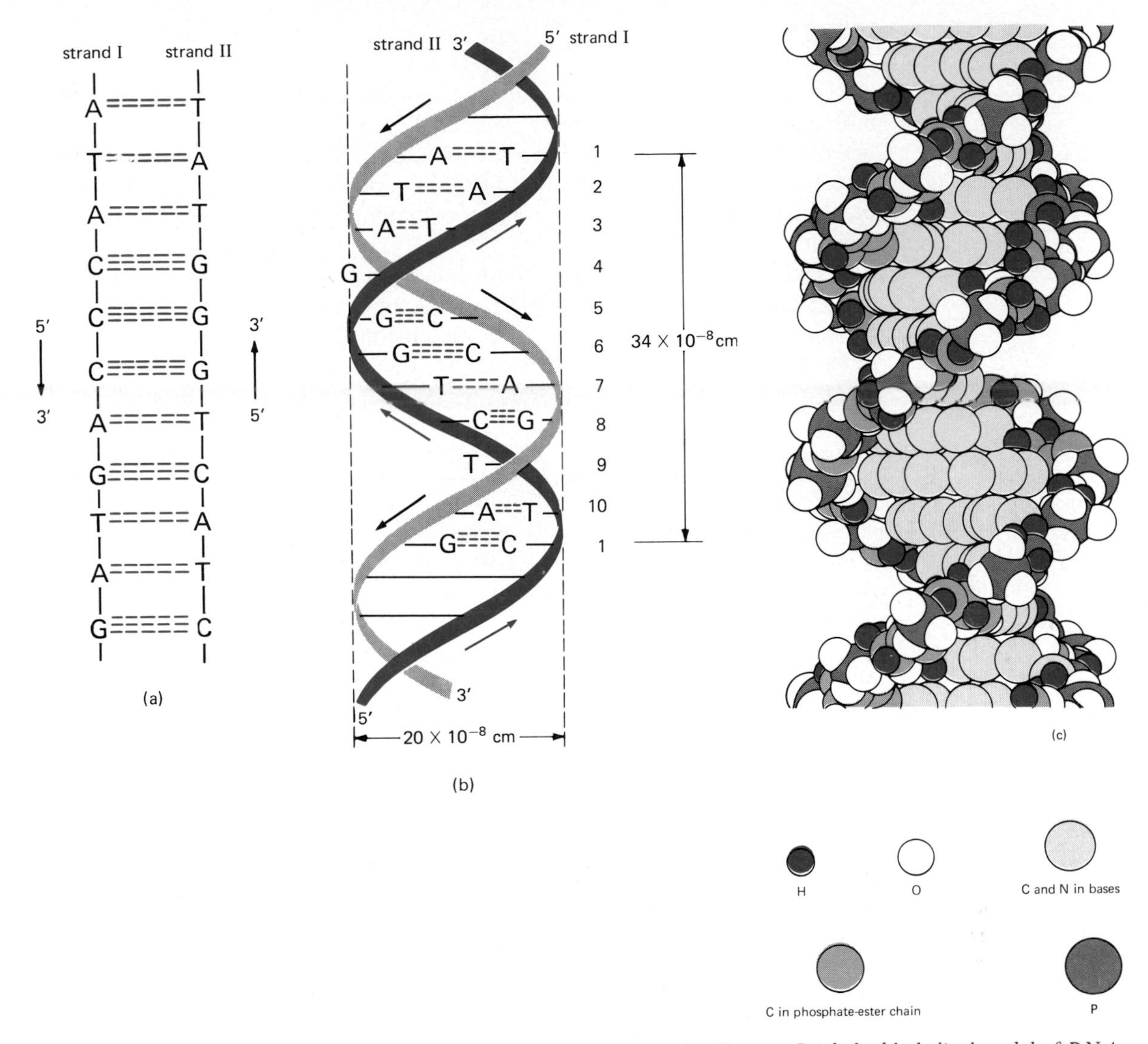

Figure 14.5 *Abbreviated representation of the Watson–Crick double-helical model of DNA. On the left are shown two complementary antiparallel polynucleotide strands and the hydrogen bonds between complementary base pairs. In the middle, the strands are twisted in a double helix of thickness 20×10^{-8} cm and a repeat distance of 34×10^{-8} cm along the axis of the double helix. There are 10 base pairs per complete turn of the helix. On the right is a space-filling model of a section of DNA double helix.*

14.5 DNA REPLICATION

At the time Watson and Crick proposed a model for the conformation of DNA, biologists had already amassed a great deal of evidence that DNA is in fact the hereditary material. Detailed studies revealed that during cell division, there is an exact copying or duplication of DNA. The challenge posed to molecular biologists was: How does the genetic material duplicate itself with such unerring fidelity?

One of the exciting things about the double helix model is that it immediately suggested how DNA might produce an exact copy of itself. The double helix consists of two parts, each the complement of the other. If the two strands separate, and each serves as a template for the construction of its own complement, then each new double strand is an exact replica of the original DNA. Because each new double-stranded DNA molecule contains one strand from the parent molecule and one newly synthesized strand, the process is called semiconservative replication (Figure 14.6).

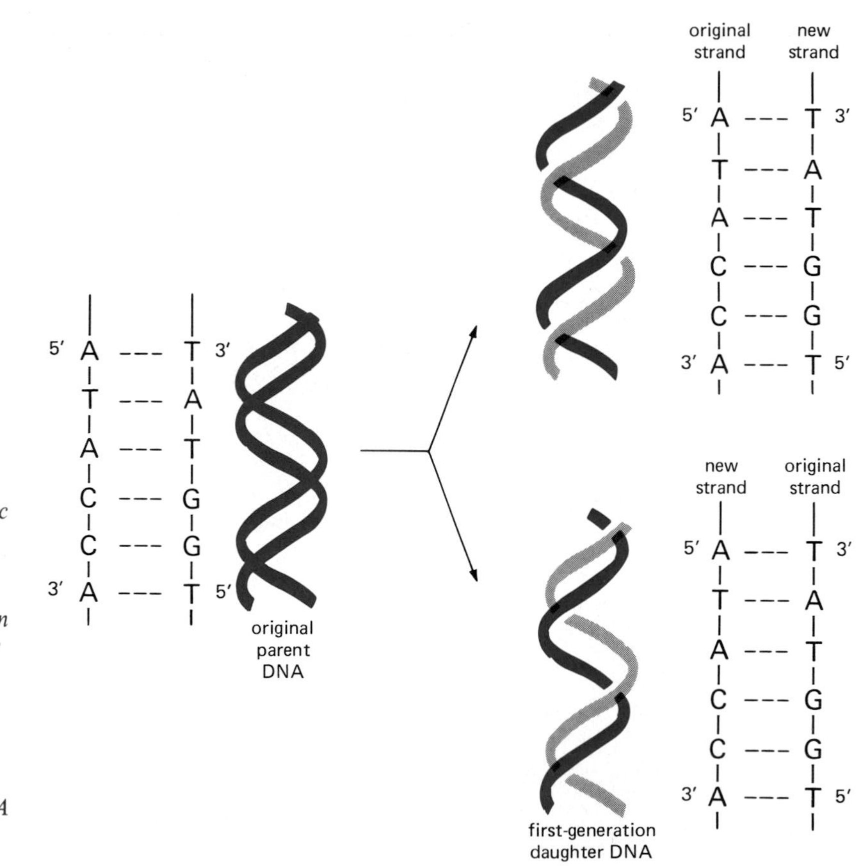

Figure 14.6 *Schematic diagram of semiconservative replication. The double helix uncoils, and each chain of the parent serves as a template for the synthesis of its complement. Each daughter DNA contains one strand from the original DNA and one newly synthesized strand.*

14.6 RIBONUCLEIC ACIDS (RNA)

Ribonucleic acids (RNAs) are similar to deoxyribonucleic acids in that they too consist of long, unbranched chains of nucleotides joined by phosphodiester bonds between the 3′-hydroxyl of one pentose and the 5′-hydroxyl of the next. Thus, their structure is much the same as that of DNA shown in Figure 14.3. However, there are three major differences in structure between RNA and DNA: (1) the pentose unit in RNA is D-ribose rather than 2-deoxy-D-ribose; (2) the pyrimidine bases in RNA are uracil and cytosine rather than thymine and cytosine; and (3) RNA is single-stranded rather than double-stranded. Following are structural formulas of D-ribose and uracil.

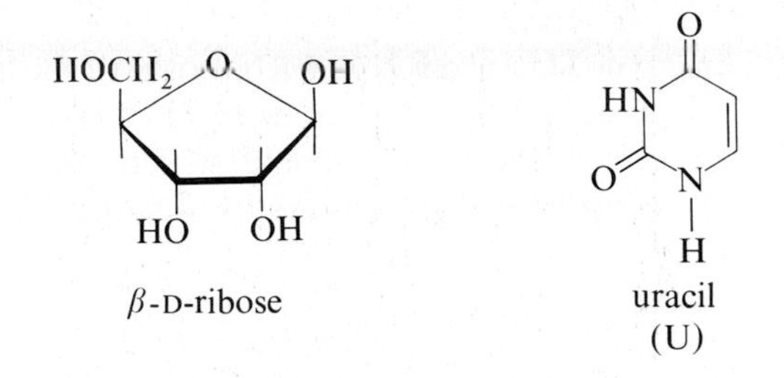

RNA is distributed throughout the cell; it is present in the nucleus, the cytoplasm, and even in subcellular particles called mitochondria. Furthermore, cells contain three types of RNA, given the names ribosomal RNA, transfer RNA, and messenger RNA. These three types of RNA differ in molecular weight and, as their descriptive names imply, they perform different functions within the cell.

Ribosomal RNA (rRNA) molecules have molecular weights of 0.5–1.0 million and comprise up to 85–90% of the total cellular ribonucleic acid. rRNA is found in the cytoplasm of the cell in subcellular particles called ribosomes, that contain about 60% rRNA and 40% protein. Under certain conditions, complete ribosomes (referred to as 70S ribosomes) can be dissociated into two subunits of unequal size, known as 50S subunits and 30S subunits (Figure 14.7). The designation S stands for Svedberg units and is a measure of

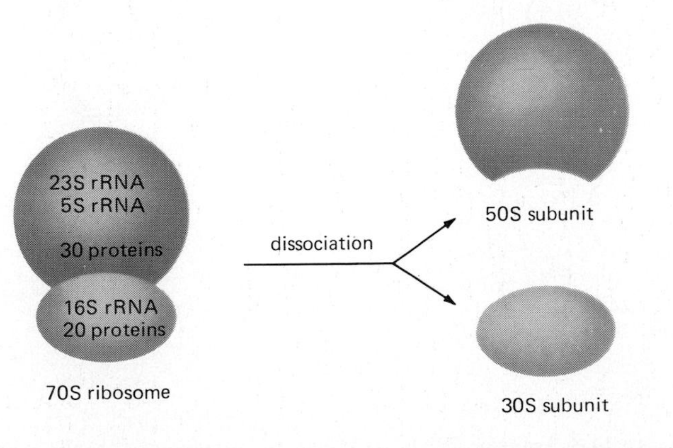

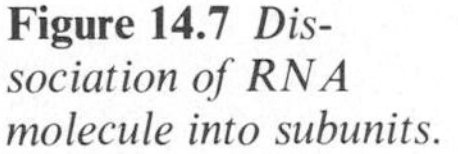
Figure 14.7 *Dissociation of RNA molecule into subunits.*

the molecular weight and compactness of the ribosomal particles. A large value of S indicates a high molecular weight and, conversely, a small value of S indicates a low molecular weight. The larger 50S subunit is about twice the size of the 30S subunit and further dissociates into a 23S and 5S subunit and approximately 30 protein molecules. The smaller 30S subunit dissociates into a single 16S subunit and about 20 different protein molecules.

Many of the proteins bound to intact 70S ribosomes have a high percentage of lysine and arginine and at the pH of the cell, the side chains of these amino acids have net positive charges. It is likely that the interaction of these positively charged amino acid side chains and negatively charged phosphate groups of RNA is an important factor stabilizing the larger 70S ribosomal particles.

Transfer RNAs (tRNAs) have the smallest molecular weight of all nucleic acids. They consist of from 75 to 80 nucleotides in a single chain that is folded into a three-dimensional structure stabilized by hydrogen bonding between complementary base pairs. The tertiary structure of tRNAs is best described

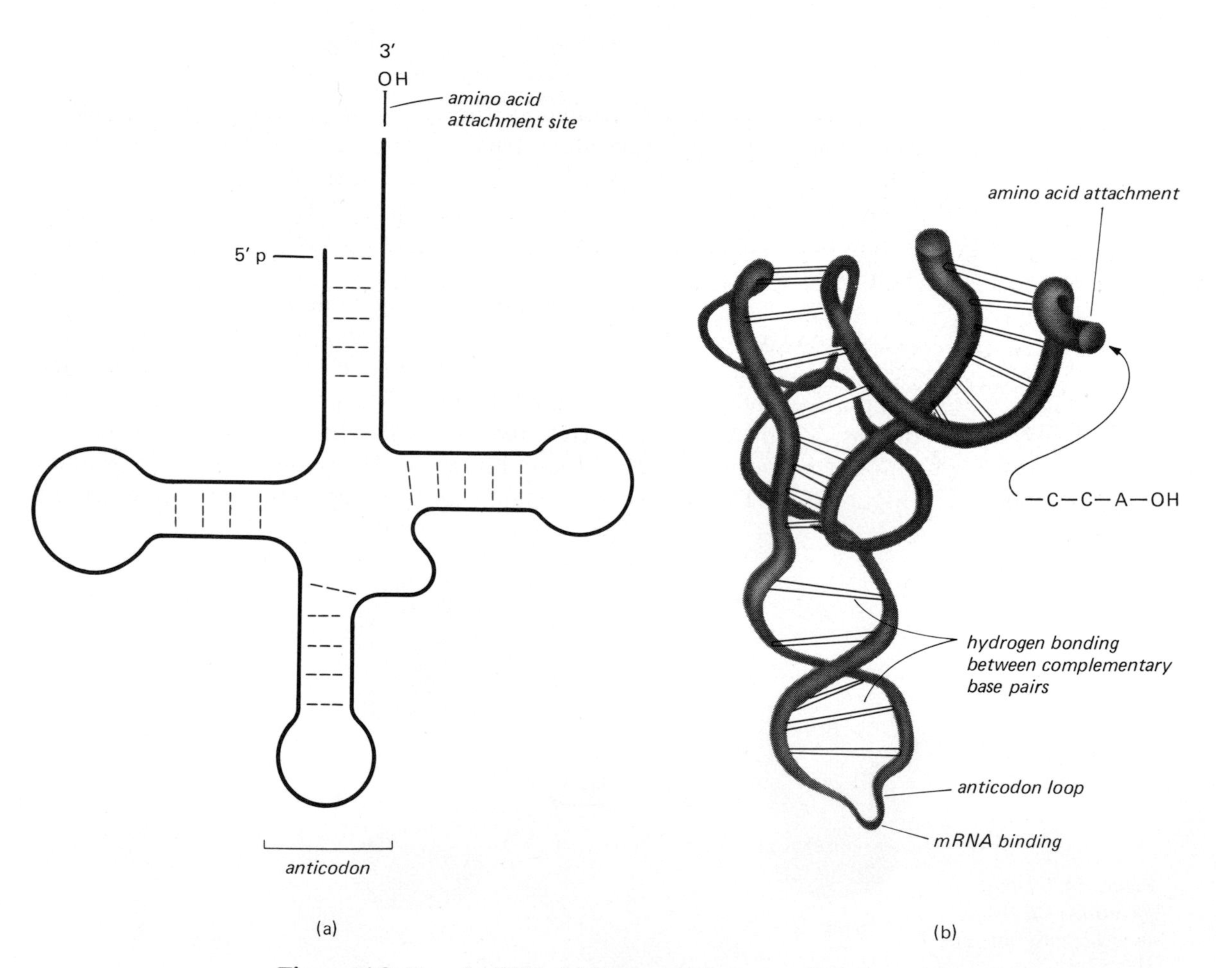

Figure 14.8 *Transfer RNA. (a) Schematic drawing and (b) three-dimensional shape.*

as L-shaped (Fig. 14.8). Nearly all tRNAs have G at the 5′ end of the chain and the sequence CCA at the 3′ end of the chain.

The function of tRNAs is to carry specific amino acids to the site of protein synthesis on the ribosome. There are some 20 tRNAs known, each specific for one of the 20 amino acids found in proteins. For transportation to the ribosome, an amino acid is joined to the 3′ end of its specific tRNA by an ester bond formed between the α-carboxyl group of the amino acid and the 3′-hydroxyl group of ribose.

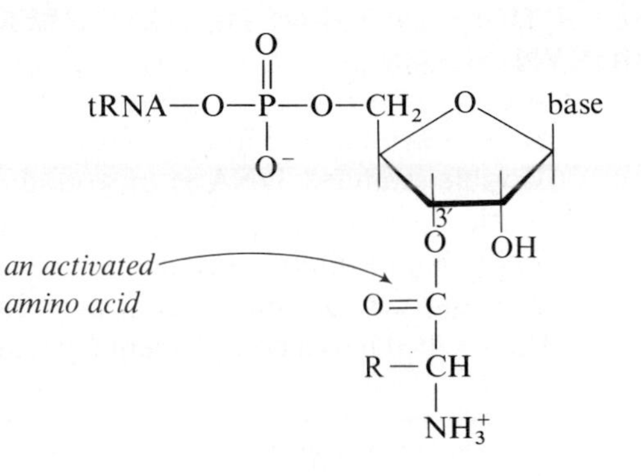

An amino acid thus bound to a tRNA is said to be "activated" because it is prepared for the synthesis of a peptide bond.

Messenger RNA (mRNA) is present in cells in relatively small amounts, and is very short lived. This RNA fraction has an average molecular weight of several hundred thousand and has a base composition much like that of the DNA of the organism from which it is isolated. mRNA is single stranded and folds back on itself to form a conformation stabilized by hydrogen bonding between complementary base pairs (Figure 14.9). It is estimated that about 50%

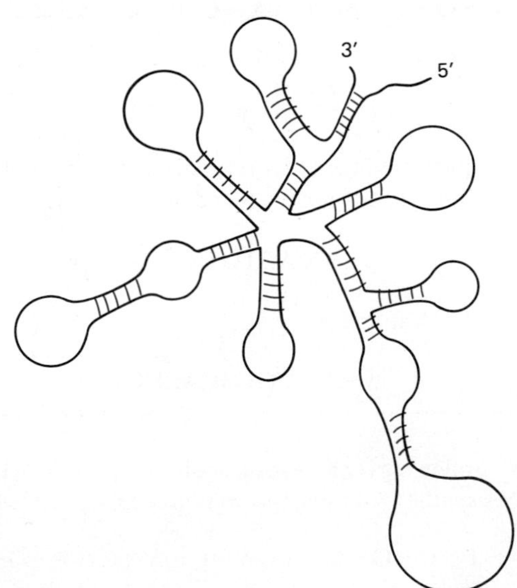

Figure 14.9 *A schematic drawing of a mRNA molecule.*

of the pyrimidine and purine bases in any given mRNA are involved in intramolecular hydrogen bonding. The name "messenger" RNA derives from the fact that this type of RNA is made in the nucleus on a DNA template and carries coded genetic information to the ribosomes for the synthesis of new proteins.

14.7 TRANSCRIPTION OF GENETIC INFORMATION: RNA BIOSYNTHESIS

RNA is synthesized from DNA in a manner similar to the replication of DNA. Double-stranded DNA is unwound and a complementary strand of RNA is synthesized beginning from the 3′ end of the DNA template. The synthesis of RNA from a DNA template is called transcription, a term that refers to the fact that genetic information contained in the sequence of bases of DNA is transcribed into a complementary sequence of bases in mRNA.

Example 14.4 Following is a base sequence from a portion of DNA. Write the sequence of bases of the RNA synthesized using this section of DNA as a template.

3′—A—G—C—C—A—T—G—T—G—A—C—C—5′

Solution RNA synthesis begins at the 3′-end of the DNA template and proceeds toward the 5′-end of the template. The complementary RNA strand is formed using the bases C, G, A, and U. Note that uracil (U) is the complement of adenine (A) on the DNA template.

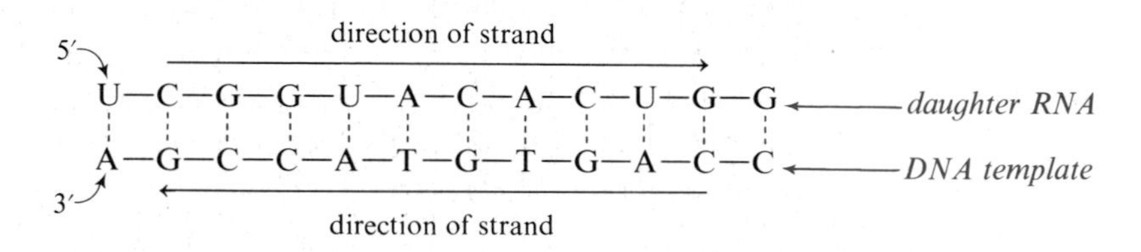

To write the base sequences of the DNA template and the complementary daughter RNA strand synthesized from it, start from the 5′ end of each. Thus, the DNA template is

5′—CCAGTGTACCGA—3′.

The complementary RNA strand is

5′—UCGGUACACUGG—3′.

PROBLEM 14.4 Following is a base sequence from a portion of DNA. Write the sequence of bases in the RNA synthesized using this section of DNA as a template.

5′—T—C—G—G—T—A—C—A—C—T—G—G—3′

14.8 THE GENETIC CODE

Given our understanding of the structure of DNA and its function as a storehouse of genetic information for the entire organism, the next questions to be asked are: In what manner is genetic information coded on DNA molecules? What is it coded for? How is this code read and expressed? In answer to the first two questions, it is now clear that the sequence of bases in DNA molecules constitutes the store of genetic information and that this sequence of bases serves to direct the synthesis of RNA and of proteins. However, the statement that the sequence of bases is the genetic information and that this sequence directs the synthesis of proteins presents a paradox. How can a molecule consisting of only four variable units (adenine, cytosine, guanine, thymine) direct the synthesis of a protein in which there are as many as 20 different units? How can a 4-letter alphabet specify the sequence of the 20-letter alphabet that occurs in proteins?

One obvious answer is that it is not one base but some combination of bases that codes for each amino acid. If the code consists of nucleoside pairs, there are $4^2 = 16$ combinations, a more extensive code, but still not extensive enough to code for 20 amino acids. If the code consists of nucleosides in groups of three, there are $4^3 = 64$ possible combinations, more than enough to specify the primary sequence of a protein. This appears to be a very simple solution to a system that must have taken eons of evolutionary trial and error to develop. Yet there is convincing evidence that nature does indeed use this simple 3-letter or triplet code to store genetic information. A triplet of nucleosides is called a codon. One of the triumphs of molecular biology is that the code has been deciphered.

The next question of course is: Which particular triplets code for each amino acid? At one time it seemed that the only hope of answering this question was to isolate a section of DNA coding for a particular protein and then compare the sequence of amino acids on the protein with the sequence of bases on DNA. However, this was not experimentally possible even as late as 1960 because the base sequences of DNAs were unknown.

Fortunately, the young biochemist Marshall Nirenberg provided a simple and very direct experimental approach to the problem. It was based on the observation that synthetic polynucleotides are able to direct polypeptide synthesis in much the same manner as mRNAs direct polypeptide synthesis. Nirenberg incubated ribosomes, amino acids, tRNAs, and the appropriate protein-synthesizing enzymes, and with only these components, there was essentially no polypeptide synthesis. However, when Nirenberg added synthetic polyuridylic acid (poly U), a polypeptide of high molecular weight was formed. The exciting result of this experiment was that poly U had served as a synthetic messenger RNA and that the polypeptide synthesized contained only phenylalanine. With this discovery, the first element of the genetic code had been deciphered. The triplet UUU codes for phenylalanine.

This same type of experiment was carried out with different polyribonucleotides. It was found that polyadenylic acid (poly A) led to the synthesis of polylysine and that polycytidylic acid (poly C) led to the synthesis of polyproline.

codon on mRNA	amino acid
UUU	phenylalanine
AAA	lysine
CCC	proline

This strategy was extended, and by 1966 all 64 codons had been deciphered (Table 14.3).

Table 14.3 The genetic code: mRNA codons and the amino acid whose incorporation each codon directs.

UUU	Phe	UCU	Ser	UAU	Tyr	UGU	Cys
UUC	Phe	UCC	Ser	UAC	Tyr	UGC	Cys
UUA	Leu	UCA	Ser	UAA	Stop	UGA	Stop
UUG	Leu	UCG	Ser	UAG	Stop	UGG	Trp
CUU	Leu	CCU	Pro	CAU	His	CGU	Arg
CUC	Leu	CCC	Pro	CAC	His	CGC	Arg
CUA	Leu	CCA	Pro	CAA	Gln	CGA	Arg
CUG	Leu	CCG	Pro	CAG	Gln	CGG	Arg
AUU	Ile	ACU	Thr	AAU	Asn	AGU	Ser
AUC	Ile	ACC	Thr	AAC	Asn	AGC	Ser
AUA	Ile	ACA	Thr	AAA	Lys	AGA	Arg
AUG	Met	ACG	Thr	AAG	Lys	AGG	Arg
GUU	Val	GCU	Ala	GAU	Asp	GGU	Gly
GUC	Val	GCC	Ala	GAC	Asp	GGC	Gly
GUA	Val	GCA	Ala	GAA	Glu	GGA	Gly
GUG	Val	GCG	Ala	GAG	Glu	GGG	Gly

A number of features of the genetic code are evident from Table 14.3.

(1) Only 61 triplets code for amino acids. The remaining three triplets (UAA, UAG, and UGA) are signals for chain termination, i.e., they are signals to the protein-synthesizing machinery of the cell that the primary structure of the protein is complete. The three chain termination triplets are indicated in Table 14.3 by Stop.

(2) The code is degenerate, which means that many amino acids are coded for by more than one triplet. If you count the number of triplets coding for each amino acid, you will find that only methionine and tryptophan are coded for by one triplet. Leucine, serine, and arginine are coded for by six triplets, and the remaining 15 amino acids are coded for by two, three, or four triplets.

(3) For the 15 amino acids coded for by two, three, or four triplets, the degeneracy is only in the last base of the triplet. In other words, in the codons for 15 amino acids, it is only the third letter of the code that varies. For example, glycine is coded by the triplets GGA, GGG, GGC, and GGU.

(4) Finally, there is no ambiguity in the code. Each triplet codes for one and only one amino acid.

We must ask one last question about the genetic code: Is the code universal—is it the same for all organisms? Every bit of experimental evidence available today from the study of viruses, bacteria, and higher animals, including humans, indicates that the code is the same for all organisms and that it is universal. Furthermore, the fact that it is the same in all these organisms means that it has been the same over billions of years of evolution.

Example 14.5 During transcription, a portion of mRNA is synthesized with the following base sequence.

5′→A—U—G—G—U—A—C—C—A—C—A—U—U—U—G—U—G—A←3′ ········

(a) Write the base sequence of the DNA from which this portion of mRNA was synthesized.

(b) Write the primary structure of the polypeptide coded for by this section of mRNA.

Solution During transcription, mRNA is synthesized from a DNA strand beginning from the 3′ end of the DNA template. The DNA strand must be complementary to the newly synthesized mRNA strand.

direction of strand

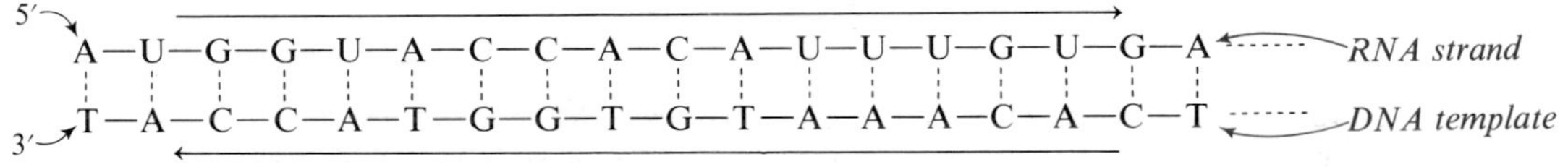

direction of strand

(b) The sequence of amino acids is shown below the mRNA strand.

A—U—G—	G—U—A	C—C—A—	C—A—U—	U—U—G—	U—G—A—
met	val	pro	his	leu	Stop

The codon UGA codes for termination of the growing polypeptide chain and therefore the sequence given in this problem codes for a pentapeptide only.

PROBLEM 14.5 The following section of DNA codes for oxytocin, a polypeptide hormone.

mRNA synthesis begins here → 3′—A—C—G—A—T—A—T—A—A—G—T—T—T—T—A—
A—C—G—G—G—A—G—A—A—C—C—A—A—C—T—5′

(a) Write the base sequence for the mRNA synthesized from this section of DNA.

(b) Given this sequence of bases in mRNA, write the amino acid sequence of oxytocin.

14.9 TRANSLATION OF GENETIC INFORMATION: PROTEIN BIOSYNTHESIS

The biosynthesis of polypeptides is usually described in terms of three major processes: initiation of the polypeptide chain, elongation of the polypeptide chain, and termination of the completed polypeptide chain. These processes along with the substances required for each are summarized in Table 14.4.

Table 14.4 Major processes in protein biosynthesis.

Process	*Substances required*
initiation	tRNA carrying N-formylmethionine, mRNA, 30S and 50S ribosomal subunits, GTP, protein-initiating factors
elongation	amino acyl tRNAs, protein elongation factors, GTP
termination	termination codon on mRNA, protein termination factors

A. INITIATION OF PROTEIN BIOSYNTHESIS

In bacteria, all polypeptide chains are initiated with the amino acid N-formylmethionine (fMet)

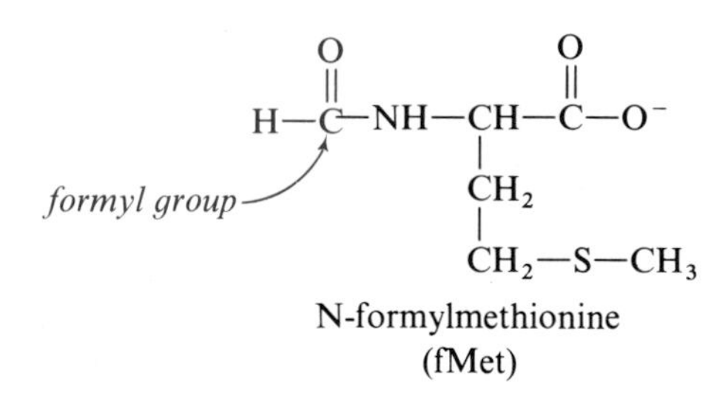

N-formylmethionine
(fMet)

Many bacterial polypeptides do have N-formylmethionine as the N-terminal amino acid. However, for most bacterial proteins, N-formylmethionine or it and several other amino acids at the N-terminal end of the polypeptide chain are cleaved to give the native polypeptide. N-formylmethionine is bound to a specific tRNA molecule given the symbol $tRNA_{fMet}$.

The first step in the initiation process is alignment of mRNA on the 30S ribosomal subunit so that the initiating codon is located at a specific site on the ribosome called the P site. The initiating codon is most commonly AUG,

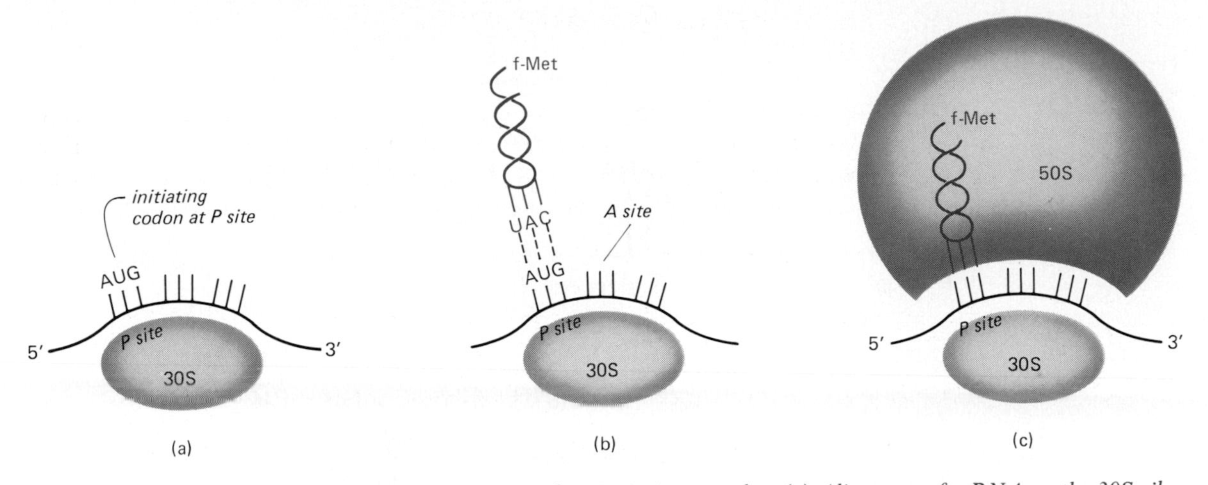

Figure 14.10 *Formation of an initiation complex. (a) Alignment of mRNA on the 30S ribosomal subunit so that AUG, the initiating codon, is located at the P site. (b) Binding of tRNA carrying N-formylmethionine to the initiating codon. (c) Association of the 50S ribosomal subunit to give the initiation complex.*

that for methionine. Next, tRNA carrying N-formylmethionine binds to the initiating codon and this complex, in turn, binds a 50S ribosomal subunit to give a unit called an initiation complex. Each of these alignments and associations of subunits requires energy and specific proteins called initiating factors. The source of energy for the formation of the initiation complex is the hydrolysis of GTP to GDP and HPO_4^{2-}. The formation of an initiation complex is shown schematically in Figure 14.10.

B. ELONGATION OF THE POLYPEPTIDE CHAIN

Elongation of a polypeptide chain consists of three steps which are repeated over and over until the entire polypeptide chain is synthesized. In the first step, a "charged" tRNA (one carrying an amino acid esterified at the 3′ end of the tRNA chain) binds to the A site of the initiating complex. The second step is formation of a peptide bond between the carboxyl group of the tRNA-bound amino acid at the P site and the amino group of the tRNA-bound amino acid at the A site (Fig. 14.11). Peptide bond formation is catalyzed by the enzyme peptidyl transferase. After formation of the new peptide bond, the tRNA bound to the P site is "empty" and the growing polypeptide chain is now attached to the tRNA bound to the A site.

The third step in the elongation cycle involves release of the "empty" tRNA from the P site and translocation of the 70S ribosomal complex by one

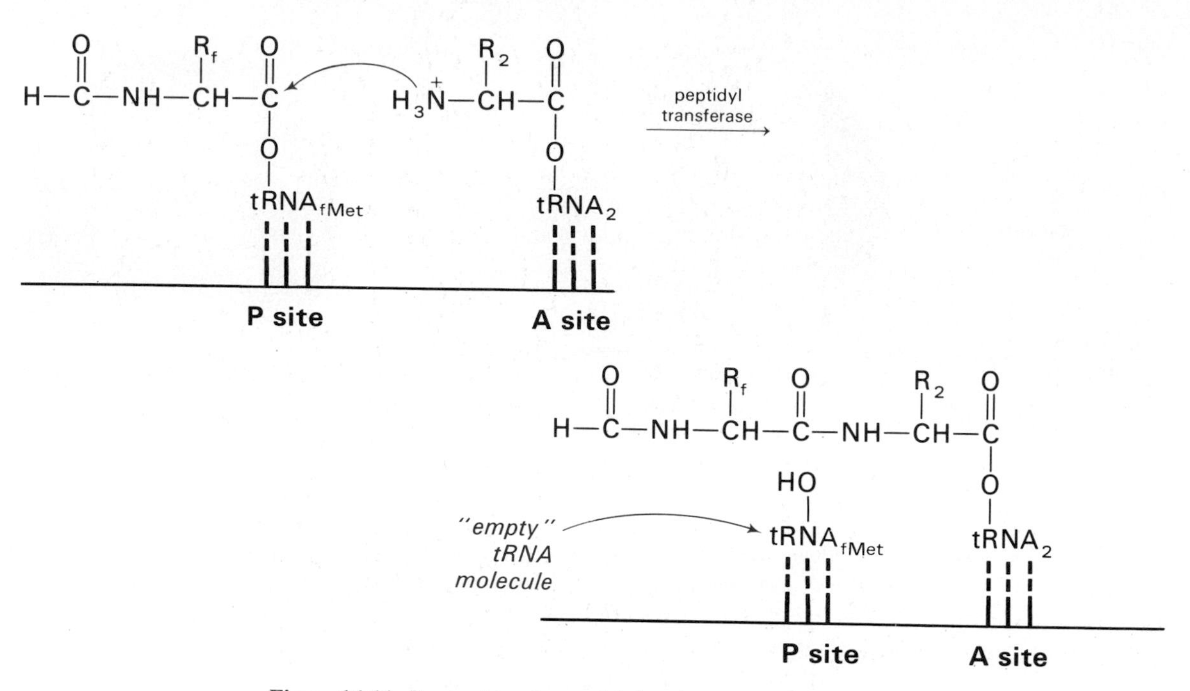

Figure 14.11 *Formation of a peptide bond.*

codon from the 5′ end of the mRNA toward the 3′ end. As a result of translocation, the tRNA carrying the growing polypeptide chain is moved from the A site to the P site. Energy for translocation is derived from the hydrolysis of GTP to GDP and HPO_4^{2-}.

The three steps in the elongation cycle are shown schematically in Figure 14.12 for the synthesis of the tripeptide of fmet-arg-phe from fmet-arg.

C. TERMINATION OF POLYPEPTIDE BIOSYNTHESIS

Polypeptide synthesis continues through the chain elongation cycle until the ribosome complex reaches a stop codon (UAA, UAG, or UGA) on mRNA. There, a specific protein called a termination factor binds to the stop codon and catalyzes hydrolysis of the completed polypeptide chain from tRNA. The "empty" ribosome then dissociates, ready for binding to another strand of mRNA, fMet-tRNA and the formation of another initiation complex.

Figure 14.13 shows several ribosome complexes moving along a single strand of mRNA and illustrates the fact that several polypeptide chains can be synthesized simultaneously from a single mRNA molecule. Figure 14.13 also illustrates the fact that as a polypeptide chain grows, it extends out from the ribosome into the cytoplasm of the cell and folds spontaneously into its native three-dimensional conformation.

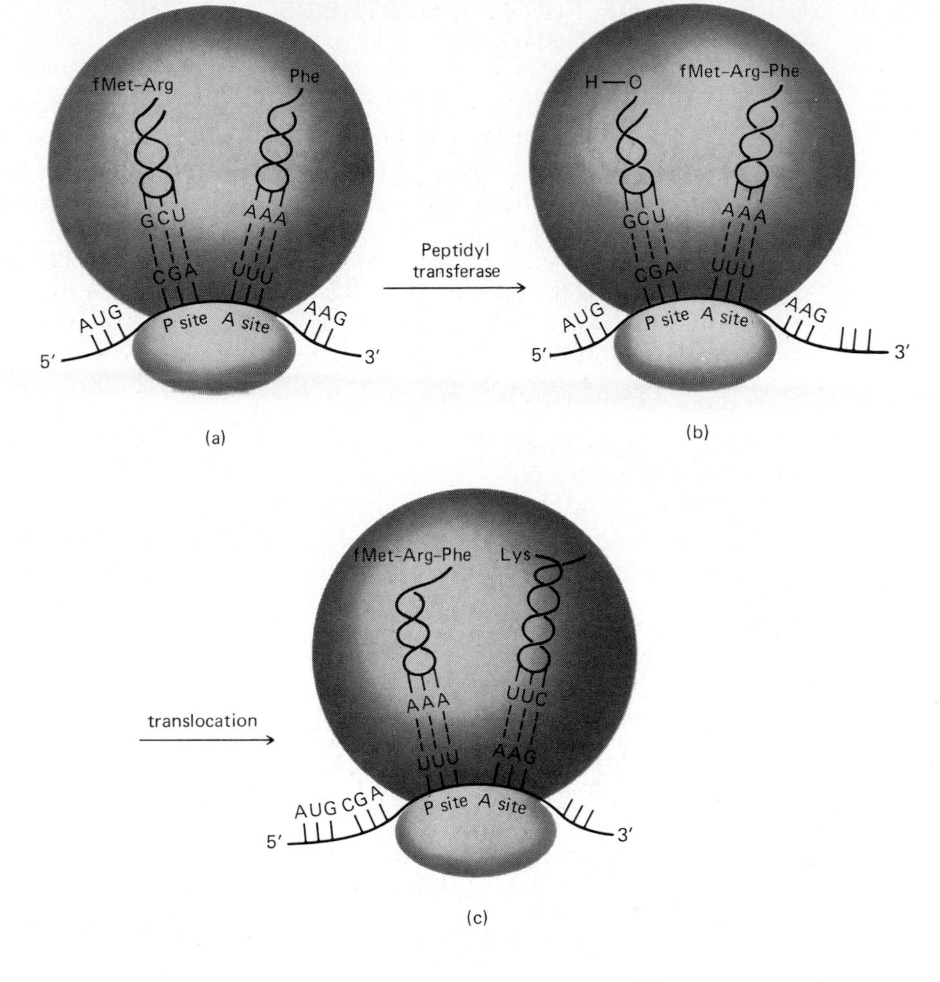

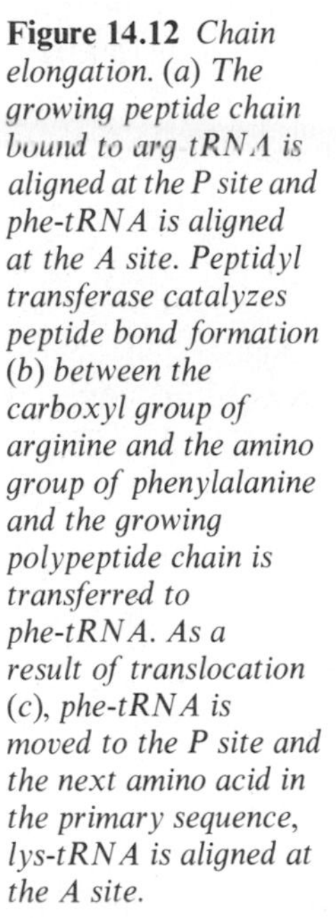

Figure 14.12 *Chain elongation. (a) The growing peptide chain bound to arg tRNA is aligned at the P site and phe-tRNA is aligned at the A site. Peptidyl transferase catalyzes peptide bond formation (b) between the carboxyl group of arginine and the amino group of phenylalanine and the growing polypeptide chain is transferred to phe-tRNA. As a result of translocation (c), phe-tRNA is moved to the P site and the next amino acid in the primary sequence, lys-tRNA is aligned at the A site.*

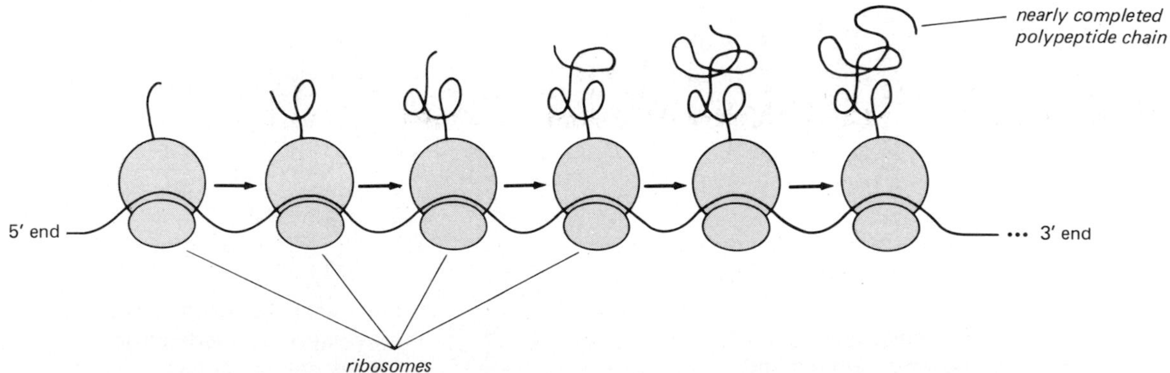

Figure 14.13 *Simultaneous elongation of several polypeptide chains on a single strand of mRNA. The growing polypeptide chains spontaneously assume the native three-dimensional conformation.*

14.10 INHIBITION OF PROTEIN SYNTHESIS AND THE ACTION OF ANTIBIOTICS

Several of the most widely used antibiotics including tetracycline, streptomycin, chloramphenicol, and puromycin (Figure 14.14) act by inhibition of protein synthesis in bacteria at the ribosomal level. Although the general process of protein synthesis we have described in the previous section operates universally,

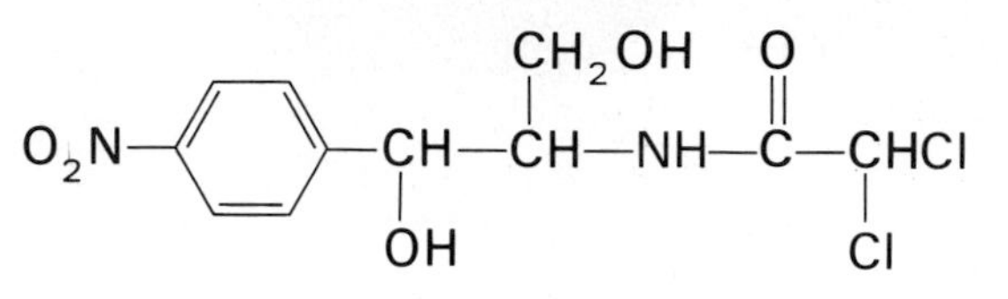

chloramphenicol (binds to the A site and inhibits binding of charged tRNAs)

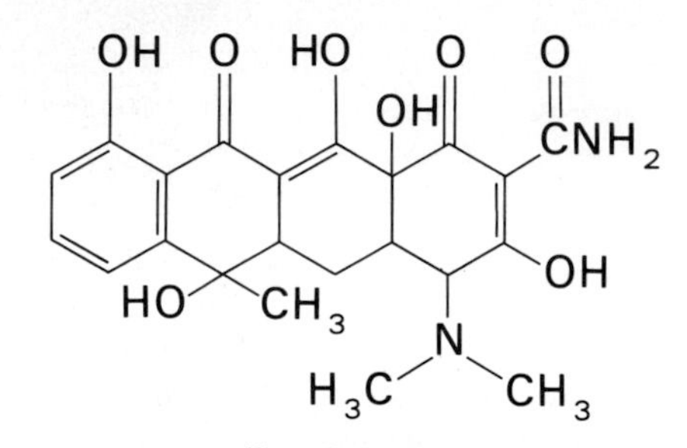

tetracycline (inhibits binding of charged tRNAs to the 30S ribosomal subunit)

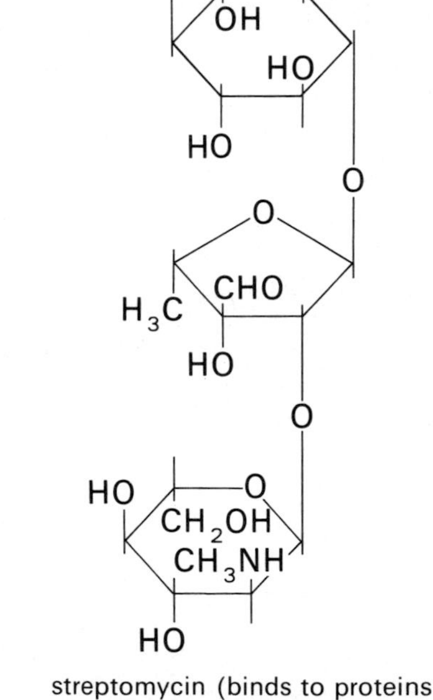

streptomycin (binds to proteins of the 30S ribosomal subunit and causes misreading of mRNA code)

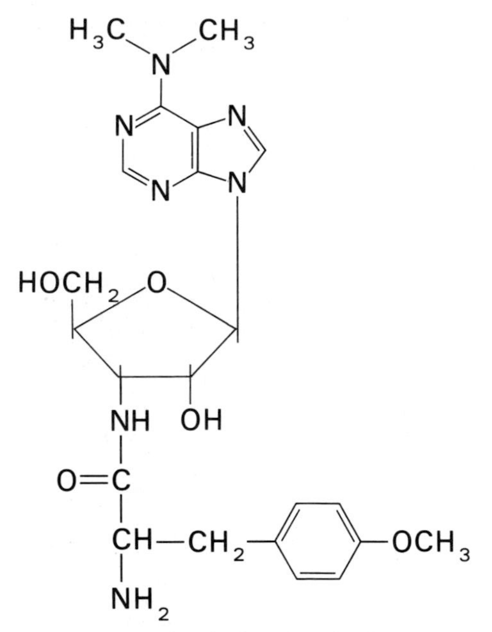

puromycin (is inserted in the growing polypeptide chain and causes premature termination of polypeptide synthesis)

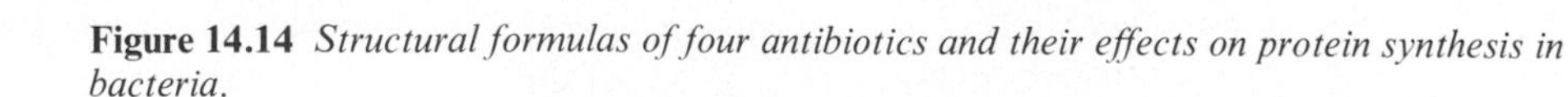

Figure 14.14 *Structural formulas of four antibiotics and their effects on protein synthesis in bacteria.*

there are differences in detail between bacteria and animals. Some of these, particularly those dealing with chain initiation and chain elongation are quite marked. Because of these differences in details of the translation process, these and other antibiotics are able to inhibit protein synthesis in bacteria but have little or no effect on host cells.

Chloramphenicol is a very broad-spectrum antibiotic, however, in some persons it causes serious, often toxic side effects. For this reason, the use of chloramphenicol is restricted largely to treatment of acute infections for which other antibiotics are ineffective or for medical reasons cannot be used.

Puromycin is a structural analog of a charged tRNA molecule (one bearing an amino acid esterified to the 3′-hydroxyl of the terminal nucleotide) and binds to the A site during the chain elongation phase of polypeptide synthesis. There, the enzyme peptidyl transferase catalyzes formation of a peptide bond between the growing polypeptide chain and the amino group of puromycin at which point further elongation of the chain ceases. Thus, puromycin causes premature termination of polypeptide synthesis. Streptomycin binds with proteins of the 30S ribosomal subunit and interferes with interactions between mRNA codons and tRNAs. This interference gives rise to errors in reading the mRNA code and results in insertion of incorrect amino acids in the growing polypeptide chain. Chloramphenicol binds specifically to the A site of the 50S ribosomal subunit and thereby prevents charged tRNAs from binding to the A site. Tetracycline prevents binding of charged tRNAs to the 30S ribosomal subunit.

PROBLEMS

14.6 Examine the structure of purine. Would you predict this molecule to be planar or puckered? To exist as a number of interconvertible conformations (as in the case of cyclohexane) or to be rigid and inflexible? Explain the basis for your answer.

14.7 An important drug in the chemotherapy of leukemia is 6-mercaptopurine, a sulfur analog of adenine. Draw the structural formula of 6-mercaptopurine.

14.8 Explain the difference in structure between a nucleoside and a nucleotide.

14.9 Name and draw structural formulas for the following. In each label the N-glycoside bond.

(a) a nucleoside composed of β-D-ribose and adenine.
(b) a nucleoside composed of β-D-ribose and uracil.
(c) a nucleoside composed of β-2-deoxy-D-ribose and cytosine.

14.10 Name and draw structural formulas for the following. Label all N-glycoside bonds, ester bonds, and anhydride bonds.

(a) dATP **(b)** ADP **(c)** dGMP **(d)** GTP

14.11 Calculate the net charge on the following at pH 7.4.

(a) ATP **(b)** 2′-deoxyadenosine **(c)** GMP

14.12 Cyclic-AMP (adenosine-3′, 5′-cyclic monophosphate), first isolated in 1959, is involved in many diverse biological processes as a regulator of metabolic and physiological activity. In it, a single phosphate group is esterified with both the 3′- and 5′-hydroxyls of adenosine. Draw the structural formula for this substance.

14.13 Following are sequences for several polynucleotides. Write structural formulas for each. Calculate the net charge on each at pH 7.4.
(a) ApGpA **(b)** pppCpT **(c)** pGpCpCpTpA

14.14 Show by structural formulas the hydrogen bonding between thymine and adenine; between uracil and adenine.

14.15 Compare and contrast the α-helix found in proteins with the double helix of DNA in the following ways.
(a) The units that repeat in the backbone of the chain.
(b) The projection in space of the backbone substituents (R— groups in the case of amino acids, purine and pyrimidine bases in the case of DNA) relative to the axis of the helix.

14.16 List the postulates of the Watson–Crick model of DNA structure. This model is based on certain experimental observations of base composition and molecular dimensions. Describe these observations and show how their model accounts for each.

14.17 Explain the role of hydrophobic interaction in stabilizing
(a) soap micelles **(b)** lipid bilayers **(c)** double-stranded DNA

14.18 What type of bond or interaction holds monomers together in
(a) proteins **(b)** nucleic acids **(c)** polysaccharides

14.19 Compare and contrast DNA and RNA in the following ways.
(a) monosaccharide units **(b)** major purine and pyrimidine bases
(c) primary structure **(d)** location in the cell
(e) function in the cell

14.20 Compare and contrast ribosomal RNA, messenger RNA, and transfer RNA as follows:
(a) molecular weight **(b)** function in protein synthesis

14.21 Draw a diagram of a mRNA-ribosome initiation complex and label the following:
(a) 30S subunit **(b)** 50S subunit
(c) 5′ and 3′ ends of mRNA

14.22 Given the DNA triplet ATC, is its complement TAG or is it GAT? Explain.

14.23 Given the following DNA sequence

5′—A—C—C—G—T—T—G—C—C—A—A—T—G—3′

(a) Write the sequence of its DNA complement.
(b) Write the sequence of its mRNA complement.

14.24 Following is a section of mRNA.

5′—A—G—G—U—C—C—C—A—G—3′

(a) What tripeptide is synthesized if the code is read from the 5′ end to the 3′ end?
(b) What tripeptide is synthesized if the code is read from the 3′ and to the 5′ end?
(c) Calculate the net charge on each tripeptide at pH 7.4.
(d) Which way is the code read in the cell and which tripeptide is synthesized?

14.25 What peptide sequences are coded for by the following mRNA sequences? (Each is written from the $5' \rightarrow 3'$ direction.)

(a) G—C—U—G—A—A—U—G—G **(b)** U—C—A—G—C—A—A—U—C
(c) G—U—C—G—A—G—G—U—G **(d)** G—C—U—U—C—U—U—A—A

14.26 Complete the following table.

DNA	*DNA complement*	*mRNA complement*	*amino acid coded for*
T—G—C			
C—A—G			
	A—C—G		
	G—T—A		
		G—U—C	
		U—G—C	
		C—A—C	

14.27 Each of the following reactions involves ammonolysis of an ester. Draw the structural formula of the amide produced in each reaction.

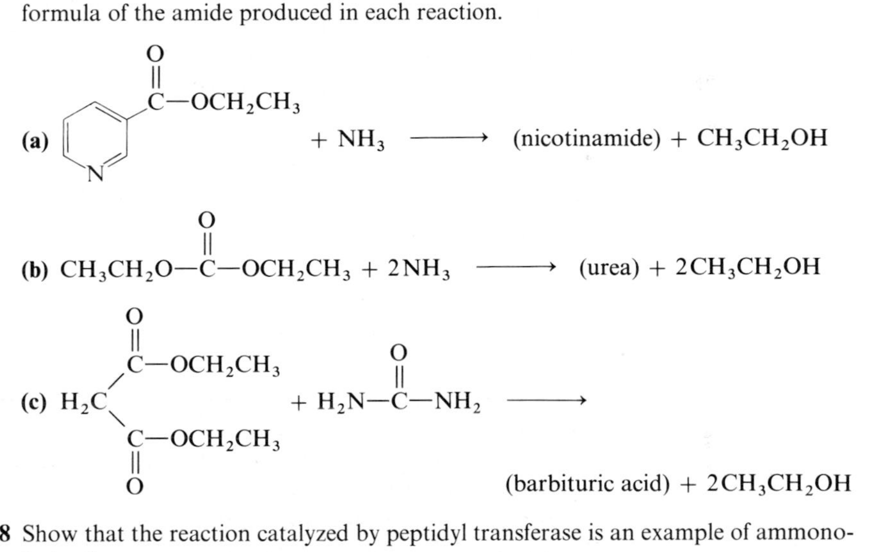

14.28 Show that the reaction catalyzed by peptidyl transferase is an example of ammonolysis of an ester.

14.29 The α-chain of human hemoglobin has 141 amino acids in a single polypeptide chain.
(a) Calculate the minimum number of bases on DNA necessary to code for the α-chain. Include in your calculation the bases necessary to specify termination of polypeptide synthesis.
(b) Calculate the length in centimeters of double-stranded DNA containing this number of bases.

14.30 In HbS, the abnormal human hemoglobin found in individuals with sickle-cell anemia (see the mini-essay "Abnormal Human Hemoglobins") glutamic acid at position 6 of the β-chain is replaced by valine.
(a) List the two codons for glutamic acid and the four codons for valine.
(b) Show that a glutamic acid codon can be converted into a valine codon by a single substitution mutation.

CHAPTER 15

Spectroscopy

The first several chapters of this text describe many of the most important functional groups in organic chemistry, the typical reactions of each, how to convert one functional group to another, and how to determine by chemical tests which functional groups a substance contains. For example, if a substance of unknown structure discharges the color of a solution of bromine in CCl_4, you should suspect immediately that it contains some kind of unsaturation, possibly an alkene or alkyne. If the same substance fails to react with 2,4-dinitrophenylhydrazine, we know that it probably does not contain an aldehyde or ketone. An understanding of the typical reactions of functional groups can help us answer other types of problems as well. For example, suppose you know that a compound is one of the following three substances.

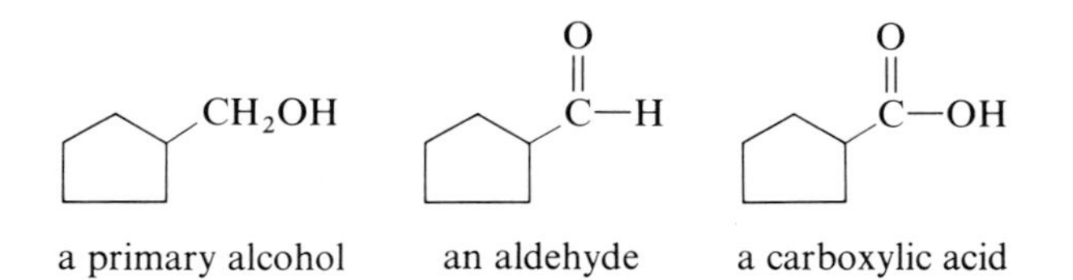

How can you tell which of the three substances you are dealing with? One is a primary alcohol, the second an aldehyde, the third a carboxylic acid, and each has its own characteristic reactions. Only the aldehyde reacts with 2,4-dinitrophenylhydrazine and only the carboxylic acid reacts with aqueous sodium bicarbonate. If the substance fails to react with either of these reagents, then it must be the alcohol. Suppose instead that you are asked to distinguish between 1-propanol and 2-propanol.

$$CH_3{-}CH_2{-}CH_2{-}OH \qquad CH_3{-}\underset{}{\overset{\displaystyle OH}{\overset{|}{C}H}}{-}CH_3$$

1-propanol   2-propanol

Each has the same molecular formula, C_3H_8O, and each contains the same functional group. These two substances can be distinguished by chemical tests, but the process is much more involved than for the first examples we examined.

The purpose of this chapter is to show how information about the presence or absence of particular functional groups can be obtained from the study of molecular spectra. As you will see, such methods have three major advantages over most chemical tests.

(1) Spectral analyses are easier and faster to do than most chemical tests.

(2) Spectral analyses generally provide far more detailed information about molecular structure.

(3) Spectral analyses are nondestructive and, if necessary, the entire sample can be recovered.

Let us begin our study of spectroscopy by first reviewing some facts about radiant energy and the interaction of energy and matter.

15.1 ELECTROMAGNETIC RADIATION

Recall from general chemistry that we can describe light (visible light, ultraviolet light, radio waves, X-rays, and so on) in a very simple way if we assume that light is a wave which travels through space. We can describe a light wave in the same terms we use to describe an ocean wave as it moves toward the shore, namely, in terms of its wavelength and its frequency.

Wavelength is the distance between consecutive crests (or troughs).

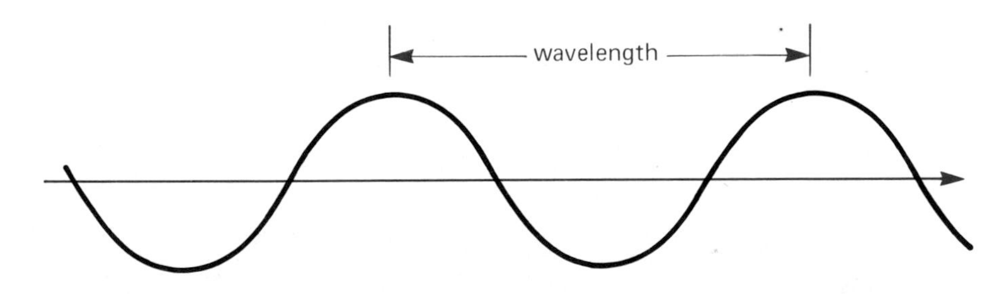

Wavelength is given the symbol λ (lambda) and is generally measured in meters (m) or fractions of meters.

$$1 \text{ centimeter (cm)} = 10^{-2} \text{ meter}$$
$$1 \text{ millimeter (mm)} = 10^{-3} \text{ meter}$$
$$1 \text{ micrometer } (\mu\text{m}) = 10^{-6} \text{ meter}$$
$$1 \text{ nanometer (nm)} = 10^{-9} \text{ meter}$$

Frequency is the number of full cycles of the wave that pass a given point in a fixed period of time. Frequency is given the symbol ν (nu) and is reported in cycles per second (cps) or Hertz (Hz; one Hz = one cps). Wavelength and frequency are inversely proportional to each other, and one can be calculated from the other using the following relationship:

$$\nu = \frac{c}{\lambda}$$

ν = frequency in Hz or cps

λ = wavelength in meters

c = velocity of light, 3×10^8 m/sec

For example, infrared radiation, or heat radiation as it is also called, has a wavelength of about 15×10^{-6} m (15 μm). The frequency of this radiation is 2×10^{13} Hz.

$$\nu = \frac{3 \times 10^{8} \text{ m/sec}}{15 \times 10^{-6} \text{ m}} = 2 \times 10^{13} \text{ Hz}$$

A third way to describe light is in terms of its energy. To scientists at the turn of the century, there were a number of puzzling experimental observations about light that could not be explained in terms of wave properties. Einstein discovered that he could explain these results if he first assumed that light has some of the properties of particles. We now call these particles of light photons. The amount of energy in a mole of photons is related to the frequency of the light by the following equation:

$$E = h\nu \qquad \begin{aligned} E &= \text{energy, in cal/mole} \\ h &= \text{Planck's constant, } 6.625 \times 10^{-27} \text{ erg-sec} \\ \nu &= \text{frequency, in Hz} \end{aligned}$$

Notice that there is a direct relationship between the frequency of a light wave and its energy—the greater the frequency, the greater the energy. Thus, ultraviolet light (frequency 10^{15} Hz) has a greater energy than infrared radiation (frequency 10^{13} Hz). The wavelengths, frequencies, and energies of the various parts of the electromagnetic spectrum are summarized in Table 15.1.

In this chapter, we will be concerned primarily with three regions of the electromagnetic spectrum: infrared light, ultraviolet-visible light, and radio waves.

Table 15.1 The electromagnetic spectrum.

wavelength (m)	3	3×10^{-2}	3×10^{-4}	3×10^{-6}	3×10^{-8}	3×10^{-10}	3×10^{-12}
frequency (Hz)	10^{8}	10^{10}	10^{12}	10^{14}	10^{16}	10^{18}	10^{20}
energy (kcal)	10^{-5}	10^{-3}	10^{-1}	10	10^{3}	10^{5}	10^{7}

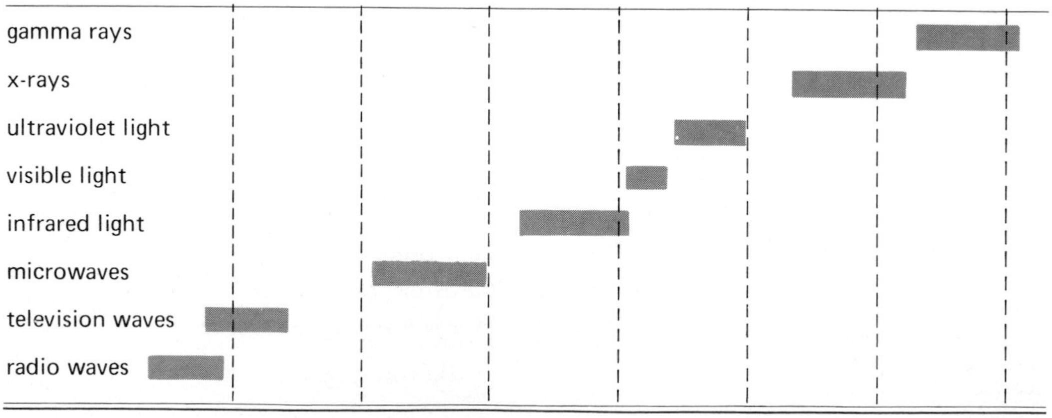

gamma rays

x-rays

ultraviolet light

visible light

infrared light

microwaves

television waves

radio waves

15.2 MOLECULAR SPECTROSCOPY

Organic molecules are surprisingly flexible structures. As we have already discussed in Chapter 2, atoms or groups of atoms rotate about covalent bonds, and a given molecule can have an almost limitless number of conformations. In addition, covalent bonds bend and stretch just as if the atoms themselves were joined by flexible springs. Furthermore, electrons within molecules can move from one electronic energy level or orbital to another. We know from experimental observations and from theories of molecular structure that these energy changes within molecules are quantized; bonds within a molecule can undergo transitions only between allowed vibrational energy levels, electrons can undergo transitions only between allowed electronic energy levels, and so on.

Organic molecules can be made to undergo a transition from energy state E_1 to a higher state, E_2, by irradiating them with electromagnetic radiation corresponding to the energy difference between states E_2 and E_1. Molecular spectroscopy is the experimental process of measuring which frequencies of radiation are absorbed by a particular molecule and then attempting to correlate these absorption patterns with details of molecular structure (Fig. 15.1).

We will study how organic molecules absorb infrared radiation, ultraviolet-visible radiation, and microwaves, and we will study what types of information these absorption patterns can give us about details of molecular structure. Specifically, we will see that:

(1) Absorption of infrared radiation causes covalent bonds within molecules to be promoted from one vibrational energy level to a higher vibrational energy level.

(2) Absorption of ultraviolet-visible radiation causes electrons within molecules to be promoted from one electronic energy level to a higher electronic energy level.

(3) Absorption of radiowaves in the presence of a magnetic field causes nuclei within molecules to be promoted from one spin energy level to a higher spin energy level.

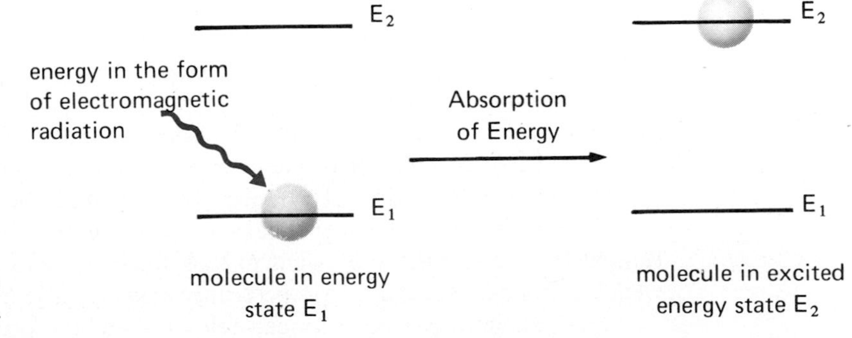

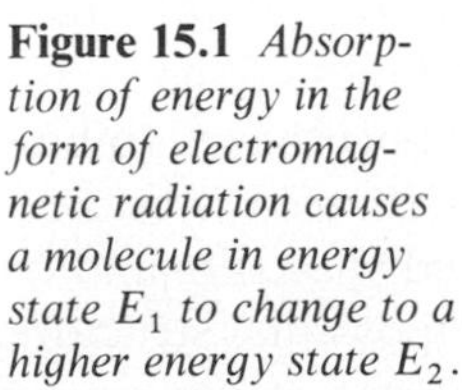

Figure 15.1 *Absorption of energy in the form of electromagnetic radiation causes a molecule in energy state E_1 to change to a higher energy state E_2.*

15.3 INFRARED SPECTROSCOPY

Wavelengths in the region of the infrared spectrum of most interest to us range from about 2.5×10^{-6} to 15×10^{-6} m. In order to simplify the reporting and tabulation of infrared information, chemists generally report absorption frequencies in either micrometers (μm) or in wavenumbers. Expressed in micrometers, this region of the infrared spectrum runs from 2.5 μm to 15 μm. Alternatively, infrared absorption frequencies may be reported in wavenumbers, that is, the number of complete wave cycles per centimeter. Wavenumbers ($\bar{\nu}$) are expressed in cm^{-1} (reciprocal centimeters) and are calculated from the wavelength by the following relationship.

$$\bar{\nu} = \frac{1}{\lambda}$$

For example, radiation of wavelength 2.5×10^{-6} m is equivalent to 4000 cm^{-1}. Expressed in wavenumbers, the infrared spectrum runs from 4000 cm^{-1} to 600 cm^{-1}. Chart paper for most infrared spectrophotometers is generally calibrated in both micrometers (μm) and reciprocal centimeters (cm^{-1}).

Virtually all organic molecules are infrared active, because radiation in this region of the spectrum corresponds to the energy required to excite the natural vibrational frequencies of covalent bonds. There are two basic types of bond vibrations: those that correspond to bond stretching and those that correspond to bond bending. In stretching, the distance between bonded atoms increases and decreases in a rhythmic manner, much as the distance between two objects connected by a spring first increases and then decreases as the spring is stretched and then relaxed. In bending, the position of the bonded atoms changes in relation to the original bond axis. You can imagine bond bending as something like the wagging motion of a dog's tail or the flapping motion of the wings of birds in flight. While infrared radiation between 4000 cm^{-1} and 600 cm^{-1} will excite both stretching and bending vibrations, we will look only at bond stretching.

An infrared spectrum is a plot of the percent of radiation transmitted through the sample versus the wavelength of the radiation. (Note that 100% transmission corresponds to 0% absorption.) Figure 15.2 is an infrared spectrum of octane, a saturated hydrocarbon.

The scale at the left of the chart paper indicates % transmittance. Notice also that the chart paper is calibrated in reciprocal centimeters (cm^{-1}) along the bottom and in micrometers (μm) along the top. The most prominent feature of this spectrum is three closely spaced peaks around 2900 cm^{-1}. It has been determined that this cluster of peaks corresponds to stretchings of the C—H bonds in the $—CH_3$ and $—CH_2—$ groups. We will not be concerned with how it has been determined which peaks correspond to which bond vibration. Rather, let us concentrate on what to look for in an infrared spectrum and on what types of information such a spectrum can give us about molecular structure.

First, an infrared spectrum can tell us about the presence or absence of

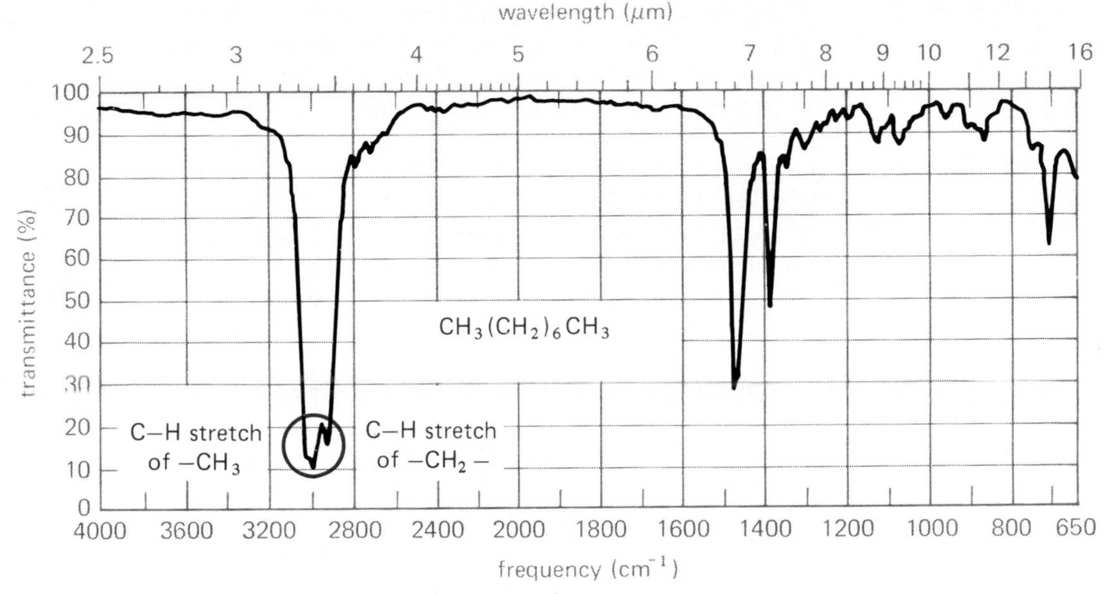

Figure 15.2 *An infrared spectrum of octane.*

particular functional groups. The stretching vibrations of each type of covalent bond (C—H, O—H, N—H, C=O, C=C, and so on) absorb infrared radiation only in certain small regions of the spectrum. Furthermore, the absorption of energy by a particular type of covalent bond is not greatly influenced by the rest of the molecule. For example, the stretching frequencies for the vibration of the carbonyl groups of aldehydes, ketones, carboxylic acids, and esters are all found within the narrow range from 1680 cm^{-1} to 1750 cm^{-1}. Characteristic infrared stretching frequencies for several types of covalent bonds and organic functional groups are given in Table 15.2.

Table 15.2 Characteristic infrared stretching frequencies.

Type of bond	*Found in*	*Frequency* (cm^{-1})
C—H	alkanes	2850–2950
=C—H	alkenes and aromatic hydrocarbons	3000–3200
O—H	alcohols, phenols	3600–3650
O—H	carboxylic acids (hydrogen bonded)	2500–3000
N—H	amines	3300–3500
C—O	esters, alcohols, ethers	1000–1300
C=O	aldehydes, ketones, carboxylic acids, esters, amides	1680–1750

Second, a comparison of infrared spectra can often tell us about structural similarities between two substances. Although each type of covalent bond has its own characteristic absorption frequencies, no two molecules have precisely the same spectrum. Although many absorption frequencies may be the same for closely related substances, almost always there are differences. In general, these differences appear in the range from 1600 cm^{-1} to 600 cm^{-1}, and accordingly, this region of the infrared spectrum is often called the "fingerprint" region. By comparing spectra, particularly in the fingerprint region, it is often possible to tell whether or not two compounds are identical. If the spectra are identical, peak for peak, then it is almost certain that the two substances are identical. If the spectra are not identical, then the two substances do not have the same molecular structure.

What should you look for as you attempt to interpret an infrared spectrum? In this course, we will not attempt any detailed analysis of an infrared spectrum, something which requires a great deal of practice. Rather, we will be concerned only with the recognition of the presence or absence of particular functional groups, as indicated by the presence or absence of particular absorption peaks in the infrared spectrum. The stretching frequencies due to O—H, C—H, N—H, C=O, and C—O are the most important for you to recognize. The relative positions of these stretching frequencies are listed in Table 15.2; they are also shown in the form of a chart (Figure 15.3).

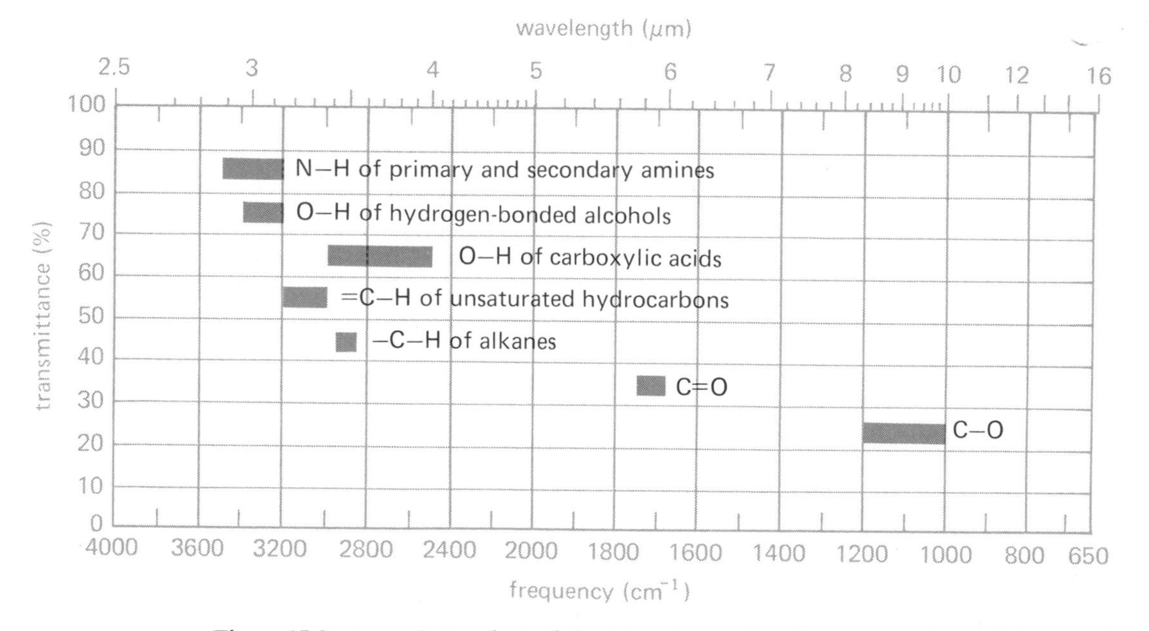

Figure 15.3 *A correlation chart of the infrared stretching frequencies of several functional groups.*

15.4 ULTRAVIOLET-VISIBLE SPECTROSCOPY

The near ultraviolet region of the electromagnetic spectrum extends from 200×10^{-9} m to 400×10^{-9} m, and the visible region extends from 400×10^{-9} m (violet light) to 700×10^{-9} m (red light). To simplify reporting of spectral information, it is common practice to report wavelengths in nanometers (nm). Reported in these units, the ranges of near ultraviolet and visible spectra are

Type of spectrum	Nanometers
near ultraviolet	200–400
visible	400–700

Prior to the adoption of the nanometer, it was common to report ultraviolet-visible spectral information in Angstroms (Å; 10Å = 1 nm) or in millimicrons (mμ; 1 mμ = 1 nm). In our discussions, we will use only the nanometer for reporting wavelengths of absorption maximum.

An ultraviolet-visible spectrum is a plot of absorbance versus wavelength. Figure 15.4 is an ultraviolet spectrum of 2,5-dimethyl-2,4-hexadiene, a conjugated diene. The spectrum consists of a single, broad absorption band between 210 and 260 nm with an absorption maximum at 241 nm. In reporting UV spectral information, it is customary to report only the position of the absorption maximum (or maxima, if there are several peaks). Thus, we would report that the absorption maximum for 2,5-dimethyl-2,4-hexadiene is 241 nm.

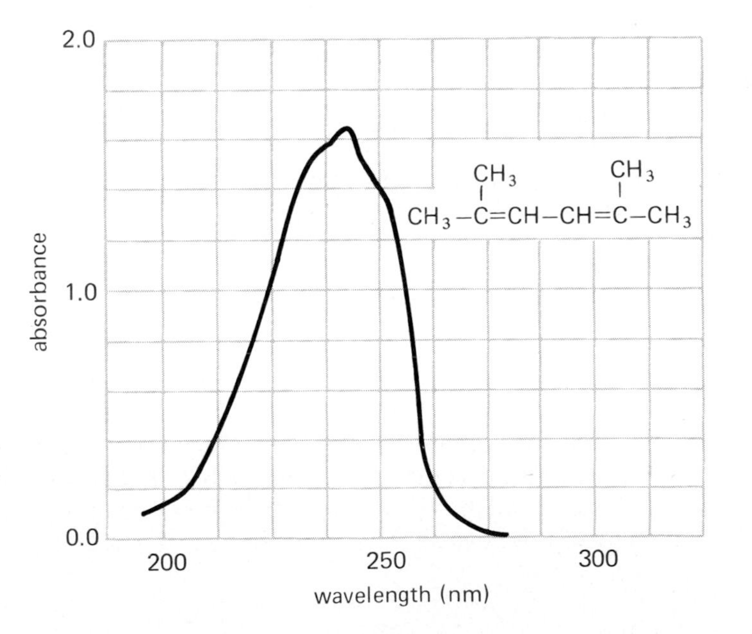

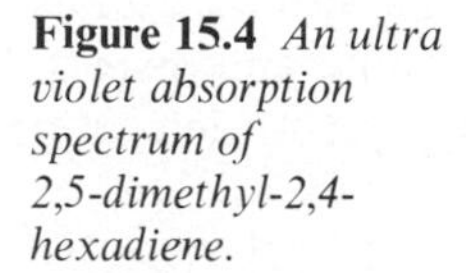

Figure 15.4 *An ultra violet absorption spectrum of 2,5-dimethyl-2,4-hexadiene.*

Absorption of ultraviolet-visible radiation is accompanied by the promotion of electrons from one energy level or orbital to a higher one. Electrons of sigma bonds are held too tightly to be affected by near ultraviolet-visible radiation. Hence, alkanes, cycloalkanes and simple alkenes show no absorption between 200 nm and 700 nm. However, pi electrons of carbonyl groups (C=O), conjugated unsaturated systems, and aromatic rings absorb this type of radiation. Butadiene itself shows an absorption maximum at 215 nm. The conjugated carbon-carbon and carbon-oxygen double bond system of 3-pentene-2-one shows an absorption maximum at 227 nm. Benzene, the simplest of the aromatic hydrocarbons, shows an absorption maximum at 257 nm.

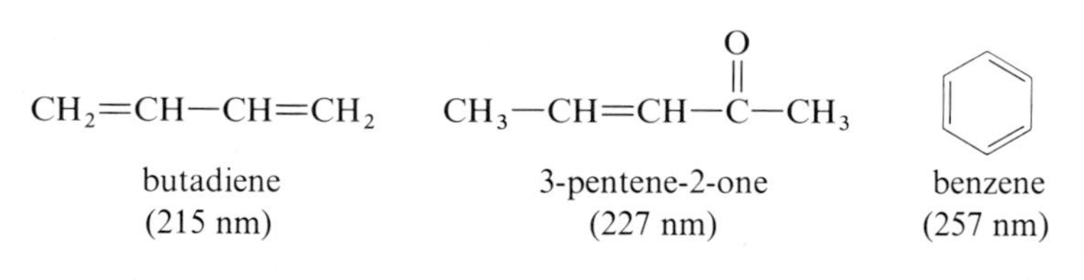

butadiene (215 nm) 3-pentene-2-one (227 nm) benzene (257 nm)

What should you look for in interpreting an ultraviolet-visible spectrum? These spectra are not as complicated as infrared spectra (compare Figures 15.2 and 15.4) or for that matter nuclear magnetic resonance spectra. For this reason, they do not provide as much information about details of molecular structure as do the other spectral methods. Our use of UV-visible spectra will be quite simple. If an organic molecule does not absorb radiation in the region 200 nm to 700 nm, we conclude that probably it does not contain any type of conjugated unsaturation. If it does absorb in this region, we immediately suspect that it contains some type of unsaturation, for example, a conjugated diene, a carbonyl group, an α,β-unsaturated carbonyl group, or an aromatic ring.

In addition to providing information about molecular structure, ultraviolet-visible spectroscopy can also be used for quantitative analysis, for there is a direct proportionality between absorbance (A), concentration of the sample (c), and the length of the light path through the sample (l). The proportionality constant which relates these three variables is called the molar absorptivity and is given the symbol ε (epsilon). The equation for this relationship is

$$A = \varepsilon \times l \times c$$

A = absorbance
ε = molar absorptivity
l = length of light path in cm
c = concentration in mole/liter

This equation, known as Beer's Law, forms the basis for the use of absorption spectroscopy, including ultraviolet-visible spectroscopy, for quantitative analysis. The molar absorptivity is a constant for any given compound and once the value of this constant has been determined, it can then be used to determine the concentration of the substance in solution. For several examples of quantitative absorption spectroscopy, read the mini-essay, "Clinical Chemistry—The Search for Specificity."

15.5 NUCLEAR MAGNETIC RESONANCE SPECTROSCOPY

Nuclear magnetic resonance spectroscopy involves absorption of electromagnetic radiation in the radiofrequency region of the spectrum, and, as you have already learned from the information in Table 15.1, the energy of this radiation is very small. Radiowaves of frequency 60×10^6 Hz (60 Megahertz or 60 MHz) have a wavelength of 5 meters and an energy of slightly less than 0.01 cal/mole. Absorption of radiowaves is accompanied by a special type of nuclear transition, and, for this reason we call this type of spectroscopy nuclear magnetic resonance (NMR) spectroscopy.

The nuclei of certain elements behave as if they were spinning charges. Any spinning charge creates a magnetic field (Figure 15.5a) and in effect behaves as if it were a tiny bar magnet (Figure 15.5b).

Of the three isotopes most common to organic compounds (^{1}H, ^{12}C, and ^{16}O), only the proton behaves in this manner. Therefore, proton magnetic resonance (PMR) spectroscopy is the only type of nuclear magnetic resonance we will study. We should note, however, that isotopes of other elements including ^{13}C, ^{15}N, ^{19}F, and ^{31}P also behave as if they were spinning charges, and the magnetic resonance spectroscopy of these nuclei is a field of active and expanding research at the present time.

When a hydrogen nucleus in an organic molecule is placed in a strong magnetic field, there are only two allowed orientations for its magnetic field: with the applied field or against it (Figure 15.6). In the lower energy state, E_1, the nuclear magnet is aligned with the applied magnetic field. If energy is supplied in the form of radiowaves of exactly the right frequency, radiation is absorbed and the spinning nuclear magnet flips and become aligned against

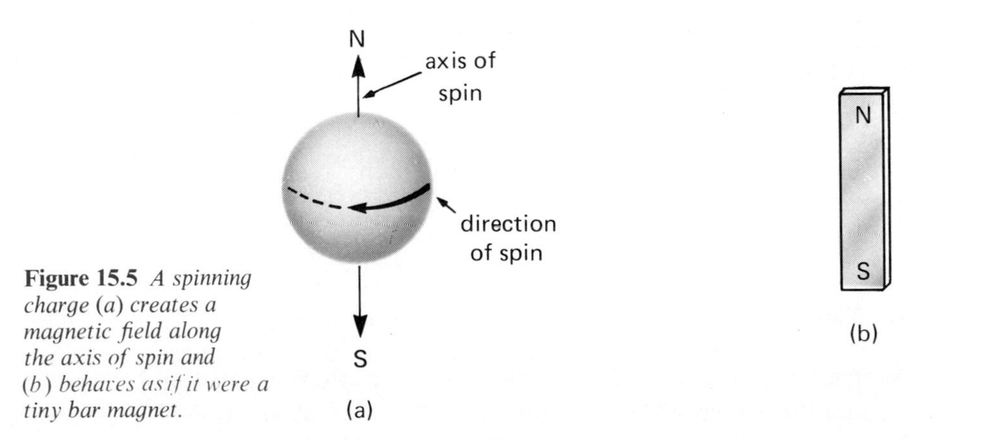

Figure 15.5 *A spinning charge (a) creates a magnetic field along the axis of spin and (b) behaves as if it were a tiny bar magnet.*

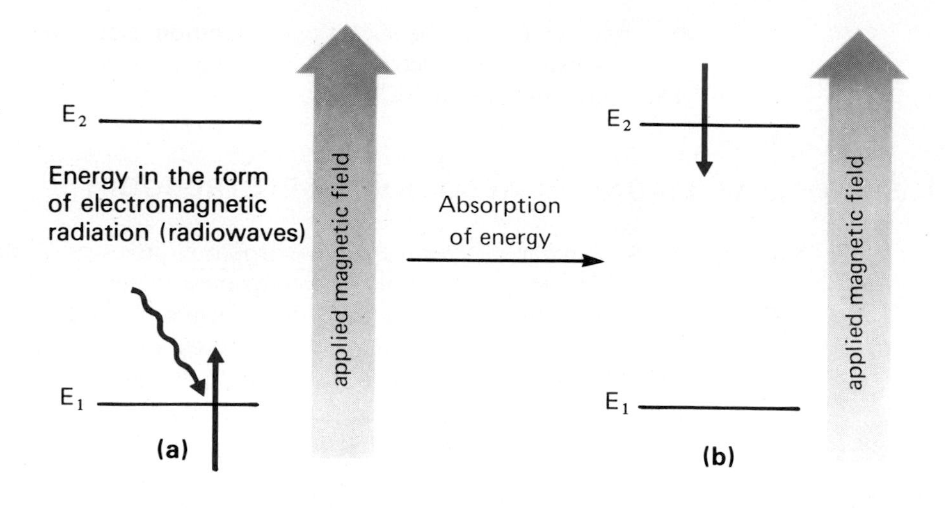

Figure 15.6 *Orientation of a hydrogen nucleus in an applied magnetic field. (a) Nucleus in the lower energy state, (b) nucleus in the higher energy state. Absorption of electromagnetic radiation causes a transition from the lower to the higher energy state.*

the applied magnetic field in the higher energy state, E_2. The most common commercially available nuclear magnetic resonance spectrometers generate magnetic fields of approximately 14,000 Gauss. By way of comparison, the magnetic field of the earth is only about 0.5 Gauss. Given this applied field strength, radiofrequency radiation in the range of 60 MHz is required to cause transitions of hydrogen nuclei from the lower to the higher energy state. One type of NMR spectrometer supplies radiowaves of frequency 60 MHz and then gradually increases the strength of the applied magnetic field. As the field strength increases, protons within a molecule absorb radiation and produce an NMR signal. An NMR spectrum, then, is a plot of the strength of the applied magnetic field versus the intensity of the absorption signal for the protons within a molecule.

What should you look for as you attempt to analyze an NMR spectrum? We will be concerned with three things:

(1) the position of each signal in the spectrum;
(2) the relative areas of the various signals;
(3) the splitting pattern of each signal.

Let us look at each of these features separately, and learn what types of information each can give us about molecular structure.

A. THE CHEMICAL SHIFT

At this point, it might seem that protons are protons, and all protons within a molecule should absorb at exactly the same point in the NMR spectrum. Fortunately, this is not the case, for the position at which any given proton

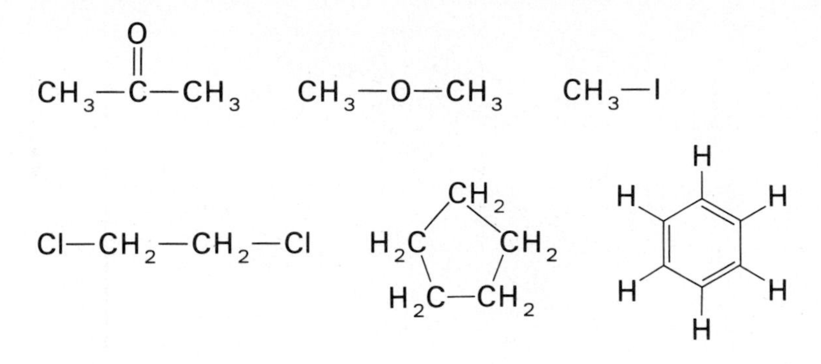

Figure 15.7 *Molecules showing only a single absorption signal in the NMR spectrum.*

absorbs depends on (1) the strength of the applied magnetic field and (2) the proton's immediate electronic environment. Electrons in the immediate environment of a proton (bonding as well as nonbonding electrons, those in sigma bonds as well as those in pi bonds) serve to shield a proton very slightly from the external magnetic field, altering the position at which it absorbs in the NMR spectrum. All chemically equivalent protons, that is, all those with identical electronic environments, absorb at the same position in the NMR spectrum. Conversely, protons with different chemical environments absorb at different positions in the NMR spectrum. Each of the molecules shown in Figure 15.7 shows only a single absorption signal in the NMR spectrum because, in each, all protons are chemically equivalent.

Each of the molecules shown in Figure 15.8 shows more than one signal in the NMR spectrum because, in each, there are two or more different sets of chemically equivalent protons.

In reporting where the various chemically equivalent protons within a molecule absorb, it is common practice to select a reference signal and then to report other signals in terms of how far each is shifted from the reference signal. This shift is called the chemical shift and is given the symbol δ (delta). The most generally used reference standard is the signal due to the twelve

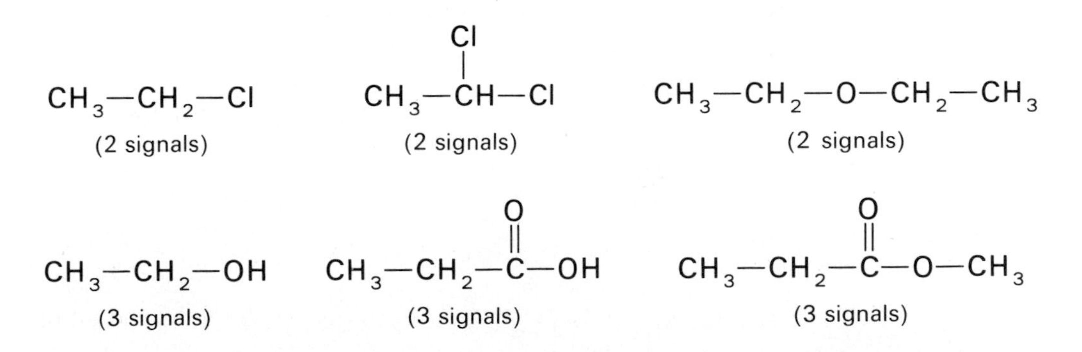

Figure 15.8 *Molecules showing two or more absorption signals in the NMR spectrum.*

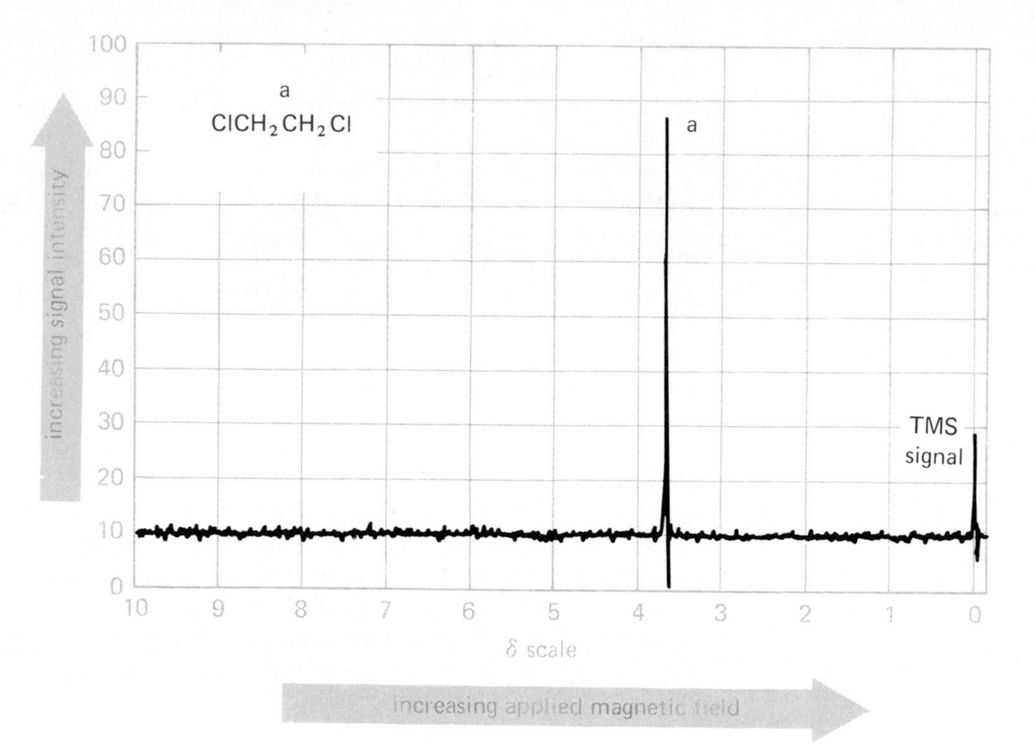

Figure 15.9 *An NMR spectrum of 1,2-dichloroethane*, $ClCH_2CH_2Cl$.

Table 15.3 Chemical shifts of several types of protons, relative to TMS.

Type of proton	*Chemical shift relative to TMS*	*Type of proton*	*Chemical shift relative to TMS*
$C-CH_3$	0.9–1.0	$O=\underset{\vert}{C}-H$	9.0–10.0
$C-CH_2-C$	1.20–1.40	(benzene ring)—H	6.5–8.3
$C-\underset{\underset{C}{\vert}}{CH}-C$	1.40–1.60		
$(C)_2C=C(H)(C)$	4.5–5.5	$-O-CH_3$	3.2–3.3
		$-O-CH_2-C$	3.2–3.8
$O=\underset{\vert}{C}-CH_3$	2.0–2.8	$-O-H$	2.0–5.0
$O=\underset{\vert}{C}-CH_2-$	2.2–2.5	$-\overset{\overset{O}{\Vert}}{C}-O-H$	10.0–13.0

chemically equivalent protons in tetramethylsilane, $(CH_3)_4Si$, abbreviated TMS On the delta scale, the peak for TMS is set at zero. Other signals are given positive delta values depending on how far they are shifted to the left on the chart paper from TMS. Figure 15.9 is the NMR spectrum of 1,2-dichloroethane. Note that there are two signals in this spectrum. The signal of low intensity on the far right at $\delta = 0$ is that of the TMS reference standard. The signal at $\delta = 3.7$ is a single sharp peak due to the four chemically equivalent protons of 1,2-dichloroethane. Delta values for the types of protons we will encounter fall within the range 0 to 13 and are summarized in Table 15.3.

B. RELATIVE SIGNAL AREAS

As we have shown in the previous section, the number of signals in an NMR spectrum corresponds to the number of sets of chemically equivalent protons within the molecule. In addition, an NMR spectrum also gives us information about the number of chemically equivalent protons there are in each set for the relative intensities of the various signals (as measured by the relative areas) are proportional to the number of protons giving rise to each signal. For benzyl acetate (Figure 15.10) the areas of the three signals are in the ratio of 3:2:5 corresponding to the three protons of the $-CH_3$ group, the two protons of the $-CH_2-$ group, and the five protons of the $-C_6H_5$ group, respectively.

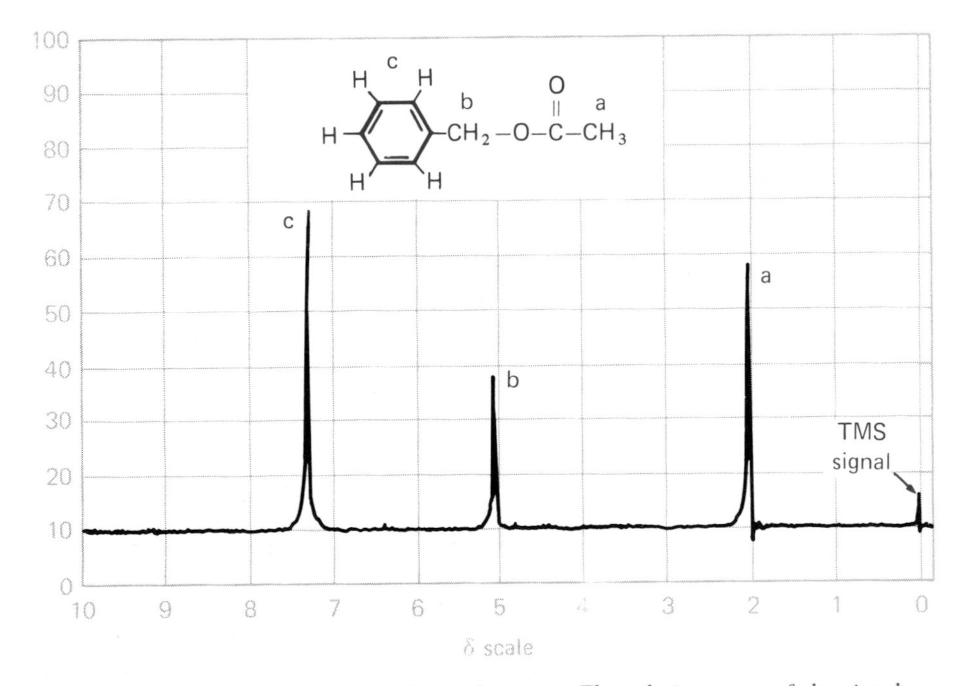

Figure 15.10 *An NMR spectrum of benzyl acetate. The relative areas of the signals at $\delta = 2.1$, 5.1, and 7.3 are in the ratio of 3:2:5.*

C THE SPLITTING PATTERN

A third kind of information can be derived from the splitting pattern of each NMR signal. Consider, for example, the NMR spectrum of 1,1-dichloroethane (Figure 15.11). Note that this molecule is an isomer of 1,2-dichloroethane, a substance whose NMR spectrum we saw in Figure 15.9.

According to what we have said so far, you would predict two signals with relative areas in the ratio 3:1, corresponding to the three protons of the $—CH_3$ group and the single proton of the $—CHCl_2$ group. Notice from the spectrum that there are in fact two signals, but neither is a single sharp peak. The signal at $\delta = 2.1$ is split into two closely spaced peaks (a doublet) and the signal at $\delta = 5.8$ is split into four closely spaced peaks (a quartet). The areas of the doublet and quartet are in the ratio 3:1, as we would have predicted they should be. But, why is the first signal split into a doublet and the second into a quartet? We can account for this splitting if we remember that protons themselves behave as tiny bar magnets. Therefore, any given proton within a molecule feels the effect not only of the external magnetic field but also that of the magnetic field generated by neighboring protons. We can predict the signal splitting due to nonchemically equivalent proton neighbors by the so-called $n + 1$ rule. According to this rule, if a proton has n nonchemically equivalent proton neighbors, then its NMR signal will be split into $n + 1$ peaks. If we apply this rule to the spectrum of 1,1-dichloroethane, we would predict that the $—CH_3$ signal should be split into a doublet by the one proton

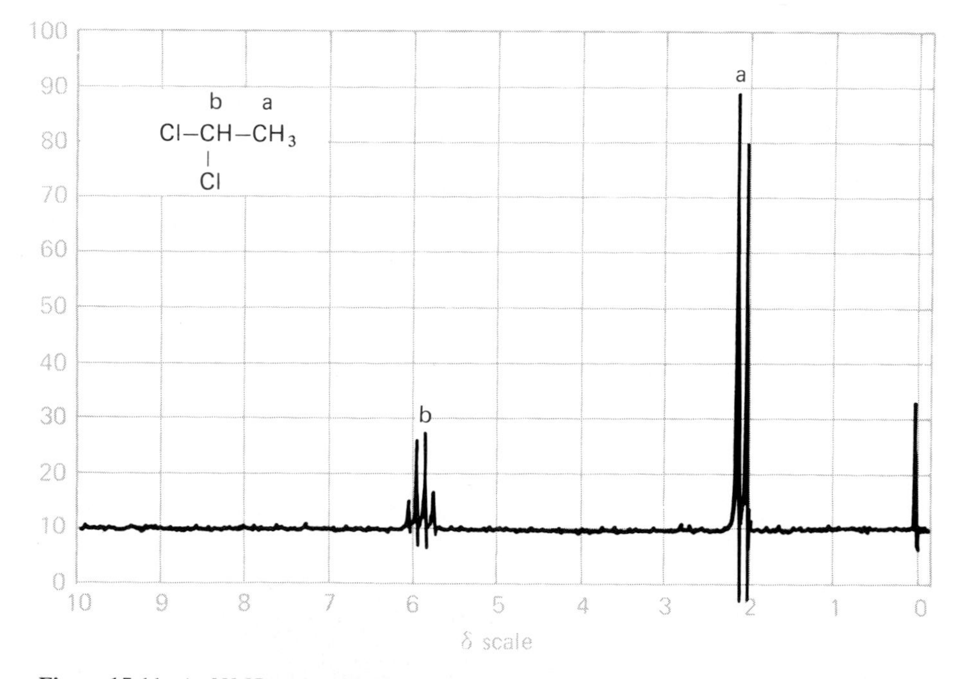

Figure 15.11 *An NMR spectrum of 1,1-dichloroethane, CH_3CHCl_2. The spectrum consists of two signals, a doublet at $\delta = 2.1$ and a quartet at $\delta = 5.8$.*

neighbor. The $—CHCl_2$ signal should be split into a quartet by the three hydrogen neighbors. This is exactly what is observed (Figure 15.8).

It is important to keep in mind that chemically equivalent protons do not split each other. For example, the NMR spectrum of 1,2-dichloroethane (Figure 15.9) consists of only a single, sharp peak. All four protons within this molecule are chemically equivalent; therefore, no splitting of the signal occurs.

In summary, an NMR spectrum can provide us with three types of information about molecular structure.

(1) The number of signals in the NMR spectrum and the chemical shift of each tell us the number of different types of chemically equivalent protons within the molecule.

(2) The relative areas of the various signals tell us how many protons there are of each chemically equivalent type.

(3) The splitting pattern of each signal tells us something about the number of neighboring protons in the environment of each type of chemically equivalent proton.

PROBLEMS

15.1 Arrange ultraviolet-visible, infrared, and radiofrequency radiation in order of
(a) increasing wavelength **(b)** increasing frequency
(c) increasing energy

15.2 Absorption of electromagnetic radiation by a molecule is accompanied by an increase in the internal energy, that is, by transitions between one energy level and another within the molecule. What types of molecular transitions are brought about by the absorption of
(a) ultraviolet-visible radiation **(b)** infrared radiation
(c) radiofrequency radiation

15.3 Define the term *molecular spectroscopy*.

15.4 **(a)** Name the unit in which ultraviolet-visible spectral information is most commonly reported. What is the numerical relationship between this unit and the Angstrom?
(b) Name two units in which infrared spectral information is commonly reported. What is the numerical relationship between these two units?
(c) Name one unit in which nuclear magnetic resonance spectral information is commonly reported.

15.5 State whether you would predict the following molecules to absorb radiation in the ultraviolet-visible region of the spectrum (between 200 nm and 700 nm).

(a) H_2O **(b)** CH_3CH_2OH

(c) $CH_2{=}CH{-}CH_2{-}CH_2{-}CH{=}CH_2$

(d) $CH_3{-}CH{=}CH{-}CH{=}CH{-}CH_3$

(e) $CH_3{-}CH_2{-}\overset{\overset{\displaystyle O}{\|}}{C}{-}CH_2{-}CH_3$ **(f)** $CH_3{-}CH_2{-}\overset{\overset{\displaystyle O}{\|}}{C}{-}CH{=}CH_2$

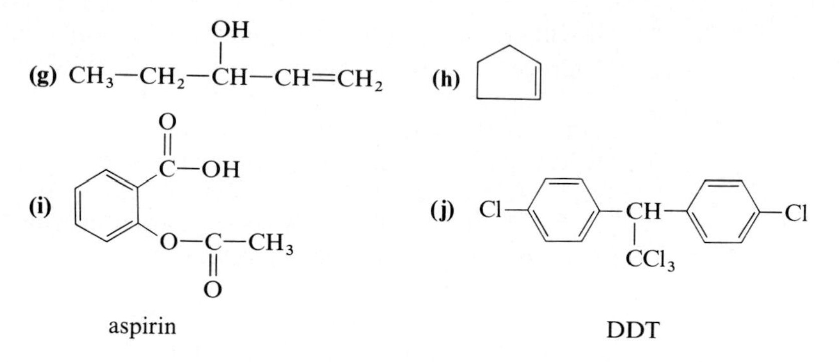

aspirin

DDT

15.6 For the following pairs of compounds, (i) name the functional groups present, and (ii) list one major feature that will appear in the infrared spectrum of the first but not the second molecule. Your answer to part (ii) should state what type of bond vibration is responsible for the spectral feature you have listed and its approximate position in the spectrum.

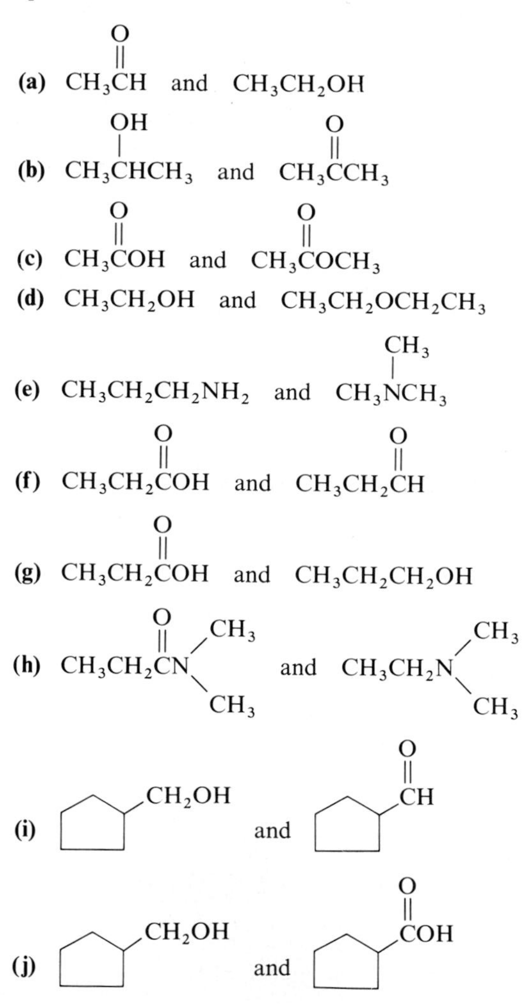

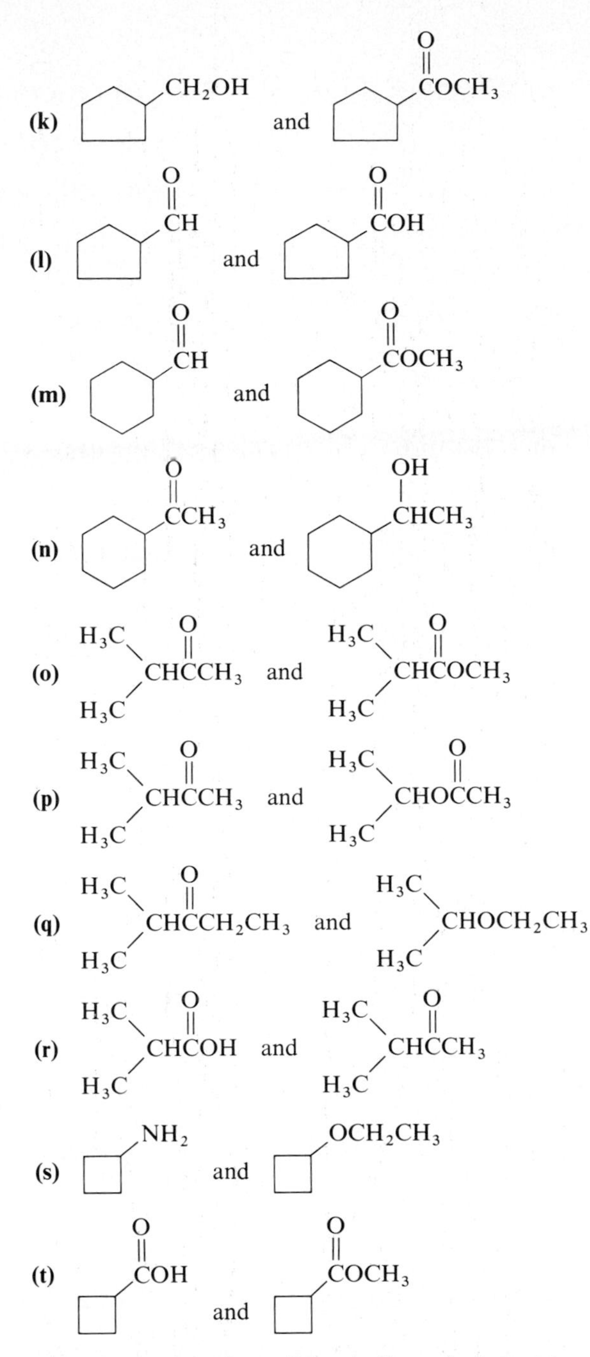

15.7 State the number of sets of chemically equivalent protons in the following molecules, and the number of protons in each set.

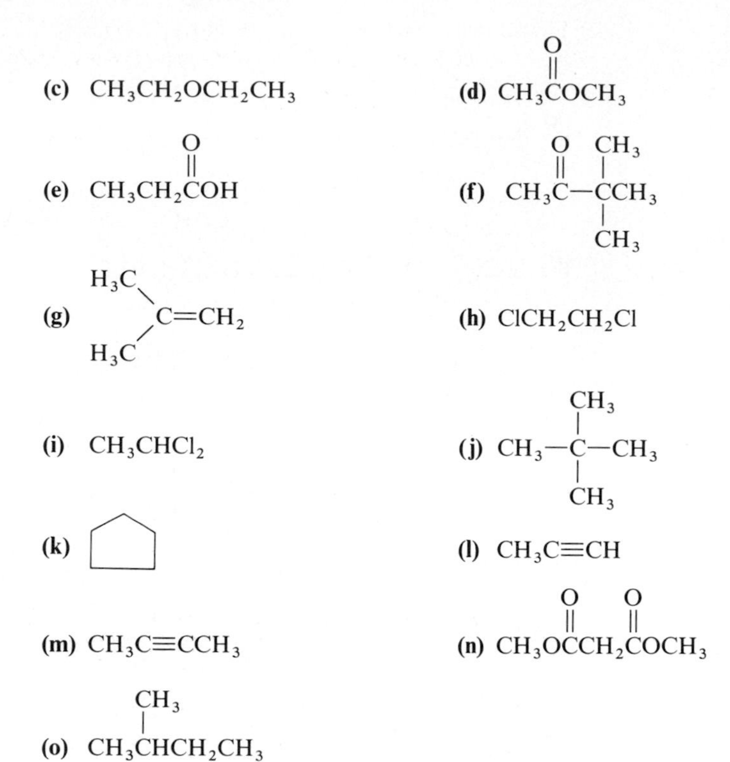

15.8 Which of the compounds in Problem 15.7 shows only a single peak in its NMR spectrum?

15.9 Use the $n + 1$ rule to predict the splitting patterns of each set of chemically equivalent protons in the following molecules.

(a) $CH_3CH_2OCH_2CH_3$

(b) 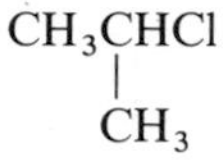 CH_3CHCl (with CH_3 on the CH)

(c) CH_3CHCl_2

(d) $ClCH_2CHCl_2$

(e) 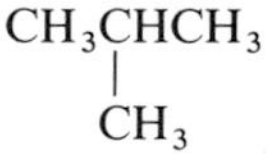 CH_3CHCH_3 (with CH_3 on the CH)

(f) 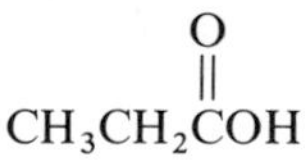 $CH_3CH_2\overset{O}{\overset{\|}{C}}OH$

(g) 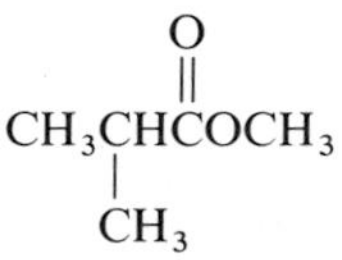 $CH_3CH\overset{O}{\overset{\|}{C}}OCH_3$ (with CH_3 on the CH)

(h) 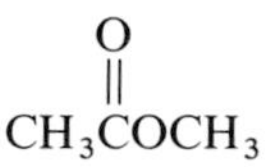 $CH_3\overset{O}{\overset{\|}{C}}OCH_3$

(i) 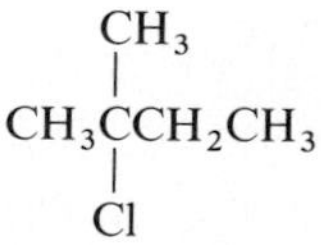 $CH_3CCH_2CH_3$ (with CH_3 above and Cl below the C)

15.10 Following are pairs of structural isomers. For each pair, state one major feature in the NMR spectrum which will allow you to distinguish the first isomer from the second.

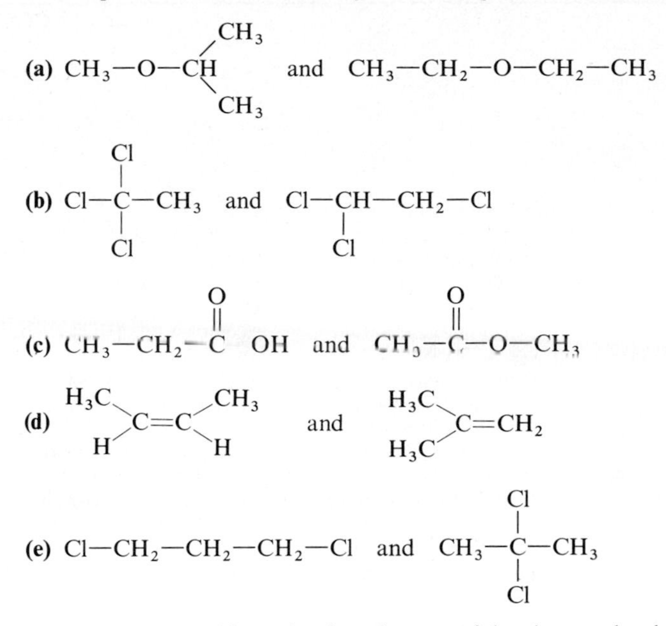

15.11 Draw the structural formula of a substance of the given molecular formula that will show only a single absorption signal in its NMR spectrum.

(a) C_5H_{12} **(b)** C_5H_{10} **(c)** C_3H_6O
(d) $C_3H_6Cl_2$ **(e)** C_2H_6O **(f)** C_4H_8
(g) $C_2H_2Cl_4$ **(h)** C_4H_9Cl **(i)** C_6H_{12}

15.12 Following are pairs of structural isomers. The members of each pair have the same number of signals in the NMR spectrum and the same splitting pattern of each signal. However, these isomers can be distinguished by chemical shifts. Use the information in Table 15.2 to show how the chemical shifts of the underlined protons enable you to distinguish between these isomers.

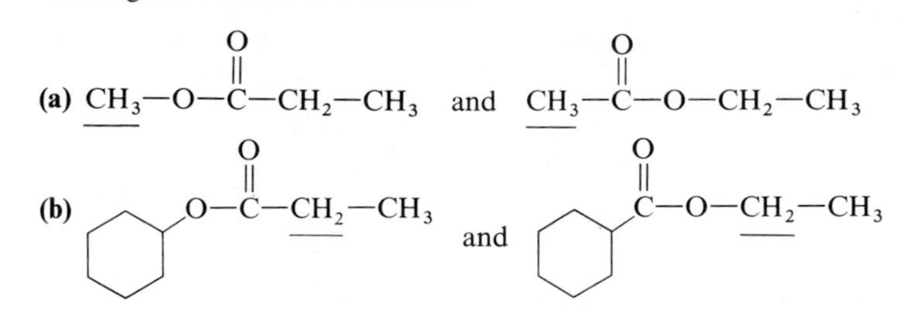

INDEX

Characteristic infrared stretching frequencies.

Type of bond	*Found in*	*Frequency* (cm^{-1})
C—H	alkanes	2850–2950
=C—H	alkenes and aromatic hydrocarbons	3000–3200
O—H	alcohols, phenols	3600–3650
O—H	carboxylic acids (hydrogen bonded)	2500–3000
N—H	amines	3300–3500
C—O	esters, alcohols, ethers	1000–1300
C=O	aldehydes, ketones, carboxylic acids, esters, amides	1680–1750

Chemical shifts of several types of protons, relative to TMS.

Type of proton	*Chemical shift relative to TMS*	*Type of proton*	*Chemical shift relative to TMS*
C—CH_3	0.9–1.0	O=C(—)—H	9.0–10.0
C—CH_2—C	1.20–1.40	aromatic H (benzene ring—H)	6.5–8.3
C—CH(—C)—C	1.40–1.60		
C_2C=C(H)C	4.5–5.5	—O—CH_3	3.2–3.3
		—O—CH_2—C	3.2–3.8
O=C(—)—CH_3	2.0–2.8	—O—H	2.0–5.0
O=C(—)—CH_2—	2.2–2.5	—C(=O)—O—H	10.0–13.0